Computing and Systems Technology Division 2018

Core Programming Area at the 2018 AIChE Annual Meeting

Pittsburgh, Pennsylvania, USA
28 October - 2 November 2018

ISBN: 978-1-5108-7623-1

Computing and Systems Technology Division 2018

Core Programming Area at the 2018 AIChE Annual Meeting

Pittsburgh, Pennsylvania, USA
28 October - 2 November 2018

AIChE
The Global Home of Chemical Engineers

Printed from e-media with permission by:

Curran Associates, Inc.
57 Morehouse Lane
Red Hook, NY 12571

Printed by Curran Associates, Inc. (2019)

AIChE
120 Wall Street, FL 23
New York, NY 10005-4020

Phone: (800) 242-4363
Fax: (203) 775-5177

www.aiche.org

Additional copies of this publication are available from:

Curran Associates, Inc.
57 Morehouse Lane
Red Hook, NY 12571 USA
Phone: 845-758-0400
Fax: 845-758-2633
Email: curran@proceedings.com
Web: www.proceedings.com

TABLE OF CONTENTS

Some abstracts in this book refer to images that were not made available by the publisher

OPTIMAL CLEANING SCHEDULING AND CONTROL OF HEAT EXCHANGER NETWORKS: PROBLEM FORMULATION AND SOLUTION STRATEGY.

Federico Lozano Santamaria, Imperial College, London, UK
Sandro Macchietto, Imperial College, London, UK

Abstract

In crude oil processing heat exchanger fouling is a major problem because it reduces the thermal and hydraulic performance of the units which has a large effect in the energy efficiency and operational cost. Fouling mitigation alternatives are mandatory and two of the most common and effective are: flow distribution control in heat exchanger networks, and periodic cleanings. These two mitigation alternatives can be defined individually by solving two distinct optimization problems, an optimal control problem (NLP) for the optimal flow distribution, and an optimal scheduling problem (MINLP) for the cleaning scheduling. However, there are important interactions between these two mitigation strategies that are neglected when addressed independently, while solving the integrated scheduling and control problem has economic benefits, but it is highly challenging. Here we present a general, flexible and accurate formulation to model heat exchanger networks under fouling, and an efficient solution strategy to tackle the optimal control and optimal scheduling problems simultaneoulsy. The formulation is able to capture all the important interactions of the operation, and includes all variables for the various decisions. The integrated control and scheduling problem is reformulated as a MPCC (mathematical problem with complementarity constraints) and then solved using a NLP solver based on interior point methods. A case study of industrial significance is used to illustrate the advantages of the formulation, the efficiency of the solution, and the benefits of considering the control and scheduling optimization problems at the same decision level.

Introduction

Petroleum refining processes supply a large number of the world chemical commodities and energy demand. There is a strong push towards more efficient operation due to tight market requirements, competition with alternative energies and environmental regulations. The industry must be able to adapt to those changes while maintaining a steady and profitable operation, therefore all sources of inefficiencies have to be minimized. In these processes there are still many sources of inefficiencies that can be addressed to improve refining operations performance, both from an economic and an environmental perspective, and given the large quantities of crude oil processed by refineries the impact of any kind of inefficiency during their operation is significant. Among the energy inefficiencies, fouling of process units has a significant impact on energy recovery because the deposition of unwanted material over process surfaces significantly compromises the heat transfer of many process units. This particularly impacts the front-end heat exchanger network and energy recovery in the crude distillation unit (CDU) section, which processes essentially the entire world crude oil. Considering only the economic implications for the operation of the refinery a loss of 1 MW in energy recovery caused by fouling has an additional operational cost of $ 0.3 M/year [1].

Fouling in heat exchangers, a slow dynamic process, reduces their thermal and hydraulic performance significantly, therefore mitigation alternatives are needed to recover the performance of the network and to maintain a profitable and safe operation. Two important and common mitigation alternatives are: the control of flow distribution in the network (by means of by-passes and flow split between parallel units), and the periodic cleaning of certain the affected units. Both mitigation alternatives can be determined by the formulation and solution of a dynamic optimization problem over a long operating horizon (from months to a few years). Usually the two alternatives have been analyzed separately, as in [2] where the optimal control of the desalter is defined, and in [3] where the cleaning schedule for a network is defined using heuristic algorithms. The integration of scheduling and control for this problem is challenging because of the different time scales involved (of the order of hours for the control problem, of months or years , for the scheduling problem) and the large number of continuous and binary decision variables. Moreover, the solution of the optimal scheduling problem on its own has very challenging features such as the combinatorial nature due to a large number of binary variables, the existence of a great number of feasible solutions with the same or similar objective function values, and the discontinuities and nonlinearities of the prediction models [4].

An accurate solution to the fouling problem from the operational perspective is needed to define the optimal operation of the network (flow control and cleaning scheduling). A rigorous mathematical formulation of the fouling problem and the heat transfer in the network, adequate to describe the dynamics of the operation was presented in [7]. This is composed by nonlinear models that capture all the important interactions of the variables in the network, and includes all the binary decision variables associated with the cleaning scheduling problem. This formulation captures both the control and the scheduling problems. Some preliminary aspects of the solution strategy and results were presented in [8], showing that it the simultaneous problem can be solved efficiently using deterministic optimization algorithms. Here, a summary of the model formulation is presented, together with some further details of the solution strategy and their application to an industrial case study, with consideration of the performance achieved.

Mathematical formulation and solution strategy

The heat exchanger network is modelled as a multigraph where the nodes are: exchangers, sources, sinks, mixers, and splitters. Mass balances and energy balances are applied to each node of the network, and only in the heat exchangers two streams interact to recover energy. To model the heat transfer in the exchangers the P-NTU model is used considering the differences in geometry and configuration of each exchanger. The connections between the nodes are defined as arcs, which correspond to the streams in the network, characterized by a flow rate, temperature, and physical properties. Another important aspect to consider in the formulation is the fouling model. Here, the Ebert-Panchal model is used and deposit ageing (defined here in terms of the fresh, 'gel'-like and aged, 'coke'-like mass fractions in the deposit) is considered because of the high temperature of the hot-end of the preheat train. These two models are differential equations, Eq. 1 – 2, that capture how the fouling thermal resistance changes as a function of the operating conditions of the exchanger. To account for changes of the conditions between the inlet and outlet of the exchanger an average of fouling and ageing rates between the two extremes is used to integrate the differential equations. Although the model is lumped in the axial direction of the exchangers, it considers the radial effects on the heat transfer by solving analytically the heat transfer equations on the different layers: tube side, fresh deposit, aged deposit, tube wall, and shell side (assuming thermal resistance in series).

$$\frac{dR_{f,i,t}}{dt} = \alpha_i Re_{i,t}^{-0.66} Pr_{i,t}^{-0.33} \exp\left(-\frac{E_{f,i}}{RT_{f,i,t}}\right) - \gamma_i \tau_{w,i,t}, \quad \forall t \in Time\backslash\{0\}, i \in HEX \qquad \text{Eq. 1}$$

$$\frac{dx_{g,i,t}}{dt} = -A_{a,i} \exp\left(-\frac{E_{a,i}}{RT_{dc,i,t}}\right) x_{g,i,t}, \quad \forall t \in Time\backslash\{0\}, i \in HEX \qquad \text{Eq. 2}$$

To discretize the differential equations the time horizon is divided in periods of variable length (continuous time approach), and each period is discretized using orthogonal collocation in finite elements to improve the accuracy of the integration. This approach is not only efficient for solving optimal control problems with a small number of discretization points, but it also defines a formulation for the scheduling problem with fewer binary variables than those of a discrete time formulation [7]. A binary decision is assigned to each period and to each exchanger to define the cleanings, and the cleaning time is defined by the starting time of the period.

The decision variables of the problem are: the throughput, the flow rate through parallel branches, and the definition of the cleanings. The first two correspond to continuous decision variables in the optimal control problem, while the last one correspond to binary variables in the optimal cleaning scheduling problem that define which exchanger to clean and when to clean it. Note that the length of the time periods is also a decision variable that is related to both the control problem and the scheduling problem. These binary decision variables are the complicating variables of the problem. They are used as switches to indicate changes in the operating mode and are involved in almost all of the problem constraints in the form of disjunctions.

The binary variables used to model the disjunctions or operating modes are listed in Table 1.

Table 1. List and definition of disjunctions

Constraint	Nature	Modes	
		Operating	**Cleaning**
Fouling	Differential equation	Rate given by Eq. 1	No fouling (rate equal 0)
Ageing	Differential equation	Rate given by Eq. 2	No ageing (rate equal 0)
Pressure drop in exchanger	Algebraic equation	Pressure drop model	No pressure drop (ΔP=0)
Heat transfer in exchanger	Algebraic equation	Heat transfer model	No heat transfer (Q=0)
Flow rates	Variables	There is flow through HEX	There is NO flow through HEX
Streams temperature	Variables	Given by heat transfer model	No change in temperature of the HEX streams

Those disjunctions are defined as nonlinear functions, and the switch from one state to the other is modelled by multiplying the binary variables and the nonlinear term of the constraint. This approach is preferred instead of a BigM formulation of the disjunctions because the latter leads to a formulation that is too loose and the solution of which will take a prohibited computational time using standard MINLP solvers. For example, with a BigM formulation of the disjunctions and using a standard branch and bound algorithm it was not possible to solve the simultaneous optimal control and cleaning scheduling problem of a simple network of two exchangers: after a computational time of 72 h the optimality gap was still 79%.

The solution approach adopted for the integrated problem of optimal control and cleaning scheduling of heat exchanger networks is to reformulate the MINLP as a mathematical program with complementarity constraints (MPCC). In this approach the binary variables are relaxed as bounded continuous variables, slack variables are introduced to the problem, and complementarity constraints between these two kinds of variables are defined. These constraints between two variables force one to be at its bound and the other to be free, being an alternative way to model the disjunctions defined by an OR operator. Eq. 3 shows the definition of the decision variables (y) associated with the cleaning scheduling, the slack variables, and the complementarity constraints. Although this reformulation reduces the complexity of the problem by removing the binary variables, it introduces other difficulties such as additional nonlinearities and nonconvexities, and the no satisfaction of constraints qualifications, for instance LICQ [5]. These characteristics make the problem difficult to solve to optimality with current NLP solvers.

$$\begin{array}{c} 0 \leq y_{i,t,c} \perp s_{a,i,t,c} \geq 0 \\ 0 \leq (1 - y_{i,t,c}) \perp s_{b,i,t,c} \geq 0 \\ 0 \leq s_{a,i,t,c}, s_{b,i,t,c} \leq 1 \\ s_{a,i,t,c} + s_{b,i,t,c} = 1 \end{array}, \quad \forall i \in HEX, t \in Time, c \in Cleaning\ Types \qquad \text{Eq. 3}$$

The complementarity constraints are therefore reformulated in terms of inequalities by introducing an $\epsilon > 0$ parameter, and this reformulation acts as a relaxation of the problem. Eq. 4 is the definition of the complementarity constraints using this regularization ϵ approach, and the solution of the original MINLP problem is achieved solving a number of problems sequentially as $\epsilon \rightarrow 0$. The final solution is accepted with a specific tolerance and the variables y satisfy its binary nature at that point.

$$y_{i,t,c}(s_{a,i,t,c}) \leq \epsilon, \quad \forall i \in HEX, t \in Time, c \in Cleaning\ Types \qquad \text{Eq. 4}$$

Finally, the objective of the problem is defined as the minimization of the total operating cost, formed by the loss of throughput, the energy cost, the environmental cost, and the cleaning cost (Eq. 5). In addition, to the model constraints mentioned, the optimization problem is also subjected to the constraints that arise from the time discretization to define the starting time of the cleanings, the connectivity of the network, the scheduling constraints (e.g. sequence of cleanings, simultaneous cleanings, maximum number of cleanings per exchanger), practical operating constraints (e.g. limits on temperatures, furnace duty, pressure drop and split ratio limits), and the disjunctions defined using nonlinear relations and complementarity constraints.

$$\begin{aligned} \min_{m,y} z = {} & P_{kg} \int_0^{t_f} (-m_{production\ ,t}) dt + P_{fuel} \int_0^{t_f} \frac{Q_f}{\eta} dt + P_{CO_2} m_{CO_2} \int_0^{t_f} \frac{Q_f}{\eta} dt \\ & + \sum_{i \in HEX} \sum_{c \in Cleaning\ Type} \sum_{t \in Time} P_{Cl,i,c}\, y_{t,i,c} \end{aligned} \qquad \text{Eq. 5}$$

Case studies and numerical results

The case study analyzed here is the hot end of the preheat train of a refinery. Figure 1 shows the structure of the network which is composed by 9 exchangers, with two parallel branches which interact through the hot residue stream (RS). All the exchangers are of shell and tube type, and more details about their specifications and normal operating conditions can be found in [6]. The operation horizon is 4 years, and the fouling parameters and deposit roughness are estimated from temperature, pressure, and flow rate measurements. The prediction error of the main variables is within acceptable levels (average errors: 0.2°C for temperature, and 0.5 MW for heat duty) The model was fitted against a more detail distributed model for heat exchanger networks under fouling, which has good prediction capabilities tested and validated against real plant data [6]. In this case study the control degree of freedom is the split fraction of the crude through the parallel branches of the network, and the scheduling degrees of freedom are the starting time, the assignment, and the sequence of the cleanings.

Two scenarios are considered for this case study: 1) only optimal cleanings scheduling fixing the flow split at 50%/50%, and 2) simultaneous optimal control and optimal cleaning scheduling.

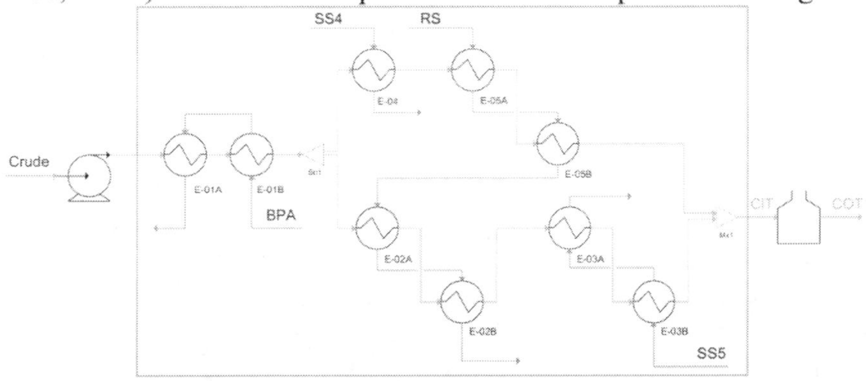

Figure 1. Heat exchanger network of the preheat train for the case study.

The number of periods used in the discretization of the time domain is varied from 10 to 40, to analyze how the numerical complexity of the problem scales up, and how this parameter affects the optimal solution. Figure 2 shows that the size of the problem, in terms of number of variables and constraints, increases linearly with the number of periods, and that there is not a significant difference in problem size between the optimal scheduling problem and the integrated optimal scheduling and control problem. Also, the number of binary variables increases very rapidly with the number of periods and given the combinatorial nature of the problem, the number of feasible scenarios and cleaning schedules increase exponentially. Although the number of continuous variables is around a thousand times larger than that of binary variables, current NLP solvers can handle problems of that size and the complexity of finding a solution comes from the introduction of binary variables.

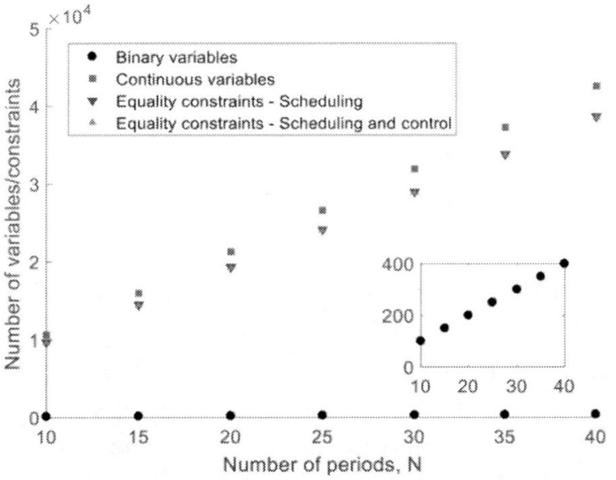

Figure 2. Problem size of the case study as a function of the number of periods (N).

The number of periods used in the discretization of the time horizon also has an effect on the optimal solution of the problem. Figure 3 shows how the minimum total cost, the total number of cleanings, and the computational time change with the number of periods for i) the optimal cleaning scheduling problem, with fixed flow split and ii) the integrated optimal control and scheduling problem. In Figure 3a it is observed that the total operational cost decreases with the number of periods for both problems, as a finer discretization allows more cleanings to take place in the same time horizon. However, for the integrated scheduling and control problem the number of cleanings does not increase significantly after N=30, while for optimal scheduling only the total optimal number of cleanings keeps increasing almost linearly. Note that for N < 30 the number of cleanings of both scenarios is similar and after this value, the two trends diverge. This is observed because up until N =25 there are not enough periods to define all the cleanings required. Another significant difference between the two scenarios is the total operating cost which is always larger for the case of optimal cleaning scheduling with fixed flow split. The additional degree of freedom from the control layer allows reducing the operating cost by $ 2 M (on average), although the minimum operating cost is found at N =30 for the case of integrated scheduling and control. For larger values of N the total cost then increases, but not significantly. Finally, from a computational effort perspective there is not a significant difference between optimising only the schedule, and the integrated schedule and control: the computational time required to find a solution using the MPCC reformulation is similar for both scenarios.

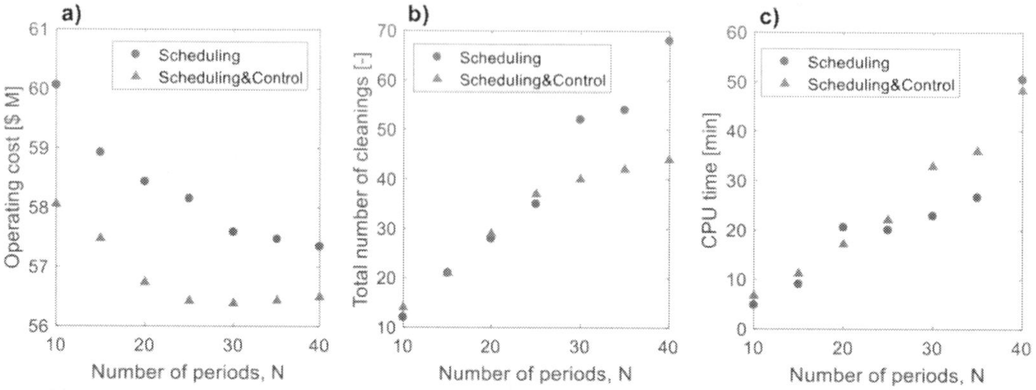

Figure 3. Optimal solution of the case study as a function of the number of periods.

The optimal solution of the two scenarios at N=30 is chosen to analyze in more detail the synergy between the scheduling and control problems. Figure 4 shows the evolution of the furnace duty during the four year operation horizon for four scenarios (with labels in italics as in Figure 4): *No mitigation* (no cleanings and flow split fixed at 50-50%), optimal *Control*, optimal cleaning *Scheduling* with fixed flow split at 50-50%, and simultaneous optimal *Scheduling and Control*. In the *No mitigation* scenario, it is observed that the efficiency of the network decreases rapidly and in less than one year, the furnace duty has reached 90% of its final value. For all the other scenarios, considering at least one mitigation strategy, there is a reduction in energy consumption. Even for the case of optimal *Control,* which is preventing and not eliminating fouling, there is a significant improvement in the operation. For the other cases that introduce cleanings there are periods of high energy consumption (during the cleanings), and periods where the duty is low (after the cleanings). This periodic and discontinuous behavior results in a good economic and energetic trade-off for the operation. Finally, when control elements are introduced into the schedule optimization problem, the frequency of the cleanings and the total number of cleanings is reduced, and the length of operating periods between cleanings is increased, relative to the of optimal *Scheduling* case with fixed flow split. In addition, comparing the optimal cleaning *Scheduling* with fixed flow split with that one with optimal *Scheduling and Control*, in the latter the furnace duty peaks during cleanings are smaller, and the energy recovered after the cleanings is higher. This shows that changing the flow rate through the network can improve the operation by taking advantage of the units with a higher capacity to transfer heat.

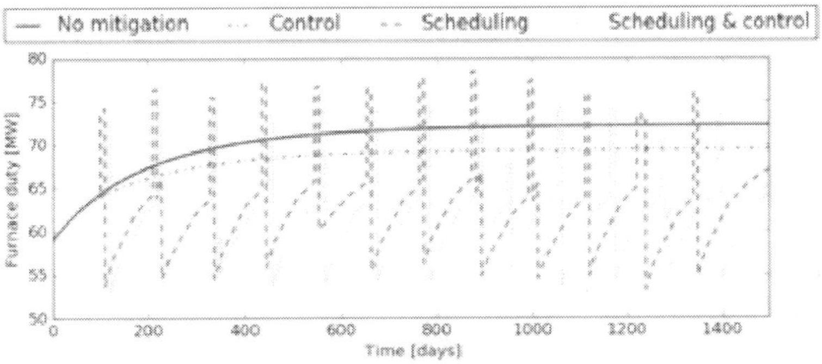

Figure 4. Furnace duty of the case study for different operating scenarios.

Figure 5 shows the profiles of the optimal solution of the integrated *Scheduling and Control*. The optimal cleaning schedule exhibits an almost period behavior in which E01A/B and E04 are cleaned more frequently as they present the highest fouling rate and the highest economic trade-off for energy recovery. On the other side, the optimal flow split (Figure 5b) changes dynamically to increase the energy recovered during and after the cleanings. The trends show that during a cleaning of E04 which is on the upper branch of the network, the greatest fraction of the flow is sent to the opposite branch, and just after this exchanger is back in service the flow through the branches is shifted in the opposite direction. This dynamic behavior allows to best exploit the heat recovery capacity of all the exchangers available at a specific time considering their respective fouling levels Finally, the simultaneous solution of this problem using a continuous time formulation and a MPCC reformulation takes 35 min of CPU time which is acceptably low considering the large size of the problem and the number of binary variables, and that such optimisations may be carried out in practice at no more than hourly frequency.

a) b)

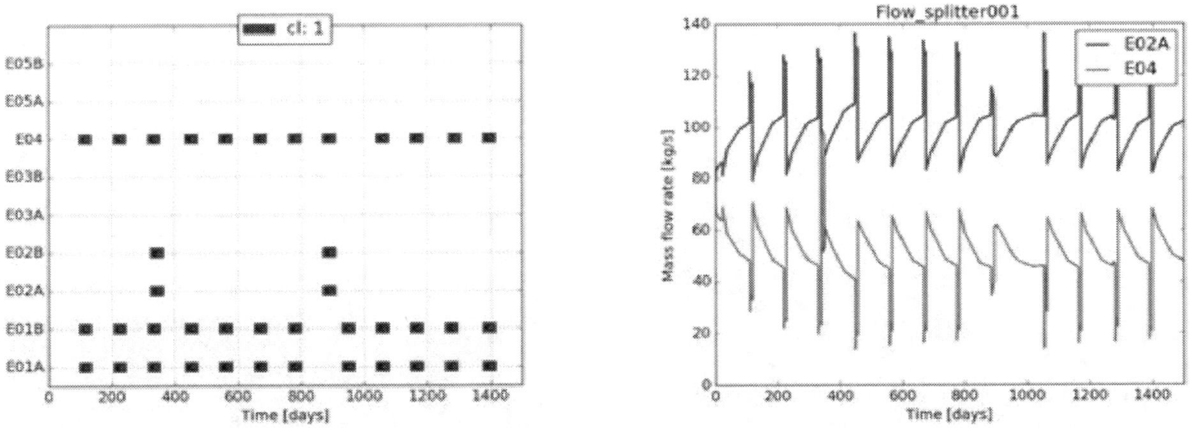

Figure 5. Optimal solution of the simultaneous scheduling and control problem. a) optimal cleaning scheduling, b) optimal flow control for the parallel branches.

Conclusions and perspectives

Defining the cleaning scheduling and the flow distribution of heat exchanger networks are two important fouling mitigation strategies, and they present synergetic effects. The simultaneous solution of these problems is challenging, but it can be achieved efficiently using the formulation and strategy presented here. The continuous time representation allows reducing the number of binary variables in the problem, and the MPCC reformulation allows solving the integrated problem efficiently. These two factors enable the solution of problems of industrial interest in a reasonable computational time (less than one hour). Exploiting the interactions of the control and scheduling mitigation strategies improves the operation of the network and results in significant additional increase in energy recovery and reductions in operating cost.

References

1. Andersson E, Tahric I (2018) Enhanced energy recovery for crude preheating. Pet. Technol. quarterly. Refin. gas Process. petrochemicals
2. Ishiyama EM, Heins A V, Paterson WR, Spinelli L, Wilson DI (2010) Scheduling cleaning in a crude oil preheat train subject to fouling: Incorporating desalter control. Appl Therm Eng 30:1852–1862
3. Deka D, Datta D (2017) Multi-objective optimization of the scheduling of a heat exchanger network under milk fouling. Knowledge-Based Syst 121:71–82
4. Georgiadis MC, Papageorgiou LG, Macchietto S (2000) Optimal Cleaning Policies in Heat Exchanger Networks under Rapid Fouling. Ind Eng Chem Res 39:441–454
5. Biegler LT (2010) Nonlinear Programming. Concepts, Algorithms and Applications to chemical process. SIAM, Society for Industrial and Applied Mathematics, Philadelphi
6. Coletti F, Macchietto S (2011) Refinery Pre-Heat Train Network Simulation Undergoing Fouling: Assessment of Energy Efficiency and Carbon Emissions. Heat Transf Eng 32:228–236
7. Lozano Santamaria F, Macchietto S (2018) INTEGRATION OF OPTIMAL CLEANING SCHEDULING AND CONTROL OF HEAT EXCHANGER NETWORKS UNDERGOING FOULING: MODEL AND FORMULATION. Ind Eng Chem Res. doi: 10.1021/acs.iecr.8b01701
8. Santamaria FL, Macchietto S (2018) Integration of Optimal Cleaning Scheduling and Flow Split Control for Crude Oil Fouling Mitigation in the Operation of Refinery Heat Exchanger Networks. Comput Aided Chem Eng 44:1087–1092

538692 Control of a Heating, Ventilation and Air Conditioning (HVAC) System Using Decentralized Extremum Seeking

Abstract:

HVAC equipment are installed in residential, commercial and industrial buildings to meet the temperature requirements that provide maximum comfort for occupants and to maintain satisfactory indoor air quality. In meeting these requirements, a considerable amount of energy is utilized. It was reported that about 37% of the overall energy consumed in the province of Ontario, Canada in 2014 was from the building sector. Improvements in the design of buildings and the monitoring of human activities and behaviours have been suggested as ways to increase energy savings.

HVAC systems are nonlinear, complex and highly integrated. The malfunctioning of any of its components or units can affect its performance and result in energy wastage. The complexity of an HVAC system makes it very challenging to effectively describe it using a mathematical model. Also, with the technological advancement of HVAC equipment, there is a need to use highly effective and efficient control techniques to minimize energy consumption.

This paper looks at the application of decentralized extremum seeking control to heating, ventilation and air conditioning (HVAC) systems in residential, commercial and industrial buildings. The air conditioning system considered comprises two rooftop units that each provide cool air to two zones. The compressor, fan and expansion valve of each rooftop unit are controlled by three inner loop proportional-integral (PI) controllers to meet specified control requirements. The objective is to determine the optimal supply air temperature set-point for each rooftop unit that minimizes the overall power consumption of the units. In addition, each set-point must satisfy the control objectives of the three inner loop PI controllers. To tackle this problem, a decentralized proportional-integral extremum seeking control technique that avoids the need for communication between the units is employed. A simulation result is included to show the effectiveness of this technique.

526774 Subsystem Decomposition of Process Networks for Simultaneous Distributed State Estimation and Control

Abstract:

A process network comprises interconnected operating units that are coupled through material, energy and information flows. Advanced control strategies are typically used for these process networks to ensure operational safety and profitability and to meet the regulations on environmental sustainability [1, 2]. Distributed control has been recognized as a very promising advanced control framework for such process networks [1, 3]. Within the distributed control framework, a process network is decomposed into subsystems, and then a local controller is developed for each subsystem [4, 5]. The local controllers communicate with each other to coordinate their actions [3]. The distributed control framework inherits the structural flexibility of the decentralized control and provides much improved control performance. Most research efforts on distributed control (especially distributed model predictive control (DMPC)) have been devoted to the algorithm development and the analysis of feasibility, optimality and stability. However, the fundamental problem of subsystem decomposition has received relatively much less attention.

Subsystem decomposition may affect the performance of a distributed predictive control system significantly. The community detection concept originating from network theory provides a very promising way to address the subsystem decomposition problem [4, 6]. By means of the measure of modularity [7], community-based approaches have been proposed to find distributed control structures where the subsystems are made well-decoupled [8, 4]. It is worth mentioning that the existing methods require a full state feedback. However, full state measurements can be difficult to obtain online for many applications. One solution is to incorporate distributed state estimation and distributed control in one integration, such that distributed control can be implemented based on output measurements. From the perspective of implementation, maintenance and communication, it is more favorable if the local state estimators and local controllers are designed based on the same subsystem decomposition. However, a systematic approach to achieve this objective is not yet available.

In this work, we focus on subsystem decomposition of nonlinear process networks for simultaneous distributed state estimation and distributed control. To achieve this goal, we propose a systematic approach based on the concept of community structure detection. We resort to the measure of modularity to quantitatively assess the quality of different community structures. Specifically, the state, manipulated input and measured output variables of a process are taken into account and are viewed as vertices in a network. The way to construct a directed graph containing all the vertices and the corresponding adjacency matrix is defined. An implementation procedure based on approximate optimization of modularity is developed, such that subsystem models for simultaneous distributed state estimation and distributed control can be established by allocating vertices into communities based on modularity. Three chemical process examples of different complexities are used to illustrate the effectiveness and applicability of the proposed approach. While there have been several heuristic approaches on subsystem decomposition for distributed control (e.g., [8, 4]), to the best of our knowledge, the approach proposed in this work is the first one that enables a distributed scheme to require much less information from the process; that is, only output-feedback information is needed for distributed control design and implementation. Also, it is not a trivial extension of the existing approaches. Technical contributions of this work include the following aspects:

- an adjacency matrix which incorporates state, manipulated input, and measured output variables is defined;
- bidirectional edges are considered to characterize the connectivity between state and measured output variables and are used for constructing the direct graph and the adjacency matrix;
- a simple yet efficient method is proposed to calculate the adjacency matrix;
- a method for initializing the community structure is proposed to better handle the constraints on the subsystem structure;
- a systematic procedure is proposed based on the fast folding algorithm [9] to decompose the entire process network into subsystems of which the number is user-specified.

References

[1] P. D. Christofides, R. Scattolini, D. Munoz de la Pena, and J. Liu. Distributed model predictive control: A tutorial review and future research directions. *Computers and Chemical Engineering*, 51:21-41, 2013.

[2] R. Scattolini. Architectures for distributed and hierarchical model predictive control - A review. *Journal of Process Control*, 19:723-731, 2009.

[3] P. Daoutidis, M. Zachar, and S. S. Jogwar. Sustainability and process control: A survey and perspective. *Journal of Process Control*, 44:184-206, 2016.

[4] W. Tang and P. Daoutidis. Network decomposition for distributed control through community detection in input-output bipartite graphs. *Journal of Process Control*, 64:7-14, 2018.

[5] P. Daoutidis, W. Tang, and S. S. Jogwar. Decomposing complex plants for distributed control: Perspectives from network theory. *Computers and Chemical Engineering*, 2017. in press, doi:10.1016/j.compchemeng.2017.10.015.

[6] D. B. Pourkargar, A. Almansoori, and P. Daoutidis. Impact of decomposition on distributed model predictive control: A process network case study. *Industrial and Engineering Chemistry Research*, 56(34):9606-9616, 2017.

[7] M. E. J. Newman and M. Girvan. Finding and evaluating community structure in networks. *Physical Review E*, 69(2):026113, 2004.

[8] S. S. Jogwar and P. Daoutidis. Community-based synthesis of distributed control architectures for integrated process networks. *Chemical Engineering Science*, 172:434-443, 2017.

[9] V. D. Blondel, J.-L. Guillaume, R. Lambiotte, and E. Lefebvre. Fast unfolding of communities in large networks. *Journal of Statistical Mechanics: Theory and Experiment*, 2008(10):P10008, 2008.

527683 Distributed Model Predictive Control Based on NLP Sensitivity

Abstract:

Distributed Model Predictive Control Based on NLP Sensitivity

Tianyu Yu [a], Jun Zhao [a], Zuhua Xu [a], Xi Chen [a], Lorenz T. Biegler [b]

[a] State Key Laboratory of Industrial Control Technology, College of Control Science and Engineering, Zhejiang University, Hangzhou, 310027, Zhejiang, P. R. China

[b] Department of Chemical Engineering, Carnegie Mellon University, 5000 Forbes Avenue, Pittsburgh, PA 15213, USA

Abstract

Model predictive control (MPC) has been widely used in the control of industrial processes over the past thirty years. It can handle constraints effectively and account for system interactions explicitly. Centralized and decentralized MPC are two common strategies in large-scale systems. In centralized scheme, a single controller is designed to calculate all the manipulated variables, which may encounter heavy computational burden and reliability issues for a large scale system. In decentralized scheme, the plant is divided into several subsystems and each one is controlled by a MPC controller without communication with the other controllers. The decentralized scheme may lead to poor performance due to the ignorance of subsystem interactions. Distributed MPC (DMPC) is developed as a compromise in which several controllers are used to control the whole process and the communication among controllers is applied to improve control performance. DMPC can be classified into two types, the cooperative DMPC and the non-cooperative DMPC. Their main difference is the objective function used in the optimization problem. The cooperative DMPC uses a global cost function while the other uses several local cost functions[1,2]. For implementation of DMPC, one key difficulty lies in the online solution of the optimization, especially for strongly nonlinear systems.

To speed up the real-time computation, a novel cooperative DMPC algorithm is proposed with NLP sensitivity in this paper. A distributed optimization structure is built with all the subsystemsâ€™ optimal inputs being evaluated in parallel. For a single MPC controller, at each sampling time, the future state at the next step can be calculated using its nominal model with its current state and control action. Therefore, the predicted optimization problem can be solved in advance. If this problem can be solved within one step, the predicted control action is available at the next step. This strategy can remove the computational delay and preserve the stability properties of the controller. After that, a fast approximate solution at the next step can be computed by taking a perturbed Newton step when a new state is obtained[3]. This concept can be extended to distributed scheme. In this paper, a cooperative DMPC strategy is used based on [4], in which many problems are solved in parallel. In background, optimal solutions of different subsystems

can be obtained in each iteration. After that, these solutions are transmitted to each other, and the iteration continues until stopping criteria is met. When new states are available, the approximate optimal solutions of subsystems can be calculated using the same updating method. In this way, fast feedback is obtained by analyzing the parametric property of the problems. A case study is presented to demonstrate how the algorithm works efficiently.

Keywords: Distributed Control, Predictive Control, NLP Sensitivity

References:

[1] P. D. Christofides, R. Scattolini, D. M. de la Peña, J. Liu, Distributed model predictive control: A tutorial review and future research directions, Computer and Chemical Engineering 51 (2013) 21-41.

[2] M. J. Tippett, J. Bao, Distributed Control of Chemical Process Networks, International Journal of Automation and Computing 12(4) (2015) 368-381.

[3] V. M. Zavala, L. T. Biegler, The advanced-step NMPC controller: Optimality, stability and robustness, Automatica 45 (2009) 86-93.

[4] B. T. Stewart, A. N. Venkat, J. B. Rawlings, S. J. Wright, G. Pannocchia, Cooperative distributed model predictive control, Systems & Control Letters 59(2010) 460-469.

532197 Event-Triggered Model-Based Control and Identification of Networked Process Systems

Abstract:

The problem of designing distributed and supervisory control systems for process networks has been the subject of significant research work in process control. In addition to handling the dynamic coupling and interactions between the constituent subsystems, the management and optimization of information flow and communication between the plant subsystems are becoming increasingly important considerations in the control problem formulation, especially in view of the growing reliance on networked sensor and control systems in plant operations and the increased emphasis placed on smart plant operations [1].

In a previous work [2], the integration of control and communication was addressed in the context of plant-wide control, where a quasi-decentralized model-based networked control structure that enforces closed-loop stability with minimal cross communication was developed. Within each local control system, a set of predictive models were used, in conjunction with the local state measurements, to generate the local control action at times when communication between the plant subsystems was suspended, and the states of the models were updated when communication was permitted at discrete times. In doing so, a minimum communication rate that guarantees closed-loop stability could be determined and it was shown to depend strongly on the size of the plant-model mismatch, with a larger mismatch requiring a higher rate of information transfer between the plant subsystems.

Due to the fixed periodic nature of the communication strategy that this approach utilizes, the local controllers may be irresponsive to time-varying external disturbances, or they may establish communication at unnecessary times. Moreover, this approach may not be the optimal strategy when the plant requires all sensors to access the communication medium simultaneously, which may lead to practical issues such as data losses and transmission delays. One way to overcome these limitations is to use event-triggered communication strategies (e.g., see [3], [4],[5]) in which the local control systems transmit their state measurements to their neighbors only when an event triggers the need for such communication. These events are typically defined based on the desired closed-loop stability and performance characteristics. This approach minimizes network utilization and enables the controllers to be more responsive to time-varying external disturbances. This approach can also reduce the energy expenditures associated with wireless networks.

While a model-based control strategy generally helps reduce the control system's reliance on the communication medium, and therefore reduces its vulnerability to sensor failures and communication outages, the use of a single model with fixed parameters may limit the achievable savings in network resource utilization, especially when plant operating conditions experience variations and drift over time. Such variations (e.g., due to catalyst deactivation in reactors and fouling in heat exchangers) are unavoidable and invariably lead to process parameters variations which exacerbate the mismatch between the existing model and the actual plant, ultimately leading to increased communication levels to maintain closed-loop stability. One way to address this problem is by constantly updating the model parameters using model identification techniques (e.g., [6],[7]) so as to try to keep plant-model mismatch to a minimum.

In this work, we present a methodological framework for the integration of model-based control and model identification in the context of networked process systems subject to external perturbations, process parameter variations and communication constraints. The framework aims to maintain closed-loop stability in the presence of plant-model mismatch while simultaneously minimizing network utilization. Initially, a networked model-based controller with an event-triggered communication logic is designed and implemented. During operation, model state updates are triggered when a certain state-dependent stability threshold on the model estimation error is breached. The threshold is obtained using Lyapunov techniques and is explicitly characterized in terms of the model uncertainty and the controller design parameters. Meanwhile, a moving-horizon error detection scheme is used to continuously track the overall network utilization and stability. When a significant plant-model mismatch is detected, based on a predefined threshold, model re-identification is carried out based on the available state and input data, and the model parameters are updated with the consideration of system stability together with network utilization. A new closed-loop stability threshold is then obtained based on the newly identified model to trigger future communications between the plant subsystems. A chemical plant example is used to illustrate the implementation of the developed methodology.

References:

[1] Christofides PD, Davis JF, El-Farra NH, Clark D, Harris KRD, Gipson JN. Smart plant operations: Vision, progress and challenges. AIChE Journal. 2007;53(11):2734–2741.

[2] Sun Y, El-Farra NH. Quasi-decentralized model-based networked control of process systems. Computers & Chemical Engineering. 2008;32(9):2016–2029.

[3] Sun Y, El-Farra NH. Quasi-decentralized networked process control using an adaptive communication policy. In Proceedings of American Control Conference, 2010;2841-2846.

[4] Garcia E, Antsaklis PJ. Model-based event-triggered control for systems with quantization and time-varying network delays. Automatic Control, IEEE Transactions on. 2013;58(2):422–434.

[5] Xue D, El-Farra NH. Networked Event-Triggered Control of Spatially Distributed Processes using a Dual-Mode Communication Strategy. In: Proceedings of 54th IEEE Conference on Decision and Control, 2015; 1899-1904.

[6] Favoreel W, Moor BD, Overschee PV. Subspace state space system identification for industrial processes. Journal of Process Control. 2000;10:149–155.

[7] Ljung L. System Identification. Prentice Hall. 1999.

535155 Coordination of Distributed MPC Systems with Closed-Loop Prediction Approximation in Dynamic Real-Time Optimization (DRTO)

Abstract:

Industrial process control systems typically follow a hierarchical structure. Within this structure, dynamic real-time optimization (DRTO) can be used a supervisory layer to compute set-point trajectories for an underlying regulatory control system based on economic optimization and a dynamic process model. It addresses the shortcomings of traditional steady-state RTO systems since it does not require the plant to be at steady state prior to execution. Tosukhowong et al. presented a DRTO formulation that handles slow dynamics of a system based on a linear dynamic model [1], while a DRTO formulation proposed by Kadam and Marquardt uses a sensitivity technique and estimation of uncertain parameters [2]. Open-loop models of the process are generally adopted by the DRTO formulation discussed above without considering the dynamic effects of the underlying regulatory controller. However, DRTO formulations based on the closed-loop prediction of the plant under constrained MPC have recently been developed [3, 4].

Using a single, monolithic MPC system is imprudent for large-scale industrial process systems, which instead favor a distributed MPC architecture that takes the modularity of a large process into consideration and provide additional flexibility [5]. Thorough reviews on the architecture and formulations of distributed MPCs have been conducted in [6, 7, 8]. The coordination scheme proposed by Camponogara et al. allows communication among MPC subsystems either once or multiple times before implementation of the control actions [9]. The distributed Lyapunov-based MPC schemes developed by Liu et al. [10] adopt a sequential one-directional and iterative bi-directional approach respectively, assuming state feedback for all controllers. The cooperative scheme proposed by Stewart et al. [11] for distributed MPCs ensures the closed-loop stability of the system upon termination of iterations for all subsystems.

In previous work, a closed-loop DRTO formulation was developed for the coordination of distributed MPC systems by assigning computed set-point trajectories based on the prediction of future interactions between the plant dynamics and distributed MPCs [12]. This closed-loop formulation naturally results in a multilevel optimization problem whose primary optimization problem computes the reference trajectories for each MPC subsystem based on a full plant dynamic model under an economic objective, and whose MPC optimization sub-problems compute optimal control actions to be applied to the full plant dynamic model based on partial models each of which contains model equations relevant to its associated subsystem. Such a multilevel optimization problem can be transformed into a single-level optimization problem by applying the Karush-Kuhn-Tucker (KKT) optimality conditions to transform the MPC optimization sub-problems into sets of algebraic constraints embedded in the primary optimization problem, resulting in a single mathematical program with complementarity constraints (MPCC). No direct exchange of information is needed between local controllers, and communication is achieved through set-point trajectories generated by the DRTO formulation for the lower level MPCs.

This work extends the use of the DRTO formulation for dynamic coordination of distributed MPC systems [12], but addresses the issue of computational complexity for systems with a long MPC control horizon or DRTO prediction horizon which can lead to large DRTO optimization problems, which is further exacerbated with increasing plant scale and complexity. Two techniques are applied for approximating the predicted closed-loop response of the plant, developed originally for closed-loop prediction under centralized MPC [3]. Relevant linear and nonlinear case studies are conducted to demonstrate the capability of target tracking and economic optimization by the DRTO formulation. Results show that the closed-loop DRTO formulation is able to coordinate distributed MPCs to achieve an optimal economic transition or the desired target value rapidly in a computationally efficient manner and without significant performance loss when compared to the counterpart with rigorous closed-loop prediction.

References

[1] T. Tosukhowong, J. Lee, J. Lee and J. Lu, "An introduction to a dynamic plant-wide optimization strategy for an integrated plant," *Comp. Chem. Eng.* , vol. 29, no. 1, pp. 199-208, 2004.

[2] J. Kadam and W. Marquardt, "Sensitivity-based solution updates in closed-loop dynamic optimization," in *Proceedings of the 7th International Conference on the Dynamics and Control of Process Systems - DYCOPS 7, Boston, Massachusetts, USA*, Boston, 2004.

[3] M. Z. Jamaludin and C. L. E. Swartz, "Approximation of Closed-loop Prediction for Dynamic Real-time Optimization Calculations," *Computers and Chemical Engineering,* vol. 103, pp. 23-38, 2017.

[4] M. Z. Jamaludin and C. L. E. Swartz, "Dynamic Real-time Optimization with Closed-loop Prediction," *AIChE Journal,* Accepted Author Manuscript. doi:10.1002/aic.15752, 2017.

[5] G. Pannocchia, "Distributed Model Predictive Control"," in *Encyclopedia of Systems and Control*, London, Springer London, 2013, pp. 1-9.

[6] R. Scattolini, "Architectures for distributed and hierarchical Model Predictive Control â€" A review," *J. Process Control,* vol. 19, no. 5, pp. 723-731, 2009.

[7] P. Christofides, R. Scattolini, D. MuÃ±oz de la PeÃ±a and J. Liu, "Distributed model predictive control: A tutorial review and future research directions," *Comp. Chem. Eng.,* vol. 51, pp. 21-41, 2013.

[8] J. Rawlings and B. Stewart, "Coordinating multiple optimization-based controllers: New opportunities and challenges," *J. Process Control,* vol. 18, no. 9, pp. 839-845, 2008.

[9] E. Camponogara, D. Jia, B. Krogh and S. Talukdar, "Distributed model predictive control," *IEEE Control Systems,* vol. 22, no. 1, pp. 44-52, 2002.

[10] J. Liu, X. Chen, D. Muñoz de la Peña and P. Christofides, "Sequential and iterative architectures for distributed model predictive control of nonlinear process systems," *AIChE J.,* vol. 56, no. 8, pp. 2137-2149, 2010.

B. Stewart, A. Venkat, J. Rawlings, S. J. Wright and G. Pannocchia, "Cooperative distributed model predictive

[11] control," *Syst. Control Lett.*, vol. 59, no. 8, pp. 460-469, 2010.

[12] Li, H., Swartz, C.L., 2017. Coordination of distributed mpc systems through dynamic real-time optimization with closed-loop prediction, in: Espua, A., Graells, M., Puigjaner, L. 42 (Eds.), 27th European Symposium on Computer Aided Process Engineering. Elsevier. Volume 40 of Computer Aided Chemical Engineering, pp. 1603 â€" 1608.

537915 Data-Based Sequential Design of Decentralized PID Controllers

Abstract:

The contemporary process industries need to adhere to stringent product specifications and environmental regulations, thus making high control performance more important than ever. With incessant technological advancements, the processes have become more complex and hence difficult to model, which is essential to the conventional model-based design methods. On the other hand, the availability of a plethora of process data has recently led to the development of data-based controller design methods, wherein controller design is carried out without resorting to a process model. One such method is the Virtual-Reference-Feedback-Tuning (VRFT) method, which uses a desired reference model to formulate an optimization problem based on process input-output data (Campi et al., 2002). The key idea is to achieve a pre-specified closed-loop response via the design of a desired reference model, which is incorporated in the aforementioned optimization problem, and its optimal solution determines the controller's parameters.

Though the VRFT design method has been studied extensively, the existing design methods have been predominantly developed for single-input and single-output (SISO) control systems. However, most industrial chemical processes deal with multiple controlled and manipulated variables referred to as multi-input and multi-output (MIMO) system. In this work, decentralized PID controllers are considered as they are commonly used in process industries. For a decentralized control system, process interaction is inherent and causes difficulties for feedback controller design. As a result, the achievable control performance of a decentralized control system is inevitably degraded or even becomes unstable.

Despite few results being available for decentralized controller design under the VRFT design framework, there have been some notable attempts on the subject nonetheless (Nakamoto, 2004; Campestrini et al., 2016; Da Silva et al., 2016). However, these methods are independent design methods, which can design decentralized controllers in a straightforward manner at the cost of being a conservative design approach. In this work, a design alternative termed sequential design method is adopted as it innately takes into account the process interactions and can utilize the information about the controller designed in the previous step. This gives sequential design the advantage of being less conservative than independent design methods.

Owing to the aforementioned reasons, a sequential design procedure in the z-domain has been employed in this work under the VRFT design framework. The rudimentary mechanism is to treat multiple control loops as a sequence of single loops. For instance, a 2x2-control system can be tuned via the following steps:

- Tune one of the loops based on open loop data
- Using the controller parameters obtained in step 1, close the first loop and tune the second loop with the second loop open
- Using the controller parameters obtained in step 2, close the second loop and tune the first loop with the first loop open
- Repeat 2 and 3 till the controller parameters for both loops converge to a fixed value

Additionally, instead of the typically used first-order or critically damped reference model, a second-order reference model has been used in this work. In order to evaluate the performance of the proposed method, simulation results of controlling Wood and Berry distillation column (Vu and Lee, 2010) were used in a comparative study with benchmark methods including model-based Dahlinâ€™s controller and existing VRFT reference models. The results show that the proposed method provides better or comparable performance to that of the benchmark decentralized controllers.

References

Campestrini, L., D. Eckhard, L. A. Chía and E. Boeira (2016). Unbiased MIMO VRFT with application to process control. Journal of Process Control 39(Supplement C): 35-49.

Campi, M. C., A. Lecchini and S. M. Savaresi (2002). Virtual reference feedback tuning: a direct method for the design of feedback controllers. Automatica 38(8): 1337-1346.

Da Silva, G. R. G., L. Campestrini and A. S. Bazanella (2016). Multivariable VRFT: an approach for systems with non-minimum phase transmission zeros. Control Applications (CCA), 2016 IEEE Conference on, IEEE. pp 1324-1329.

Nakamoto, M. (2004). An application of the virtual reference feedback tuning for an MIMO process. SICE 2004 Annual Conference. pp 2208-2213 vol. 2203.

Vu, T. N. L. and M. Lee (2010). Independent design of multi-loop PI/PID controllers for interacting multivariable processes. Journal of Process Control 20(8): 922-933.

528077 Distributed Estimation and Nonlinear Model Predictive Control of a Benzene Chlorination Process

Abstract:

Model predictive control (MPC) is a prevalent control strategy inherently tailored to accommodate optimal performance criteria and handle constraints, which are two major concerns of the control of processes in the chemical industry [1]. The philosophy underlying MPC design is to cast the control problem in the form of a repeated constrained dynamic optimization problem that computes a sequence of future manipulated inputs at each sampling time [2]. Consequently, the applicability of MPC hinges on the real-time solvability of the underlying optimization problem. This is a challenge when applying a centralized control architecture to industrial scale process systems that require solving a large-scale constrained nonlinear dynamic optimization in real-time [3]. An alternative approach to overcome the computational challenges of centralized model predictive control (CMPC) is to decompose the original optimization-based control problem into a number of smaller control problems. This idea results in a distributed model predictive control (DMPC) architecture comprised of local controllers with some level of cooperation and communication [4], [5].

The decomposition of the system, i.e. deciding on the number of subsystems and identifying how the manipulated inputs and output variables distribute among the network of subsystems is a key factor for DMPC implementation. From a network theory perspective, this problem can be posed as identifying weakly-connected subsystems whereby the variables of each subsystem are strongly connected. The impact of system decomposition on the closed-loop performance and the computational requirements of employing DMPC has been investigated by considering different levels of communication and cooperation between local controllers, levels of system uncertainty, dynamic optimization platforms, and operating points for the benchmark reactor-separator process network [6], [7]. The identified optimal distributed architecture by the community detection method has been then employed to address the DMPC problem for a class of real chemical process networks [8]. Such studies focused on the distributed control problem assuming full state information at each sampling time in order to isolate and investigate the effects of decomposition on the controller design problem itself. However, such an assumption cannot be generally invoked in many practical cases, so the use of a state estimation method is required.

Among the several available methods for state estimation, moving horizon estimation (MHE) has attracted a lot of attention since it can be formulated as a similar constrained dynamic optimization problem [9], [10]. The MHE provides an optimal state estimation compatible with the system measurements at the past sampling times [11]. However, solving such an optimization-based estimation problem at each sampling time remains computationally challenging when applying a centralized architecture to large-scale process networks [12]. Any delay in such computations will directly affect the closed-loop performance since the distributed controllers require the initial values of all state variables at each sampling time. Like DMPC, an alternative approach to overcome the computational challenges of centralized estimation is the system decomposition and a distributed moving horizon estimation (DMHE) structure consisting of local estimators with some level of cooperation and communication. Extending the optimal decomposition of process networks to the combined control and estimation problem is the subject of this work.

Specifically, we provide a framework to combine iterative DMPC and DMHE to address the output-feedback control problem for a benzene chlorination plant. Such a process consists of three tubular reactors and two flash separators. A series of two elementary reactions take place in the reactors to produce mono and dichlorobenzene where monochlorobenzene is the desired product [13]. The first flash column separates the unreacted benzene in the product stream and recycles it to the fresh feed stream to be reused in the reactors. For each tubular reactor, we have five nonlinear parabolic partial differential equations (PPDEs) to describe the spatiotemporal dynamics of temperature and the concentrations of benzene, chlorine, mono, and dichlorobenzene. Galerkinâ€™s method is employed to derive low-dimensional reduced order models in the form of ordinary differential equations (ODEs) which can approximate the dominant dynamics of the governing PPDEs [14]. The resulting ODEs are added to the set of algebraic equations and ODEs that describe the temperature and concentrations in the flash separators and are used as the basis for the combined DMPC and DMHE design.

The maximum temperature in the tubular reactors and the monochlorobenzene concentration at the outlet streams of the reactors and separators are considered as the controlled outputs. We can only measure the temperature at the outlet streams of the reactors and inside the flash separators and the total concentration of chlorobenzene (monochlorobenzene + dichlorobenzene) at the outlet streams of the reactors and flash separators. By considering these measured outputs, we require estimating the temperature and concentrations of all components in the reactors and separators at each sampling time by DMHE. Then the estimated state variables are used by DMPC as the initial values of the predictive model. Since the sets of controlled and measured outputs are different in this case study, we obtain different optimal decompositions for the control and estimation problems by applying the community detection method. Both optimal decompositions result in two subsystems, recommending a distributed estimation/control structure of two local estimators/controllers, which can communicate over the network. The closed-loop performance and the average computation time are evaluated using the detailed simulations. We additionally compare the results with those of centralized and fully decentralized control synthesis. The proposed distributed output feedback control design based on the optimal decomposition which minimizes the interactions between the distributed local estimators and controllers, is shown to enable closed-loop performance close to that of the centralized, while reducing the computational effort significantly.

[1] J. Rawlings, D. Mayne and M. Diehl, Model predictive control: Theory, computation, and design, Madison: Nob Hill Publishing, 2017.

[2] D. Mayne, "Model predictive control: Recent developments and future promise," *Automatica,* vol. 50(12), 2967-2986, 2014.

[3] L. Biegler, Nonlinear Programming: Concepts, Algorithms and Applications to Chemical Processes, Philadelphia: SIAM, 2010.

[4] P.D. Christofides, R. Scattolini, D. Munoz de la Pena and J. Liu, "Distributed model predictive control: A tutorial review and future research directions," *Computers and Chemical Engineering,* 51, 21-41, 2013.

[5] R. Scattolini, "Architectures for distributed and hierarchical model predictive control: A review," *Journal of Process Control,* 19, 723-731, 2009.

[6] D.B. Pourkargar, A. Almansoori and P. Daoutidis, "The impact of decomposition on distributed," *Industrial and Engineering Chemistry,* 56(34), 9606-9616, 2017.

[7] D.B. Pourkargar, A. Almansoori and P. Daoutidis, "Comprehensive study of decomposition effects on distributed output tracking of an integrated process over a wide operating range," *Chemical Engineering Research and Design,* under review, 2018.

[8] M. Moharir, D. Pourkargar, A. Almansoori and P. Daoutidis, "Distributed model predictive control of an amine gas sweetening plant," *Industrial and Engineering Chemistry Research,* under review, 2018.

[9] J. Rawlings and B. Bakshi, "Particle filtering and moving horizon estimation," *Computers and Chemical Engineering,* 30(10-12), 1529-1541, 2006.

[10] D. Copp and J. Hespanha, "Simultaneous nonlinear model predictive control and state estimation," *Automatica,* 77, 143-154, 2017.

[11] C. Rao, J. Rawlings and D. Mayne, "Constrained state estimation for nonlinear discrete-time systems: Stability and moving horizon approximations," *IEEE Transactions on Automatic Control,* 48(2), 246-258, 2003.

[12] X. Yin and J. Liu, "Distributed moving horizon state estimation of two-time-scale nonlinear systems," *Automatica,* 79, 152-161, 2017.

[13] A. Kokossis and C. Floudas, "Synthesis of isothermal reactor-separator-recycle systems," *Chemical Engineering Science,* 46(5/6), 1361-1383, 1991.

[14] P.D. Christofides, Nonlinear and robust control of PDE systems: Methods and applications to transport-reaction processes, Boston: 2001.

531795 Pynumero: Python Numerical Optimization

Abstract:

Numerical optimization has proven to be an efficient tool for characterizing, designing, and operating chemical processes. Efficient optimization algorithms are typically complex code bases written in low-level compiled programming languages. However, this complexity also impedes development of new algorithms, requiring significant programming and software engineering expertise and a steep learning curve. This is especially true when developing decomposition approaches to solve large-scale structured problems in parallel.

To address and mitigate these challenges, we present PyNumero, a Python package for numerical optimization that provides a high-level programming framework for rapid development of nonlinear optimization algorithms. The package gives access to all high-level features of the Python programming language without making large sacrifices on computational performance. It combines the capabilities of the modeling language Pyomo with efficient libraries like the AMPL Solver Library (ASL), the Harwell Subroutine Library (HSL), the Message Passing Interface (MPI) and NumPy/SciPy. This combination makes PyNumero an excellent tool to develop numerical optimization algorithms that are interfaced with AMPL and Pyomo. Furthermore, PyNumero performs all linear algebra operations in compiled code, and is designed to avoid marshalling of data between the C and Python environments, allowing for high-level development of algorithms without a significant sacrifice in performance. We show that an implementation of the sequential quadratic programing algorithm in PyNumero has comparable solution times with the state of the art solver IPOPT when solving a dynamic optimization problem with 100K variables and constraints. The overhead from the python interface to ASL and HSL only increases the solution time by 10%.

In this presentation we describe the software design of PyNumero and demonstrate its capabilities through a series examples from different mathematical programming applications.

521795 Graph-Based Modeling Abstractions and Computational Tools for Complex Systems

Abstract:

This talk presents advances in graph-based modeling abstractions and computational tools for complex systems [1]. Under the proposed graph-based abstraction, a complex system is represented as a collection of algebraic optimization models that live on nodes and that are connected through edges representing linking constraints between algebraic models. The graph abstraction can be exploited to create nested/hierarchical graphs in which a node can be a graph model itself. These features can be exploited to model coupled heterogeneous and multi-scale networks (e.g., supply chains, biological networks, and infrastructure networks). The graph abstraction also enables the modeler to perform partitioning, aggregation, and community detection tasks, which are commonly used in the design of decomposition strategies [2]. The graph abstraction also facilitates model processing (e.g., computation of derivatives) and analysis (e.g., visualization and data management) as well as the development of interfaces with existing standard and structured solvers [3,4].

We also show that a graph-based abstraction can be naturally be extended to represent computational workflows. A computational workflow is a virtual graph in which a collection of computational tasks live in nodes and edges represent communication links between tasks. This abstraction generalizes other modeling paradigms such as discrete-event and agent-based simulation, naturally accommodates computational algorithms, and enables the simulation of synchronous and asynchronous computing environments [5]. A computational workflow can be used to simulate the performance of algorithms such as Benders decomposition, decentralized control architectures, markets, and swarm robotic systems under communication and decision-making delays and failures [6,7,8,9]. We discuss an implementation of a graph-based modeling platform in Julia, that we call Plasmo.jl.

[1] Jalving, J., Abhyankar, S., Kim, K.,Hereld, M., Zavala, V. M. A Graph-Based Computational Framework for Simulation and Optimization of Coupled Infrastructure Networks. To Appear in IET Generation, Transmission & Distribution, 2016.

[2] Tang, W., Allman, A., Pourkargar, D. B., & Daoutidis, P. (2017). Optimal decomposition for distributed optimization in nonlinear model predictive control through community detection. Computers & Chemical Engineering, 111, 43–54

[3] N. Chiang, C. G. Petra, and V. M. Zavala. Structured nonconvex optimization of large-scale energy systems using PIPS-NLP. In Proc. of the 18th Power Systems Computation Conference (PSCC), Wroclaw, Poland, 2014.

[4] K. Kim and V. M. Zavala. Algorithmic innovations and software for the dual decomposition method applied to stochastic mixed-integer programs. Optimization Online, 2015

[5] Cassandras, C. G., & Lafortune, S. Introduction to Discrete Event Systems, (2008).

[6] Ruszczyński, Andrzej P. Nonlinear optimization. Vol. 13. Princeton university press, 2006.

[7] J. B.Rawlings and D. Mayne. Model predictive control: theory and design. Nob Hill Publishing, 2009.

[8] Lopes, Y. K., Trenkwalder, S. M., Leal, A. B., Dodd, T. J., & Groß, R. (2016). Supervisory control theory applied to swarm robotics. Swarm Intelligence, 10(1), 65–97.

[9] Scott, W. Naomi L. (2017). Optimal evasive strategies for groups of interacting agents with motion constraints, (January), 1–12.

528251 Plasmoalgorithms, a Collection of Decomposition Algorithms for Graph-Based Problem Representations

Abstract:

Mathematical programming has been successfully applied to solve decision-making problems in a variety of applications and industries, such as petroleum, chemical, pharmaceutical, and consumer products[1]. The success of the first experiences in these applications has continuously motivated the need for solving larger, and more complex problems. Furthermore, the recent focus on optimization under uncertainty has increased the problem sizes by at least one order of magnitude. At the same time solvers have improved significantly over the years[2], while computing hardware has not increased in speed but in number of parallel cores available. To keep improving and take advantage of this scenario the advancement of decomposition algorithms is necessary.

Currently, implementing and testing decomposition algorithms is a common and time consuming task in Process Systems Engineering (PSE) research. Lagrangean decomposition[3], Benders decomposition[4], and derived algorithms have been the choice for solving large-scale problems. The implementations of these algorithms are usually tailored to the application, there is an absence of general purpose implementations. A few attempts in this direction include the Benders decomposition implementation in CPLEX since version 12.7[5], and DSP[6], an implementation of dual decomposition for solving stochastic programming problems with integer recourse. The first one is a commercial code, while the second one is a complex code written in C++, making them hard to extend and improve.

We present PlasmoAlgorithms, a collection of general decomposition algorithms implemented in Julia programming language, using Plasmo[7] graphs as an input. Plasmo is a Julia package that allows the generation of graphs of models, where each node represents an independent model, and each edge represents a linking constraint. Since the Julia programming language is easy to code, and efficient, PlasmoAlgorithms is a good candidate to become a meeting place for the PSE and Operations Research communities to implement state-of-the-art algorithms, and for testing of novel ideas.

PlasmoAlgorithms currently includes implementations of multilevel Benders decomposition, Lagrangean decomposition, and Cross decomposition. The Benders implementation uses a multicuts, in which an independent cut is generated by each subproblem; and also includes cuts suitable for mixed-integer subproblems. The Lagrangean decomposition implementation includes several options for updating the multipliers, including a probing subgradient method, in which the bound of the candidate step is first probed to decide the length of the step taken. The Cross decomposition implementation follows the algorithm proposed by Mitra et al[8]. All the code is open source, and the implementations are modular enough to allow for easy extension of the algorithms.

References

[1] Grossmann, I.E., 2012. Advances in mathematical programming models for enterprise-wide optimization. *Computers & Chemical Engineering*, *47*, pp.2-18.

[2] Bonami, P., KilinÃ§, M. and Linderoth, J., 2012. Algorithms and software for convex mixed integer nonlinear programs. In *Mixed integer nonlinear programming* (pp. 1-39). Springer, New York, NY.

[3] You, F. and Grossmann, I.E., 2013. Multicut Benders decomposition algorithm for process supply chain planning under uncertainty. *Annals of Operations Research*, *210*(1), pp.191-211.

[4] Guignard, M. and Kim, S., 1987. Lagrangean decomposition: A model yielding stronger Lagrangean bounds. *Mathematical programming*, *39*(2), pp.215-228.

[5] CPLEX, IBM ILOG, 12.7, Userâ€™s Manual for CPLEX, 2016.

[6] Kim, K. and Zavala, V.M., 2017. Algorithmic innovations and software for the dual decomposition method applied to stochastic mixed-integer programs. *Mathematical Programming Computation*, pp.1-42.

[7] Jalving, J., Abhyankar, S., Kim, K., Hereld, M. and Zavala, V.M., 2017. A graph-based computational framework for simulation and optimisation of coupled infrastructure networks. *IET Generation, Transmission & Distribution*, *11*(12), pp.3163-3176.

[8] Mitra, S., Garcia-Herreros, P. and Grossmann, I.E., 2014. A novel cross-decomposition multi-cut scheme for two-stage stochastic programming. In *Computer Aided Chemical Engineering* (Vol. 33, pp. 241-246). Elsevier.

538215 Component Based Development of Application Specific Computer-Aided Tools

Abstract:

Chemical engineering is in constant evolution and access to computational resources are changing the way chemical engineering problems are addressed. Process simulation, which emerged in the 1960s, has become one of the great success stories in the use of computing in the chemical industry. For instance, steady-state simulation has largely reduced experimentation and pilot plant testing in process development for commodity chemicals. Parallel with the development of new process-product development technologies, associated computer aided tools are needed for faster dissemination. However, the current unit-operation based process simulators, although having great success and wide applications, are not able to satisfy the demands for new technologies (multiscale, multipurpose, new concepts leading to new algorithms, visualization, etc.). Therefore, the demand for a new class of computer-aided software has increased, especially for new areas of chemical engineering in which there are no standard software tools. For example, computer tools are developed to target and accelerate the half life of product innovation (time to market) due to competitive global economy and increasing consumer demands.

This work focuses on a generic component-based architecture for development of tailor made computer-aided tools for specific application objectives. The architecture is based on â€œcomponentsâ€, which can be model-objects that are stored in a library (developed through tools like MoT [1]); collection of property models; databases; numerical solvers; design-algorithms (work-flows) etc. The main idea of these components is to incorporate a plug and play framework, where for a specific purpose, after the work-flow and associated data-flow are established, a specific computer-aided tool can be configured without too much additional programming work. That is, after implementing through an interface, the work-flow and data-flow associated with the new tool, retrieve and connect the necessary components for each step of the work-flow. This architecture has been used to develop various computer-aided tools like ProCAPD (computer aided product design [2]), ProCAFD (computer aided flowsheet design [3]), ProCACD (computer aided controller design), ProCAPE (computer aided property estimation), ProCAMT (computer aided modelling toolbox).

The developed product simulator (ProCAPD) is analogous to a process simulator, that is, it helps make the product design and development easier and faster and provides for a unified and consistent product documentation. However, unlike process simulators, ProCAPD can also be used directly for design of chemicals-based products, such as, single molecule products, formulations, blends, emulsions and devices. To design this chemical product simulator (ProCAPD), the components that must be selected are databases (chemicals, solvents, active ingredients, aroma, color-agents etc); model objects (properties, product performance, etc.); solvers (product attributes, blend compositions, environmental impact, etc.); design algorithms (single molecules, blends, formulations, emulsions, devices). Similarly based on this generic component architecture a computer-aided flowsheet design tool (ProCAFD), capable of enumerating the entire feasible search space for processing routes, analyzing and determining the most sustainable process, is developed. The components required for ProCAFD are databases (chemicals, solvents, process-groups as opposed to unit operations, reactions); model objects (pure and mixture properties, flowsheet property, phase equilibrium, driving force etc.); calculation tools (solvent attributes, mixture compositions, environmental impact, etc.); design algorithms (operational design, solvent design, separation synthesis etc.). Tools like ProCAPE and ProCAMT are more problem specific and require smaller number of components, for example, ProCAPE requires database of chemicals and properties, models for pure component and mixture properties, property calculation options, graphics for phase diagrams and property analysis, and, model parameter estimation toolbox as the main components. The application range of this tool depends on the property models available in the model library. In ProCAMT, the main components are a model generator, a model analyzer, a model solver (collection of numerical solvers) and a model reuser. This tool serves to create model objects for other tools.

This presentation will focus on the architecture, components and the generic main user interface through which different components are connected based on the identified data and work flows. Tailor-made tool development based on this component-based architecture will be illustrated to emphasize the power of true "plug and play".

References:

[1] M. Fedorova, G. Tolksdorf, S. Fillinger, G. Wozny, M. Sales-Cruz, G. Sin, R. Gani, 2015. Development of Computer Aided Modelling Templates for Model Re-use in Chemical and Biochemical Process and Product Design: Import and export of models. Computer Aided Chemical Engineering, 37, 953-958

[2] A.K. Tula, J. Bottlaender, M.R. Eden, R. Gani, 2017, A computer-aided software-tool for sustainable process synthesis-intensification, Computers Chem Eng, 105, 74-95.

[3] S. Kalakul, M.R. Eden, R. Gani, 2017, The Chemical Product Simulator â€“ ProCAPD, Computer Aided Chemical Engineering, 40, 979-984.

538298 SPICE: A Computer-Aided Platform for Simultaneous Process Synthesis and Intensification

Abstract:

Process intensification improves the process performance and brings technical innovation with improved controllability, sustainability and energy efficiency [1-3]. Process synthesis identifies the best process alternatives under several design constraints to generate desired products from available raw materials, while giving insights on optimal unit operation type and their interconnections [4-7]. Several computer-aided tools and software frameworks already exist for integrating the process synthesis methods and simulation packages to automatically generate the desired process flowsheets [8-11].

In this work, we describe the update on SPICE (Systematic Process Intensification of Chemical Enterprises) computational framework to involve process intensification within process synthesis framework. We have recently proposed block superstructure to address process synthesis for process intensification and formulated a corresponding mixed-integer nonlinear optimization (MINLP) model [1,5]. This MINLP model serves as the basis for SPICE. Specifically, the proposed SPICE framework enables: 1) collection of model input data on feeds, products and materials through a user-friendly interface; 2) determination of block superstructure size via a mixed integer linear (MILP) optimization model; 3) generation of optimal block configurations by solving the MINLP model via commercial solvers; 4) conversion of obtained block configuration to equipment-based process flowsheets. With these features, SPICE framework automates the generation and optimization of process flowsheets including intensified alternatives without postulating their existence *a priori*. The fundamental MINLP model embedded in SPICE involves general mass and energy conservations, product demand and purity requirements, reaction and separation modeling formulations, and material selection constraints.

Additionally, we will discuss an MILP-based method for determining the block superstructure size with full connectivity. We interpret each block as node and flow connections in block superstructure as path segments connecting these nodes. The proposed MILP model involves constraints on the identification of initial and destination block for each path, path propagation for achieving node connectivity and elimination of forbidden path overlaps. Through the SPICE framework, we investigate several process synthesis and intensification examples including methanol production process and several reactive and hybrid separation processes. The results show promise towards identifying novel intensified options within process synthesis framework without postulating flowsheet alternatives *a priori*. With these case studies, the versatile capabilities of the software framework are demonstrated in terms of data collection and transferring, as well as automatic generation and optimization of process flowsheets.

References:

[1] Demirel, S. E., Li, J., Hasan, M. M. F. Systematic process intensification using building blocks. *Computers & Chemical Engineering*, 2017, 105: 2–38.

[2] Stankiewicz, A.I., Moulijn, J.A., 2000. Process intensification: transforming chemical engineering. *Chemical Engineering Process*, 2000, 96(1): 22-34.

[3] Reay, D., Ramshaw, C., Harvey, A. Process intensification: engineering for efficiency, sustainability and flexibility. Waltham, MA: Butterworth-Heinemann, 2013.

[4] Chen Q, Grossmann I. Recent developments and challenges in optimization-based process synthesis. *Annual review of chemical and biomolecular engineering*. 2017, 8: 249–283.

[5] Li, J., Demirel, S.E., Hasan, M.M.F. Process Synthesis using Block Superstructure with Automated Flowsheet Generation and Optimization. AIChE Journal, 2018, under review.

[6] Li J., Demirel S.E., Hasan M.M.F. Process Integration using Block Superstructure. *Industrial & Engineering Chemistry Research*, 2018, 57: 4377–4398.

[7] Li J., Demirel S.E., Hasan M.M.F. Fuel Gas Network Synthesis Using Block Superstructure. *Processes*. 2018, 6(3): 23.

[8] Kravanja Z, Grossmann IE. PROSYN-an automated topology and parameter process synthesizer. *Computers & Chemical engineering*. 1993, 17: S87–S94.

[9] Gani R, Hytoft G, Jaksland C, Jensen AK. An integrated computer aided system for integrated design of chemical processes. *Computers & Chemical Engineering*. 1997, 21(10):1135–1146.

[10] Friedman Z, Ingalls J, Siirola JD, Watson JP. Block-oriented modeling of superstructure optimization problems. *Computers & Chemical Engineering*. 2013, 57:10–23.

[11] Tula AK, Babi DK, Bottlaender J, Eden MR, Gani R. A computer-aided software-tool for sustainable process synthesis-intensification. Computers & Chemical Engineering. 2017, 105:74–95.

526364 Data Driven Modeling in Alamo: Feature Selection and Non-Parametric Modeling Applications

Abstract:

ALAMO is a computational methodology designed to learn algebraic models from data. The models produce via this approach constitute a linear combination of parametric transformations of input variables. Recent developments comparing variants of the lasso and best subset selection [1] have motivated the inclusion of additional heuristics to identify optimally predictive models. The linear model selection methodology utilized by ALAMO is explained and demonstrated across a diverse set of benchmark problems. The problem of feature engineering, or the identification of promising transformations of input variables for inclusion in linear models, is directly addressed by the ALAMO methodology. Diverse sets of algebraic features are applied to optimization benchmark problems in order to identify accurate models that are tailored for optimization with equation oriented solvers.

When physical interpretability of a model is not prioritized, it is common to use non-parametric regression techniques to provide smooth interpolations of available data. ALAMO is capable of using non-parametric transformations of points in the domain of interest as features for linear model selection. Examples utilizing Gaussian radial basis functions are explored in order to identify optimal interpolative models. Constrained regression techniques used by ALAMO can force these non-parametric models to obey insights enforced by the modeler, enhancing their ability to model accurately in extrapolative domains.

[1] Hastie, Trevor, Robert Tibshirani, and Ryan J. Tibshirani. "Extended Comparisons of Best Subset Selection, Forward Stepwise Selection, and the Lasso." arXiv preprint arXiv:1707.08692 (2017)

519756 Recent Advances in the EaGO Platform: Global and Robust Optimization in Julia

Abstract:

Parametric uncertainty in process systems modeling and simulation is ever present. Robust optimization has significant applications in design for process safety, control of water-treatment facilities, and the economic assessments of novel projects [1,2]. However, the currently available software tools (ROME, YAMLP), address only narrowly structured robust optimization problems and efforts to develop generalized solvers for parametric uncertainty remain in a nascent state [3,4,5]. For general problems, highly-specialized bounding routines may be necessary [6]. While numerous options exist for constructing meta-optimization algorithms (e.g. Pyomo [7]) easy manipulation of low-level solution routines and propagation of information from these low-level routines to meta-algorithms is still quite difficult.

In 2018, the EAGO optimization platform was introduced to address this gap [8 By introducing, a low-level global/robust optimization platform in Julia, we're able to reduce the overhead to develop and implement algorithms addressing these novel applications [9]. Our hope is that this will act as a development platform to complementing existing commercial optimization packages. Prior work on EAGO focused on the development of McCormick relaxation libraries [10,11,12], a global optimizer for explicit & implicit functions [6,13], and meta-algorithms for solving semi-infinite programs [5,6]. New features updated this year include:

Additional relaxation schemes for nonconvex NLP problems such as alpha-BB and interval methods [14].

- New library of domain reduction subroutines [15].
- Improved handling for sparse systems including GPGPU-based enhancements to existing global and robust solution routines.
- Incorporation of mixed-integer programming routines into existing robust solvers [16]
- Improved benchmarking framework for use with standard libraries [17].

Other advances in manipulation techniques for directed-acyclic graphs will be discussed. Demonstrations of the above improvements will be given with comparisons to prior techniques as appropriate. Multiple examples will illustrate EAGO's utility for solving systems that commonly arise in model-based system engineering applications. Additionally, a series of industrially-relevant examples in process systems engineering will be given to illustrate use cases and performance aspects of the new EAGO features.

[1] Ben-Tal A.; El Ghaoui L.; Nemirovski A. Robust optimization. Princeton University Press, 2009.

[2] Li, P., and Geletu, A. Recent Developments in Computational Approaches to Optimization under Uncertainty and Application in Process Systems Engineering. ChemBioEng Reviews, 1(4): 170-190 2014.

[3] Joel Goh. Distributionally Robust Optimization and its Tractable Approximations with ROME. *ISMP*, Aug

2009, Chicago.

[4] Lofberg, Johan. "YALMIP: A toolbox for modeling and optimization in MATLAB." *Computer Aided Control Systems Design, 2004 IEEE International Symposium on*. IEEE, 2004.

[5] Mitsos, A. Global Optimization of Semi-Infinite Programs via Restriction of the Right-Hand Side. *Optimization*. 60:10-1,1291-1308.

[6] Stuber, M.D., and Barton, P.I. Semi-Infinite Optimization with Implicit Functions. *Ind. Eng. Chem. Res.*, 54:307-317, 2015.

[7] Hart WE, Laird C, Watson J-P, Woodruff DL. 2012. *Pyomo - Optimization Modeling in Python*, Springer US, Boston, MA. doi:10.1007/978-1-4614-3226-5.

[8] Wilhelm, M. and Stuber, M. Easy Advanced Global Optimization (EAGO): An Open-Source Platform for Robust and Global Optimization in Julia. AIChE Annual Meeting 2017.

[9] Bezanson, J. et al. Julia: A fresh approach to numerical computing. *SIAM Review*. 59, 65-98, 2017.

[10] Scott, J.K., Stuber, M.D., and Barton, P.I. Generalized McCormick Relaxations. *J Global Optim*, 51:569-606, 2011.

[11] Mitsos, A., Chachuat, B., and Barton, P.I. McCormick-Based Relaxations of Algorithms. SIAM J. Optim. 20(2): 573-601.

[12] Khan, K.A., Watson, H.A., and Barton, P.I. Differentiable McCormick Relaxations. *J Global Optim*, 67(4):687-729, 2017.

[13] Stuber, M.D., Scott, J.K., and P.I. Barton. Convex and Concave Relaxations of Implicit Functions. *Optimization Methods and Software*. 30(3), 424-460, 2014.

[14] Adjiman, Claire S., et al. A global optimization method, αBB, for general twice-differentiable constrained NLPs—I. Theoretical advances. *Computers & Chemical Engineering* 22.9 (1998): 1137-1158.
[15] Puranik, Y. and N. V. Sahinidis. Domain reduction techniques for global NLP and MINLP optimization. Constraints, 22: 338-376 (2017).

[16] Floudas, Christodoulos A. *Nonlinear and mixed-integer optimization: fundamentals and applications*. Oxford University Press, 1995.

[17] O. Shcherbina, et al. Benchmarking Global Optimization and Constraint Satisfaction Codes, In *Global Optimization and Constraint Satisfaction*, Bliek, Ch., Jermann, Ch. and Neumaier, A. Springer, Berlin 2003, 212-222.

531025 A Sensitivity-Based Nonlinear Model Predictive Control and State-Estimation Framework in Python

Abstract:

With the current software and computer capabilities the implementation of advanced techniques for nonlinear model predictive control enables users to create more sophisticated solution strategies, and apply them to solve complex chemical and energy process control problems. Increased popularity of open-source modelling platforms, like Pyomo, and the robustness of nonlinear optimization solvers, like Ipopt, make it possible to construct generalized frameworks for meta-algorithms such as Nonlinear Model Predictive Control (NMPC) and Moving Horizon Estimation (MHE).

In this work we provide details of a object-oriented framework based on Python and Pyomo for running NMPC and MHE. Variants of these approaches based on parametric sensitivity computations have also been implemented using a new sensitivity computation software. The framework accepts as an input a user-declared (and initialized) differential algebraic model which then is automatically augmented and discretized to form the entirety of the controller and/or state estimation scheme. Because of the intrinsic computational cost of the solution of the NLP problems for control and state-estimation, standard NMPC requires delay on the online control action. However, this work demonstrates that the delay can be circumvented by implementing a parametric sensitivity update of NLPs solved in background for predicted parametric information (i.e. state value for control, measurement for state-estimation), which is far less computationally expensive than the online solution of the NLPs. For this purpose we also introduce `*k_aug*`, the sensitivity computation software that handles the relevant post-optimality computations, like parametric sensitivity matrices or reduced Hessians, both required for the control and state estimation frameworks.

We demonstrate the capabilities and limitations of the framework by solving a number of case studies, including a distillation column sequence, and a Bubbling Fluidized Bed reactor for CO2 capture. For both control problems it is assumed there is no knowledge of the full state of the plant, so state-estimation must be used. When the NMPC approach for control is coupled with MHE for state-estimation the results show favorable performance of the sensitivity-based approaches compared to the ideal case with no online computational delay. In this context the bulk of the online computational delay with sensitivity for control and state-estimation is an order of magnitude *less* than the ideal case for the bubbling fluidized example. As a result, the proposed framework might serve as an excellent foundation for more complex dynamic optimization formulations like economic NMPC, parameter-estimation, etc.

535345 Multi-Period Design and Planning of Centralized and Distributed Manufacturing Networks

Abstract:

Distributed and modular manufacturing has arisen as a promising option for supply chain networks in which the logistics are the main bottleneck. However, economies of scale will always favor large scale centralized production. Therefore, there is a need for a general framework that can support this decision, taking into account the potential tradeoffs [1]. In this paper, we address the multi-period design and planning of manufacturing networks considering the option of centralized and/or distributed facilities. This problem involves the selection of which facilities to build in each time period, and their location in the continuous 2-dimensional space, in order to meet demand and maximize profits.

The problem is formulated as a version of the continuous facility location and allocation problem with limited capacity, also known as the Capacitated Multi-facility Weber Problem (CMWP) [2]. The objective of this type of problem is to determine locations in continuous 2-dimensional space for opening new facilities based on their maximum capacity and the given coordinates of the suppliers or customers. We propose an extension of the original CMWP that considers fixed cost for opening new facilities, fixed transportation costs, and two sets of fixed-location points: suppliers i and customers j [3], and investment decisions in different time periods. The model is formulated as a multi-period nonlinear Generalized Disjunctive Programming (GDP), and reformulated as a multi-period nonconvex Mixed-Integer Nonlinear Programming (MINLP).

We develop a bilevel decomposition algorithm that consists of decomposing the problem into a master problem and a subproblem. The master problem is based on a relaxation of the nonconvex MINLP, which yields an MILP that predicts the selection of facilities and their links to suppliers and customers, as well as a lower bound on the cost of problem. The subproblem corresponds to a nonconvex NLP of reduced dimensionality that results from fixing the binary variables in the MINLP problem, according to the binary variables predicted in the MILP master problem. Based on the bounding properties of their subproblems, Îµ-convergence is proved for this algorithm.

The applicability of the proposed multi-period model and solution framework is illustrated with a biomass supply chain case study. The results show that the algorithm is more effective at finding global optimal solutions than general purpose global optimization solvers tested.

[1] Lara, C.L., Grossmann, I.E.: Global Optimization for a Continuous Location-Allocation Model for Centralized and Distributed Manufacturing. Proc. of.26th European Symposium on Computer Aided Process Engineering, June 2016, PortoroÅ¾, Slovenia.

[2] Brimberg, J., Hansen, P., Mladonovic, N., Salhi, S.: A survey of solution methods for the continuous location allocation problem. International Journal of Operational Research 5(1), 1â€"12 (2008).

[3] Lara, C.L., Trespalacios, F., & Grossmann, I.E.: Global Optimization Algorithm for Capacitated Multi-facility Continuous Location Allocation Problems, Journal of Global Optimization, 2018. https://doi.org/10.1007/s10898-018-0621-6.

525298 A Stochastic Game Theoretic Framework for Optimization of Decentralized Supply Chains Under Uncertainty

Abstract:

The management of supply chains is a complex decision-making process involving multiple stakeholders, each of which only controls part of the supply chain, and may pursue different objectives, thus leading to compromised solutions [1]. Nevertheless, most existing studies on optimal design and operations of supply chains rely on centralized optimization models [2-5]. Consequently, the optimal solutions of centralized models can be suboptimal or even infeasible in a decentralized, multi-stakeholder supply chain. To address this research challenge, multiple game theoretic models are developed to explicitly account for the performance of multi-stakeholder systems. Examples include optimization models for cooperative multi-enterprise supply chains based on the generalized Nash bargaining solution approach [6-9]. On the other hand, optimization models integrating Stackelberg game and Nash-equilibrium are proposed for noncooperative supply chain optimization [10-14]. However, these models assume perfect information sharing among different stakeholders. In other words, the resulting optimal decisions are based on deterministic information. In practice, various types of uncertainties, such as price and productivity fluctuation, exist concerning the decision-making processes of stakeholders. These uncertainties may significantly influence the rational behaviors of stakeholders. Therefore, it remains a research challenge to simultaneously consider decentralized features of multi-stakeholder supply chains and incorporate uncertainty in the noncooperative stakeholders' optimal decision-making process for supply chain design and operations. To fill this knowledge gap, it is necessary to develop a holistic game theoretic model of multi-stakeholder decentralized supply chains that captures the influences of uncertainty on stakeholders' optimal decisions in a systematical way.

In this work, we propose a novel modeling framework to investigate the influences of uncertainty in decentralized optimization of supply chains. This modeling framework integrates the Stackelberg game with stochastic programming approach into a holistic two-stage stochastic game theoretic model. Specifically, this modeling framework allows consideration of one leader and multiple followers. Following the sequence of decision making process, decision variables for both the leader and the followers are classified into design decisions that must be made "here-and-now" and operational decisions that are postponed to a "wait-and-see" mode after the realization of uncertainties. As a result, both types of stakeholders interact with each other to determine their optimal design strategies at the first stage. After uncertainties from both the leader and the followers are realized, all stakeholders then determine their operational strategies as "recourse" decisions of the uncertainty information based on their previous design decisions. Following the stochastic programming approach, uncertainties are depicted with discrete scenarios with known probabilities. The objectives of the leader and the followers are to maximize their own expected net present value (NPV). The resulting problem is formulated as a two-stage stochastic mixed-integer bilevel programming (MIBP) problem. The applicability of the proposed modeling framework is demonstrated with a large-scale application to Marcellus shale gas supply chains. By solving this optimization problem, we find that the expected NPV for the leader is $68.9 MM. The three followers corresponding to three processing plants are expected to achieve $2.57 MM, $3.39 MM, and $1.21 MM NPVs, respectively. To demonstrate the advantage of this two-stage stochastic MIBP model, we further compare the optimal results obtained in the proposed two-stage stochastic game theoretic model with those of deterministic game theoretic models with different expectation of stakeholders. Based on the optimization results, we concluded that stakeholders tended to choose more conservative strategies when considering uncertainties in the optimization of decentralized supply chains. Although the conservatism might affect the overall performance of stakeholders, it effectively hedged against the risk of extreme cases when stakeholders wrongly anticipated others' performances.

References

[1] G. Cachon and S. Netessine, "Game Theory in Supply Chain Analysis," in *Handbook of Quantitative Supply Chain Analysis*. vol. 74, D. Simchi-Levi, S. D. Wu, and Z.-J. Shen, Eds., ed: Springer US, 2004, pp. 13-65.

[2] I. E. Grossmann, "Challenges in the new millennium: product discovery and design, enterprise and supply chain optimization, global life cycle assessment," *Computers & Chemical Engineering,* vol. 29, pp. 29-39, 2004.

[3] L. G. Papageorgiou, "Supply chain optimisation for the process industries: Advances and opportunities," *Computers & Chemical Engineering,* vol. 33, pp. 1931-1938, 2009.

[4] D. Yue, F. You, and S. W. Snyder, "Biomass-to-bioenergy and biofuel supply chain optimization: Overview, key issues and challenges," *Computers & Chemical Engineering,* vol. 66, pp. 36-56, 2014.

[5] D. Garcia and F. You, "Supply chain design and optimization: Challenges and opportunities," *Computers & Chemical Engineering,* vol. 81, pp. 153-170, 2015.

[6] J. Gjerdrum, N. Shah, and L. G. Papageorgiou, "Transfer Prices for Multienterprise Supply Chain Optimization," *Industrial & Engineering Chemistry Research,* vol. 40, pp. 1650-1660, 2001.

[7] J. Gjerdrum, N. Shah, and L. G. Papageorgiou, "Fair transfer price and inventory holding policies in two-enterprise supply chains," *European Journal of Operational Research,* vol. 143, pp. 582-599, 2002.

[8] D. Zhang, N. J. Samsatli, A. D. Hawkes, D. J. L. Brett, N. Shah, and L. G. Papageorgiou, "Fair electricity transfer price and unit capacity selection for microgrids," *Energy Economics,* vol. 36, pp. 581-593, 2013.

[9] D. Yue and F. You, "Fair profit allocation in supply chain optimization with transfer price and revenue sharing: MINLP model and algorithm for cellulosic biofuel supply chains," *AIChE Journal,* vol. 60, pp. 3211-3229, 2014.

[10] M. A. Zamarripa, A. M. Aguirre, C. A. Méndez, and A. Espuña, "Improving supply chain planning in a competitive environment," *Computers & Chemical Engineering,* vol. 42, pp. 178-188, 2012.

[11] D. Yue and F. You, "Game-theoretic modeling and optimization of multi-echelon supply chain design and operation under Stackelberg game and market equilibrium," *Computers & Chemical Engineering,* vol. 71, pp. 347-361, 2014.

[12] J. Gao and F. You, "Game theory approach to optimal design of shale gas supply chains with consideration of economics and life cycle greenhouse gas emissions," *AIChE Journal,* vol. 63, pp. 2671-2693, 2017.

[13] J. Gao and F. You, "Economic and Environmental Life Cycle Optimization of Noncooperative Supply Chains and Product Systems: Modeling Framework, Mixed-Integer Bilevel Fractional Programming Algorithm, and Shale Gas Application," *ACS Sustainable Chemistry & Engineering,* vol. 5, pp. 3362-3381, 2017.

[14] D. Yue and F. You, "Stackelberg-game-based modeling and optimization for supply chain design and operations: A mixed integer bilevel programming framework," *Computers & Chemical Engineering,* vol. 102,

pp. 81-95, 2017.

524234 A Column Generation Approach to Multiscale Capacity Planning for Continuous Power-Intensive Processes

Abstract:

In light of highly time-sensitive electricity prices in deregulated electricity markets, power-intensive industries have the increasing need to shift load to low-price time periods in order to maintain cost-competitive operations. An industrial plant's ability to shift load is primarily given by its operational efficiency and dynamic behavior, as well as its production and storage capacities. Therefore, careful capacity planning is required for assessing the value of additional flexibility and planning for expected long-term changes in demand. Here, the main challenge lies in the simultaneous consideration of decisions at two different time scales, namely the long-term decisions related to the capacity expansion strategy and the short-term operational decisions.

The multiscale capacity planning problem for continuous power-intensive processes was first introduced by Mitra et al. (2014). The objective is to find the optimal long-term capacity expansion plan, which involves adding or modifying process units and storage capacity over the course of multiple years. The resulting mixed-integer linear program (MILP) is very large in size and hence difficult to solve.

In this work, we apply a column generation approach to solve an extended deterministic version of the model proposed by Mitra et al. (2014). The Dantzig-Wolfe decomposition takes advantage of the block-diagonal structure of the problem and allows us to obtain the solution by solving multiple independent MILPs, one per season, in an iterative fashion. We demonstrate the effectiveness of the proposed algorithm in an extensive computational study as well as in a real-world industrial case provided by Praxair.

References

Mitra, S., Pinto, J. M., & Grossmann, I. E. (2014). Optimal multi-scale capacity planning for power-intensive continuous processes under time-sensitive electricity prices and demand uncertainty. Part I: Modeling. *Computers and Chemical Engineering*, *65*, 89â€“101.

528612 Solving Robust Vehicle Routing Via a Branch-Price-and-Cut Approach

Abstract:

The Vehicle Routing Problem (VRP) is one of the most highly studied combinatorial optimization problems in the area of Supply Chains and Logistics [1]. It aims to determine a set of optimal routes for vehicles to serve customers such that all known demands are satisfied and in a way such that all vehicle capacities are respected. The Branch-Price-and-Cut (BP&C) method has been gradually accepted as the state-of-the-art algorithm to tackle deterministic VRP and its variants. The BP&C algorithm is sophisticated in a way that the linear relaxation at every node in the branch-and-bound tree is solved via Column Generation (CG) and cutting planes are added to strengthen the relaxation. In our VRP context, columns denote feasible routes, and they are generated by solving a Shortest Path Problem with Resource Constraints (SPPRC) via a labeling algorithm [2]. It has been shown that Rounded Capacity Inequalities (RCI) [3] and limited-memory subset row cuts [4] can effectively tighten the linear model in the deterministic setting.

Recently, VRP under demand uncertainty [5][6] has attracted considerable attention. It is motivated by the fact that customer demands are often not readily available at the time the distributor needs to commit to some routing plan. In this work, we combine Robust Optimization (RO) [7] and BP&C in order to develop an effective approach for robust VRP under demand uncertainty. When extending BP&C for a robust VRP, the optimal routing design has to be robust feasible. There are two different perspectives for how to achieve this. The first and most intuitive way is to generate deterministically feasible routes via solving a standard SPPRC and enforce robust feasibility via robust RCI [6]. This approach is very generic and can be ideally applied under any uncertainty set selection. Another approach is to directly generate robust feasible columns. This necessitates an efficient algorithm to exactly solve a robust SPPRC, which can be a tough proposition, since this problem is in general strongly NP-hard [8]. Recently, Pessoa ***et al.*** [9] showed that for some special structure uncertainty sets (e.g., Γ uncertainty set [10]), robust columns can be generated via solving a sequence of polynomially many deterministic SPPRC.

In this work, we first evaluate the two approaches and compare their performances using several well-known uncertainty sets, including the Γ set and budget set [6] among others. The comparison studies are mainly focused on the computational performance and theoretical limitations to different uncertainty sets. In addition, we consider combining the two techniques together and generating robust feasible columns in the CG stage as well as separating robust cuts in the cut generation stage to further tighten the linear relaxation.

References:

[1] Toth, P., & Vigo, D. (Eds.). (2014). Vehicle routing: problems, methods, and applications. ***Society for Industrial and Applied Mathematics***.

[2] Irnich, S., Desaulniers, G., & others. (2005). Shortest path problems with resource constraints. ***Column Generation***, 33–65.

[3] Fukasawa, R., Lysgaard, J., Poggi de Aragão, M., Reis, M., Uchoa, E., & Werneck, R. (2004). Robust branch-and-cut-and-price for the capacitated vehicle routing problem. ***Integer programming and combinatorial optimization***, 11-17.

[4] Pecin, D. G. (2014). Exact Algorithms for the Capacitated Vehicle Routing Problem (***Doctoral dissertation***, PUC-Rio).

[5] Sungur, I., Ordónez, F., & Dessouky, M. (2008). A robust optimization approach for the capacitated vehicle routing problem with demand uncertainty. ***IIE Transactions***, 40(5), 509-523.

[6] Gounaris, C. E., Wiesemann, W., & Floudas, C. A. (2013). Demand Uncertainty The Robust Capacitated Vehicle Routing Problem Under Demand Uncertainty. ***Operations Research***, ***61***(3), 677–693.

[7] Ben-Tal, A., El Ghaoui, L., & Nemirovski, A. (2009). Robust optimization. ***Princeton University Press***.

[8] Alves Pessoa, A., Di Puglia Pugliese, L., Guerriero, F. and Poss, M., 2015. Robust constrained shortest path problems under budgeted uncertainty. ***Networks***, ***66***(2), pp.98-111.

[9] Pessoa, A., Poss, M., Sadykov, R. and Vanderbeck, F., 2018, June. Solving the robust CVRP under demand uncertainty. In ***ODYSSEUS 2018-7th International Workshop on Freight Transportation and Logistics***.

[10] Bertsimas, D. and Sim, M., 2003. Robust discrete optimization and network flows. ***Mathematical programming***, ***98***(1-3), pp.49-71.

533204 Autologous Cancer Therapies:How Can We Handle the Complexity of the Supply Chain?

Abstract:

Autologous Cancer Therapies:

How can we handle the complexity of the supply chain?

Maria M. Papathanasiou, Nilay Shah

Dept. of Chemical Engineering, Centre for Process Systems Engineering (CPSE), Imperial College London SW7 2AZ, London, U.K

In August 2017, the U.S. Food and Drug Administration made a historic approval of the first autologous, cell-based cancer therapy, changing the future of cancer therapies. Novartis' Kymriah, is an autologous Chimeric Antigen Receptor (CAR) T-cell therapy for B-cell acute lymphoblastic leukaemia (ALL) and is the first therapy to open the road towards innovative cancer therapies. Following that Kite's Yescarta was approved by FDA in October 2017, indicated for the treatment of adult patients with relapsed or refractory large B-cell lymphoma after two or more lines of systemic therapy. Due to their highly promising results [1], the CAR T-cells are currently experiencing a tremendous growth with more than 200 clinical trials in place globally (data as presented in ClinicalTrials.gov).

Unlike common pharmaceutical products, those therapies are of autologous nature, suggesting 1:1 business and/or manufacturing models, where the drug is manufactured based on the patient's own cells (Figure 1) [2]. The latter increases the complexity both of the production process, as well as of the reimbursement strategies that need to be established. Such products are of highly sensitive nature, requiring special handling, while their manufacturing process involves a plethora of steps that need to be well-coordinated. Moreover, due to their high specificity and complex lifecycle, CAR T-cell therapies are characterized by high manufacturing costs that are reflected by excessive end-prices for the patients (approximately $400.000 per patient). Consequently, CAR T-cells are currently considered elite-type products that cannot be accessed by all patient groups and are very challenging to reimburse.

Figure 1 Current CAR T-cell process/distribution steps along with some of the key bottlenecks and challenges.

Apart from the afore-mentioned challenges .CAR T-cell therapies are currently following a hand-to-hand logistics model that will prove to be insufficient once they become available to bigger patient populations. Therefore, the pharmaceutical industry is urged by the regulators to establish robust supply chain models that will be able to support full-scale commercialization as those therapies become available.

In this work, we focus on the investigation, design and comparison of alternative supply chain models that will be able to ensure: (i) lower costs, (ii) return time optimization, (iii) scalability and (iv) risk mitigation in cases of supply shortages and/or high demand. We present and discuss alternative distribution solutions and their impact on the cost and regulations associate with those therapies, as well as challenges associated to adverse events (e.g. extreme weather conditions). We focus on decisions related to: (i) number of clinical sites, (ii) number of manufacturing sites, as well as (iii) location of sites. Moreover, we investigate the establishment of intermediate collection sites to serve as de-bottlenecking solutions between the clinical and manufacturing sites. The impact of all presented alternatives on the Net Present Value (NPV) of the therapy is assessed, along with potential challenges related to the associated regulatory environment.

Acknowledgments

Funding from the UK Engineering & Physical Sciences Research Council (EPSRC) for the Future Targeted Healthcare Manufacturing Hub hosted at University College London with UK university partners is gratefully acknowledged (Grant Reference: EP/P006485/1). Financial and in-kind support from the consortium of industrial users and sector organisations is also acknowledged.

References

[1] S. L. Maude *et al.*, "Tisagenlecleucel in Children and Young Adults with B-Cell Lymphoblastic Leukemia," *N. Engl. J. Med.*, vol. 378, no. 5, pp. 439–448, 2018.

[2] B. L. Levine, J. Miskin, K. Wonnacott, and C. Keir, "Global Manufacturing of CAR T Cell Therapy," *Mol. Ther. - Methods Clin. Dev.*, vol. 4, no. March, pp. 92–101, 2017.

532375 Surrogate-Based Derivative-Free Optimization of a Multi-Enterprise Supply Chain Simulation

Abstract:

Simulations are widely used to capture the details of a complex physical or scientific phenomenon. Optimizing over these simulations is thus a problem that is often encountered and is of growing interest. As these simulations are computationally expensive or have no available closed form, this problem is often nontrivial. To address this, a number of derivative-free optimization (DFO) algorithms have been developed in the last couple of decades [1][2]. These algorithms have shown a great potential to be applied for practical problems. Efficient use of surrogate models plays an important role towards the success of many of these algorithms [3]. Some of the algorithms utilize a surrogate model for approximations in small local regions whereas other algorithms use it for global approximation over the entire feasible domain. As the search progresses, quality of the sample set is not maintained globally because of concentrated local searches. As a result, factors such as local variability in the output response may lead to a poor surrogate model that exhibits high variance and poor prediction accuracy. This highlights the need for improved surrogate modeling strategy for DFO.

In this work, an agent-based simulation model is considered for a supply chain network consisting of multiple enterprises. Agent-based models are popular in the area of the supply chain because of their ability to model complex interactions between entities. The problem is to find optimal inventory allocation such that the total cost of the supply chain is minimized. This problem has been shown to be highly nonconvex in the presence of complex interaction mechanisms such as auctions[4]. The problem becomes more difficult as problem size increases with the size of the supply chain. Moreover, the total cost of the supply chain is a combination of several components such as transport cost, inventory cost, production cost, and backorder cost. These individual components display a local variability in the objective function with respect to warehouse capacity. As a result, the output has multiple closely placed local optima. Because of these variations, and because of the quality of sample design, this problem becomes difficult for DFO.

Finally, this work proposes a DFO framework based on Gaussian process regression models with noise variance estimation. The noise variance or nugget parameter is estimated along with other model hyperparameters by maximizing likelihood estimate. Starting from a space filling design, this algorithm relies on expected improvement based adaptive sampling strategy to perform a global search. A model based local search is triggered when promising samples are found during the global search. A comparison of proposed framework with competitive DFO algorithms is presented on test problems and on a large supply chain problem of up to 30 dimensions.

References

[1] A. Conn, K. Scheinberg, and L. Vicente, *Introduction to Derivative-Free Optimization*. Society for Industrial and Applied Mathematics, 2009.

[2] L. M. Rios and N. V. Sahinidis, "Derivative-free optimization: A review of algorithms and comparison of software implementations," *J. Glob. Optim.*, vol. 56, no. 3, pp. 1247–1293, 2013.

[3] A. Bhosekar and M. Ierapetritou, "Advances in surrogate based modeling, feasibility analysis, and optimization: A review," *Comput. Chem. Eng.*, vol. 108, pp. 250–267, 2018.

[4] N. Sahay and M. Ierapetritou, "Multienterprise supply chain: Simulation and optimization," *AIChE J.*, vol. 62, no. 9, pp. 3392–3403, 2016.

527248 Strategic Time Window Assignment in Vehicle Routing Operations

Abstract:

The delivery (or pickup) of goods within a scheduled time window is widespread in several real world distribution networks. In many industries, these time windows are mutually agreed upon by the distributor and customer through long-term (e.g., annual) delivery contracts. From the customer's point of view, this is crucial for efficient inventory management and scheduling of personnel to process the delivery. From the distributor's point of view, this is crucial for reducing the variability across repetitive deliveries and can expose efficiencies that add up to significant cost savings. Examples of applications where such operations are typical include, among others, retail distribution [1], vendor-managed inventory routing [2] and courier services [3].

Once a time window has been agreed upon and communicated to the customer, the distributor must attempt to meet it on an operational (e.g., daily) basis as well as possible. This is done by solving a Vehicle Routing Problem with Time Windows (VRPTW) [4] to determine a delivery schedule that adheres to the agreed time windows. The assigned time windows strongly influence the structure of feasible schedules and, hence, the daily incurred distribution costs. Therefore, a natural choice is to assign time windows based on the arrival times at customer locations in the optimal (i.e., minimum cost) vehicle routing schedule computed under nominal conditions. However, this seemingly optimal decision may become highly suboptimal in the presence of operational uncertainty, particularly in situations that deviate from the nominal case. For example, the demand volume of a customer typically fluctuates per delivery. Similarly, travel times vary on a day-to-day basis (e.g., because of unpredictable traffic conditions). The true values of these parameters become known only on the day of delivery before the vehicles are dispatched. This makes the strategic assignment of time windows a non-trivial task.

In this work, we study the Time Window Assignment Vehicle Routing Problem (TWAVRP), which consists of strategically assigning long-term delivery time windows to customers. Feasible time windows may either be of pre-specified width within some exogenous windows (e.g., representing operating hours or government regulations) or selected from a set of *a priori* windows (e.g., AM/PM). A finite set of scenarios, each describing a possible realization of the uncertainty, is also assumed to be given with known probability of occurrence. This information is used to formulate a two-stage stochastic program, in which the first-stage decisions are to assign time windows, while the second-stage decisions are to design vehicle routing schedules satisfying the assigned time windows, one for each of the scenarios. The objective is to minimize the total routing costs, averaged over the postulated scenarios.

Our first contribution is to show that the two-stage stochastic programming formulation of the TWAVRP can be reduced to a variant of a large class of problems, popularly known as Consistent Vehicle Routing Problems (ConVRP) [5]. The goal of the ConVRP is to design minimum cost vehicle routes over a finite, multi-day horizon which remain "consistent" over time. We formally prove that the endogenous time windows in the TWAVRP serve to satisfy the arrival-time consistency requirement of the ConVRP [6], which requires that every customer be visited at roughly the same time whenever service is requested.

Our second contribution is to leverage the above stated result to develop a new, efficient scenario decomposition algorithm, to solve the TWAVRP. We demonstrate that our algorithm strongly outperforms all existing solution methods [1, 7, 8], and furthermore, that it is both modular and scalable. From a modeling viewpoint, it can accommodate general scenario-based models of uncertainty for several routing-specific parameters as well as continuous or discrete sets of feasible time window assignments. From an algorithmic viewpoint, it can be easily parallelized, can utilize any available vehicle routing solver in a black box fashion, and can be readily modified as a heuristic for large-scale instances. Notably, we show that our algorithm strongly outperforms all previous state-of-the-art methods, solving for the first time 54 previously open instances. Among other features, our algorithm also incorporates a new class of path-based disjunctions for the TWAVRP; in view of the relationship between the TWAVRP and ConVRP, these disjunctions are new for the ConVRP as well.

Our third contribution is to conduct computational experiments with a parallel implementation of our algorithm to solve instances consisting up to fifteen scenarios, representing a five-fold increase compared to existing methods. Using our algorithm, and via out-of-sample simulations, we numerically quantify the tradeoff between computational tractability and expected cost savings when considering more scenarios of future uncertainty during strategic time window assignment. Our business insights are that long-term costs are expected to decrease by modeling more scenarios of future uncertainty but the marginal benefits rapidly diminish as a function of the number of postulated scenarios. In other words, a few scenarios are sufficient to obtain time window assignments that would incur lower long-term costs compared to those that would be obtained by completely ignoring uncertainty.

References:

[1] R. Spliet, and A.F. Gabor (2015). The Time Window Assignment Vehicle Routing Problem. *Transportation Science*, 49(4):721—731.
[2] C. Zhang, G. Nemhauser, J. Sokol, M.-S. Cheon, and D. Papageorgiou (2015). Robust Inventory Routing with Flexible Time Window Allocation. Available at *Optimization Online* (2015/01/4744).
[3] O. Jabali, R. Leus, T. van Woensel, and T. de Kok (2015). Self-imposed Time Windows in Vehicle Routing Problems. *OR Spectrum*, 37(2):331—352.
[4] G. Desaulniers, O.B.G. Madsen, and S. Ropke (2014). Chapter 5: The Vehicle Routing Problem with Time Windows. In *Vehicle Routing: Problems, Methods and Applications*, P. Toth, and D. Vigo (editors), 119—159, Society for Industrial and Applied Mathematics, Philadelphia, PA.
[5] C. Gröer, B. Golden and E. Wasil (2009). The Consistent Vehicle Routing Problem. *Manufacturing & Service Operations Management*, 11(4):630—643.
[6] A. Subramanyam, and C.E. Gounaris (2018). A Decomposition Algorithm for the Consistent Traveling Salesman Problem with Vehicle Idling. *Transportation Science*, 52(2):386—401.
[7] R. Spliet, and G. Desaulniers (2015). The Discrete Time Window Assignment Vehicle Routing Problem. *European Journal of Operational Research*, 244(2):379—391.
[8] K. Dalmeijer, and R. Spliet (2018). A Branch-and-cut Algorithm for the Time Window Assignment Vehicle Routing Problem. *Computers & Operations Research*, 89:140—152.

534647 Time Window Based Berth and Yard Allocation Planning of Container Vessels

Abstract:

Transshipment of container vessels has been playing a significant role in international sea freight transport. Over the past several years, a reduction in the productivity of port has been observed due to inefficient usage of the available resources. Amongst all the operations, the most important is an optimal berth allocation of the ship and yard allocation for the containers. The berth location of ships and the yard location of containers are inter-dependent on each other. Also container vessels might arrive earlier or later than their nominal arrival time due to all kinds of uncertainties during port operations. Conventional berth allocation and planning models have difficulties to handle the uncertainty of vessel arrival while constructing a nominal berth plan or nominal berth allocation. However, in reality, the delay of a single vessel might interrupt the entire schedule and make it very difficult or even impossible to recover. And terminal operators strive to optimize the berth and yard allocation planning so as to minimize the travel distance of all containers. Because long travel distance not only causes high cost of carriers but also has a potential threat to the loading and discharging operations at the quayside. So, it is crucial to consider these two planning problems in an integrated manner.

Mathematical optimization has been used to improve the operational capability of port terminals using various realistic constraints. An integrated model of berth and yard template planning is developed for transshipment hubs [1]. But a weighted sum of berth-side and yard-side cost is utilized in the objective, and yard crane contention during the unloading activities has not been considered in the model. A joint planning problem for berth and yard allocation in transshipment terminals has also been developed using multi-cluster stacking strategy [2]. However, the two objectives are linearly combined by assigning each a weight which disregards that they are of different units. Existing publications are either focused on planning problem without considering realistic flexible arrival and departure time windows or using fixed maximal deviation from nominal arrival time [3].

The goal of this study is to optimally use the resources on the quay. For instance, making the most of quay cranes and minimizing total carrier transportation cost. A time window-based berth and yard allocation planning problem has been developed. Flexible arrival and departure time windows have to be taken into account specifically as control variables. We consider the simultaneous berth allocation and yard planning problem at a tactical level. This implies that certain decisions have already been made at a strategic level. We also assume that an optimal layout of the yard has been determined. And the number of storage blocks as well as their sizes and locations are known. We allow transshipment not only from vessel to storage block, but also from storage block to vessel. And containers can only be loaded or discharged when the vessel is berthing in the terminal.

The problem consists of two parts, first is the berth allocation planning model, which we called BAP model. BAP model is a MIP model and the objective function is to minimize total operational cost while satisfying all the transportation demand and operation constraints. The second part is the berth and yard allocation planning model, which we called BAYAP model. The berthing and departure time of each vessel are results of BAP model, which based on the expected amount of containers to be handled and the necessary quay and crane capacity to do so. These results can serve as an input to the following BAYAP model. The BAYAP model is a MINLP model and the objective function is to minimize total carrier transportation cost. Commercial solvers such as CPLEX, BARON, and ANTIGONE have been employed to obtain the optimal solution of the developed planning models. After solving the BAYAP model, we can get the optimal schedule of containers transshipment. Computational results of the case study demonstrate the efficacy of the developed planning model.

References

1. Zhen, L., E.P. Chew, and L.H. Lee, *An Integrated Model for Berth Template and Yard Template Planning in Transshipment Hubs.* Transportation Science, 2011. **45**(4): p. 483-504.
2. Tao, Y. and C.-Y. Lee, *Joint planning of berth and yard allocation in transshipment terminals using multi-cluster stacking strategy.* Transportation Research Part E: Logistics and Transportation Review, 2015. **83**: p. 34-50.
3. Hendriks, M.P.M., E. Lefeber, and J.T. Udding, *Simultaneous berth allocation and yard planning at tactical level.* OR Spectrum, 2013. **35**(2): p. 441-456.

535327 Verifying Performance Specifications for Dynamic Processes Under Uncertainty Using Backward Reachability Analysis

Abstract:

A new algorithm is presented for efficiently characterizing the set of uncertain inputs to a dynamic process for which the process is guaranteed to satisfy a given set of performance specifications. These specifications are generally described as a combination of 'goal sets' that the system states must eventually reach or remain in indefinitely (e.g., attaining a desired level of conversion or purity), and 'obstacles' that must be avoided (e.g., operating regions that are unsafe or otherwise detrimental to the process or product). Characterizing the set of inputs for which these specifications are guaranteed to be satisfied provides a robust operating envelope for the process, and is important in a number of applications where one or more of the following properties hold: (i) large input variability is unavoidable, (ii) critical product specifications are difficult or impossible to measure online, and therefore must be ensured by operating within a precomputed envelope, and/or (iii) even temporary violations of a specification can contaminate or otherwise destroy the product, and are therefore intolerable. For example, all of these qualities occur in various pharmaceutical and biopharmaceutical processes, which has led to widespread interest in the concept of 'Quality-by-Design' (QbD). In the QbD paradigm, the set of process inputs for which all critical product quality attributes can be ensured is known as the 'design space', and algorithms for efficiently computing it are of critical importance. Other emerging applications where robust performance verification is critical include remote modular systems, such as wellhead gas purification systems, remote power systems, and advanced manufacturing systems using automated industrial robotics.

To address the formal verification of such processes, we will present an algorithm that provides guaranteed inner and outer approximations of the set of inputs that lead to satisfaction of all performance specifications with certainty. This problem is commonly referred to as 'backward reachability analysis,' and is solved in the most basic approach by embedding forward reachability calculations in a branch-and-bound framework (forward reachability analysis aims to bound the set of model outputs that are achievable with given set of inputs). In our algorithm, we apply this standard branch-and-bound approach with several novel advancements. First, we use advanced forward reachability algorithms recently developed in our group, which are able to achieve much tighter bounds than other algorithms of similar complexity in many important application areas by exploiting model redundancy and centered-form enclosures. Second, we use custom dynamic constraint propagation techniques based on the specified goal and obstacle sets to accelerate convergence of the algorithm. Our final algorithm will be demonstrated using case studies in design space construction, and will be compared against state-of-the-art algorithms from the literature in terms of both accuracy and efficiency.

521998 An Embedded Model Predictive Controller for a Medical Oxygen Concentrator Device

Abstract:

Medical Oxygen Concentrators (MOCs) are commericial devices which produce 90% oxygen (O_2) from ambient air using Rapid Pressure Swing Adsorption (RPSA) technology. RPSA processes are cyclic, and use materials (most commonly zeolites) which selectively adsorb one species in a gas mixture. The MOC market is large, estimated to be more than $1.8 billion in 2017, with many manufacturers trying to reduce the size and weight of ever smaller MOC devices. We recently designed and built a novel, single bed MOC device which features a single adsorber column concentrically inserted into a product storage tank. The single bed design reduces the overall size and weight of the MOC device and does not require the synchronization of multiple adsorbent beds found in other MOC devices. Operating RPSA systems is highly complex due to fast (<10 sec) total cycle times, flow reversals, non-steady dynamic behavior and other effects, which makes controlling MOC devices both challenging and necessary. As MOC devices become more portable, they are also subject to changing operating conditions which further complicate their design and operation. In our previous work, we demonstrated that the single-bed design poses a multivariable control problem, and we developed a Model Predictive Control (MPC) algorithm in simulation which controls the RPSA process by manipulating the cycle step durations. The MPC features a linear model identified from a data-driven sub-space identification procedure and PseudoRandom Binary Sequence (PRBS) input signals as well as a convex quadratic optimization program with linear inequality constraints. The advantage of using a data-driven model identification technique is that the same procedure used in simulation can be repeated on the prototype unit, thus eliminating potential modeling errors in the controller formulation.

In this work, we present the implementation of this multivariable MPC algorithm on a lab-scale MOC prototype using Raspberry Pi® hardware. Python code and custom circuitry is used to both operate the RPSA cycle and solve the MPC quadratic program in real-time on a cycle-to-cycle basis. The model identification procedure uses measured output data, carefully designed PRBS-type input signals and the sub-space identification algorithm n4sid to generate the linear model used in the MPC. Implementation challenges such as hardware requirements, sensor dynamics, measurement error and task synchronization and imposed safeguards are discussed in detail. The closed-loop MOC-MPC system is able to operate reliably at the desired performance by operating the RPSA cycle and rejecting process disturbances. Several disturbance rejection and set point tracking case are presented to evaluate the MPC performance.

534112 Online Scheduling Design Principles

Abstract:

It is now widely recognized that unforeseen disruptions or arrival of new information can make the current schedule of chemical production facility suboptimal or even infeasible, thus motivating the need for online (re)scheduling (Subramanian et al., 2012; Gupta and Maravelias, 2016). However, while the need for online scheduling is recognized, there are a number of open questions regarding the design of online scheduling algorithms. Three pertinent, but for the most part unexplored, aspects of an online scheduling algorithm are (1) the look-ahead or prediction horizon length, (2) the frequency of rescheduling, and (3) the incorporation of feedback through re-optimization (Gupta et al., 2016).

The goal of this work is to develop the principles and a methodology that would allow us to address the following fundamental question: Given a production environment (i.e., resource availability and recipes) and the distributions of the disturbances, what are the best rescheduling frequency and prediction horizon?

We first study how demand uncertainty affects the quality of the implemented schedule. For example, does the time at which uncertainty is observed affect solution quality? How does demand variability affect solution quality? Do different demand patterns, characterized by, for example, the frequency of orders, pose any particular challenges? In addition, how do network characteristics (e.g. shared intermediate materials, availability of buffer production capacity), aggravate or diminish the above effect?

Thereafter, we identify the key factors that affect the selection of online scheduling parameters, namely, production load, demand uncertainty, and process network characteristics. Second, for each factor, we define an â€œauxiliary parameterâ€ to quantify it. In some cases, defining this auxiliary parameter is straightforward (e.g., use standard deviation to describe demand variability) but in some cases it is far from trivial. For example, quantifying the load of a facility, subject to a demand pattern, depends on a series of factors and there is no explicit expression to calculate it. We present procedures to calculate all these auxiliary parameters. Third, we present methods to determine the online scheduling parameters (e.g., horizon of optimization problem and re-optimization frequency) from the instance-specific auxiliary parameters. Finally, to test our framework, we present results for various networks.

Interestingly, the work presented herein can be viewed as the counter part of controller design for dynamic systems. Given a production system (rather than a dynamic process) described in terms of the aforementioned auxiliary variables (rather than time constants(s)), we decide the parameters of the online scheduler (rather than controller). To our knowledge, the proposed framework is the first systematic attempt to design the online scheduling algorithm, and thus takes us one-step forward, towards synthesizing a general strategy to obtain high-quality closed-loop schedules.

References

Subramanian, K.; Rawlings J.B.; Maravelias, C.T. A state-space model for chemical production scheduling. *Computers & Chemical Engineering* 47 (2012): 97-110.

Gupta, D.; Maravelias, C.T. On deterministic online scheduling: Major considerations, paradoxes and remedies. *Computers & Chemical Engineering*, 94 (2016), 312â€“330.

Gupta, D.; Maravelias, C.T.; Wassick, J.M. From rescheduling to online scheduling. *Chemical Engineering Research and Design*, 116 (2016), 83-97.

524709 Demand Response-Oriented Modeling and Production Scheduling Optimization for Chlor-Aklali Processes

Abstract:

The recent years have witnessed a significant increase in the amount of electricity generated from renewable resources, such as wind and solar power plants. These resources have significant environmental benefits but present the challenge of significant variations in generation rate, both over the course of the day and from season to season. Energy storage can, in principle, be used to smooth these fluctuations, but remains prohibitively expensive at the grid level. Thus, incorporating renewable resources in the power grid calls for the development of new operating strategies. Demand-side management or demand-response (DR) is one of the most widespread approaches to balancing power generation with power demand. In the DR context, customers are encouraged (via time-varying price signals or other economic incentives) to curtail their consumption during peak grid demand times. Such strategies are particularly amenable for industrial implementation, given that industrial electricity use is not driven by human needs and behaviors as is the case for residential demand.

Chlor-alkali production is one of the most electricity-intensive chemical processes, historically absorbing about two percent of the electricity generated in the United States [1,2]. The end products of this process, chlorine and caustic soda, are highly valuable chemicals that support other major industries like alumina smelting, organic chemicals, building and manufacturing (through the production of plastics and polymers etc.). The North American chlor-alkali market was valued at USD 13.68 billion in 2015 [3].

The majority of chlor-alkali production relies on brine electrolysis [4, 5]. Given the high electricity demand on this process, important research efforts have been targeted at improving the process efficiency, including the development of energy efficient membranes [6] and sophisticated electrodes [7]. Incipient efforts aimed at the intensification of the process have also been reported [8]. Industrial demand response, as described above, remains an alternative solution approach to resolve the high energy cost/operating cost challenge [2, 4, 9].

Demand response by means of fluctuating production rates, overproducing products during off-peak demand hours when electricity is typically cheap and storing excess product to supplement reduced production rate when electricity demand peaks (and thus, prices) have been studied for electricity-intensive chemical processes including air-separation [10-12] and aluminum smelting [13]. However, our literature survey revealed a relative paucity of studies rigorously covering the DR operation of brine electrolysis processes, in spite of the fact that this strategy is currently applied in practice.

Motivated by the above, in the present paper, we investigate the provision of demand response services by brine electrolysis plants, focusing on engagement in in electricity markets with fast dynamics (markets where prices are only known a few minutes before the operating period and prices change every few minutes). Such engagements in short-term markets have been shown to be highly profitable [14, 15].

The dynamics of chemical processes are typically slow with dominant time constants in the order of hours. Consequently, scheduling plant operations for providing demand response in fast paced electricity markets must explicitly account for process dynamics [10, 11, 16]. In order to consider both longer process time constants and fast market dynamics, we present a demand response-oriented modeling framework for brine electrolysis, paying specific attention to the dynamics of relevant variables and phenomena such as the cell temperature and current-voltage relationship. Additionally, we develop a production scheduling formulation for the provision of demand response by the chlor-alkali industry in fast-paced electricity markets. We evaluate the financial benefits and discuss potential process limitations to the provision of fast demand response. Our work show that the chlor-alkali industry is capable of safely and profitably providing significant load reductions in fast-changing electricity markets.

References

1. Annual Energy Outlook, 2017. U.S. Energy Information Administration, pp. 122. Accessed online https://www.eia.gov/outlooks/aeo/pdf/0383(2017).pdf, on April 7, 2018.
2. A. Kent, Ed., Riegel's handbook of industrial chemistry, 10th ed. Springer, 2003.
3. North America Chlor-alkali Market Analysis, 2018 â€“ 2024. Grand View Research Inc. Accessed online https://www.grandviewresearch.com/industry-analysis/north-america-chlor-alkali-market, on April 7, 2018.
4. F. O'Brien, T. V. Bommaraju, and F. Hine, Handbook of Chlor-Alkali Technology: Volume I: Fundamentals, Volume II: Brine Treatment and Cell Operation, Volume III: Facility Design and Product Handling, Volume IV: Operations, Volume V: Corrosion, Environmental Issues, and Future Developments. Springer Science & Business Media, 2007, vol. 1.
5. Chemical economics handbook. IHS market. Accessed online https://ihsmarkit.com/products/chlorine-sodium-chemical-economics-handbook.html on April 7, 2018.
6. Gelb, Process for generating chlorine and caustic soda using a membrane electrolysis cell coupled to a membrane alkaline fuel cell, Mar. 3 1987, U.S. Patent 4,647,351.
7. Morimoto, K. Suzuki, T. Matsubara, and N. Yoshida, Oxygen reduction electrode in brine electrolysis," Electrochimica acta, vol. 45, no. 25, pp. 4257-4262, 2000.
8. Granados Mendoza, Intensification of the chlor-alkali process using a rotor-stator spinning disc membrane electrochemical reactor, Eindhoven University of Technology, 2016.
9. Chlistuno, Advanced chlor-alkali technology, Los Alamos, NM, USA, 2005.
10. Pattison, R. C., Touretzky, C. R., Johansson, T., Harjunkoski, I., & Baldea, M. (2016). Optimal Process Operations in Fast-Changing Electricity Markets: Framework for Scheduling with Low-Order Dynamic Models and an Air Separation Application. Industrial & Engineering Chemistry Research, 55(16), 4562-4584.
11. Cao, Y., Swartz, C. L., Baldea, M., & Blouin, S. (2015). Optimization-based assessment of design limitations to air separation plant agility in demand response scenarios. Journal of Process Control, 33, 37-48.
12. T. Kelley, R. C. Pattison, R. Baldick, and M. Baldea, An efficient MILP modeling framework for demand response operation of air separation units, accepted
13. Todd, M. Caufield, B. Helms, A. P. Generating, I. M. Starke, B. Kirby, and J. Kueck, Providing reliability services through demand response: A preliminary evaluation of the demand response capabilities of alcoa inc, ORNL/TM, vol. 233, 2008.
14. I. Otashu and M. Baldea, Grid-level battery operation of chemical processes and demand-side participation in short-term electricity markets. Applied Energy, vol. 220, pp. 562-575, 2018
15. Dowling, A. W., Kumar, R., & Zavala, V. M. (2017). A multi-scale optimization framework for electricity market participation. Applied Energy, 190, 147-164.
16. Baldea, M., & Harjunkoski, I. (2014). Integrated production scheduling and process control: A systematic review. Computers & Chemical Engineering, 71, 377-390

531257 Forty Years of Computers and Chemical Engineering (1977-2017): Analysis of the Field Via Natural Language Processing Techniques

Abstract:

Recent advances in machine learning and natural language processing have produced a variety of tools for computerized mining of textual databases. However, while the mining of numerical data has enjoyed considerable attention, relatively little work has been done by chemical engineers in the analysis of textual information. In this work, we analyzed a large body of literature in chemical engineering using state-of-the-art text mining techniques. Particularly, the focus of attention is in process systems engineering and computer applications in the field of chemical engineering. Given its prominence and relevance to the field, the forty year old journal of Computers and Chemical Engineering (CACE) was investigated for our purposes.

In this work we are interested in answering following two questions. What are the main topics that have appealed to the CACE community since the journal was launched? How has interest in these topics evolved over time? These questions were addressed through topic modeling techniques, including non-negative matrix factorization and the structural topic model [1-4]. The results show that CACE covers diverse topics from process synthesis to artificial neural networks. A total of 18 research topics were identified when we maximized topic coherence [5]. Two dominant research topics that draw a great attention to the community are mathematical programming and process control. The topic prevalence analysis shows that the use of machine learning techniques can be traced back to 1990s. Additionally, we discovered that there is a close relationship between topics "artificial neural network" and "fault diagnosis." These and other findings from our analysis will be discussed in detail.

References

[1] Greene, Derek, and James P. Cross. "Exploring the political agenda of the European parliament using a dynamic topic modeling approach." Political Analysis 25.1 (2017): 77-94.

[2] Roberts, Margaret E., et al. "Structural Topic Models for Open-Ended Survey Responses." American Journal of Political Science 58.4 (2014): 1064-1082.

[3] Roberts, Margaret E., Brandon M. Stewart, and Edoardo M. Airoldi. "A model of text for experimentation in the social sciences." Journal of the American Statistical Association 111.515 (2016): 988-1003.

[4] Wang, Chong, and David M. Blei. "Variational inference in nonconjugate models." Journal of Machine Learning Research 14.Apr (2013): 1005-1031.

[5] O'Callaghan, Derek, et al. "An analysis of the coherence of descriptors in topic modeling." Expert Systems with Applications 42.13 (2015): 5645-5657.

533596 Developing Engineering Thinking during an Ill-Structured Activity for Kinetics and Reactor Design: A Technology Solution

Abstract:

The majority of core engineering science courses in the School of Chemical, Biological, and Environmental Engineering at Oregon State University utilize a studio structure [1] where large lecture sections (100 - 350 students) are interspersed with smaller studio meetings (approximately 24 students). In studios, students work together in mostly 3-person teams, facilitated by trained graduate student teaching assistants (GTAs) and undergraduate learning assistants. Studios are designed to extend students' thinking and problem-solving techniques while simultaneously reinforcing core content and developing teamwork and communication skills. In its original manifestation, Studio 1.0, the activity relied on sequestered, worksheet-based problems. These problems helped students identify and practice key conceptual and procedural knowledge, and connect that understanding to lecture, but the activities were clearly limited in developing students' ability to connect the activity to professional practice and to develop value systems corresponding to the profession.

In the Studio 2.0 reform, our intent is to shift activity and re-situate learning by engaging students in *meaningful, consequential work* that directly and clearly relates to professional practice and desired professional attitudes and behaviors. The intent for these experiences is to provide a foundation for development of students' chemical engineering identity. Rather than attempting to direct students procedurally to a "correct" solution, a Studio 2.0 memorandum might explain a situation where a company is seeking to optimize a particular process and ask students to collaboratively decide on and perform calculations to make a design recommendation. Assessment is formative and immediate, focused on whether teams are "making progress" in grappling with the task. In this framing, as learners struggle with difficult concepts and may even sometimes fail to accomplish their short-term goals, they are continually positioned as engineers seeking meaningful progress towards a viable solution, rather than students following directions to get a grade.

In this paper, we present a four-hour studio progression where students work in teams in studio on a situated problem where they must determine the rate constant for a sugar reaction and, importantly, recommend a time for manufacturing to yield a minimum conversion. We have developed a technology-based solution where the teams are provided data that is generated by simulation. Each team has two reactors to work with and the simulation generates unique results for the team. Variation is added mandating the teams apply statistical reasoning in addition to utilizing their knowledge of kinetics. For some teams, the reactors are different, for others they are the same. The technology solution prompts students to respond to a set of prompts in which they report their findings and engineering recommendations and justify them. We report on how we use these collected data to better understand the way students are making choices in this ill-structured space, how students within the same team (with the same data sets) align, and how we can interpret the results to improve the activity design. Our ultimate goal is to automate these assessment processes to provide an instructor a broad understanding of the choices students are making that supports delivery and assessment of more authentic ill-structured tasks.

[1] Koretsky, M. D. (2015). Program level curriculum reform at scale: Using studios to flip the classroom. *Chemical Engineering Education*, *49*(1), 47-57.

531450 Reading and Repetition Using an Interactive Textbook for Material and Energy Balances

Abstract:

zyBooks creates interactive textbook replacements, and a zyBook for a material and energy balances course will be discussed. Features of the interactive textbook were highlighted previously, including animations, question sets, and auto-graded homework questions. Reading analytics have been presented previously with average reading rates of 87 and 93% for 2016 and 2017, respectively (n = 100 in 2016 and 88 in 2017). Now, new research questions are asked based on large sets of web analytics generated by students using/reading the interactive textbook. The high reading rates correspond to thousands of complete animation views, and new data will cover the most popular animations from the number of students re-watching an animation. Further analysis of studentsâ€™ attempts and success when completing auto-graded homework problems within randomized numbers and content will be presented. For the 2018 cohort, over 280 auto-graded questions were assigned to the students. Questions needing the most attempts before correct will be identified and analyzed. Finally, in addition to re-watching animations, students complete many attempts *after* answering the auto-graded questions correctly and earning their grade, i.e., performing self-guided repetition.

528915 Check Your Homework with Your Phone

Abstract:

Recently, computers have been used to deliver homework, quiz, and examination problems that were traditionally paper based. These electronic problems have several advantages for the instructor as well as the students. Advantages for students include instant feedback, multiple tries, and follow-up computer-based instruction to address any learner deficiencies. Less instructor grading and difficulties with students copying solutions manuals, along with the possibility of large problem banks are some of the advantages for instructors.

However, there are several disadvantages to computer-based delivery systems. Students are normally bound to the computer and an internet connection while they are working the problem. While this can normally be accomplished, it is certainly not as flexible as a problem on a piece of paper. Another disadvantage is that engineers are normally encouraged to clearly document their work and this step is less likely to be reinforced with a computer-based homework system.

The biggest disadvantage however, may be that students are less likely to efficiently engage the problems that are presented on a computer. This is especially true with traditional engineering education problems that tend to have several pieces of information, involve physical laws, mathematical manipulation and logical reasoning. When given these types of problems on the computer, some students will attempt to do more of the steps in their heads, usually becoming frustrated, instead of transcribing the material from the computer to a piece of paper to manipulate it. Obtaining a hard copy by printing it can sometimes be difficult if the problem is not formatted for printing.

To address these issues, a problem repository has been developed and continues to be populated with engineering problems that are made to be printed. Unlike traditional paper-based problems, each student receives a different version of the problem. In addition, the printed file contains a QR Code and website where students can check their responses with their phone or computer. Context specific instructional materials are provided on the website for students who need instructional support. Besides homework problems, this system lends itself to in-class practice and group problems especially for classrooms that are not equipped with computers.

Many of these problems have been tested in several chemical engineering classes including Energy Balances, Thermodynamics, and as a curriculum review in Chemical Process Design I.

The proposed system combines the flexibility and efficiency of a paper-based system with many of the advantages of a computer-based system and allows students to get immediate feedback utilizing their cell phones. The student interface for this problem system will be demonstrated in the session.

533334 Educational Augmented Reality Tools: Development, Implementation, and Assessment of Phase I

Abstract:

Pairing teaching and learning strategies with appropriate device and technology trends can greatly enhance the motivation to learn [1]. Through years of empirical studies, research on the use of computer or Internet in education has come to understand the relationships between learnersâ€™ characteristics and technology use, and has already shown some effectiveness of technologies in promoting learning outcomes [2]. Wearable technologies equipped with Augmented Reality (AR or Mixed Reality capabilities offer a lot of promise for education and training [3]. Such technologies have features that greatly align with the expectations of new student learners. At the same time, unlocking the full potential of these emerging technologies through proper tools and application implementation is essential for realizing the intended benefits [4]. However, compared to studies of more mature technologies in education, research of AR applications in education is in an early stage, and evidence of the effects of AR on teaching and learning appears to be shallow [5].

During the past years, in the context of EduART I (Transformative Educational Experiences 2016, a Texas A&M at Qatar grant), we investigated the use of AR to enhance the instructional experience of engineering students. A team of three faculty, one laboratory staff, two students, and an external programmer developed a visual support tool as a means to improve learning of chemical engineering students engaged in a specific laboratory activity while mitigating the challenges of laboratory work. The pilot demonstration of EduART I took place October-December 2017, where around 30 students, from multiple levels, participated. The experimental evaluation showed encouraging results, 43% showed significant and marginal improvements on the learning outcomes. A tremendous impact on the learning attitude was observed while impressive satisfaction and significant cognitive validity were captured in the self-surveys. To the best of our knowledge, there are very few similar efforts in engineering education and here we share the experiences gained during the development, implementation and assessment of EduART.

537060 Preparing Chemical Engineering Students for the Digitalization of Tomorrow – Integrating Modelling across the Curriculum

Abstract:

The advent of increased computational power, availability of more data and improved algorithms is allowing industry to "digitalize" operations. In essence, this means industry is able to use models to represent the plant accurately, use large amounts of data to determine the state of the plant, and use algorithms to optimize the plant performance on-the-fly. To prepare students for such a working environment, it is important to train students in the fundamentals of modelling. At present, modelling often does not go much beyond a flowsheeting exercise, which provides little understanding of the fundamentals underlying a model, and often has become a simple enter-data-and-push-button exercise. Modelling plays an important underlying role in almost all chemical engineering areas, from reaction engineering, physical transport phenomena to thermodynamics and of course process systems engineering, because the core laws, and chemical and physical phenomena, are expressed in equations. Introducing modelling as an integral part of a curriculum is therefore not only logic, but crucial in order to prepare students for the industry of tomorrow.

The Department of Chemical Engineering at University College London has successfully implemented an Integrated Engineering Program (IEP) where modelling is integrated across the entire curriculum, rather than being covered in a single course in the junior or senior year. To support the integration, PSE has developed and provided training material, exercises and tests through the Process systems engineering Academic Teaching Highway (PATH) initiative. The material is modular in nature and covers the typical fundamental topics from a modelling perspective and are supplementary to existing course material. We will reflect on the implementation and delivery of the new curriculum over the past four years, and how the PATH material has supported the students' learning, ultimately preparing them for a digital future.

537567 Wastewater Minimization and Energy Conservation Software Developed and Used in Teaching Process Systems Engineering at Vanderbilt

Abstract:

Russell F. Dunn, Ph.D., P.E.

Professor of the Practice

Department of Chemical and Biomolecular Engineering

Vanderbilt University

Nashville, TN 37235-1604

Ph: 615-343-4268

russell.dunn@vanderbilt.edu

Scott A. Guelcher, Ph.D.

Associate Professor

Department of Chemical and Biomolecular Engineering

Vanderbilt University

Nashville, TN 37212

Ph: 615-322-9097

scott.guelcher@vanderbilt.edu

Bryan R. Beyer

Adjunct Instructor

Department of Chemical and Biomolecular Engineering

Vanderbilt University

Nashville, TN 37235-1604

Ph: 615-343-4268

bryan.r.beyer@vanderbilt.edu

Wastewater Minimization and Energy Conservation Software Developed and Used in Teaching Process Systems Engineering at Vanderbilt

Significant efforts have been devoted to the development of process integration design methodologies over the past several decades. Two specific and widely-used developments in process systems analysis are the identification of optimal heat exchange networks and water allocation networks. These systems-based tools are widely applicable and are now essential elements of undergraduate process design instruction. This presentation will highlight the development of customized heat integration and water network software at Vanderbilt through senior capstone design projects. It will then be shown how these custom software packages are used in teaching process systems engineering in our design course and in applying process systems engineering in our senior unit operations laboratory course, when coupled with pilot plant data generated in the lab.

528546 Aspen Plus® Videos for Chemical Engineering Undergraduates

Abstract:

The University of Kansas (KU) Department of Chemical and Petroleum Engineering (CPE) has introduced ASPEN Plus® in the sophomore level Material and Energy Balance (MEB) course using videos and handouts created by an undergraduate who has taken the course. The ASPEN Plus® modules teach MEB students how to create a chemical process – the hydrogenation of benzene to cyclohexane in the program, including reactors, heat exchangers, distillation columns, and multiple recycle loops. The presentation will highlight how the videos were created using Camtasia Studio 9 screen capture software, the advantages of traditional learning via handouts and more innovative learning through videos, and benefits of a "student teaching students" learning experience. The goal of the speaker (junior CPE undergraduate) is to create ASPEN Plus® training modules for each course of the KU CPE curriculum. Thus, by graduation, a library of videos will be available for students to use in all the chemical engineering courses at KU.

533479 Problem Solving Skills When Using Student-Generated Problems That Reverse Engineer YouTube Videos

Abstract:

Homework problems generally come from textbooks, and many solutions manuals are available on the Internet. Students finding solutions manuals likely focus on copying the right answer and putting little effort on learning. Complex problem solving skill are in high demand for engineers. Here, students apply course concepts by reverse engineering YouTube videos and creating new homework problems and solutions. This research is examining rigor and problem-solving skills related to the YouTube problems. A quasi-experimental, treatment/control group design is in process. Rigor of YouTube problems is examined using the NASA Task Load Index (TLX), and problem solving skill are examined using a modified version of Problem definition, representing the problem, organizing information, calculations, evaluating the solution, solution communication, and self-assessment (PROCESS). Results from the first cohort from the 2017-2018 academic year will be examined and expand upon results presented at the 2018 ASEE annual meeting.

540773 2018 Outlook for Energy: A View to 2040

Abstract:

The Outlook for Energy is ExxonMobil's global view of energy demand and supply through 2040. It also highlights the dual challenge of providing affordable energy to support prosperity while reducing environmental impacts – including the risks of climate change. We use the Outlook to help inform our long-term business strategies and investment plans.
The Outlook anticipates significant changes through 2040 across the world to boost living standards, reshape the use of energy, broaden access to abundant energy supplies, and accelerate decarbonization of the world's energy system to address the risks of climate change.

- Global energy needs will rise about 25 percent over the period to 2040
- Electricity demand will nearly double in non-OECD countries
- Worldwide electricity from solar and wind will increase about 400 percent
- Natural gas will expand its role, led by growth for electricity generation
- Growth in oil demand will be driven by commercial transportation and chemical needs
- Efficiency gains and growing use of less-carbon intensive energy sources will contribute to a nearly 45 percent decline in the carbon intensity of global GDP

Energy is fundamental to modern life, and as the world's population approaches 9 billion people in 2040, we are challenged to improve living standards everywhere. Understanding the factors that drive the world's energy needs – and likely choices to meet those needs – is the mission of the Outlook. By sharing the Outlook with the public, we hope to broaden that understanding. Energy matters to everyone, and we all play a role in shaping its future.

541543 Energy Decarbonisation Scenarios

Abstract:

With the energy sector being associated with 2/3 of the annual greenhouse gas emissions, decarbonizing the energy is a key step to reduce their level. The National Determined Contributions (NDC's) announced so far by the participating countries do not meet the objectives of the Paris-climate agreement, as far as reaching a peak in emissions in the near future or decreasing to a level that is compatible with the stated long-term targets.

Using different scenarios for energy demand and supply, options for energy sector decarbonisation are evaluated. Meeting the 2-degree target requires a major transformation in the power sector, efficiency improvements – specially in the buildings and in the industry, and a revolution in transport towards fuel and modal shifts. With a significant increase in renewable energy, system integration and optimization will also be key in reducing the cost of the energy transition. Meeting more ambitious climate goals ("well below 2 degrees") will push further the technological innovation and implementation requirements throughout the entire energy value chain. A scenario aiming at zero net-emissions by 2060 is presented, along with a comparison to the 2-degree case.

542625 Fundamental Research Needs to Advance Energy Technologies

Abstract:

The U. S. Department of Energy (DOE), Office of Science, Office of Basic Energy Sciences (BES) supports fundamental research in chemical and materials sciences to provide the foundations for new energy technologies and to support DOE missions in energy, environment, and national security. This presentation will discuss opportunities for fundamental research to impact DOE's energy mission "to catalyze the timely, material, and efficient transformation of the nation's energy system and secure U.S. leadership in energy technologies" with a focus on the way we generate, store and use energy nationally. I will provide an overview of BES strategic planning over the past decade that identified priority research directions for advancing energy applications, highlight key scientific advances in these areas, and discuss some future opportunities for modern science, particularly chemical sciences, to accelerate the transformation of the U. S. energy portfolio.

538025 On the Value of Global Optimality in Machine Learning Problems: The Case of Feature Selection for Linear Support Vector Machines

Abstract:

Numerical optimization is at the core of many contemporary machine learning algorithms. Often, the optimization problems of interest are nonconvex (e.g., training neural nets) or involve binary decisions (e.g., zero-norm feature selection). However, due to the extreme size of the problems of interest in this domain, the vast majority of optimization algorithms in common use are based on satisfying first-order optimality conditions, and hence furnish only locally optimal solutions, often with no indication of how far from global optimality they may be. In recent years, there has been increasing interest in methods with better optimality guarantees, either through increased capacity to escape from suboptimal local minima, or through convex relaxations or other tractable approximations with known theoretical relationships to the original problem (i.e., provable bounds on the difference in their optimal solutions). There have also been some attempts to apply rigorous global optimization methods, particularly in the case of problems that can be formulated as mixed-integer linear programs, although this is a very significant challenge. The tacit assumption in all of these approaches is that a global solution will necessarily perform better than a suboptimal local solution for the learning task at hand. However, this need not be the case because the performance metric of interest in most machine learning tasks (e.g., accurate classification of new data points in the future) is distinct from the objective function minimized during optimization (e.g., classification error for given training data, along with additional regularization terms). Given this fact, what is the value of achieving global optimality in such problems? When should we expect a clear correlation between suboptimality during training and degraded performance in practice? And how strong is this correlation (i.e., how hard should we be willing to try to achieve global optimality)?

In this talk, we will present our recent results investigating the value of global optimality in the context of zero-norm feature selection for linear support vector machines (SVMs). Linear SVMs aim to classify data points in a high-dimensional space into one or more clusters based on separating hyperplanes. The individual components of each data point are called features. For example, the features may represent expression levels for various genes in a cell, while the clusters might represent different disease classes. In this context, feature selection refers to the problem of selecting a specified number K of features (e.g., genes) from the full set such that classifying the data based on only those K features is as good or better than classifying the data based on any other choice of K features. This problem is motivated by the observation that classifiers based on fewer features often generalize better from the training data to new data sets, especially when there are many noisy features in the original data set. Moreover, some interesting applications call for a very aggressive reduction in the number of features used for classification. For example, in the design of biosensors that aim to infer the levels of pollutants or toxins in their environment based on the gene expression of immobilized microbes, it is impractical to accurately measure more than a small number of genes (~5).

In general, zero-norm feature selection is a very large-scale mixed-integer quadratic optimization problem (MIQP) that is difficult to solve to global optimality. However, modern MIQP solvers are capable of solving such problems for data sets of modest size. Using several standard data sets in this size range from various application areas, we tested the classification accuracy of SVMs with fixed numbers of features obtained by solving the zero-norm feature selection problem to guaranteed global optimality, as well as by using a variety of more efficient approximations that are in common use but furnish only approximate solutions. Moreover, to more directly test the correlation between suboptimality and classification accuracy, we also developed a rigorous technique for computing a sequence of progressively more suboptimal SVMs for each test problem in a controlled manner, ranging from globally optimal solutions to solutions with objective values that deviate from global optimality by up to 25%. For all of the SVMs obtained, we then evaluated their classification errors on new data using a standard five-fold validation procedure. Interestingly, our results indicate that, when a moderate to large number of features are allowed to be retained (>15%), there is very little correlation between suboptimality and classification error. In other words, SVMs that achieve objective values within 10-20% of global optimality during training performed just as well as global solutions on average when classifying new data. In contrast, when only a small number of features are allowed to be retained (~5%), as in the biosensor example mentioned above, a very clear positive correlation emerges between suboptimality during training and degraded classification performance on new data sets. Moreover, in comparisons with SVMs obtained by existing state-of-the-art feature selection algorithms, the SVMs obtained by rigorously solving the feature selection problem to global optimality were not substantially more accurate at high feature selection levels, but showed significant advantages at low feature selection levels (~5%). In this talk, we will present the data supporting these conclusions for several test cases, provide hypotheses for these surprising trends with supporting evidence, and discuss the implications of these trends for the usefulness and practicality of rigorous global optimization for feature selection.

532682 Towards Developing a Learning Based Evolutionary Assistive Paradigm for Surrogate Selection (LEAPS2)

Abstract:

Complex systems are typically studied via physical experiments, computer experiments or their combinations. Physical experiments involve conducting laboratory trials, making field observations, etc. and usually compute intensive. Besides, some experiments may not even be feasible in practice. In such cases, computer experiments may be preferred over physical experiments. These involve experimenting on a rigorous first-principles or physics-based model instead of a real system. Revolutionary advances in algorithmic and computing technologies over the last two decades have empowered researchers to incorporate greater details and accuracy into such models. However, this comes at the expense of larger model size and greater computational burden. Repetitive evaluations of such high-fidelity models for tasks such as sensitivity analysis and optimization often prove uneconomical. Commercial simulators, which are essentially black-box in nature, are often used for developing and simulating these high-fidelity models. Additionally, the development and implementation of such models demand deep domain expertise, which makes them inaccessible to users from other disciplines. Therefore, it is beneficial to replace high-fidelity models (in case of computer experiments) or real physical systems (in case of physical experiments) by computationally cheaper surrogate models that offer a simpler overall picture of the underlying system.

A surrogate model, also known as a meta-model or a response surface, is an empirical expression quantifying the relationships among the most important or relevant inputs and outputs of a system. It is computationally much cheaper alternative to a physical system or its high-fidelity model and is easily comprehensible to users with little domain knowledge. Its construction requires one to obtain the system response at several sample points via experiments or simulations. Then, one needs to choose a form and technique for the surrogate model development. The form is a mathematical relationship between the inputs and outputs, while the technique is a procedure to derive the best parameters of that form for the sampled input-output data. These two steps together yield a final model that we call the surrogate model. Henceforth, we will use "modelling technique" to mean the surrogate modelling technique, and "surrogate" to mean the final surrogate model.

The literature offers a variety of modelling techniques such as polynomial response surface models, support vector regression, kriging, radial basis functions (RBF), multivariate adaptive regression splines, and artificial neural networks. Many techniques (*e.g.* kriging and RBF) offer a variety of functional forms, hence can yield several surrogates. Therefore, selecting the best modelling technique and the final surrogate are non-trivial tasks that have largely been done based on trial and error or intuition. Therefore, we focus on the choice of the modelling technique, corresponding functional form, and the factors that impact these decisions. To this end, our work addresses the following two key questions that a typical user may ask while developing a surrogate from a given set of input-output data.

1. Which is the best surrogate for approximating my system/data set?
2. Is there a systematic, automated, user-friendly, efficient, and reliable procedure for selecting a few

surrogates that are likely to be the best?

Although several previous works have attempted to answer the above questions, most are straightforward and enumerative benchmarking studies. They numerically compare and rank the surrogates for some well-known test functions. Inferences from such studies are obviously limited and their extrapolation to unknown systems may not be reliable. Naturally, this has inspired many researchers to explore alternate approaches [1, 2]. Their key idea is to use a set of simple basis functions (*e.g.* sin, cos, exp, *etc.*), and then iteratively evolve their best mix by solving a series regression problems. In other words, this approach is more adaptive and more system-targeted than the earlier benchmarking studies. It is further enhanced by some researchers by employing a variety of more complex surrogates in place of simple analytical functions [3]. Despite their generalized appeal, all these approaches consider surrogate selection as a stand-alone task. For every system, a user needs to solve an optimization problem that guides surrogate selection. Solving such optimization problems is not only arduous but also compute-intensive; and defeats the underlying incentive for using surrogates.

In this work, we develop an evolving knowledge-based framework that defines and computes prior information from sampled data of several systems to aid surrogate selection for other systems. We call this **L**earning based **E**volutionary **A**ssistive **P**aradigm for **S**urrogate **S**election (LEAPS2) and its key features are as follows:

- It relies on the most commonly employed modelling techniques and their corresponding functional forms.
- It defines several system attributes (based on the response, gradient, and extrema) and performance metrics to characterize and quantify information about surrogate selections for various systems.
- It utilizes this prior information to derive knowledge that drives the surrogate selection for unknown systems with minimal surrogate construction efforts.
- It can perform surrogate selections over wide ranges of dimensions and sample sizes; thus, making it practically useful.
- It enables users to add new modelling techniques as well as system attributes without disturbing the existing architecture of the paradigm.

LEAPS2 relies on the concept of knowledge pyramid to select surrogates based on the input-output data alone and with minimal surrogate fitting efforts. The knowledge pyramid involves three phases: (1) Data collection, (2) Information extraction, and (3) Knowledge derivation. The development of LEAPS2 begins with the data collection phase involving the following steps:

(D1) Select 66 test functions as representative systems.

(D2) Select 25 surrogates resulting from six modelling techniques mentioned earlier.

(D3) Use Sobol sampling (QS) to generate four input-output data sets for each test function. This amounts to total 264 data sets.

(D4) Construct 25 surrogates for each input-output data set yielding 6600 surrogates in total.

This yields a rich set of system-surrogate data which can be updated dynamically as and when more data become available. We then process these data to extract information.

(I1) Compute system attributes to extract the essence of the data sets from (D3).

(I2) Compute performance metrics to quantify the quality of the surrogates.

From the information computed in steps (I1) and (I2), we quantify and embed knowledge into LEAPS2 using regression tree ensemble.

(K1) Discard correlated attributes via the attribute selection called RReliefF.

(K2) Correlate the selected system attributes with surrogate performance metrics using regression tree ensemble. This serves as the brain of LEAPS2 for future surrogate selections. The fully-grown ensemble then can be used to select the best possible surrogate out of 25 surrogates for any given data set.

The learning and evolution of LEAPS2 towards a powerful system modelling tool can occur along three dimensions, *viz.* data sets, attributes, and surrogates. Our progressive development of LEAPS2 in this work shows its evolution with respect to data sets. As we added more and more data sets progressively, LEAPS2 learnt. In other words, by progressively adding data, we demonstrate that LEAPS2 learns to improve computational efficiency and functional accuracy. Apart from data sets, the architecture of LEAPS2 enables its evolution via more attributes and surrogates. The addition of the new attributes will equip LEAPS2 for better system identification while the new surrogates will provide more options for system approximation. Overall, such learning will impart more versatility to LEAPS2 and extend its ability to handle more complex systems.

Besides its learning ability, we demonstrate the practicality of LEAPS2 by employing it to recommend surrogates for estimating the bubble and dew point temperatures of LNG. Interestingly, our assistive tool suggests a different surrogate for each temperature, and hints that DPT may be harder to approximate than BPT. Finally, we compare the performance of LEAPS2 with the metalearning-based recommendation scheme (CRS) of Cuit *et al.* [4]. CRS in contrast to LEAPS2 recommends only a surrogate modelling technique rather than a surrogate model. Therefore, a straightforward comparison is not entirely fair. However, we can still compare CRS and LEAPS2 based on their success in identifying the best modelling technique. To this end, we use the LEAPS2 to recommend the best modelling techniques for all 264 data sets. LEAPS2 identifies the true best for 236 data sets, thus achieves coefficient of success (CoS) of 0.89. CRS achieves an average CoS of 0.84 [4]. Clearly, LEAPS2 performs better than CRS in recommending modelling techniques despite its generalized philosophy.

In brief, LEAPS2 is a comprehensive, practical, and assistive data-driven surrogate selection paradigm that holds much promise.

References:

[1] Lessmann, S., Stahlbock, R., & Crone, S. F. (2006). Genetic algorithms for support vector machine model selection. In Neural Networks, 2006. IJCNN'06. International Joint Conference on (pp. 3063–3069). IEEE.

[2] Yeun, Y.-S., Ruy, W.-S., Yang, Y.-S., & Kim, N.-J. (2004). Implementing linear models in genetic programming. IEEE transactions on evolutionary computation, 8, 542–566.

[3] Goel, T., Haftka, R. T., Shyy, W., & Queipo, N. V. (2007). Ensemble of surrogates. Structural and Multidisciplinary Optimization, 33, 199–216.

[4] Cui, C., Hu, M., Weir, J. D., & Wu, T. (2016). A recommendation system for metamodeling: A meta-learning based approach. Expert Systems with Applications, 46, 33–44.

522672 Explore the Potential of Machine Learning in Building Reaction Models for Chemical Industry

Abstract:

Process optimization has been widely considered to maximize the benefit of chemical manufacturing plants, either from a technical and/or economic point of view. Machine learning techniques like neural network have been used to approximate equation-based first-principle reaction models [1] or numerical simulation models with excellent accuracy and efficiency. We have developed a standard framework to combine neural network based reaction model in process optimization model and demonstrated its excellent performance. However, unlike equation-based models and simulation models, machine learning models heavily rely on training data. This limitation requires machine learning models to be re-trained for different reactions and any major changes in the reactors. The goal of this research is to explore the potential of developing a generalized machine learning model that can cover a series of reactions and a large reactor geometry space. Success of this work will facilitate knowledge transfer among similar plants and similar reactions with even limited production data or experimental measurements. Potentially, the advanced machine learning model will help plant diagnosis and new plant design.

References

[1] C. A. O. Nascimento, R. Giudici, R. Guardani; Computers and Chemical Engineering 24 (2000) 2303â€“2314

534719 Data-Driven Optimization with Implicit Constraints: Application to an Ethane Steam Cracking Process

Abstract:

Data-driven modeling and derivative-free optimization relies on different sampling strategies that provides an initial plan for the controlled experiments on problem simulators, which is commonly known as the Design of Experiments (DoE). The goal of DoE is to provide possible candidate locations for input variables within the pre-defined box-constraints in such a way that these experiments capture a variety of system dynamics. However, the DoE is a statistical procedure and not guided by the physical information that entails an engineering process. Thus, a subset of candidate initial points generated by the DoE may result in unphysical and/or undesirable outcomes, such as an early termination of the problem simulator due to failures or solving a numerically unstable problem (stiff Ordinary Differential Equations, ODEs). This generally implies that a constraint between the optimization variables exists, in which the explicit analytical formulation of this, as a function of the input variables, is unknown to the user. As a result, the global optimization of such a system using a data-driven methodology will be hindered since the returned optimal solution may not be a feasible one.

The aim of this work is to handle implicit constraints that might exist between the input variables and guide the initial sampling strategy in such a way that the unphysical/numerically unstable combinations of input variables are filtered out, leaving only an appropriate set of samples for simulating the problem. To this end, a supervised machine learning algorithm, Support Vector Machine (SVM) is employed. The problem is formulated as nonlinear SVM classification problem where an optimal hyperplane, separating numerically stable and unstable samples, is built to implement an implicit constraint for the input space. The model is trained and cross-validated using a large set of simulated samples offline, and then incorporated in the AlgoRithms for Global Optimization of coNstrAined grey-box compUTational problems (p-ARGONAUT) [1-3] framework to assess the feasibility of the candidate points *a priori* to sample collection. This methodology has been tested on an ethane steam cracking process, where the continuity, energy and momentum balances are characterized by stiff ODEs in which our objective is to maximize the profitability of operation. The results show that SVM model provides a highly accurate implicit constraint and helps the grey-box optimizer to return feasible profitable solutions to the steam cracking optimization process.

References

[1] Boukouvala, F. & Floudas, C.A. ARGONAUT: AlgoRithms for Global Optimization of coNstrAined grey-box compUTational problems. Optimization Letters, 2017, 11:895-913.

[2] Boukouvala, F., Hasan, M.M.F. & Floudas, C.A. Global optimization of general constrained grey-box models: new method and its application to constrained PDEs for pressure swing adsorption. Journal of Global Optimization, 2017, 67:3-42.

[3] Beykal, B., Boukouvala, F., Floudas, C.A., Sorek, N., Zalavadia, H. & Gildin, E. Global Optimization of Grey-Box

Computational Systems Using Surrogate Functions and Application to Highly Constrained Oil-Field Operations. Computers & Chemical Engineering, 2018, https://doi.org/10.1016/j.compchemeng.2018.01.005

536413 Recurrent Neural Networks, Numerical Integrators and Nonlinear System Identification

Abstract:

Based on older work [1-4] we formulate the problem of identifying nonlinear ODEs and PDEs from time-dependent data as a recurrent neural net templated on established numerical simulation algorithms: initial value problem solvers, like Runge-Kutta methods for ODEs, and method of lines for PDEs.

"Unrolled" feedforward as well as recurrent architectures (with shared parameters) are used – more importantly, we consider implementations templated on implicit integrators (like the Crank-Nicolson), and explore the ability of these architectures to learn relations (and not just functions).

We couple these methods with nonlinear manifold learning techniques (giving us data-driven variables in terms of which the ODEs/PDEs are formulated) and illustrate the approach by identifying nonlinear effective PDEs arising in the study of large ensembles of heterogeneous networks. In particular, we study models of the neuronal network of the SCN (the suprachiasmatic nucleus, responsible for controlling circadian rhythms).

Finally, we note connections of the form of our learned network--with an explicitly included differential equation initial value solver--to popular recurrent network architectures of less specific form, such as the gated recurrent unit (GRU) or long short-term memory (LSTM) networks.

[1] Rico-Martínez, R., & Kevrekidis, I. G. (1995). Nonlinear system identification using neural networks: dynamics and instabilities. In A. B. Bulsari (Ed.), Neural Networks for Chemical Engineers (pp. 409–442). Elsevier.
[2] González-García, R., Rico-Martínez, R., & Kevrekidis, I. G. (1998). Identification of distributed parameter systems: A neural net based approach. Computers & Chemical Engineering, 22(98), S965–S968.
[3] Rico-Martínez, R., Kevrekidis, I. G., Kube, M. C., & Hudson, J. L. (1993). Discrete- vs continuous-time nonlinear signal processing attractors, transitions and parallel implementation issues. In American Control Conference (pp. 1475–1479).
[4] Anderson, J. S., Kevrekidis, I. G., & Rico-Martínez, R. (1996). A comparison of recurrent training algorithms for time series analysis and system identification. Computers & Chemical Engineering, 20(96), S751–S756.

537013 Data-Driven Identification of Interpretable Reduced-Order Models Using Sparse Regression

Abstract:

Reduced-order models (ROMs) can be thought of as computationally inexpensive mathematical representations, of a high-fidelity dynamical model that preserves essential behaviors and dominant effects that offer the potential for near real-time analysis. The field of reduced-order modeling is large, and new techniques are developing rapidly. Most of these model reduction methods can be classified into either subspace-based or projection-based methods. Subspace-based methods, including Subspace-based State Space System Identification (N4SID) [1], Multivariable Output Error State Space (MOESP) [2] etc., are attractive in that they provide accurate state-space models for multivariable linear systems directly from input-output data. They have also been applied for nonlinear systems with limited success provided the system is properly excited. However, the states of the identified system are void of any meaning in the physical sense and moreover, a large order is necessary to fully resolve the nonlinear dynamics. The projection or spectral-based methods involve decomposing a vector field into a set of modes and projecting the dynamics onto a low-dimensional space in which the solution lies. These methods assume a priori knowledge of the state evolution with system dynamics given by a set of partial or ordinary differential equations. Proper Orthogonal Decomposition [3], Eigensystem Realization [4] and Dynamic Mode Decomposition [5] are some of the widely used projection-based model reduction methods.

A recent breakthrough in nonlinear model identification has been triggered by the use of regression techniques to discover the governing equations of dynamical systems by balancing model complexity with accuracy [6,7]. In this work, we approach the reduced order modeling problem from the perspective of compressive sampling and sparse regression. We rely solely on time series data collected at a fixed number of spatial locations to identify parsimonious and physically interpretable ROMs. The fundamental assumption in this method is that there are only a few relevant terms that dictate the system dynamics. This is a reasonable assumption which holds for many physical systems as long as a set of appropriate basis functions is selected. Within this context, we perform nonlinear sparse regression on a large library of potential candidate functions to determine the fewest terms that most accurately represent the data. Therefore, the resulting ROMs are functions of the original system states and inputs, and the identified dynamical system is interpretable. The innovations in regression techniques have resulted in several attractive features, such as avoiding overfitting, selecting sparse models and sampling strategies to deal with large amounts of data, which can be readily applicable here. This method is not limited to linear model identification as several nonlinear terms can be included in the candidate library. Additionally, if any a priori information regarding the system is available, it can be easily incorporated into the algorithm. We demonstrate the performance of the proposed regression-based model reduction method to design a closed-loop operation of a hydraulic fracturing process, which is characterized by a system of nonlinear highly-coupled PDEs with time-dependent spatial domain [8]. In addition to achieving the desired model approximation accuracy, by introducing delayed inputs into the model the proposed algorithm readily realizes the time-delay structure which primarily characterizes a fracturing operation. We also compare the performance of the models identified using the proposed and the subspace-based MOESP methods. The results show that the proposed method is competitive with respect to the MOESP both in terms of ROM accuracy and closed-loop performance.

Literature cited:

[1] Van Overschee, P., De Moor, B., 1994. N4SID: subspace algorithms for the identification of combined deterministic-stochastic systems. Automatica, 30, 75–93.

[2] Van Overschee, P., De Moor, B., 1996. Subspace Identification for Linear Systems: Theory – Implementation – Applications. Springer, New York.

[3] Holmes, P., Lumley, J.L., Berkooz, G., 1996. Turbulence, Coherent Structures, Dynamical Systems and Symmetry. Cambridge University Press, New York.

[4] Juang, J.N. and Pappa, R.S., 1985. An eigensystem realization algorithm for modal parameter identification and model reduction. Journal of guidance, control, and dynamics, 8(5), 620-627.

[5] Schmid, P.J., 2010. Dynamic mode decomposition of numerical and experimental data. J. Fluid Mech. 656, 5–28.

[6] Bongard, J. and Lipson, H., 2007. Automated reverse engineering of nonlinear dynamical systems. Proceedings of the National Academy of Sciences, 104(24), 9943-9948.

[7] Brunton, S.L., Proctor, J.L. and Kutz, J.N., 2016. Discovering governing equations from data by sparse identification of nonlinear dynamical systems. Proceedings of the National Academy of Sciences, 113(15), 3932-3937.

[8] Economides, M.J., Nolte, K.G., 2000. Reservoir stimulation. Chichester, Wiley.

530734 A Comparison of Mathematical Optimization and Deep Reinforcement Learning for Supply Chain Materials Planning

Abstract:

This paper presents a novel approach to supply chain planning optimization through the utilization of deep reinforcement learning and compares it to modern, mathematical optimization techniques. The supply chain is modeled as a factory with multiple products and stochastic demand for each product. The deep reinforcement learning agent interacts with this environment to learn by experience, as captured in neural networks, to determine which product to produce and when to produce it to satisfy customer demand. The neural networks approximate the value of each state with respect to a value function that accounts for on time delivery, the level of inventory on hand, and the customer service level. The reinforcement learning agent learns through simulations of the manufacturing environment and demand patterns in order to maximize the reward received through the value function. In parallel, we formulate a mixed-integer program that has to make identical decisions under the same constraints. The mixed-integer program can see orders out to a certain horizon, and does not use any information from historical order patterns. We compare and contrast the performance between the two methodologies on several scheduling test examples, and suggest how to choose between them, as well as how they may be used in concert.

534066 Dimensional Analysis Based Uncertainty Quantification: Modeling of Erosion in Pipeline Transportation

Abstract:

The transport of solids in multiphase flows is common practice in energy industries due to the unavoidable extraction of solids from oil and gas bearing reservoirs either from onshore or offshore sites. The safe and efficient operation requires reliable estimates of erosion happening in the pipelines. The phenomenon that leads to erosion, especially in multiphase flow systems, is very complex and depends on many factors including the fluid and solid characteristics, the pipeline material properties, and the geometry of the flow lines. Empirical and semi-mechanistic models have been developed to quantify the erosion rate within pipelines that transport solids in both single-phase and multiphase flows. However, due to the process complexity, the discrepancies between these models' predictions and actual erosion rates can be very significant, e.g., several orders of magnitude for some operation conditions[1], and the uncertainty of model predictions is generally not provided.

For estimating uncertainty of erosion-rate prediction, particularly, the distribution of the model discrepancy, we successfully adopted Gaussian Process Modeling1,[2],[3] (GPM) in our previous studies. The model discrepancy is defined as the difference in erosion-rate between experimental measurement and model prediction. For erosion rate predictions, we employ a widely used erosion model[4]. The Gaussian Process Regression (GPR) model specified by a mean and a covariance function with optimized hyperparameters is used to estimate the distribution (mean and variance values) of model discrepancy2. The prediction of erosion model is then corrected by adding the mean value of the estimated model discrepancy. The estimated variance is used to calculate a confidence interval of the corrected erosion-rate prediction at a specified level. To assess the GPR model performance, Area Metric[5] (AM) is used, where a smaller AM value represents a more reliable prediction of the model discrepancy.

In this talk, we introduce dimensional analysis to extend the applicable range of operating conditions of the GPR model. The analysis starts by compiling all 11 physical variables (i.e. sand particle size and density, gas/liquid viscosity, density and flow rate, pipe diameter, gravity constant and surface tension) that influence the erosion process in a pipeline. Buckingham Pi analysis is employed to determine all 648 dimensionless numbers using 81 combinations of repeating variables. We then construct a set that contains the 366 unique dimensionless numbers. Considering the three fundamental dimensions (mass, length and time), from this set, the eight most relevant dimensionless numbers are identified using Automatic Relevance Determination[6] (ARD) approach, and these numbers are used to construct GPR models. The model discrepancy is normalized by the prediction from the erosion model to keep a consistent non-dimensional relationship between the inputs and output. Prior to training GPR models, the training data is clustered using an object-cluster similarity metric based iterative clustering learning approach[7] to capture different fundamental phenomena that governs erosion process in multi-phase flow. Seven clusters are obtained and for each cluster, a GPM is trained where the inputs are gas Galileo number, ratio of gas and liquid density, Reynolds number of gas, liquid and mixed flows, mixed Weber number, gas Froude number and gas Capillary number, and the output is normalized model discrepancy. A comparison with the

uncertainty predictions developed using dimensional inputs in our previous study[8] reveals that the dimensional analysis based approach noticeably improve the prediction of model uncertainty with a 30% reduction in the overall AM value. The developed GPR models can also be extended to a wide range of operating conditions with the obtained dimensionless numbers and used as a suggestion for possible improvements in erosion process modeling and experimental design.

References

[1] Dai, W. and Cremaschi, S., Quantifying Model Uncertainty in Scarce Data Regions - A Case Study of Particle Erosion in Pipelines, 2015. doi:10.1016/B978-0-444-63577-8.50147-9.

[2] Dai, W., Cremaschi, S., A Data-Mining Framework for Uncertainty Analysis in Pipeline Erosion Modeling, 2016 AIChE Annual Meeting, November 13-18, 2016, San Francisco, CA, USA.

[3] Dai, W., Cremaschi, S., Machine Learning Based Uncertainty Quantification of Erosion in Pipeline Transportation, 2017 AIChE Annual Meeting, October 29-31, 2017, Minneapolis, MN, USA.

[4] Mazumder, Q.H., Shirazi, S.A., McLaury, B.S., Shadley, J.R., and Rybicki, E.F., Development and validation of a mechanistic model to predict solid particle erosion in multiphase flow, Wear, 259 (2005), 203–207.

[5] Ferson, S., Oberkampf, W.L., and Ginzburg, L., Comput. Methods Appl. Mech. Eng., 197 (2008), 2408–2430.

[6] Rasmussen, C. E. and Williams, C. K., Gaussian Process for Machine Learning, *The MIT Press*. 2006.

[7] Cheung, Y., and Jia, H., Categorical-and-numerical-attribute data clustering based on a unified similarity metric without knowing cluster number, Pattern Recognit. 46 (2013) 2228–2238.

[8] Dai, W., Cremaschi S, Subramani H.J., Gao H., Uncertainty Analysis in Multiphase Flow Prediction in Presence of Solids. In: Foundations of Computer-Aided Process Operations, 2017.

537714 Efficient Computation of Local Sensitivity Information for Nonsmooth Process Models

Abstract:

Discrete behavior may be introduced into an otherwise continuous process model to reflect changes in thermodynamic phase or flow regime, to represent discrete design decisions, or to represent closed-loop control actions. While discrete behavior is traditionally represented in process models using integer variables or disjunctions, these representations cannot preserve continuity between different discrete modes. Nonsmooth representations, on the other hand, provide a means to preserve this continuity information, and have been recently used to model discrete phenomena such as pinch analysis for heat integration, and flash drums that can handle different thermodynamic phases.

This presentation describes a recent improved *branch-locking* method [1] for computing local sensitivity information for these nonsmooth models, for use in overarching methods for simulation or optimization. This method borrows the computational benefits of the efficient reverse mode of automatic differentiation for smooth functions, even though the reverse mode does not apply directly to nonsmooth systems. The branch-locking method uses inexpensive probing steps to temporarily â€œlockâ€ the nonsmooth model into a carefully-chosen smooth variant; this smooth model is then differentiated using standard efficient techniques. This method is accurate and automatable, and typically yields far superior computational performance to existing methods for nonsmooth sensitivity analysis. Implications and examples are discussed.

Reference

[1] KA Khan, Branch-locking AD techniques for nonsmooth composite functions and nonsmooth implicit functions, Optim. Methods Softw., in press, 2017

534303 A Novel Geometric Based Algorithm to Solve Multiparametric Programming Problems

Abstract:

Multiparametric programming (mpP) is a technique that develops explicit maps of solutions to optimization problems comprising unrealized and bounded parameters. Advancements in mpP have led to "MPC-on-a-chip" technology [1], global solutions to certain classes of bi-level optimization problems [2], and a procedure for integrated multi-scale decision making in process systems engineering [3], amongst others. However, multiparametric programming for large scale applications is still prohibitive due to the combinatorial nature of the active constraints. Current algorithms suffer from the curse of dimensionality arising from an increase in the number of optimization variables, constraints, and parameters. Improving the performance of mpP problems, primarily linear (mpLP), mixed integer linear (mpMILP), and quadratic (mpQP) is still an open question in the literature [4,5].

In this work, we propose a parametric space exploration based algorithm to address mpP limitations with increasing number of optimization variables and constraints. The algorithm exploits key properties of mpP problems where (i) critical regions are convex, (ii) active sets within a critical region remain identical, and (iii) optimization variables are affine functions of the parameters. Exploring the parametric space follows an iterative procedure of vertex identification based on properties (i) and (ii), followed by identifying critical region boundaries based on (i), and determining the exact expressions of the optimization variables as a function of the parameters based on (iii). The proposed algorithm effectively handles mpP problems with a larger number of optimization variables and constraints, while still scaling exponentially with the number of parameters. The applicability of the algorithm is demonstrated through a computational study of various mpP problem sizes and classes (mpLP, mpMILP, and mpQP), and compared against existing software [6,7].

References

[1] Pistikopoulos, E. N. Perspectives in multiparametric programming and explicit model predictive control. AIChE Journal 2009, 55 (8), 1918-1925.

[2] Avraamidou, S.; Pistikopoulos, E. N. A novel algorithm for the global solution of mixed-integer bi-level multi-follower problems - Application on the problem of Planning and Scheduling integration. European Control Conference (ECC 2018); 2018; Accepted manuscript.

[3] Burnak, B.; Katz, J.; Diangelakis, N. A.; Pistikopoulos, E. N. Simultaneous Process Scheduling and Control: A Multiparametric Programming Based Approach. Industrial & Engineering Chemistry Research 2018, 57 (11), 3963-3976.

[4] Oberdieck, R.; Diangelakis, N. A.; Pistikopoulos, E. N. Explicit Model Predictive Control: A connected-graph approach. Automatica 2017, 76, 103-112.

[5] Ahmadi-Moshkenani, P.; Johanses, T.A.; Olaru, S. Combinatorial Approach towards Multi-Parametric Quadratic Programming based on Characterizing Adjacent Critical Regions. IEEE Transactions on Automatic Control, Institute of Electrical and Electronics Engineers, In press.

[6] Oberdieck, R.; Diangelakis, N. A.; Papathanasiou, M. M.; Nascu, I.; Pistikopoulos, E. N. POP - Parametric Optimization Toolbox. Industrial & Engineering Chemistry Research 2016, 55 (33), 8979-8991.

[7] M. Herceg, M. Kvasnica, C.N. Jones, and M. Morari. Multi-Parametric Toolbox 3.0. In Proc. of the European Control Conference, pages 502–510, Zurich, Switzerland, July 17–19 2013.

526568 Bayesian Machine Learning Modeling of a Reformer Furnace Using CFD Data

Abstract:

Hydrogen is widely used as precursor across a variety of manufacturing industries, e.g., fertilizer (ammonia), petroleum refinery and methanol plants are responsible for 49%, 37% and 8% of world-wide hydrogen consumption, respectively [1]. The current demand for hydrogen is large; for instance, the annual worldwide hydrogen production in 2008 was 500 billion Nm^3 and corresponded to a global market of $40 billion, and this demand is also expected to grow at an annual rate of 5% to 10%. Therefore, solving challenges encountered in commercial-scale production of hydrogen and optimizing these production lines has become an issue of great interest to both academia and industry. Among a variety of commercial hydrogen production processes, steam methane reforming (SMR) is by far the most common. For instance, SMR was responsible for 80% to 85% of the world-wide hydrogen production in 2007 [2]. SMR is typically carried out in top-fired steam methane reforming furnaces which are the most expensive equipment in terms of the maintenance and operating costs compared to other major equipment such as the hydrotreating, prereforming, water-shift and purification units at centralized SMR-based facilities. For instance, the re-tubing cost of a furnace is 10% of the total capital investment, and the annual operating cost to procure fresh natural gas for a SMR-based hydrogen plant with a production rate of 2.7 million Nm^3 per day is $62 million [3]. In practice, the energy efficiency of the reformer is subjected to various operational disturbances; for example, a failure in the flow control valve distribution prevents the optimized reformer fuel input determined off-line from being implemented. Hence, the need for a computationally efficient high-fidelity model for the reformer controlled variables (i.e., the outer tube wall temperature (OTWT) distribution) as a function of the reformer manipulated inputs (i.e., the flow control valve distribution and the total fuel flow rate) becomes apparent. First-principles models [3] and computational fluid dynamics (CFD) models [4] of the reformer may require significant computation time if solved repeatedly as part of an optimization problem that seeks to reduce nonuniformity in the OTWT distribution, and therefore would not be suitable for designing a real-time robust tool that allows the reformer fuel input to be adjusted to account for disturbances. Data-driven modeling is an appealing alternative as data-driven models are computationally inexpensive and can have reasonable accuracy.

Motivated by the above considerations, the present work introduces a statistical-based model identification scheme that generates a computationally efficient data-driven model for the OTWT distribution as a function of the total fuel flow rate, the fuel distribution and interactions among neighboring reforming tubes from reformer data. The proposed scheme is structured to have two fully-parallelized components, namely, a prediction step and a correction step. An algorithm for the prediction step is developed from Bayesian variable selection, Bayesian model averaging, sparse nonlinear regression, reformer geometry and theories of thermal radiation so that the OTWT of each reforming tube is computed based on the total fuel flow rate and its spatial distribution inside the reformer. An algorithm for the correction step is developed from the ordinary Kriging so that the OTWT of each reforming tube is computed based on the OTWTs of the neighboring reforming tubes. Finally, the data-driven model for the OTWT of each reforming tube is formulated as the weighted average of the respective prediction and correction models. Results from the goodness-of-fit and out-of-sample prediction tests of the data-driven model for the OTWT distribution are used to demonstrate the effectiveness of the scheme proposed in this work.

[1] Balat, M., 2008. Potential importance of hydrogen as a future solution to environmental and transportation problems. International journal of hydrogen energy 33, 4013–4029.

[2] Simpson, A.P., Lutz, A.E., 2007. Exergy analysis of hydrogen production via steam methane reforming. International Journal of Hydrogen Energy 32, 4811–4820.

[3] Latham, D.A., McAuley, K.B., Peppley, B.A., Raybold, T.M., 2011. Mathematical modeling of an industrial steam-methane reformer for on-line deployment. Fuel Processing Technology 92, 1574–1586.

[4] Tran, A., Aguirre, A., Durand, H., Crose, M., Christofides, P.D., 2017b. CFD modeling of a industrial-scale steam methane reforming furnace. Chemical Engineering Science 171, 576–598.

524689 Image-Based Modeling of Fibroblasts Modifying PDGF-BB Gradient Explains Cellsâ€™ Alternated Directional Decision during Chemotaxis in a Microfluidic Maze

Abstract:

Directed fibroblast migration is central to highly proliferative processes in regenerative medicine and developmental biology, such as wound healing and embryogenesis. However, the mechanisms by which single fibroblasts affect each other's directional decisions, while chemotaxing in microscopic tissue pores, are not well understood. Here we use image-based multiscale modeling in order to explain an experimentally observed contradiction to the classical chemotaxis theory: fibroblasts alternating their directional decisions in microscopic tissue-sized confinement, instead of following the steepest chemotactic gradient. The chemotaxis experiment was performed in a microfluidic maze (see Figure 1A), which contains micron-sized paths (i.e. a long path and a short path) that the cells preferred depending on the sequence in which they entered the maze. The maze had two ports, one entrance and one exit. The entrance was connected to a cell reservoir, while the exit was connected to a chemoattractant source of 50 ng/ml of PDGF-BB. NIH-3T3 fibroblasts were allowed to migrate along the PDGF-BB gradient, towards the exit of the maze. Real-time imaging of the migration was performed using an automated Olympus IX83 microscope. Experimental data, such as directional decisions of the cells, was manually extracted from time-lapse images acquired at 15-min intervals. An image-based multiscale COMSOL model was performed in order to visualize the gradient of PDGF-BB inside of the microfluidic maze. The macroscopic scale of the model simulated the global chemoattractant gradient established in the maze, while the microscopic scale was used to represent the localized gradient disturbances caused by the presence of the cells. In the former case, we used a transient two-dimensional transport-of-diluted-species physics with two constant concentration boundary conditions applied at the maze ports. Moreover, the model took into account depletion of the PDGF-BB via natural decay. On the microscopic level, on the other hand, we modeled endocytosis of the PDGF-BB by the individual cells. The simulation results showed that the reason why the fibroblasts alternate the paths that they take through the maze can be explained by the localized PDGF-BB consumption by the preceding cell. Namely, the leading cell follows the shortest path (i.e. the steepest gradient), as expected from classical chemotaxis theory (see Figure 1B). The subsequent cell, however, sees a gradient *altered* by the leading cell ahead of it (see Figure 1C), and gets steered into the long path. Overall, these finding supports the conclusion that the presence of the individual cells in microconfinement can modify the external chemotactic cues via self-generation of local gradients. Consequently, accounting for such effects could lead to a better understanding of tissue generation in vivo, and result in more advanced engineered tissue products in vitro.

Figure 1. Image-based modeling of localized fibroblast effects on the PDGF-BB distribution within the maze. (A) Demonstration of the image-based concept. Upper half corresponds to the COMSOL simulation, with the PDGF-BB concentration scaled by the exit boundary condition concentration; while the lower half is the phase microscopy image of the fibroblasts (labeled in green) chemotaxing through the maze. (B) & (C) Simulation results of real-time modifications to the PDGF-BB gradient overlaid on the experimental microscopy images, showing the first and second cells' directional decisions at the maze bifurcation, respectively. Note that although the gradient is scaled between 0 and 100%, the actual value corresponding to the latter is chosen differently between the two timeframes in order to enhance visibility. Dashed white circle highlights the decision-making cell, while white arrow points to the higher chemoattractant gradient path chosen by the cells.

526783 Multiscale Computational Modeling of Renal Intercellular Cross-Talk at the Onset of Diabetic Kidney Disease

Abstract:

Diabetic kidney disease (DKD) is one of the most severe complications of diabetes and is the primary cause for kidney failure. In the early stages of DKD, kidney tissue damage is primarily focused in glomeruli, which are bundles of capillaries within the filtration units (nephrons) of the kidney where blood is filtered to produce urine. Significant damage to glomeruli occurs before the leakage of proteins through glomeruli into urine. This damage is detectable by non-invasive measurements in a condition called proteinuria, which is a clinical marker of DKD. Continued damage ultimately leads to kidney failure. Some of these renal tissues including podocyte cells are terminally differentiated and cannot repair themselves. Thus, it is critical to slow the rate of progression of DKD before irreversible injury and subsequent nephron loss occurs.

Hyperglycemia-induced metabolic dysregulation is known to initiate DKD through complex biochemical signaling pathways. The connections between the various molecular mechanisms that lead to glomerular injury in DKD are not completely understood. A glomerulus consists of a bundle of capillaries and a scaffold of cells and extracellular domains that hold the bundle together. The scaffold includes the mesangium, podocytes, and endothelial cells. The mesangium forms the interior of the capillary cluster of a glomerulus. Podocyte cells surround the glomerular capillaries with an interlocking network of long projections or foot processes with small slits between them for filtration. The endothelial cells line the capillaries and are covered by a cell-surface layer called the glycocalyx. Additionally, the glomerular basement membrane (GBM) is a specialized extracellular matrix that exists between the endothelial cells and podocytes. Experimental studies have shown different biochemical pathways are triggered in hyperglycemic or diabetic conditions that alter these cells leading to glomerular injury mainly in four ways: podocyte cell loss, mesangial expansion, GBM thickening, and endothelial glycocalyx thinning. The roles in glomerular damage of cross-talk between these cell types have been suggested in the literature; however, there is as yet no way to predict the systemic impacts of the cross-talk throughout entire glomeruli in DKD. The objective of this work is to create a computational model to simulate the biochemical and physical interactions between glucose, cell signaling pathways, and various cell types within glomeruli at the onset of DKD in advance of detection of proteinuria. The hypothesis is that simultaneous concerted interactions between glucose-stimulated signaling pathways in multiple cell types within the glomeruli and the cross-talk between cell types are needed in order for hyperglycemia to progress to diabetic proteinuria. Such a computational model can provide insight into understanding complex networks of the physiological and biochemical processes that may lead to emergent behavior in DKD.

Podocytes express a local renin-angiotensin system (RAS) that is altered in diabetes. RAS is a network of biochemical reactions involving multiple enzymes and hormones. Angiotensin II (ANG II) is one of the crucial hormones in the RAS and is responsible for regulating blood pressure. Studies have shown that ANG II is upregulated in diabetic conditions and triggers irreversible glomerular injury by causing mesangium expansion, GBM thickening, shrinking of the glycocalyx, podocyte apoptosis and detachment. This deterioration of the glomerular filtration barrier leads to leakage of proteins into the urine and eventually, kidney failure. Angiotensin converting enzyme (ACE) is responsible for conversion of ANG I to ANG II. ACE inhibitors are widely used pharmaceuticals that inhibit the production of ANG II. However, these drugs are not well characterized for use in CKD. The progression of CKD could be slowed by controlling the ANG II levels to prevent irreversible podocyte loss. We have developed an ODE-based PK/PD model of glucose-stimulated RAS dynamics in podocytes using MATLAB. The model predicts the e ects of di erent dosages of pharmaceutical drugs on the concentrations of ANG II in local podocyte RAS. Glucose dependency is added to the model through enzymatic parameters that can help predict ANG II for di erent glucose levels. The model is parametrized for two pharmaceutical drugs in the cases of normal and impaired renal function. The model can also take patient specific glucose dynamics data as a model input to predict ANG II levels and required drug dosage to suppress elevated ANG II levels inside of podocytes.

Another scale of the computational model is developed for the convection and diffusion of chemical species through the porous mesangium matrix. This model can be applied to hyperglycemic conditions to study the effects of glucose-stimulated elevated ANG II levels and transport rates on the different pathophysiological processes including GBM thickening and mesangial matrix density.

The biochemical mechanisms and their effects in the podocytes and mesangium can be incorporated together in a hybrid continuous and discrete agent-based model to connect the biochemical models to the physical effects on the glomerular cell types. This model is developed to produce a anatomically-scaled geometry of the cross section of the glomerular filtration barrier composed of the podocytes, mesangium, capillarie,s and GBM in Compucell3D to simulate the interactions among the kidney cells. Compucell3D uses a cellular Potts model to model the tissue environment and computes an energy associated with cell properties. User-created Python scripts are also used to augment the behavior of the cells and add functionality to the simulation. While Compucell3D does have an embedded partial differential equation (PDE) solver, the method is forward Euler explicit method, which is not suited for simulating convection. The present work has tailored Compucell3D to allow for the arbitrary execution of any Python code, including the package FiPy, which is an extremely flexible PDE solver library. In our model, FiPy will be used to solve the diffusion and convection of biochemical reaction networks in different cell types (podocytes and mesangium) and predict the overall effect of hyperglycemia-stimulated cellular cross-talk on glomerular injury. The agent-based part of the model includes features related to cell migration, growth, death, and adhesion.

The overall objective of this work is to connect these models focused individually on podocytes and on the mesangium to build a multiscale computational model that can improve the understanding of the mechanisms governing physiological effects of various chemicals, e.g., metabolites (glucose), hormones (ANG II), and pharmaceutical drugs, on kidney damage as a complication of diabetes. By detecting DKD at the onset through conditions identified by the model, treatments could be administered to slow the progression before significant damage occurs.

530361 Controllability of the Influenza Virus-Host Protein-Protein Interaction Network: Engineering Insights into Host-Virus Interactions

Abstract:

Controllability of the Influenza Virus-Host Protein-Protein Interaction Network: Engineering Insights into Host-Virus Interactions

Emily Ackerman, University of Pittsburgh

Influenza virus is a major threat to public health each year. The World Health Organization estimates that there are 3-5 million severe cases of seasonal influenza worldwide yearly, with 290,000-650,000 cases resulting in death[1]. Yearly seasonal outbreaks are especially harmful to at-risk populations such as children and pregnant women, the elderly, and the immunocompromised[2,3]. Additionally, there is a continual threat of a pandemic, with the most recent being the 2009 H1N1 outbreak. As such, there is a high demand for effective treatment strategies and novel methods for antiviral drug target discovery.

Understanding which cellular components are key in enabling viruses to interrupt the normal activity of healthy cells provides a major advantage in the design of novel disease treatments[4,5]. Systems biology methods aim to investigate this problem through the application of mathematical and computational based methods. While kinetic modeling approaches detail and quantify cellular events in relevant pathways, the effectiveness of the approach can be limited by data availability and lack of interaction mechanism knowledge. Alternatively, protein-protein interaction (PPI) data is used to explore the same biological information in a convenient, highly simplified network representation[6]. Though the method lacks the quantitative nature of ODE modeling, the generality of cell interactions opposed to specific pathway analysis allows for the study of PPI relationships on a whole cell level. Previous work with PPI networks have identified disease-associated genes and predicted drug targets using established network topology (or centrality) measures such as degree and betweenness[7–9]. However, there is limited work describing the ways in which PPI networks are differently regulated when comparing changing cell states.

In classic control theory, controllability is the idea that a system can be driven to any final state in state space given an appropriate external input in finite time[10]. A PPI network represents a cell system of protein states. During infection, viral proteins manipulate host proteins to control the system and drive it to a final infected state which performs virus replication tasks. While PPI systems are too large to directly calculate the controllability matrix, two methods of network controllability analysis exist which use bipartite graphs to simplify the associated calculations. One analysis determines if it is more or less difficult to control the network in the absence of each protein[11]. Ultimately, this analysis calculates the minimum set of nodes needed to control the network with or without a selected protein. This set is non-unique; which leads to the second analysis which places each protein in the broad context of its importance to all configurations of network control[12]. Prior to this work, only the first method has been applied to a human PPI network[13]. These analyses identify key proteins being manipulated in the externally controlled, infected cell system.

Here, we complete a dual controllability analysis of two PPI networks: a host PPI network (HIN) and an integrated virus-host PPI network (VIN). This design allows for the comparison of changes between the healthy and infected cell states to identify proteins that display changes to the regulatory behavior of the network. Further, the analysis compares the characteristics of influenza A virus (IAV) interacting host proteins and "driver" proteins (proteins which must be manipulated for the system to be fully controlled, a non-unique set) to all proteins of the network to determine if these specialized groups are enriched for virus replication host factors; i.e. human proteins required for influenza virus replication.

An observed increase in the basic topological measurements of degree and betweenness after the addition of viral interactions to the network is proof of the wide-reaching effects to cellular behavior during infection. This effect is especially strong for proteins that are both IAV interacting and driver proteins. The controllability analyses determine that it is more difficult to control the overall network in the absence of IAV interacting host proteins, and that they are well integrated into the control structure without requiring their own manipulation for total network control. This suggests that viruses prefer to interact with host proteins that offer the best advantages in controlling total system behavior. Conversely, controllability results predict that the absence of driver proteins makes it easier to control the network, meaning that they act as a barrier to control of the system. Over 75% of driver proteins are able to act as non-driver nodes given certain control configurations, opening up the idea of viral escape routes: the idea that virus proteins could use alternative pathways under pressure and maintain system control.

The first controllability classification does not identify proteins of interest in the contrast between the healthy HIN and infected VIN, despite known changes to the cell system during infection. However, the second controllability classification identifies a set of 24 proteins which exhibit notable changes between the HIN and VIN systems. Together, they are predicted as indicators of infected state regulatory roles, with the most prominent being protein arginine methyltransferase 5 (Entrez ID: 10419) with a VIN betweenness of 250 times its HIN value. 50% of the 24 proteins are identified as having interferon regulating roles and an IPA[14] analysis shows that they are centered around NF-kB, which has been previously identified as inhibited by influenza virus[15,16]. Though the set of 24 proteins are not significantly enriched in the results of six siRNA screening studies for host factors of influenza A replication, one protein, heterogeneous nuclear riboprotein A0 (Entrez ID: 10949), is triple validated. This rare observation suggests that controllability methods hold power for the prediction of disease host factors.

By analyzing the control structure of the cell system, we are able to predict of the protein components which allow an influenza A infection to take hold. Disruption of these methods of viral control may prove to be an effective way of disrupting the infection as a whole. As such, the identification of these key disease proteins advances the prediction of possible antiviral drug targets. Additionally, this approach enables the study of alternative pathways for maintaining viral replication which could lead to drug resistance. A window into the control behavior and robustness of influenza infection could position us one step ahead of both yearly, seasonal outbreaks as well as the potential of pandemic infection.

1. WHO. Influenza (Seasonal) Fact Sheet. (2017).

2. Mertz, D. *et al.* Populations at risk for severe or complicated influenza illness: systematic review and meta-analysis. *BMJ* **347,** f5061--f5061 (2013).

3. Rodríguez-Baño, J., Paño-Pardo, J. R., Múñez Rubio, E. & Segura Porta, F. Pregnancy, obesity and other risk factors for complications in influenza A(H1N1) pdm09 infection. *Enferm. Infecc. Microbiol. Clin.* **30,** 32–37 (2012).

4. Klipp, E. & Liebermeister, W. Mathematical modeling of intracellular signaling pathways. *BMC Neuroscience* **7,** (2006).

5. Schoeberl, B., Eichler-Jonsson, C., Gilles, E. D. & Muüller, G. Computational modeling of the dynamics of the MAP kinase cascade activated by surface and internalized EGF receptors. *Nat. Biotechnol.* **20,** 370–375

(2002).

6. Cho, D.-Y., Kim, Y.-A. & Przytycka, T. M. Chapter 5: Network Biology Approach to Complex Diseases. *PLoS Comput. Biol.* **8,** e1002820 (2012).

7. Jonsson, P. F. & Bates, P. A. Global topological features of cancer proteins in the human interactome. *Bioinformatics* **22,** 2291–2297 (2006).

8. Hase, T., Tanaka, H., Suzuki, Y., Nakagawa, S. & Kitano, H. Structure of protein interaction networks and their implications on drug design. *PLoS Comput. Biol.* **5,** (2009).

9. Yildirim, M. a, Goh, K.-I., Cusick, M. E., Barabási, A.-L. & Vidal, M. Drug-target network. *Nat. Biotechnol.* **25,** 1119–26 (2007).

10. Lin, C. T. Structural Controllability. *IEEE Trans. Automat. Contr.* **19,** 201–208 (1974).

11. Liu, Y. Y., Slotine, J. J. & Barabási, A. L. Controllability of complex networks. *Nature* **473,** 167–173 (2011).

12. Jia, T. & Barabási, A. L. Control capacity and a random sampling method in exploring controllability of complex networks. *Sci. Rep.* **3,** (2013).

13. Vinayagam, A. *et al.* Controllability analysis of the directed human protein interaction network identifies disease genes and drug targets. *Proc. Natl. Acad. Sci.* **113,** 4976–4981 (2016).

14. Krämer, A., Green, J., Pollard, J. & Tugendreich, S. Causal analysis approaches in ingenuity pathway analysis. *Bioinformatics* **30,** 523–530 (2014).

15. Kumar, N., Xin, Z.-T., Liang, Y., Ly, H. & Liang, Y. NF-kappaB signaling differentially regulates influenza virus RNA synthesis. *J. Virol.* **82,** 9880–9 (2008).

16. Ludwig, S. & Planz, O. Influenza viruses and the NF-κB signaling pathway - Towards a novel concept of antiviral therapy. *Biological Chemistry* **389,** 1307–1312 (2008).

528569 Novel Approaches for the Integration of Supply Chain Planning and Scheduling

Abstract:

Supply chain decisions are hierarchically organized in strategic, tactical, and operational decisions. In practice, these planning processes are frequently conducted independently with minimal exchange of information between them, thereby leading to suboptimal solutions. Achieving a better coordination between these processes allows companies to capture benefits that are currently out of their reach and improve the communication among their functional areas[1]. Optimization methods for the integration of strategic and tactical decisions are well established[2]. Nevertheless, the integration of tactical and operational decisions has not been solved[3],[4]. There are at least three shortcomings that need to be addressed to create integrated planning and scheduling systems that can be used in real-world applications: 1) the models proposed in previous work are too simplistic, assuming a mid-term planning level setting inventory targets for the scheduling; 2) they aim to solve the scheduling model in the horizon of the planning model, and 3) there are no efficient algorithms capable of handling industrial-sized problems in reasonable amount of time.

In this work, we present an integrated formulation of short term multisite planning with batch scheduling using the UOPSS modeling framework[5]. The planning level considers an (s,S) inventory policy for all inventories in the system[6], and distinguishes between make-to-order and make-to-stock products[7]. The scheduling level receives planned orders, corresponding to a mix of inventory replenishment orders and customer orders, to determine the optimal production schedule. The scheduling objective includes minimization of priority-dependent backlog and changeovers. To address the difference in timescales we present an objective scaling approach. In the proposed scaling method the objective function of the model with the largest horizon (planning) is downscaled to make it comparable to the objective of the model of the shorter horizon (scheduling). The method is evaluated through simulation, and compared with traditional integration methods: full-space solution, and representative day approach. Results on industrially sized problems clearly indicate significantly improved solutions with the proposed approach. To address the computational complexity added by the integration of planning and scheduling, Benders decomposition with mixed-integer subproblems inspired by the stochastic programming literature[8],[9], and cross decomposition[10] methods, have also been developed to effectively address the integration problem.

References

[1] Brunaud, B. and Grossmann, I.E., 2017. Perspectives in multilevel decision-making in the process industry. *Frontiers of Engineering Management*, *4*(3), pp.256-270.

[2] Barbosa-Póvoa, A.P., 2012. Progresses and challenges in process industry supply chains optimization. *Current Opinion in Chemical Engineering*, *1*(4), pp.446-452.

[3] Maravelias, C.T. and Sung, C., 2009. Integration of production planning and scheduling: Overview, challenges and opportunities. *Computers & Chemical Engineering*, *33*(12), pp.1919-1930.

[4] Garcia, D.J. and You, F., 2015. Supply chain design and optimization: Challenges and opportunities. *Computers & Chemical Engineering*, *81*, pp.153-170.

[5] Zyngier, D. and Kelly, J.D., 2012, June. UOPSS: a new paradigm for modeling production planning & scheduling systems. In *Symposium on Computer Aided Process Engineering* (Vol. 17, p. 20).

[6] Brunaud, B., Lainez-Aguirre, J.M., Pinto, J.M. and Grossmann, I.E., 2017. Mixed-integer Models for Simultaneous Optimization of Inventory Policies and Supply Chain Planning. In *Computer Aided Chemical Engineering* (Vol. 40, pp. 1255-1260).

[7] Soman, C.A., van Donk, D.P. and Gaalman, G.J., 2007. Capacitated planning and scheduling for combined make-to-order and make-to-stock production in the food industry: An illustrative case study. *International Journal of Production Economics*, *108*(1-2), pp.191-199.

[8] Zou, J., Ahmed, S. and Sun, X.A., 2017. Stochastic dual dynamic integer programming. *Mathematical Programming*, pp.1-42.

[9] Gade, D., Küçükyavuz, S. and Sen, S., 2014. Decomposition algorithms with parametric gomory cuts for two-stage stochastic integer programs. *Mathematical Programming*, *144*(1-2), pp.39-64.

[10] Mitra, S., Garcia-Herreros, P. and Grossmann, I.E., 2014. A novel cross-decomposition multi-cut scheme for two-stage stochastic programming. In *Computer Aided Chemical Engineering* (Vol. 33, pp. 241-246).

529298 Optimal Demand Response Operation of an Industrial Air Separation Unit Using Data-Driven, Scheduling-Relevant Dynamic Models

Abstract:

Renewable sources now make up a significant portion of the electricity generation portfolio, and the intermittency of such sources, coupled with market deregulation, has resulted in electricity prices that fluctuate over short time intervals (one hour or less). These time-sensitive prices effect demand-response behaviors, implicitly motivating consumers to lower electricity consumption during times of peak power demand, when electricity prices are typically also at a maximum. Chemical processes, in particular, are a natural candidate to benefit from such demand-side modulation of electricity use, as production rates (and electricity demand) can be increased during off-peak hours, and excess product can be stored and used to supplement lowered production rates during times of peak electricity demand [1]. In addition to lowering production costs, such demand-response (DR) modulation also has a stabilizing effect on the electricity grid, as electricity consumption is shifted away from times of peak demand to times of peak renewable-based generation, when electricity prices are low.

Air separation units (ASUs) use large amounts of electricity to separate air into purified gases via cryogenic distillation, with refrigeration cycles driven by large electric compressors. ASUs are a prime candidate for engagement in DR programs. Nevertheless, modulating air separation production rates in response to electricity prices involves operating in a transient regime, a clear deviation from the steady-state operational paradigms for which these processes are usually designed [1]. Additionally, dynamic models of air separation plants are usually nonlinear, high-dimensional, and ill-conditioned, making computation of optimal production schedules using dynamic models intractable in a practical amount of time [2]. In previous work [3], we took a data-driven approach, using historical operating data to build scale-bridging models (SBMs), or low-order models of the scheduling-relevant dynamics of a process. We showed that these models reduced the computation time required for the aforementioned scheduling calculations and that air separation units can improve their profitability by overproducing when electricity prices are high and storing oxygen, nitrogen, and argon in the form of cryogenic liquids.

In this work, we apply our framework for scheduling under dynamic constraints to an industrial ASU operating under Model Predictive Control, using historical industrial data to model closed-loop dynamics of the process. The process is a three-column ASU producing nitrogen, oxygen and argon, and we identify scale-bridging models of Hammerstein-Wiener form, that relate power demand to production rate targets and measurable disturbances (e.g., ambient conditions). We then solve scheduling problems using a sequential solver available in gPROMS, showing significant economic benefits compared to operation at a constant operating point. We also examine the increased benefits from scheduling a small network of multiple plants connected via a pipeline, which are subject to different time-varying electricity price trajectories, and we carry out sensitivity analyses to determine the effects of inaccuracies in electricity price forecasts on the economic benefits of the optimal production schedules.

[1] M Baldea and I Harjunkoski. Integrated production scheduling and process control: A systematic review. *Comp. Chem. Eng.*, 71:377-390, 2014.

[2] LS Dias, RC Pattison, C Tsay, M Baldea, and MG Ierapetritou. Multi-resolution model of an industrial hydrogen plant for plantwide operational optimization with non-uniform steam-methane reformer temperature field. *Comp. Chem. Eng.*, 107:271-283, 2017.

[3] RC Pattison, CR Touretzky, T Johansson, I Harjunkoski, M Baldea. Optimal process operations in fast-changing electricity markets: framework for scheduling with low-order dynamic models and an air separation application. *Ind. Eng. Chem. Res.*, 55(16):4562-4584, 2016.

534448 Integration of Planning, Scheduling and Control Using Feasibility Analysis and Surrogate Models

Abstract:

Optimal decision-making in a process industry is fundamental in order to improve performance and guarantee efficient operation in highly competitive and global markets. Process operations decision-making problems range across different scales, and include control, operational, tactical and strategic decisions. Traditionally, these decisions are considered in a hierarchical fashion and addressed individually by different areas within an organization. However, remarkable benefits can be achieved though the proper integration of the decision-making process. The academic community has made significant progress towards this goal [1, 2]. Nevertheless, the available methodologies and tools do not fully meet industrial needs, and more practical and efficient methods to coordinate the different levels of decisions are needed.

In this work, a novel framework for the integration of planning, scheduling and control problems is presented. The first step is to design the control level simulation of the system in consideration. The simulation consists of iterations between the solution of a Model Predictive Control problem and the implementation of control actions in a detailed dynamic model of the system. The systems of interest commonly involve highly nonlinear models, and the sample time of the MPC problem is usually in the order of minutes or seconds. Therefore, transmitting the closed-loop behavior of the simulation in a time scale that is relevant to the scheduling problem is challenging, and it is achieved through the use of surrogate models. Once surrogates are built to effectively capture the closed-loop behavior of the control-level simulations, they can be incorporated as constraints in the scheduling problem.

To enable the complete vertical integration of the decision making process, we then investigate how the scheduling problem can be posed as a feasibility problem [3, 4], where feasible production targets are defined as a function of the initial state of the system, the scheduling horizon, and the given network. By reformulating the scheduling problem as a feasibility problem, surrogate models and classification methods can be used to define the feasible space, following the work by Wang and Ierapetritou [5]. We discuss how the feasible space of the integrated scheduling and control problem is expected to change when compared to the individual scheduling problem. An explicit function that provides the feasible production targets and associated production costs is obtained, and the explicit models can be incorporated as constraints in the planning problem.

Finally, the planning problem is posed as a nonlinear problem and solved in a moving-horizon fashion in order to address uncertainties and disruptions associated to the scheduling and control levels. Several case studies are presented to illustrate the proposed approach and provide comparisons with the hierarchical decision-making process.

1. Dias, L.S. and M.G. Ierapetritou, *From process control to supply chain management: An overview of integrated decision making strategies*. Computers & Chemical Engineering, 2017. **106**(Supplement C): p. 826-835.
2. Grossmann, I., *Enterpriseâ€wide optimization: A new frontier in process systems engineering*. AIChE Journal, 2005. **51**(7): p. 1846-1857.
3. Sung, C. and C.T. Maravelias, *An attainable region approach for production planning of multiproduct processes*. AIChE Journal, 2007. **53**(5): p. 1298-1315.
4. Sung, C. and C.T. Maravelias, *A projection-based method for production planning of multiproduct facilities*. AIChE Journal, 2009. **55**(10): p. 2614-2630.
5. Wang, Z. and M. Ierapetritou, *A novel feasibility analysis method for black-box processes using a radial basis function adaptive sampling approach*. AIChE Journal, 2017. **63**(2): p. 532-550.

528443 Decode: Detection of Communities for Optimization Decomposition

Abstract:

Optimization is a mathematical tool that has become pervasive in process systems engineering for making decisions using mathematical models. In chemical engineering applications, these models often contain integer variables and nonconvex constraints which make the e ort required to solve the problem scale poorly with problem size. For such large or difficult to solve optimization problems, a common strategy is to apply a decomposition, whereby the original problem is broken into a series of subproblems with smaller size and reduced computational complexity. Once a decomposition is obtained, there exist many decomposition solution methods that may be applied, the most common of which include a Bender's decomposition for problems with complicating variables, or a Lagrangean decomposition of problems with complicating constraints. Typically, subproblems are generated in an ad hoc manner governed by the user's best intuition about the structure of the problem and the type of solution method they wish to employ.

In contrast to the ad hoc approaches typically used, here we present a systematic method for generating decompositions for a generic optimization problem and a software package allowing for the implementation of this method. We call this method and software package DeCODe: the Detection of Communities for Optimization Decomposition. As the name suggests, the method uses communities as the basis for finding subproblems which can be solved e ciently by decomposition solution methods [1]. Community detection is a concept that has been used widely in network theory to find groups within a network which have high modularity; that is, groups which have a statistically high amount of intra-group connectivity and therefore a statistically low amount of inter-group connectivity. To apply this method to an optimization problem, we develop a novel bipartite variable-constraint graph where nodes representing a variable or constraint are connected if the variable appears in the constraint [2]. The communities found in this network represent subproblems in a decomposition which have low connectivity, or coupling, between them, reducing the e ort needed by the decomposition solution method. The software package implements the methods described above with minimal input required from the user: only the constraint equations and an initial guess for the variables are required. The software then uses its embedded functions to generate the variable-constraint graph, calculate modularity of different partitions, detect communities, and return which variables or constraints belong to which subproblem. The software package is currently implemented in MATLAB, with plans to extend its use to the Julia programming language in the future.

To document the ability of our software to generate e cient decompositions even when intuition about the problem structure does not exist, we apply it to several problems from the MINLP test library which cannot be solved e ciently without decomposition. In doing so, we demonstrate that the decompositions generated allow for the problem to be solved faster than it could be without a decomposition in cases where the solver for the original problem has trouble improving both the upper bound (best found) solution and lower bound (best possible) solution. We also apply our method to several problems from the literature, such as our previous work which found the optimal design of a combined microgrid and biorefinery system [3], to document cases where intuition matches the decomposition found by our method, and cases where our method finds different decompositions which may admit a more efficient solution. In doing so, we show that our method for finding decompositions is amenable to many of the different decomposition solution methods used by others.

[1] Allman, A., Tang., W., Daoutidis, P. Towards a generic algorithm for identifying high-quality decompositions of optimization problems. Comput. Aided Chem. Eng. 2018, Accepted.
[2] Tang, W., Allman, A., Pourkargar, D.B., Daoutidis, P. Optimal decomposition for distributed optimization in nonlinear model predictive control through community detection. Comput. Chem. Eng. (111). 2018, pp. 43-54.
[3] Allman, A., Daoutidis, P. Optimal design of synergistic renewable fuel and power systems. Renew. Energy (100). 2017, pp. 78-89.

535033 Identification of Optimal Dopant Patterns in a Doped Perovskite Oxygen Carrier

Abstract:

Perovskites have been shown to exhibit fast and tunable oxygen release, making them good candidates for oxygen carriers in chemical looping combustion (CLC) [1,2,3,4]. The introduction of dopant atoms in the lattice can have a significant impact on both the lattice structure and the reducibility of the perovskite. As our system of interest, we consider a $BaFeO_3$ perovskite doped with indium that has been shown to exhibit oxygen release at lower temperatures than the undoped counterpart, potentially enabling lower temperature operation of the CLC process. The presence of indium in the lattice is predicted to increase the strength of bonds to neighboring oxygen atoms while bonds to oxygen further away are weakened. Due to the combinatorial number of ways that dopant could be placed in the lattice, it is not obvious which arrangement of dopant would lead to the best release of oxygen.

To that end, this work aims to formulate and solve a mathematical optimization model to identify doped perovskite lattices with optimal properties. To translate the search for a highly reducible perovskite into a mathematical optimization formulation, we first quantify the impact of dopant on the reducibility of the perovskite via a suitable metric, namely the oxygen excess energy, which is defined as the energy difference between the original lattice and the lattice with a particular oxygen removed. This quantity can be calculated accurately by Hubbard corrected density functional theory given a supercell with a particular arrangement of dopant around an oxygen atom in question [5,6]. In order to explore the impact that dopant has on the oxygen excess energy, a motif of relevant locations was defined and the set of possible configurations of dopant in a motif were enumerated. This resulted in 74 rotationally-unique ways to place dopant in the ten nearest sites near a given oxygen. The oxygen excess energy for each of these motifs was evaluated at T=0K and then additional free energy excess energy calculations were performed at process conditions.

Given a complete set of evaluations, we selected a subset of motifs with the lowest oxygen excess energy to consider as "target" motifs to pack in an optimal material. A mathematical optimization model was formulated to relate the placement of dopant atoms in a supercell of the perovskite lattice to the number of oxygen sites that achieved one of the target motifs. Using binary variables and indicator constraints, we were able to model the problem as a mixed integer linear program. This modelling approach is generic in the sense that additional constraints, such as bounds on dopant concentration, fabricability, and stability of the material can be enforced to further restrict the search to materials of interest. In addition to the basic formulation, we derive several approaches to symmetry breaking that are able to improve the tractability of the model. A key advantage of casting our problem as a mathematical optimization model is the ability to rigorously guarantee the exploration of the entire search space and the ability to collect a list of the top designs for evaluation offline against secondary criteria.

Our optimal solutions demonstrate that a significant fraction of oxygen sites can display a target motif. By comparing the quality of our optimal designs against the baseline qualities of designs formed by random dopant placement, we show that nanostructured designs can significantly outperform unstructured materials. Furthermore, we can use our model of randomly doped perovskites to explain trends in the experimentally observed temperature of oxygen release. While the precise placement of dopant in the lattice may be currently impractical, the identification of ideal dopant patterns can serve to guide future development of synthesis methods and to establish rigorous upper bounds on the performance of doped perovskite materials.

References

[1] Adanez J., Abad A., Garcia-Labiano F., Gayan P., Diego L. F. De, "Progress in chemical-looping combustion and reforming technologies," *Progress in Energy and Combustion Science*, 38(2):215–282, 2012.

[2] Rydén M., Lyngfelt A., Mattisson T., Chen D., Holmen A., Bjørgum E., "Novel oxygen-carrier materials for chemical-looping combustion and chemical-looping reforming; $La_xSr_{1-x}Fe_yCo_{1-y}O_{3-d}$ perovskites and mixed-metal oxides of NiO, Fe_2O_3 and Mn_3O_4," *International Journal of Greenhouse Gas Control*, 2(1):21–36, 2008.

[3] Lekse J. W., Natesakhawat S., Alfonso D., Matranga C., "An experimental and computational investigation of the oxygen storage properties of $BaLnFe_2O_{5+d}$ and $BaLnCo_2O_{5+d}$ (Ln = La, Y) perovskites," *Journal of Materials Chemistry A*, 2(7):2397–2404, 2014.

[4] Lu Y., Zhao H., Cheng X., Jia Y., Du X., Fang M., Du Z., Zheng K., Świerczek K., "Investigation of In-doped $BaFeO_{3-\delta}$ perovskite-type oxygen permeable membranes," *Journal of Materials Chemistry A*, 3(11):6202–6214, 2015.

[5] Anisimov V., Aryasetiawan F., Lichtenstein A. I., "First-principles calculations of the electronic structure and spectra of strongly correlated systems: the LDA+U method," *Journal of Physics: Condensed Matter*, 9:767–808, 1997.

[6] Kresse G., Furthmüller J., "Efficiency of ab initio total energy calculations for metals and semiconductors using a plane wave basis set," *Computational Materials Science*, 6(1):15, 1996.

537954 Unipopt: Univariate Projection-Based Optimization without Derivatives

Abstract:

Many design and optimization problems are challenging black-box problems where, due to various reasons, it is not possible to obtain reliable derivative information [1-6]. This has led to the growing importance of derivative-free optimization (DFO). Despite developments in the past, the scale of the problems that can be solved by DFO solvers are relatively small compared to exact gradient-based methods. They also often converge to local optima [7], which may be prohibitive for many applications. In this work, we present a novel derivative-free framework UNIPOPT (UNIvariate Projection-based OPTimization) that is based on projecting all the samples onto a univariate space defined by summation of the decision variables [8]. The projection of the samples results in a point-to-set mapping of the projected variable to multiple function values. We illustrate that a univariate function exists on this map such that its minima is the same as that of the original function. The UNIPOPT framework is based on two key algorithms: (i) EPIC (Envelope PredIctor and Corrector) that identifies the points on this univariate function that is the lower envelope of the point-to-set map, and (ii) LEO (Lower Envelope Optimization) that uses these samples identified by EPIC to optimize the lower envelope.

We have extensively tested the UNIPOPT framework using many test problems from literature. For instance, we applied it to solve 393 bound-constrained black-box problems comprising of 232 nonconvex smooth problems and 161 convex nonsmooth problems taken from Rios and Sahinidis [7]. Its performance has been compared to other DFO solvers such as BOBYQA, ORBIT, SNOBFIT and IMFIL. For the test suite of nonconvex smooth problems, UNIPOPT finds solutions within 1% of the global minima for 10â€"65% more problems compared to these solvers. For convex nonsmooth problems, UNIPOPT can solve 20-30% more problems to 1% global optimality. In terms of convergence, we show that UNIPOPT converges to local minima when certain conditions are satisfied. The UNIPOPT framework is further extended to handle constrained black-box problems. Specifically, we will present the performance of the algorithm against other DFO solvers that can handle constraints for 92 numerical problems taken from GlobalLib. UNIPOPT finds solutions within 1% of the global optimality for 20-52% more problems than those solved by the other solvers.

References:

[1] Agarwal, A., Biegler, L. T. and Zitney, S. E. A superstructure-based optimal synthesis of PSA cycles for post-combustion CO_2 capture. *AIChE Journal*, 56(7):1813-1828, 2010.

[2] Caballero, J. A. and Grossmann, I. E. An algorithm for the use of surrogate models in modular flowsheet optimization. *AIChE Journal*, 54(10):2633-2650, 2008.

[3] Nuchitprasittichai, A. and Cremaschi, S. Optimization of CO_2 capture process with aqueous amines using response surface methodology. *Computers & Chemical Engineering*, 35 (8):1521-1531, 2011.

[4] Jia, Z., Davis, E., Muzzio, F. J. and Ierapetritou, M. G. Predictive modeling for pharmaceutical processes using kriging and response surface. *Journal of Pharmaceutical Innovation*, 4(4):174-186, 2009.

[5] Beykal, B., Boukouvala, F., Floudas, C. A. and Pistikopoulos, E. N. Optimal design of energy systems using constrained grey-box multi-objective optimization, *Computers and Chemical Engineering*, 2018, https://doi.org/10.1016/j.compchemeng.2018.02.017.

[6] Arora, A., Bajaj, I., Iyer, S. S. and Hasan, M. M. F. Optimal synthesis of periodic sorption enhanced reaction processes with application to hydrogen production. *Computers and Chemical Engineering*, 115, 89-111, 2018.

[7] Rios, L. M. and Sahinidis, N. V. Derivative-free optimization: a review of algorithms and comparison of software implementations. *Journal of Global Optimization*, 56(3):1247â€"1293, 2013.

[8] Bajaj, I. and Hasan, M. M. F. UNIPOPT: Univariate projection-based optimization without derivatives. *Under Review*.

[9] Fiacco, A. V. Introduction to sensitivity and stability analysis in nonlinear programming. *Academic Press, New York*, 1984.

521738 Hierarchical MPC Schemes for Periodic Systems Using Stochastic Programming Techniques

Abstract:

Model Predictive Control (MPC) is a powerful planning framework that can handle inequality constraints, multivariable interactions, and accommodate different kinds of cost functions and disturbances [1, 2]. A well-known challenge arising in MPC is the computational tractability associated with the length of the planning horizon and with the time resolution of the state and control policies. These tractability issues are often encountered in energy system applications that exhibit phenomena and disturbances emanating at multiple timescales. For instance, long horizons are often required to respond to low-frequency (e.g., seasonal) supply/demand variations, peak electricity costs (e.g., demand charges) while fine time resolutions are needed to tackle high-frequency supply/demand variations (e.g., from wind/solar supply) and to participate in real-time electricity markets [3-5]. Tractability issues are currently handled by using ad-hoc receding horizon (RH) approximations, which are practical but do not provide optimality guarantees [6,7].

Hierarchical MPC schemes [8, 9] have recently been proposed to handle multiple scales and achieve stability. These schemes, however, do not provide optimality guarantees in the sense that the computed policies match those of the long-horizon optimal control problem of interest. The
hierarchical scheme proposed in [10] uses adjoint information obtained from a long-term and coarse MPC controller to guide a short-term MPC controller operating at fine time resolutions. Computational experiments are provided to demonstrate that this approach can achieve optimality but no guarantees are given. Moreover, such an approach requires smoothness and continuity of the adjoint profiles, which is not guaranteed in general applications. The hierarchical scheme proposed in this work relies on the observation that, if the optimal policy of an infinite horizon optimal control problem is periodic (or can be approximated with a periodic policy), the problem can be cast as a stochastic programming (SP) problem. Periodicity is a property that is commonly observed in systems driven by exogenous factors with strong periodic components (e.g., energy demands and prices) [11-13]. Under the SP abstraction, the periodic states are interpreted as design variables and operational policies over the periods are interpreted as recourse variables.

We have recently observed that the SP representation provides a mechanism to construct hierarchical MPC schemes in which a long-term (supervisory) MPC controller provides periodic targets to guide a short-term MPC controller [14]. Under nominal conditions with perfect forecasts, we have shown that the hierarchical scheme delivers an optimal policy for the infinite horizon problem. For the more practical case of imperfect forecasts, the hierarchical scheme needs to re-compute periodic targets. While this can certainly be done using an RH scheme (e.g., computes targets by anticipating multiple future periods), such an approach would not provide optimality guarantees. In fact, to the best of our knowledge, no RH scheme currently exists that can provide optimal policies in the presence of imperfect forecast information.

The key contribution of this work is the observation that, under a periodic setting, one can derive retroactive optimization schemes that accumulate real historical disturbance information to asymptotically deliver optimal targets. We argue that this retroactive design principle offers a key advantage over proactive RH schemes (which rely only on forecast information). The targets obtained with a retroactive scheme are used to guide a low-level MPC controller operating at fine time resolutions within the periods. In the case of linear systems, one can derive a specialized retroactive scheme by using incremental cutting-plane (CP) techniques [15,16]. The SP setting also reveals strategies to construct retroactive schemes for general nonlinear systems and to derive metrics to monitor optimality. We demonstrate the concepts using an application in buildings with electrochemical storage.
[1] J.B. Rawlings and D.Q. Mayne. Model Predictive Control: Theory and Design. Nob Hill Publishing, Madison, WI, 2009.

[2] S. Qin and T. Bagdwell. A survey of industrial model predictive control technology. Control Engineering Practice, 1:733-764, 2003.

[3] Jingran Ma, Joe Qin, Timothy Salsbury, and Peng Xu. Demand reduction in building energy systems based on economic model predictive
control. Chemical Engineering Science, 67(1):92-100, 2012.

[4] Ranjeet Kumar, Michael J Wenzel, Matthew J Ellis, Mohammad N ElBsat, Kirk H Drees, and Victor M Zavala. A stochastic model predictive control framework for stationary battery systems. IEEE Transactions on Power Systems, 2018.

[5] James E. Braun. Reducing energy costs and peak electrical demand through optimal control of building thermal storage. In ASHRAE
Transactions, number pt 2, pages 876-888, 1990.

[6] Michael J Risbeck, Christos T Maravelias, James B Rawlings, and Robert D Turney. A mixed-integer linear programming model for realtime
cost optimization of building heating, ventilation, and air conditioning equipment. Energy and Buildings, 142:220-235, 2017.

[7] Nishith R Patel, Michael J Risbeck, James B Rawlings, Michael J Wenzel, and Robert D Turney. Distributed economic model predictive control
for large-scale building temperature regulation. In American Control Conference (ACC), 2016, pages 895-900. IEEE, 2016.

[8] Riccardo Scattolini and Patrizio Colaneri. Hierarchical model predictive control. In Decision and Control, 2007 46th IEEE Conference on, pages 4803-4808. IEEE, 2007.

[9] Bruno Picasso, Daniele De Vito, Riccardo Scattolini, and Patrizio Colaneri. An mpc approach to the design of two-layer hierarchical
control systems. Automatica, 46(5):823-831, 2010.

[10] Victor M Zavala. New architectures for hierarchical predictive control. IFAC-PapersOnLine, 49(7):43-48, 2016.

[11] Rui Huang, Eranda Harinath, and Lorenz T Biegler. Lyapunov stability of economically oriented nmpc for cyclic processes. Journal of Process Control, 21(4):501-509, 2011.

[12] Kaushik Subramanian, James B Rawlings, and Christos T Maravelias. Economic model predictive control for inventory management in supply chains. Computers & Chemical Engineering, 64:71-80, 2014.

[13] Michael J Risbeck, Christos T Maravelias, James B Rawlings, and Robert D Turney. Cost optimization of combined building heating/

cooling equipment via mixed-integer linear programming. In American Control Conference (ACC), 2015, pages 1689-1694. IEEE, 2015.

[14] Ranjeet Kumar, Michael J. Wenzel, Matthew J. Ellis, Mohammad N. ElBsat, Kirk H. Drees, and Victor M. Zavala. Handling long horizons
in mpc: A stochastic programming approach. In American Control Conference (ACC), 2018. IEEE, 2018. To appear.

[15] Julia L. Higle and Suvrajeet Sen. Stochastic decomposition: An algorithm for two-stage linear programs with recourse. Mathematics of Operations Research, 16(3):650-669, 1991.

[16] Julia L Higle and Suvrajeet Sen. Stochastic decomposition: a statistical method for large scale stochastic linear programming, volume 8.
Springer Science & Business Media, 2013.

520106 Concurrent Canonical Variate Analysis for Process Operating Condition Deviations and Dynamic Anomalies Monitoring

Abstract:

Data-driven process monitoring methods are widely applied in industry for fault detection to improve process safety and product quality (e.g., [1]). Conventional process monitoring methods rely on the assumption that the characteristics of the data variations are relatively unchanged at normal operating conditions. However, variations in operating conditions are likely and common, which can result in high false alarm rates. Recursive process monitoring methods update the model with respect to the changing normal operating conditions, but cannot effectively distinguish between a slow drifting fault and a changing normal operating condition (e.g., [2]). The quality monitoring method resolves part of the problem, in the sense that the system can tell whether a deviation affects the quality variables (e.g., [3]). However, the quality monitoring method fails to trigger the alarm when dynamic process anomalies are present. To address these issues, we extend the concurrent canonical correlation analysis of Zhu et al. [4] for static quality monitoring to accommodate system dynamics. Concurrent canonical variate analysis (CCVA) is developed with two groups of monitoring indices to monitor quality-relevant fault and system dynamic anomaly. A real fault is considered as either the deviations of the quality-relevant variables or the abnormal temporal dynamic behaviors. Due to the capability of CVA on dealing with the dynamic data, the proposed approach has higher accuracy compared to other conventional quality monitoring methods, and the corresponding two proposed groups of monitoring indices can distinguish whether the disturbance in operating conditions is normal or a fault, which yields a complete description of the system. The Tennessee Eastman process is used to illustrate the effectiveness of CCVA-based monitoring method.

[1] L.H. Chiang, E.L. Russell, and R.D. Braatz (2000) Fault Detection and Diagnosis in Industrial Systems. Springer Verlag, London, United Kingdom.

[2] W. Li, H.H. Yue, S. Valle-Cervantes, and S.J. Qin (2000) Recursive PCA for adaptive process monitoring. Journal of Process Control, 10, 471-486.

[3] J.F. MacGregor, C. Jaeckle, C. Kiparissides, and M. Koutoudi (1994) Process monitoring and diagnosis by multiblock PLS methods. AIChE Journal, 40(5), 826-838.

[4] Q. Zhu, Q. Liu, and S.J. Qin (2016) Concurrent canonical correlation analysis modeling for quality-relevant monitoring. IFAC-PapersOnLine, 49(7), 1044-1049.

A combined graphical / algebraic method for model reduction and analysis of chemical reaction networks: application to atomic layer deposition process

Hossein Salami, Aisha Alobaid, Raymond Adomaitis

Department of Chemical and Biomolecular Engineering,
University of Maryland College Park

In many cases, the study of chemical engineering systems involves mathematical modeling of a potentially complex reaction network (RN) consisting numerous species and reactions. One example of such systems is atomic layer deposition (ALD), a thin-film manufacturing technique based on sequential reactions between gaseous precursors and a solid substrate to conformally deposit thin-film with accurately controlled composition. An ALD process starts with precursor adsorption on the substrate surface and proceeds with surface reactions by different mechanisms such as dissociation, transition state formation, ligand exchange, and densification; this reaction sequence ends with desorption and ultimately evacuation of reaction by-products from the reactor [1]. These reactions may occur with significantly different rates. For example in many cases the initial adsorption of precursor molecules on surface sites is barrierless and has a very short timescale; likewise equilibrium reactions result in pseudo-steady state conditions. Some ligand exchange surface reactions, on the other hand, may have a large energy barrier making them finite-rate processes.

One path to formulating the molar-balance system of equations for the species present in the RN system is to represent each equilibrium reaction in form of a finite-rate process by introducing an artificial time constant ($1/\varepsilon$), where ε can later be set to 0 to recover the original equilibrium relationship [1,2]. Due to interconnected nature of the reactions in the network, the system of differential equations obtained from writing molar balances in this manner sometimes produces a singularly perturbed system in non-standard form, meaning ε cannot be directly set to 0 to recover each equilibrium process; by doing so, information regarding the dynamics of a subset of chemical species would be lost. Moreover, if the number of independent reactions in the RN under study is less than number of species, the RN is expected to have invariant states giving rise to additional algebraic constraints. These issues can be systematically addressed by finding a transformation of species molar amounts (original variables) to a new state space in a process that identifies redundant dynamic modes (invariants) and decouples reactions with different time scales to produce a well-posed differential algebraic system of equations (DAE) upon setting $\varepsilon \to 0$ [1,2,3].

We recently proposed a graphical approach to identifying invariant states associated with an RN [4]. Our approach makes use of species-reaction (SR) graphs first introduced by Craciun and Feinberg [5]. We have shown that a RN's invariant states correspond to paths on the SR graph and that these paths can be generated by a relatively simple set of rules. In this work, we introduce a computational procedure to facilitate identification of these paths. This procedure follows the mathematical algorithm proposed by Schuster and Hofer [6] and is based on constructing a series of tableaus based on the stoichiometric matrix constructed from RN. The invariant states obtained from this graphical / algebraic method are all semi-positive (nonzero and non-trivial linear combination of species molar amounts); the semi-positive properties of the invariants will be shown to be crucial to interpreting the physical significance of each invariant state; furthermore, the properties of the invariants themselves also can be used to identify potential defects in the RN structure. We will show that this procedure can be modified to not just obtain invariant states, but also to decouple reactions with different time scales to complete the transformation described earlier. Extension of our algorithmic and graphical techniques to the study of reaction variants also will be presented.

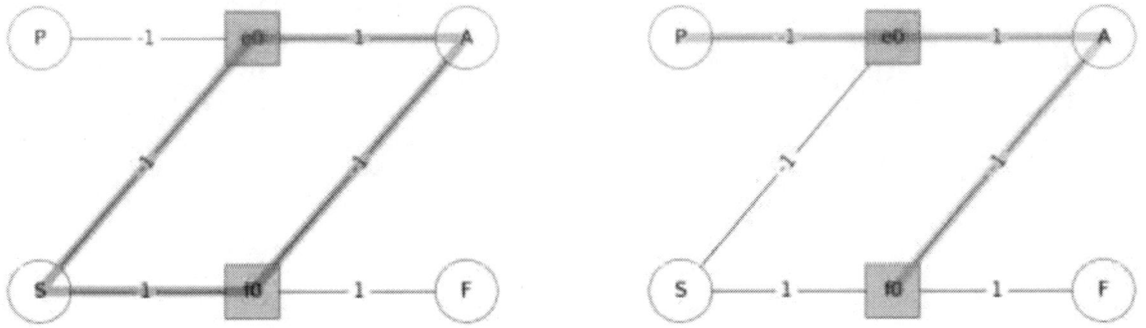

Fig1. SR graph associated with RN of P+S→A (e0), A→S+F (f0) giving a simplified picture of one ALD half-reaction (precursor P, site S, adsorbed species A and film material F) and possessing 2 invariant states. One of the invariants, associated with conservation of surface sites, can be formulated from the cyclic path shown in blue (left). Variant state associated with the effect of finite rate reaction f0 can be obtained from the P to A path shown in red (right). The node corresponding to equilibrium reaction e0 is surrounded by two species nodes with different stoichiometric signs, resulting in removal of e0 effect in the formulated variant state and leaving the f0 node in decoupled form: $\dot{P} + \dot{A} = -f0$

References

[1] Adomaitis R., 2016. J Vac. Sci. Technol. A 34, 01A104.

[2] Daoutidis, P., 2015. Surveys in diferential-algebraic equations II., Springer International Publishing, New York, 69.

[3] Rodrigues, D., Srinivasan, S., Billeter, J., Bonvin, D., 2015. Comput. Chem. Eng. 73, 23.

[4] Salami, H., Ramakrishnasubramanian, K., Adomaitis, R., 2017. Phys. Status Solidi B 1700091.

[5] Craciun, G., Feinberg, M., 2006. SIAM J. Appl. Math. 66(4), 1321.

[6] Schuster, S., Hofer, T., 1991. J. Chem. Soc. Faraday Trans. 87(16), 2561.

537317 Optimal Solar Cell Configuration Under Partially Shaded Conditions

Abstract:

The objective of this work is to develop computational tools to identify optimal configurations when integrating a large number of PV cells under shady or faulty conditions. Monte Carlo simulation is used to simulate a PV cell module under partially shaded conditions and to find the optimal number of bypass diodes needed to minimize performance losses. Both event-driven Monte Carlo (MC) simulation and Multi-Attribute Utility Theory (MAUT) are used to evaluate the effect of the number of bypass diodes on the PV module maximum power point under uncertain shading conditions [1]. The number of shaded cells and their position in the PV module are considered random variables, and the model is evaluated under different shading levels.

Figure 1: P-V curve for one shaded cell (50% shading) at different bypass diodes [1]

Figure 2: MAUT total value function (50% shading) [1]

References:

[1] Alobaid, A. and R. A. Adomaitis, (2018). Monte Carlo Simulation for Optimal Solar Cell Configuration. *Proceedings of PSE 2018.*

533454 Multi-Objective Optimization of Cchp Systems Using Particle Swarm Algorithms

Abstract:

The development of the global economy and the growth of population have brought about an ever-increasing demand for energy consumption. At the same time, related studies have pointed out that generally there are more than one-third of global energy consumption comes from building energy supply systems. Since the excellent performance on energy saving and emission reduction, the combined cooling heating and power (CCHP) system has been widely used to supply energy in residential houses, offices, shopping malls, hotels and other buildings or regions. However, in the actual design planning process, there is a lack of scientific and rational optimization methods and management strategies for many CCHP systems.

According to the characteristics of energy use from specific buildings, most CCHP systems may operate following the electric load (FEL), the thermal load (FTL) or the hybrid electric-thermal load (FHL). In this study, thermal modeling and optimal design of a typical CCHP system for a designated office building was conducted using the multi-objective particle swarm optimization (MOPSO) algorithm. The cooling capacity of electrical and absorption chillers as well as electric cooling ratio were optimized as the design variables. The Pareto principle was applied to evaluate the system indicators under different operate strategies, including the energy, economy, and environmental parameters simultaneously. It was observed that the improvement of energy efficiency would inevitably lead to the increase of investment. The performance of CCHP system under different operate strategies during a year was also compared. Finally, the results of this study would provide a certain technical reference for related planning and design.

526596 HPC Modeling and Simulation of Mass Transport in Wavy Falling Liquid Films

Abstract:

Corresponding author. E-mail address: hengyi@mail.sysu.edu.cn (Yi Heng)

It is known that the wavy falling film exhibits better heat and mass transfer performance. However, the underlying kinetic phenomena are still not fully understood. In recent years, different modeling and simulation studies for local hydrodynamics and mass transport of falling film have received great attention in both academia and industry. A typical way for calculating the fluid dynamics in a falling film is to solve the Navier-Stokes equations (NSE) in the liquid and gas phases simultaneously and use volume-of-fluid (VOF) method to capture the gas-liquid interface. To the best of our knowledge, only a few direct numerical simulation (DNS) results related to the modeling of two-phase falling films are available in the literature. Xu et al. [1] used a hydrodynamic parameter to characterize the scalar transport across the interface and gave an empirical relation. Albert et al. [2] employed a two-scalar approach to simulate the absorption of oxygen into wavy film and identified the enhancement mechanism of mass transfer related to local Sherwood number and hydrodynamic phenomena. Hu et al. [3] found out that the concentration of CO_2 absorbed in the film is closely related to the vorticity and defined a correlation coefficient to characterize the relationship between them. Besides the VOF method, other methods such as the Arbitrary-Lagrangian-Eulerian approach [4] or level set method [5, 6] can also be used to simulate the mass transport if fluid dynamics is apriori known.

However, DNS approaches usually require significant computational effort that is in general not applicable in practice. An alternative way, which can reduce computational effort, is to apply a simplified single-phase model, i.e., by solving the NSE only in the liquid phase with a free boundary at the two-phase interface. Sisoev et al. [7] used a simplified Navier-Stokes model derived by Shkadov to study the gas absorption into a wavy falling film. Bandi et al. [8] utilized a reduced model, i.e., the long-wave expansion developed by Balakotaiah et al. [9, 10], to simulate the absorption of wavy falling film and validated the results with 2-D experimental data obtained by the PLIL measurement method. Wylock and Scheid [11] reduced the 2-D governing equations into a 1-D second-order weighted integral boundary layer (WIBL) model and obtained similar hydrodynamic with the 2-D model when Reynolds number is less than 100.

In this work, instead of using the simplified model to reduce computational effort, the DNS of the falling film is investigated by means of high performance computing (HPC) techniques on the supercomputer. The long-term goal is to establish an efficient computing platform that can solve large-scale forward and inverse problems arising in falling film on an industrial scale (decimeters to meters) at high calculation precision. As a first step, a series of 2-D simulations for a laminar, Newtonian and incompressible water film flowing down a vertical flat plane are investigated in this work. Ansys Fluent 18.0 is adopted to solve the NSE with the VOF method [12] employed to capture the gas-liquid interface and the continuum surface force (CSF) model [13] to account for the surface tension. To simulate the oxygen transport in the film, the user-defined scalar (UDS) function is used to calculate the convection-diffusion equation. The inlet water velocity is assumed as a well-accepted parabolic Nusselt velocity profile with a sinusoidal perturbation to generate wavy film. The rest of the computation domain is assumed to be filled with air. The oxygen concentration at the gas-liquid interface is assumed to be equal to the equilibrium concentration, i.e. 0.28 mol/m^3 according to Henry s law [8]. Homogeneous Neumann conditions are applied at the remaining boundaries.

A grid independence analysis is carried out at different grid levels. The grid size of 0.02*0.04 mm^2, in the height and flow direction, respectively, is an appropriate choice to balance the computational effort and solution accuracy (cf. the red line in Figure 1). A time step size of 3*10^{-5} s is applied to ensure the numerical stability.

Figure 1 A representative example of simulated instantaneous film thickness at t=0.3 s using three different levels of grids

Furthermore, based on the proposed computational model, scalability tests on the supercomputing platform (24 cores per node) are carried out (cf. Figure 2). Without loss of generality, the results for Re=41 with a simulation time of 0.1s are presented.

Figure 2 (left) Strong scalability test (right) Weak scalability test

In the strong scalability analysis, a fixed-size problem of 1.4*960 mm^2 is computed by using a series of increasing number of cores/nodes. Figure 2(left) shows that for a problem of such size the speed-up ratio can be increased up to 2 nodes (48 cores). As more nodes are used the communication overhead across the nodes becomes the choke point and limits the possibility of further speed-up. In the weak scalability analysis, the problem size is set to be proportional to the CPU cores, i.e. the computational domain of 1.4*60 mm^2 for 3 cores, 1.4*120 mm^2 for 6 cores, 1.4*240 mm^2 for 12 cores, 1.4*480 mm^2 for 24 cores (1 node), 1.4*960 mm^2 for 48 cores (2 nodes), and 1.4*1920 mm^2 grid for 96 cores (4 nodes). Figure 2(right) shows that the speed-up ratio can be increased by using the multi-node parallel computing if the problem size is sufficiently large. In this case, the influence of communication overhead can be diluted.

Based on the proposed HPC computational strategy, a series of simulations for a range of Re numbers are performed. The waves are induced by a perturbation of frequency f=12 in a computational domain of 0.25 m. The simulated velocity and concentration distributions in quasi-steady state for 4 representative Re numbers are shown in Figure 3 and Figure 4, respectively. The perturbation in the inlet generates solitary waves followed by several capillary waves periodically. Similar to the patterns of high concentrations locally found in other literature [8,9,10], it is also observed that the velocity magnitude in the wave crest is higher than other zones, which results in intensified convection mass transfer and high concentration in the solitary wave.

Figure 3 Instantaneous velocity distributions at t=1s for different Re(tenfold scaling of the Y axis is used for better view)

Figure 4 Instantaneous concentration distributions at t=1s for different Re(tenfold scaling of the Y axis is used for better view)

In summary, a 2-D wavy laminar falling film model was proposed for the simulation of the oxygen absorption into water. Through parallel computing by the supercomputing platform, it has high potential to solve large-scale forward problems at a practically reasonable cost. Moreover, the separability of the problem allows to perform batch processing at different operating conditions. In the future work, it is expected to prepare big data generated by large-scale simulations on falling film by using the proposed HPC computational strategy, so as to meet the future industrial requirements. Future work will also include the development of optimal experiment design techniques for the mass transport in falling films by solving arising large-scale inverse problems. The forward modeling approach with high efficiency and precision proposed in this work is considered as an important step toward this long-term modeling goal.

References

[1] Z.F. Xu, B.C. Khoo, N.E. Wijeysundera, Mass transfer across the falling film: Simulations and experiments, Chemical Engineering Science 63(9) (2008) 2559-2575.

[2] C. Albert, H. Marschall, D. Bothe, Direct Numerical Simulation of interfacial mass transfer into falling films, International Journal of Heat and Mass Transfer 69 (2014) 343-357.

[3] J. Hu, X. Yang, J. Yu, G. Dai, Numerical simulation of carbon dioxide (CO2) absorption and interfacial mass transfer across vertically wavy falling film, Chemical Engineering Science 116 (2014) 243-253.

[4] P.S. Weber, H. Marschall, D. Bothe, Highly accurate two-phase species transfer based on ALE Interface Tracking, International Journal of Heat and Mass Transfer 104 (2017) 759-773.

[5] P. Chasanis, A. Lautenschleger, E.Y. Kenig, Numerical investigation of carbon dioxide absorption in a falling-film micro-contactor, Chemical Engineering Science 65(3) (2010) 1125-1133.

[6] C. Yang, Z.-S. Mao, Numerical simulation of interphase mass transfer with the level set approach, Chemical Engineering Science 60(10) (2005) 2643-2660.

[7] G.A. Sisoev, O.K. Matar, C.J. Lawrence, Absorption of gas into a wavy falling film, Chemical Engineering Science 60(3) (2005) 827-838.

[8] P. Bandi, M. Modigell, S. Groß, A. Reusken, L. Zhang, Y. Heng, W. Marquardt, A. Mhamdi, On reduced modeling of mass transport in wavy falling films, AIChE Journal (2018), In Press.

[9] C.E. Meza, V. Balakotaiah, Modeling and experimental studies of large amplitude waves on vertically falling films, Chemical Engineering Science 63(19) (2008) 4704-4734.

[10] R.R. Mudunuri, V. Balakotaiah, Solitary waves on thin falling films in the very low forcing frequency limit, AIChE Journal 52(12) (2006) 3995-4003.

[11] C. Wylock, B. Scheid, Two-dimensional modeling of an absorbing falling film in its development zone, AIChE Journal 63(10) (2017) 4370-4378.

[12] C.W. Hirt, B.D. Nichols, Volume of fluid (VOF) method for the dynamics of free boundaries, Journal of Computational Physics 39(1) (1981) 201-225.

[13] J.U. Brackbill, D.B. Kothe, C. Zemach, A continuum method for modeling surface tension, Journal of Computational Physics 100(2) (1992) 335-354.

530288 Autotuning with Derivative-Free Optimization

Abstract:

As the landscape of high performance computing evolves, there is a growing need to design software that is optimal on a variety of system architectures. To address this need, algorithms are designed with tunable parameters to allow for performance portability. However, identifying optimal tuning parameters requires solving a large global optimization problem without an explicit algebraic function. Past research has been conducted to automatically determine good tuning parameters through the use of hill climbing algorithms [1], Nelder-mead simplex algorithms [5], genetic algorithms [2], and simulated annealing [4].

In this work we investigate the benefits of autotuning with derivative-free optimization (DFO). While most of the search techniques implemented in autotuners utilize local DFO solvers or heuristic based methods, by employing model-based and global DFO solvers [6], we are able to quickly find high quality tuning parameters. We compare our proposed methodology against state-of-the-art autotuning techniques such as OpenTuner [1], Active Harmony [7], and Starchart [3]. We carry out experiments on dense linear algebra kernels that are essential to the performance of numerous algorithms. In this work, we tune GPU algorithms which have a more complex parameter space than traditional CPU algorithms. We demonstrate the ability of our approach to tune any system by tuning a collection of algorithms on different GPU architectures.

References

[1] J. Ansel, S. Kamil, K. Veeramachaneni, J. Ragan-Kelley, J. Bosboom, U. O'Reilly, and S. Amarasinghe. Opentuner: An extensible framework for program autotuning. In Parallel Architecture and Compilation Techniques (PACT), 2014 23rd International Conference on, pages 303–315, 2014.

[2] D. E. Goldberg. Genetic Algorithms in Search, Optimization and Machine Learning. Addison-Wesley, 1989.

[3] W. Jia, E. Garza, K. Shaw, and M. Martonosi. GPU performance and power tuning using regression trees. ACM Transactions on Architecture and Code Optimization (TACO), 12:13, 2015.

[4] S. Kirkpatrick, C. D. Gelatt, and M. P. Vecchi. Optimization by simulated annealing. Science, 220:671-680, 1983.

[5] J. A. Nelder and R. Mead. A simplex method for function minimization. Computer Journal, 7:308–313, 1965.

[6] L. M. Rios and N. V. Sahinidis. Derivative-free optimization: A review of algorithms and comparison of software implementations. *Journal of Global Optimization*, 56:1247-1293, 2013.

[7] C. Ţăpuş, I. Chung, and J. Hollingsworth. Active harmony: Towards automated performance tuning. In

Proceedings of the 2002 ACM/IEEE conference on Supercomputing, pages 1–11, 2002.

538037 An Ontology-Based Automated Generation of Training Scenarios: Development of Process Safety Rule Engine

Abstract:

The preconditions of the usefulness of an operator safety training model in large plants are the versatility and accuracy of operational procedures, obtained by detailed analysis of the various types of risks associated with the operation, and the systematic representation of knowledge.

Existing scenario-based training system were not flexible, it can be focused on implantable memories due to they are limited, and the efficiency of knowledge acquisition is low owing to uniformity process and no differences trainees. Therefore, automatic generation of training scenarios provides a variety of training situations and opportunities when mounted on a VR-based training system, and can provide optimal step-by-step intelligent training tailored needs and abilities of individual trainees. This requires the development of events in the training domain, simulation and the representation of knowledge about goals and behaviors. Ontologies enable coherent processes, devices, operations, reasoning and modeling, sharing and reuse of knowledge.

In this study, we use the ISO 15926 an international standard for plant data integration, management and the process ontology of OntoCAPE to generate a knowledge node based on the HAZOP analysis of the RDS process of the petrochemical plant. Ontology on the results of causation for 22 items of action items due to process safety and operational impact. And automatic generation of safety training scenarios is implemented by using inference modeling of SWRL (Semantic Web Rule Language) based on OWL-DL. It is expected that it will be possible to develop a swift automatic generation system by increasing the recyclability of scenario inference engine compared with the automatic generation based on XML extraction + HTN planning of the proposed ontology.

538167 Construction of a Semi-Stochastic Intracellular Signaling Model Via Global Sensitivity Analysis and Probability Density Estimation

Abstract:

Biological models often describe the population-average dynamics of signaling pathways through a deterministic modeling approach. However, recent single-cell studies have revealed that cells from a clonal population exhibit a large degree of cell-to-cell variability [1]. The cell-to-cell variability is believed to originate from intrinsic (stochastic chemical kinetics) and extrinsic (different initial conditions between cells) sources. Previous studies have applied stochastic modeling methodologies such as Gillespie's algorithm [2] to simulate the single-cell dynamics. However, this method is often computationally expensive; this is particularly problematic as many biological models have a large number of states and parameters, which inevitably leads to the formidable computational cost.

As an alternative to the stochastic simulation framework, a semi-stochastic modeling approach has been proposed [3], which uses a deterministic model with model parameters that have distributions. More specifically, a pre-specified probability density function (PDF) of the model parameters is used to generate different parameter values, which are subsequently used in the deterministic model to simulate corresponding distinct signaling dynamics. This approach allows simulation of cell-to-cell variability with a manageable computational cost. However, the PDF of the model parameters is usually unknown *a priori* as values of the model parameters are difficult to measure experimentally; therefore, the PDF needs to be inferred from measurements.

In this study, a sequential approach that consists of global sensitivity analysis and probability density estimation is proposed to systematically estimate the PDF of model parameters. First, a sampling-based global sensitivity analysis method [4] is implemented to identify a set of model parameters that impact model outputs most significantly. Next, the PDF of the identified parameters is estimated by minimizing the difference between the measured and predicted output probability densities, which are approximated through particle filtering [5] and kernel density estimation [6]. Through the proposed methodology, the parameter PDF of a dynamic model can be inferred accurately, which can be used to construct a semi-stochastic model to identify the source of heterogeneity and to quantify their magnitude in the single-cell dynamics. As a test case, the proposed methodology is applied to estimate the PDF of the NFκB signaling pathway model [7], which includes about 150 model parameters, to validate the capability of the proposed approach.

[1] Mitchell, S.; Hoffmann, A. Identifying noise sources governing cell-to-cell variability. *Curr. Opin. Syst. Biol.*, **2018**, *8*, 39-45.

[2] Gillespie, D.T. Exact stochastic simulation of coupled chemical reactions. *J. Phys. Chem.*, **1977**, *82*, 2340–2361.

[3]Hasenauer, J.; Waldherr, S.; Doszczak, M.; Scheurich, P.; Radde, N.; Allgöwer, F. Analysis of heterogeneous cell populations: A density-based modeling and identification framework. *J. Process Control*, **2011**, *21*, 1417–1425.

[4] Chu, Y.; Hahn, J. Parameter set selection for estimation of nonlinear dynamic systems. *AIChE J.*, **2007**, *53*, 2858-2870

[5] Rawlings, J. B.; Bakshi, B. R. Particle filtering and moving horizon estimation. *Comp. & Chem. Eng.*, **2006**, *30*, 1529-1541.

[6] Silverman, B.W. Density estimation for statistics and data analysis. Chapman & Hall/CPC: London, 1986.

[7] Lee, D.; Ding, Y.; Jayaraman, A.; Kwon, J.S. Mathematical Modeling and Parameter Estimation of Intracellular Signaling Pathway: Application to LPS-induced NFκB Activation and TNFα Production in Macrophages, *Processes*, **2018**, *6*, 21.

538800 Mechanical Perturbation Approach for TreatingÂ Cardiac Arrhythmias

Abstract:

Cardiac arrhythmias such as atrial fibrillation (AF) and ventricular fibrillation (VF) are characterized by irregular electrical activity, which lead to asynchronous contraction and may lead to sudden cardiac death. AF and VF can be treated by the application of an electrical shock, which is referred to as electrical defibrillation therapy . This electrical stimulus depolarizes a large amount of the heart muscle cells and has the effect of stopping any propagation of the impulse and terminating reentrant activations. If successful, normal heart beat, which is referred to as sinus rhythm, is restored. This procedure, called defibrillation, has been successfully applied to a human for the first time in 1947. The electrical treatments, which have remained the most reliable approach to terminate AF and VF, may cause a damage to the heart due to delivery of high electrical current.

The electrical waves propagate through the cardiac tissue and initiate contraction in human heart. In its turn, contraction causes cardiac tissue deformation which feeds back on the wave propagation and affects electrophysiological properties via mechano-electric feedback (MEF) [1]. Cardiac excitation and contraction of the heart can be linked by an electromechanical model of cardiac tissue. MEF is used in the literature to describe the influence of mechanical deformation of cardiac tissue on the electrophysiology. Mechanisms of MEF which include immediate influence on the action potential through stretch-activated channels, has been studied in the clinical community for well over a century and has been found to have to have anti- and pro-arrhythmic effects.

Recently, a novel mechanical perturbation algorithm to suppress cardiac alternans [2,3], which is an alternation in the action potential duration, was presented. This mechanical perturbation is reflected in cardiac excitation through the mechanisms of MEF. In this study, we explore this approach in the treatment of VF where MEF plays also a role in restoring normal electrical activity. Our control algorithm considers a modified version of error based feedback control that is implemented using mechanical perturbation control strategy and applied sequentially in subregions of cardiac tissue. To this end, a realistic electromechanical model of cardiac tissue is employed to explore the feasibility of terminating VF at cellular and tissue levels.
The cardiac bidomain model of action potential propagation in tissue is presented. The bidomain model consists of a system of coupled parabolic and elliptic PDEs for two potentials in the cardiac muscle, coupled with a nonlinear system of ODEs describing the ionic currents flowing across the cardiac membrane. The Luo-Rudy 1 [4] model is used to represent the electrophysiological properties of the heart, while for the active contractile tension model, the Niederer-Hunter-Smith [5] model is used to represent mechanical activity. The exponential passive elasticity [6] model is employed to describe passive mechanical behaviour of the myocardium.

1 - M.J. Lab, Mechanoelectric feedback (transduction) in heart: concepts and implications, Cardiovasc.Res, vol.32, p. 3-14, 1996.

2 - A. Hazim, Y. Belhamadia, S. Dubljevic, Control of cardiac alternans in an electromechanical model of cardiac tissue, Computers in Biology and Medicine 63 (2015) 108-117.

3 - A. Hazim, Y. Belhamadia, S. Dubljevic, Mechanical perturbation control of cardiac alternates, Physical Review E, accepted

4 - C. Luo and Y. Rudy, A model of the ventricular cardiac action potential. Depolarization, repolarization, and their interaction, Circ. Res. 68, 1501 (1991).

5 - S. A. Niederer, P. J. Hunter, and N. P. Smith, A quantitative analysis of cardiac myocyte relaxation: a simulation study, J Biophysics, vol. 90, p. 1697-722, 2006.

6 - J. M. Guccione, A. D. McCulloch, L. K. Waldman, Passive material properties of intact ventricular myocardium determined from a cylindrical model, J. Biomech. Eng. 113 (1) (1991) 42-55.

514702 The Impact of Glottis Opening on Drug Aerosol Delivery in a Subject-Specific Lung-Airway Model: A Numerical Study

Abstract:

Computational fluid-particle dynamics (CFPD) simulations have been widely used to investigate transport phenomena of airflow and inhaled aerosols in human pulmonary routes. Compared to *in-vitro* and *in-vivo* methods, CFPD models are non-invasive, time-saving, and cost-effective. Properly validated, they can provide high-resolution results of and new physical insight to the transport and deposition of inhalable drugs or toxicants in human respiratory systems. Existing CFPD studies assume that glottis openings are identical between the respiratory system geometries reconstructed from CT/MRI images and the real scenarios when subjects draw drug aerosols from inhalers. However, this assumption is not physiologically correct and hence will influence the accuracy of simulation results of aerosol transport and deposition. As a result, the changing glottis geometry significantly influences the airflow structure and inhaled particle deposition in tracheobronchial airways. Thus, it is necessary to investigate such glottis-opening effects on aerosol transport and deposition in human upper respiratory systems. It will provide patient-specific guidance for operating the drug inhalation process, i.e., breathing patterns, inhaler activation timing, inhaler type/configuration and administered dosage, with the overall goal of best drug delivery efficacy. In this study, the impact of glottis opening on aerosol transport is investigated in a human upper lung model, i.e., from mouth to Generation 8 (G8). Specifically, using an experimentally validated Euler-Lagrange CFPD model, the transport and deposition of drug-aerosol particles on the micro/nanoscale were performed with three different glottis openings, i.e., 86.39 mm^2, 163.47 mm^2, and 253.00 mm^2. At different inhalation flowrates, significant influences were discovered among different glottis openings regarding airflow structures, i.e., laryngeal jet and turbulence intensity, as well as particle transport characteristics and local deposition values. Results support the importance of physiological correctness of glottis openings associated with realistic drug inhalation patterns. Furthermore, the numerical results provide physical insight and preliminary data for fluid-structure interaction (FSI) modeling of the real-time glottis motion during drug inhalation, as part of future inter-subject variability studies.

536888 Investigating LecA Binding Mechanisms with a Cellular Membrane Containing Multiple Types of Receptors Via Kinetic Monte Carlo Simulation

Abstract:

Pseudomonas aeruginosa is a widespread bacterium that causes various acute and chronic infections such as pneumonia and meningitis [1]. LecA protein is one of the adhesins that mediate *P. aeruginosa* attachment to a host cell membrane, leading to the development of the infection [2]. LecA is a homotetramer that consists of four identical binding units. The four binding sites are grouped into two pairs that are aligned in opposite directions. Hence, one LecA can bind to maximum two glycolipid receptors at a time on a cell membrane. Moreover, in contrast to the classic antigen-antibody interaction, LecA can simultaneously bind to various types of glycolipids with different affinities. These two distinct characteristics, multivalency and semi-specificity, make the LecA recognition principle inherently different from classic bimolecular interactions (e.g. antibody-antigen binding).

Our recent study has experimentally demonstrated that mixing different glycolipid receptors on cell membranes could results in the alternation of LecA binding capacity, avidity, and kinetics, which influence the downstream biological processes. However, no theoretical model can interpret the complex interplay between LecA and multiple glycolipids. Here, we aim to develop a kinetic Monte Carlo model (kMC) to simulate the kinetics of LecA binding to a cell membrane containing two types of receptors. The kMC framework is selected as it can track the temporal evolution of the surface configuration, which is important in the overall binding kinetics on a cell membrane [4]. Specifically, we represent a cell membrane as square lattices, where LecA receptors reside and are allowed to migrate; on the other hand, LecA are allowed to attach and detach from the simulation lattices during the simulation. The overall binding kinetics are represented by a few discrete events in the proposed kMC: (a) the attachment of a LecA protein (b) the detachment of a membrane-bound LecA protein (c) surface reactions between membrane-bound LecA proteins with receptors (d) receptor migration on a membrane. During the simulation, reaction rates of these events, which depend on the surface configuration, are calculated to randomly select one event to be performed at a time until the pre-specified end time. For the computational efficiency, the receptor migration event is decoupled as the rate of the receptor migration is overwhelmingly large [5]. We apply this methodology to systematically analyze the overall binding process and identify the most important factors. The kMC simulation not only supports our experimental results, but also offers a fundamental mechanism to predict the complex glycobiology processes on cell membrane surface.

[1] Grishin, A.V.; Krivozubov, M. S.; Karyagina A.S.; Gintsburg, A.L. *Acta Naturae*, **2015**, *7*, 29-41.

[2] Chi, E.; Mehl, T.; Nunn, D.; Lory, S. Interaction of Pseudomonas aeruginosa with A549 pneumocyte cells. *Infect. Immun.* **1991**, *59*, 822-828.

[3] Blanchard, B; Nurisso, A; Hollville, E; Tétaud, C; Wiels, J; Pokorná, M; Wimmerová, M; Varrot, A; Imberty, A.

Structural basis of the preferential binding for globo-series glycosphingolipids displayed by Pseudomonas aeruginosa lectin I. *J. Mol. Bio.*, **2008**, *383*, 837-853.

[4] Mayawala, K.; Vlachos, D. G.; Edwards, J. S. Spatial modeling of dimerization reaction dynamics in the plasma membrane: Monte Carlo vs. continuum differential equations, *Biophys. J.*, **2006**, *121*, 194-208.

[5] Crose, M.; Kwon, J. S.; Nayhouse, M.; Ni, D.Christofides, P.D. On Operation of PECVD of Thin Film Solar Cells. *Chem. Eng. Sci.*, **2015**, *136*, 50-61.

[6] Krishnan, P.; Singla, A.; Lee, C.A.; Weatherston, J.D.; Worstell, N.C.; Wu, H.-J. Hetero-multivalent binding of cholera toxin subunit B with glycolipid mixtures. *Colloids Surf. B Biointerfaces*, **2017**, *160*, 281-288.

521141 Rigorous Parameter Estimation for Model Validation in Oncological Systems

Abstract:

The application of quantitative and formal methods from the physical sciences, engineering, and mathematics to physiological systems, provides an immense opportunity for novel and high-impact discovery. Specifically, such approaches in cancer research may enable a fundamental understanding of the limitations of conventional therapies, uncover novel treatment strategies, and someday help optimize therapy for individual patient outcomes. Recently, the normalization of tumor blood vessels has been proposed to improve the the delivery of nanomedicines [1, 2, 3, 4]. However, enhancing the delivery of nanomedicines is a multi-faceted engineering problem, and so a model-based systems engineering approach is proposed to better understand the underlying physical phenomena and complex relationships of the biological system.

Characterizing the fundamental transport properties of nanomedicines in tumors is paramount to developing highly-effective targeted drug delivery cancer therapies. For this purpose, a physics-based modeling approach must be taken. Baxter and Jain originally developed a one-dimensional spherical tumor model that describes vascular and transvascular transport [5, 6, 7, 8]. Further, a two-dimensional percolation-based tumor vasculature network model has also been established, that describes more details about the transmural coupling of fluid flow in the tumors [1, 9]. To enhance the predictive capabilities of models and provide confidence in their utility for the model-based approach for drug and therapy development, formal methods for model validation are required.

In this work, we propose using deterministic global optimization to solve the parameter estimation problem and provide a rigorous quantitative foundation for model discrimination. This approach requires solving a nonconvex optimization problem constrained by the physics-based tumor transport model as a partial-differential equation (PDE). A simulation-based feasible-path approach is taken and the PDE-constrained optimization problem is reformulated as a box-constrained nonlinear program with implicit functions embedded [10]. The objective function is the sum of squared error between the experimental data and the PDE model prediction.

In this study, we utilize a new experimental dataset of the dynamic transport of liposomes in solid tumors in mice. The data is in the form of the intensity profile over the spatial domain of the tumor for 1h and 24h discrete time points. Using distance-based anomaly detection, the data is filtered to eliminate outliers and obtain better results from solving the parameter estimation problem. We select the 1-D Baxter and Jain model for this study because the intensity data exists over the entire tumor domain consistent with this model [5, 6, 7, 8]. We apply a partial-factorial design of experiments approach to estimate the relative global sensitivities of the model parameters and identify the important parameters for consideration in the parameter estimation problem. The problem is solved using the open-source EAGO package[11] for global optimization with implicit functions in Julia [12].

Our work seeks to enhance the practicability and predictive capabilities of tumor transport models using data-driven model validation and rigorous methods in global optimization for stronger model-based systems engineering

approaches in cancer research. The information obtained through this approach aides in the development of better models and provides deeper insight into the physical behavior of molecular transport in tumors to guide drug development and delivery.

References

[1] V. P. Chauhan, T. Stylianopoulos, J. D. Martin, Z. Popovic, O. Chen, W. S. Kamoun, M. G. Bawendi, D. Fukumura, and R. K. Jain, "Normalization of tumour blood vessels improves the delivery of nanomedicines in a size-dependent manner," *Nat Nano*, vol. 7, pp. 383-388, June 2012.
[2] S. M. Tolaney, Y. Boucher, D. G. Duda, J. D. Martin, G. Seano, M. Ancukiewicz, W. T. Barry, S. Goel, J. Lahdenrata, S. J. Isako, et al., "Role of vascular density and normalization in response to neoadjuvant bevacizumab and chemotherapy in breast cancer patients," *Proceedings of the National Academy of Sciences*, vol. 112, no. 46, pp. 14325-14330, 2015.
[3] R. K. Jain, R. T. Tong, and L. L. Munn, "Effect of vascular normalization by antiangiogenic therapy on interstitial hypertension, peritumor edema, and lymphatic metastasis: insights from a mathematical model," *Cancer research*, vol. 67, no. 6, pp. 2729-2735, 2007.
[4] C. Polydorou, F. Mpekris, P. Papageorgis, C. Voutouri, and T. Stylianopoulos, "Pirfenidone normalizes the tumor microenvironment to improve chemotherapy," *Oncotarget*, vol. 8, no. 15, p. 24506, 2017.
[5] L. T. Baxter and R. K. Jain, "Transport of fluid and macromolecules in tumors. i. role of interstitial pressure and convection," *Microvascular research*, vol. 37, no. 1, pp. 77-104, 1989.
[6] L. T. Baxter and R. K. Jain, "Transport of fluid and macromolecules in tumors. ii. role of heterogeneous perfusion and lymphatics," *Microvascular research*, vol. 40, no. 2, pp. 246-263, 1990.
[7] L. T. Baxter and R. K. Jain, "Transport of fluid and macromolecules in tumors: iii. role of binding and metabolism," *Microvascular research*, vol. 41, no. 1, pp. 5-23, 1991.
[8] L. T. Baxter and R. K. Jain, "Transport of fluid and macromolecules in tumors. iv. a microscopic model of the perivascular distribution," *Microvascular research*, vol. 41, no. 2, pp. 252-272, 1991.
[9] J. W. Baish, P. A. Netti, and R. K. Jain, "Transmural coupling of fluid ow in microcirculatory network and interstitium in tumors," *Microvascular research*, vol. 53, no. 2, pp. 128-141, 1997.
[10] M. D. Stuber, J. K. Scott, and P. I. Barton, "Convex and concave relaxations of implicit functions," *Optimization Methods and Software*, vol. 30, no. 3, pp. 424-460, 2015.
[11] M. Wilhelm and M. D. Stuber, "Easy advanced global optimization (EAGO): An open-source platform for robust and global optimization in Julia," in *AIChE Annual Meeting 2017*, Minneapolis, MN, Oct 31, 2017.
[12] J. Bezanson, A. Edelman, S. Karpinski, and V. B. Shah, "Julia: A fresh approach to numerical computing," *SIAM review*, vol. 59, no. 1, pp. 65-98, 2017.

530358 An MCMC-Based Approach to Inferring Cell Counts in Diseased Tissue

Abstract:

Motivation

Currently, gene expression data is highly underutilized. Gene expression data is primarily used to rapidly probe the genome to discover gene activity associated with specific phenotypes or used to infer gene regulatory networks active within pure cell cultures. However, gene expression data developed from complex tissue contains dynamic information of changes in intracellular signaling and changes in the cellular makeup of the sample itself. Engineers and scientists must develop methods to maximally exploit biological data for model construction.

We and others have proposed that gene expression data can be used to quantify the dynamics of immune cells during disease. The immune system coordinates the activity of diverse immune cells and intracellular signaling pathways to identify and protect against disease. Aberrant immune responses have been associated with different influenza infection outcomes [1]. However, a systems-level understanding of a healthy immune response is needed to improve treatment of severe infections. In this work, we present a novel algorithm for inferring cell counts from influenza-infected lung gene expression data and compare the performance of our algorithm to the current state-of-the-art inference techniques. Lastly, we discuss how current algorithms can be integrated into mathematical model development protocols.

Background

Transcriptional profiling of diseased tissues and cells has resulted in large, publicly-available gene expression datasets. Though changes in gene transcript levels in tissue might be due to both gene regulation and changes in cell populations, most research has focused on identifying genes differentially expressed between phenotypes using traditional statistical tests or on pathway activity inferred using a variety of methods. Inferring changes in cellular demographics remains a challenge and no method to date has focused on temporal (e.g. time-point) samples.

Multiple deconvolution algorithms have been developed to predict cellular demographics through the use of gene expression data. Basic deconvolution postulates that expression profiles of cell mixtures are linear combinations of the expression profiles of pure cells. Thus, linear regression tools are used to infer cell proportions in a given sample. Examples include: modified linear least-squares regression (modified LLSR) [2], cell-type identification by estimating relative subsets of RNA transcripts with support vector regression (CIBERSORT) [3], and digital cell quantification with elastic net regulation (DCQ) [4]. These tools are designed to de-convolute one sample at a time. Therefore to estimate dynamics of cells, they need to be run for each timepoint to generate predictions. Furthermore, the predictions can only be accurately determined if all cell types of the sample and their pure-cell expression data are known *a priori*.

Another approach is to apply dynamic clustering to identify co-regulated modules (groups) of genes and then apply bioinformatics approaches to associate each module with specific cell populations, e.g. cell type enrichment (CTen) [5]. Different from the former deconvolution approach, CTen implements only one computation for time-series data. Although it determines relative fold changes as opposed to cell proportions, CTen requires less prior knowledge of sample composition.

The reported demonstrations of these deconvolution algorithms are mostly conducted by artificial datasets or cell mixtures with simple compositions. When applied to complex and time-course data, we found the accuracy of these algorithms was insufficient for several key immune cell types (including CD8+ T Cells, CD4+ T cells, B Cells, and lung resident macrophages). For modified LLSR, CIBERSORT or DCQ, they are lacking consideration of dependency and contiguity among time points. Instead CTen takes into account the time course profile of each gene. It was tested with the same temporal data and accurately predicted dynamic changes of some immune cell types (showing high correlation for lung resident macrophages). While CTen showed a lower accuracy for a few other immune cell subsets, which might be due to its gene marker database. Based on these results, we have developed a novel approach to cell count inference that better predicts cell quantity change in tissue across time using temporal transcriptomic data.

Methods

We constructed a novel cell signature database from data available from ImmGen [6]. Expression intensities of transcripts are averaged across replicates and mapped to associated genes. To reduce computation complexity, genes with small variation across the cell type library were removed. To find genes that best distinguish a certain cell type, we establish an optimization problem which seeks a set of K genes to minimize the correlation coefficient between this cell type and all others. We use MCMC to identify multiple solutions that satisfactorily reduce the cost function. In doing so, we create a database of cell signatures in which each cell type has 3 cell signature sets. The cell signature sets allow us to provide multiple estimates of the cell counts and provide a confidence score.

To evaluate the MCMC-based inference algorithm, differentially expressed genes (DEG) are obtained from microarray data of influenza-infected mouse lung tissue [7]. Cell enrichment of the DEGs is detected in comparison with the cell signature database by Fisherâ€™s exact test and log fold changes of immune cell counts are assumed to correspond with log fold changes of mean expression profiles of the cell markers. For comparison, we applied the same lung tissue dataset to 4 common algorithms: modified LLSR, CIBERSORT, DCQ and CTen. Algorithm performance is measured by the sum of squared errors.

Conclusions and Discussion

Here we present a new approach for inference of immune cell population changes across time. Compared with other popular algorithms, our method demonstrated improved prediction accuracy of the log fold change of several common immune cells. This improved quantification of cell count changes in samples may help computational biologists with characterizing immune response and/or disease pathology, and mining disease transcriptomics. Future work will be to integrate this inference approach with other biological data to promote tissue-level systems inference. Furthermore, applications of our model are beyond inference. The cell signature database we generated contains unique information for each immune cell population. Study on these signatures, combining with predictions of dynamic changes for key immune cells in disease, may provide new insights on potential biomarkers for disease diagnosis.

Reference

1. Cilloniz, C., et al., *Lethal influenza virus infection in macaques is associated with early dysregulation of inflammatory related genes.* PLoS Pathog, 2009. **5**(10): p. e1000604.
2. Abbas, A.R., et al., *Deconvolution of blood microarray data identifies cellular activation patterns in systemic lupus erythematosus.* PLoS One, 2009. **4**(7): p. e6098.
3. Newman, A.M., et al., *Robust enumeration of cell subsets from tissue expression profiles.* Nat Methods, 2015. **12**(5): p. 453-7.
4. Altboum, Z., et al., *Digital cell quantification identifies global immune cell dynamics during influenza infection.* Mol Syst Biol, 2014. **10**: p. 720.
5. Shoemaker, J.E., et al., *CTen: a web-based platform for identifying enriched cell types from heterogeneous microarray data.* BMC Genomics, 2012. **13**: p. 460.

6. Heng, T.S., M.W. Painter, and C. Immunological Genome Project, *The Immunological Genome Project: networks of gene expression in immune cells.* Nat Immunol, 2008. **9**(10): p. 1091-4.
7. Shoemaker, J.E., et al., *An Ultrasensitive Mechanism Regulates Influenza Virus-Induced Inflammation.* PLoS Pathog, 2015. **11**(6): p. e1004856.

530522 Optimization in Cancer Therapeutics: Model Integration for Tumor Dynamics and Myelosuppression to Predict Chemotherapy Dosing Profiles

Abstract:

Optimization in Cancer Therapeutics: Model Integration for Tumor Dynamics and Myelosuppression to Predict Chemotherapy Dosing Profiles

Ian C. Dunn[1], and Kirti M. Yenkie[1,*]

[1]Department of Chemical Engineering, Henry M. Rowan College of Engineering,

Rowan University, NJ - 08028, USA

(Tel: 856-256-5375; e-mail: yenkie@rowan.edu)

Abstract

In 2016, cancer was reported to be the second most common cause of death in the United States and according to 2017 projections, there will be around 1.7 million new cancer diagnoses and 700 thousand cancer deaths [1]. Fig. 1 shows the estimate by National Cancer Institute's (NCI) Surveillance, Epidemiology, and End Results (SEER) program, and by the Centers for Disease Control and Prevention's (CDC) National Program of Cancer Registries (NPCR) on the incidence of new cancer cases for 2018 in the United States.

Fig. 1. Estimated new cancer cases for 2018 in the United States, categorized by gender and cancer incidence sites [2].

By analyzing these reported numbers, it can be inferred that the occurrence of a lumped cancerous mass, as observed in prostrate, breast, colon, bladder, kidney, etc., is quite high and deserves attention. In such cancers, the tumors grow much faster as compared to normal cells and spread to other parts of the body. However, tumor growth can be controlled as well as the undesired cells can be killed using Chemotherapy, the most established and widely used anti-cancer treatment [3]. However, the medicines and drugs involved also affect the healthy cells which decrease patient immunity and lead to multiple side-effects.

One of the major side-effects of chemotherapeutic drugs is myelosuppression [4], [5], a condition when the bone marrow activity is decreased, resulting in fewer red blood cells, white blood cells, and platelets. The WBCs, also known as leukocytes, are responsible for protecting the body against infectious disease and foreign invaders. The immunity of a person depends on their blood count, which should be maintained within a certain threshold, thus rendering them as a dose-limiting factor during treatment. Additionally, it is not impractical to measure a patient's WBC concentration during the course of treatment. Effective chemotherapy treatment must simultaneously balance treatment efficacy and toxicity to the patient; herein lies an optimization problem that may be addressed with optimal control theory [6], [7].

In phase-I, a simplified model to simulate tumor growth dynamics and white blood cell concentrations during chemotherapy was developed. Parameter estimation was conducted using tumor growth data from literature [8]. This model was evaluated via the methods of maximum-principle and discretized non-linear programming (NLP) to obtain an optimal drug-concentration profile that minimized the tumor volume while keeping the white blood cell concentration above a certain threshold [9].

Fig. 2. Combined model showing two compartments C1 (blood plasma) and C2 (tumor) [adpated from 8]. **Leukocytes are present in C1 and their structured model is shown below with five compartments: proliferative cells (Prol), three transit compartments (T1, T2, and T3), and circulating blood cells (Circ)** [adpated from 10].

In phase II of this work, this tumor dynamics model was expanded to include pharmacokinetic compartment models to simulate the concentration of the anticancer agent in different parts of the body. Additionally, a more complex model of myelosuppression (leukocyte dynamics in the blood plasma (C1) compartment) with demonstrated accuracy [10] was integrated with the existing model to quantitate hematological toxicities induced by the anticancer agent (see Fig. 2). This combined model was evaluated via maximum principle and discretized NLP methods to gain insights into optimal dosage regimens for several anticancer agents (e.g. CPT-11 for the treatment of colon cancer).

This model consists of drug-specific and patient-specific parameters. A drug-specific parameter accounts for the efficacy of the drug. A patient-specific parameter accounts for how quickly the drug is eliminated from the body. In theory, the modeling and optimization strategy developed in this work will not only be robust enough to be applied to multiple types of chemotherapies but also capable of being personalized for the individual patient. This work will contribute to the design of reliable, personalized tools to optimize chemotherapy treatment and improve patient outcomes.

Keywords: Model formulation, Myelosuppression, Chemotherapy, Pharmacokinetics, Personalized Medicine, Optimal control

References:

[1] A. B. Ryerson *et al.*, *Cancer*, vol. 122, no. 9, pp. 1312–1337, May 2016.

[2] R. L. Siegel, K. D. Miller, and A. Jemal, *CA: A Cancer Journal for Clinicians*, vol. 66, no. 1, pp. 7–30, Jan.

2016.

[3] M. Engelhart, D. Lebiedz, and S. Sager, *Mathematical Biosciences*, vol. 229, no. 1, pp. 123–134, Jan. 2011.

[4] S. Nanda, H. Moore, and S. Lenhart, *Mathematical Biosciences*, vol. 210, no. 1, pp. 143–156, Nov. 2007.

[5] A. L. Quartino, L. E. Friberg, and M. O. Karlsson, *Invest New Drugs*, vol. 30, no. 2, pp. 833–845, Apr. 2012.

[6] F. S. Lobato, V. S. Machado, and V. Steffen, *Computer Methods and Programs in Biomedicine*, vol. 131, pp. 51–61, Jul. 2016.

[7] K. M. Yenkie and U. Diwekar, *Journal of theoretical biology*, vol. 355, pp. 219–228, 2014.

[8] M. M. Hadjiandreou and G. D. Mitsis, *IEEE Transactions on Biomedical Engineering*, vol. 61, no. 2, pp. 415–425, Feb. 2014.

[9] K. M. Yenkie and U. M. Diwekar, *Computers & Chemical Engineering*, vol. 71, pp. 708–714, 2014.

[10] L. E. Friberg, A. Henningsson, H. Maas, L. Nguyen, and M. O. Karlsson, *J. Clin. Oncol.*, vol. 20, no. 24, pp. 4713–4721, Dec. 2002.

534939 Modeling Heat Transfer Using an Integral Equation Approach Via Green's Function: Application to Cancerous Tumor Undergoing Hyperthermia Treatment

Abstract:

In recent years, studies have been performed in which cancer tissues are exposed to high temperatures in order to damage/kill cancer cells. This treatment is referred to as hyperthermia, and the goal of this work is to develop a model that describes the heat transfer and temperature profile throughout a cancerous tumor undergoing hyperthermia. The model consists of two portions: linear and nonlinear aspects. The linear aspects of this transient model involve the accumulation and conductive transport terms of the model, while the nonlinear aspect of the model involves the heat sources term resulting from applied hyperthermia treatment. The resulting model equation is a second-order, non-homogeneous partial differential equation and is challenging to solve using standard methods used in engineering. Instead, a Green's function approach is employed which allows the problem to become decoupled and solved through the use of integral equations and associated eigenvalue problems. This approach also allows for a certain "flexibility" through which the generation function can be manipulated without altering the Green's function, thus allowing for multiple scenarios to be modeled efficiently. Another possible implication of this work is utilizing the information provided by the temperature profile in order to predict the killing rate of cancerous tissues (and possibly even the killing/healing rate of healthy tissues) which can be described using an F-kill ratio described by Pascal, et al. 2013. Fundamental aspects and illustration will be described in the contribution.

References:

1. Pascal, Jennifer, et al. "Mechanistic Patient-Specific Predictive Correlation of Tumor Drug Response with Microenvironment and Perfusion Measurements." Proceedings of the National Academy of Sciences, vol. 110, no. 35, Aug. 2013, p. 14266, doi:10.1073/pnas.1300619110.

535894 A Modeling Framework to Characterize Kinetics, Efficacy and Toxicity of Hydroxyurea Based Treatment of Individual Sickle Cell Disease Patients

Abstract:

Hydroxyurea (HU) is the current FDA approved drug for sickle cell disease (SCD) treatment. Though several studies have indicated that the drug is effective in ameliorating disease pathophysiology, its use is limited due to myelosuppression and uncertainties associated with its long-term safety. Use of hydroxyurea presents major challenges to clinicians due to wide interpatient variability in pharmacokinetic-pharmacodynamic (PK-PD) profiles, cytotoxicity, lack of an effective biomarker to predict treatment efficacy and consequently lack of an optimal dosing regimen for individual patients.

A semi-mechanistic compartmental PK model using ODEs is developed that consists of three compartments: gastrointestinal tract, plasma and tissue where the drug gets metabolized to nitric oxide and its derivatives. Fetal hemoglobin (HbF) and mean cell volume (MCV) of red blood cells (RBCs) have been clinically used as biomarkers to indicate treatment efficacy. To capture dynamics of HbF, a signaling model using ODEs, based on the hypothesis that HU activates HbF through the NO-cGMP pathway, is formulated. To quantify toxicity and change in MCV of RBCs, erythropoiesis and leukopoiesis processes are modeled where stem cells undergo biochemical and morphological changes to produce fully functional erythrocytes and leukocytes in the blood circulation. The model is framed using a population balance modeling approach where all the cellular properties are lumped into two internal coordinates, cell maturation age and cell volume, then solved using the method of characteristics. The model predicts the volume distribution of RBCs and concentrations of leukocytes and erythrocytes of SCD patients with time under treatment with hydroxyurea. The PK model is integrated with the signaling model and the hematopoiesis model to characterize kinetics with both efficacy and toxicity. Furthermore, sensitive parameters are obtained using a global sensitivity analysis approach to obtain an individual patient's model. Model parameters are obtained after fitting the model predictions to an individual patient PK-PD variables i.e., drug concentration in the plasma, MCV, HbF, and blood cell count. Once the individual patient response is predicted, the optimal dose is obtained using model predictive control that keeps the model predictions within target values. Another major issue which plagues the entire treatment process is patients' non-compliance where they do not take the drug. It is difficult to differentiate non-compliance from treatment inefficacy because non-compliant patients might mislead clinicians into believing that treatment at the current dose level is insufficient. In this case, the dose might be increased but once the patient starts complying, the excess dose might prove to be harmful. Non-compliant patients are modeled and simulated to study the effect of missing the dose on patient PK-PD variables and based on that, an optimal dose for a non-compliant patient is calculated from the integrated model.

Hydroxyurea shows the potential for improvements in the health of patients with sickle cell disease, however, if it is not administered in the right amount, it leads to myelosuppression. Through our integrated PK-PD model, we seek to predict the individual patient trajectory and answer questions as to why some of the patients do not respond, why some of the patients respond well for the same amount of dose, and how to differentiate non-compliant patients from poor responders.

536068 Customized Robust Optimal Dosage Determination in the Face of Uncertainties for IVF Practices

Abstract:

In Vitro Fertilization (IVF) treatment, one of the most advanced and highly recommended procedure for treating infertility in women; remains experimental even with recent advancements in technology. The increasing levels of infertility in women per year, uncertain outcome, and the high cost of this procedure; necessitate the need to develop a patient-specific treatment protocol. IVF comprises of four stages; namely, Superovulation, Egg-retrieval, Insemination/Fertilization and Embryo transfer. Superovulation stage involves stimulation and growth of multiple follicles or oocytes through externally injected hormones which makes it a highly controllable process.

In this work, we used the data from 50 IVF patients and developed customized models for drug dosage prediction. Based on this, the analysis is conducted to find patient dependent and independent parameters in the IVF model for customized dosage profiles in the given data. In real time scenario, data for two days is required to model which poses character uncertainties.

We derived robust optimal drug dosage profiles in the face of uncertainties using optimal control. Further, our analysis predicts starting optimal dosage profiles for diverse age distributions.

The resulting customized dosage profiles of the externally injected hormones differ significantly in comparison to the prescribed dosage and show a better final day follicle size distribution, lower overall hormonal dosages, and less testing. The dosage profile is robust in the presence of uncertainties.

525650 Metabolites from Blood Samples of Pregnant Mothers Predict Autism Risk

Abstract:

Autism Spectrum Disorder (ASD) is a neurodevelopmental disorder characterized by difficulties in social interaction, verbal and nonverbal communication, and repetitive behaviors [1]. ASD has a high prevalence as approximately one in every 68 children in the United States are estimated to be diagnosed with ASD [1]. This prevalence has significant economic impacts and a recent study found an annual national economic burden of $268 billion resulting from ASD [2].

It is generally acknowledged that the earlier ASD is diagnosed in life, the sooner support services can start, and the better the general outcome will be. However, ASD is currently a clinical diagnosis as no lab test exists. The current average age of diagnosis is 4 years old, but the desired diagnosis would be 18-24 months [3]. One promising approach has found that there are significant differences in measurements of the metabolites of the Folate One-Carbon Metabolism (FOCM) and Transsulfuration (TS) pathways between children with ASD and their typically developing (TD) peers [4]. These differences have the potential for the development of a test to support ASD diagnosis. However, the key question is how early differences could be detected that point towards an ASD diagnosis in the future. Ideally a diagnosis could be made even prior to birth, however, it is unclear if this is possible. This work investigates this point and focuses on metabolite measurements of the FOCM/TS pathways of the mothers during pregnancy as the earliest possible time point of detection.

For this investigation, blood samples were taken at three different time points throughout the pregnancy, during the 1st trimester, 2nd trimester, and 3rd trimester. The samples came from two groups of pregnant mothers: mothers who have already had a child with ASD (High Risk) and those who have not (Control). The data from these mothers are labeled as High Risk and Control because the probability of having a child with ASD across the general population is 1.5% [1], but the probability increases to 18.7% once the mother has already had a child diagnosed with ASD [5]. Furthermore, the High Risk group is divided into two subgroups based upon an evaluation of the children at three years of age. One subgroup consists of the High Risk mothers whose children have been diagnosed with ASD at age three, while the other subgroup contains mothers from the High Risk group whose children do not have an ASD diagnosis by age three. It should be noted that implementation of this study involved tracking of the mothers for the entire pregnancy as well as following up with the mothers at age three of the child to determine if an ASD diagnosis was made, which resulted in a very long and non-trivial to conduct clinical study.

The data collected from this clinical trial were analyzed using multivariate statistical analysis. Specifically, Fisher Discriminant Analysis (FDA) [6] was used to separate the different groups of data. The results of this study show that it is not possible to predict if a child will be diagnosed with ASD from the FOCM/TS metabolites of the mother during pregnancy. However, the results indicate that it is possible to predict with reasonably high probability (misclassification errors of approximately 10%) if a mother falls into the High Risk or the Control group. These results are important insofar as no test for predicting ASD risk exists so far and our results indicate if the ASD risk is the same as the one found in the general population (Control group) or is increased by more than an order of magnitude (High Risk group). While more studies will be needed to replicate these results, the initial findings are promising that the probability of having a child with ASD can be predicted during pregnancy based on the multivariate analysis of FOCM/TS metabolites.

[1] "Facts About ASDs," *CDC - Facts about Autism Spectrum Disorders - NCBDDD*. [Online]. Available: http://www.cdc.gov/ncbddd/autism/facts.html. [Accessed: 22-May-2017].

[2] J. P. Leigh and J. Du, "Brief Report: Forecasting the Economic Burden of Autism in 2015 and 2025 in the United States," *J. Autism Dev. Disord.*, vol. 45, no. 12, pp. 4135–4139, Dec. 2015.

[3] Autism and Developmental Disabilities Monitoring Network Surveillance Year 2008 Principal Investigators and Centers for Disease Control and Prevention, "Prevalence of autism spectrum disorders--Autism and Developmental Disabilities Monitoring Network, 14 sites, United States, 2008," *Morb. Mortal. Wkly. Rep. Surveill. Summ. Wash. DC 2002*, vol. 61, no. 3, pp. 1–19, Mar. 2012.

[4] D. P. Howsmon, U. Kruger, S. Melnyk, S. J. James, and J. Hahn, "Classification and adaptive behavior prediction of children with autism spectrum disorder based upon multivariate data analysis of markers of oxidative stress and DNA methylation," *PLOS Comput. Biol.*, vol. 13, no. 3, p. e1005385, Mar. 2017.

[5] S. Ozonoff *et al.*, "Recurrence Risk for Autism Spectrum Disorders: A Baby Siblings Research Consortium Study," *Pediatrics*, vol. 128, no. 3, pp. e488–e495, Sep. 2011.

[6] R. Fisher, "The Use of Multiple Measurements in Taxonomic Problems," *Ann. Eugen.*, vol. 7, no. 2, pp. 179–188, 1936.

520101 Review and Comparative Study of Nonlinear PCA Fault Detection Methods

Abstract:

In the process and manufacturing industries, safe operations and high quality product are assured by process monitoring systems (e.g., [1]). The frontline of any process monitoring system is fault detection, which is the detection of anomalous behavior. Most fault detection methods implemented in industrial systems are data-driven methods, that is, methods based on models constructed entirely or nearly entirely from process data. The data-driven fault detection methods most used in the early chemical industry were one-variable-at-a-time control charts, namely, Shewart, CUSUM, and EWMA charts. The past many decades have seen major growth in the application of multivariate statistical techniques, which can handle correlations between the data associated with different variables (aka â€œmultivariable correlationsâ€). The multivariate statistical technique most widely used in industry is principal component analysis (PCA), which handles high-dimensional, highly correlated, and noisy data [1].

Part of the success of PCA has been that (i) the sampling times used in many applications have been long, which minimizes temporal correlations within the data that would violate one of the assumptions underlying PCA, and (ii) most continuous processes have input-output relationships that are very close to linear, which is another assumption underlying PCA. Some industrial applications have been using sampling times that are fast enough that the data have significant temporal correlations. It is well known that the temporal correlations are well addressed by data-driven fault detection methods based on canonical variate analysis (e.g., [1]), and such methods have been applied in industrial practice for some years now.

This presentation considers the second limitation of PCA, which is that PCA can give poor fault detection performance for processes that have nonlinear input-output behavior. To overcome the linear restriction, different nonlinear extensions of PCA have been proposed, going back to early work by Thomas McAvoy, Jong Min Lee, and others (e.g., [2,3]). The literature on nonlinear PCA has grown substantially over the past twenty years, both within and outside of the chemical process systems community, and this presentation provides a thorough systematic review and comparative study. Our main interest is in the characteristics of nonlinear PCA methods relevant for their effective application for data-driven fault detection. The most widely studied nonlinear PCA methods for fault detection, including neural network PCA [2] and kernel PCA [3], are compared with PCA in several case studies that collectively illustrate significant limitations and strengths of various nonlinear PCA methods. The case studies are (1) a simple nonlinear example, (2) a realistic simulation of a full chemical plant commonly used in the literature, and (3) a realistic simulation of a highly nonlinear pharmaceutical manufacturing facility. In addition to pointing out some common misconceptions in the literature, the case studies are used to provide general guidelines on when to use nonlinear PCA, how to use nonlinear PCA effectively, and how to interpret the results for the purposes of fault detection of nonlinear systems.

[1] L.H. Chiang, E.L. Russell, and R.D. Braatz (2000) Fault Detection and Diagnosis in Industrial Systems. Springer Science & Business Media.

[2] D. Dong and T.J. McAvoy (1996) Nonlinear principal component analysis based on principal curves and neural networks. Computer and Chemical Engineering, 20(1), 65-78.

[3] J.M. Lee, C. Yoo, and S.W. Choi (2004) Nonlinear process monitoring using kernel principal component analysis. Chemical Engineering Science, 59(1), 223-234.

534370 Development of the Texas a&M Superfund Research Program Computational Platform for Data Integration, Visualization, and Analysis

Abstract:

The National Institute of Environmental Health Sciences (NIEHS) Superfund Research Program (SRP) aims to support university-based multidisciplinary research on human health and environmental issues related to hazardous substances and pollutants [1]. The Texas A&M Superfund Research Program comprehensively evaluate the complexities of hazardous chemical exposures, potential adverse health impacts, and potential hazards of exposures to complex mixtures through a number of multi-disciplinary projects and cores [2]. One of the essential components of the Texas A&M Superfund Research Center is the Data Science Core, which serves as the basis for translating the data produced by the multi-disciplinary research projects into useful knowledge for the community via data collection, quality control, analysis, and model generation. The four main projects of the TAMU Superfund Research Center mainly focuses on (i) understanding dynamic exposure pathways under the conditions of environmental emergencies, (ii) designing novel broad-acting sorption materials for reducing bioavailability of contaminants, (iii) studying in vitro and in vivo hazard, kinetics and inter-individual variability of responses to chemical mixtures, and (iv) developing in vitro multiplex single-cell assays to detect endocrine disruption potential of mixtures. Using the data generated by the projects, the overall goal is to develop comprehensive tools and models to address the environmental and health effects of exposure to chemical mixtures during environmental emergency-related contamination events. In this presentation, we demonstrate the Texas A&M Superfund Research Program computational platform, which houses and integrates large-scale, diverse data sets generated across the Center, and provides basic visualization service to facilitate interpretation, monitors the data quality, and finally implements variety of state-of-the-art data analytics tools for knowledge extraction [3]. Specifically, we use both supervised (i.e. classification, regression analysis) and unsupervised learning algorithms (i.e hierarchical, k-means, spectral clustering analysis) along with dimensionality reduction techniques to create decision support models and tools. The platform is aimed to facilitate effective integration and collaboration across the Center.

References

[1] National Institute of Environmental Health Sciences. Superfund Research Program (2018). https://www.niehs.nih.gov/research/supported/centers/srp/index.cfm (accessed 9 April 2018

[2] TAMU Superfund Research Center (2017). https://superfund.tamu.edu/ (accessed 9 April 2018).

[3] Onel, M.; Beykal, B.; Wang, M.; Grimm, F.A.; Zhou, L.; Wright, F.A.; Phillips, T.D.; Rusyn, I.; Pistikopoulos, E.N. Optimal Chemical Grouping and Sorbent Material Design by Data Analysis, Modeling and Dimensionality Reduction Techniques. Proceedings of the 28th European Symposium on Computer Aided Process Engineering, 2018, Graz, Austria (Accepted).

534782 Advanced Data Analytics for Process-Shop Base+Delta Sub-Model Estimation in Planning and Scheduling Decision-Making

Abstract:

Typically, process data used for planning and scheduling is outdated or not properly integrated with the current state of the production, generating inconsistencies in the prediction of product streams in a chain-wide transformation throughout the manufacturing network [1]. Therefore, estimating accurate process transformation formulas to propagate feed and process characteristics into yields and properties of product streams is one of the major challenges facing any integrated plant when making planning and scheduling decisions. Due to uncertainties in proper transformation formula, feed quality, measured data, etc., high-fidelity modeling becomes unrealistic. For better process-shop predictions on their yields and properties, we propose new data analytics techniques combining reconciliation and regression into an estimation strategy.

This better process transformation modeling can be integrated in a continuous cycle to provide process feedback, aiming to remove the gap between the model prediction and the actual values in the plant. Thus, the use of data and information measured within automated systems leads to greater precision and higher model reliability, especially when using techniques of parameter estimation and data reconciliation for a better quality or integrity of the plant information [2]. A case study involving a reactor and a tank in a production chain was considered by Kelly and Zyngier [3], in which the gap between the modeled and the actual inventory or level of the tank can be reduced to zero by using parameter feedback from the process. According to the authors, the proposed example points out that differences between the planned and the actual values of the process do not necessarily result from task execution problems or measurement errors, since the absence of feedback makes it impossible to discern the difference between an inadequate model and problems in implementation of the operations. Franzoi et al. [4] recently introduced a complete crude-oil refinery blend scheduling problem and applied both data reconciliation techniques and parameter feedback to effectively optimize the complex process system. The authors solved the feedback strategy within an iterative mixed-integer linear and nonlinear programming (MILP-NLP) decomposition and highlighted the importance of properly integrating data to the decision automation core to cope with uncertainties, to reduce inaccuracies and to close the gap between predictions and productions.

In the proposed data analytics, by doing constrained and weighted least squares to fit better base+delta sub-models using data reconciliation and regression techniques, the reconciliation forces the yields to add up to the unity (or 100%) and the regression fits base and delta coefficients simultaneously across all yields. In this proposition, the method has the following protocol: 1) independent variable bases are taken as the averages/means from the plant trials, test runs, survey days or sample experiments; 2) the estimation of dependent yield variables is ran one-at-a-time to get the minimum sum-of-squares of residuals also known as the standard-error; 3) the one-at-a-time yield objective function values are divided by the number of observations or samples minus one in order to properly compute the base and delta variances by compensating for the nominal number of degrees-of-freedom. This also equalize or balance the objective function so that all dependent variables or output responses are weighted equally when their base and delta coefficients are estimated simultaneously. In addition to the base and delta coefficient variances, which easily translate into 95% statistical confidence-intervals, gross-error detection statistics for each yield (and property) deviation from its measured and actual values is also performed. It should be noted that when estimating the property base+deltaâ€™s sub-models, as there is no overall consistency relationship with the yields summing to unity, we simply regress each property separately.

[1] MENEZES, B. C.; KELLY, J. D.; GROSSMANN, I. E.; VAZACOPOULOS, A. Generalized Capital Investment Planning of Oil-Refineries using MILP and Sequence-Dependent Setups. Computers & Chemical Engineering, v. 80, p. 140-154, 2015.

[2] KELLY, J. D.; ZYNGIER, D. Unit-operation nonlinear modeling for planning and scheduling applications. Optimization and Engineering, p. 1-22, 2016.

[3] KELLY, J. D.; ZYNGIER D. Continuously improve the performance of planning and scheduling models with parameter feedback. In: Proceedings of the foundations of computer-aided process operations (FOCAPO), 2008.

[4] FRANZOI, R.E.; MENEZES, B.C.; KELLY, J.D.; GUT, J.W. Effective scheduling of complex process-shops using on-line parameter feedback in crude-oil refineries, Process Systems Engineering, 2018.

529692 A Segmentation Approach for Oscillation Characterization

Abstract:

Oscillations are common in a wide range of systems. Natural phenomena such as brain activity [1], earthquake [2], solar corona [3] and variations at the atmospheric boundary layer [4] are examples of fields where studying the underlying oscillation patterns can lead to important insights. Oscillations can be observed as a symptom of system reconfiguration, **e.g.**, predator-prey systems [5]. Recently, it has been shown that high frequency oscillations in electroencephalogram [6] (EEG) can be used as a biomarker for epilepsy detection. A study of heartbeat data [7] revealed that patients with severe congestive heart failure exhibit intermittent low frequency oscillations compared to healthy subjects.

Given importance of oscillations across disciplines, several techniques have been developed to detect oscillations in data using time, frequency and time-frequency analysis techniques. A straightforward time domain analysis technique is visual inspection for presence of oscillations. Although simple, this makes the analysis subjective. The empirical mode decomposition [8] (a widely used time domain technique) can be used to reduce the effect of noise, and has been employed to analyze intermittent and elliptic oscillations at the atmospheric boundary layer. However, intermittent oscillation in data leads to presence of several modes in a single IMF, referred to as mode mixing. Moreover, a single mode of oscillation can also appear in multiple IMFs which makes mode mixing a serious concern for reliable detection of intermittent oscillations.

On the other hand, several frequency domain techniques exist that can be used for detection of oscillations. The most common frequency domain technique is the Fourier transform which transforms the time domain signal to frequency domain in which the power content at different frequencies can be examined to detect oscillations. However, phenomena such as spectral leakage make it difficult to correctly identify the frequencies of oscillations. An extension to Fourier Transform is the Short Time Fourier Transform for signals with time varying frequency that restricts the calculation of Fourier transform to a moving window of finite length. However, this approach suffers from the problem of fixed window size and hence fixed time and frequency resolution. The wavelet transforms use waveforms with certain desired characteristics (such as orthogonality) called (mother and daughter) wavelets as basis functions. However wavelet based techniques suffer from the fact that they lack clear definition of sinusoidal oscillations and there is a need to select appropriate mother wavelet for desired feature extraction.

Despite their limitations, the above techniques and many more have been extensively used for detection of oscillations in diverse applications. However, there is a lack of a common framework that can accommodate several important features in data such as non-stationarity, intermittent oscillation, measurement noise and multiple oscillations. In this paper, we report an approach that addresses all of the above stated problems. The proposed framework combines both time and frequency information for better oscillation characterization. A key segmentation idea that allows the localization of the oscillating regimes using template matching is used to overcome many of the difficulties with both time and frequency based oscillation characterization approaches. We show that the proposed approached is robust to non-stationarity, intermittency and noise while characterizing oscillations with multiple modes. The power of this approach is demonstrated using data from various disciplines.

References:

[1] P. Fries, D. Nikolić, and W. Singer, "The gamma cycle," ***Trends Neurosci.***, vol. 30, no. 7, pp. 309–316, 2007.

[2] H. Zhang, C. Thurber, and C. Rowe, "Automatic P-wave arrival detection and picking with multiscale wavelet analysis for single-component recordings," ***Bull. Seismol. Soc. Am.***, vol. 93, no. 5, pp. 1904–1912, 2003.

[3] A. Mendoza-bricen and R. Erde, "Intermittent coronal loop oscillations by random energy releases ´," pp. 722–731, 2006.

[4] J. K. Lundquist, "Intermittent and Elliptical Inertial Oscillations in the Atmospheric Boundary Layer," ***J. Atmos. Sci.***, vol. 60, no. 1997, pp. 2661–2673, 2003.

[5] A. Mougi and Y. Iwasa, "Evolution towards oscillation or stability in a predator--prey system," ***Proc. R. Soc. London B Biol. Sci.***, p. rspb20100691, 2010.

[6] J. Röschke, J. Fell, and K. Mann, "Non-linear dynamics of alpha and theta rhythm: Correlation dimensions and Lyapunov exponents from healthy subject's spontaneous EEG," ***Int. J. Psychophysiol.***, vol. 26, no. 1–3, pp. 251–261, 1997.

[7] A. L. Goldberger, D. R. Rigney, J. Mietus, E. M. Antman, and S. Greenwald, "Nonlinear dynamics in sudden cardiac death syndrome: heartrate oscillations and bifurcations.," ***Experientia***, vol. 44, no. 11–12, pp. 983–987, 1988.

[8] D. Kim and H. Oh, "EMD: A Package for Empirical Mode Decomposition and Hilbert Spectrum," ***R J.***, vol. 1, no. 1, pp. 40–46, 2009.

537392 A Dead Time Compensation Approach for State Estimation of Sampled-Data Systems in the Presence of Large Measurement Delays

Abstract:

State estimation of continuous-time dynamical systems in the presence of sampled and delayed measurements (e.g., chemical processes, networked control systems) has caught researchers' attention over the last decade [1-3]. In addition to fast-sampled measurements, it is crucial to incorporate slow-sampled measurements in the observer design framework to make the entire system observable as well as improve the estimation accuracy, despite their low sampling rate and large output delay. Moreover, different sampling rates and different sizes of delays should be handled in the observer design.

In this paper, a multi-rate sampled-data observer design is adopted as a starting point before considering the presence of possible measurement delays. The multi-rate observer design (see [4] for linear systems and [5] for nonlinear systems) is based on an available continuous-time observer design coupled with multiple, asynchronous inter-sample predictors for the sampled measurements. Each predictor generates an estimate of a sampled output between consecutive measurements and gets reinitialized once the associated, most-recent measurement becomes available. Sufficient conditions were derived to guarantee stability property of the error dynamics by using Lyapunov's second method and the vector small-gain theorem [6], respectively.

To handle measurement delays, we propose a model-based dead time compensation algorithm and show that the stability property of the multi-rate observer will be preserved under arbitrarily large measurement delays. The algorithm is described as follows: once a delayed measurement is obtained, the state estimates at its sampling instant can be re-calculated by using the multi-rate sampled-data observer. Then the process model can be utilized to predict the state at current time, given the delayed estimates, in the same spirit as in Smith-predictor methods. During the compensation, any available measurement can be treated as a delay-free output and the compensator will get reinitialized at every sampling instant where the measurement is available. By integrating the observer and compensator equations, corrected estimates can be obtained at current time which will subsequently be used in the inter-sample prediction during the time when there is no sampled measurement available. This algorithm inherits all the nice properties (i.e., stability and robustness) of a multi-rate sampled-data observer. Moreover, the convergence property is not affected by perturbations in the size of each measurement delay. An industrial gas-phase polyethylene reactor example will be revisited to illustrate the applicability of the proposed method in both linearized system and nonlinear system.

References:

[1] Ahmed-Ali, T., Karafyllis, I., & Lamnabhi-Lagarrigue, F. (2013). Global exponential sampled-data observers for nonlinear systems with delayed measurements. *Systems & Control Letters*, 62(7), 539-549.

[2] Khosravian, A., Trumpf, J., Mahony, R., & Hamel, T. (2015). Recursive attitude estimation in the presence of multi-rate and multi-delay vector measurements. *Proceedings of the American Control Conference*, 3199-3205.

[3] Shen, Y., Zhang, D., & Xia, X. (2017). Continuous observer design for a class of multi-output nonlinear systems with multi-rate sampled and delayed output measurements. *Automatica*, 75, 127-132.

[4] Ling, C., & Kravaris, C. (2017). Multi rate observer design for process monitoring using asynchronous inter sample output predictions. *AIChE Journal*, 63(8), 3384-3394.

[5] Ling, C. and Kravaris, C. (2017). Multi-rate sampled-data observers based on a continuous-time design. *Proceedings of the 56th Conference on Decision and Control*, 3664-3669.

[6] Karafyllis, I., & Jiang, Z. P. (2011). A vector small-gain theorem for general non-linear control systems. *IMA Journal of Mathematical Control and Information*, 28(3), 309-344.

538393 Iterative Fault Isolation for Integrated Chemical Systems Based on Approximate Linear Model Inversion

Abstract:

Chemical process integration by material and energy recycle is a common practice in the chemical industry, but fault detection and isolation have been challenging due to the complexity of the integrated system. In this study, we use the approximate inversion of strictly proper linear time-invariant subsystems to isolate the errors on an integrated plant.

The framework involves structuralized state space representation of the integrated system, a dynamic simulation of each subsystem (which may include process units as well as operating valves), an algorithm that simultaneously recovers the state and input variables of each subsystem based on output, as well as an iterative estimation of state and disturbance dynamics. For a strictly proper system where the feedthrough D matrix is absent, this improved algorithm presents opportunities to provide insights of the dynamics when the overall plant transfer function, G, is uncontrollable, or when the disturbance transfer function, G_d, has a large norm. In addition, observable variables are reconciled by an optimization formulation whose objective function is minimizing the norm of the error vector between two or more linked subsystems, and its constraints are determined by physical operations (e.g. valve percent opening, etc.) This methodology is fundamentally different from the data-driven fault detection methods, which neglects the process knowledge. Moreover, it improves the previous methods by creating an essentially â€œidealâ€ control of the simpler sub-models instead of the entire plant.

Finally, a numerical simulation of a chemical process network demonstrates the effectiveness of the methodology. We hope to create a real-time structure such that a fault can be detected as soon as it occurs. In future work, such a structure can have decentralized controllers to avoid the error from propagating downstream in the system and restoring the system back to normal operation, through either an active control action or a passive signal correction.

535633 Hypothesis-Driven Data-Based Modeling to Study the Effect of Specialization on Hospital Performance

Abstract:

< style="padding-left: 5pt; text-align: justify;">

Fig. 1. The proposed path diagram linking specialization (Is) and total charge (TOTCHG) through direct, mediation and moderation paths.

References:

1. M. Hartman, A. Martin, P. McDonnell, A. Catlin, and N. H. E. A. Team, "National health spending in 2007: slower drug spending contributes to lowest rate of overall growth since 1998," *Health Aff.*, vol. 28, pp. 246–261, 2009.

2. M. J. Pesch and R. G. Schroeder, "Measuring factory focus: an empirical study," *Prod. Oper. Manag.*, vol. 5, pp. 234–254, 1996.

3. J. R. Edwards and L. S. Lambert, "Methods for integrating moderation and mediation: a general analytical framework using moderated path analysis.," *Psychol. methods*, vol. 12, 2007.

4. D. F. Alwin and R. M. Hauser, "The decomposition of effects in path analysis," *Am. Sociol. Rev.*, pp. 37–47, 1975.

5. D. Delen, A. Oztekin, and L. Tomak, "An analytic approach to better understanding and management of coronary surgeries," *Decis. Support Syst.*, vol. 52, pp. 698–705, 2012.

6. S. R. Eastaugh, "Hospital specialization and cost efficiency: benefits of trimming product lines," *J. Healthc. Manag.*, vol. 37, 1992.

537163 Dynamic Mode Decomposition Based Model Reduction for Control of Moving Boundary Problems via Approximate Dynamic Programming

Abstract:

Dynamic Mode Decomposition Based Model Reduction for Control of Moving Boundary Problems via Approximate Dynamic Programming

Mohammed Saad Faizan Bangi[1, 2], Harwinder Singh Sidhu[1, 2], Prashanth Siddhamshetty [1, 2],

Joseph Sang-Il Kwon [1, 2]

[1] Artie McFerrin Department of Chemical Engineering, Texas A&M University, College Station, TX 77845 USA

[2] Texas A&M Energy Institute, Texas A&M University, College Station, TX 77845 USA

There are a large number of industrial control problems which can be characterized by highly nonlinear distributed parameter systems (DPS) with moving boundaries, such as catalytic reactors [1] and chemical vapor decomposition [2]. In general, mathematical models of these processes are derived from first-principles and consist of nonlinear parabolic partial differential equations (PDEs) with time-varying spatial domains. The literature over the last three decades is affluent with the development in the modeling and nonlinear control of parabolic PDEs and has mainly focused on systems with time-invariant spatial domains [3]. Despite the progress on nonlinear control of parabolic PDE systems with time-invariant spatial domains, few investigations are available on control and estimation of parabolic PDE systems with time-varying spatial domains [4].

In the past 15 years, owing to its ability to handle a large-scale multivariable system with both input and state constraints, significant efforts have been made to apply model predictive control (MPC) to systems modeled by nonlinear PDEs [5]. Despite the popularity of this approach, there are two drawbacks of the MPC approach. First, the online computational load to calculate the optimal control moves at each sampling time can be significant and demand the use of long prediction/control horizons to ensure satisfactory performance. The second is its insufficiency to handle the uncertainty, which is an inherent feature while solving the real-time optimization problems in chemical plants.

Alternatively, both the issues of MPC formulation can be addressed by the approximate dynamic programming (ADP) approach [6]. In this approach, an improved control policy is derived after starting with suboptimal control policies, while circumventing the 'curse-of-dimensionality' of the traditional dynamic programming (DP) approach. Function approximators such as k-nearest neighbor or artificial neural networks are used to obtain the map between the 'cost-to-go' values and system states, and then Bellman equation is solved offline with the approximator in an iterative manner. However, it is not straightforward to apply ADP to systems characterized by nonlinear PDEs with time-varying spatial domains because of the high computational efforts required to derive the 'cost-to-go' function. Therefore, it is necessary to develop reduced-order models (ROMs) for the application of ADP to such systems.

The primary objective of this work is two-fold. First, we develop a temporally local ROM that reproduces the essential features of the underlying parabolic PDE system with time-varying spatial domains. This is accomplished by partitioning the temporal domain into multiple temporal subdomains using global optimum search (GOS) framework [7], and then applying dynamic mode decomposition (DMD) technique [8] within each cluster. Second, we employ the developed temporally local ROM in ADP based controller. In application, we will apply the proposed ADP based controller to several parabolic PDE systems with moving boundaries.

[1] Bremer, J., Goyal, P., Feng, L., Benner, P. and Sundmacher, K., 2017. POD-DEIM for efficient reduction of a dynamic 2D catalytic reactor model. *Computers & Chemical Engineering*, *106*, pp.777-784.

[2] Baker, J. and Christofides, P.D., 1999. Output feedback control of parabolic PDE systems with nonlinear spatial differential operators. *Industrial & Engineering Chemistry Research*, *38*(11), pp.4372-4380.

[3] Christofides, P.D., 2012. *Nonlinear and robust control of PDE systems: Methods and applications to transport-reaction processes.* Springer Science & Business Media.

[4] Armaou, A. and Christofides, P.D., 2001. Robust control of parabolic PDE systems with time-dependent spatial domains. *Automatica*, *37*(1), pp.61-69.

[5] Dufour, P., Touré, Y., Blanc, D. and Laurent, P., 2003. On nonlinear distributed parameter model predictive control strategy: on-line calculation time reduction and application to an experimental drying process. *Computers & Chemical engineering*, *27*(11), pp.1533-1542.

[6] Lee, J.H. and Lee, J.M., 2006. Approximate dynamic programming based approach to process control and scheduling. *Computers & Chemical engineering*, *30*(10-12), pp.1603-1618.

[7] Tan, M.P., Broach, J.R. and Floudas, C.A., 2007. A novel clustering approach and prediction of optimal number of clusters: global optimum search with enhanced positioning. *Journal of Global Optimization*, *39*(3), pp.323-346.

[8] Schmid, P.J., 2010. Dynamic mode decomposition of numerical and experimental data. *Journal of fluid mechanics*, *656*, pp.5-28.

537642 Data-Driven Optimization for Process Intensification Governed By High-Fidelity Models

Abstract:

Process intensification merges processes with different aims to reduce the cost and energy consumption. When the processes are modeled explicitly using algebraic equations, standard local and global solvers can be used for optimization [1]. However, many processes may not have an available algebraic model, or it is insufficient to allow for detailed optimization. We are especially interested in process intensification problems where the process is modeled by high-fidelity models such as those given by set of Nonlinear Algebraic and Partial Differential Equations (NAPDE). One of the approaches studied in the literature involves discretizing the NAPDE model to convert it into a large-scale nonlinear programming problem [2]. However, the level of discretization needs to be moderate to keep the model tractable. An alternate strategy is to use simulation data for optimization. Since the simulations are generally computationally expensive [3-6], it critical that the optima (preferably global) is obtained using few evaluations.

Despite several advancements made in data-driven optimization, many algorithms require that a feasible initial guess is provided. Furthermore, handling hard constraints that leads to simulation failure remains a challenge. In this work, we propose a two-phase algorithm [7] comprising of feasibility and optimization phases. Feasibility phase finds a feasible point if the initial point is infeasible and then moves to the optimization phase that aims to find the optimum point. Both the phases involve iteratively developing surrogate models for each of the black-box objective and constraints and then optimizing the subproblems in a trust-region framework. A novel optimization-based sampling strategy is also proposed that can handle hard constraints.

The two-phase algorithm is applied to optimize two process intensification case studies with governing model given by NAPDE. The performance of the algorithm is compared with NOMAD [8] and COBYLA [9]. The first is a carbon capture case study that utilizes CO_2 from flue gas and natural gas to produce syngas [5]. The resulting optimization problem is a grey-box problem with 10 decision variables, black-box objective function, 7 black-box constraints and 2 hard constraints. For the case of minimizing cost of production of syngas, the two-phase algorithm and COBYLA gives the optimum objective function value as $109.57 and $115.75, respectively. The respective required number of simulations are 681 and 364. On the other hand, NOMAD could not find a feasible point. When the algorithm is applied to maximize CO_2 utilization, the two-phase algorithm, COBYLA and NOMAD yields CO_2 utilization of 99.71%, 99.6% and 97.59%, respectively. The number of evaluations required by the two-phase algorithm, COBYLA and NOMAD are 332, 206 and 1248, respectively. The second case study is the design and synthesis of sorption enhanced steam methane reforming to produce H_2 using natural gas [6]. The resulting optimization problem is a grey-box problem with 10 variables, black-box objective function, 2 black-box constraints and 11 hard constraints. The two-phase algorithm leads to an optimal design of the process that yields 95% pure hydrogen with 10.86% lower cost compared to existing small-scale technologies. The performance of the algorithm will also be compared to other solvers.

References:

[1] Demirel, S. E., Li, J. and Hasan, M. M. F. Systematic process intensification using building blocks. *Computers and Chemical Engineering*, 105:2-38, 2017.

[2] Kawajiri, Y. and Biegler, L. T. Optimization strategies for simulated moving bed and powerfeed processes. *AIChE Journal*, 52(4):1343-1350, 2006.

[3] Caballero, J. A. and Grossmann, I. E. An algorithm for the use of surrogate models in modular flowsheet optimization. *AIChE Journal*, 54(10):2633-2650, 2008.

[4] Beykal, B., Boukouvala, F., Floudas, C. A. and Pistikopoulos, E. N. Optimal design of energy systems using constrained grey-box multi-objective optimization, *Computers and Chemical Engineering*, 2018, https://doi.org/10.1016/j.compchemeng.2018.02.017.

[5] Iyer, S. S., Bajaj, I., Balasubramanian, P. and Hasan, M. M. F. Integrated carbon capture and conversion to produce syngas: Novel process design, intensification, and optimization. *Industrial & Engineering Chemistry Research*, 56(30), 8622-8648, 2017.

[6] Arora, A., Bajaj, I., Iyer, S. S. and Hasan, M. M. F. Optimal synthesis of periodic sorption enhanced reaction processes with application to hydrogen production. *Computers and Chemical Engineering*, 115, 89-111, 2018.

[7] Bajaj, I., Iyer, S. S. and Hasan, M. M. F. A trust region-based two phase algorithm for constrained black-box and grey-box optimization with infeasible initial point. *Computers and Chemical Engineering*, 2017, https://doi.org/10.1016/j.compchemeng.2017.12.011.

[8] Abramson, M. A., Audet, C., Couture, G., Dennis Jr, J. E., Le Digabel, S. and Tribes, C. The NOMAD project, 2011.

[9] Powell, M. J. D. A direct search optimization method that models the objective and constraint functions by linear interpolation. In *Advances in optimization and numerical analysis*, pages 51-67, Springer, 1994.

538071 Constrained Least Square Parameter Identification Algorithms for Dual-Rate Systems with Inter-Sample Output Estimation

Abstract:

Multi-rate systems, where inputs and/or outputs have different sampling rates, are encountered in many chemical and biological engineering applications. Typically, this occurs in systems where multiple variables are measured. Process variables such as temperature, pressure and pH are measured more frequently, at times nearly continuously, while other variables, such as species concentrations in a reaction mixture, are measured less frequently. The simplest way to handle multi-rate systems is to neglect excess data from fast sampling signals and synchronize the signals with the slowest sampling rate. With this method, great amount of data will be discarded, the data that may contain crucial information regarding system dynamics. Therefore, some effort has been devoted to modeling, analysis and identification of multi-rate systems to take advantage of the rich information available in the experimental databases. In all the techniques proposed, multi-rate systems are generally simplified to multiple dual rate systems where the slower sampling rate is a positive integer () multiple of the faster sampling rate. Polynomial transformation technique is used to derive a dual rate model from a single rate model. Conventional identification algorithms for multiple input single output (MISO) dual rate systems are unconstrained, unstable, and slow to converge. Recursive data driven models are developed for dual rate systems here by using a modified constrained least square (MCLS) algorithm for parameter identification. Appropriate parameter constraints are imposed in parameter estimation algorithms and stability of these is examined. The MCLS algorithm-based models - dual rate for parameter identification and single rate for output prediction - allow for output prediction at frequency of faster sampled inputs with inter-sample output estimation, a beneficial feature. The MLCS algorithm is guaranteed to be stable, enables more accurate parameter identification with better convergence rate, and results in better prediction vis-a-vis two conventional dual rate models, identification algorithms for which are unstable. The beneficial features of the MLCS algorithm are illustrated with two examples, one being a simple discrete time process with a single slower sampled output and another being a fed-batch mammalian cell culture with three slower sampled outputs, concentrations of glucose, glutamine and monoclonal antibody. Faster prediction of the slower sampled output is achieved using a single rate-ARX prediction model, the parameters in the model being related to parameters identified for the dual rate model for the output. The more frequent output prediction vis-a-vis output sampling will be of substantial utility in optimization and control of processes. For mammalian cell culture, glucose and glutamine feed rates are considered as inputs. The data required for parameter estimation are generated from simulated fed-batch experiments using a well-tested first principles model. The predictions of the less frequently sampled concentrations of glucose, glutamine, and monoclonal antibody track very well the data for these, with the errors for reasonable prediction horizons considered being limited to 10% or less. The prediction accuracy can be increased further if data from prior experiments with dynamic similarities are available. The models can be used reliably for model predictive control.

537605 Global Optimization of a Class of Black-Box Problems with Bounded Hessian

Abstract:

Data-driven optimization is a practical alternative to optimize problems for a class of problems with unknown algebraic form of the objective function and/or constraints [1-5]. Trust-region based algorithms [6-7] have been proposed in the literature that guarantees convergence to first-order critical point if the model is full linear. However, global optimization of black-box problems is a challenging problem. In this work, we are interested in deterministic global optimization of bound constrained black-box problems.

Constructing a valid underestimator is a major impediment for applying branch and bound algorithm on black-box problems. Note that the lower bound at a node is calculated by minimizing the relaxed problem. Therefore, an underestimator whose minima can be found by performing finite number of evaluations is desirable for it to be used in global optimization of black-box problems.

To this end, edge-concave underestimator is an attractive alternative [8] because its minima lie at one of the corner points. To determine the underestimator, the upper bound of the diagonal elements of the hessian of the original function are needed. Therefore, if we assume to know the maximum of the upper bound of the diagonal elements of the hessian matrix, we can evaluate the underestimator at different x values. When we evaluate the edge-concave underestimator at all the corner points, then the minimum of these points is a valid lower bound of the original function.

In this work, we present a global optimization scheme based on a branch and bound framework. At each node, a valid lower bound is found using the scheme from previous paragraph and BOBYQA [9] is used to find the upper bound. The algorithm is implemented on problems listed in [8 (Table 3)] and all the problems converge to Îµ-global optimality. For many problems, the number of function evaluations and the nodes explored are economical. For instance, for problem 1 (n=5), the number of nodes explored and function evaluations to obtain global minima are 8 and 240 respectively. For problem 4 (n=5), the number of nodes explored and required function evaluations are 44 and 970 respectively. The algorithm will be implemented on several problems and compared with other DFO solvers to illustrate its effectiveness.

References:

[1] Liu, J., Ploskas, N. and Sahinidis, N. V. Tuning BARON using derivative-free optimization algorithms. *Journal of Global Optimization,* 1-27, 2018, https://doi.org/10.1007/s10898-018-0640-3.

[2] Fernandes, F. A. Optimization of fischer-tropsch synthesis using neural networks. *Chemical Engineering & Technology*, 29(4):449-453, 2006.

[3] Palmer, K. and Realff, M. Metamodeling approach to optimization of steady-state flowsheet simulations: Model generation. *Chemical Engineering Research and Design*, 80(7):760-772, 2002.

[4] Iyer, S. S., Bajaj, I., Balasubramanian, P. and Hasan, M. M. F. Integrated carbon capture and conversion to produce syngas: Novel process design, intensification, and optimization. *Industrial & Engineering Chemistry Research,* 56(30), 8622-8648, 2017.

[5] Arora, A., Bajaj, I., Iyer, S. S. and Hasan, M. M. F. Optimal synthesis of periodic sorption enhanced reaction processes with application to hydrogen production. *Computers and Chemical Engineering,* 115, 89-111, 2018.

[6] Conn, A. R., Scheinberg, K. and Vicente, L. N. Global convergence of general derivative-free trust-region algorithms to first-and second-order critical points. *SIAM Journal on Optimization*, 20(1):387-415, 2009.

[7] Eason, J. P. and Biegler, L. T. A trust region filter method for glass box/black box optimization. *AIChE Journal*, 62(9):3124-3136, 2016.

[8] Hasan, M. M. F. An edge-concave underestimator for the global optimization of twice-differentiable nonconvex problems. *Journal of Global Optimization,* 1-18, 2018 https://doi.org/10.1007/s10898-018-0646-x.

[9] Powell, M. J. D. The BOBYQA algorithm for bound constrained optimization without derivatives. Cambridge, *NA Report NA2009/06*, University of Cambridge, Cambridge, 2009.

538990 Ontology Engineering Approach to Support Process of Model Integration

Abstract:

Process modelling and simulation is a vital tool to plan, evaluate, assess, and develop different alternatives for the design of products and processes. The complexity of problems as well as heterogeneity of modelling methods make process modelling and simulation challenging, time consuming and often tedious process requiring a wide range of expertise. Inconsistencies in model development are the main cause for redundant work. Models remain implicit to the engineers who have built them, which further limits the potential of reusability. The only model integration framework in use, the CAPE-OPEN, addresses the issue of standardisation of interfaces to enable interoperability between simulator software components from different sources. It is the framework built around a middleware, the Common Object Request Broker Architecture (CORBA) that hosts communication between unit operations defined for a specific function and the process modelling environments. The standard specification is defined as a property package which is needed for a thermodynamic or physical property calculation. The interoperability of models, such as model selection, parameter identification, and experimental work is enabled through the connection related to the unit operations and physical properties. It is not necessary to match all parameters in order to facilitate Input-Output (I-O) matching. However, the shortcoming of the CAPE-OPEN is in the need for identifying key variables for each unit operation.

In this paper a new approach for model integration is proposed which builds upon the CAPE-OPEN framework and proposes the use of ontology and replaces the object bus with more flexible semantic repository (Koo, Trokanas and Cecelja, 2017). Models are described by Semantic Web Services (SWS) using Ontology Web Service Description (OWL-S) as an enabler of web services through service discovery, selection, composition, and execution stages (Figure 1). The discovery stage allows formation of an integrated model through matching requests from a public repository(ies). The best match that satisfy the requestor's functionality is selected in the model selection stage. The model composition stage then formulates the chain of integrated models and execution of integrated model takes place during the execution stage. This paper focuses on the matching parameters related to the domain of process system engineering, with emphasis placed on the role of physical properties and unit operation. Each model representing a device (e.g. unit operations, flowsheets, subflowsheets etc.) is semantically described in domain ontology including domain assumptions and descriptions of the functionality of the model. The domain ontology guides the process of registering models and instantiation of ontology through ontology parsing, which makes the model and data explicit and sharing terminology in domain ontology improves consistency. The devices in a process are connected via streams that transmit information through multiple inlets and outlets from one unit to the other. The connection between devices are described in ontology by introducing the concept of 'ports' and 'connections'. The 'ports' generally describe inlets and outlets of devices and three different types of streams are distinguished: material, energy, and information, which are further described by objective properties. The 'connections' are the object that is responsible for establishing a link between two ports, which contains information regarding methods, types, quantities, and units of streams.

539298 Prototype Study for Monitoring Flare Performance

Abstract:

Flares are a key source of volatile organic chemicals and particulate matter in petrochemical plants. Automating the measurement of these flare emissions offers plants the ability to reduce the dependence on human operators and enhance environmental compliance efforts. In industrial settings, plant operators often neglect the existing flare automation because it is just as easy to control flaring behavior through visual confirmation; integrating this data into automated flare control systems will allow for better flare control. In this paper, a soft sensor is developed that can predict emissions using this visual data with still images. A natural gas burnerâ€™s carbon monoxide and methane emissions are predicted using this deep neural network.

Abstract:

An Artificial Neural Network Approach for the Identification of Stochastic Models of Travelling Traders' Exchange Process

C. Huang[1], *P. M. Piccione*[2], *F. Cattani*[3], *F. Galvanin*[1,*]

[1] *Department of Chemical Engineering, University College London*

Gower Street, London WC1E 6BT (UK)

[2] *Process Studies Group, Technology and Engineering, Syngenta, Breitenloh 5, CH-4333 Münchwilen, Switzerland*

[3] *Process Studies Group, Technology & Engineering, Syngenta, Jealott's Hill International Research Centre,*

Bracknell, Berkshire, RG42 6EY (UK)

* *f.galvanin@ucl.ac.uk*

The Travelling Traders' Exchange Process [1] (TTEP) is a stochastic process formulated to study the nature and insights of the dispersion and redistribution of a physical property (money) as a consequence of random sequentially-occurring exchange events between traders. The TTEP models are developed to study how the money distribution to traders will evolve over processing time and the properties of steady state that the system will eventually achieve given different model constructions. The analysis of stochastic model realisations obtained from the simulation of TTEP models is essential for process understanding, control and optimisation.

Stochastic simulations have been executed using a simple 1-D TTEP model in which operating and initial conditions are established including money transfer coefficients, trader population and initial money distribution. Simulation results at steady state in terms of money distributions to traders involve distinct output patterns that are influenced by the values of operating variables in the model. Figure 1 presents three possible money distribution patterns (A, B and C) observed at steady state. The major challenge is to select appropriate regression models to analyse the evolution of money distribution in time and to represent and classify the distribution patterns obtained from TTEP stochastic simulations.

Artificial Neural Networks [2] (ANNs) are versatile, self-adaptive and scalable tools to tackle large and highly complex data analysis tasks. Depending on the amount of data available for network training, ANNs are likely to outperform other Machine Learning techniques on even complex tasks [3]. It is interesting to link this powerful tool to the problem of identifying the nature of the outcomes of TTEP stochastic simulations. ANNs enable elucidating the connection between inputs and outputs to the process which is otherwise extremely complex to analyse in the case of stochastic systems.

Conventionally, distribution patterns are analysed via classical regression techniques after the simulation terminates and the distribution pattern has become an observable output [4]. This manual manipulation presents efficiency and precision issues: (i) The pre-set simulation time must be large enough as to insure that steady state has been reached; (ii) The selected regression models may have less estimate precision in regression analysis.

The objective of this work is to develop a computational framework to recognise and classify money distributions captured in TTEP stochastic simulation processes using ANN techniques so as to assist the regression model identification procedure.

To validate the neural network, firstly, the network layers are established and neurons in each layer are initialised by random weights and biases. Then, a set of training data are produced to give network training so as to adjust the values of weights and biases of neurons in network layers.

After training, the network will be able to recognise and classify the data configuration captured at each sampling point and correctly tell if the system achieves steady state and needs to commence further investigation. With well-trained ANNs, we may move to the stochastic model investigation section. The entire model investigation process comprises three steps that are presented by Figure 2. In the first step stochastic simulations are executed using TTEP model with given input variables. After the simulation process, samples of money distribution are collected at different time points and a database will be built and sent to the ANN for analysis. In the second step, the sample is recognised by the ANN, verifying that steady state has been detected and distribution pattern of the sample has been classified into classes (A, B, C, etc.). The process then moves to the next step. In the third step, the corresponding regression model is selected based on the category classified in step 2 to drive the choice of the most suitable regression model in the identification procedure. The integration of simulation and identification facilitates automatic investigation on the TTEP stochastic model.

The development of ANNs from stochastic models provides an efficient way to identify the nature and insights of stochastic models with high precision and accuracy. Future work aims at extending the applicability of the ANNs approach to TTEP models with higher complexity regarding model structure and money exchange mechanism, so that the trained ANNs will be capable of identifying data configurations and distribution patterns independently from the stochastic model structure.

References

[1] C. Huang, P. M. Piccione, F. Cattani and F. Galvanin, "A two-layer identification strategy for the development of stochastic models of the travelling traders' exchange problem," in *27th European Symposium on Computer Aided Process Engineering*, Barcelona, 2017.

[2] S. Samarasinghe, Neural Networks for Applied Sciences and Engineering, Boca Raton: Auerbach Publications, 2006.

[3] A. Géron, Hands-on Machine Learning with Scikit-Learn and TensorFlow, Sebastopol: O'Reilly Media, 2017.

[4] D. M. Bates and D. G. Watts, Nonlinear Regression Analysis and Its Applications, New York: Wiley, 2007.

530484 Process Data Analytics Using Deep-Learning Based Methods

Abstract:

In chemical manufacturing plants, numerous types of data are accessible, which could be process operational data (historical or real-time), process design and product quality data, economic and environmental (including process safety, waste emission and health impact) data. Effective knowledge extraction from data has always been a very challenging task, especially the data needed for a type of study is huge. Analyzing a high volume of data of different types is frequently difficult to perform by traditional learning methods.

In recent years, various deep learning methods have been introduced for knowledge extraction. Deep learning is a class of machine learning techniques developed based on learning data representations[1], as opposed to task-specific feature engineering. Effective deep learning methods can extract features automatically from data, which results in much better performance.

In this paper, we introduce our recent investigation on the prediction of catalyst age of a Liquid Phase Methanol (LPMEOH) process using a recurrent neural network (RNN) technique. Facing the existence of massive time series data involving exogenous parameters, we adopt a powerful technique to construct a dual-stage attention-based recurrent neural network (DA-RNN)[2]. This method consists of an encoder with an input attention mechanism, which can adaptively select the most relevant input features, and a decoder with a temporal attention mechanism that can capture the long-term temporal dependencies of a time series. The case study will demonstrate the methodological efficacy, which shows the outstanding prediction performance as compared with traditional learning methods.

[1] Bengio Y, Courville A, Vincent P (2013). Representation learning: A review and new perspectives. Pattern Analysis and Machine Intelligence, IEEE Transactions on 35(8):1798â€"1828. doi:10.1109/TPAMI.2013.50.

[2] Yao Qin, Dongjin Song, Haifeng Chen, Wei Cheng, Geoff Jiang, Garrison Cottrell (2017). A Dual-Stage Attention-Based Recurrent Neural Network for Time Series Prediction. Proc. 27th Intl. Joint Conf. on Artificial Intelligence (IJCAI).

---

* All correspondence should be addressed to Prof. Yinlun Huang (Phone: 313-577-3771; E-mail: yhuang@wayne.edu).

533376 Development of an Efficient Control for Smr Using Rigorous Modelling Techniques, to Improve Plant Performance, Stability, and Reliability during Feed Disturbances

Abstract:

As the gas reserves are depleting in the world, existing reforming plants are open to use natural gas from different sources. Different natural gas sources offer different compositions in the gas. This wide range of feed natural gas composition has significant effect on the performance of the steam reformers. In order to operate the reforming process at optimum conditions, precise control of reformer temperature, and steam intake are very crucial. Disturbances in the feed gas create instability in the reforming process. Furthermore, some steam reformers are operated with carbon dioxide added to the reformer feed, to have a lower hydrogen to carbon mole ratio, as required for downstream methanol or plasticizer alcohols processes. These dry gas combined steam reformers are operated at higher temperatures and lower steam to carbon ratios than that of pure steam reformers. Hence, frequent feed composition disturbances imply greater in-stability in the dry gas combined steam reformers in terms of reformer performance, reformer temperature, methane slip, fuel gas consumption and excess air in reformer box.

The existing control philosophy of the reformers cannot cope with the feed disturbances and can be effected by big swings in the reformer temperature, methane slip, fuel gas and steam & carbon dioxide flows in case of dry gas combined steam reformer. These disturbances may also have adverse effects on the plant efficiency, capacity, safety and reliability, and the plant integrity. In order to have a stable operation and cope with these disturbances, most of the existing plants are opting for advanced process control possibilities like Aspen DMC+ etc

In the paper, a new feed forward control strategy is developed using dynamic simulation of steam reformer and shift reactor sections using ASPEN DYNAMICS that can control the steam reforming process efficiently and effectively for a large composition variations. The dynamic simulation includes the all the key closed control loops of the reformer and shift reactors. The basic methodology of the new control strategy is to control the steam reforming key parameters like reformer temperature, stream flow, process air, and Fuel gas flow, based on the carbon count of the natural gas in a feed forward way. Additionally, a small correction to the individual manipulators is done in a feedback control philosophy. This feed-forward & feedback control philosophy enables the operating plant to react to the disturbances quickly and efficiently. Hence, the plant performance disorder is kept minimum and huge swings in the parameters are avoided during the feed disturbances. Also, with this new control strategy, the variations in the plant parameters are reduced and stable and steady production of the hydrogen can save the energy about 0.5 GJ/ton of H_2 produced from the reforming process. More than the energy, the process operates more steadily and downstream processes that depend on the reforming process operate steadily as well. This developed strategy is tested and implemented and has improved the plant reliability and integrity.

525775 Leveraging Open Source, Big Data and the Cloud for Chemical Process Control

Abstract:

Traditionally process controls have been a highly specialized area of engineering, requiring not only expensive hardware but also experts to implement it effectively. Over the last decade advances in hardware, software and integration have made highly advanced process control methodologies available to researchers for relatively little upfront investment. Control and automation tasks that used to require tens of thousands of dollars of investment could now be accomplished with a hundred-dollar microcontroller and an internet connection. However, both industry and academia have been slow to adapt to these changes and embrace all the new functionality afforded to them. Recent developments in electronics and software have greatly reduced the cost and complexity of control systems, from less expensive sensors to easily being able to store petabyte levels of data in the cloud and look for complex multivariate trends automatically. By embracing modern trends in both open-source platforms and cloud computing researchers and organizations will be able to greatly enhance their level of control, data collection, agglomeration and analytical capabilities, depth of insight into their systems, and ultimately process safety and efficiency.

The primary capability of a control system is to be able to input data from sensors and act upon this data. On the sensor front, cost reductions in silicon manufacturing have resulted in a significant cost reduction for high-precision thermocouples, pressure transducers, level sensors, accelerometers, thermal cameras and other common components. Hardware to interpolate and process data from these sensors has also become not only much cheaper but also much more capable, robust and easy to implement. Consider for example a scenario where you need to collect temperature data from each tray of a distillation column. Before you would need to run endless wires, have a large cabinet with a thousand-dollar thermocouple reader, and constantly be concerned that a loose wire could cripple your entire system. Today you could accomplish the same task using a set of Arduino boards networked through Zigbee, each with a 31855 thermocouple reader chip. Now all you need to run a single low-voltage DC power cable, and the failure of a single node would no longer impact the entire system. It is also much easier to add redundancy, augment your system for future expansion, and remotely monitor your process, not to mention significant cost savings. Overall recent developments in hardware have significantly decreased the cost and complexity of control system installations.

Another aspect that has significantly decreased the cost and increased the capability of controller instillations is the development of single-board computers. Ranging from the Raspberry Pi to the TK1 Jetson, the possibilities offered by these boards are hard to have imagined even several years ago. On the lower end these computers have roughly the processing power of a top-of-the-line desktop from the late 90s, for $30 and the size of pack of chewing gum. On the upper end these boards can train convolutional neural networks and process data at TFLOP/s-levels (that is one billion floating-point arithmetic operations per second). This allows for a lot of the analytical functions of a system to be moved away from centralized cabinets and off to the edge, close to sensors and actuators. Additionally, this will enable much quicker troubleshooting and modification without having to shut down the entire system. Most importantly though, this opens up the possibility of control algorithms that were not previously possible at a laboratory scale.

Instead of traditional PID loops, engineers can implement much more complicated non-linear schemes thanks to the extra processing power. The scaling down of powerful computers opens up many new doors for controls and automation.

The final recent development that allows for a paradigm shift in control systems design is big data and cloud computing. Over the last several years it has become very simple to stream large amounts of sensor and process data to an off-site server. It is not unfeasible to record a thousand sensor values once a second and store this data for years. In fact, storing 30 billion data points would only cost you about $4 in cloud services. This enables a really interesting a new ability, combing through this data and looking for non-linear trends between parts of the system that one would never intuitively think effect each other. For example, how does ambient humidity effect the temperature distribution in your columns due to condensation or how do slight frequency changes in the AC power your receive from the grid change the flowrate of your pumps. Discovering these trends will enable engineers to adapt their control algorithms for them, and reduce the overall number of process deviations.

533568 A Synthesis Framework for Structure Constrained Thermally Coupled Distillation Sequences Including Divided Wall Columns

Abstract:

Abstract

The optimality of distillation sequence decides the energy consumption of multicomponent zeotropic mixture separation directly[1]. Thermally coupled structure can effectively eliminate the backmixing of intermediate components, which is proved to be one of the most promising energy saving configurations. It is reported that in some cases thermally coupled sequences can decrease energy consumption for more than 30 percents and reduce total annualized cost for nearly 20 percents compared with that of conventional ones[2-4]. However, most of the highly thermally coupled structures can't be utilized in chemical industry due to the difficulties caused by equipment manufacture, automatic control and practical operation[5]. Besides, distillation sequence with feasible thermally coupled structure, like divided wall columns (DWC), is usually optimized through enumeration method which can't be adopted to zeotropic mixture separation with more than five components[6-7].

Therefore, a novel synthesis framework for thermally coupled distillation sequences with structure constraints is proposed. To obtain structure constraint equations, a systematic methodology for qualitatively analyzing the complexity of thermally coupled distillation sequences and a strategy for developing structure constraints are given. The concepts of couple degree and maximum section number of mass loop to qualitatively describe the complexity of system as well as the mechanism to classify the reason of thermally coupled structure generation are proposed. Based on these, a series of structure constraints are generated considering technologies applied in industry (DWC). With these constraints, a mathematic model for thermally coupled distillation synthesis based on state-task network superstructure is developed. A five components zeotropic mixture case study is carried out demonstrating that the existence, position and number of thermally coupled structure is related to the feed condition and investment cost. Besides, a six components zeotropic mixture case study is also carried out, the resulting sequence can decrease energy consumption for 30.8 and 21.1 percents and reduce total annualized cost for 21.5 and 16.2 percents respectively compared with that of conventional ones and thermally coupled sharp split ones, demonstrating the feasibility of the proposed methodology in large scale zeotropic mixture separation.

Reference

[1] Zou X, Cui Y, Dong H, Wang J, Grossmann I E. Optimal design of complex distillation system for multicomponent zeotropic separations[J]. Chemical Engineering Science, 2012, 75:133-143.

[2] Caballero J A, Grossmann I E. Synthesis of complex thermally coupled distillation systems including divided wall columns[J]. Aiche Journal, 2013, 59(4):1139-1159.

[3] Caballero J A, Grossmann I E. Design of distillation sequences: from conventional to fully thermally coupled distillation systems[J]. Computers & Chemical Engineering, 2004, 28(11):2307-2329.

[4] Asprion N, Kaibel G. Dividing wall columns: Fundamentals and recent advances[J]. Chemical Engineering & Processing Process Intensification, 2010, 49(2):139-146.

[5] Shah V H, Agrawal R. A matrix method for multicomponent distillation sequences[J]. Aiche Journal, 2010, 56(7):1759-1775.

[6] Luceroâ€Robles E. Petlyuk Columns in Multicomponent Distillation Trains: Effect of Their Location on the Separation of Hydrocarbon Mixtures[J]. Chemical Engineering & Technology, 2016, 39(12):2207-2216.

[7] Tututi-Avila S, DomÃnguez-DÃaz L A, Medina-Herrera N, et al. Dividing-Wall Columns: Design and Control of a Kaibel and a Satellite Distillation Column for BTX separation[J]. Chemical Engineering & Processing Process Intensification, 2017, 114:1-15.

527804 A Novel Robust Kalman Filter Algorithm Using Incremental PID Controller for Model Uncertainties

Abstract:

Kalman filter has been recognized as a powerful tool for many control problems. Its performance mainly depends on the accuracy of its dynamic model. In employing the filter in real problem, various sources of uncertainty actually should be considered such as unmodeled dynamics, parameter variations, model reduction, and linearization.

It is inevitable to model such inherent inaccuracy in practical applications. It is therefore difficult for conventional Kalman filter to provide satisfactory performance for such cases with significant model uncertainties. To tackle this difficulty, several approaches tried to modify the conventional Kalman filter algorithm in terms of robust estimation and filtering over the past decades.

Alternately, a simple but powerful method is introduced by incorporating incremental PID controller structure to the standard Kalman filter algorithm in this paper. The calculated estimation error defined as the difference between estimation values and measurements is used to correct the state estimates based on proportional, integral, and derivative terms. It makes errors due to state estimates to be much smaller compared to the standard approach which only uses estimation error as a corrective term.

To illustrate the applicability of the proposed method, two case studies are presented for Single-Input Single-Output (SISO) and Multiple-Input Multiple-Output (MIMO) systems. The performance comparison of proposed algorithm versus standard Kalman filter algorithm was conducted by calculating the mean square error (MSE) of the estimation error in each case. The proposed algorithm showed the better tracking performance than standard Kalman filter. The current works are under way to employ the proposed improved Kalman filter in actual control applications.

537131 Simultaneous Uncertainty Reduction and Control of Hydraulic Fracturing

Abstract:

Accurate characterization of reservoir properties is of central importance to achieve a desired fracture geometry during a hydraulic fracturing process. However, the estimation of spatially varying geological properties in hydraulic fracturing is inherently ill-posed due to a limited number of measurements. One way to address this is to perform parametrization that reduces the dimensionality of reservoir properties (for example, spatially varying Youngâ€™s modulus profiles). In our previous work, we proposed an integrated framework that combines proper orthogonal decomposition (POD) with a data assimilation technique, ensemble Kalman filter (EnKF) [2], to estimate the parameter values in the reduced low-dimensional subspace [3]. Usually, the process and measurement noise information within the EnKF is assumed to be known and independent of inputs. However, unlike other industrial applications, for hydraulic fracturing processes that use microseismic monitoring technology, the occurrence of a measurement depends on the fluid injection rate into the fracture, which is an input to the process [4]. To be more precise, higher flow rates are likely to trigger a higher number of microseismic events as a result of increased stress caused by higher flow rates. A larger number of almost simultaneous microseismic events can effectively reduce measurement errors. This suggests the use of new control approaches based on the input dependent measurement error covariance to help reduce state and output estimation errors [5].

In this work, we explore the idea of using a controller for the purpose of simultaneous uncertainty reduction and set-point tracking of a nonlinear hydraulic fracturing process. More specifically, we propose to design a model-based feedback control system that will help in accurately identifying the spatially varying geological properties by reducing the measurement uncertainty, while at the same time accomplishing the original control tasks. Here, the objective function fed to the controller consists of both the estimation error covariance (uncertainty reduction) as well as the deviation from the desired target value (set-point tracking). The proposed closed-loop system is applied to a simulated model of a hydraulic fracturing process with synthetic spatially varying geological parameters.

Literature cited:

[1] Holmes, P., Lumley, J.L., Berkooz, G., 1996. Turbulence, Coherent Structures, Dynamical Systems and Symmetry. Cambridge University Press, New York.

[2] Evensen, G., 1994. Sequential data assimilation with a non-linear quasi-geostrophic model using Monte Carlo methods to forecast error statistics. Journal of Geophysical Research, 99, 10143â€"10162.

[3] Narasingam, A., Siddhamshetty, P., Kwon, J. S., 2018. Handling spatial heterogeneity in reservoir parameters using proper orthogonal decomposition based ensemble Kalman filter for model-based feedback control of hydraulic fracturing. Industrial & Engineering Chemistry Research, 57(11), 3977-3989.

[4] Maxwell, S. C., Rutledge, J., Jones, R., Fehler, M., 2010. Petroleum reservoir characterization using downhole microseismic monitoring. Geophysics, 75(5), 129â€"137.

[5] Sun, Z., Gu, Q., Dykstra, J., 2016. Uncertainty reduction of hydraulic fracturing process. In American Control Conference (ACC). 2135-2141.

538413 Non-Linear Model Predictive Control of Module Temperature in Photovoltaic System

Abstract:

With the growing demand in energy and environmental concerns, a paradigm shift towards renewable energy sources is imminent. Among the existing renewable resources, solar technology is the most promising one, because of its abundance, relatively low installation cost and ease of scale-up in the plant capacity. One of the prominent technologies based on solar energy is built on the photovoltaic (PV) cell, which transforms the solar energy either into electrical or heat energy. A PV module is a large combination of cells that are connected either in series or parallel in order to meet the power demand. The efficiency of the PV module is highly dependent on module temperature which is sensitive to solar irradiance and ambient temperature. The effect of module temperature on voltage and power is shown in Figure 1.

PIC

Figure 1: Effect of temperature variation on voltage and power

As evident from Figure 1, the efficiency of PV module decreases with increase in module temperature, which is significantly influenced by the intermittent nature of insolation and ambient temperature. Since it is not possible to adjust the levels of these disturbances, a cooling system is necessary to control the module temperature [2]. The success of such a cooling system critically rests on the development of an accurate model that explains the effect of solar irradiance and ambient temperature on the module temperature.

Ilhan [3] proposed a spiral heat exchanger placed on the surface of the panel in order to provide active cooling. Teo [4] proposed an active air cooling system attached to the back of the PV panel. The few works that appear in the reported literature control the module temperature by use of ON/OFF controller for the coolant system. On the other hand, the use of well-established model-predictive control (MPC) framework for improving the power efficiency is quite attractive.

In this work, we first present a revised nonlinear model of PV/thermal and coolant system that explains the dependency of insolation and ambient temperature on the module temperature. Subsequently a nonlinear model predictive control (NL-MPC) strategy is developed which determines the optimal coolant flow rate so as to control the panel temperature in the presence of operating constraints.

PIC

Figure 2: A schematic structure of NL-MPC

It may be noted that an NL-MPC strategy was implemented by Eduardo [1] for a solar trough, where the objective is to control the temperature of the oil flow in order to improve the solar thermal plant efficiency.

The proposed NL-MPC architecture for the PV cell is shown in Figure 2. Simulation results show (to be reported in the full-length paper) that the proposed NL-MPC for the PV module results in a significant improvement in PV/thermal system efficiency. References

[1] Eduardo F. Camacho and A. J. G. Model predictive control in solar trough plants: A review. IFAC-PapersOnLine, 48(23):278 – 285, 2015. 5th IFAC Conference on Nonlinear Model Predictive Control NMPC 2015.

[2] Dong-J. Kim, Dae H. Kim, Sujala Bhattarai, and Jae-H. Oh. Simulation and model validation of the surface cooling system for improving the power of a photovoltaic module. Journal of Solar Energy Engineering, 133(4):041012, 2011.

[3] lhan C., Ali E. G., Hsamettin D., and Bahri A. Cooling of a photovoltaic module with temperature controlled solar collector. Energy and Buildings, 72:96 – 101, 2014.

[4] H.G. Teo, P.S. Lee, and M.N.A. Hawlader. An active cooling system for photovoltaic modules. Applied Energy, 90(1):309 – 315, 2012. Energy Solutions for a Sustainable World, Special Issue of International Conference of Applied Energy, ICA2010, April 21-23, 2010, Singapore.

525208 Discrete-Time Nonlinear Observer-Based Globally Linearizing Control of a PEM Fuel Cell

Abstract:

Exponential increase of energy demand and continuous depletion of fossil fuels have forced us to search and utilize the alternate energy sources [1]. Accordingly, among many renewable energy devices, the fuel cell is chosen as one of right candidate for clean and sustainable future development. Again, amongst various fuel cells, the proton exchange membrane (PEM) fuel cell is preferred mostly due to its low operating temperature, high power density, fast start-up and, more importantly zero emission and less operating noise [2]. Because of these notable advantages, its practical applications are widely found in transport, residential and, stationary and portable applications [3]. With this, we have started working towards modelling and automation of the PEM fuel cell for its stable operation. In early stage of PEM fuel cell blooming, there are several steady state and linearized models reported. But these models may not produce satisfactory response if the operating conditions slightly deviate from its base condition. Later, a few nonlinear dynamic control oriented models have been developed [4,5]. Again, regarding control of PEM fuel cell, apart from conventional controllers, nonlinear control works have been found in open literature [6-9]. The PEM fuel cell utilizes the reactants, namely hydrogen and oxygen to convert the chemical energy into electrical energy using platinum as a catalyst. As per the energy requirement (load demand), the reactants flow are to be altered accordingly to avoid irregularities (i.e., cell breathing issue, membrane damage and voltage degradation) during the operation. As the PEM fuel cell is inherited with fluid-heat-electrochemistry interacting complex nonlinear dynamics, we need to use the nonlinear controller for its efficient operation. In this work, a discrete-time nonlinear globally linearizing control (GLC) system is developed (Fig. 1) for the PEM fuel cell [10]. The GLC control structure includes a nonlinear transformer, an adaptive state observer (ASO) and a dual-loop PI controller. The ASO is formulated based on shortcut model (i.e., only two measured component balance equations) to estimate the unknown state information as per the GLC requirement. To reduce the resulting process/model mismatch in the observer, a polynomial equation based dynamic tuning parameter is used. Finally, the hybrid GLC-ASO shows its supremacy in tracking the set point as well as rejecting the disturbances compared to the conventional dual-loop PI controller. *Keywords*: PEM fuel cell; modelling and simulation; adaptive state observer; discrete-time globally linearizing controller; conventional dual-loop PI controller

Description: C:\Users\pdc@fuelcell\Desktop\GLC-ASO.jpg

Fig. 1. GLC-ASO control system of PEMFC.

Reference [1] Bhattacharyya S. C. Influence of India's transformation on residential energy demand. Applied Energy, 143, pp. 228-237, 2015. [2] Larminie J. Dicks A. Fuel cell systems explained, John Wiley & Sons Ltd, West Sussex, England, 2001. [3] Panik F. Fuel cells for vehicle applications in cars-bringing the future closer. Journal of Power Sources, 71, 36-8, 1998. [4] Xue X., Tang J., Smirnova A., England R. and Sammes N. System level lumped parameter dynamic modelling of PEM fuel cell. Journal of Power Sources, 133, pp. 188-204, 2004. [5] Li Q., Chena W., Wang Y., Jia J. and Han M. Nonlinear robust control of proton exchange membrane fuel cell by state feedback exact linearization. Journal of Power Sources, 194, pp. 338-48, 2009. [6] Matraji I., Laghrouche S., Jemei S. and Wack M. Robust control of the PEM fuel cell air-feed system via sub-optimal second order sliding mode. Applied Energy, 104, 945-57, 2013. [7] Pukrushpan J., Stefanopoulou A.G. and Peng H. Control of fuel cell breathing, IEEE Control Systems Magazine, 24, pp. 30-46, 2004. [8] Arce A., del Real A. J., Bordons C. and Ramirez D. R. Real-time implementation of a constrained MPC for efficient airflow control in a PEM fuel cell, IEEE Transaction on Industrial Electrononics, 57, pp. 1892-1905, 2010. [9] Na W. K. and Gou B. Feedback-linearization-based nonlinear control for PEM Fuel cells, IEEE Transaction on Energy Conversion, 23, 179-90, 2008. [10] Sankar K. and Jana A. K. Dynamics and estimator-based nonlinear control of a PEM fuel cell. IEEE Transaction on Control Systems Technology, accepted, 2017.

532817 Robust Model Predictive Control for Smart Grid Integrated with Solar Power and Energy Storage System Under Regular and Abnormal Loads

Abstract:

In this work, a scenario-based robust model predictive control (MPC) integrated with a load estimator is proposed to minimize the power loss and stabilize voltage fluctuation when photovoltaic (PV) cells and energy storage system (ESS) are connected to the grid.

Due to the high level installation of rooftop photovoltaic cells in California, the solar power becomes an important renewable energy source injected into the grid. However, the rapid growth of capacity and penetration of the PV may lead to serious fluctuation of voltage in the grid and thereby degrade the power quality of the distribution line. Moreover, the collection of solar energy is dependent on the weather and solar irradiation. Thus it may not be available when unexpected power demand occurs. In the meanwhile, it is worthwhile to note that ESS such as Lithium battery has progressed significantly in terms of the capacity, charging/discharging rate and life time. The combination of PV and ESS may provide an attractive scheme to meet energy demands and stabilize voltage in a smart grid. Nevertheless, the resulting integrated system contains fast dynamics from the PV inverter and slow dynamics from the Lithium battery. Thus how to coordinate, control and optimize the fast electrical components and slow chemical processes become an important research topic.

In the controller design, the power demand data from Southern California Edison (SCE) and California Independent System Operator (CAISO) is used to build different scenarios for residential/commercial/industrial regions, respectively. Then a scenario-based robust MPC is developed to coordinate PV and ESS, reduce power loss and compensate the abnormal power load. Both battery and dynamic load models are carefully selected to simulate the real process and then integrated into the DistFlow equations [1] to describe the power of the distribution line. Directly solving MPC with such complex multi-scenario model is very time consuming. Therefore, a fast algorithm is proposed to find a high quality solution within reasonable time.

To formulate a closed-loop control scheme, an optimization-based power load estimator is also developed to deliver states information to MPC. This
estimator is able to reconstruct the real/reactive power load according to the noise-corrupted measurement data. The accuracy of this estimator can be quantified and guaranteed especially when high power load happens. Such closed-loop control framework enables the grid to quickly recover from undesired disturbances (i.e. abnormal loads) by effectively changing the real and reactive powers provided by PV cells and ESS.

Reference
[1] M. E. Baran and F. F. Wu, Optimal capacitor placement on radial distribution systems,â€ IEEE Trans. Power Del., vol. 4, no. 1, 725â€"734, Jan. 1989.

528587 Feedback Predictive Control Versus Model Predictive Control for Automatically Controlling Blood Glucose Concentration

Abstract:

Type 1 diabetes is the condition that features the loss of endogenous insulin production ability. Artificial pancreas (AP) is the system that can automatically control glucose level by determining insulin infusion rate via control algorithms. Model predictive control (MPC) is a widely used control algorithm in AP for the automatic control of glucose concentration. However, it has drawbacks due to long time lags associated with inputs (e.g. carbs and insulin) and inaccurate predictions over long prediction horizons. The proposed algorithm, feedback predictive control (FBPC), has the potential to avoid those issues to achieve fully closed-loop control of glucose and significantly reduce the variations of glucose around its target.

In this work, FBPC is first demonstrated in a first order plus dead time (FOPDT) process to explicitly showcase its superiority over MPC mathematically. Then the results of FBPC is compared to MPC for the 30 *in-silico* subjects in a â€œhead-to-headâ€ comparison with all things being equal except for the differences in the control algorithms. Lastly, to further evaluate the capability of handling unmeasured disturbances in control of glucose, meals are treated as unmeasured disturbances for first five (5) cases of virtual subjects.

For the standard deviation of glucose about its mean (Stdev), FBPC was smaller than MPC for all 30 cases. On the average, relative to FBPC, the Stdev for MPC was 33% higher. With meals as unmeasured disturbances, the Stdev was 133% larger for MPC relative to FBPC. For FBPC, its Stdev was less than 10% larger for unmeasured meals versus measured.

The proposed FBPC algorithm has the ability to produce significantly tighter variability of glucose about its target and has the capability in handling unmeasured disturbances. Thus, FBPC should be considered as an alternative to MPC for fully automatic control of glucose.

534599 Dual Control Framework with Multistep Ahead Prediction Model

Abstract:

In this work, a dual control framework, which is able to simultaneously track setpoint and online re-identify process model, is presented. Moreover, a new optimization algorithm is integrated into this framework to minimize the multistep prediction error for the auto regressive with exogenous inputs (ARX) model.

Though the model predictive control (MPC) has shown great capabilities to achieve the optimal solution for setpoint tracking, its real performance may degrade rapidly due to model mismatch and parameters drift. To overcome this drawback, the dual control scheme, in which control and model identification are solved in parallel, has been received great attentions. In such scheme, the persistent excitation (PE) condition becomes significantly important, because the identification from closed-loop data without such condition will lead to biased model. However, how to strike a balance between PE condition and setpoint tracking is still outstanding because these two objectives are naturally conflicted. An analytical solution is hardly developed [1] and numerical optimization should be used. However, the resulting problem is usually non-convex [2]. Therefore it can be very complex and difficult to implement for real time regulation.

Our first contribution is the development of a tractable two-step formulation to decide the control input such that the information from input/output data sequence is maximized while the stability of the process is still guaranteed. The resulting optimization is relatively easy to solve. Our second contribution is the development of a simple optimization scheme for ARX model identification by explicitly minimizing the multistep prediction error. Hence, the resulting process model is very suitable for process prediction for a long time horizon.

Reference

[1] J. Sternby, A simple dual control problem with an analytical solution, IEEE Transactions on Automatic Control 21 (6) (1976) 840-844.

[2] H. Genceli, M. Nikolaou, New approach to constrained predictive control with simultaneous model identification, AIChE Journal 42 (1996) 2857â€“2868.

533585 Black Box Operation Optimization for Temperature Control of Basic Oxygen Furnace Process

Abstract:

The basic oxygen furnace(BOF)steelmaking is an important process of the iron and steel production. To ensure the quality of molten steel, the main task of this process is to remove impurities from molten iron by blowing oxygen. Continuous oxidation reactions of chemical components occur during the oxygen blowing process. This is a heat dissipation process which leads to the increase in the temperature of molten steel in the furnace. And it makes it is necessary to control the temperature of the molten in this process to meet production requirements and ensure production safety. Thus, in this paper, according to the history data, a black box operation optimization (BBOO) model with a box constraint is extracted to realize temperature control for the endpoint of BOF steelmaking. At the same time, in the model, we also take production materials saving into account. To solve the BBOO problem, we proposed a box constrained derivative free optimization algorithm. Finally, numerical experiments are performed.

533587 Optimal Control of Bof Steelmaking with Considering Energy Consumption

Abstract:

The basic oxygen furnace(BOF) steelmaking is a high-energy consumption industrial production process. In actual production, controlling temperature and carbon content at the end of process is of great significance. Thus, the literature on carbon content and temperature prediction is extensive. The works that Considering energy consumption of this process are scarce. In this paper, based on the mechanism, the system equation is established, and then set up a optimal control model, of which the deviation of temperature, carbon content, and energy consumption are considered. Finite difference method and Lasserre relaxation are used to transform the problem to Semidefinite programming(SDP), then solving this problem by ADMM algorithm. The model proposed in this paper is designed to accurately describe the relationship between the quality requirement of molten steel (temperature, carbon content) and the amount of materials, so as to make the production plan quickly and reduce the energy consumption as much as possible.

538901 Machine Learning Techniques for Model Identification from Historical Data for Control

Abstract:

Model predictive controllers (MPC) rely on system model to identify optimal input profile. Data collection for identifying such models is the most time consuming step in implementation of MPC in industries [1]. Storing operational data using process historian is a common practice in industries. In general usage of such data is restricted to performance assessment, fault detection etc. Historical data is not used for model identification due to the presence of multiple time scales, presence of long time disturbance effects and correlated inputs etc. Within the time period for which the historical data is collected, there will be intervals of time where the data segment is informative enough to be used for modeling exercise. Identifying and marking these segments is a complex problem. Peretzki et al [2], proposed a method for data segmentation based on variance of the variables, condition number of the information matrix and the estimated parameters using Laguerre's filter being non-zero. They also provided a quality measure for the data. In this paper we propose a new data segmentation technique that can identify regions of informative data from historical data. An interval-halving technique that can handle multivariate data is used for this purpose [3]. Machine learning techniques are used to identify regions of data with similar features to address the presence of disturbance effects. Self-learning and model update after online implementation of such models will also be discussed.

References

[1] H. Genceli and M. Nikolaou, "New Approach to Constrained Predictive Control with Simultaneous Model Identification," vol. 42, no. 10, pp. 2857–2868, 1996.

[2] D. Peretzki, A. J. Isaksson, A. C. Bittencourt, and K. Forsman, "Data mining of historic data for process identification," *AIChE Annu. Meet.*, 2011.

[3] L. Das, B. Srinivasan, and R. Rengaswamy, "Multivariate Control Loop Performance Assessment With Hurst Exponent and Mahalanobis Distance," *IEEE Trans. Control Syst. Technol.*, vol. 24, no. 3, pp. 1067–1074, 2016.

537136 Modeling and Control of Proppant Distribution of Multi-Stage Hydraulic Fracturing in Horizontal Wells

Abstract:

In shale rock formations, multiple hydraulic fracturing treatments with slick water (i.e. a low-viscosity fluid) are generally used to stimulate the recovery of natural gas and oil. The slick water can create massive hydraulic fractures, but it is inefficient at suspending proppant due to the low-viscosity, resulting in the creation of significant proppant bed at the bottom of the fracture, which will influence the subsequent proppant transport mechanism inside the fracture. Currently available hydraulic fracture simulators mainly focused on simulating simultaneously propagating multiple fractures considering stress shadow effects without considering this proppant transport phenomena [1, 2]. As of now, the simulation of proppant transport has mainly focused on simple fracture geometry, i.e. a single, planar fracture [3, 4]. Even though some efforts have been made to simulate the proppant transport using the Eulerian-Eulerian method [5], dense proppant transport and proppant bed formation were not considered. Recently, the Eulerian-Lagrangian method was used to model particle movement [6]. However, the approach is highly computationally expensive to be incorporated into simultaneously propagating multiple fractures model. Motivated by these considerations, the proposed research will first address the development of a new high-fidelity model to describe the fracture propagation by explicitly accounting for stress shadow effects as well as proppant transport to fluid conductivity.

The hydrocarbon production through these simultaneously propagating multiple fracture networks is strongly dependent on the proppant distribution within the fractures, because they will create conductive channels through which natural oil and gas can be easily transported. Therefore, the proppant distribution at the end of pumping in simultaneously propagating multiple fractures must be regulated by developing model-based pumping schedules. In this work, utilizing the new high-fidelity model, we will focus on introducing a new model-based control algorithm to compute online pumping schedules that compensate for stress shadow effects; the computed pumping schedule will be able to generate uniform proppant distribution in simultaneously propagating multiple fractures.

References:

[1] Wu, K., & Olson, J. E. (2015). Simultaneous multifracture treatments: fully coupled fluid flow and fracture mechanics for horizontal wells. *SPE journal*, 20(02), 337-346.

[2] Peirce, A., & Detournay, E. (2008). An implicit level set method for modeling hydraulically driven fractures. *Computer Methods in Applied Mechanics and Engineering*, 197(33-40), 2858-2885.

[3] Siddhamshetty, P., Yang, S., & Kwon, J. S. I. (2017). Modeling of hydraulic fracturing and designing of online pumping schedules to achieve uniform proppant concentration in conventional oil reservoirs. *Computers & Chemical Engineering. URL: https://doi.org/10.1016/j.compchemeng.2017.10.032.*

[4] Siddhamshetty, P., Kwon, J. S. I., Liu, S., & Valkó, P. P. (2017). Feedback control of proppant bank heights during hydraulic fracturing for enhanced productivity in shale formations. *AIChE Journal*, 64(05), 1638-1650.

[5] Liu, Y. (2006). Settling and hydrodynamics retardation of proppants in hydraulic fractures. The University of Texas at Austin, Ph.D. thesis.

[6] Wu, C., Yi, S., & Sharma, M.M. (2017). Proppant distribution among multiple perforation clusters in a horizontal wellbore. *Presented at SPE Hydraulic Fracturing Technology Conference and Exhibition,* The Woodlands, Texas, USA.

535855 A Monte Carlo Simulation Study to Evaluate the Limits of Prediction Accuracy for Blood Glucose Concentration

Abstract:

Predictions of blood glucose concentration is an important aspect of glucose management for people with type 1 diabetes. Especially in applications of artificial pancreas (AP), systems that can automatically determine insulin infusion rate with the frequency up to every minute, model-based control algorithms utilize predictions of glucose extensively. Model predictive control (MPC), as one of the most widely used control algorithms in this application, requires long prediction horizons into the future (e.g. over 5 hours into the future) due to long inputs dynamics (i.e. carbs and insulin).

This study used data collected from 11 subjects with type 1 diabetes under free-living conditions (i.e. no restrictions on the life styles of participating subjects) with noises added to mimic sensor noise/measurement errors during the study. The autoregressive (AR) models, widely used in AP applications, were used to model predictions of glucose up to one (1) hour into the future. First part of the study was to examine the deterioration of prediction accuracy as predication horizon increases. Then the effects of noises were evaluated. Lastly, an adaptive modeling methodology was evaluated against fixed model parameterization.

In terms of the average absolute deviation (AAD), indicating the model deviations, the model performance deteriorates almost linearly as prediction horizon increases. The AAD reached over 35 mg/dL, and the correlation coefficient between measured (treated as true glucose concentrations) and predicted glucose (rfit) fell to 0.74, with prediction of one hour into the future. The AAD also increases as the noise added increases. However, as standard deviation of added white noise increases from 5 to 15 mg/dL, the AAD for one hour prediction horizon only increased by 1.1%. For the adaptive modeling methodology, the AAD and rfit improved from mean AAD of 34.41 mg/dL to 33.32 mg/dL against fixed model parameterization.

The study shows prediction accuracy deteriorates mainly due to long prediction horizons other than sensor noises. Additionally, the adaptive modeling methodology improves model performance. However, the limits of prediction accuracy are mainly bound by the length of prediction horizons. Control algorithms that avoid the use of long prediction horizons may be preferable options for this AP application.

528780 Dynamic Optimization of Natural Gas Network with Rigorous Thermodynamics

Abstract:

Pipelines are widely used in the transportation of natural gas, and their optimal operation requires accurate models. In particular, accurate thermodynamic properties are important for the dynamic models, as the gas goes through significant changes of thermodynamic properties across time and space.

Gas thermodynamic properties, like gas compressibility factor (Z-factor), are influenced by internal properties like gas composition, and environmental conditions like temperature and pressure. Previous work has applied experiment results in the form of charts for reference. The major drawback of these reference charts is that they only provide rough estimations and cannot be used directly in the mathematical optimization models. In the last few decades, several methods for gas thermodynamic properties have been proposed and accurate approximations can be obtained. These methods solve implicit nonlinear equations, and their direct implementation would make the dynamic model prohibitively expensive to solve. As a result, many researchers have simplified the natural gas model by assuming the natural gas to be an ideal gas, which inevitably introduces errors to the models.

In this study, we develop and compare different thermodynamic property methods for the Z-factor to find the most accurate approximation for the existing dynamic transmission model. Since the direct implementation of the existing thermodynamic calculating methods are difficult to solve, reduced order models(ROM) are used to replace the complex thermodynamic equations. Here, the ROM method can help accelerate the computation but may lead to inaccurate approximations to the original model. To guarantee convergence to an optimum, a trust-region based optimization framework is required. However, ROMs are computational tractable for problems with less than a few dozen variables. While in the dynamic pipeline network, we need to calculate Z-factors at every spatial and temporal discretization point, and the number of variables grows rapidly as the resolution of the model increases. Therefore, the ROM-based trust-region method is overwhelmed by the dynamic model. To address this challenge, we propose a distributed trust region method, which builds separate trust regions at every discretization point. In this way, each trust region only needs to handle a small number of ROM variables. When the resolution discretization increases, only the number of trust regions increases, while the size of the ROMS remains the same for every trust region.

This approach is incorporated within Pyomo, an optimization modeling platform and solver interface. This platform works as well as any other algebraic modeling languages such as GAMS and AMPL in terms of model formulation and analysis. Moreover, it contains a rich set of meta-solvers developed by Process Systems Engineering community. Finally, Pyomoâ€™s modeling objects are embedded within Python, which makes it flexible to implement user-defined optimization framework.

To demonstrate this ROM-based trust region approach, we consider a dynamic natural gas pipeline network with multiple compressors and present significant improvements in the efficiency of the optimization strategy and the quality of solutions.

537488 Modelling and MPC Design of Mineral Column Flotation Process

Abstract:

Column Flotation is a commonly used separation method in the mineral processing industry, it can effectively separate useful minerals from ore with low grade and complex mineral composition and has been widely used in industry for decades. Since it was first used in the commercial application of mineral separation in 1980s [1], the modelling and/or control of the column flotation processes have attracted the attention of many researchers and gradually become a popular research field. Its main purpose is to ensure maximum recovery of minerals while ensuring products (concentrates) grade, the optimization of the flotation process through control effects could improve recovery and product grades for greater benefits [2].

The column flotation process is a highly nonlinear distributed parameter system (DPS) consisting of gas, water and solid three-phases flows, with multiple inputs and various important parameters are highly interrelated, there are also various sub-process occur during the flotation process. After many years of research and development, flotation process is still not fully understood and production is still relatively inefficient [3].

Motivated by the above, in this work, a three-phase dynamic model and model predictive control design of column flotation process is developed. Based on the process description and conservation of mass, the overall system is described by a set of nonlinear coupled hyperbolic PDEs and ODEs of interface, froth and collection regions, these three subsystems are connected through the boundaries. The steady-state profiles are utilized to linearize the original nonlinear model. The structured preserving Cayley-Tustin time discretization transformation [4] is utilized for the discrete controller design which renders discrete infinite dimensional system without spatial discretization or any model reduction. The model predictive controller is designed by solving the optimization problem to minimization the open-loop objective function at sampling time K [5]. By adding constraints to future inputs, outputs or state variables, the constraints can be explicitly expressed in an online solving quadratic programming. The controller performance to keep the output at the steady state within the constraint range by adjusting the input is demonstrated by simulation studies.

[1] Finch, J. A., and G. S. Dobby. "Column flotation" (1990).

[2] McKee, D. J. "Automatic flotation control-a review of 20 years of effort." *Minerals Engineering* 4.7-11 (1991): 653-666.

[3] Shean, B. J., and J. J. Cilliers. "A review of froth flotation control." *International Journal of Mineral Processing* 100.3-4 (2011): 57-71.

[4] Xu, Qingqing, and Stevan Dubljevic. "Linear model predictive control for transportâ€reaction processes." *AIChE Journal* 63.7 (2017): 2644-2659.

[5] Muske, Kenneth R., and James B. Rawlings. "Model predictive control with linear models." *AIChE Journal* 39.2 (1993): 262-287.

521931 Multiple Phase Shifted Chirp Signals for Rapid Impedance Estimation: Applications in Diagnosis of Electrochemical Systems

Abstract:

Multiple phase shifted chirp signals for rapid impedance estimation: Applications in diagnosis of electrochemical systems

Resmi Suresh[1], Sathish Swaminathan[1] and Raghunathan Rengaswamy[1*]

[1]Department of Chemical Engineering, IIT Madras, India

Email address: raghur@iitm.ac.in

ABSTRACT

A new approach that uses multiple large bandwidth and short duration chirp signals for rapid impedance measurement of electrochemical systems is proposed in this work. The existing techniques for impedance measurements are not suited for non-invasive, on-line diagnosis of electrochemical systems as they are time consuming and require the system under test to be idle during measurements[1]. In the proposed method, chirp signals in tandem with a definition of instantaneous frequency are used in the development of the impedance plots. Unlike the standard multi-sine EIS techniques, where at each frequency transients have to die down before measurements are made, in the proposed approach, only the transient data is used in the analysis. An interesting result from our approach is that while the impedance plots that are generated are approximate, qualitatively (and to a large extent quantitatively) they match the true electrochemical impedance spectroscopy plot (EIS plot). However, the impedance plots are generated using very short duration signals. As the impedance developed using the proposed technique mirrors the EIS plot, this method can be used as an alternative to EIS for various applications like fault diagnosis and optimization studies[1]. The proposed method improves on the work presented in Bullecks[2] where a single chirp signal was used for impedance estimation. Experimental and simulation studies using the proposed technique (as shown in Figure 1) are included in this work.

Figure 1: Comparison of the impedance profiles obtained using dual chirp analysis with that of the theoretical impedance

REFERENCES

1. Yuan, Xiaozi, Haijiang Wang, Jian Colin Sun, and Jiujun Zhang. "AC impedance technique in PEM fuel cell diagnosis—A review." International Journal of Hydrogen Energy 32, no. 17 (2007): 4365-4380. 2. Bullecks, Brian, Resmi Suresh, and Raghunathan Rengaswamy. "Rapid impedance measurement using chirp signals for electrochemical system analysis." Computers & Chemical Engineering 106 (2017): 421-436.

534862 Development of Advanced Model-Based Controllers for Optimal Load-Following Operation of the Supercritical Pulverized Coal Power Plants

Abstract:

High penetration of variable renewable generators into the electricity grid has resulted in changing operating requirements of traditional power generation plants, particularly in super- and sub-critical pulverized coal plants. A number of these coal-fired plants were originally designed to operate under base-load condition, but are now being forced to load follow. This leads to low efficiency, high operation and maintenance (O&M) costs, and potential increases in environmental emissions. Optimal operation during load following is necessary for the commercial viability of these plants.

High-fidelity, plant-wide, dynamic modeling of supercritical pulverized coal (SCPC) plants is critical for studying optimal load-following operation, but is lacking in the existing literature. It is desired that the model is accurate during load-following operation when significant changes in the operating conditions do occur. Under sliding-pressure operation during load-following, the boiler can transition to subcritical condition. Due to high nonlinearity in the physical properties of steam and due to the heat transfer characteristics under part-load operation, the temperature profile in the boiler can change considerably. Thus a model that can capture these characteristics during load-following operation is desired. Furthermore, due to the high heat and mass integration in the SCPC plant, it is critical to consider the dynamics of the entire plant.

Dynamic models can be highly useful for developing efficient control strategies for SCPC plants [1-2]. However, due to the considerable differences in the time scales of the key phenomena and due to the transport delay, development of efficient controllers under load-following operation becomes challenging. Uncertainties in the heat transfer model in the boiler due to the time-varying fly ash deposit on the tubes and due to the complex multi-dimensional convective, radiative and conductive hear transfer can lead to poor performance of the model-based controllers.

In this work, a plant-wide dynamic model of a 640 MW SCPC power plant is developed following the cost and performance baseline studies conducted by the National Energy Technology Laboratory [3]. Using the Aspen Custom Modeler software, high-fidelity process models are developed for the once-through pulverized coal boiler, the steam turbine system with consideration for fixed and sliding pressure operation, the surface condenser, the selective catalytic reduction unit, and the heat exchangers in the feed-water heating section. These custom models are compiled into the complete plant-wide SCPC steady-state Aspen Plus and Aspen Plus Dynamics models considering steam extractions and other auxiliary operations.

The steady-state SCPC model is validated against the NETL Baseline Report. A regulatory control layer is designed for disturbance rejection and plant stabilization. A supervisory control layer is first developed following the industry-standard coordinated control technique. Finally, both linear and nonlinear model predictive controllers under state and rate-of-change constraints are developed. The state and rate-of-change constraints help to ensure that the creep and fatigue damages due to load-following are reduced. Several load-following scenarios are evaluated using the dynamic SCPC model. The presentation will include comparisons of control performance by studying the tradeoff between the efficiency, ramp rate, state, rate of change, and environmental emission.

Bibliography

[1] J.-Z. Liu, S. Yan, D.-L. Zeng, Y. Hu and Y. Lv, "A dynamic model used for controller design of a coal fired once-through boiler-turbine unit," *Energy,* no. 93, pp. 2069-2078, 2015.

[2] W. Shinohara and D. E. Koditschek, "A Simplified Model Base Supercritical Power Plant Controller," in *Proceedings of the 35th Conference on Decision and Control*, Kobe, Japan, 1996.

[3] National Energy Technology Laboratory (NETL), "Cost and Performance Baseline for Fossil Energy Plants Volume 1a: Bituminous COal (PC) and Natural Gas to Electricity," U.S. DOE/NETL, 2015.

527843 Recurrent Neural Network-Based Model Predictive Control for Continuous Pharmaceutical Manufacturing

Abstract:

In the recent past, the pharmaceutical industry has witnessed exponential growth in momentum in transforming operations away from the conventional batch processes [1, 2] towards continuous manufacturing [3, 4]. This fundamental shift is necessary to allow this industry to effectively address key challenges such as increasing margin, achieving superior Critical Quality Attributes (CQAs), extending manufacturing range (e.g., for Active Pharmaceutical Ingredients that are not amenable to conventional batch processes) while reducing cost, waste, footprint, lead-times, in the face of ever-increasing regulatory requirements globally.

With the aforementioned developments in continuous manufacturing technologies and shift towards Quality by Design paradigms, there is increasing demand for more advanced model identification and process control strategies in the pharmaceutical industry. Model Predictive Control (MPC) has represented a step-change in the way process industries approach and deploy advanced process control solutions at scales ranging from local unit operations to plant-wide control. In similar fashion, for continuous pharmaceutical manufacturing, Model Predictive Control (MPC) is considered to be a key technology that enables this vision by being able to rigorously address technical requirements such as the control of highly non-linear reactions (e.g., reversible reactions, side reactions, etc.) in the presence of tight constraints.

Within industry, empirical, data-driven models are oftentimes used in MPC, as first-principles models, whilst desirable in some situations, are costly to obtain. This may be especially so in pharmaceutical manufacturing and may not be necessary nor practical. In the context of data-driven approaches, while it is generally recognized that Neural Networks (NNs) have broad-ranging capabilities of approximating non-linear functions, their usage as part of MPC has not found wide applicability.

Our contribution is in demonstrating the applicability of a special class of NNs, termed Recurrent Neural Networks (RNNs), for MPC applications for continuous pharmaceutical manufacturing applications. In particular, RNNs possess inherent state-space structures that naturally lend themselves well for time-series modeling for predictive control. While MPC has been proposed and investigated in the control of CQAs in continuous pharmaceutical manufacturing [5], it has rarely been applied towards the control of the reactor, which is the heart of the process. RNNs in the context of closed-loop MPC has also been studied in but not for Continuous Pharmaceutical Manufacturing.

We study a single, Multi-Input Multi-Output CSTR example (per [6], [7]) which experiences input multiplicity due to reversible kinetics ($A \leftrightarrow R \leftrightarrow S$). The optimal set-point has been carefully selected to optimize production against excessive downstream separations cost. This set-point is close to a point of inflexion where the system gain changes in sign with respect to reactor temperature. We show how a deep RNN can be learned and present associated closed-loop performance results for two scenarios that require the RNN-based Non-Linear Model Predictive Controller to move from either side of the inflexion point towards the desired set-point. As a result of input multiplicity, it is noted that a single linear controller is not expected to work well for the problem of concern [8]. We show favorable results of the RNN-MPC controller when compared against a Non-Linear MPC benchmark that uses the true plant model for control in terms of closed-loop performance and computational time.

Reference:

[1] Glasnov T. Continuous- flow chemistry in the research laboratory. Springer; 1st ed. 2016 edition (June 2, 2016); 2016.

[2] Poechlauer P, Colberg J, Fisher E, Jansen M, Johnson MD, Koenig SG, et al. Pharmaceutical Roundtable Study Demonstrates the Value of Continuous Manufacturing in the Design of Greener Processes. Organic Process Research and Development 2013; 17(12):1472–1478.

[3] Lakerveld R, Benyahia B, Heider PL, Zhang H, Wolfe A, Testa CJ, et al. The Application of an Automated Control Strategy for an Integrated Continuous Pharmaceutical Pilot Plant. Organic Process Research and Development 2015; 19(9):1088– 1100.

[4] Schaber SD, Gerogiorgis DI, Ramachandran R, Evans JMB, Barton PI, Trout BL. Economic Analysis of Integrated Continuous and Batch Pharmaceutical Manufacturing: A Case Study. Industrial & Engineering Chemistry Research 2011 9; 50(17):10083–10092.

[5] Mesbah A, Paulson JA, Lakerveld R, Braatz RD. Model Predictive Control of an Integrated Continuous Pharmaceutical Manufacturing Pilot Plant. Organic Process Research & Development 2017 6; 21(6):844–854. http://pubs.acs.org/ doi/10.1021/acs.oprd.7b00058.

[6] Seki H, Ooyama S, Ogawa M. Nonlinear Model Predictive Control Using Sucessive Linearization - Application to Chemical Reactors. Trans of the Society of Instrument and Control Engineers 2004; E-3(1):66–72.

[7] Koppel LB. Input multiplicities in nonlinear, multivariable control systems. AIChE Journal 1982; 28(6):935–945.

526770 Smart Constrained Model Predictive Control

Abstract:

Process control of chemical plants needs to address a number of objectives including safety, environmental regulations, product quality, energy efficiency, profitability, etc. In the control system design, these objectives are usually translated into three types of control objectives: setpoint tracking, constraint control or zone control and economic optimization. The different control objectives are either handled separately at different levels of a hierarchical framework or altogether in an integrated framework. An important yet challenging task in practice is to obtain desired closed-loop performance via controller parameter tuning, which can be overwhelmingly difficult due to lack of transparency in the tuning process.

Model predictive control (MPC) has been the ideal platform for integrated control system design due to its ability to handle constraints and incorporate economic considerations. The literature is rich with theories for conventional setpoint tracking MPC [1]. The past decade has seen an increasing academic interest in economic model predictive control (EMPC) [2] which integrates economic objectives into regulatory setpoint tracking control. On the contrary, zone control has received less attention. In the MPC framework, zone control is usually dealt with by the so-called soft constraint technique [3-5]. As its name suggests, soft constraint is often dismissed as a trick to avoid feasibility issue with hard constraint, and is addressed separately from the stability analysis of setpoint tracking MPC or economic MPC. To the best of our knowledge, only a few MPC frameworks explicitly handle zone tracking objectives [6-8] with guaranteed stability.

In this work, we propose a smart constrained MPC framework which addresses setpoint tracking, zone control and economic optimization in a unified framework. A zone tracking stage cost which penalizes weighted l1 norm and squared l2 norm distance to the target zone is incorporated into the existing EMPC framework to form a multi-objective optimization problem. We provide sufficient conditions for asymptotic stability of the optimal steady state and characterize exact penalty for the zone tracking cost which prioritizes the zone tracking objective over the economic objective. Moreover, we propose an algorithm to modify the target zone based on the economic performance and reachability of the optimal steady state in the target zone. The modified target zone is constructed as an invariant subset of the original target zone in which closed-loop transient economic performance is guaranteed. EMPC with modified target zone effectively decouples the zone tracking and economic objectives and enjoys a simplified and more transparent parameter tuning procedure. Finally, two numerical examples are investigated which reveal the intrinsic difficulties in parameter tuning for EMPC with zone tracking and demonstrate the efficacy of the proposed approach.

References

[1] Mayne, D.Q.; Rawlings, J.B.; Rao, C.V.; Scokaert, P.O.M. Constrained model predictive control: stability and optimality. Automatica **2000**, 36, 789–814.

[2] Ellis, M.; Durand, H.; Christofides, P.D. A tutorial review of economic model predictive control methods.

Journal of Process Control **2014**.

[3] Scokaert, P.; Rawlings, J.B. Feasibility issues in linear model predictive control. AIChE Journal **1999**, 45, 1649–1659.

[4] Kerrigan, E.C.; Maciejowski, J.M. Soft constraints and exact penalty functions in model predictive control. Proc. UKACC International Conference Control, **2000**.

[5] De Oliveira, N.M.C.; Biegler, L.T. Constraint handing and stability properties of model-predictive control. AIChE journal **1994**, 40, 1138–1155.

[6] Ferramosca, A.; Limon, D.; González, A.H.; Odloak, D.; Camacho, E. MPC for tracking zone regions. Journal of Process Control **2010**, 20, 506–516.

[7] Lez, A.H.G.; Marchetti, J.L.; Odloak, D. Robust model predictive control with zone control. IET Control Theory & Applications **2009**, 3, 121–135.

[8] González, A.H.; Odloak, D. A stable MPC with zone control. Journal of Process Control **2009**, 19, 110–122.

538792 Simultaneous Scheduling of Refinery Manufacturing and Pipeline-Based Multi-Oil Product Distribution

Abstract:

Refinery manufacturing and product distribution systems are highly correlated in the entire refinery supply chain. If they were handled separately, it may cause imbalances or discrepancies between the production and distribution sectors, and thus the loss of the profit margin across the refinery supply chain. On the contrary, the optimal coordination between these two systems could benefit not only the industrial profitability, but also the reduction of both management risks. For instances, based on stable oil products from refineries, inventories of distribution center could be kept in a reasonable lower level to reduce the distribution cost; meanwhile, based on a reliable demand from distribution center, the refinery could conduct a need-based production to avoid unnecessary overproduction. Therefore, refinery manufacturing and product distribution should be and could be optimally integrated if at any possibility. Certainly, such simultaneous scheduling work present a grand challenge to refiners.

Many published studies have addressed refinery production and pipeline distribution system separately, while only a few works have addressed two sectors simultaneously. Existing publications either focused on the strategic and tactical decision level for the upstream and midstream (Azadeh et al., 2017), or sequentially and simply considered the scheduling problem with a discrete framework (Guyonnet et al., 2009, Tong et al., 2012). However, there is still lack of the systematic and integrated study dealing with a detailed continuous-time and continuous-slot scheduling with long-distance pipeline system. In this work, a continuous-time based integrated MINLP model is developed to optimize the scheduling of both refinery manufacturing operation and multi-oil product distribution. The general objective is to maximize the total profit; meanwhile, operation and product specifications, inventory limits, transportation constraints and production demands have to be satisfied. The scope of this work includes two main parts: refinery and long-distance pipeline distribution system. The refinery includes crude distillation, reforming, cracking, hydrotreating, blending, gas processing, and sulfur recovering facilities. The long-distance pipeline distribution system connecting the refinery plant and multiple depots, which transport multi-oil product to different marketing areas. Each depot contains various oil-product storage tanks and mixing tanks dedicated to hold pure products and mixtures respectively. A lot of realistic features such as feed and product blending, multiple depots receiving oil product simultaneously, brine settling time, and interface mixing between two kinds of products inside pipeline have been modelled or constrained. Overall, the simultaneous scheduling model could provide optimal solutions such as processing stream property and flowrate for unit operations, yield and inventory of various petrochemical products, product movements inside pipeline, as well as oil product discharging rate to multiple depot tanks based on consumer demands. The efficacy of the developed integrated scheduling model is demonstrated by a detailed case study.

References

Azadeh, A., Shafiee, F., Yazdanparast, R., Heydari J., and Keshvarparast, A., *Optimum Integrated Design of Crude Oil Supply Chain by a Unique Mixed Integer Nonlinear Programming Model.* Industrial & Engineering Chemistry Research, 2017. **56**: 5734â"5746.

Guyonnet, P., Grant, F.H., Bagajewicz, M.J., *Integrated Model for Refinery Planning, Oil Procuring, and Product Distribution.* Industrial & Engineering Chemistry Research. 2009, **48**: 463â€"482

Tong, K., Feng, Y., Rong, G., *Integrated Model for Refinery Production and Pipeline System Scheduling.* Proceedings of the 22nd European Symposium on Computer Aided Process Engineering, 17 - 20 June 2012, London.

532025 Single & Multi-Objective Optimizations Using Parallelized Process Simulators

Abstract:

Equation-oriented (EO) optimization methods are difficult to apply to rigorous thermodynamic methods, discontinuous variables, or systems where the number of equations is not constant. Despite these challenges, EO methods are still quite commonly used for process optimization, due in part to a perception that EO methods are faster than process simulation-based methods.

By integrating parallel processing techniques into modern process simulation and optimization, rapid results can be achieved comparable to the speed of more traditional EO modeling methods. These speeds can be attained even while using first-principle models, without risking validity of the model through too many simplifying assumptions. In this presentation, several process examples will be considered to explore the advantages of using simulators for optimization projects.

526765 Development of a Spatio-Temporal Multi-Objective Optimisation Model for Multi-Product Oil Palm Value Chains

Abstract:

Oil palm is a versatile and high-yielding crop that can be used to produce a wide variety of products, including oleochemicals, energy and materials [1]. The oil palm fruit, called fresh fruit bunch (FFB), can be milled to produce crude palm oil (CPO) and kernel palm oil (KPO), which are the main feedstocks for the production of oleochemicals and biodiesel. The resulting â€œwasteâ€ streams from the oil production can be further processed into valuable products. The palm oil mill effluent (POME) can be fed into anaerobic digesters to produce biomethane, which can be used to generate heat and electricity. The empty fruit bunches (EFB), on the other hand, can be converted to: syngas via gasification; bio-oil via pyrolysis; bioethanol via lignocellulosic fermentation; fibre mat via mechanical treatment; composts and other valuable products.

In this conference, we will present a spatio-temporal mixed integer linear programming (MILP) model that can simultaneously determine the design and operation of multi-product value chains [2-3] for oil palm. A value chain is a set of activities required to convert raw materials into products and services, including pre-processing and pre-treatment, conversion to intermediates and final products, logistics, inventory and waste management [4]. The model is used to examine and optimise different scenarios in the oil palm sectors in Peninsular Malaysia. In order to capture the spatial-dependencies of the problem, such as the candidate locations for oil palm plantations and processing facilities, and location of demands, Peninsular Malaysia is represented as a grid of 50 km squares. The candidate locations for oil palm plantations are modelled in GIS in order to exclude forests, peatlands and other land areas whose utilisation may result in emissions of stored CO_2. The model considers a long planning horizon, out to 2050, in order to model the staged investment in and retirement of technologies. The model determines interdependent decisions such as where to locate the plantations and processing facilities, when to invest in them and what size/capacity; what products to produce, how to transport and store resources, centralised or distributed production, among others. Different objectives are considered such as maximisation of profit and minimisation of GHG emissions. Pareto sets are generated in order to determine optimal solutions that represent a balance between economic gain and environmental protection.

References:

[1] A. Abdulrazik, M. Elsholkami, A. Elkamel, L. Simon (2017). Multi-products productions from Malaysian oil palm empty fruit bunch (EFB): Analyzing economic potentials from the optimal biomass supply chain. Journal of Cleaner Production, 168, 131-148, 2017.

[2] S. Samsatli, N.J. Samsatli (2017). A multi-objective MILP model for the design and operation of future integrated multi-vector energy networks capturing detailed spatio-temporal dependencies. Applied Energy. DOI:

10.1016/j.apenergy.2017.09.055.

[3] S. Samsatli, N.J. Samsatli, N. Shah (2015). BVCM: a comprehensive and flexible toolkit for whole-system biomass value chain analysis and optimisation - mathematical formulation. Applied Energy, 147, 131-160.

[4] S.M. Jarvis, S. Samsatli (2018). Technologies and infrastructures underpinning future CO_2 value chains: a comprehensive review and comparative analysis. Renewable & Sustainable Energy Reviews, 85, 46-68.

522743 Optimal Synthesis of Reaction Networks for the Manufacture of Benzaldehyde from Toluene Via the P-Graph Methodology

Abstract:

Herein, the optimal and near-optimal reaction networks yielding benzaldehyde from toluene are algorithmically synthesized from a set of reactions of industrial interest. This is accomplished by the judicious adaptation of the principles and methodology of process-network synthesis (PNS) based on P-graphs; thus, the problem explored in this work is termed reaction-network synthesis (RNS) accordingly. In fact, RNS is the first step in optimizing chemical processes via PNS whose design is guided by the P-graph framework. At the outset, the feasible reaction networks are selected from the set of all plausible reactions producing benzaldehyde from toluene available in the literature; this selection is effected by ranking the plausible reactions by their industrial significance. Moreover, the reactions selected are stoichiometrically balanced and depicted by P-graphs, thereby graphically representing the reacting species, i.e., reactants and products. Subsequently, the superstructure formed by combining the P-graphs of the feasible reactions is generated by resorting to algorithm maximal-structure generation (MSG). In addition, the combinatorially and stoichiometrically feasible reaction networks are determined from the superstructure via algorithm solution-structure generation (SSG) and integer linear programming (LP). Finally, the optimal and near-optimal reaction networks are identified from those stoichiometrically feasible by means of an objective function defined in terms of the profit that could be obtained from the reactions. Furthermore, the objective function embeds the reaction networksâ€™ environmental impact by means of suitable metrics, such as explosiveness and aquatic-toxicity potential.

538718 Simulation Approach for Natural Gas Sweetening Using Mixed Amines

Abstract:

Global warming is a major threat to peopleâ€™s and animalâ€™s lives due to the increase of greenhouse gases emissions, especially CO_2 emission. This critical issue is expensive and consume large amount of energy. The purpose of this term project is to extant an inclusive review and illustrative comparison of different amine solvents and amine mixtures available and suitable for removal of acid gas from natural gas the production specifications. A base case of amine base CO_2 removal process from a real plant is used to create a steady state simulation using Aspen HYSYSÂ® V9.0 process design program that can practically predict the CO_2 capture mechanism from the natural gas using DGA, MDEA+DEA and MDEA+PZ solvents. Adding piperazine improved the solubility of CO_2 in low CO_2 partial pressure associated to pure MDEA and DGA. The CO_2 loading increased with increasing CO_2 partial pressure, decreasing temperature, and increasing solvents concentration. In this study, the design models are tested for all the possible key parameters that can affect the performance of the process. Since, achieving the product specification with minimum possible energy requirement is the goal of any process plant, hence, the effect of operational parameters is investigated in terms of CO_2 content in the sweet gas and reboiler duty. Energy and cost requirements have been investigated for more rescannable comparative analysis

532202 Impact of Biomass Densification on the Overall Economics of Renewable Gasoline and Diesel Production

Abstract:

The feedstock to make renewable gasoline and diesel (RGD) fuels via integrated fast pyrolysis and hydrodeoxygenation (IFPH) pathway must be procured from energy crops, forest and crop residues, and forest and organic wastes to avoid potential impacts on food and water security with the bioenergy production. The low bulk density of these renewable feedstocks results in large shipping volumes that create significant logistical challenges. The requirement of huge transportation volumes for bioenergy production can be reduced with the densification of renewable feedstocks before they are processed to make RGD fuels. However, the addition of costs and a change in overall mass and energy yields of RGD fuel production with the biomass densification can influence the overall economic performance of RGD fuel production via IFPH pathway. In this study, we investigated the impact of biomass densification on the overall economic performance of RGD fuel production through the comparison of the techno-economic studies of eight scenarios. In each scenario, either non-densified feedstock (switchgrass/wood chips) or densified feedstock in the form of pellets, torrefied pellets and pyrolysis oil is shipped to a central facility where non-densified and densified feedstock is converted to RGD fuel via (IFPH) pathway. It is found that the densification reduces the transportation cost of renewable feedstock by at least 20%. However, the requirement of capital and operating costs for the renewable feedstock densification increases the minimum selling price of RGD fuel by at least 3% as compared to that of RGD fuel made from non-densified feedstock. This result indicates that the densification of renewable feedstock via pelletization, torrefaction, and pyrolysis is not an economically viable option to produce RGD fuel via the IFPH pathway when there is no competition for procuring renewable feedstock from a land surrounding the biorefinery. Therefore, the influences of feedstock competition and capital cost scaling factor on the impact of distributed biomass densification economics will also be discussed.

529412 A Novel Conceptual Design for Simultaneous Production of Biodiesel and Glycerol Carbonate from Soybean Oil

Abstract:

Biodiesel, a viable, non-toxic and biodegradable alternative to the conventional crude diesel, has prompted growing interest worldwide. Glycerol is the by-product of biodiesel production plant, the rapid increase in demand and production of biodiesel has resulted in excessive glycerol supply in the market. Converting the cheap glycerol into the high-value glycerol carbonate (GC) is a potential solution to the current worldwide excessive glycerol supply. GC, a green substitute for important petro-derivative compounds as ethylene carbonate or propylene carbonate, offers useful applications as a novel component of gas separation membranes, surfactant component, and a multipurpose green solvent for several types of materials. In this paper, a novel process for simultaneously producing biodiesel and GC using soybean oil as the main feedstock has been developed, the annual production capacities of 35.4 kilotonnes/yr of biodiesel and 4.4 kilotonnes/yr of GC are achieved. First, the simultaneous production of the biodiesel and GC has been investigated and simulated, glycerol from the biodiesel production is used as the feedstock for the GC production, while methanol, which is the by-product of GC production, can be supplied as the feedstock of the biodiesel production. Second, the operating conditions and configurations of distillation columns of GC production are optimized via thermal analysis, significant amounts of cooling/heating duties in condensers/reboilers of distillation columns are saved. Next, a mathematic model for heat integration of process streams has been built to minimize total utility costs, and thus a heat exchange network (HEN) is designed based on the optimal solution. Finally, economic evaluation for the simultaneous production of biodiesel and GC has been performed. The studied results show that the simultaneous production of biodiesel and GC is profitable, with a net present value (NPV) of $28,720,000 and a discounted cash flow return rate (DCFRR) of 60%.

537979 Spry Drying System Modelation for Orange (*Citrus sinensis*) Juice Drying Using Open Foam

Abstract:

Drying is a method for food preservation that consists in extracting the water to decrease the water activity and delay degradation reactions. Conventional drying processes involve high temperatures and long residence time, this can decrease the quality of the food in terms of its appearance and organoleptic properties, also, it can lose much of its nutritional value. Spray drying is a method that uses moderate pressure and temperature conditions. It is used to dry thermally sensitive foods, with a solids content lower than 50%. The product obtained is a powder material with a low moisture content and a uniform particle size.

There are 6 variables that control the spray drying phenomenon of mass and heat transfer: 1) air and feedstock flow ratio 2) temperature of the air stream 3) temperature of the feedstock stream 4) atomizer speed 5) type of carryer agent and 6) concentration of the carryer agent. This amount of variables complicate the search of an optimal point of operation, doing the experimental work arduous and expensive. For this reason, the popularity of computational tools like computational fluid dynamics (CFD) has grown in past years, as this tool allows system behavior evaluation for diverse operating conditions with low investment in resources

In this work the drying of a Colombian native orange variety (Citrus sinensis) is evaluated using spray drying at laboratory scale. The operational variables of the process were obtained and validated through computer models for the design of an industrial scale spray drying process, in order to obtain a powder that conserves more than 90% of the nutrients and good organoleptic properties that allow new market opportunities (mainly exportations) to the farmers who grow the fruit.

533064 Tackling the Challenges/Limitations Posed By Heat Exchanger Network in Work-Heat Exchange Network Synthesis

Abstract:

The recent concerns of global warming and exhausting natural resources have encouraged global efforts for increased energy efficiency and conservation. As industry is one of the major energy consumer, tools and methods are being developed and applied to reduce the overall industrial energy consumption. In process industries, heat and work are the two main forms of energy consumption. Heat integration is a well-established tool in process industries to reduce the overall utility consumption. These principles of heat integration can also be extended to work integration. Furthermore, considerable energy savings can be achieved by simultaneously integrating both forms of energy (work and heat) requirements of the process streams.

There are novel techniques available to incorporate the heat integration model in process synthesis and optimization. Duran and Grossmann [1] developed the Pinch Location method for obtaining the minimum utilities within process flowsheet optimization. Kong, et al. [2] developed a generalized transshipment model for unclassified hot/cold streams for simultaneous process synthesis and heat integration. However, the focus of these heat integration models is to obtain the minimum utility consumption rather than designing the network.

In work-heat exchange network synthesis (WHENS), a superstructure for designing heat exchanger network (HEN) is incorporated. This embedded HEN superstructure is of importance. The overall mathematical complexity of the synthesis might increase with the complexity of this HEN superstructure. As a result, simpler HEN superstructures with simplified assumptions are used to reduce the complexity and computation time of the model. These superstructures have some limitations associated with it, which reflects in the final result of WHENS. This may eliminate many industry relevant promising configurations and operating conditions. For example, some of the commonly used superstructures allow utility exchangers only at the end of superstructure. Furthermore, some of the assumptions in the HEN model such as isothermal mixing might eliminate simpler and optimal configurations. One more important challenge is that the identity of the process streams as hot/cold is not known beforehand in WHENS. Hence, it is difficult to classify the streams and enumerate the matches in the HEN superstructure. Thus, novel methods and strategies are required to tackle these challenges and obtain a reasonable solution for the entire problem.

In this work, we consider superstructure based simultaneous WHENS. We have developed a novel framework for work-heat integration in which we do not fix the identity of the streams as hot or cold a priori in heat integration [3]. Furthermore, we used zone-based pressure-temperature dependent properties, allowing liquid, vapor, and 2-phase streams. However, this mixed-integer non-linear programming (MINLP) model uses stage-wise superstructure [4] with isothermal mixing assumption for heat integration. For a case study of C3 splitter, we observe that the reboiler has a utility exchanger in series rather than the preferred and more practical parallel configuration. Thus, we evaluate the effect of the HEN superstructure and model assumptions on such embedded process synthesis and implement novel strategies for a case study of C3 splitter. Overall, we improve the HEN superstructure and relax certain assumptions to improve the result of WHENS.

[1] M. A. Duran and I. E. Grossmann, "Simultaneous optimization and heat integration of chemical processes," *AIChE Journal*, vol. 32, pp. 123-138, 1986.

[2] L. Kong, V. Avadiappan, K. Huang, and C. T. Maravelias, "Simultaneous chemical process synthesis and heat integration with unclassified hot/cold process streams," *Computers & Chemical Engineering*, vol. 101, pp. 210-225, 2017.

[3] S. K. Nair, H. N. Rao, and I. A. Karimi, "Framework for workâ€heat exchange network synthesis (WHENS)," *AIChE Journal*.

[4] T. F. Yee and I. E. Grossmann, "Simultaneous optimization models for heat integration--II. Heat exchanger network synthesis," *Computers and Chemical Engineering*, vol. 14, pp. 1165-1184, 1990.

536656 Uncertainty Analysis Including Safety, Environmental and Economic Performance of Chemical Processes

Abstract:

Traditionally, major decisions for chemical process design have been based on economic and technical analysis. This approach is usually carried out considering average values to key data such as feedstock and product prices, capital and operating costs, as well as process related variables, such as reactor performance or equipment efficiency. The incorporation of uncertainty at the process design stage aims to take explicitly into account uncertainty sources and associated variations of economic and process conditions to provide better solutions as compared to the traditional use of single design variables based on nominal values. In this work, we develop a systematic integrated framework to include uncertainty at the design stage of a chemical process. Furthermore, the typical techno-economic analysis is expanded to include safety and sustainable factors to generate a more comprehensive analysis of the process potential. Within the proposed design framework, an insightful assessment of uncertaintyâ€™s effect on the safety, environmental and economic characteristics of the process are carried out by generating ranges for the performance indicators, supported by probabilistic methodologies. To illustrate the proposed approach, cases of study related to processes to produce ethylene and methanol are considered, where uncertainty is incorporated to the three performance indicators using Monte Carlo techniques. Two variables that affect the three indicators were considered under uncertainty, namely raw materials prices and compressor efficiencies in the methanol synthesis loop for the methanol process, and raw material prices and catalyst yield for the ethylene processes. Detailed performance profiles are reported, and risk-reward zones and inherent trade-offs are highlighted as part of the identification of profitable design options. Also, as a special feature of the approach, implications on key safety properties of the process are examined concurrently.

524133 Design of Carbon-Hydrogen-Oxygen Symbiosis Networks with CO_2 Monetization and Footprint Constraints

Abstract:

Carbon dioxide (CO_2) has significant potential as a chemical feedstock as it is an abundant source of carbon atoms, entails zero or negative acquisition costs and utilization reduces environmental impact. However, significant challenges remain as the use of this feedstock involves unfavorable thermodynamics in reaction and separation, material transportation and end use application. Opportunities exist to overcome these limitations especially within eco-industrial parks which involve a group of facilities cooperating by exchanging materials and energy. An important class of EIPs, Carbon-Hydrogen-Oxygen SYmbiosis Networks (CHOSYN), has been recently introduced to integrate hydrocarbon streams from multiple plants while allowing for chemical conversion (in addition to the conventional exchange, separation, and allocation) through a multiscale targeting approach [1-3]. The objective of this work is to address the problem of designing a CHOSYN with the conversion of CO_2 into value-added chemicals while maintaining carbon-footprint constraints for the whole system. The design and/or retrofitting of CO_2 capture systems is dependent on situational conditions such as source purity, source flowrate, end use specification and availability of waste heat. Previous works addressed the selection and design of CO_2 capture systems and the allocation of CO_2 through optimization frameworks [4]. This study focuses on the synthesis and screening of CO_2 monetization reactions and the development of a simplified and systematic procedure for the selection and design of CO_2 separation processes to connect CO_2 source to CO_2 sink. Each separation process possesses unique characteristics that are applicable to a given source of CO_2 and a sink that will accept a purified CO_2 containing stream. A search tree is developed to guide the selection and design of the separation system given key performance indicators and economic data. Cost correlations are developed from surveyed academic literature and industrial implementations of CO_2 separation systems. Pre-synthesis studies are carried out to identify optimality conditions for various reaction and separation systems. Carbon-footprint constraints are added based on life-cycle analysis of the used sources and energy associated with the various reaction and separation steps. This systematic procedure captures significant interactions between surrounding CO_2 sources and CO_2 sinks while reducing the difficulty of selection and design of the CO_2 capture process. A case study is developed and solved to demonstrate the application of this new method.

References:

[1] Noureldin, M.M.B. and M.M. El-Halwagi, Synthesis of C-H-O Symbiosis Networks. AIChE Journal, 61(4), 1242-1262 (2015)

[2] El-Halwagi, M. M., "A Shortcut Approach to the Multi-Scale Atomic Targeting and Design of C-H-O Symbiosis Networks", Process Integration and Optimization for Sustainability (DOI: 10.1007/s41660-016-0001-y), 1(1), 3-13 (2017)

[3] Topolski, K., M. M. Noureldin, F. T. Eljack, and M. M. El-Halwagi, "An Anchor-Tenant Approach to the Synthesis of Carbon-Hydrogen-Oxygen Symbiosis Networks", Comp. Chem. Eng. (doi.org/10.1016/j.compchemeng.2018.02.024 in press, 2018)

[4] Al-Mohannadi, D. M., and P. Linke, On the systematic carbon integration of industrial parks for climate footprint reduction. *Journal of Cleaner Production*, *112*, 4053-4064 (2016)

528576 Integrating Mass and Heat in the Synthesis of Carbon-Hydrogen-Oxygen Symbiosis Networks

Abstract:

The concept of eco-industrial parks (EIPs) integrating mass and energy among several adjacent plants offers industries synergistic opportunities that can lead to enhanced profit, reduced pollution, and increased conservation of natural resources. A special class of EIPs, Carbon-Hydrogen-Oxygen SYmbiosis Networks (CHOSYNs), has been introduced for the integration hydrocarbon processing plants [1, 2]. The synthesis of CHOSYNs establishes multi-scale benchmarks for the design by tracking the atomic fluxes of carbon, hydrogen and oxygen atoms to develop minimum usage of raw materials and by including reaction pathways that can combine the atoms in a meaningful manner to produce value-added products and intermediates. Recently, the synthesis of CHOSYNs has been expanded upon by introducing "anchor-tenant" models [3]. In this approach, first-to-build and/or existing plants are defined as "Anchors". The potential plants to invite, build and integrate with Anchors are termed as "Tenants". Past CHOSYN studies have focused on increasing resource efficiency through mass integration. Opportunities exist to further increase resource efficiency by incorporating heat integration into the design of CHOSYNs. This work addresses the design of CHOSYN through the inclusion of mass and heat integration in the benchmarking and synthesis of the network configurations. The design of CHOSYNs will be further evaluated by introducing environmental constraints to determine the trade-off between profit and emissions. Alternative heat integration solutions considering intra-plant and inter-plant heat integration are explored. The anticipated outcome of this work are increased savings via the synergistic integration of streams and species, conservation of material and energy, and reduction in environmental emissions. A case study is developed to demonstrate the benefits of this methodology.

References

[1] Noureldin, M.M.B. and M.M. El-Halwagi, Synthesis of C-H-O Symbiosis Networks. AIChE Journal, 61(4), 1242-1262 (2015)

[2] El-Halwagi, M. M., "A Shortcut Approach to the Multi-Scale Atomic Targeting and Design of C-H-O Symbiosis Networks", Process Integration and Optimization for Sustainability (DOI: 10.1007/s41660-016-0001-y), 1(1), 3-13 (2017)

[3] Topolski, K., M. M. Noureldin, F. T. Eljack, and M. M. El-Halwagi, "An Anchor-Tenant Approach to the Synthesis of Carbon-Hydrogen-Oxygen Symbiosis Networks", Comp. Chem. Eng. (doi.org/10.1016/j.compchemeng.2018.02.024 in press, 2018)

535772 Optimal Design of PHAs Plants with Alternative Substrates

Abstract:

In this work we proposed a mixed integer nonlinear programming (MINLP) model for the optimal design of a poly(hydroxyalkanoate)s (PHAs) biorefinery. PHAs are a family of biodegradable polyesters synthesized by certain microorganisms as an energy reserve. PHAs mechanical and thermal properties together with its biodegradable abilities make them promising candidates for sustainable polymer production (Dietrich et al., 2017). The proposed PHAs production process present different technologies embedded within a superstructure for the three main stages: raw materials pretreatment, biosynthesis and biopolymers extraction and purification.

Different potential raw materials are considered as carbon source for pretreatment stage, namely, glycerol, starch and sugarcane. Crude glycerol (glycerol 60.05 wt %, methanol 22.59 wt %, water 10.00 wt %, ashes 2.80 wt %, sodium methoxide 2.62 wt %, soaps 1.94 wt %) can be fed to the biosynthesis stage or it can be purified to improve PHA productivity (Ramos et al., 2017). If glycerol purification step is selected, methanol can be sold as a co-product after distillation. Also selling purified glycerol is set as a possible alternative (García Prieto et al., 2017). Three potential processes are considered for starch obtention. The first one includes starch production process from corn, the second option involves the acquisition of corn starch from market and the third one contemplates the possibility of using cassava starch. In this point, the superstructure considers the possibility of obtaining glucose through the liquefaction and saccharification of starch, to be used as a substrate for the microbial growth in the PHAs production step. Furthermore, a sugarcane-based process for sucrose production is included in the propose superstructure. Sugarcane bagasse, an important residue from sugar industries, can be processed for thermal and electrical energy co-generation. Sugarcane juice (mainly sucrose) can be used as a carbon source for PHAs bioproduction. As alternatives to the mentioned process, the possibility of buying processed sucrose o sugarcane molasses are included in the model. Concerning the biosynthesis, the technologies involved in this stage are highly dependent of the selected substrate. They include a sterilization step and two bioreactors, one for biomass growth and the other one for biopolymer accumulation. Finally, four extraction and purification alternatives are considered in the PHA production process: use of enzymes, solvent, surfactant and NaOCl, or surfactant and chelate.

The proposed superstructure is formulated as an MINLP problem and implemented in GAMS (Brooke et al., 2013) for net present value (NPV) maximization. Model equality constraints include mass and energy balances, yield equations and detailed capital cost for process equipment, while inequality constraints include process and product specifications and operating bounds on process units. The resulting MINLP model for the production of 10,000 t/y of PHA has 8,249 continuous variables, 25 discrete variables, 7,456 constraints and it is solved using DICOPT (CONOPT and CPEX) (Grossmann et al., 2003) in a CPU time of 14,625 s. The optimal configuration includes the use of sugarcane as raw material for the production of PHA employing the enzymatic extraction and purification method. The value of the objective function results NPV= 75.01 MM$. The biopolymer production cost is 3.02 $/kg, which is in concordance to worldwide PHA production cost (nearly 3 $/kg) reported by Koller et al. (2012). The energy consumption for the PHA production is 22,56 MJ/kg, which results similar to other PHAs plants presented by Lopez-Arenas et al. (2017) and Akiyama et al. (2003). Also, other profitability indexes are calculated. In this sense, considering a discount rate of 10 % and a project lifetime of 15 years, the return on investment (ROI) of 22.36 %, the payback period (PP) of 3 years, and the internal rate of return (IRR) of 52.53 % indicate the economic profitability of the project. Finally, we performed a sensitivity analysis in order to point out the technological aspects that could be improved to achieve a higher profit on the biorefinery.

References

Akiyama, M.; Tsuge, T.; Doi, Y. Environmental life cycle comparison of polyhydroxyalkanoates produced from renewable carbon resources by bacterial fermentation. Polymer Degradation and Stability. 2003, 80, 183-194.

Brooke A.; Kendrick D.; Meeraus A.; Raman R. GAMS- A Userâ€™s Guide, Washington, DC, USA. 2013.

Dietrich, K.; Dumont, M. J.; Del Rio, L. F.; Orsat, V. Producing PHAs in the bioeconomy - Towards a sustainable bioplastic. Sustainable Production and Consumption. 2017, 9, 58-70.

GarcÃa Prieto, C. V.; Ramos, F. D.; Estrada, V.; Villar, M. A.; Diaz, M. S. Optimization of an integrated algae-based biorefinery for the production of biodiesel, astaxanthin and PHB. Energy. 2017, 139, 1159-1172.

Koller, M.; Salerno, A.; Tuffner, P.; Koinigg, M.; BÃ¶chzelt, H.; Schober, S.; Pieber, S.; Schnitzer, H.; Mittelbach, M.; Braunegg, G. Characteristics and potential of micro algal cultivation strategies: a review. Journal of cleaner production. 2012, 37, 377-388.

Lopez-Arenas, T.; Gonzalez-Contreras, M.; Anaya-Reza, O.; Sales-Cruz, M. Analysis of the fermentation strategy and its impact on the economics of the production process of PHB (polyhydroxybutyrate). Computers and Chemical Engineering. 2017, 107, 140-150.

Ramos, F. D.; Villar, M. A.; Diaz, M. S. Optimal Design of poly(3-hydroxybutyrate) production using alternative carbon sources. Computer Aided Chemical Engineering. 2017, 40, 877-882.

535794 Optimizing Energy System Design Using a Parallel Tabu Search Algorithm

Abstract:

The goal of this work is to develop novel parallel computing strategies for the solution of multi-objective energy production problems. The Tabu search algorithm provides a hierarchical approach, and the efficacy of the method derives from the ability of the algorithm to navigate to and escape from local minima. In this project, the Tabu search method is applied to Fazlollahi et al.â€™s [1] model for a multi-objective optimization problem involving energy design, which simplifies a potentially complex energy system by classifying energy conversion technologies into six subgroups which can then be optimized. The search employs the methods of Lin et al. [2] to minimize both environmental effects and cost, producing Pareto optimality curves which mimic those produced by the evolutionary algorithm originally utilized by the modelâ€™s author. This algorithm was implemented in the Python programming language. Many problems in terms of design and scheduling of energy systems can be formulated as MINLPâ€™s, while objectives are formulated for cost and environmental impact. In this work, we apply results on both single and multiple processors to show the parallel efficiency of the approach.

References

[1] Fazlollahi, S., Mandel, P., Becker, G., Marechal, F. (2012). Methods for Multi-objective Investment and Operating Optimization of Complex Energy Systems. *Energy, 45, 12-22.*

[2] Lin, B., Chavali, S., Camarda, K., Miller, D.C. (2004). Computer-aided molecular design using Tabu search. *Computers and Chemical Engineering, 29, 337-347.*

538331 Computer-Aided Tools for Process and Product Design

Abstract:

Computer-aided methods and tools for current and future productâ€"process design and development need to manage problems requiring efficient handling of models, data, and knowledge from different sources and at different times and size scales. In this work, the development and features of such tools are presented for both process and product design.

ProCAPD is new product design simulator based on systematic model-based framework for product design and evaluation [1]. ProCAPD is the one and only software that like the well-known process simulators offer options for chemical product evaluation (simulation of their performance). However, unlike the process simulators, also provide options for chemical product synthesis, design and analysis. It also allows access to vast amounts of data that could be useful for chemical substitution. ProCAPD offers the following options:

Database with search engine (chemicals, solvents, active ingredients, aroma, color-agents, etc.), Product evaluation (estimation of a wide range of properties used to evaluate product performance), Single species product design (database search, generate-test approach, mathematical programming), Multi-species product design templates (blends, formulated liquid products, formulated emulsions), Modelling tool-box and New product template design.

ProCAFD is a computer aided tool [2,3] capable of enumerating the entire feasible search space for processing routes, analyzing and determining the most sustainable process. On the highest level of the tool, it has 6 different component structures, which are connected to the main user interface, which interact to provide sufficient information to solve the process synthesis-design problem. They are: 1) Database library for storing, retrieval and addition of information; 2) Flowsheet property model library, which estimates the properties of the entire flowsheet that can be used to benchmark the alternatives; 3) Methods library to solve sub-problems or to generate data required to solve the synthesis problem; 4) Tools library, which has wrapper classes for integrating different in-house tools, which are used to generate results at various levels of the problem solution; 5) Simulation tools library that facilitates the wrapper classes required to interact with simulation packages like Aspen Plus, ProII etc. 6) Graph interface to interact with user.

The frameworks requires integration of different tools, methods, models and database libraries to solve the process-product design problems. The computer aided tools are completely built in visual C# environment, using the concepts of object oriented programming. The tools are developed, keeping in mind the present and future integration needs to further expand the capabilities of the tool.

References:

[1] S. Kalakul, M.R. Eden, R. Gani, 2017, The Chemical Product Simulator â€" ProCAPD, Computer Aided Chemical Engineering, 40, 979-984.

[2] A.K. Tula, M.R. Eden, R. Gani, 2015, Process synthesis, design and analysis using a process-group contribution method. Computers and Chemical Engineering, 81, 245-259.

[3] A.K. Tula, J. Bottlaender, M.R. Eden, R. Gani, 2017, A computer-aided software-tool for sustainable process synthesis-intensification, Computers and Chemical Engineering, 105, 74-95.

528671 Optimal Design of Gas Supply Chains Including Shale with Economic and Environmental Criteria

Abstract:

The recent advances in extraction technologies and the discovery of enormous shale reserves around the world have generated a great interest in exploiting of this resource to satisfy the current increasing energy demand in addition to the low greenhouse gas emissions associated to this fossil fuel. However, shale gas industry is surrounded by several challenges such as the sustainable planification of new gas supply chains considering all the aspects involved in shale gas production (which is relevant mainly for the regions where the shale gas production is not currently carried out but it is planned to start in the short term). Hence, this work proposes the holistic design of a gas supply chain through an optimization approach considering economic and environmental targets in shale gas production. Additionally, the model formulation considers that the fixed gas demands of the markets can be met by shale gas in conjunction with conventional gas as well as the possible importation of gas. Then, the methodology requires unit costs and environmental indicators for each source of gas including its transportation route. Other important aspect accounted by the proposed approach is the optimal selection of the fracking fluid with the purpose to evaluate the performance of a set of fluids that have been efficiently tested in this task. Also, the mathematical model is configured to consider the processing hubs installation for shale gas once it is produced; these types of plants are commonly employed to eliminate undesired residues and even to compress the gas.

Moreover, the economic objective consists in maximizing the total annual profit, which is composed by the gas selling in markets and shale and conventional gas production costs (including the fracking and processing costs), the importation of gas, as well as the transportation costs and the capital costs generated by the required processing hubs. Whereas the environmental aspect deals with the minimization of the global warming, human health and eutrophication categories with the purpose to find a supply chain pathway with the lowest impact It is worth to mention that these indicators were evaluated using the Tool for Reduction and Assessment of Chemicals and Other Environmental Impacts (*TRACI*) technique.

Finally, an example for the case of Mexico is implemented to evaluate the proposed methodology, it should be noticed that Mexico has not yet exploited its shale reserves on a large scale.

536625 Multi-Scale Simultaneous Parameter Estimation in Rate-Based Processes

Abstract:

Unbiased and consistent estimation of model parameters using noisy experimental data is an essential requirement for developing predictive models. While there has been considerable progress in this area in the last few decades, optimal estimation of model parameters is still challenging for reactive solvent-based processes. Simultaneous mass and heat transfer coupled with fast chemical reactions that cannot be distinguished from each other make the parameter estimation problem for these systems considerably more difficult. The typical approach in the literature to circumvent this problem is to develop these models sequentially by designing experiments that attempt to isolate a specific mechanism. For example, a diffusivity model is first developed considering a surrogate system where no reaction takes place. Then, data from a bench scale wetted wall column (WWC) is used to develop models for the mass transfer coefficients and reaction kinetics, where the previously established diffusivity model is assumed to be valid. Finally, an interfacial area model is developed for a given packing based on pilot scale absorber and regenerator data. Another approach is to obtain the mass transfer coefficient model using experimental data from a nonreactive system in the packed tower. Then data for the actual, reactive system can be used to develop the interfacial area submodel.

The traditional sequential approach implicitly assumes that the diffusivity and mass transfer coefficient models obtained from different types of equipment are valid for each type of equipment and system (e.g., that the WWC column and non-reactive system are valid for the reactive system in a packed column). However, the hydrodynamics, liquid and gas velocities, loading of the solvent and operating temperatures can be very different. Also, significant differences in density, viscosity, and surface tension between the reactive and non-reactive systems can affect the wettability and flow characteristics of the fluids, hence the interfacial area. Furthermore, the mass transfer for electrolyte systems is affected by the ionic species present in the solution, ion-molecule interactions, and ion mobility. Therefore, the parameters that are estimated for a given scale or a surrogate system may not be necessarily optimal for the system. Thus, the error in parameter estimation from one step gets propagated to the next step.

In this work, a novel simultaneous parameter estimation technique is developed where parameters for mass transfer, diffusivity, interfacial area, and reaction kinetics are optimally estimated leveraging noisy, experimental data from multiple scales (mm-scale to m-scale) and operating conditions using high-fidelity process models (Chinen et al., 2018). The proposed approach leads to a large-scale optimization problem consisting of a large number of parameters and millions of linear and nonlinear equality and inequality constraints. The optimization problem is solved using the Institute for the Design of Advanced Energy Systems' (IDAES) computational framework that facilitates large-scale parameter estimation at multiple scales embedded in complex process models and property submodels while utilizing data from various sources and scales (Eslick et al., 2018). The software platform provides modular process modeling capabilities using the Pyomo algebraic modeling language (Hart et al., 2011) and the PySP module in Pyomo for stochastic programming and parameter estimation (Watson et al., 2012). Our previous work focused on the parameter estimation problem for a MEA-H_2O-CO_2 system where the data from the WWC and pilot plant were used (Eslick et al., 2018). The Non-random-two-liquid (NRTL) model was implemented to model the vapor-liquid equilibrium at the interface of the liquid/gas films. Even though the WWC and packed tower models were rigorous in that work, a simple mass transfer model was used to describe the CO_2 and H_2O transport across the liquid and gas films. In particular, it was assumed that the change in the concentration driving force is linear across the films. Chemical reactions in the liquid film were not explicitly considered in that model, rather an enhancement factor was used to capture the enhancement in mass transfer due to the chemical reactions. However, due to the fast ionic reactions in the liquid film of the MEA-H_2O-CO_2 system, complex thermodynamics of this electrolyte system, and interactions between the molecular and ionic species, the CO_2 concentration profile in the liquid film is highly nonlinear. Therefore, the parameter estimates under linear assumption can be inaccurate. In the updated model, simultaneous mass transfer and chemical reactions are considered in the liquid film. Multicomponent mass transfer in the liquid and gas films is modeled by the Maxwell-Stefan equations. Furthermore, a rigorous electrolyte NRTL model is implemented to model the vapor-liquid equilibrium at the interface of the liquid/gas films. Experimental data sets are also expanded to include the data from the WWC, bench-scale, and pilot-scale systems extending the richness of the experimental data considerably. These modifications increased the size of the optimization problem substantially. Model reformulations are carried out and novel solution strategies are developed to solve the parameter estimation problem within reasonable time. Various approaches investigated in this work includes direct simultaneous optimization, decomposition via PySP and an approach using cross-validation and statistical inference. It is observed that the variance in the parameter estimates can be substantially improved by including the data from multi-scale systems simultaneously.

A. S. Chinen, J. C. Morgan, B. Omell, D. Bhattacharyya, C. Tong, D. C. Miller, 2018, Development of a gold-standard model for solvent-based CO2 capture. Part 1: hydraulic and mass transfer models and their uncertainty quantification, (internal review)

J. C. Eslick, P. T. Akula, D. Bhattacharyya, D. C. Miller, 2018, Simultaneous Parameter Estimation in Reactive-Solvent-Based Processes, PSE 2018 conference San Diego.

J.-P. Watson, D. L. Woodruff and W. E. Hart (2012). PySP: modeling and solving stochastic programs in Python. Mathematical Programming Computation **4**(2): 109-149

W. E. Hart, J.-P. Watson and D. L. Woodruff (2011). Pyomo: modeling and solving mathematical programs in Python. Mathematical Programming Computation **3**(3): 219-
260

536707 Development of a One-Dimensional Bubbling Fluidized Bed Model for a Coal-Fed Chemical Looping Combustion Fuel Reactor

Abstract:

Chemical looping combustion (CLC) is an advanced combustion technology which has been receiving increasing attention in the research community. In CLC, a solid oxygen carrier (OC) circulates between an air reactor and a fuel reactor. In the fuel reactor, the OC undergoes a reduction reaction as it oxidizes the fuel to produce a flue gas (mainly H_2O and CO_2); with the reduced OC then transferred to the air reactor where it is oxidized by O_2 in air, before being recirculated back to the fuel reactor to complete the loop.

CLC provides some advantages for the generation of low carbon energy, as the CLC process produces a CO_2 rich flue gas without the additional energy and capital costs associated with post-combustion capture from conventional fossil-based power plants, thus potentially leading to lower costs of electricity in comparison to conventional capture processes.

Past CLC research primarily focused on gaseous fuels; however, in the last decade there has been an increasing emphasis on adapting chemical looping technology for the direct combustion of solid fuels such as coal and biomass. The direct CLC of solid fuels avoids the extra processing step associated with the indirect CLC of solid fuels where gasification of the solid fuel to syngas (mainly CO_2 and H_2) is done before the CLC of the resulting syngas. Though the potential for direct CLC of solid fuels like coal have been successfully demonstrated in several pilot scale demonstration facilities, there have been no commercial scale plants to date (Adanez et al. 2018).

One way to accelerate the development of CLC technology is through the use of mathematical models. These models can provide a representation of the physical phenomena that might be difficult or expensive to measure in physical models. Furthermore, these models can be used as part of plant-scale modeling studies to identify optimal design and operating conditions for further investigation in bench and pilot-scale experiments.This work presents the development of a steady-state, one-dimensional, bubbling fluidized bed model of a coal-fed CLC fuel reactor that is suitable for both simulation and optimization type studies. Mass and energy balances of the gas and solid phases in the model are developed from first principles, while semi-empirical correlations are used for hydrodynamic, mass and heat transfer computations. The methodology used for this model development effort is adapted from prior work on gas-fired CLC (Okoli et al. 2018), with changes made to the reaction and hydrodynamic sub-models to account for the use of a solid fuel. The resulting model consists of a set of differential algebraic equations, and accounts for the gas and solid phase interactions in the reactor, such as the devolatilization and gasification reactions of coal, as well as the reactions between the OC and the evolved gases from the coal reactions. Furthermore, the model considers the mass and heat diffusion phenomena occurring in the bubble, cloud-wake and emulsion regions of the reactor. The output of the model includes the axial profiles of the pressure, velocities, concentrations, flows and temperatures in the reactor.

The modelâ€™s capabilities are demonstrated for the simulation of different coal-fed CLC fuel reactor configurations with an iron-based OC. In addition, the impacts of key design variables on the reactorâ€™s performance are also evaluated.

The model is built using the Institute for the Design of Advanced Energy Systems (IDAES) open-source equation-oriented process systems engineering framework which enables the rapid development and optimization of next generation advanced energy systems (Miller et al. 2018). Its modular nature allows for easy reuse and integration of different models and solvers, thus allowing the quick assembly of flowsheets for large scale simulation and optimization studies. The IDAES platform utilizes Pyomo (Hart et al. 2017); a Python based algebraic modeling language and incorporates the object-oriented features of Python.

References

J. Adanez, A. Abad, T. Mendiara, P. Gayan, L.F. de Diego, F. Garcia-Labiano, 2018. Chemical looping combustion of solid fuels, Progress in Energy and Combustion Science, 65, 6 - 66.

W.E. Hart, C. D. Laird, J. P. Watson, D. L. Woodruff, G. A. Hackebeil, B. L. Nicholson and J. D. Siirola, 2017, Pyomo â€" Optimization Modeling in Python, 2nd ed., Springer, Vol. 67.

D.C. Miller, J.D. Siirola, D.A. Agarwal, A.P. Burgard, A. Lee, J.C. Eslick, B.L. Nicholson, C.D. Laird, L.T. Biegler, D. Bhattacharyya, N.V. Sahinidis, I.E. Grossmann, C.E. Gounaris, and D. Gunter, 2018. Next Generation Multi-Scale Process Systems Engineering Framework, *Proceedings of the 13th International Symposium on Process Systems Engineering*, San Diego, California, USA.

C.O. Okoli, A. Lee, A.P. Burgard, D.C. Miller, 2018. A fluidized bed process model of a chemical looping combustion fuel reactor, *Proceedings of the 13th International Symposium on Process Systems Engineering*, San Diego, California, USA.

523249 A New Optimization-Based Computer-Aided Molecular and Mixture Design (OptCAMD) Framework

Abstract:

Abstract

Chemical engineering has been expanding its focus from process-centered products such as ethanol and ammonia to chemical-based products such as mosquito repellent sprays and blended fuels [1]. Modern society requires many chemical-based products for its survival, such as fuels, plastics, detergents, medicines and foods [2]. The sustainability of modern society depends on the continuous availability of these products. Nowadays, chemical-based product design and development has become a key topic in chemical engineering since high value-added products, green chemistry and product-process sustainability have started to receive increasing focus in research and development. Various requirements have to be satisfied for different product design problems, which are hard to be determined from limited number of experiments, considering the cost of time and money. At the same time, potential products may be missed with empirical and knowledge-based design methods. Therefore, it is important to develop a systematic way to determine the optimal products based on the appropriate product attributes that need to be converted to property constraints. The design of molecular products is the foundation of all chemical-based product design problems. Even with established molecular design methods, the mathematical formulation and solution of the design problem for a wide range of chemical-based products is still a tough problem, considering the diversity of products and their requirements, huge number of isomers, and the availability of models for the prediction of the needed properties. Therefore, it is necessary to developed new molecular and mixture product design tools.

In this paper, an optimization-based framework for molecular and mixture product design is presented. The molecular product design problem is decomposed into preliminary design, isomer generation and evaluation and verification steps in this framework. In the preliminary design step, an optimization-based mathematical programming model is established to generate feasible sets of molecular groups. In the isomer generation step, a SMILES-based isomer generation strategy is proposed to generate all isomers. In the evaluation and verification step, the properties of the isomers are verified using database and software with rigorous property model, to evaluate the availability of the generated molecules and accuracy of the properties, and further requirements including unconventional properties and process constraints are then evaluated. In the case of mixture design, a MILP/MINLP problem is formulated to find the optimal mixture (chemical identity and their composition) that satisfy a set of desired target properties. Typical examples, are tailor-made fuel blends. The design work-flows for molecular design and mixture design have been implemented into a chemical-based product design tool â€œOptCAMDâ€, which is available as an option in ProCAPD [3], the chemical product design tool. Several design case studies are presented to illustrate the application of OptCAMD involving molecular and mixture design problems.

References

[1] Zhang, L., Fung, K. Y., Zhang, X., Fung, H. K., & Ng, K. M. (2017). An integrated framework for designing formulated products. *Computers & Chemical Engineering*, 107, 61-76.

[2] Zhang, L., Babi, D. K., & Gani, R. (2016). New vistas in chemical product and process design. *Annual Review of Chemical and Biomolecular Engineering*, 7, 557-582.

[3} Kalalul, S., Zhang, L., Fang, Z., Choudhury, H. A., Intikhab, S., Elbashir, N., Eden, M. R., Gani, R. (2018). Computer-aided chemical product design â€“ ProCAPD & tailor-made blended products. *Computers & Chemical Engineering*. DOI: 10.1016/j.compchemeng.2018.03.029.

528023 The Optimization of Integrated Energy System Under Uncertainty Based on Genetic Algorithm

Abstract:

The integrated energy system is a multi-energy flow coupling system which includes the energy production, transportation, distribution, conversion and storage processes. It has greatly improved the efficiency since the system achieves cascade utilization of energy. The cold, heat and electricity can be transformed into each other in the system for the peak load shifting. In order to solve the increasingly serious resources and environmental problems, many countries regard the integrated energy system as the future development strategy. However, the design of the integrated energy system are quite complex because of the obvious volatility and uncertainty of the renewable energy and energy load.

In this work, the mathematical model of integrated energy system was established which included wind power generation, solar power generation and energy storage technology. According to the energy demand characteristics and the renewable energy resources conditions, the uncertain parameters of the model were presented based on historical data. In this paper, a database which contains various device parameters such as the price and equipment efficiency was set up, concerning the current development status of related energy equipment. Furthermore, an corresponding equations were also given to describe the relationship between equipment capacity and these parameters. Moreover, the genetic algorithm was improved to optimize the system design parameters based on the uncertain parameters. Finally, the optimized design of the integrated energy system was obtained which contains the system composition and equipment parameters

536247 Application of Sequential Design of Experiments (SDoE) to a MEA-Based CO_2 Capture Pilot Plant

Abstract:

The U.S. DoEâ€™s Carbon Capture Simulation for Industry Impact ($CCSI^2$) is a partnership among national laboratories, industry, and academic institutions, focused on the application and enhancement of a suite of computational tools and models developed in its predecessor project, the Carbon Capture Simulation Initiative (CCSI). The CCSI Toolset was developed to accelerate the development and deployment of novel carbon capture technologies in part by reducing the risk associated with process scale-up through the use of fundamental predictive models with full uncertainty quantification (UQ).

As a part of CCSI, a modeling framework for CO_2 capture solvent systems was developed with the capability to quantify uncertainty in submodel (e.g. physical properties, mass transfer, kinetics) parameters. In this work, this model is used to identify optimal test conditions through a sequential design of experiments (SDoE) implemented in a pilot plant. Due to the expense of pilot-scale testing, it is important to allocate resources to maximize learning during the test period. In a typical experimental design, test cases are chosen using a space-filling approach that does not consider the process output space or leverage the collected process data to update the test plan. The SDoE approach, however, incorporates the process model and UQ to optimally select experiments, to achieve a specific objective or goal for the test campaign. The data collected are incorporated into a Bayesian framework in which model uncertainty can be quantified and with that knowledge used to improve the model by reducing its parametric uncertainty. A new set of experiments are then selected using the updated model, thus utilizing the knowledge gained in the previous test runs.

The objective function of the proposed SDoE in this work seeks to minimize the cost of CO_2 capture and considers both capital and operating costs as well as thermodynamic improvements. Moreover, the CO_2 capture percentage is treated as an optimization variable in this objective function, whereas this is often considered fixed at a typical level (85-90%) in many campaigns. To reduce the computational expense of the SDoE procedure, multiple-input/multiple-output surrogate models are developed for the absorber and the stripper columns. These reduced models accurately predict the output variables of interest over the operating and parametric spaces of interest, effectively reproducing the rigorous column model results within 5% error for most cases. The SDoE will be implemented at Norwayâ€™s Technology Centre Mongstadâ€™s 12 MWe scale pilot facility in an upcoming test campaign.

541781 Process Synthesis and Simultaneous Minimization of Inherent Risk

Abstract:

Although process synthesis has received significant attention during last decades, only a few studies on process synthesis with simultaneous consideration of inherent safety have been performed. The safety of processes and their subsystems, however, is among the most important considerations for obtaining operable process plants. The safety can be efficiently enhanced if it is considered at the early stages of the design, as implementation of safety principles and measures during early decision making is both more effective and less costly. The aim of this study is to develop a risk assessment method by which it would be possible to obtain an appropriate trade-off between process economy and inherent risk. A mixed-integer nonlinear programming (MINLP) model has thus been formulated where the inherent risk expressed by monetary cost terms is imbedded in the objective function defined as yearly profit or net present value. The inherent risk is defined as an annual expected cost by multiplication of frequency of failure and estimated money losses due to expected severity of consequences. The developed approach is illustrated by two examples, i) the synthesis of a heat exchanger network and ii) the synthesis of an overall process scheme. The results indicate that the simultaneous approach enables obtaining designs which are significantly safer and yet still economically viable.

532153 Quantitative Risk Assessment of Soft Sensor Predictions Using Fast PDF Estimation

Abstract:

Soft sensors (also referred to as virtual sensors) are algorithms that provide estimates of unmeasurable quantities, based on mechanistic/empirical models and on experimental measurements of other measurable quantities. More specifically, consider a process including both measurable and unmeasurable variables, and assume we can build a parametric model of such process. The rationale of conventional soft sensors involves two steps in series: (I) calculation of the optimal values of the most important parameters of the process model from prior and/or current data and (II) use of the resulting model to compute estimates of some key unmeasurable process variables.

Virtual sensors find application in many different industrial sectors, including but not limited to the chemical and pharmaceutical industry, and can be utilized for both monitoring and closed-loop control purposes. In chemical plants, they are responsible for most of the inferential composition measurements, e.g. top/bottom tray composition in distillation columns[1], and may as well be used to detect fouling and catalyst deactivation issues. In the pharmaceutical sector, soft sensors are equally important because they allow not only estimation of key unmeasurable process variables, e.g. API (Active Pharmaceutical Ingredient) temperature in freeze-drying processes[2], but also indirect evaluation of essential product quality indicators, e.g. dosage API loading or tablet tensile strength.

Conventional virtual sensors only provide point estimates of unmeasurable process variables, possibly complemented with minimal statistical information. Typically, the latter consists of the variance on the estimates of the unmeasurable variables of interest, computed under restrictive assumptions: (I) use of a linear/linearized version of the full-fledged process model and (II) presence of random uncorrelated errors on the measurements of the (measurable) process states. Although these conventional virtual sensors often perform satisfactorily in many process applications, pharmaceutical systems may require more advanced soft sensors, which provide extensive statistical information on the estimates of the unmeasurable process variables of interest. This type of information may be used for risk analysis purposes and, specifically, for quantitative risk assessment.

Figure 1: Simplified architecture of the new type of soft sensor proposed in this work.

This paper reports on the development of a new type of soft sensor, which allows estimation of the marginal PDFs (Probability Density Function) of unmeasurable process variables, at a reasonable computational cost. This new type of sensor relies on a combination of dynamic data reconciliation techniques[3] and fast PDF estimation strategies[4] (see Figure 1). Specifically, first, all the available measurements of the manipulated variables, of the disturbances and of the (measurable) states of the process are used to compute the parameters of the process model as well as the reconciled trajectories of manipulated variables and disturbances. Then, these intermediate results are sent to a fast PDF

estimation algorithm, which computes the joint PDF of the model parameters. Finally, this joint PDF is converted into the marginal distributions of the unmeasurable process variables of interest, using numerical integration techniques (

) and/or a combination of ad-hoc mapping strategies and dynamic simulation ( ). Note that, although this new type of virtual sensor makes use of parallel computing techniques, it may not be suitable for closed-loop control purposes. On the other hand, it is valuable for monitoring purposes, allows rapid identification of off spec product in continuous manufacturing lines, and provides very useful information for risk analysis/assessment.

The new type of soft sensor, proposed in this work, is demonstrated on a pilot scale plant for continuous production of tablets, which includes multiple feeders and blenders, a roller compactor and a tablet press. More specifically, we show how this new sensor can be employed to estimate the marginal PDFs of several important unmeasurable process variables, at an acceptable computational cost.

References

1. Ahmadreza G., Mehdi S., Gholamabbas S., 2015, Soft Sensor Development for Distillation Columns Using Fuzzy C-Means and the Recursive Finite Newton Algorithm with Support Vector Regression (RFN-SVR), Industrial & Engineering Chemistry Research, 54, 12031-12039.

2. Bosca S., Barresi A.A., Fissore D., 2014, Use of soft sensors to monitor a pharmaceuticals freeze-drying process in vials, Pharmaceutical Development and Technology, 19, 148-159.

3. Martinez Prata D., Schwaab M., Lima E.L., Pinto J.C., 2009, Nonlinear dynamic data reconciliation and parameter estimation through particle swarm optimization: Application for an industrial polypropylene reactor, Chemical Engineering Science, 64, 3953-3967.

4. Rossi F., Mockus L., Manenti F., Reklaitis G., expected 2018. Assessment of accuracy and computational efficiency of different strategies for estimation of probability distributions applied to ODE/DAE systems. Computer Aided Chemical Engineering, accepted.

536724 Development of Artificial Lift Infrastructure Plan Under Endogenous and Exogenous Uncertainties

Abstract:

Development of Artificial Lift Infrastructure Plan under Endogenous and Exogenous Uncertainties

Zuo Zeng, Selen Cremaschi

Department of Chemical Engineering, Auburn University, Auburn, AL 36849, USA

selen-cremaschi@auburn.edu

Shale gas has become a significant source of natural gas in the United States. The U.S. Energy Information Administration (EIA) estimates that in 2017, shale gas provided over 60% of U.S. dry natural gas production compared to its 1% share in 2000 [1]. Because large amounts of fluids are injected to the shale formation during the fracturing process, shale gas wells often require deliquification to unload the well relatively quickly, generally within their first or second year of production [2]. Artificial lift methods (ALMs) are used in horizontal shale-gas-producing wells to lift the accumulated fluids in the well and to help sustain well performance. Artificial lift methods can be divided into two types: passive and active. Passive systems, such as velocity strings, plunger lift and foam lift, mainly naturally carry and remove liquid from the wellbore because sufficient energy remains in the reservoir. The active systems, such as sucker rod pump, well head compressors and gas lift, add energy to the system, and are generally implemented once the reservoir pressure drops below the wellhead pressure [3].

The artificial lift infrastructure plan (ALIP) includes which ALM should be installed, when exactly the selected ALM should be installed and how long the installed ALM should be operated to maximize the net present value (NPV) of the well. Selecting an appropriate ALM for gas well deliquification mainly depends on the well geometry, operating conditions, and well performance. Constructing an artificial lift infrastructure plan is complicated due to uncertainties in economic factors, such as natural gas price, and in production rates of the well.

In this work, we present a multistage stochastic programming model to develop artificial lift infrastructure plans (ALIPs) for shale-gas producing horizontal wells. The model discretize the planning horizon into equal time periods (corresponding to one month) $t = 1, 2, 3 \ldots T$, where period t starts at time t-1 and ends at time t. The decision variables are $w_{i,p,s}$ and $z_{i,t,s}$, which track the installation (p T) and removal t T time periods of ALM i I under scenario s S, respectively. One more binary variable, $y_{i,t,p,s}$, is introduced, which is equal to one if ALM i is installed on period p and uninstalled on period t in scenario s. The sequencing constraints ensure that each ALM is installed and uninstalled at most once, only one ALM is installed and uninstalled at any given month, and each ALM is uninstalled only after its installation time. Each ALM has defined design and operational limitations such as the ones specified in the Artificial Lift R&D Council guidelines [5] and in typical attribute tables [6]. These limitations are incorporated as technical limitation constraints in the model.

The model incorporates the impact of exogenous natural gas price uncertainty, which is independent of decisions, and of endogenous uncertainty associated with the production rate of the well, which depends on the selected ALM. We assume that natural gas price follows a multiplicative binomial process over discrete periods [7], and that its rate of change (ROC) has two possible outcomes $\{H^P, L^P\}$ with equal probabilities. Figure 1 represents the scenario tree for natural gas prices. The natural gas price at initial stage is given by the parameter P_1. We assume that the price uncertainty is realized every four months, resulting in a total of $st = T/4$ stages for the exogenous uncertain parameter realizations, where T is length of the planning horizon in months. Further, we assume that the two outcomes of ROC H^P and L^P are constant throughout the planning horizon and equal to each other. We define $_{st}$ as the random variable associated with natural gas price at stage st $T' = \{1,2,\ldots,T/4\}$.

Figure 1: Natural gas price scenario tree

The production rate of a horizontal well is described using a hyperbolic curve [5]. At first, the well is producing naturally, represented with the solid curve in Figure 2. After liquid loading, the production from the well becomes unstable, and is halted $(t = p - 1)$ unless an appropriate ALM is installed to stabilize it. Dashed line in Figure 2 represents a potential production rate after ALM i is installed at time p. We define the ratio of the production rate at time p (with ALM i installed) to the production rate at p-1 (the production right before ALM i installation) as the flowrate change ratio, Qrc_i, of ALM i.

Figure 2. Production decline curve when ALM i is installed.

The flowrate change ratio is the endogenous uncertain parameter of the model. Let ξ_i represent the random variable associated with the endogenous uncertain parameter Qrc_i. We assume that the random variable ξ_i has two outcomes with equal probabilities, i.e., $\Omega_i = \{H^{Qcr}, L^{Qcr}\}$. The full scenario set for the MSSP model is constructed using the Cartesian product of outcomes of endogenous and exogenous uncertain parameters, and it is defined as $S := \{\times_{st\ \ T'}\ \ st\} \times \{\times_{i\ \ I}\ \xi_i\}$.

We apply the proposed MSSP model to develop ALIPs for two wells in Woodford shale field. Well I has high average production conditions with a gas rate of 800 thousand standard cubic feet per day (Mscf/D) and a liquid rate of 520 BPD. Well II has a low liquid flowrate and a gas production rate of 200 Mscf/D. We consider 12, 16, 20, and 24-month planning horizons for both wells. We first solve the deterministic versions of the problems constructed using the expected values of all uncertain parameters. The optimal solution of Well I recommends Sucker Rod Pump (SRP) method for all planning horizons. For Well II, which has a low liquid flowrate, the optimal ALM is SRP if the planning horizon is 12, 16 and 20 months, and Well Head Compression (WHC) for the planning horizon of 24 months. The optimal solutions to the deterministic problems reveal that different ALMs may be installed for different planning horizons indicating a trade-off between flowrate change ratios and low operating costs among ALMs. The MSSP models have 64, 128, 256 and 516 scenarios for 12, 16, 20, and 24-month planning horizons. The deterministic equivalent of the MSSP models (which are mixed integer linear programming models) cannot be solved to optimality for planning horizon longer than 12 months within 36000 CPUs. In 12-month planning horizon, the optimal solution for Well I recommends to install SRP at the beginning and install the Electrical Submersible Pump (ESP) at month 7 for all scenarios. To obtain feasible solutions for all instances, we apply the generalized knapsack problem based decomposition algorithm (GKDA) [9]. The GKDA decomposes both time periods and scenarios of the original MSSP into a series of knapsack sub-problems and solves these problems at appropriate decision points of the planning horizon. The GKDA yields feasible solutions within 60 CPUs for all instances. For example, for the 12-month planning horizon, the GKDA selects SRP at initial time period and selects ESP at month 7, which is the optimal solution of the MSSP.

References

1. https://www.eia.gov/tools/faqs/faq.php?id=907&t=8.

2. Bondurant, A. B., Dotson, B. D., and Oyewole, P. O. (2007). "Getting the Last Gasp: Deliquification of Challenging Gas Wells," SPE paper 11651 presented at The International Petroleum Technology Conference, Dubai.

3. Oyewole, P. O., Lea, J. F. (2008). Artificial-lift selection strategy for the life of a gas well with some liquid production. SPE Annual Technical Conference and Exhibition held in Denver

4. Zeng, Z., & Cremaschi, S. (2017). Artificial lift infrastructure planning for shale gas producing horizontal wells. Proceedings of the FOCAPO/CPC, Tuscan, AZ, USA, 8-12.

5. http://beta.alrdc.com/recommendations/gas%20well%20deliquification/artificial%20lift%20selection%20---%20new%20version.htm

6. Weatherford. (2013). Introduction to Artificial Lift Systems. <http://www.alrdc.com/recommendations/Gas%20Well%20Deliquification/artificial%20lift%20selection/WFT%20Intro%20to%20Artificial%20Lift%20Systems.pdf> (accessed 10 April 2018.)

7. Cox, J. C., Ross, S. A., & Rubinstein, M. (1979). Option pricing: A simplified approach. Journal of financial Economics, 7(3), 229-263.

8. M. J. Fetkovich, E. J. Fetkovich, and M. D. Fetkovich, 1996, Useful Concepts for Decline Curve Forecasting, Reserve Estimation, and Analysis. Society of Petroleum Engineers 11(01): 13-22.

9. Zeng, Z., & Cremaschi, S. (2018). A generalized knapsack-problem based decomposition heuristic for solving multistage stochastic programs with endogenous and/or exogenous uncertainties. *Industrial & Engineering Chemistry Research* (In review).

535154 Bayesian Design of Experiments for Fault Detection and Isolation

Abstract:

Fault detection and isolation (FDI) is a challenging process in modern cyber-physical systems (CPS), which are often characterized by complexity and corresponding parametric and environmental uncertainty.[1] When feasible, optimal test design is necessary for the FDI of these systems, due to the limitations in maintenance budget, time, and the size of datasets available. In systems with capability for down-time or relaxed performance requirements, optimal tests can be designed by manipulating the system input trajectories with the objective to maximize the information extracted from available system sensors.[2] However, due to the deterministic nature of CPS models and the finiteness of the inputs selected to execute FDI tests, model-based active FDI approaches are vulnerable to uncertainty and error.[3] In most approaches, FDI test designs vary with the parameter uncertainty space. For instance, approaches that utilize metrics of the Fisher Information Matrix or Shannon Information are subject to assumptions with respect to the neighborhood around which numerical derivatives expressing parametric sensitivities are calculated. In the majority of methods, these derivatives are, by design, calculated locally, assuming that they hold accurately for the entirety of the parameter space.[4] The latter, cannot be proven and, in fact, it is not correct for certain classes of problems (e.g., system models that are non-linear with respect to their parameters). To address this issue, this work proposes the use of Bayesian approaches for the design of FDI tests, capable of maximizing test information for the entire model parameter space.

The implications of assuming no dependence of the information extracted from a test on the parameter estimates and the advantages of Bayesian approaches will be illustrated in various implementations of the benchmark 3-tank system model.[5] We focus this discussion and presentation on the dependence of the system sensitivity matrix on the assumption about the neighborhood around which the sensitivities are estimated using the system model. We showcase that depending on the type of system (non-linearities involved) and the number of errors and uncertainty, assumptions about the effect of our anticipation of fault severity can lead to unique or multiple optimal designs. The latter leads to uncertainty in the decision process for the execution of one set of tests, and lack of certification for the detectability of faults, using model-/sensitivity-based approaches.

Acknowledgment

This work was sponsored by the UTC Institute for Advanced Systems Engineering (UTC-IASE) of the University of Connecticut and the United Technologies Corporation. Any opinions expressed herein are those of the authors and do not represent those of the sponsor.

Literature

1. Venkatasubramanian, V., Rengaswamy, R., Yin, K. & Kavuri, S. N. A review of process fault detection and diagnosis. *Comput. Chem. Eng.* **27,** 293–311 (2003).

2. Franceschini, G. & Macchietto, S. Model-based design of experiments for parameter precision : State of the art. **63,** 4846–4872 (2008).

3. Venkatasubramanian, V., Rengaswamy, R., Kavuri, S. N. & Yin, K. A review of process fault detection and diagnosis Part III: Process history based methods. *Comput. Chem. Eng.* **27,** 293–311 (2003).

4. Palmer, K. A., Hale, W. T., Such, K. D., Shea, B. R. & Bollas, G. M. Optimal design of tests for heat exchanger fouling identification. *Appl. Therm. Eng.* **95,** 382–393 (2016).

5. Mesbah, A., Streif, S., Findeisen, R. & Braatz, R. D. *Active fault diagnosis for nonlinear systems with probabilistic uncertainties. IFAC Proceedings Volumes (IFAC-PapersOnline)* **19,** (IFAC, 2014).

539475 Design of Optimal Multistage Heat Exchange Networks

Abstract:

Design of Optimal Multistage Heat Exchange Networks

N.N. Ziyatdinov, I.I. Emelyanov, A.N. Bezrukov, and D.V. Kubanov

Kazan National Research Technological University, Kazan, Russia

nnziat@yandex.ru

Development of an optimal heat exchange system is a complex combinatorial problem with a large number of dimensions due to multiple heat sources and outputs in technological units. The studies in the last several decades resulted in the development of a series of approaches and methods of heat integration which provide significant decrease in the value of the sum of capital and energy costs. We offer a series of modifications of a classic decomposition approach of the mathematic programming which is related to the "assignment" problem. It should be considered as a multilevel iteration procedure with the method of fixing of intermediate variables. The lower level activities are the search and optimization of structurally interconnected units which cool the "hot" flows and heat the "cold" flows to the given temperatures. To formalize the heat exchange optimal economy problem, each pair of technological flows is offered to be related to a certain superstructure of heat exchangers with a set of structural subunits – elementary blocks of a heat exchange system [1]. An elementary block is a structural subunit allowing transfer of a certain given amount of heat to the j-th "cold" flow and the extraction of a certain given amount of heat from the i-th "hot flow". The problem of optimization of elementary blocks of a heat exchange system can be solved by nonlinear mathematic programming models with the limitations to an admissible force of a heat exchange process.

The second solution level of the discrete linear programming "assignment" problem is finding the single stage heat exchange system structure. A single stage heat-exchange system as a system with a single j-th "cold" flow assigned to the each i-th "hot" flow. The optimal structures of elementary blocks of a heat exchange system and control variables are found for a set of such systems. The input data for such a problem are the matrix of economic evaluations obtained at the lower level. A multistage heat exchange system can be represented by a sequence of single stage heat exchange systems with each i-th "hot" flow able to transfer its thermal energy to several "cold" flows, while each j-th "cold" flow is able to accept thermal energy several "hot" flows. To solve the problem of synthesis of optimal multistage heat exchange systems, we offer to develop an iteration procedure with such search variables as the amount of heat taken from the i-th "hot" flow and transferred to the j-th "cold" flow at the each stage. The third level will be dedicated to the optimization of the heat exchange system with the structure obtained at the previous step. The result will be the determination of new approximations by the search variables.

Let's install splitters at a heat exchange system inputs at each i-th "hot" and j-th "cold" flows and mixers at its output. We will obtain a heat exchange system with the division of process flows. The optimal synthesis problem statement will in this case be reduced to the previous one by the structural modification and expansion of a single stage heat exchange system. The "assignment" problem is offered to be supplemented by the estimations including or excluding the recuperating heat exchange for the each set of the l-th part of the i-th "hot" flow and the k-th part of the j-th "cold" flow. Such an approach will obviously increase the heat transfer moving force.

The efficiency of the proposed method is compared to the results obtained with SYNHEAT software [2] by the method of integral synthesis of heat exchange systems.

[1] Ziyatdinov N.N., Ostrovskii G.M., Emel'yanov I.I. Designing a Heat-Exchange System upon the Reconstruction and Synthesis of Optimal Systems of Distillation Columns // Theoretical Foundations of Chemical Engineering. 2016. V. 50. № 2. P. 178.

[2] Ponce-Ortega, J. Optimal synthesis of heat exchanger networks involving isothermal process streams / J. Ponce-Ortega, A. Jimenez-Gutierrez, I. Grossmann // Comput. Chem. Eng. 2008. № 32. P. 1918.

535102 Experimental and Numerical Investigation to Develop the Ultrasound Assisted Oxidative Desulfurization (UAOD) Process in a New Continuous-Flow System

Abstract:

Ultrasound assisted oxidative desulfurization (UAOD) is a promising technology, which can result in ultra-low sulfur fuels in order to reduce the environmental crisis. Most of the UAOD studies have been executed in the batch-mode systems. Comparing to ultrasonic-assisted process in batch-mode, the ultrasonic-assisted process in continuous mode of operation seems to be more beneficial because of its several advantages. Moreover, development of the different ultrasonic reactors in continuous flow causes generation of the wide potential in the area of sonochemistry which can help to introduce into industry. Hence, cylindrical continuous-flow system was introduced. In this system injection below the horn tip was employed as an efficient strategy to intensify the reaction rate and minimization of the catalyst consumption. Indeed, one of the decisive factors for increasing the efficiency of reactions in two-phase systems is to increase the contact area between the phases to attain more efficient exposure of the reagents. Since the maximum energy dissipation rate takes place near the horn tip owing to the much fluctuating component of the velocity, this region can be considered as "active zone". In addition, detection of more hydroxyl radicals (.OH) generation near the radiating surface rather than other zones demonstrated that this region also could be chemically active zones, as well. As a result, this zone could be opted as the location of introducing the feed and/or oxidant/catalyst into the reactor in which the maximum energy is dissipated. Diesel fuel as the oil phase with sulfur content of 1550 ppmw and an appropriate mixture of hydrogen peroxide and formic acid as the aqueous phase were used. Significant improvement on the sulfur removal was observed specially in lower sonication time in the case of dispersion method in comparison with the conventional contact between two phases. Furthermore the effect of nozzle number was examined. Ultimately, the flow pattern induced by ultrasonic device, and also injection of the aqueous phase were analyzed quantitatively and qualitatively by capturing the sequential images. A hydrodynamic parameter is introduced to evaluate the relative strength of the ultrasonic jet versus aqueous jet in the injection strategy system.

Combination of experimental and simulation methods can be applied to understand a process. Besides, the elaboration of the ultrasound reactor characterizations can be identified by simulation outcomes. Numerical simulation could be used as an efficient tool to investigate the hydrodynamic behavior and other related phenomena. It can be beneficial to provide some insights regarding the complex fluid behavior in the sonoreactor, which might be difficult to obtain with the experimental methods.

In the present study, a computational fluid dynamic (CFD) model was developed to investigate the hydrodynamic behavior and flow characteristics in a sonoreactor. The CFD results indicate that the physical and chemical effects associated with the ultrasonic field can contribute to the enhancement of the reaction and sulfur removal rates. However, the physical effects are predominant as compared to the chemical effects. Indeed, homogenous mixing and fine micro-emulsification as a consequence of the physical effects lead to increase the interfacial area and mass transfer rate between the aqueous and oil phases.

531831 Model-Based Analysis and Optimization of a Semi-Lean MBC Process for Natural Gas Sweetening

Abstract:

Over the past decade, a great deal of research has been devoted to membrane contactors (MBC) for CO2 absorption in natural gas sweetening applications. One key advantage of MBC over conventional absorption towers is their large intensification potential. In this work, the MBC mathematical model presented in Quek et al. [1], is used to carry out global sensitivity analysis and process optimization studies using the gPROMS ProcessBuilder. Global sensitivity analysis is used to determine the membrane characteristics and operating conditions that have the largest impact on CO2 removal performance and costs. Next, the standalone MBC unit model is integrated with gPROMS ProcessBuilder standard unit operations model libraries to model an MBC-based flowsheet of a semi-lean process for natural gas sweetening (Fig. 1). This flowsheet is optimized with the objective of minimizing the annualized cost, subject to a given CO2 specification. The optimization is carried out by applying a simple direct search in Matlab using the gO:Matlab interface.

Fig. 1. Semi-lean MBC process flowsheet.

References:

[1] V. C. Quek, N. Shah, and B. Chachuat, Modeling for design and operation of high-pressure membrane contactors in natural gas sweetening, Chem. Eng. Res. Des. 132:1005-1019, 2018.

538592 Process Intensification of Hydrogen Production Systems

Abstract:

Integrated gasification combined cycle (IGCC) is considered a promising technology for producing clean, affordable and secure power, and this technology has potential for further technological developments and improvements for carbon capture within efficiency. In the conventional IGCC plant, the water gas shift reaction (WGSR) is typically carried out in a sequence of two reactors (high temperature-WGSR and low temperature-WGSR), with inter-stage cooling, to overcome both kinetic and equilibrium limitations for the enrichment of H_2. However, carbon capture and storage process causes energy efficiency penalties since a significant amount of energy is needed to separate CO_2 from flue gas and to compress CO_2 for transport/storage. According to National Energy Technology Laboratory (NETL), the efficiency of IGCC (with CO_2 capture) decreases approximately 15-20% compared to IGCC without CO_2 capture. For these reasons, alternative technologies are required to eliminate these penalties and ensure competitive aspects of power plants with CO_2 capture. The process is therefore well-suited for the application of novel intensification concepts, integrating reaction and separation into a single unit. Adsorption and membranes are the several techniques to accomplish such intensification for the WGSR, as part of a multifunctional reactor for simultaneous production, separation, and purification of hydrogen from syngas mixtures. The primary objective of this work is to develop an advanced multi-scale mathematical model and demonstrate design/intensification of reaction/separation systems for the hydrogen production. The study on various hydrogen production processes (conventional high temperature shift reactor-HTSR/low temperature shift reactor-LTSR, proposed HTSR/membrane separation-MS/LTSR/MS and proposed low temperature membrane reactors-LTMRs) is carried out in terms of design and intensification concepts.

540280 Heat Integration and Controllability Analysis of Heat Exchanger Networks

Abstract:

Research Interests: Process Engineering

Teaching Interests: Process Dynamics and Control

The objective of this research is to present a methodological framework for designing heat exchanger networks (HENs), which best addresses heat recovery, economics and controllability. The proposed framework formulates a systematic approach consisting of a series of simple design steps. The steps include heat integration techniques such that the design can achieve its energy recovery targets: a detailed cost analysis to minimize both capital and utility costs of the design, and steady-state controllability measures to keep the design controllable. A heat exchanger network case study was used to test the proposed framework, and the results were compared with previous works in the literature. Pinch and Superstructure heat integration methods were applied to the case study; both designs achieved the required heat recovery of the system, however, the Superstructure design showed lower costs than the Pinch design. Both heat integration methods were also compared in terms of their inner loop interactions by performing Relative Gain Array (RGA) and Singular Value Decomposition (SVD) analyses.

541959 An Open Source Process Simulation Environment on Python for Automated Preliminary Techno-Economic Analysis

Abstract:

Research Interests:

Software development of process simulation components and platforms. Biofuel and bioproduct production. Technological development pathways based on quantitative sustainable design.

Teaching Interests:

Any undergraduate chemical engineering course and senior design. Biofuel and bioproduct production. Working with CAPE-OPEN platforms.

Abstract:

A fast and flexible open source simulation environment on python is being developed for automated preliminary technoeconomic analysis. The development and evaluation of a production process is a complex task that involves minimizing both cost and environmental impact while satisfying design specifications and product requirements. Therefore, tailoring a design to study how different feed compositions and design parameters impact its sustainability is a laborious task. To automate this task, the simulation environment allows for easy implementation of simplifying assumptions that lie within marginal error in the final cost evaluation. Additionally, an efficient recycle converging algorithm is employed whereby mass balances are solved without the need for any rigorous calculation. After each outer iteration of a rigorous simulation loop, an inner loop based on mass balances and collected simulation data is converged. The simulation environment maintains accuracy in rigorous thermodynamic calculations with the aid of standard CAPE-OPEN thermodynamic packages and the Chemical Engineering Design Library on python. The validity of the simulation environment to correctly predict production costs will be tested by reevaluating a design for the co-production of ethanol and biodiesel from lipid cane. The automation capabilities will be tested by varying design parameters in a Monte Carlo framework and performing multivariable sensitivity analysis. The overarching goal of the python simulation environment is to better prioritize research and development by evaluating how possible technological innovations can impact the sustainability of a production process. This open source simulation environment would allow third party designs and unit operations to be easily shared without any economic barriers. Researchers would benefit with the capability to test the implications of their technological innovations.

533937 Optimization-Based Retrofit of a Cryogenic Air Separation Unit for Flexible Operation

Abstract:

The increasing advancement and use of intermittent renewable energies results in fast-changing electricity prices that motivate flexible operation of industrial processes for demand side management (e.g., [1,2]). Previous literature has identified a considerable potential for air separation units (ASU) to increase their economic profitability by demand side management [3-5], since ASUs are large electrical energy consumers due to necessary compression of air for cryogenic separation and on-demand supply of gaseous products via pipelines.

Recent publications for dynamic operation of ASUs have mainly focused on either advanced process control strategies for existing plants (e.g., [6]) or on long-term scheduling of production cycles considering storage and vaporization of products (e.g., [3,7]). They found that adding further production equipment is economically desirable for cases with high utilization of the plant. However, they did not consider the transition time between operating modes in particular. Cao et al. [8] assessed design limitations to the dynamic behavior of the high-pressure nitrogen column in response to demand fluctuations and proposed the introduction of liquid nitrogen as an additional reflux to improve the plant's agility.

In this work, we propose a methodology for the redesign of the ASU process that considers the process dynamics, i.e., its transient behavior. This process design problem results in a large-scale dynamic optimization problem due to dynamic balances on the ASU columns' trays. In order to improve the computational efficiency, the rigorous dynamic stage-wise model is reduced using the collocation-based approach. Herein, dynamic balance equations are only considered for few trays, which reduces the number of differential equations and variables. The remaining trays are coupled using Lagrange polynomials. Then, a characteristic load change of the plant is optimized using a single-shooting dynamic optimization algorithm. In a post-optimal sensitivity analysis, we further identify bottlenecks that limit the plant's agility and propose new design modifications to overcome those limitations. The results show that resizing of column equipment, e.g., the volume of column sumps, and additional liquid buffer tanks can enhance process agility.

[1] Ghobeity, Amin & Mitsos Alexander 2010. Optimal time-dependent operation of seawater reverse osmosis. Desalination, 263, 76-88.

[2] Zhang, Qi, & Grossmann, Ignacio E. 2016. Enterprise-wide optimization for industrial demand side management: Fundamentals, advances, and perspectives. Chemical Engineering Research and Design, 116, 114–131.

[3] Cao, Yanan, Swartz, Christopher L.E., & Flores-Cerrillo, Jesus. 2017. Preemptive dynamic operation of cryogenic air separation units. AIChE Journal, 24(5), 467.

[4] Pattison, Richard C., & Baldea, Michael. 2014. Optimal Design of Air Separation Plants with Variable Electricity Pricing. Pages 393–398 of: Proceedings of the 8th International Conference on Foundations of Computer-Aided Process Design. Computer Aided Chemical Engineering, vol. 34. Elsevier.

[5] Zhang, Qi. 2016. Enterprise-Wide Optimization for Industrial Demand Side Management. Dissertation, Carnegie Mellon University, Pittsburgh, Pennsylvania.

[6] Xu, Zuhua, Zhao, Jun, Chen, Xi, Shao, Zhijiang, Qian, Jixin, Zhu, Lingyu, Zhou, Zhiyong, & Qin, Haizhong. 2011. Automatic load change system of cryogenic air separation process. Separation and Purification Technology, 81(3), 451–465.

[7] Pattison, Richard C., Touretzky, Cara R., Johansson, Ted, Harjunkoski, Iiro, & Baldea, Michael. 2016. Optimal Process Operations in Fast-Changing Electricity Markets: Framework for Scheduling with Low-Order Dynamic Models and an Air Separation Application. Industrial & Engineering Chemistry Research, 55(16), 4562–4584.

[8] Cao, Yanan, Swartz, Christopher L.E., Baldea, Michael, & Blouin, Stéphane. 2015. Optimization-based assessment of design limitations to air separation plant agility in demand response scenarios. Journal of Process Control, 33, 37–48.

535215 Planuling: A Hybrid Planning and Scheduling Optimization to Schedule Slow and Plan Fast Processes

Abstract:

Planuling is our term for combining planning and scheduling together within the same future time-horizon. It is a portmanteau of planning and scheduling and may also be classified as hybrid planning and scheduling given that within the same time-horizon some units, machines or equipment are planned and others are scheduled. This typically occurs when units with long run-lengths or time-constants are optimized together with units with short run-lengths whereby determining a greatest common factor would result in many small time-periods. This also occurs when numerically integrating ordinary differential equations (ODE) and is the phenomenon known as stiff-systems. Our novel approach to reduce the number of time-periods and hence decrease the time to find good solutions is to schedule the slow processes and to plan the fast processes with the understanding or assumption that planning the short processes makes sense or is appropriate. These types of situations are common in industry and occur when large amounts of materials are produced in bulk operations and then need to be packed into smaller amounts where the packing processes are much faster than the bulking or batching operations. A specific example can also be found in oil-refining when crude-oils are distilled and converted (refined) into many diverse intermediate products in slower major processing units but are blended into finished products such as gasoline, diesel and heating oil in faster blending operations. From a mathematical formulation perspective, we essentially allow the planuling model to contain both big-buckets or planning time-periods (with a pile of production orders to be disaggregated) and small-buckets or scheduling time-periods. Big-buckets are used for the fast unit-operations and small-buckets are used for the slow unit-operations. Other examples of planuling other than bulking to packing or refining to blending is smelting to casting, fermenting to purifying, rolling to sheeting and batching to dispensing.

Figure 1 depicts a sample flowsheet problem with a batch-process unit (BatchUnit), a continuous-process unit (PackUnit) and three storage units (Tank1, Tank2, Tank3) drawn using the unit-operation-port-state superstructure (UOPSS) [1], [2] and [3].

Figure 1. Planuling UOPSS Flowsheet.

Based on the above planuling discussion, the greatest common denominator is 1-hour and would require 24 times the matrix elements compared to our current implementation using 120 24-hour time-periods and enhancing the modeling to include our new planuling formulation. The current matrix has approximately 11000 rows, 8000 columns of which 4000 are binary, 30000 non-zeros. If we used a 1-hour time-period this would require 120 * 24 = 2880 time-periods where the larger matrix would have 264000 rows, 192000 columns, 96000 binaries and 720000 non-zeros. Obviously, planuling is able to drastically reduce the size of the problem provided that planning and not scheduling of the fast unit-operations is sufficient for the decision-making at hand.

Our new concept of planuling is definitely novel and to our knowledge has not been presented previously. It is a pragmatic procedure in the sense that for some faster units, they can be planned instead of being scheduled with little loss in decision-making accuracy or capability. It has the benefit of requiring significantly less matrix elements and will translate into quicker solutions for what would previously be intractable or insolvable APS problems. If necessary, as used in the example below, a second level of scheduling can be performed to find the schedule of the fast processes defined as planning inside the planuling optimization. In this case, the total amount from the planning time-step containing the pile of production orders are disaggregated within the faster or shorter time-grid scheduling.

Figure 2 below is the Gantt chart of the flowsheet found in Figure 1 configured in IMPL's IML file and computed using GUROBI in about 3-seconds using 1-day time-periods. There is a 3-day past horizon where Batch2 starts 2-days before the start of the 120-day future time-horizon. The batch-sizes and batch-times of the BatchUnit for Batch1 are

2000 m^3 and 2-days, Batch2 are 4000 m^3 and 4-days and Batch3 are 5000 m^3 and 5-days. The three tanks or pools each have a capacity of 4000 m^3 and open inventories of 2000 m^3 each. The PackUnit has six operations Batch1Pack1, Batch1Pack2, Batch2Pack1, Batch2Pack2, Batch3Pack1 and Batch3Pack2 with charge-sizes of 1200, 1200, 2400, 2400, 2400 and 2400 m^3 respectively. These operations have a minimum run-length of 11-hours and must not exceed 24-hours of continuous operation i.e., they have a maximum run-length of 24-hours.

Figure 2. Gantt Chart with 3-day and 120-day Past and Future Horizons.

[1] Kelly, J.D., "Production modeling for multimodal operations", Chemical Engineering Progress, February, 44, (2004).

[2] Kelly, J.D., "The unit-operation-stock superstructure (UOSS) and the quantity-logic-quality paradigm (QLQP) for production scheduling in the process industries", In: MISTA 2005 Conference Proceedings, 327, (2005).

[3] Zyngier, D., Kelly, J.D., "UOPSS: a new paradigm for modeling production planning and scheduling systems", ESCAPE 22, June, (2012).

520193 GoNDEF: A New Exact Method to Generate All Non-Dominated Points of Multi-Objective Mixed-Integer Linear Programs

Abstract:

Most real-life decision problems involve more than a single criterion that are often conflicting with each other. This category of problems are called multi-criteria/multi-objective optimization problems (MCOP/MOOP). One of the most important tasks for MOOPs is finding the non-dominated (ND) points in the objective space or efficient solutions in the decision space. A ND point results in objective function values that cannot be improved without worsening another objective function. In this paper, we present a novel method to generate the set of entire ND points for a multi-objective mixed-integer linear program (MOMILP). The Generator of ND and Efficient Frontier for MOMILPs (GoNDEF) finds the ND points represented as points, line segments, and facets consisting of all types of the ND points. First, GoNDEF finds integer values for integer variables that generate the ND points. Then, our method fixes integer variables to specific values. The resulting problem is a multi-objective linear program (MOLP). This MOLP has its own set of ND points. A subset of this set establishes a subset of the ND points set of the MOMILP. We present an extensive theoretical analysis of the GoNDEF and illustrate its effectiveness on a set of instance problems.

525525 Data-Driven Robust Optimization with Principal Component Analysis and Kernel Smoothing

Abstract:

Optimization applications abound in many areas of science and engineering. In real practice, some parameters involved in optimization problems are subject to uncertainty [1]. The issue of uncertainty could render the solution of a deterministic optimization problem suboptimal or even infeasible. Therefore, optimization under uncertainty has attracted tremendous attention from both academia and industry. A wide array of mathematical programming techniques, such as stochastic programming and robust optimization, have been proposed [2]. The key idea of stochastic programming is to model the randomness in uncertain parameters with probability distributions. However, this approach requires accurate information on probability distribution. Furthermore, the resulting optimization problem is usually computationally challenging. By introducing â€œwait-and-seeâ€ decisions, adaptive robust optimization (ARO) approach well models the sequential nature of decision-making processes, and ameliorates the conservatism issue [3]. In recent years, ARO has demonstrated various applications, such as process scheduling [4, 5], energy systems [6], and supply chain optimization [7].

Despite the burgeoning popularity of ARO, conventional ARO models have some limitations. They fall short of leveraging uncertainty data to facilitate the decision-making processes. Moreover, their adopted uncertainty sets fail to account for the correlations or asymmetry among uncertainties. With the explosion in uncertainty data and great stride in machine learning, data-driven optimization is becoming an active research area [8-10]. Therefore, the research objective of our work is to overcome the mentioned two limitations by proposing a data-driven asymmetric uncertainty set that effectively captures correlations.

In this work, we propose a novel data-driven robust optimization framework using principal component analysis (PCA) in conjunction with kernel smoothing methods. The PCA technique is employed to acquire the principal components via eigenvalue decomposition of the covariance matrix. This way, the latent uncorrelated uncertainties behind observed uncertainty data are effectively identified. We then project uncertainty data points onto each principal component. To truthfully capture the probability distribution of latent uncertainty, nonparametric density estimation approaches are applied to the projected uncertainty data. Based on forward and backward deviation vectors [11], we further develop a data-driven polyhedron uncertainty set, in which quantile functions are used to describe confidence intervals. A notable merit of the proposed uncertainty set is that it flexibly adapts to the intrinsic structure and complexity of uncertainty data. Based upon this uncertainty set, we propose a data-driven robust optimization framework, which includes both a data-driven static robust optimization model and a data-driven ARO model. The proposed framework not only mitigates the conservatism issue of robust optimization but also enjoys considerable computational benefits. For data-driven static robust optimization, we derive a computationally tractable robust counterpart to facilitate the solution and guarantee constraint feasibility. For data-driven ARO, a decomposition-based algorithm is developed to solve the resulting multi-level optimization problem. To demonstrate the effectiveness of the proposed framework, we present three applications on optimization under uncertainty, including model predictive control (MPC), production scheduling of multipurpose batch processes, and strategic planning of process networks.

References

[1] E. N. Pistikopoulos, "Uncertainty in process design and operations," *Computers & Chemical Engineering*, vol. 19, pp. 553-563, 1995.

[2] I. E. Grossmann, R. M. Apap, B. A. Calfa, P. García-Herreros, and Q. Zhang, "Recent advances in mathematical programming techniques for the optimization of process systems under uncertainty," *Computers & Chemical Engineering*, vol. 91, pp. 3-14, 2016.

[3] A. Ben-Tal, A. Goryashko, E. Guslitzer, and A. Nemirovski, "Adjustable robust solutions of uncertain linear programs," *Mathematical Programming*, vol. 99, pp. 351-376, 2004.

[4] Q. Zhang, M. F. Morari, I. E. Grossmann, A. Sundaramoorthy, and J. M. Pinto, "An adjustable robust optimization approach to scheduling of continuous industrial processes providing interruptible load," *Computers & Chemical Engineering*, vol. 86, pp. 106-119, 2016.

[5] H. Shi and F. You, "A computational framework and solution algorithms for two-stage adaptive robust scheduling of batch manufacturing processes under uncertainty," *AIChE Journal*, vol. 62, pp. 687-703, 2016.

[6] J. Gong, D. J. Garcia, and F. You, "Unraveling optimal biomass processing routes from bioconversion product and process networks under uncertainty: An adaptive robust optimization approach," *ACS Sustainable Chemistry & Engineering*, vol. 4, pp. 3160-3173, 2016.

[7] D. Yue and F. You, "Optimal supply chain design and operations under multi-scale uncertainties: Nested stochastic robust optimization modeling framework and solution algorithm," *AIChE Journal*, vol. 62, pp. 3041-3055, 2016.

[8] C. Shang, X. Huang, and F. You, "Data-driven robust optimization based on kernel learning," *Computers & Chemical Engineering*, vol. 106, pp. 464-479, 2017.

[9] C. Ning and F. You, "Data-driven adaptive nested robust optimization: General modeling framework and efficient computational algorithm for decision making under uncertainty," *AIChE Journal*, vol. 63, pp. 3790-3817, 2017.

[10] C. Ning and F. You, "A data-driven multistage adaptive robust optimization framework for planning and scheduling under uncertainty," *AIChE Journal*, vol. 63, pp. 4343â€"4369, 2017.

[11] X. Chen, M. Sim, and P. Sun, "A robust optimization perspective on stochastic programming," *Operations Research*, vol. 55, pp. 1058-1071, 2007.

Validation of CFD Prediction Accuracy of VOC Generation Rate for Cost-Effective Design of VOC Recovery Equipment

Xidong Hu, Manabu Kumagami, Shaoxiang Qian, Nobuhiro Yamada, Masahiro Kawasaki, Takashi Iitsuka and Syuichi Oguro, JGC Corporation, Yokohama, Japan

1. Introduction

Volatile organic compounds (VOC) emissions from storage facilities significantly affect the environment and human health. The storage and transport of crude oil and gasoline produce emissions of VOC [1]. To minimize the VOC emissions from storage tanks, the emitted VOC vapor is usually recovered by recovery equipment. In designing the VOC recovery equipment, the maximum instantaneous VOC discharge rate and vapor composition discharged from tanks are crucial parameters for selecting the recovery process and deciding the capacity of recovery equipment [2]. Although several quantitative estimation methods have been developed, few methods can predict the change of breathing VOC amount with time considering tank types, tank size, liquid level, property of the storage liquids and local weather conditions.

In view of this, a semi-empirical model, which is based on analogy between mass transfer and heat transfer on a flat plate, has been proposed in the present research and incorporated into the CFD code to more properly predict the discharge rate of VOC under higher ambient temperature and condensation rate of its vapor under lower ambient temperature for a fixed-roof gasoline tank. The present research aims to verify the CFD prediction accuracy of VOC evaporation and condensation rate by comparison with the experimental data from both the laboratory experiments and literature.

2. Evaporation and Condensation Model

As described above, a semi-empirical model based on analogy between mass transfer and heat transfer on a flat plate is proposed. The actual liquid surface in the tank is quiescent and is modelled as a flat plate shown in Figure 1. In simulating the evaporation and condensation at the interface, mass and heat transfer exists between the gas and liquid phases. Similarly, mass and heat transfer also exists near the inner wall surface of the tank when condensation occurs.

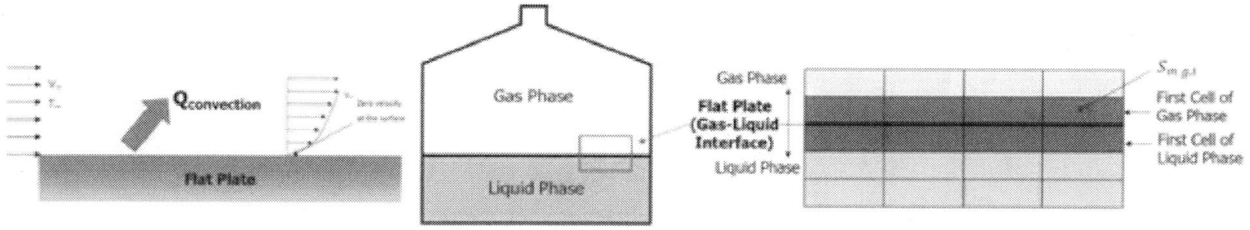

(a) Flat Plate Model (b) Product Storage Tank (c) Partially Enlarged View of Liquid-Gas Interface

Figure 1 Schematic Diagram of Flat Plate Model Application to Gas-Liquid Interface

The mass transfer rate due to evaporation and condensation is calculated as follows:

$$S_{m\,g,i} = ka\rho_g(Y_{g,i}^* - Y_{g,i}) \tag{1}$$

where k is mass transfer coefficient, a is the amount of surface area per unit volume of each cell at the interface, ρ_g is the gas density, $Y_{g,i}^*$ is the mass fraction of i component at equilibrium state and $Y_{g,i}$ is the local mass fraction of i component for each cell at the interface. When $Y_{g,i}^*$ is smaller than $Y_{g,i}$, $S_{m\,g,i}$ is positive and evaporation will occur. While $Y_{g,i}^*$ is larger than $Y_{g,i}$, $S_{m\,g,i}$ is negative and condensation will occur.

The mass transfer coefficient k can be evaluated, based on the analogy between mass transfer and heat transfer. The following empirical correlations are applicable for evaluation of forced convective heat transfer on a flat plate:

$$Nu = 0.664Re^{1/2}Pr^{1/3} \quad Re \le 15000 \tag{2}$$

$$Nu = 0.037Re^{4/5}Pr^{1/3} \quad Re > 15000 \tag{3}$$

where Nu is the Nusselt number, Re is the Reynolds number and Pr is the Prandtl number. Based on the analogy between mass transfer and heat transfer, the following empirical correlations are also valid for forced convective mass transfer on a flat plate:

$$Sh = 0.664Re^{1/2}Sc^{1/3} \quad Re \le 15000 \tag{4}$$

$$Sh = 0.037Re^{4/5}Sc^{1/3} \quad Re > 15000 \tag{5}$$

where Sh is the Sherwood number and Sc is the Schmidt number. The mass transfer coefficient k can be calculated from the Sherwood number (Sh):

$$k = \frac{ShD}{L} \tag{6}$$

where D is mass diffusivity and L is a characteristic length. In the proposed model, the turbulence length scale l in the gas phase is used to calculate the characteristic length L in equation (6). Thus, the L can be written as below:

$$L = l = C_\mu^{3/4}\frac{k_g^{3/2}}{\varepsilon_g} \tag{7}$$

where k_g is the kinetic energy of gas phase, ε_g is the turbulence dissipation rate of gas phase and C_μ is the model constant.

Simultaneously, the heat transfer rate due to evaporation and condensation at the interface is calculated using:

$$S_{hg,i} = (\Delta H_i + \Delta_{vap} H_i) S_{m\,g,i} \tag{8}$$

where ΔH_i is the sensible heat of i component and $\Delta_{vap} H_i$ is the latent heat of i component accompanied with the mass transfer rate $S_{m\,g,i}$.

The inter-phase mass and heat transfer are modeled using mass source terms in multicomponent transport equations and heat source term in energy equation in the CFD code.

3. Experimental Data for CFD Validation

The proposed semi-empirical model has been verified by comparing its predictions with the experimental data from both our laboratory experiments and literature.

3.1 Laboratory-Scale Experiment Data

Laboratory-scale experiments were carried out in a 2-liter tank, as shown in Figure 2. The tank was immersed in a constant temperature water bath and the surface temperature of tank was continuously measured by temperature sensors on the tank outer surface. The experiments for two cases were performed as shown in Table 1. Case 01 is for single component liquid (n-C_6H_{14} 100 vol%) and Case 02 is for multi-component liquid (n-C_6H_{14} 50 vol% and 2,3-Dimethylbutane 50 vol%). The tank was filled to half height level and purged by nitrogen to remove the air before heating. The tank surface temperature was heated up to 65 degC for Case 01 and 54 degC for Case 02, as shown in Figure 3. The discharged VOC vapor is sampled using the glass syringes, and then, its composition is analyzed using the gas chromatography analyzer. The experimental results of instantaneous VOC discharge flow rate of n-C_6H_{14} for Case 01 and Case 02 are shown in Figure 4 (a) and Figure 4 (b) respectively.

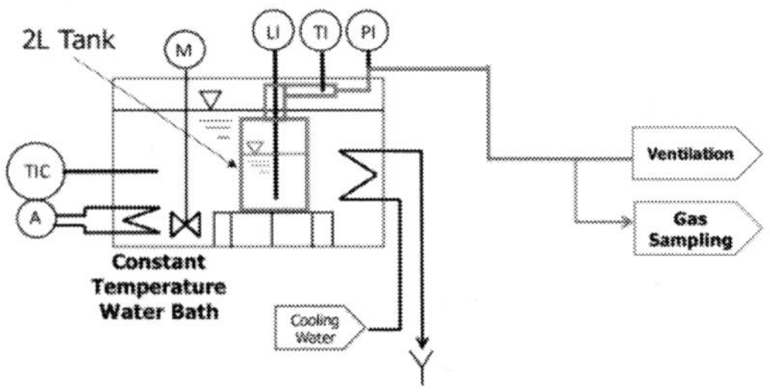

Figure 2 Experimental Apparatus

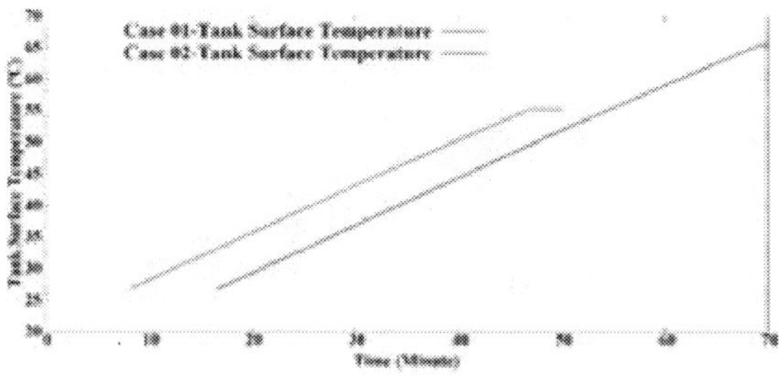

Figure 3 Heating Curve of Tank Surface

Table 1 Cases of Laboratory-Scale Experiment

Cases	Liquid Component
Case 01	n-C_6H_{14} (100 vol%)
Case 02	n-C_6H_{14} (50 vol%), 2,3-C_6H_{14} (50 vol%)
Note: 23-C_6H_{14} is 2,3-Dimethylbutane	

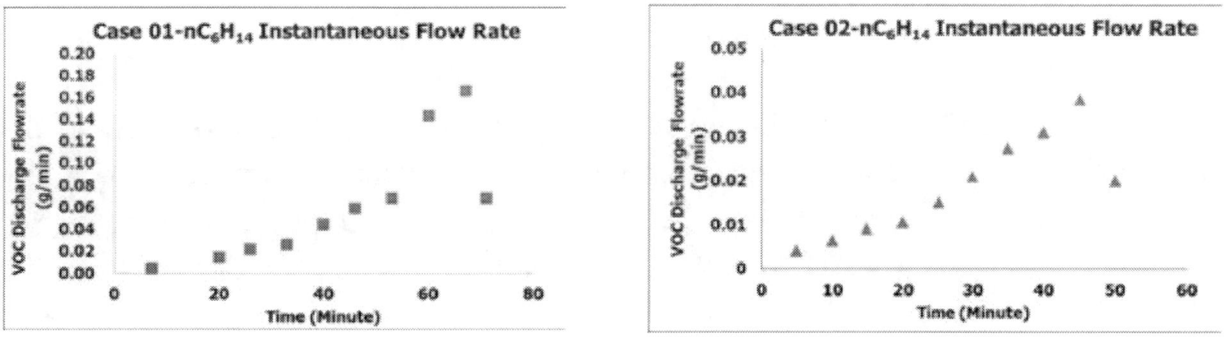

(a) Case 01 (b) Case 02

Figure 4 Variation of Instantaneous Flow Rate of n-C_6H_{14} with Time

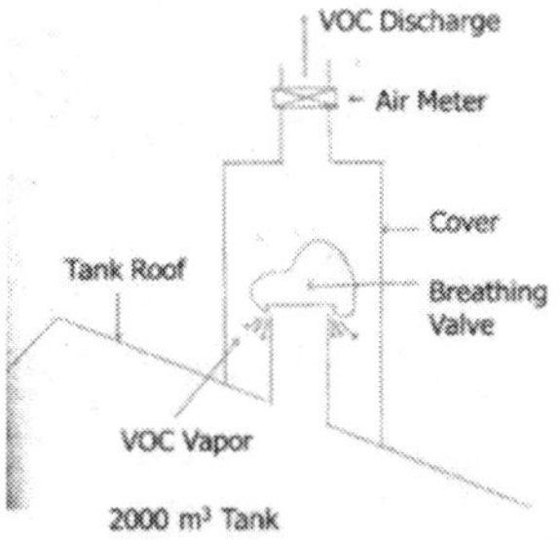

Figure 5 VOC Measurement for a 2000 m³ tank

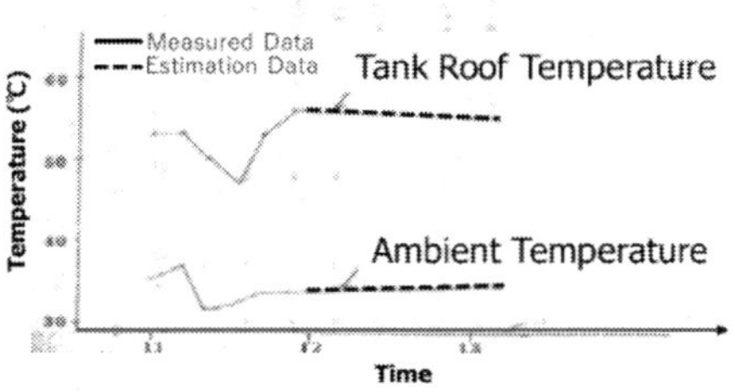

Figure 6 Variation of Ambient and Roof Temperature with Time

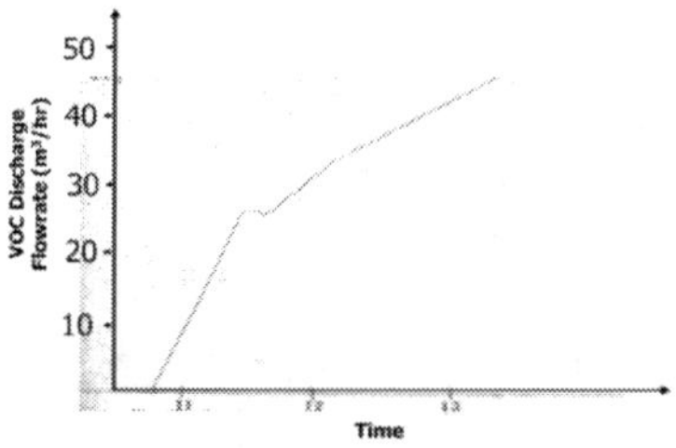

Figure 7 Variation of Measured VOC Flow Rate with Time

3.2 Industry-Scale Experiment Data

To further validate CFD predictions based on the proposed model, the VOC emission data of industry-scale tanks in literature are investigated. Few past studies provided the complete set of data including variation of VOC discharge flow rate with time, temperature profile, tank geometry and liquid properties, which make the reproduction study difficult. Agency for Natural Resources and Energy of Japan Ministry of Economy, Trade and Industry performed the VOC measurement for a 2000 m^3 tank in 1975 and summarized the measurement results into a report.

The VOC measurement was carried out by installing a cover and flow meter at the top of the breathing valve of a 2000 m^3 gasoline tank shown in Figure 5. The variation of ambient and roof temperature with time was measured and recorded, as shown in Figure 6. The measured instantaneous flow rate of discharged VOC vapor with time is indicated in Figure 7.

4. CFD Validation Model and Study Conditions

4.1 CFD Study for a Laboratory-Scale Tank

The CFD study using the same conditions as the laboratory-scale experiment is firstly performed. As shown in Figure 8, one quarter of the tank geometry is used to save the computational time. Same as the laboratory-scale experimental cases of Table 1, the CFD simulations for two cases are performed, which includes single component case (Case 01) and multi-component case (Case 02). The physical properties as a function of temperature used in the simulation are cited from the data of National Institute of Standards and Technology (NIST) Chemistry WebBook.

The main numerical methods used are summarized in Table 2. A transient analysis is applied to obtain variation of the VOC discharge amount with time when the temperature of tank outer surface is raised according to the heating curves in Figure 3.

4.2 CFD Study for an Industry-Scale Tank

Based on the available data indicated in section 3.2, a simulation for a 2000 m^3 industry-scale tank is performed as Case 03 using the same simulation approach as the laboratory-scale tank study. The geometry and main dimensions are shown in Figure 9. The variation of temperature distribution on the roof and side wall with time is applied as boundary conditions shown in Figure 10. The major components of the gasoline liquid used in the simulation are shown in Table 3.

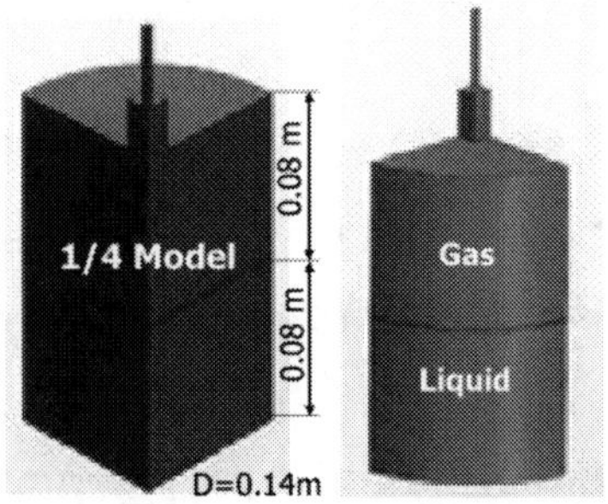

Figure 8 Geometry of Laboratory-Scale Tank for Case 01 and Case 02

Table 2 Main Numerical Methods

Item	Details
CFD Code	STAR-CCM+ 12.02
Turbulence Model and	Realizable $k-\varepsilon$
Simulation Mode	Transient
Gas Model	Ideal Gas Multi-Component
Discretization Scheme	Second Order

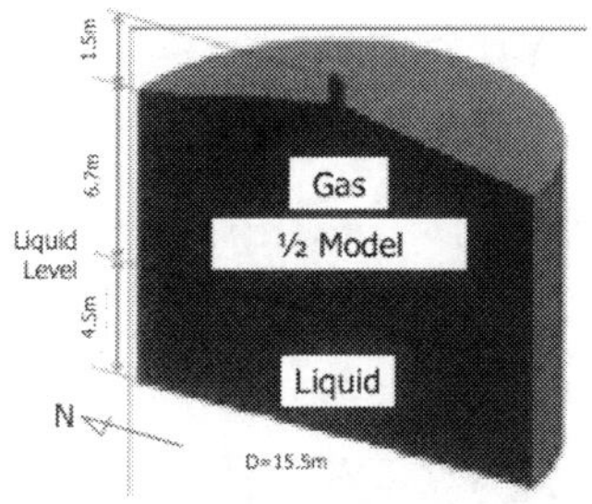

Figure 9 Geometry of Industry-Scale Tank

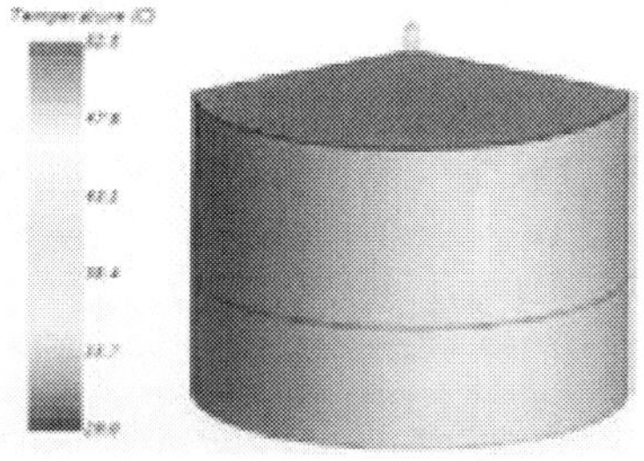

Figure 10 Initial Tank Surface Temperature

Table 3 CFD Study Case of Industry-Scale Tank Study

Cases	Liquid Component
Case 03	i-C_5H_{12} (22.8 vol%) n-C_5H_{12} (23.0 vol%) n-C_6H_{14} (54.2 vol%)

5. CFD Study Results and Discussion

5.1 CFD Study for a Laboratory-Scale Tank

The CFD study results of Case 01 and Case 02 are discussed here. Figure 11 shows the instantaneous CFD results of temperature, flow pattern and mass fraction distribution for Case 01. A higher temperature is observed in the gas phase compared with the liquid phase because of the specific heat difference between gas and liquid and stronger natural convective flow in gas. The temperature difference in gas phase shown in Figure 11 (a) is relatively small, while the temperature difference in liquid phase is big. As shown in Figure 11 (b), the upward flow near the side wall is observed both in the gas phase and liquid phase because of the high temperature on the outer surface wall. The upward flow near the side wall of the liquid phase reaches the liquid surface and flows toward the tank center. Thus, the shear stress of the liquid drags the gas near the interface flowing toward the tank center. The Figure 11 (c) shows that C_6H_{14} component evaporates at the interface, and then, diffuses toward the tank outlet.

Variation of the cumulative VOC discharge amount with time is compared between the CFD results and experimental data for Case 01 and Case 02 in Figure 12. It can be found that the CFD predicted VOC discharge flowrate agrees with the experimental data very well.

5.2 CFD Study for an Industry-Scale Tank

The CFD results for Case 03 are discussed here. As shown in Figure 13 (a), a high gas temperature near the roof is observed because of the high roof temperature shown in Figure 10. In addition, the temperature difference between the south wall and north wall leads to an upward flow near the south wall and downward flow near the north wall, as shown in Figure 13 (b).

Identically to the laboratory-scale tank study, variation of CFD predicted cumulative VOC discharge amount with time is compared with the limited measurement data from literature as shown in Figure 14. The CFD predicted results are close to the literature data well, although there is some difference. The reason for the observed difference is probably because the equilibrium state initial conditions used in the CFD study is different from the actual conditions in the experiment to some extent.

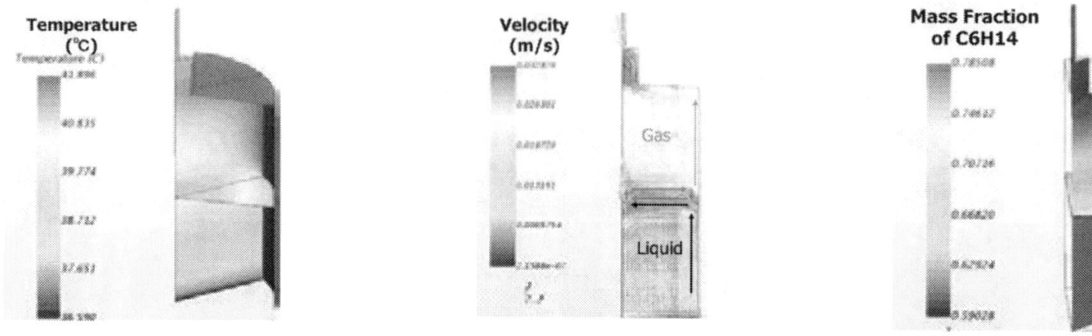

(a) Temperature Distribution (b) Flow Pattern (c) Mass Fraction of C_6H_{14}

Figure 11 Instantaneous CFD Results for Case 01

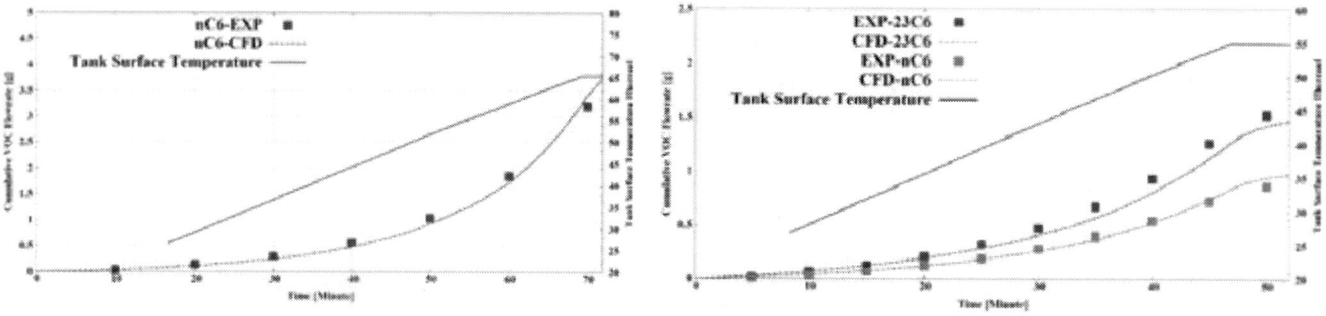

(a) Case 01 (b) Case 02

Figure 12 Comparison of CFD-Predicted Cumulative VOC Amount at Tank Outlet for Case 01 and Case 02 with Experimental Results

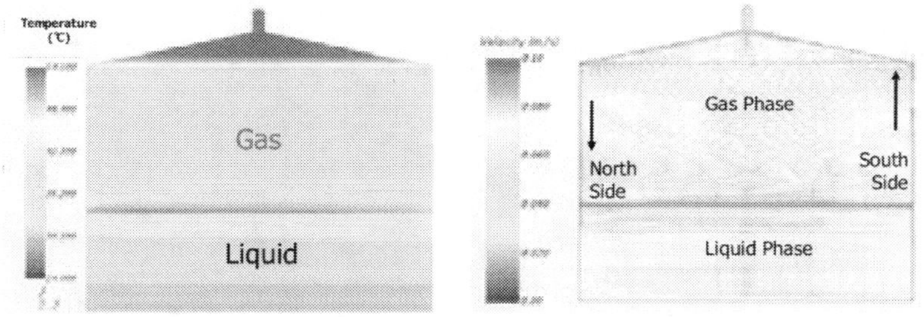

(a) Temperature Distribution (b) Flow Pattern

Figure 13 Distribution of Temperature and Velocity Vectors on the Vertical Cross-section

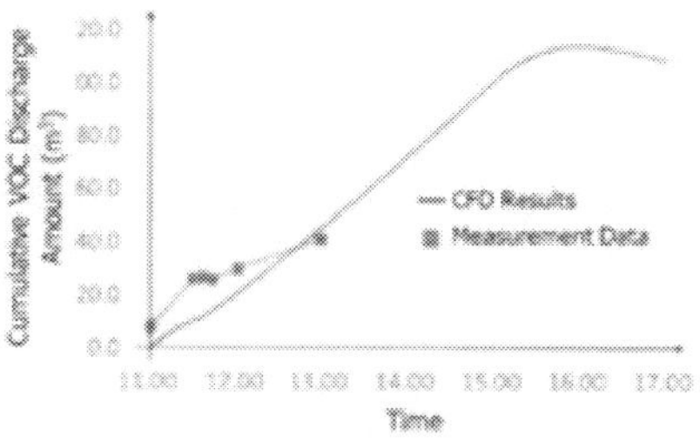

Figure 14 Comparison of Cumulative VOC Amount for Case 03 between CFD and Experiment Results

6. Conclusions

A semi-empirical model, which is based on analogy between mass transfer and heat transfer on a flat plate, has been proposed and incorporated into the CFD code to more properly predict the evaporation rate and condensation rate of VOC vapor. The analogy-based approach considers the variance of evaporation and condensation rate induced by time-dependent temperature on the tank outer surface based on the local weather data.

Laboratory-scale experiments were conducted in a 2-liter tank with single component (n-C_6H_{14}) and multi-component liquids (n-C_6H_{14} and 2,3-Dimethylbutane) for CFD validation. In addition, the experimental data of a 2000 m^3 industry-scale gasoline tank from literature was also used for CFD validation.

Transient CFD analyses were performed to validate the CFD prediction accuracy of VOC generation rate. The CFD analysis results show that the proposed semi-empirical model can predict the VOC discharge rate very well, compared with the experimental results. Therefore, it is expected cost-effective design of VOC recovery equipment can be realized using this approach.

References

1. Mark A. DeLuchi (1993), "Emissions from the Production, Storage, and Transport of Crude Oil and Gasoline", Air & Waste Management Association, Vol 43, November, pp. 1486-1495.
2. Tamaddoni M., et al. (2014), "Experimental Study of the VOC Emitted from Crude Oil Tankers", Process Safety and Environmental Protection, 92, pp 929-934.

525266 Process Design Quality for the Success of Industrial R&D at SABIC

Abstract:

Process systems engineering plays a key role in the industrial R&D at SABIC. Good process design work is critical, as it is the bridge connecting lab experimental results with the final economics of the entire technology. Rules and guidelines were defined from detailed process simulation to develop a high level road map of plant design to ensure a high design quality. Optimization and statistical analysis were applied in the Process Design work, which not only improved the design quality, but also revealed bottlenecks for technology development.

Two examples will be given for the demonstration of Process Design Quality at SABIC. One is to ensure data quality by using statistical analysis, and the other one is an example about a technology optimization from the systems point of view.

526855 A Multi-Objective MILP Model for Spatio-Temporal Design and Operation of Multi-Product Oil Palm Value Chains

Abstract:

The economic potential of oil palm is realised due to its abundance, cheapness and versatility in producing different oleochemicals and energy products [1]. The oil palm fruit or fresh fruit bunch (FFB) can be processed to produce crude palm oil (CPO) and kernel palm oil (KPO). These can be converted into oleochemicals and energy products in which side products such as palm oil mill effluent (POME) and empty fruit bunches (EFB) can be converted further to valuable products. Palm-oil derived products are useful in different applications such as in power generation, agriculture and polymer industries. Considering a wide range of applications for oil palm, it is important to consider techniques for design and operation of palm oil value chains to maximise its economic and environmental benefits.

In this conference, a spatio-temporal mixed integer linear program (MILP) model for planning and designing multi-product value chains [2-3] for oil palm will be presented. A value chain includes all activities from pretreatment of raw materials to distribution of final products and waste management [4]. Due to the presence of different options in each stage of the value chain in oil palm industry, a mathematical model is necessary to gain important insights for large-scale planning. Spatial and temporal decision factors are considered by the model to optimise different scenarios in oil palm sectors. Spatial factors include identifying locations for oil palm plantations and processing facilities and location of demand. Temporal factors are considered in order to optimise the schedule of different processing and transportation operations. The model gives important insights on other aspects such as how much production is needed and how it will be allocated and transported. These decisions will depend on decision makerâ€™s objectives including profit maximisation and GHG emissions minimisation. A case study using Peninsular Malaysian oil palm sectors is used to illustrate the model, which is implemented in AIMMS and solved using the CPLEX solver. In order to capture the spatial-dependencies of the problem, Peninsular Malaysia is represented as a grid of 50 km squares and candidate locations for oil palm plantation are modelled in GIS. The planning horizon for the problem is up to 2050 in order to illustrate the modelâ€™s capability to provide insights about the long-term evolution of the system, optimising the investments in technologies and infrastructures required in each decade.

References:

[1] A. Abdulrazik, M. Elsholkami, A. Elkamel, L. Simon (2017). Multi-products productions from Malaysian oil palm empty fruit bunch (EFB): Analyzing economic potentials from the optimal biomass supply chain, Journal of Cleaner Production, 168, 131-148, 2017.

[2] S. Samsatli, N.J. Samsatli (2017). A multi-objective MILP model for the design and operation of future integrated multi-vector energy networks capturing detailed spatio-temporal dependencies. Applied Energy. DOI:

10.1016/j.apenergy.2017.09.055.

[3] S. Samsatli, N.J. Samsatli, N. Shah (2015). BVCM: a comprehensive and flexible toolkit for whole-system biomass value chain analysis and optimisation - mathematical formulation. Applied Energy, 147, 131-160.

[4] S.M. Jarvis, S. Samsatli (2018). Technologies and infrastructures underpinning future CO_2 value chains: a comprehensive review and comparative analysis. Renewable & Sustainable Energy Reviews, 85, 46-68.

529617 Numerical Analysis of Effect of Diaphragm Structure Based on Thermo-Electro-Magneto-Hydrodynamics Coupling Model in Magnesium Electrolysis cell

Abstract:

Magnesium production is one of the most energy-intensive industrial processes. Like the Hall-Herault process of aluminum production, The core reactor is the electrolysis cell and its electrolysis efficiency is a crucial factor that profoundly affects the amount of energy consumed. Thus the energy consumption and current efficiency are crucially important indexes.

In the newly designed electrolysis process, magnesium oxide is dissolved in a rare earth chloride-containing electrolyte and electrolyzed to produce magnesium and oxygen gas similar to the production of aluminum [1] and lithium [2] in electrolytic cells. Considerable research efforts are reported on the use of commercial software to simulate flow field[3,4], thermoelectric field [5–7] and magnetohydrodynamic models [8] in magnesium electrolysis cells.

To improve the circulation of the electrolyte, a baffle plate were set between the electrolysis department and collect department. The ratio of the quantity of magnesium droplets in the metal separating compartment during the primary circulation to the whole quantity of magnesium droplets generated at the cathode is defined as the primary separation rate of magnesium droplets (PSR)[10]. In order to improve the electrolysis efficiency, the cell needs to increase its primary separation rate of magnesium droplets.

This work concerns the optimization of the electrolysis cell based on a three-dimensional thermo-electro-magneto-hydrodynamics coupling model. The flow field of mathematical model was validated by the cold model experiments of PIV in the previous research[9]. The structure of the diaphragm and buffle plate were studied to improve the PSR. The PSR could increase from 18% to 40% after optimizing the structure and position of the diaphragm and buffle plate.

1. Tessier J, Duchesne C, Tarcy GP, Gauthier C, Dufour G. 2012 Multivariate Analysis and Monitoring of the Performance of Aluminum Reduction Cells. ***Ind. Eng. Chem. Res.*** **51**, 1311–1323. (doi:10.1021/ie201258b)
2. Oliaii E, Désilets M, Lantagne G. 2017 Numerical analysis of the effect of structural and operational parameters on electric and concentration fields of a lithium electrolysis cell. ***J. Appl. Electrochem.*** , 1–16. (doi:10.1007/s10800-017-1073-2)
3. Liu C-L, Sun Z, Lu G-M, Song X-F, Yu J-G. 2015 Experimental and numerical investigation of two-phase flow patterns in magnesium electrolysis cell with non-uniform current density distribution. ***Can. J. Chem. Eng.*** **93**, 565–579. (doi:10.1002/cjce.22135)
4. Sun Z, Zhang H, Li P, Li B, Lu G, Yu J. 2009 Modeling and simulation of the flow field in the electrolysis of magnesium. ***JOM*** **61**, 29–33. (doi:10.1007/s11837-009-0066-y)
5. Sun Z, Liu C, Lu G, Song X, Sun S, Sun Y, Yu J. 2011 Coupled thermoelectric model and effects of current fluctuation on thermal balance in magnesium electrolysis cell. ***Energy and Fuels*** **25**, 2655–2663. (doi:10.1021/ef2004333)
6. Sun Z, Cai L, Liu C, Lu G, Yu J. 2017 Analysis for effects of electrolyte level on energy consumption in magnesium electrolysis by finite element method. ***Can. J. Chem. Eng.*** **95**, 648–655. (doi:10.1002/cjce.22708)
7. Sun Z, Cai L, Ni H, Lu G-M, Yu J-G. 2018 Coupled electro-thermal field in a high current electrolysis cell or liquid metal batteries. ***R. Soc. Open Sci.*** **5**, 171–179. (doi:10.1098/rsos.171309)
8. Sun Z, Li P, Lu GM, Li B, Wang J, Yu JG. 2010 Effect of Electromagnetic Field on Three-Phase Flow Behavior. ***Ind. Eng. Chem. Res.*** **49**, 10798–10803. (doi:Doi 10.1021/Ie100513w)
9. Liu C, Sun Z, Lu G, Song X, Yu J. 2015 3D and 2D experimental views on the flow field of gas-evolving electrode cold model for electrolysis magnesium. ***Flow Meas. Instrum.*** **45**, 415–420. (doi:10.1016/j.flowmeasinst.2015.07.011)
10. Liu C, Zhao Q-W, Sun Z, Lu G, Yu J. 2018 Analysis of Magnesium Droplets Characteristics and Separation Performance in a Magnesium Electrolysis Cell Based on Multiphysical Modeling. ***Arab. J. Sci. Eng.*** (doi:10.1007/s13369-018-3148-8)

536024 Integrated Design and Control of Intensified Membrane-Based Hydrogen Production Via Methane Steam Reforming

Abstract:

Significant effort has been invested in order to improve chemical process economic efficiency through process design that aims to optimize economic criteria (e.g., total annualized cost). Chemical process design aims at determining the design point, both structural and operational, at which the process can fulfil the product specifications and meet safety requirements, whereas minimizing both equipment and operational cost. Process synthesis, flowsheet configuration, unit interconnections followed by the calculation of equipment sizing and nominal operating conditions are among the steps of the traditional chemical process design procedure. Usually, the evaluation of alternative process flowsheets is performed within an optimisation framework subject to safety regulations and environmental constraints, which utilizes engineering knowledge and experience in order to define a meaningful design space by eliminating bad possible choices.

Nevertheless, process systems under real operating environments are affected by exogenous disturbances, intrinsic process uncertainties, and switching between different operating conditions. As a result, a control system is employed that compensates the effects of disturbances and uncertainties in order to satisfy product specifications. However, in order to compensate for those disturbances additional resources are needed resulting to increased actual dynamic operational cost of the process. The traditional approach is to design the chemical process followed by the design of an efficient automatic control system, which is essential to achieve an acceptable dynamic performance. In chemical processes there are cases, where the achievable dynamic performance cannot be improved unless design limitations are waived, despite the employment of sophisticated control strategies. It is therefore, quite important to incorporate process operability as an additional criterion during process design in order to ensure the achievement of the desirable dynamic performance. Integrated process and control system design approaches have reached a certain degree of maturity in terms of methods and employed techniques [1-3]. However, the current work investigates a process of significant degree of complexity representative of actual process systems and an increased degree of intensification.

Compared to the conventional method of hydrogen production via methane steam reforming, the employment of a palladium-based membrane reactor allows for the simultaneous separation of hydrogen from the produced synthesis gas mixture. Hence, in integrated membrane steam reforming reactors a higher methane conversion can be achieved at much lower reactor temperature than in conventional reactors. The intensified process therefore results to a highly complex system. The interactions among the reforming reactions mechanism, the convective and molecular diffusion of the reacting mixture species, the diffusion of hydrogen through the membrane, and the thermal effects due to reaction and heat exchange with an external heat source must be then taken into consideration for the optimal design of a highly performing system [4].

This work focuses on the development and implementation of an integrated process design and control framework for the highly intensified process of hydrogen production in a membrane-based reactor via methane steam reforming.

More specifically, the framework accounts for the optimal selection of process system design variables (material, process units and their interconnections, and operating conditions) along with the control structure configuration (selection of manipulated and control variables) for a given control algorithm. The design problem is stated as a mixed integer constrained non-linear dynamic optimization problem with an objective function that incorporates both investment and operational costs along with closed-loop dynamic performance criteria. Rigorous dynamic, one-dimensional, nonlinear, pseudo-homogeneous mathematical models of the tubular membrane reactor equipped flowsheets, which consists of mass, energy and momentum balances, validated utilizing experimental data, are used for the modelling of the system. Additionally, a linear model predictive control scheme (MPC) [5] is employed for each alternative flowsheet and in order to evaluate the closed-loop dynamic performance of the process under the presence of multiple disturbances. The disturbance scenario to be simulated consists of a regulatory problem, where the controller has to counteract the effect of multiple simultaneous disturbances, and a servo problem, where the controller has to keep the output close to a given reference input. Simultaneous disturbances associated with the direction of maximum variability for the system are used for the evaluation of the control performance. The direction of maximum variability is estimated by performing a singular value decomposition of the sensitivity matrix, which incorporates variation of process states and eigenvalues with respect to disturbance perturbation around its nominal values. The utilization of the information included in this matrix is essential towards the definition of the direction of maximum variability which is utilized in order to effectively determine the dynamic performance of each flowsheet under consideration by minimizing computational cost (evaluation of the dynamic performance under the worst case instead of every possible disturbance scenario). Given the set of possible design variables, control problem and algorithm and predefined possible disturbance scenarios, a simulated annealing algorithm is utilized for the solution of this highly complex non-linear dynamic optimization problem. After the initialization of the problem, at every iteration of the optimization algorithm, where a new set of design variable is properly chosen, a steady state problem for the calculation of the equipment and operation cost is solved. Successively, the dynamic simulation along with the solution of the control algorithm is performed in order to evaluate the dynamic performance at each design point. This step includes the linearization of the non-linear dynamic model (to be utilized in the MPC) and the calculation of the direction of maximum variability for the determination of the disturbances.

The proposed developments are illustrated in case studies considering the comparative assessment of three alternative flowsheeet design configurations of hydrogen production process via methane steam reforming utilizing membrane-based hydrogen separation. The first flowsheet utilizes an integrated multi-tubular palladium-based membrane reactor which results to a highly intensified process. The second flowsheet utilizes a common tubular reactor connected in series with a tubular palladium based separation module. Several successive reaction-separation steps are used to fully describe the process. The resulting process is less intensified compared to the first flowsheet configuration but its performance (both dynamic and steady state) is limited due to thermodynamic limitations. The last flowsheet under consideration utilizes the successive reactor-separator configuration, as described in the second flowsheet, but with two reaction-separation modules connected in series, where recycle of several streams is also taken into account. In all flowsheets additional units (heat exchangers, flowrate splitters and mixers) are utilized. Optimal designs ensure efficient operation while being able to perform better under dynamic transition compared to the designs obtained by the traditional successive approach of process design followed by the controller design. Additionally, the integrated membrane reactor scheme exhibits superior performance than the cascaded systems due to its ability to better handle the investigated disturbance scenarios. This ability is attributed to the simultaneous hydrogen separation from the traction zone which makes the membrane reactor design more flexible compared to the cascaded designs.

Acknowledgements

Cited References

[1] Diangelakis, N.A., Burnak, B., Katz, J., Pistikopoulos, N. Process Design and Control Optimization: A Simultaneous Approach by Multi-Parametric Programming. AIChE, 2017, 63(11), 4827-4846.

[2] Ricardez Sandoval, L.A.; Budman, H.M.; Douglas, P.L. Simultaneous design and control of processes under uncertainty: A robust modeling approach. J Process Control. 2008, 18, 735-752.

[3] Sanchez Sanchez, K.B.; Ricardez Sandoval, L.A. Simultaneous Design and Control under Uncertainty Using

Model Predictive Control. Ind. E ng. Chem. Res. 2013, 52, 4815-4833.

[4] Kyriakides, A.-S.; Rodríguez-García, L.; Voutetakis, S.; Ipsakis, D.; Seferlis, P.; Papadopoulou, S. Enhancement of pure hydrogen production through the use of membrane reactor. Int. J. Hydrogen Energy. 2014, 39(9), 4749-4760.

[5] Kyriakides, A.-S.; Seferlis, P.; Voutetakis, S.; Papadopoulou, S. Model Predictive Control for Hydrogen Production in a Membrane Methane Steam Reforming Reactor. Chem. Eng. Trans. 2016, 52, 991-996.

537551 Scale up Design Optimization of Pressure Swing Adsorption Processes for Gas Separation

Abstract:

The scale up design technology of pressure swing adsorption (PSA) processes for gas separation is introduced. The technology includes the optimization method (Ko et al., 2005) of PSA processes at cyclic steady state (CSS) and the mathematical model equations (Ko, 2016a). A four-bed twelve operating step PSA process were adopted and the optimal size of the adsorption bed (the inside diameter and packing bed length) and operating conditions (adsorption pressure, desorption pressure, adsorption step time, etc) were conveniently determined using the technology. Also, the trends of purity and recovery of the purified product gas were investigated depending on the design and operating conditions, and the design optimization results up to a large commercial size were verified very well by comparing with the design data of a PSA vendor company in Korea (Ko, 2016b). Finally, a simple and convenient rule of thumb for the scale up bed sizing is also suggested.

537617 Scale up Design Optimization of a Membrane Module for Gas Separation

Abstract:

This work introduces the development of dynamic simulation model of a gas separation membrane by deriving a mass balance and energy balance equations (Ko, 2017a) and a mathematical based scale up design method. The adopted membrane module is a hollow-fiber membrane with counter-current flow to recover methane from coalbed methane (CBM). The bore sides within the fibers receive the feed flow and emit the retentate flow, which is the purified CBM gas. The permeate flow from the shell side is discharged as a waste stream. The simulation results agree well with lab scale experimental data provided by a membrane vendor company in Korea. To perform the scale up design of membrane modules, the number of fibers as an integer variable should be determined properly using an optimization solver. But if the integer variable is included in the optimization problem, the computation load is too big for the problem to be solved reasonably. Hence this study introduces the new optimization method to efficiently solve the problem (Ko, 2017b), and the scale up design optimizations were successfully performed up to a commercial plant scale (500Nm^3/h of feed flow rate). As a conclusion, the optimal design conditions, i.e., the effective length, the number of fibers, and the inside shell diameter, are easily obtained from a small laboratory scale to a large commercial plant scale by employing the developed optimization algorithm and simulation model.

Solving Real-World Natural Gas Gathering Systems

Abstract ID: 538343
Author(s): Russell Burnett, **Michael Sellers**, and Charles C. Solvason, Bryan Research & Engineering, Bryan, TX
Session Date: Monday, October 29, 2018
Session Time: 3:30 PM - 5:00 PM

Natural gas gathering systems can be very large and complex. They can join hundreds of wells and include hundreds of miles of pipelines. Consequently, the effort required to build an accurate and reliable model for a large network in a process simulator is substantial, perhaps even prohibitive.

This paper presents an automated method of generating models for natural gas gathering systems directly from geographic information systems (GIS) data. Once the network inlet and/or outlets are specified, all network connectivity is resolved automatically, with easy tools available to help the user resolve regions of uncertainty in the source data. Where no elevation information is available, elevation profiles are generated from knowledge of the underlying terrain.

Once built, the model is then capable of solving the entire network system simultaneously based solely on physical input parameters (temperature, pressure, flow rate and composition). No special numerical relationships are needed. Further, the model can then be optimized for various considerations, such as limiting capital and operating costs. A case study will be presented to highlight the features of the developed method using the process simulator ProMax®.

536927 On the Temporal Evolution of the Material Stress Profile in a Supercritical Pulverized Coal Boiler Under Load-Following Operation

Abstract:

As more variable renewable generators of energy are penetrating the power grid, coal-fired power plants are increasingly forced to follow the load. Under load-following conditions, supercritical pulverized coal (SCPC) boilers experience considerable changes in their thermal and pressure profiles. Operation at higher temperatures than the design condition causes creep while thermal cycling leads to fatigue. Combined material creep-fatigue damage under cycling operation can eventually lead to the failure of the boiler system. For understanding the stress evolution in an SCPC boiler, a dynamic model of the boiler that can accurately estimate the temperature and pressure profiles in the boiler during transient operation is highly desired. Under sliding pressure operation, not only can the region where the water transitions to supercritical steam change, the boiler can also eventually slide into the subcritical phase under low-load conditions.

Current dynamic models for SCPC boilers mostly use a lumped-parameter approach. This approach fails to accurately capture the local transients of variables since there are highly nonlinear changes in physical properties and heat transfer characteristics near the critical point for water [1]. Effects of the burner configuration [2] and tube layout and design [3] on the boiler transients are also mostly neglected. Furthermore, there is hardly any work that has studied the stress profile in SCPC boilers under load-following operation.

In this work, a rigorous and modular dynamic process model for a once-through SCPC boiler is developed in Aspen Custom Modeler. Separate one-dimensional, distributed-parameter sub-models are considered for the different sections of the boiler (e.g. economizer, final superheater, platen superheater, water wall). It should be noted that due to the complicated flow paths of the water/steam and flue gas through the boiler, it is important to model the entire boiler with all its components to capture the transients accurately. Within each sub-model, rigorous property calculations are carried out. The heat transfer model considers various heat transfer mechanisms, and also accounts for the change in the heat transfer characteristics due to the change in the hydrodynamics and thermal properties of the steam under supercritical and subcritical conditions. This high-fidelity dynamic boiler model provides spatial and temporal changes in the thermal and pressure profiles in the boiler.

A material damage model is developed and integrated with the boiler dynamic model. The damage model considers the combined effect of creep and fatigue degradation phenomena. The creep model is developed with due consideration of the material properties, working load, plastic collapse load, and the yield stress. The fatigue damage model is developed by considering the principal strains at a given fatigue cycle with due consideration of the wall temperature variation along the tube, the temperature gradient along the tube wall and the variation of internal pressure. The integrated creep-fatigue damage model is used to study the dynamic evolution of the spatial stress profile under different load-following scenarios. The model suggests optimal load-following strategies that can significantly reduce material damage due to load-following.

References

[1] O. Palmqvist, *Dynamic Modelling of Heat Transfer Processes in a Supercritical Steam Power Plant,* Göteborg, Sweden: Chalmers University of Technology, 2012.

[2] D. Rakopoulos, I. Avagianos, D. Almpanidis, N. Nikolopoulos and P. Grammelis, "Dynamic Modeling of a Utility Once-Through Pulverized-Fuel Steam Generator," *Journal of Energy Engineering,* vol. 143, no. 4, 2017.

[3] K. Yamamoto, H. Suganuma, K. Domoto, Y. Yamasaki, Y. Kanemaki and H. Nakaharai, "Design Technology for Supercritical Sliding Pressure Operation Vertical Water Wall Boilers - First report: History of Practical Application and Introduction of Enhanced Rifled Tube," *Mitsubishi Heavy Industries Technicla Review,* vol. 50, no. 3, pp. 59-68, 2013.

537697 Optimization of Single-Well CO_2 Injection for Enhancement of Tight Oil Production

Abstract:

Optimization of single-well CO_2 injection for enhancement of tight oil production

CO_2 injection is a common approach to enhance recovery in conventional reservoirs, but recent efforts to apply this technique in tight oil wells have shown limited incremental oil recovery. A novel idea to address the problem employs a single horizontal well configuration, as proposed by Zhu et al. (2015) for secondary recovery in a tight formation. The proposed configuration uses alternately distributed injection and production fractures along a horizontal well, to create an array of plane-to-plane displacements in a single well. Figure 1 illustrates a portion of the envisioned well.

Figure 1: Single well injection and production

In this work we propose a dual tubing system in the wellbore, with one tubing for production and the other for injection. The injection and production sections would be separated with packers. Figure 2 shows a diagram of a completion configuration with these features. The nature of the hydraulic communication between injection and production fractures is essential to successful well optimization. The proposed optimization considers production and injection sensors and controls for each fracture with capability to shut off flow from any fracture with excessive CO_2 injection or production.

Figure 2: Completion configuration

In this research, we are replacing the capacitance-resistance model (CRM) (Yousef, et al., 2015) suited to water injection with a process control-relevant model, capable of representing compressible flow in the reservoir. Then we are using this model to forecast displacements along the well and anticipate what the injection and production sensors would record under prescribed total well injection rate and production pressure controls.

The initially described well configuration likely has more sensors and/or controls than necessary. Once we understand optimization with sensors and controls on every fracture, we will study how to reduce the sensor and control requirements, for complexity reduction without compromising recovery. The ultimate goal is a cost-effective well configuration likely to greatly increase recovery factors currently common in tight oil wells.

References:

Yousef, A. A., Gentil, P. H., Jensen, J. L., & Lake, L. W. (2005, January 1). A Capacitance Model to Infer Interwell Connectivity from Production and Injection Rate Fluctuations. Society of Petroleum Engineers. doi:10.2118/95322-MS

Zhu, P., Balhoff, M. T., & Mohanty, K. K. (2015, September 28). Simulation of Fracture-to-Fracture Gas Injection in an Oil-Rich Shale. Society of Petroleum Engineers. doi:10.2118/175131-MS

526389 A Novel Methodology to Optimize the Operation of Combined Cooling Heat and Power Systems

Abstract:

This work presents a novel methodology to optimize the operation of CCHP systems. We introduce a novel performance curve (NPC) which enables determinations of optimal operating strategies and show that previously operating strategies are the special cases of our methodology. The PGU overall optimum partial loads for demands below and above the CCHP operating curve are determined based on the optimization criteria and system characteristics. The proposed methodology is flexible and adaptable; it accounts for energy prices, carbon dioxide emissions (CDE), primary energy consumption (PEC) factors and load variations with the weather conditions. The operating of the CCHP system is analyzed for small hotel buildings in San Francisco and Miami having different energy demand profiles. These locations have different energy prices, CDE and PEC factors. The proposed methodology leads to the best operation when compared to following electrical load and following thermal load strategies based on operating cost, carbon dioxide emissions, primary energy consumption and a combination.

535765 Simultaneous Crude Procurement Planning and Movement Scheduling for Petroleum Refineries

Abstract:

The petroleum refinery production is a very complicated process, since huge volumes of crudes should be continuously handled through a highly integrated network of extremely intensive units. Nowadays, the refining process can be reliable and efficient with the aid of mature techniques, including advanced process control and real-time optimization. As comparison, crude procurement and crude supply are still vulnerable to tough missions and various uncertainties. Thereby, they will unavoidably affect the profitability and operating feasibility of refineries. The crude procurement cost is comparable to the operating cost in quantity, taking up a significant proportion of the overall cost [1]. On the other hand, the crude market has become highly volatile since the last two decades [2]. Thus, improper crude procurement in terms of amount and timing will inevitably reduce the profit margin of refinery, or even cause huge operating problems. Note that after the crude procurement has been planned, operations of different crude movement units including vessels, berths, storage tanks, pipelines and charging tanks have also to be well scheduled for a smooth and stable crude transfer, which involves complex scheduling problems. In previous studies, these issues were used to be addressed sequentially, providing separate crude purchase plans and crude movement schedules [3,4]. Some works studied a partial combination of crude purchasing and blending operations [5,6]. However, the two crucial elements have not been simultaneously considered in a full scope of crude supply chain.

In this study, a novel MINLP model with continuous-time representation has been developed with simultaneous consideration of crude procurement planning and crude movement scheduling. Through this model, the optimal procurement plan and transfer schedule can be efficiently determined towards the minimal overall cost. Optimal decisions on crude purchasing (i.e., types, volumes, and timing) and transfer operations (i.e., unloading, storing, blending, and charging) will be simultaneously and harmoniously obtained to accomplish economic and operable crude movement for petroleum refineries. In addition, realistic purchasing constraints, such as crude characteristics and shipping time, and process requirements, for instance tank inventory and blending specs have also been taken into account, so that the model applicability can be guaranteed. The efficacy and performance of the developed model has been tested and demonstrated through various case studies.

References

[1] Ji, X., Huang, S., & Grossmann, I., (2015) Integrated operational and financial hedging for risk management in crude oil procurement. Industrial & Engineering Chemistry Research, 54, 9191-9201.

[2] Chen, R., Deng, T., Huang, S., & Qin, R., (2015) Optimal crude oil procurement under fluctuating price in an oil refinery. European Journal of Operational Research, 245, 438-445.

[3] Kallestrup, K., Lynge, L., Akkerman, R., & Oddsdottir, T., (2014) Decision support in hierarchical planning systems: The case of procurement planning in oil refining industries. Decision Support Systems, 68, 49-63.

[4] Zhang, S., & Xu, Q., (2015) Refinery continuous-time crude scheduling with consideration of long-distance pipeline transportation. Computer and Chemical Engineering, 75, 74-94.

[5] Oddsdottir, T., Grunow, M., & Akkerman, R., (2013) Procurement planning in oil refining industries considering blending operations. Computer and Chemical Engineering, 58, 1-13.

[6] Zhang, J., Wen, Y., & Xu, Q., (2012) Simultaneous optimization of crude oil blending and purchase planning with delivery uncertainty consideration. Industrial & Engineering Chemistry Research, 51, 8453-8464.

536795 Dynamic Production Planning and Scheduling for an Chemical Plant

Abstract:

Chemical/petrochemical industries are facing increasingly challenges to sustain their profitability nowadays due to volatile raw material and product markets, increasingly rigorous environmental and safety regulations, and global competitions. As the key player of chemical industry, the olefin plant is the most important sector for the entire value-added chemical product chain. An olefin plant employs multiple cracking furnaces in parallel to convert various hydrocarbon feed stocks to smaller hydrocarbon molecules, mostly ethylene and propylene. The continuous operational performance of cracking furnaces gradually decays because of coke formation in the reaction coils, which requires each furnace to be periodically shut down for decoking. Thus, production and decoking scheduling of the cracking furnace system is an importance and challenging task for olefin plants. In the previous studies, olefin plant scheduling works were almost exclusively aiming at maximizing the total profit of the furnace system based on the available raw material types and quantities, performance of each furnace, as well as the market prices of raw materials and products. The front-end feedstock procurement planning and the back-end production response to product market have been simplified or just neglected.

In this study, a systematic methodology for the optimization of the entire supply chain of olefin plants have been developed, which covers feedstock purchasing and storage management, olefin production, and productivity control and response to short and long-term product markets. It couples the production planning for both front-end feedstock procurement and back-end production response to product market, as well as the olefin plant scheduling for the optimal olefin product manufacturing. The planning solution optimally determines how much of each major product to produce with respect to time; when and how much of each of feedstock to buy; and solutions of feedstock inventory management. The scheduling solution will determines the detailed production information such as batch processing numbers of each furnace during the given scheduling time, feed allocation, starting and ending time of each batch of each furnace, and the decoking sequence. Such dynamic planning and scheduling results will be simultaneously obtained through a novel MILP (mixed-integer linear programming) model. A case study with real olefin plant data is employed to demonstrate the efficacy of our development,

523461 Sustainable Strategic Planning for a National Natural Gas Energy System Accounting for Unconventional Sources

Abstract:

This work presents the interaction of the different forms of extraction of natural gas either by unconventional methods such as gas shale, offshore extraction and by conventional means of extraction of associated natural gas, where different technologies of enhanced oil & gas recovery such as the injection of heat and the injection of carbon dioxide are taken into account with the intention of improving the extraction of natural gas remaining from the wells; the extraction of non-associated natural gas, as well as importation is also considered. Traditionally, only economic aspects have been contemplated for the design of such processes, this has generated important safety problems when the processes are already in operation. Recently, the inherent safety aspect of processes has been involved since the design stage, particularly for the physical layout of industrial facilities. However, safety from the point of view of occupational health of personnel and people who live near to the facilities has not been adequately addressed, this can be assessed through an analysis of occupational health in these processes; that is, how harmful a process is compared to all processes considered, determining which parts of the process and which chemicals are hazardous to the health of workers who are in constant contact through the skin, eyes, inhalation during their work hours. Furthermore, the environment is also important, a consideration of CO_2 emissions is included to evaluate the impact in the surrounding environment. In addition, the costs associated with the entire process are taken into account, from extraction to distribution of natural gas.

Finally, the objective function consists in maximizing profit which involves the sale of petroleum and natural gas, and the different capital and operational costs; the minimization of the total fresh water used in gas shale extraction, conventional extraction of natural gas associated and non-associated, and the minimizing of total emissions in gas shale, offshore extraction, in conventional extraction of natural gas associated and non-associated, and the emissions generated during importation. The proposed model is a Mixed Integer Linear Programming problem and it was implemented in the software GAMS, where the solver DICOPT was used to solve a case study from Mexico. Results show that there is very important to consider the interaction between different ways of extraction due to the recent increase of energy demand in the world, but also is important to take into account the occupational health of workers, as well as the environment.

526643 A Multi-Objective MILP Model for Planning, Design and Operation of Biomass Supply Chains â€" Capturing the Trade-Offs within the Food-Energy-Water-Environment Nexus

Abstract:

Biomass is a strategic resource for the production of fuels, energy, chemicals and other valuable materials. Demands for biomass are increasing due to many governments subsidising its use to meet renewable energy and GHG emissions targets. These growing demands can cause greater competition with food production and have profound effects on the environment and ecosystems. Producing biomass requires large areas of land, water for irrigation and often fertilisers and pesticides, which can leach into bodies of water. Large and rapid changes in land use (especially deforestation) degrade biodiversity. Effective management of biomass supply chains is crucial to ensure its production and utilisation are sustainable, do not contribute to global warming and preserve the delicate balance in the food-energy-water-environment nexus.

A spatio-temporal multi-objective optimisation model, based on mixed integer programming, was developed for biomass value chains that can maximise synergies and minimise conflicts within the food-energy-water-environment nexus. The model considers a planning horizon, from 2010 to 2050, in order to model the investment in and retirement of technologies and land use changes, as well as seasonal time steps in order to capture the seasonal production and storage of biomass. The model accounts for spatial dependencies of the problem at a 50 km level in order to model the variation of biomass yield potentials across different locations, suitable siting of biomass plantations and processing facilities, transport of biomass and products etc. Land use is modelled using GIS and different constraints were developed, according to the suitability of the land for cultivation and harvesting and social and environmental criteria, in order to limit the use of land for bioenergy. Different types of biomass are modelled including food crops, energy crops and forestry resources. Different technologies for the production of different types of biofuel, as well heat and electricity, are modelled at different scales. The model optimises a large number of interdependent decisions such as: what biomass to grow, where and when to grow them; what energy services/products to generate (e.g. heat, biofuels, electricity, chemicals); what processing technologies to use, where to locate them, when to invest in them; centralised or distributed processing; whether or not to densify/pre-process biomass before transportation; how to transport biomass and distribute products to customers; and how to manage inventory (as the production of most types of biomass is seasonal and not available all year round), in order to obtain the greatest benefit with the lowest impact on the nexus. In this conference, the MILP model will be presented along with the efficient and robust value chains that are synergistic with the nexus, identified using the model.

References:

[1] S. Samsatli, N.J. Samsatli (2018). A multi-objective MILP model for the design and operation of future integrated multi-vector energy networks capturing detailed spatio-temporal dependencies. Applied Energy. DOI: 10.1016/j.apenergy.2017.09.055.

[2] S. Samsatli, N.J. Samsatli, N. Shah (2015). BVCM: a comprehensive and flexible toolkit for whole-system biomass value chain analysis and optimisation - mathematical formulation. Applied Energy, 147, pp. 131-160.

[3] S.M. Jarvis, S. Samsatli (2018). Technologies and infrastructures underpinning future CO2 value chains: a comprehensive review and comparative analysis. Renewable & Sustainable Energy Reviews, 85, 46-68.

[4] S. Samsatli, N.J. Samsatli (2015). A general spatio-temporal model of energy systems with a detailed account of transport and storage. Computers and Chemical Engineering, 80, 155-176.

537295 Data-Driven Multi-Period Planning Model and Global Optimization for Entire Petroleum and Petrochemical Operations

Abstract:

Over the past few decades, the petroleum and petrochemical industry has made significant contribution to the economic development and life quality improvement. The industry converts and upgrades crude oil into a range of useful products including liquefied petroleum gas (LPG), kerosene, naphtha, gasoline, jet fuel, fuel oil and many precursors for commodity chemicals such as benzene, toluene, and xylenes (mainly p-xylenes) via different reactions and separations. However, the petroleum and petrochemical industry is facing great challenges these days from global competitions due to the declined crude quality, fluctuation of crude prices and product demands, as well as stricter environmental regulations. The difficulty in making effective planning decisions has driven the petroleum and petrochemical industry to make use of cutting-edge techniques, especially advanced mixed-integer mathematical programming techniques.

Integrated petroleum and petrochemical operations include both refinery and chemical production operations, such as crude oil blending and processing, production unit operations, and product blending and distribution [1-2]. An integrated petroleum and petrochemical operational planning involve simultaneous optimization of crude blending, selection of units and their production modes, flow connections between production units, and pooling and blending operations to satisfy quality requirements of production units, intermediates, and final products. Mathematical modelling of production units, pooling and blending operations often introduces bilinear, trilinear terms, exponential and higher order terms, which result in a non-convex mixed integer nonlinear programming problem (MINLP).

The refinery planning problem has received considerable attention since the introduction of linear programming in the 1950s. The research focused on developing different models and algorithms to solve large-scale industrial problems, leading to the development of commercial software such as RPMS (Refinery and Petrochemical Modeling System) [3], PIMS (Process Industry Modeling System) [4], and GRTMPS (Haverly Systems) [5]. While the commercially available software can be extended to model and optimize integrated petroleum and petrochemical plants, inaccuracy caused by non-rigorous linear models for product yield and property prediction in processing units may reduce the overall profitability or sacrifice product quality. Additionally, Îµ-global optimality cannot be guaranteed. On the other hand, nonlinear models have been proposed for product yield and property prediction in processing units during refinery planning without ensuring Îµ- global optimality [6-9]. A comprehensive review of refinery planning can be found in [1-2].

All the above efforts were made for solely the refinery planning without integrating petrochemical operations. However, modern refineries are often integrated with petrochemical operations to improve the site efficiency, and consequently the profit margin. Recently, Li et al [10] developed a data-driven single-period planning model for integrated petroleum and petrochemical operations. They covered in detailed how the data was initially collected and processed to build the basis for the model. The postulation of yield and outlet properties was done thoroughly by using a swing-cut approach to model crude distillation unit (CDU) and linear, quadratic, bilinear or polynomial models, or the combination of them to model the secondary unit operations such as hydrotreating units, hydrocracking units, and fluidized catalytic cracking unit. This comprehensive model has also accounted for parallel unit selection and operation mode selection. However, the operation mode selection was done using a lumping method, which may lead to the incorrect prediction of product properties, affecting the yield and property prediction for the downstream units.

In this work, a multi-period MINLP planning model is developed using the data-driven prediction correlations of product yields and properties in processing units from Li et al. [10]. The planning horizon is divided into multiple periods based on profiles of production modes and product demands. In each period, only one production mode is allowed for units with multiple operating modes such as the CDU and hydrocracking units. The objective is to maximize the total profit including revenues from product sale minus raw material and operating cost. The resulting multi-period planning model is a large-scale nonconvex mixed-integer nonlinear programming model. To solve the model to Îµ-global optimality, two-stage solution approach is proposed where a feasibility problem is first constructed by introducing slack variables to the yield and property prediction correlations in the original MINLP model and changing the objective to minimization of these slack variables. This feasibility model is solved using ANTIGONE/GAMS [11], whose solution indicates whether the original problem is feasible or not. If all slack variables are very close to or exactly zero, the original problem is feasible. Then, all slack variables are fixed to their values obtained from the feasibility problem, another MINLP optimisation problem having the same objective function as the original MINLP model and the same constraints as the feasibility problem is constructed and solved to Îµ-global optimality using ANTIGONE/GAMS[11]. Two case studies from a real petroleum and petrochemical site in China are used to illustrate the capability of the proposed multi-period planning model and postulated solution approach. Comparisons were also made between the results generated from the proposed multi-period planning model and that from the single-period planning model. The results demonstrate that the proposed multi-period planning model could generate much higher profit than that of single-period planning model, with more accurate yield and property predictions for those processing units with different operating modes.

References

[1] Shah, N. K.; Li, Zu; Ierapetritou, M. G. Petroleum Refining Operations: Key Issues, Advances, and Opportunities, *Industrial and Engineering Chemistry Research*, 2011, *50*, 1161-1170.

[2] Khor, Cheng Seong, Varvarezos, Dimitrios, Petroleum Refinery Optimization, Optim Eng, DOI 10.1007/s11081-016-9338-x.

[3] RPMS (Refinery and Petrochemical Modeling Systems): A System Description; Bonner and Moore: Houston, TX, 1979.

[4] Aspen PIMS System Reference (v7.2), Aspen Technology Inc.; Burlington, MA, 2010.

[5] GRTMPS (Haverly Systems), http://www.haverly.com/main-products/13-products/9-grtmps.

[6] Pinto, J. M.; Moro, L. F. L. A Planning Model for Petroleum Refineries. *Braz. J. Chem. Eng.* 2000, *17*, 575-585.

[7] Li, W. K.; Hui, C. W.; Li, A. X. Integrating CDU, FCC, and Product Blending Models into Refinery Planning. *Computers and Chemical Engineering* 2005, *29*, 2010-2028.

[8] Alhajri, I.; Elkamel, A.; Albahri, T.; Douglas, P. L. A Nonlinear Programming Model for Refinery Planning and Optimization with Rigorous Process Models and Product Quality Specifications. International Journal of Oil, Gas, and Coal Technology 2008, 1, 283-307.

[9] Alattas, A. M.; Grossmann, I. E.; Palou-Rivera, I. Integration of Nonlinear Crude Distillation Unit Models in Refinery Planning Optimization. *Industrial and Engineering Chemistry Research* 2011, *50*, 6860-6870.

[10] Li, J.; Boukouvala, F., Floudas, C. A., Zhao, B., Du, G., Su, X., Liu, H., Data-Driven Mathematical Modelling and Global Optimization Framework for Entire Petrochemical Planning Operations, AIChE Journal, 62, 3020-3040.

[11] Misener, R.; Floudas, C. A. ANTIGONE: Algorithms for continuous/Integer Global Optimization of Nonlinear Equations. Journal of Global Optimization, Journal of Global Optimization, 2014, 59, 503-526.

544816 CESMII Overview

Abstract:

The Clean Energy Smart Manufacturing Innovation Institute (CESMII), established by the US Department of Energy, is part of a network of institutes under the Manufacturing USA Initiative with a goal to secure the future of manufacturing in the U.S. through innovation, collaboration and education. CESMIIâ€™s mission is to radically democratize and accelerate the development and adoption of Smart Manufacturing (SM) technologies (sensing, control, data analytics, predictive modeling and platforms) to become the driving, sustainable engine that delivers energy efficiency and real-time business improvements in U.S. manufacturing. CESMII integrated strategies to achieve this mission include business practice innovation, enabling technologies, workforce development and a collaborative Smart Manufacturing Platformâ„¢ for industrial applications. Haresh Malkani, Chief Technology Officer â€“ CESMII, will provide an overview of the institute, its technology focus areas and recent progress.

544817 CESMII Roadmap Projects – Advancing CESMII's Mission

Abstract:

CESMII is about to kick off a portfolio of 10 innovative R&D roadmap projects covering a range of technologies including sensing, process control, data analytics, modeling, as well as workforce development, to address significant challenges faced by the manufacturing community. The projects cut across manufacturing types (discrete, hybrid, process), industry verticals (steel, cement, chemical, aerospace), and scale of focus (unit process, plant, enterprise). Miguel Corcio, Director - Technology Programs - CESMII, we will provide brief overview of the portfolio during this session.

546992 Panel Session: Democratization of Smart Manufacturing (SM)

Abstract:

Industrie 4.0, Smart Manufacturing, Industrial Internet of Things, Big Data, Cloud Computing are no longer just buzz words. They have become a reality and every year more manufacturers are getting started on this journey. But it takes investment in technology and talent to do so. Does it mean that this reality is only within the reach of the large and mighty companies? What can small and medium manufacturers do to be in this game? How do we make technology and expertise accessible for them cost effectively? Join our panelists representing both industry and academia in this healthy dialogue.

Panelists:

- Larry Megan – Praxair
- Sergio Butkewitsch-Choze – Arconic
- Panagiotis Christofides – UCLA
- B. Wayne Bequette – Rensselaer Polytechnic Institute

Moderator:

- Jim Davis, Vice Provost, Information Technology and Chief Academic Technology Officer, UCLA

532505 Enhancing Relaxations for Nonconvex Mixed-Integer Quadratically-Constrained Quadratic Programs

Abstract:

Enhancing Relaxations for Nonconvex Mixed-Integer Quadratically Constrained Quadratic Programs

Mixed-integer quadratically-constrained quadratic programs (MIQCQPs), and related classes of problems such as quadratically-constrained quadratic programs (QCQPs) and quadratic programs (QPs) arise in a wide variety of engineering and scientific applications including pooling and blending in process networks [12], facility location [23], production planning [8], quadratic assignment [9] and portfolio optimization [15]. Due to their practical importance, these classes of problems have been extensively studied in the mathematical programming literature. They are typically solved using spatial branch-and-bound algorithms, whose efficiency primarily depends on the quality of the relaxations constructed to bound the problem.

Commonly used relaxations for these type of problems can be classified into four groups. The first group involves relaxations obtained via standard factorable programming techniques whereby each quadratic function is decomposed iteratively by introducing intermediate variables and constraints for each nonconvex term, until each intermediate expression can be outer-approximated by its convex and/or concave envelopes [10, 20, 22]. While these techniques are relatively simple to implement, they typically result in large relaxation gaps. The second group is given by semi-definite programming (SDP) relaxations. This approach has received significant attention in recent years [1, 5, 6, 21]. While SDP relaxations often provide strong bounds, they are computationally expensive to solve, which limits their use within general purpose branch-and-bound global solvers. The third group relies on the reformulation-linearization techniques (RLT) [17–19]. Though these techniques typically yield tight bounds, the resulting relaxations grow quickly in size, and thus can become expensive to solve. The fourth group consists of factorable relaxations augmented by adding various classes of valid inequalities such as SDP-based cuts [7, 16], RTL-based cuts [24, 25], facets of the envelopes of edge-concave and multilinear subexpressions [2, 3, 11, 13], and mixed-integer cuts [4]. These augmented relaxations provide stronger bounds than standard factorable relaxations while remaining computationally inexpensive to solve. Due to its practicality, this class of relaxations is employed in most general-purpose global optimization packages.

In this work, we show how certain facets of the Boolean Quadric Polytope [14] can be utilized in order to enhance factorable relaxations of MIQCQPs constructed by global optimization solvers. To illustrate the benefits of this approach, we propose a computational implementation which integrates these facets as cutting planes into the branch-and-reduce global solver BARON. Our implementation consists of several algorithmic developments including tools for the identification of substructures in a graph representation of the problem, as well as efficient cut generation and separation algorithms which are embedded at every node of the branch-and-bound tree. We demonstrate the impact of the proposed implementation by conducting an extensive computational study on a large collection of problems selected from publicly available test sets.

References:

[1] Anstreicher, K. M., Semidefinite programming versus the reformulation-linearization technique for nonconvex quadratically constrained quadratic programming, Journal of Global Optimization, 43:471–484, 2009.

[2] Bao, X., Khajavirad, A., Sahinidis, N.V., and Tawarmalani, M., Global optimization of nonconvex problems with multilinear intermediates. Mathematical Programming Computation, 7:1–37, 2015.

[3] Bao, X., Sahinidis, N.V., and Tawarmalani, M., Multiterm polyhedral relaxations for nonconvex quadratically constrained quadratic programs. Optimization Methods and Software, 24:485–504, 2009.

[4] Bonami, P., Günlük, O., and Linderoth, J. Globally solving nonconvex quadratic programming problems with box constraints via integer programming methods, Mathematical Programming Computation, 2018, https://doi.org/10.1007/s12532-018-0133-x

[5] Buchheim, C. and Wiegele, A. Semidefinite relaxations for non-convex quadratic mixed-integer programming, Mathematical Programming, 141:435–452, 2013.

[6] Burer, S., Vandenbussche, D., A finite branch-and-bound algorithm for nonconvex quadratic programming via semidefinite relaxations. Mathematical Programming, 112:259–282, 2008.

[7] Dong, H., Relaxing nonconvex quadratic functions by multiple adaptive diagonal perturbations. SIAM Journal on Optimization, 26:1962–1985, 2016.

[8] Liu, M.L. and Sahinidis, N.V., Process planning in a fuzzy environment, European Journal of Operational Research, 100:142–169, 1997.

[9] Loiola, E.M., De Abreu, N.M.M., Boaventura-Netto, P.O., Hahn, P., and Querido, T., A survey for the quadratic assignment problem, European Journal of Operational Research, 176:657–690, 2007.

[10] McCormick, G. P., Computability of global solutions to factorable nonconvex programs: Part I—Convex underestimating problems, Mathematical Programming, 10:147–175, 1976.

[11] Misener, R. and Floudas, C.A., Global optimization of mixed-integer quadratically-constrained quadratic programs (MIQCQP) through piecewise-linear and edge-concave relaxations, Mathematical Programming, 136:155–182, 2012.

[12] Misener, R., Gounaris, C.E., and Floudas, C.A., Mathematical modeling and global optimization of large-scale extended pooling problems with the (EPA) complex emissions constraints, Computers and Chemical Engineering, 34:1432–1456, 2010.

[13] Misener, R., Smadbeck, J.B, and Floudas, C.A., Dynamically generated cutting planes for mixed-integer quadratically constrained quadratic programs and their incorporation into GloMIQO 2, Optimization Methods and Software, 30:215–249, 2015.

[14] Padberg, M. The boolean quadric polytope: Some characteristics, facets and relatives. Mathematical Programming, 45:139–172, 1989.

[15] Rios, L. and Sahinidis, N.V., Portfolio optimization for wealth-dependent risk preferences, Annals of Operations Research, 177:63–90, 2010.

[16] Saxena A., Bonami, P., and Lee, J., Convex relaxations of mixed integer quadratically constrained programs: Projected formulations, Mathematical Programming, 130:359–413, 2011.

[17] Sherali, H.D. and Adams, W.P., A hierarchy of relaxations between the continuous and convex hull representations for zero-one programming problems, SIAM Journal on Discrete Mathematics, 3:411–430, 1990.

[18] Sherali, H.D., and Alameddine, A., A new reformulation-linearization technique for bilinear programming problems, Journal of Global Optimization, 2:379–410, 1992.

[19] Sherali, H.D. and Tuncbilek, C.H., A reformulation–convexification approach for solving nonconvex quadratic programming problems, Journal of Global Optimization, 7:1–31, 1995.

[20] Sherali, H. D. and Wang, H., Global optimization of nonconvex factorable programming problems, Mathematical Programming, 89:459–478, 2001.

[21] Shor, N.Z., Quadratic optimization problems, Soviet Journal of Computer and Systems Sciences, 25:1–11, 1987.

[22] Tawarmalani, M. and Sahinidis, N. V., Global optimization of mixed-integer nonlinear programs: A theoretical and computational study. Mathematical Programming, 99:563–591, 2004.

[23] Xie, W. and Sahinidis, N.V., A branch-and-bound algorithm for the continuous facility layout problem, Computers and Chemical Engineering, 32:1016–1028, 2008.

[24] Zorn, K. and Sahinidis, N.V., Global optimization of general non-convex problems with intermediate bilinear substructures, Optimization Methods and Software, 29:442–462, 2013.

[25] Zorn, K. and Sahinidis, N.V., Global optimization of general nonconvex problems with intermediate polynomial substructures, Journal of Global Optimization, 59:673–693, 2014.

531123 Advances in Decomposition Strategies for Non-Convex Nonlinear Programs

Abstract:

Optimization plays an important role in a variety of areas, including process design, operation, and control. Large-scale nonconvex optimization problems that commonly arise in these areas need to be solved efficiently. These problems are characterized by having a large number of equations and variables and can become computationally intractable, exceeding the computational capabilities of a serial algorithm on a single workstation. However, these problems often have a structure that can be exploited by decomposition algorithms.

In this presentation, we describe two decomposition strategies for solving large-scale nonconvex structured optimization problems. In the first strategy, we decompose all scale-dependent operations of a serial algorithm to achieve parallelization. We focus on the parallelization of the interior-point (IP) algorithm and use the Schur-complement to decompose the solution of the KKT system in each iteration of the IP algorithm. The second decomposition strategy partitions the large-scale optimization problem into subproblems and coordinates their solutions to find the overall optimal solution. For this second type of decomposition, we focus on studying the Alternating Direction Method of Multipliers (ADMM). We present the pros and cons of both approaches and propose acceleration techniques to improve both.

We demonstrate these two decomposition strategies on several nonconvex stochastic programming and dynamic optimization problems. We also study the convergence of ADMM on nonconvex problems, and present connections between ADMM and the classical augmented Lagrangian (AL). We summarize our results in terms of a Lyapunov function and primal and dual feasibility metric and demonstrate that performing multiple coordination steps at each ADMM iteration approaches the behavior of the AL method.

529958 Online Generation Via Offline Selection of Strong Linear Cuts from a Semidefinite Programming Relaxation

Abstract:

Deterministic global optimization is highly relevant solving quadratically-constrained quadratic optimisation (QCQP) problems that arise in a variety of engineering applications [1]. In a branch-and-cut framework, cutting planes are known to significantly expedite the solution for certain classes of QCQP problems [2-4].

Convex, particularly semidefinite programming (SDP), relaxations for non-convex continuous quadratic optimization can provide tighter bounds than the reformulation-linearization technique [5]. But using SDP relaxations directly in branch and cut is impeded by a lack of warm starting and inefficiency when combined with linear cut classes, e.g. the McCormick bilinear envelopes [6] or the reformulation-linearization technique [7]. State-of-the-art deterministic global optimization solvers use linear relaxations [4, 8-12], so SDP technology is effectively unavailable.

Previous work in developing SDP-based cutting planes has shown that these cuts can be very effective for certain classes of QCQP [13, 14]. But it's still unclear *when* to generate these relatively expensive cuts, i.e. when to solve an SDP problem with the goal of generating a cutting plane. We seek an inexpensive oracle that will tell us when it is relevant to generate a cut and over which variables.

We present a general framework based on machine learning for a strong linear outer-approximation that can retain most tightness of such SDP relaxations, in the form of few strong low-dimensional linear cuts selected offline. The cut selection complexity is taken offline by using a neural network estimator (trained before installing solver software) as a selection device for the strongest cuts. Once the neural network estimator selects the appropriate subsets of variables, we generate hyperplanes cutting off the current objective based on negative eigenvalues [15, 16].

We present results of our method on several classes of non-convex application problems from the engineering literature. We also test the strategy on generic QCQP and integrate into the global optimization solver ANTIGONE [17].

References

[1] Boukouvala F, Misener R, Floudas CA. Global optimization advances in mixed-integer nonlinear programming, MINLP, and constrained derivative-free optimization, CDFO. *Eur J Oper Res*. 252(3): 701-727, **2016**.

[2] Bao X, Sahinidis NV, Tawarmalani M. Multiterm polyhedral relaxations for nonconvex, quadratically constrained quadratic programs. *Optim Met Softw*. 24(4-5): 485-504, **2009**.

[3] Misener R, Smadbeck JB, Floudas CA. Dynamically generated cutting planes for mixed-integer quadratically constrained quadratic programs and their incorporation into GloMIQO 2. *Optim Met Softw*. 30(1): 215-49, **2015**.

[4] Bonami P, Günlük O, Linderoth J. Solving box-constrained nonconvex quadratic programs. *Optimization online*. **2016**.

[5] Anstreicher KM. Semidefinite programming versus the reformulation-linearization technique for nonconvex quadratically constrained quadratic programming. *J Glob Optim*. 43(2-3): 471 – 484, **2009**.

[6] McCormick GP. Computability of global solutions to factorable nonconvex programs: Part 1-convex underestimating problems. *Math Program*. 10: 147–175, **1976**.

[7] Sherali HD, Adams WP. *A Reformulation-Linearization Technique for Solving Discrete and Continuous Nonconvex Problems*, Nonconvex Optimization and its Applications, Kluwer Academic Publishers, Dordrecht, The Netherlands, **1999**.

[8] Tawarmalani M, Sahinidis NV. A polyhedral branch-and-cut approach to global optimization. *Math Program*. 103: 225-249, **2005**.

[9] Belotti P, Lee J, Liberti L, Margot F, Wächter A. Branching and bounds tightening techniques for non-convex MINLP. *Optim Met Softw*. 24: 597-634, **2009**.

[10] Misener R, Floudas CA. Global optimization of mixed-integer quadratically-constrained quadratic programs (MIQCQP) through piecewise-linear and edge-concave relaxations. *Math Program B*. 136: 155-182, **2012**.

[11] Berthold T, Heinz S, Vigerske S. Extending a CIP framework to solve MIQCPs, in Mixed Integer Nonlinear Programming, In Lee, Leyffer (ed) *Mixed Integer Nonlinear Program-ming, The IMA Volumes in Mathematics and its Applications*. 154: 427-444, **2012**.

[12] Misener R, Floudas CA, GloMIQO: Global mixed-integer quadratic optimizer. *J Glob Optim*. 57: 3-50, **2013**.

[13] Saxena A, Bonami P, Lee J. Convex relaxations of non-convex mixed integer quadratically constrained programs: Extended formulations, *Math Program*. 124: 383-411, **2010**.

[14] Saxena A, Bonami P, Lee J. Convex relaxations of non-convex mixed integer quadratically constrained programs: Projected formulations. *Math Program*. 130: 359-413, **2011**.

[15] Qualizza A, Belotti P, Margot, F. Linear programming relaxations of quadratically constrained quadratic programs. In Lee, Leyffer (ed) *Mixed Integer Nonlinear Programming, The IMA Volumes in Mathematics and its Applications*, 154: 407-426, **2012**.

[16] Sherali HD, Fraticelli BMP. Enhancing RLT relaxations via a new class of semidefinite cuts.*J Glob Optim*. 22(1-4): 233-261, **2002**.

[17] Misener R, Floudas CA, ANTIGONE: Algorithms for continuous / integer global optimization of nonlinear equations, *J Glob Optim*. 59(2-3): 503-526, **2014**.

537734 New Underestimator and Branching Scheme for the Global Optimization of General Nonconvex Problems

Abstract:

A deterministic global optimization approach, when minimizing a nonconvex problem, is based on three key steps: (i) generating lower bounds using valid underestimators, (ii) obtaining feasible/local solutions and use as upper bounds, and (iii) branching variables to create new nodes for progressively updating the lower and upper bounds until they converge to Îµ-global optimality. While local solvers can be used for upper bounds, the selection of the underestimator and branching scheme are central to the effectiveness of a branch-and-bound (B&B) algorithm for faster convergence. Underestimators are usually obtained by relaxing the original problem such that the relaxation is convex, thereby ensuring that the solution of the convex problem is a valid lower bound. While convex envelopes are the tightest convex underestimators, currently these are available only for a handful of functions with explicit forms. Furthermore, the relaxation of a highly nonconvex problem using general convex underestimators may not be tight, since the relaxation often depends on how close the original problem is to convexity. In terms of branching, the feasible space is generally divided into rectangular partitions by dividing along one of the edges. Different strategies on selecting the branching variable and branch point could lead to different computational performances, especially if the number of variables is large.

Our contribution is two-fold. First, instead of using convex underestimators, we propose to exploit a new class of underestimators that is edge-concave [1]. The underestimator is constructed by subtracting a positive quadratic expression such that all non-edge-concavities in the original function is overpowered by the added expression. While the edge-concave underestimator is nonlinear, the linear facets of its vertex polyhedral convex envelope [2] leads to a linear programming (LP)-based relaxation of the original nonconvex problem. We will present some theoretical results on this new class of underestimators and compare the performance of the LP relaxation with relaxations obtained by convex underestimators such as Î±BB and its variants. Second, we propose a novel branching technique based on a univariate projection of the original multi-dimensional space into a single auxiliary space [3]. The idea is based on the fact that when all function values are projected onto a univariate space t defined by summation of the decision variables x, a point-to-set map can be generated and a univariate function exists on this map such that its global minima is the same as that of the original function. This enables us to devise a novel branching scheme only on t rather than on x in a B&B framework We will present separate results showing the efficacy of the proposed relaxation and branching schemes. Furthermore, results will be presented for the cases where we employ the proposed LP relaxation and the t-branching scheme together in a B&B framework for fast convergence to Îµ-global optimality.

References:

[1] Hasan, M. M. F. An edge-concave underestimator for the global optimization of twice-differentiable nonconvex problems. *Journal of Global Optimization*, 1-18, 2018. https://doi.org/10.1007/s10898-018-0646-x.

[2] Tardella, F.: On the existence of polyhedral convex envelopes. In: Floudas, C.A., Pardalos, P.M. (eds.) Frontiers in

Global Optimization, pp. 563â€"573. Kluwer Academic Publishers, Dordrecht (2003).

[3] Bajaj, I. and Hasan, M. M. F. UNIPOPT: Univariate projection-based optimization without derivatives, *Under Review*.

529503 On Piecewise Under- and over-Estimators of Fractional Terms

Abstract:

Fractional terms are encountered in a wide variety of applications including heat exchange networks, multicomponent distillation sequences, etc. The nonconvexity of fractional terms poses a challenge for local solvers, which can get trapped in a local optimal solution. In contrast, global solvers implement spatial branch-and-bound, which finds the global optimum by iteratively refining the search domain. In this process, the global solvers use bounds on the optimum, which are derived by replacing the fractional terms with a convex outer-approximation. The standard relaxation process involves replacing each fractional term with an auxiliary variable, which is then overestimated by its concave envelope [1 â€" 4] and underestimated by either an outer-linearization of the bilinear term [4], a quadratic/linear underestimator [1,3], and/or its convex envelope [3]. However, empirical computational evidence suggests that, for many practical models that involve fractional terms, particularly if they also involve combinatorial choices, the state-of-the-art solvers may either fail to converge or do so rather slowly. In contrast, MI(L/SOC)P solvers have improved drastically, over the last two decades, and can now handle reasonably large instances. Here, we explore ways to use the MI(L/SOC)P solvers to solve hard (MI)NLP instances involving fractional terms to near optimality, which is not possible with current global solvers.

We consider various ways of constructing piecewise under- and over-estimators of fractional terms. These estimators are obtained using outer-linearization of bilinear terms [1 â€" 4], quadratic/linear underestimator [1,3], and outer-approximation followed by outer-linearization, an approach recently proposed in [5]. Since these relaxations rely heavily on bounds, we introduce binary variables, and use the incremental cost formulation [6] along with the reformulation-linearization technique [7] to develop piece-wise relaxations. This step differs from the standard approach of constructing IP relaxations via piecewise relaxations of bilinear terms obtained after cross-multiplying the fractional term. We show that using the proposed IP relaxations we are able to solve, for the first time, to near optimality, an MINLP that identifies optimal multicomponent distillation sequences. The fractional terms in this formulation arise from Underwood equations and the combinatorial choices model the structural specification of the distillation configuration. We present computational results, in the context of this application, comparing the efficacy of various estimators for the fractional terms.

1. Zamora, J. M., & Grossmann, I. E. (1998). A global MINLP optimization algorithm for the synthesis of heat exchanger networks with no stream splits. Computers & Chemical Engineering, 22(3), 367-384.
2. Zamora, J. M., & Grossmann, I. E. (1999). A branch and contract algorithm for problems with concave univariate, bilinear and linear fractional terms. Journal of Global Optimization, 14(3), 217-249.
3. Tawarmalani, M., & Sahinidis, N. V. (2001). Semidefinite relaxations of fractional programs via novel convexification techniques. Journal of Global Optimization, 20(2), 133-154.
4. Tawarmalani, M., & Sahinidis, N. V. (2002). Convexification and global optimization in continuous and mixed-integer nonlinear programming: theory, algorithms, software, and applications (Vol. 65). Springer Science & Business Media.
5. He, T., & Tawarmalani, M. On relaxing composite functions exploiting inner function structure: polynomially equivalent formulations. Working paper.
6. Dantzig, G. B. (1960). On the significance of solving linear programming problems with some integer variables. Econometrica, Journal of the Econometric Society, 30-44.

519764 Quadratic Underestimators of Differentiable Mccormick Relaxations for Deterministic Global Optimization

Abstract:

Global optimization is extremely important in process systems engineering due to the nonconvexity of commonly-encountered models. Deterministic global solvers are comprised of a spatial branch-and-bound algorithm in conjunction with a means of calculating and tightening valid bounds, and creating valid convex relaxations of the original problem on relevant subdomains [1,2]. Typically, upper bounds can be determined by solving locally the original nonlinear program (NLP) to feasibility [3]. The computation of the lower bound has been the proven challenge. While the auxiliary variable methods provide a valid approach to computing lower bounds, the use of generalized McCormick relaxations may have significant benefits: they allow for the naive relaxation of algorithmically-defined functions and a potentially smaller decision space for the global optimization problem [4-7].

Lower-bounding problems obtained through standard or multivariate McCormick relaxations are potentially nonsmooth NLPs necessitating the use of nonsmooth solvers such as the proximal bundle method, PBUN [3,6-9]. Nonsmooth solvers are typically restricted to affine constraints and are generally not as robust as commercial local solvers such as IPOPT [10]. Mitsos et al. makes use of affine relaxations of McCormick relaxations of the objective function and constraints in order to generate linear programs [7]. Khan et al. developed differentiable McCormick relaxations which can be used to form smooth NLPs that can be readily addressed by local solvers [11]. However, these potentially-tighter relaxations are not without a cost. LP relaxations only require a single function and constraint evaluation to generate the lower-bounding problem. While the number of function evaluations for commercial solvers vary greatly depending on the problem [7,9].

In this work, we introduce a novel manner of generating nonlinear differentiable lower bounds based on differentiable McCormick relaxations. We utilize these new relaxations to generate convex quadratic programs (QP) and quadratically-constrained quadratic programs (QCQP) [12]. Each of these subproblems requires a single function (gradient and hessian) evaluation to generate a small QP or QCQP. These relaxations are often tighter than affine relaxations of similar functions, require fewer original function evaluations than full NLP subproblems, and can be readily solved using commercial solvers such as Gurobi [13].

We will introduce the theory behind these QP/QCQP relaxations. Practical aspects of generating these relaxed problems, such as the varying automatic differentiation schemes, will be reviewed [14]. Lastly, the performance of this relaxation scheme will be compared to nonsmooth NLP, smooth NLP, and LP relaxations within a global optimization routine using literature examples and problems taken from the COCONUT library [15].

[1] Horst, R. and Tuy, H., A. *Global Optimization: Deterministic Approaches*. Springer Berlin Heidelberg. 2013.

[2] Floudas, Christodoulos A., and Chrysanthos E. Gounaris. "A review of recent advances in global optimization." *Journal of Global Optimization* 45.1 (2009): 3-38.

[3] Stuber, M.D., Scott, J.K., and P.I. Barton. Convex and Concave Relaxations of Implicit Functions. *Optimization Methods and Software*. 30(3), 424-460, 2014.

[4] Tawarmalani, M. and N. V. Sahinidis, A polyhedral branch-and-cut approach to global optimization, Mathematical Programming, 103(2), 225-249, 2005.

[5] Sahinidis, N. V., BARON 14.4.0: Global Optimization of Mixed-Integer Nonlinear Programs, User's manual, 2014.

[6] Scott, J.K., Stuber, M.D., and Barton, P.I. Generalized McCormick Relaxations. *J Global Optim*, 51:569-606, 2011.

[7] Mitsos, A., Chachuat, B., and Barton, P.I. McCormick-Based Relaxations of Algorithms. SIAM J. Optim. 20(2): 573-601.

[8] Tsoukalas, A. and Mitsos, A. Mutlivariate McCormick Relaxations. *J Global Optim*, 59:633-662, 2014.

[9] L. Luksan and J. Vlcek. Algorithms for non-differentiable optimization. ACM

Transactions on Mathematical Software, 27(2):193â€"213, June 2001.

[10] Waechter, A.W., Biegler, L.T., On the Implementation of a Primal-Dual Interior Point Filter Line Search Algorithm for Large-Scale Nonlinear Programming. Mathematical Programming 106(1): 25-57, 2006.

[11] Khan, K.A., Watson, H.A., and Barton, P.I. Differentiable McCormick Relaxations. *J Global Optim*, 67(4):687-729, 2017.

[12] Boyd, Stephen, and Lieven Vandenberghe. *Convex optimization*. Cambridge University Press, 2004. [13] Gurobi Optimization, Inc. Gurobi Optimizer Reference Manual. http://www.gurobi.com. 2016

[14] Griewank, A. *Evaluating derivatives: Principles and techniques of algorithmic differentiation* (2. rev. ed.). Philadelphia: Soc. for Industrial and Applied Mathematics. 2009.

[15] O. Shcherbina, et al. Benchmarking Global Optimization and Constraint Satisfaction Codes, In *Global Optimization and Constraint Satisfaction*, Bliek, Ch., Jermann, Ch. and Neumaier, A. Springer, Berlin 2003, 212-222.

528773 A Novel Branching Scheme for Problems with Reverse Convex Quadratic Constraints and Its Application to Packing Problems

Abstract:

A quadratic constraint f(x) â‰¥ 0 is called reverse convex if f(x) is convex [1]. Reverse convex quadratic constraints are abundant in mathematical models for many industrial applications such as cutting and packing problems (CPP) [2] and wireless sensor network localization [3] among others. Though there have been impressive improvements in the general purpose global solvers for the past decades, solving problems with reverse convex quadratic constraints to Îµ-optimality is still a very challenging task. In this work, we propose a novel branching scheme in order to tackle this type of constraint and we compare our approach with global solvers via computational studies on cutting and packing instances.

Unlike a traditional, general purpose global solver, which must generically relax a nonconvex constraint and tighten that relaxation via branching on the domains of the native variables [4], our novel approach recognizes the physical meaning of the reverse convex quadratic constraint and proposes a direct branching on nonconvex constraints themselves. The inspiration emanates from two observations. First, branching on nonconvex constraints is enforcing their feasibility in a more direct way and therefore should be more effective in cutting off infeasible space. Second, in the context of specific applications such as CPP, due to a large number of these constraints and their close relationship, feasibility-based tightening techniques can be exploited in a more sophisticated fashion if the constraints feasible domain is explicitly split.

The CPP aims to find a packing of shapes that minimizes the amount of material needed to carve them out of a stock. This minimization of waste can be of utmost economic importance in certain large industrial sectors, wherein a minuscule reduction in the amount of stock material used would result in significant monetary savings across the whole industry. In this work, we focus on two types of CPP: circle packing problems (CP) [5] and irregular shape packing problems (ISP, also known as nesting problems) [2]. For CP and ISP, relevant studies in the CPP community have mostly focused on the development of heuristics to obtain large-scale packings. While heuristic methods are practically valuable for the generation of good solutions [6], they lack any guarantees of optimality and cannot rigorously quantify the quality of packings produced. The challenge of solving CP and ISP to global optimality results from the fact that enforcing no overlap between two shapes leads to certain nonconvexities that are still numerically very challenging for even the state-of-the-art global optimization solvers.

In this work, we focus on tackling reverse convex quadratic constraints and propose a novel branching scheme. In order to evaluate our approachâ€™s effectiveness and compare its performance with deterministic global solvers, we conduct comprehensive computational studies on a suite of circle packing instances and irregular shape packing instances [7]. Preliminary results indicate that our new branching method is competitive to global solvers including BARON [8], COUENNE [9] and GloMIQO [10], LINDOGlobal [11] and SCIP [12].

References:

[1] Hillestad, R.J. and Jacobsen, S.E., 1980. Reverse convex programming. *Applied Mathematics and Optimization*, *6*(1), pp.63-78.

[2] Wäscher, G., Haußner, H. and Schumann, H., 2007. An improved typology of cutting and packing problems. *European journal of operational research*, *183*(3), pp.1109-1130.

[3] Biswas, P. and Ye, Y., 2004, April. Semidefinite programming for ad hoc wireless sensor network localization. In *Proceedings of the 3rd international symposium on Information processing in sensor networks* (pp. 46-54). ACM.

[4] Floudas, C.A. and Gounaris, C.E., 2009. A review of recent advances in global optimization. *Journal of Global Optimization*, *45*(1), pp.3-38.

[5] Castillo, I., Kampas, F.J. and Pintér, J.D., 2008. Solving circle packing problems by global optimization: numerical results and industrial applications. *European Journal of Operational Research*, *191*(3), pp.786-802.

[6] Hopper, E. and Turton, B.C., 2001. A review of the application of meta-heuristic algorithms to 2D strip packing problems. *Artificial Intelligence Review*, *16*(4), pp.257-300.

[7] Wang, A., Hanselman, C.L. and Gounaris, C.E., 2018. A customized branch-and-bound approach for irregular shape nesting. *Journal of Global Optimization*, pp.1-21.

[8] Tawarmalani, M. and Sahinidis, N.V., 2005. A polyhedral branch-and-cut approach to global optimization. *Mathematical Programming*, *103*(2), pp.225-249.

[9] Belotti, P., Lee, J., Liberti, L., Margot, F. and Wächter, A., 2009. Branching and bounds tighteningtechniques for non-convex MINLP. *Optimization Methods & Software*, *24*(4-5), pp.597-634.

[10] Misener, R. and Floudas, C.A., 2013. GloMIQO: Global mixed-integer quadratic optimizer. *Journal of Global Optimization*, *57*(1), pp.3-50.

[11] Lindo Systems Inc.: LINDOGlobal 11.0 (2017).

[12] Achterberg, T., 2009. SCIP: solving constraint integer programs. *Mathematical Programming Computation*, *1*(1), pp.1-41.

536749 Subtangent-Based Approaches for Optimization of Parametric Process Systems

Abstract:

Several established methods for deterministic global optimization typically compute useful global bounding information by constructing and minimizing convex underestimators of parametric process models. Convex underestimators (and analogous concave overestimators) are also useful in their own right for reachable-set methods, design centering, and constructing useful simpler approximations of complicated process models.

This presentation examines linear â€œsubtangentsâ€ of nonlinear convex underestimators of process models, constructed from the original underestimators using subgradient evaluations. Subtangents are weaker relaxations than the original underestimators, yet are straightforward to analyze and generally inexpensive to compute. It is shown that, under mild assumptions, subtangents inherit second-order convergence properties from the original underestimators, and are therefore guaranteed to be useful in applications in their own right [3]. This result is strengthened further if the original underestimators are sufficiently smooth, as is true of the differentiable McCormick relaxations [2] and the Î ±BB relaxations [1]. These results are applied to yield powerful bounding methods involving bundles of subtangents, which are particularly effective in cases where evaluating the original convex underestimators is computationally expensive. Implications and examples are discussed.

References

[1] CS Adjiman, S Dallwig, CA Floudas, and A Neumaier, A global optimization method, Î±BB, for general twice-differentiable constrained NLPs: I. Theoretical advances, Comput. Chem. Engng, 22:1137-1158, 1998.

[2] KA Khan, HAJ Watson, and PI Barton, Differentiable McCormick relaxations, J. Glob. Optim., 67:687-729, 2017.

[3] KA Khan, Subtangent-based approaches for dynamic set propagation, submitted, 2018.

516566 Perspectives on the Control of Advanced Manufacturing Systems

Abstract:

Several countries and companies in recent years have become interested in advanced manufacturing, which is the highly efficient, effective, and reliable manufacturing of complex products to satisfy tight specifications. Such complex products typically are composite solid materials manufactured in over a dozen unit operations that require precise control over a very wide range of time and length scales, with many of the unit operations involving one or more solids. An example of a complex product is a pharmaceutical tablet that requires tight control of molecular purity, crystal structure, crystal size distribution, and the long-term chemical stability of the tablet when the active pharmaceutical ingredient is compressed with excipients [e.g., see Refs. 1-2 and citations therein].

Advanced manufacturing systems have many characteristics that break existing control theories, which include (1) high to infinite state dimension, (2) probabilistic parameter uncertainties, (3) time delays, (4) unstable zero dynamics, (5) actuator, state, and output constraints, (6) noise and disturbances, and (7) phenomena described by combinations of algebraic, ordinary differential, partial differential, and integral equations (that is, generalizations of descriptor/singular systems). While control theories have been developed for systems that have some of these characteristics, attempting to address all of these characteristics results in computational costs that are too high to be implemented, a lack of theoretical guarantees, and/or conservatism.

This presentation provides some perspectives on the analysis and design of control systems that satisfy all of the characteristics of advanced manufacturing systems. Model predictive control (MPC) formulations are reviewed that have the flexibility to handle dynamical systems with these characteristics by employing polynomial chaos theory and projections. The MPC formulations have low sensitivities to model uncertainties and low online computational cost, and have been demonstrated in realistic simulations of advanced manufacturing systems. Some promising directions are proposed for future research.

References

1. R. Lakerveld, P. L. Heider, K. D. Jensen, R. D. Braatz, K. F. Jensen, A. S. Myerson, and B. L. Trout. End-to-end continuous manufacturing: Integration of unit operations. In Continuous Manufacturing of Pharmaceuticals, edited by P. Kleinebudde, J. Khinnast, and J. Rantanen, Wiley, New York, Chapter 13, pages 447-483, 2017.
2. A. T. Myerson, M. Krumme, M. Nasr, H. Thomas, and R. D. Braatz. Control systems engineering in continuous pharmaceutical processing. Journal of Pharmaceutical Sciences, 104(3):832-839, 2015.

532292 Investigating the Impacts of Time-Varying Operation on Equipment Fidelity

Abstract:

Economic model predictive control [1] is an optimization-based control design that has received significant attention in recent years due to its ability to compute control actions which explicitly optimize the process economics and may induce time-varying operation of the process states. Despite many reports of the benefits of EMPC with respect to an instantaneous measure of process profit rooted in the dynamics of the chemical process system (e.g., [2]), an avenue that has not been explored is whether the time-varying operating policies computed by an EMPC may impact process equipment fidelity through mechanisms such as fatigue, and if so, whether this would impact capital costs and still make EMPC a competitive control design from an economics standpoint. The power plant cycling literature suggests that time-varying operation of processes can result in costs, such as increased maintenance costs [3], which need to be considered in the design of an EMPC.

In this work, we begin investigating the types of issues that equipment may face under time-varying operation of a nonlinear process utilizing EMPC by performing the closed-loop computational fluid dynamics modeling of a steam methane reforming tube from [4] but coupling this simulation with finite element analysis (FEA) to analyze how transients induced by controllers impact stresses in the reforming tube walls. We begin by investigating the differences between the stresses in the walls at steady-state and under the transients involved when the process is initialized off steady-state but driven to a steady-state by a classical (i.e., proportional-integral) control design that enforces steady-state operation. We then analyze the closed-loop process under EMPC and seek to better understand how differences in the stresses in the walls are related to the control designs with the purpose of proposing EMPC designs with constraints which encourage the EMPC to choose control actions which are economically-optimal but do not cause unnecessarily significant damage to the process equipment.

[1] M. Ellis, H. Durand and P. D. Christofides, "A tutorial review of economic model predictive control methods," *Journal of Process Control*, 24, 1156-1178, 2014.

[2] M. Ellis and P. D. Christofides, "Optimal time-varying operation of nonlinear process systems with economic model predictive control," *Industrial & Engineering Chemistry Research*, 53, 4991-5001, 2014.

[3] N. Kumar, S. Paterson, K. Coleman, C. Lee, D. Agan and S. Lefton, "Power plant cycling measure – evaluating historical cycling to model future grid operations," In *IEEE Power Energy Society General Meeting*, pages 1-5, Vancouver, British Columbia, Canada, 2013.

[4] L. Lao, A. Aguirre, A. Tran, Z. Wu, H. Durand and P. D. Christofides, "CFD modeling and control of a steam methane reforming reactor," *Chemical Engineering Science*, 148, 78-92, 2016.

532872 Generalized Chaos Expansions with Arbitrary Multivariate Probability Measures: Applications in Closed-Loop Performance Verification for Stochastic Dynamic Systems

Abstract:

Modern control design methods, such as model predictive control (MPC) and its many variants including robust & stochastic MPC, often use a control-relevant model and a control-relevant uncertainty description (e.g., finite number of scenarios) that is amenable to online computations [1]. However, in many practical applications, there can be significant mismatch between the "real system" behavior and the model internally used in the controller. A complex *high-fidelity* model of the true system can sometimes be developed using first principles models, which more accurately predicts the system response over a significantly larger range of length- and time-scales. High-fidelity models can then be used for performance verification as well as for shaping closed-loop performance via tuning the many different changeable parameters (e.g., objective function, uncertainty description, prediction horizon, etc.) of the "robust" model-based controller [2].

In many engineering applications, however, even the high-fidelity models will have significant sources of uncertainty. This fact has prompted the development of broad range of tools for uncertainty quantification (UQ), which attempts to systematically understand the impact of these errors on the model predictions [3]. Therefore, the goal of this work is to develop an efficient approach for UQ in the context of closed-loop performance verification of nonlinear systems with *arbitrary* probability uncertainties. In particular, we focus on solving the "forward problem" of propagating uncertainty in unknown parameters on key quantities of interests (QoIs) that are directly related to closed-loop performance metrics (e.g., objective or constraint violations).

A huge variety of methods have been developed for stochastic UQ each with their own set of advantages and disadvantages. Our particular choice of methods is motived by two key features that are important in the context of closed-loop performance verification: (1) the closed-loop system is expensive to simulate for every realization of the uncertain model parameters due to the fact that an embedded optimization problem must be solved at sample time and (2) the uncertain parameters can have an *arbitrary* probability distribution. This latter point is motivated by the fact that the parameter distribution will often be defined as the solution to a Bayesian inversion problem, which is known to lead to highly correlated/dependent random variables. Monte Carlo Sampling (MCS) is the standard approach for solving this type of problem and has been applied for tuning backoff constraints in MPC [4]; however, the estimated statistics of the QoIs are known to converge very slowly with respect to number of parameter realizations meaning a large number of realizations is required for accurate predictions. A recently developed alternative to MCS, broadly termed generalized polynomial chaos (gPC) [5], has become one of the most widely used methods for stochastic UQ. The gPC framework can be thought of as a spectral representation (i.e., basis function expansion) in random space wherein the basis is chosen to be a set of polynomials that are orthogonal with respect to a user-specified *germ* distribution to simplify computations. However, gPC and its variants require the germ to have *independent* random elements. Thus, in order to apply gPC, the dependent random parameters must be transformed to any chosen independent germ. Not only is defining this transformation difficult (or impossible) in many practical cases, but the use of this transformation can introduce additional nonlinearities that result in a big drop in the expansion's rate of convergence

The main contribution of this work is to further generalize the concept of gPC so that the germ can have a truly arbitrary multivariate distribution. This notion is very similar to that of so-called arbitrary PC (aPC) [6]; however, it differs in the sense that the basis functions do not have to be polynomials and the independence assumption is completely removed. This proposed concept of a generalized chaos expansion (GCE) involves two key steps: (1) defining an orthonormal basis (ONB) with respect to arbitrary probability measures and (2) determining the expansion coefficients. Thus, this work also presents an efficient algorithm for numerically solving these two problems. The proposed method only requires *moments* of the germ distribution to be known so that the full distribution does not need to be known in closed-form. Owing to the orthogonal projection theorem, the latter problem of determining the expansion coefficients can be defined exactly as a multivariate integral with respect to the chosen arbitrary measure. This represents the computational bottleneck in this setting due to the expensiveness of the closed-loop simulations. Utilizing concepts from moment-matching optimization [7], we can develop a "quasi-optimal" cubature rule to accurately approximate these integrals with a small number of simulations, which substantially reduces the computational cost of the method.

The proposed tractable stochastic UQ method is demonstrated in simulations on a benchmark continuously-stirred tank reactor (CSTR) that is run in closed-loop with a robust "scenario-tree" (or multi-stage) MPC algorithm. It is shown that the proposed method can guarantee satisfaction of distributional state chance constraints through proper determination of controller parameters (e.g., the MPC constraint backoff in this case). Additionally, the proposed method is compared to gPC wherein we show that it can achieve higher accuracy with lower-order expansions while requiring an order-of-magnitude fewer closed-loop simulations (resulting in huge computational savings), owing to the fact that the germ transformation has been avoided.

References

[1] D. Q. Mayne, "MPC: Recent developments and future promise," Automatica, vol. 50, pp. 2967–2986, 2014.

[2] J. A. Paulson and A. Mesbah. "Arbitrary Polynomial Chaos for Quantification of General Probabilistic Uncertainties: Shaping Closed-loop Behavior of Nonlinear Systems," submitted to the 57th IEEE Conference on Decision and Control (under review).

[3] D. Xiu. "Fast numerical methods for stochastic computations: a review." *Communications in computational physics* 5.2-4 (2009): 242-272.

[4] H. Jang, J. H. Lee, and L. T. Biegler. "A robust NMPC scheme for semi-batch polymerization reactors." *IFAC-PapersOnLine* 49.7 (2016): 37-42.

[5] D. Xiu and G. E. Karniadakis, "The Wiener-Askey polynomial chaos for stochastic differential equations," SIAM Journal of Scientific Computation, vol. 24, pp. 619–644, 2002.

[6] J. A. Paulson, E. A. Buehler, and A. Mesbah, "Arbitrary polynomial chaos for uncertainty propagation of correlated random variables in dynamic systems," IFAC-PapersOnLine, vol. 50, pp. 3548–3553, 2017.

[7] E. K. Ryu and S. P. Boyd, "Extensions of Gauss quadrature via linear programming," Foundations of Computational Mathematics, vol. 15, pp. 953–971, 2015.

535747 Constrained Control Lyapunov Function Construction Via Approximation of Static Hamilton-Jacobi-Bellman Equations

Abstract:

The control of complex chemical processes is challenged by the presence of model nonlinearities and input constraints. As such, Lyapunov functions are often used to determine how to stabilize such systems, but only if the states lie within the chosen Lyapunov function's stability region. If it lies outside, the Lyapunov function provides no information about the stabilizing control, so the controller may fail to avert catastrophic destabilization even when it is possible to do so. Thus, designing controllers based on Lyapunov functions with large stability regions is an important safety consideration. Therefore, in this conference presentation, we consider the problem of constructing Lyapunov functions for nonlinear input-constrained systems with the largest possible stability regions by solving the associated Hamilton-Jacobi-Bellman PDE. To solve this equation, we employ a finite difference approximation and novel boundary conditions based on a recently-developed algorithmic construction of the boundary of the system's null controllable region, which efficiently determines all reachable states. Furthermore, since even smooth H-J-B PDEs are observed to contain discontinuities, the artificial viscosity perturbation method is used to improve the quality of the approximation. The sub-problem of determining the optimal input at each node is reduced to finding the roots of a certain nonlinear polynomial. Lastly, we illustrate the results via simulation examples.

536108 On Multiparametric/Explicit NMPC for Quadratically Constrained Problems

Abstract:

Providing exact solutions to the multiparametric nonlinear programming problem (mpNLP) has been proven a challenging task. Even though there has been encouraging work [1-3], most of the major subsequent efforts have focused on providing approximate solutions to the problem. For example, [4], an outer-approximation of the mpNLP is created through the linearization of the nonlinear terms of the objective function and the constraints. In [5,6] a quadratic approximation to the objective function and linear approximations to the constraints are obtained and the mpNLP is approximated by a mpQP, where an explicit solution can be found. An algorithm for the solution of nonlinear parametric optimization of polynomial functions subject to polynomial constraints based on cylindrical algebraic decomposition was presented in [7]. More recently, [8] proposed the decomposition of a mpMINLP into a series of mpQPs using quadratic approximations of the objective function, while [9,10] focused on multiparametric polynomial programming (mpPP) and based on symbolic calculations the analytical solution of the optimality conditions is obtained.

We present the expansion of the Basic Sensitivity Theorem to a second order Taylor expansion approach and the implications to explicit model predictive control of quadratically constrained systems. The expansion enables the derivation of an algorithm for the analytical solution of convex multiparametric quadratically constraint programming (mpQCQP) problems and explicit quadratically constrained NMPC problems. We derive the analytical parametric expressions of the control actions for a quadratically constrained MPC problem and its corresponding critical regions. We also expand the case to problems with non-convex quadratic constraints. We finally show the piecewise non-linear form of the solution and closed-loop validation of the results.

References

[1] Fiacco, A., Sensitivity analysis for nonlinear programming using penalty methods (1976) Mathematical Programming, 10(1), 287-311.

[2] Fiacco, A., Sensitivity analysis and stability analysis in nonlinear programming (1983).

[3] Fiacco, A., Kyparisis, J., Convexity and concavity properties of the optimal value function in parametric nonlinear programming (1986) Journal of optimization theory and applications, 48(1), 95-126.

[4] Dua, V., Pistikopoulos, E. N., Algorithms for the solution of multiparametric mixed-integer nonlinear optimization problems (1999) Industrial and Engineering Chemistry Research, 38(10), 3976-3987.

[5] Johansen, T., On multi-parametric nonlinear programming and explicit nonlinear model predictive control (2002) Volume 3, 2768-2773.

[6] Grancharova, A., Johansen, T., Tondel, P., Computational aspects of approximate explicit nonlinear model predictive control (2007) Lecture Notes in Control and Information Sciences, 358(1), 181-192.

[7] Fotiou, I., Rostalski, P., Parrilo, P., Morari, M., Parametric optimization and optimal control using algebraic geometry methods (2006) International Journal of Control, 79(11), 1340-1358.

[8] Dominguez, L., Pistikopoulos, E.N. A quadratic approximation-based algorithm for the solution of multiparametric mixed-integer nonlinear programming problems (2013) AIChE Journal, 59(2), 483-495.

[9] Dua, V. Mixed integer polynomial programming (2015) Computers and Chemical Engineering, 72, 387-394.

[10] Charitopoulos, V., Dua, V., Explicit model predictive control of hybrid systems and multiparametric mixed integer polynomial programming (2016) *AIChE Journal*, 62(9), 3441-3460.

536121 Explicit Model Predictive Control Using Nonlinear Intrinsic Variables

Abstract:

Model predictive control (MPC) is among the most successful approaches for process control and has become a *de facto* standard across the process industries. However, there remain some applications for which MPC is impractical due to the demand that an optimization problem is solved at every time step. For systems with high-dimensional state spaces or highly nonlinear dynamics, solving this optimization problem is prohibitive. "Explicit" MPC eliminates this demand by using an explicit formulation of a control law that is approximately equivalent to the typical online MPC controller, either by finding an analytical solution to the optimization problem (only available in some rare circumstances), or by interpolating via a "lookup table" of precomputed solutions (only valid for limited operating conditions, and unfeasible for high-dimensional problems). In this work, we present a new explicit MPC formulation using recent advances in data mining, and especially manifold learning, to build effective interpolation functions between the system states, process outputs, and the MPC-determined control policies. Because all variables of interest (states, outputs, control policies) lie on the same low dimensional embedding space, we can determine any one of these variables from either of the other two. We use the diffusion maps algorithm to discover such low dimensional embeddings, and then develop functions to interpolate between the embedding space, system states, process outputs, and control policies. Following this procedure, we can substantially expand the class of systems for which explicit MPC methods are practical.

First, we will provide a background on manifold learning, and particularly the diffusion maps algorithm. Then, we will demonstrate how to use diffusion maps to uncover the embedding space on which lies all the variables of the closed loop MPC system. Next, we will use geometric harmonics to construct interpolation functions between the different variables of interest, and show that these functions can be used online as feedback control laws. To demonstrate our approach, we will present three example systems: (1) a nonisothermal CSTR, (2) a mechanistic model of an optogenetic circuit (an emerging technology from synthetic biology), and (3) a PDE with low-dimensional long-term dynamics. Finally, we will compare our approach using diffusion maps to other data-driven methods for explicit MPC, including neural networks and deep Gaussian processes.

538014 Securing Process Control Systems Using Dynamic Watermarking

Abstract:

In this work, we address the problem of securing a process control system. Specifically, we consider a multiple-input-multiple-output process control system, an arbitrary subset of whose sensors could be "malicious." A malicious sensor need not report the observations that it measures truthfully to the controller. Rather, they could distort the measurements in an arbitrary fashion to achieve any objective that they may have, such as degrading the system performance or destabilizing the closed loop. In this work, we examine the efficacy of dynamic watermarking, developed in the prior work [1] in the context of securing linear stochastic systems. The fundamental idea of dynamic watermarking is for the actuators in the system to superimpose a small random noise, referred to as the private excitation, on the pre-specified control input, and check if the measurements reported by the sensors are appropriately correlated with the private excitation. If not, the actuators can conclude that the output signal has been made unauthorized alterations, and thereby declare the presence of malicious sensors in the system.

As an initial effort to experimentally demonstrate that dynamic watermarking is effective in securing a nonlinear process control system, a representative coupled water tank system is considered. Specifically, the coupled water tank system consists of a single pump with two tanks, referred to as the upper-level tank and the lower-level tank. The pump feeds water to the upper-level tank, and its outflow is introduced to the lower-level tank. Each tank is instrumented with a pressure sensor which measures the water level in the tank, and feeds it back to the controller. The controller's objective is to maintain the water level at a given set-point by controlling the voltage across the pump.

We emulate a scenario wherein an adversary compromises the pressure sensors in the system and reports erroneous values of the water level to the controller. This eventually leads to the tanks over flowing. We then implement dynamic watermarking in this system to check whether or not the measurements reported by the pressure sensors are manipulated, and show how it can detect malicious sensors, if they are present, within a short span of time. Once malicious sensors are detected, the controller can abstain from closing the loop around the measurements reported by them and prevent the tanks from overflowing.

The demonstrated efficacy of dynamic watermarking in securing this simple system, together with the observation that it can be implemented on legacy control systems with minimal modifications to the control logic, will build confidence in dynamic watermarking and make it attractive for widespread adoption by the industry for securing process control systems.

References:

[1] B Satchidanandan, PR KumarDynamic. Watermarking: Active defense of networked cyberâ€"physical systems, Proceedings of the IEEE, 2017, 105 (2), 219-240.

538582 Nonlinear Manifold Learning of Nucleosome Dynamics from Molecular Simulation

Abstract:

The identification of important dynamical motions remains a challenge in molecular simulations of complex systems. Here, we use a nonlinear manifold learning technique known as the diffusion map to extract key dynamical modes from molecular simulations of the nucleosome, a complex biomolecular system consisting of 147 base pairs of DNA wrapped around a disc-shaped histone octamer. We demonstrate that diffusion maps are effective at extracting motions consistent with those previously found through a detailed free energy analysis (e.g. DNA translocation along the histone complex). We also show that diffusion maps can reveal more subtle features, including the formation of looping conformations, in which DNA bulges away from the histone complex, as well as motions involved in DNA breathing, where DNA spontaneously unwraps from the histone octamer.

536589 Machine Learning Predicts Functional Classes of Family 7 Glycoside Hydrolases with High Accuracy

Abstract:

Glycoside hydrolases (GH) are a class of enzymes that catalyze the hydrolysis of glycosidic bonds in saccharides. They are utilized in industries, such as the biofuel and textile industries, for enzymatic degradation and reorganization of saccharides. GHs are presently classified into 152 families based on sequence identity. Family 7 glycoside hydrolases (GH7s) are predominantly found in fungi and are often the largest composition by mass of the secretomes of cellulolytic fungi. In the biofuel industry, GH7s are the primary components of the enzymatic cocktails used in cellulose degradation. GH7s fall into one of two classes: cellobiohydrolases (CBHs) or endoglucanases (EGs). GH7 CBHs hydrolyze cellulose processively, i.e. they carry out multiple catalytic steps without dissociating from the substrate. GH7 EGs, on the other hand, are non-processive and dissociate from the substrate after hydrolyzing a glycosidic bond. Processive GH7s (CBHs) have become a focus of research because they provide the greatest hydrolytic potential in enzymatic cellulose degradation. As many of the known GH7 sequences have not yet been classified in terms of activity, we have set out to develop a predictive approach for classifying GH7 activity. We first retrieved a large and diverse set of 1,521 GH7 sequences from the genomic databases. The functional classes (i.e. CBH or EG) are reported for only about 30% of these GH7s. We trained multiple machine learning classifiers (decision tree, SVM, naïve Bayes and logistic regression) using known structural differences between GH7 CBHs and EGs as features. We determined, using Monte Carlo cross validation, that the overall accuracy of the machine learning classifiers ranges from 95 to 97%, suggesting that GH7 functional class can be readily predicted from sequence alone.

534508 Implementation and Automation of a Hierarchical Graph Based Approach for Extracting Coarse-Grain Mapping Operators

Abstract:

Coarse-grain (CG) molecular dynamics (MD) simulation allows us to overcome the limitations of all-atom (AA) MD simulation in terms of length and time scales. While CG-MD has been successfully applied to various systems like protein folding, there is still a need for a general guideline to determine the choice of CG mapping operators. Recently we have proposed a technique of incorporating multiple CG representations of a molecule within a single hierarchical graph. With the ability to extract valid CG mapping operators from the graph and relevant thermodynamic property like entropy driving the selection of one among the various mapping operators extractable from the hierarchical graph, this technique is a step towards automating the selection of CG mapping operators. As a proof of concept we have built a hierarchical graph for methanol which encodes all symmetry preserving CG mapping operators. We also demonstrated the feasibility of using the graph for automated CG mapping operator selection by using the uniform entropy flattening method to select CG mapping operators for methanol. In this talk, we will describe the implementation of the algorithm along with an automated pipeline framework to create a database of benchmark CG simulations. The discussion will include different thermodynamic measures to compare the CG and the AA performances. The large data set of CG simulations will aid in the development of standardized and scalable CG mapping strategy that is system independent.

521801 Machine Learning Algorithms for High-Throughtput Chemical Sensing Using Liquid-Crystals

Abstract:

Security, health, and environmental monitoring requires the development of chemical sensing technologies that can be used in-situ and with limited equipment and human intervention. The impact of such technologies when deployed at a large scale can be significant; for instance, the U.S. Department of Energy Savannah River laboratory analyzes over 40,000 groundwater samples per year at a cost of $1,000 per sample ($40 million per year) [1]. Liquid crystals (LCs) are fluid phases with preferred molecular orientations, a so-called director, that undergo surface-driven ordering transitions in the presence of chemical species such as organophosphonates [2, 3, 4, 5, 6], chlorine, ammonia, and hydrogen sulfide [7]. The optical characteristics (features) of the LC transitions can be tailored and exploited to design chemical sensors. For instance, LCs can be designed to assume homeotropic (perpendicular) orientations on surfaces decorated with metal salts [2, 4, 8]. Chemical species that diffuse into the LCs and bind more strongly to the metal cations than the LC functional groups will trigger a transition of the LCs orientation from homeotropic to planar (see Figure 1) [9, 10, 11]. It is possible to manipulate the selectivity and response characteristics (e.g., dynamics) of the LC by tuning the binding energies of the LC functional groups (e.g., nitrile and pyridine groups) to the surface (e.g., Fe+3, La3+) [9, 10]. For instance, ordering transitions of LC sensors fabricated using a nematic LC called 4-cyano- 4'-penthylbiphenil (5CB) and surfaces presenting aluminum perchlorate salts have been studied in [12, 13, 14].

In this work, we present a machine learning (ML) framework to optimize the specificity and speed of LC-based chemical sensors. Specifically, we demonstrate that ML techniques can uncover valuable feature information from surface-driven LC orientational transitions triggered by the presence of different gas-phase analytes (and the corresponding optical responses) and can exploit such feature information to train accurate and automatic classifiers. We demonstrate the utility of the framework by designing an experimental LC system that exhibits similar optical responses to a stream of nitrogen containing either 10 ppmv dimethyl-methylphosphonate (DMMP) or 30% relative humidity (RH). The ML framework is used to process and classify thousands of images (optical micrographs) collected during the LC responses and we show that classification (sensing) accuracies of over 99% can be achieved. For the same experimental system, we demonstrate that traditional feature information used in characterizing LC responses (such as average brightness) can only achieve sensing accuracies of 60%. We also find that high accuracies can be achieved by using time snapshots collected early in the LC response, thus providing the ability to create fast sensors. We also show that the ML framework can be used to systematically analyze the quality of information embedded in LC responses and to filter out noise that arises from imperfect LC designs and from sample variations. We evaluate a range of classifiers and feature extraction methods and conclude that linear support vector machines are preferred and that high accuracies can only be achieved by simultaneously exploiting multiple sources of feature information.

References

[1] Clifford K Ho, Alex Robinson, David R Miller, and Mary J Davis. Overview of sensors and needs for environmental monitoring. Sensors, 5(1):4–37, 2005.

[2] K-L Yang, Katie Cadwell, and Nicholas L Abbott. Contact printing of metal ions onto carboxylate-terminated self-assembled monolayers. Advanced Materials, 15(21):1819–1823, 2003.

[3] Kun-Lin Yang, Katie Cadwell, and Nicholas L Abbott. Mechanistic study of the anchoring behavior of liquid crystals supported on metal salts and their orientational responses to dimethyl methylphosphonate. the Journal of Physical Chemistry B, 108(52):20180–20186, 2004.

[4] Katie D Cadwell, Mahriah E Alf, and Nicholas L Abbott. Infrared spectroscopy of competitive interactions between liquid crystals, metal salts, and dimethyl methylphosphonate at surfaces. The Journal of Physical Chemistry B, 110(51):26081–26088, 2006.

[5] Maricar L Bungabong, Peng Bin Ong, and Kun-Lin Yang. Using copper perchlorate doped liquid crystals for the detection of organophosphonate vapor. Sensors and Actuators B: Chemical, 148(2):420–426, 2010.

[6] Santanu Kumar Pal, Claribel Acevedo-Velez, Jacob T Hunter, and Nicholas L Abbott. Effects of divalent ligand interactions on surface-induced ordering of liquid crystals. Chemistry of Materials, 22(19):5474–5482, 2010.

[7] Sheila E Robinson, Bart A Grinwald, Laura L Bremer, Kurt A Kupcho, Bharat R Acharya, and Patrick D Owens. A liquid crystal-based passive badge for personal monitoring of exposure to hydrogen sulfide. Journal of occupational and environmental hygiene, 11(11):741–750, 2014.

[8] Jacob T Hunter and Nicholas L Abbott. Adsorbate-induced anchoring transitions of liquid crystals on surfaces presenting metal salts with mixed anions. ACS applied materials & interfaces, 6(4):2362–2369, 2014.

[9] Tibor Szilv´asi, Luke T Roling, Huaizhe Yu, Prabin Rai, Sangwook Choi, Robert J Twieg, Manos Mavrikakis, and Nicholas L Abbott. Design of chemoresponsive liquid crystals through integration of computational chemistry and experimental studies. Chemistry of Materials, 29(8):3563– 3571, 2017.

[10] Huaizhe Yu, Tibor Szilv´asi, Prabin Rai, Robert J Twieg, Manos Mavrikakis, and Nicholas L Abbott. Computational chemistry-guided design of selective chemoresponsive liquid crystals using pyridine and pyrimidine functional groups. Advanced Functional Materials, 2018.

[11] Luke T Roling, Jessica Scaranto, Jeffrey A Herron, Huaizhe Yu, Sangwook Choi, Nicholas L Abbott, and Manos Mavrikakis. Towards first-principles molecular design of liquid crystalbased chemoresponsive systems. Nature communications, 7:13338, 2016.

[12] Daniel S Miller, Rebecca J Carlton, Peter C Mushenheim, and Nicholas L Abbott. Introduction to optical methods for characterizing liquid crystals at interfaces. Langmuir, 29(10):3154–3169, 2013.

[13] Alireza Hassanzadeh and Robert G Lindquist. Liquid crystal sensor microchip. IEEE Sensors Journal, 12(5):1536–1544, 2012.

[14] Rahul R Shah and Nicholas L Abbott. Orientational transitions of liquid crystals driven by binding of organoamines to carboxylic acids presented at surfaces with nanometer-scale topography. Langmuir, 19(2):275–284, 2003.

533502 Materials Informatics for Process Optimization: Case Studies Using P3HT and PP Composites

Abstract:

The traditional paradigm in materials process optimization is energy-intensive and time consuming. Generally, single factor, trial-and-error based experimentation is used to screen a wide array of complex processing conditions. This generates a vastly multidimensional design space that is impossible to fully characterize. Furthermore, results are difficult to reproduce because of a lack of standardization due to the exploratory nature of the field. As a result, it is challenging to develop global, robust process-structure-property relationships to generate novel material formulations. Instead, informatics based methodologies can leverage existing knowledge from the literature to guide high-throughput screening of promising candidate experimental conditions. In this work two case studies of applying materials informatics methodologies to polymer systems will be presented.

In the first study, the impact of processing conditions on the organic field effect mobility of the semiconducting polymer, poly(3-hexylthiophene) (P3HT), will be presented. A database of over 200 devices was created to identify significant trends and to identify an optimal operating space. Processing conditions were sorted and filtered to identify a standard device, with reported mobility values that span two orders of magnitude. Furthermore, both integer and categorical processing conditions resulting in mobility values exceeding 0.1 cm^2/V-s were identified to reduce the dimensionality of the design space. This enabled the generation of new hypotheses for future high-throughput experimentation.

In the second study, a database with over 140 entries quantifying the development of polypropylene and talc composites for high strength materials was created. However, this database suffered from inconsistent reporting of processing conditions. Physical based models, for example relating melt flow index to molecular weight, were proposed to help fill in missing data. Varying thresholds of Youngâ€™s Modulus were applied to identify the span of processing conditions that result in high strength composites. In both studies quantified analysis of all processing conditions has identified key design spaces to enable high through-put experimentation.

537478 Learning Many-Body Molecular Interactions from Machine Learning

Abstract:

The accurate representation of multidimensional potential energy surfaces is a necessary ingredient for realistic computer simulations of molecular systems. It has recently been shown that chemical and spectroscopic accuracy can be achieved with analytical potential energy functions (PEFs) rigorously derived from many-body expansions. In this contribution, we demonstrate the equivalence between permutationally invariant polynomials, neural networks, and Gaussian approximation potentials in representing lower-order many-body terms of molecular interactions in aqueous systems from the gas to the condensed phase. Besides demonstrating the synergy between high-quality electronic structure data and machine-learning techniques for developing transferable PEFs, our results show that machine learning provides a powerful tool for learning the underlying physics of many-body molecular interactions.

531504 Combining Machine Learning and Evolutionary Computing for Accelerating Materials Design

Abstract:

Machine learning models have been progressively adopted by the materials science community to accelerate design of materials with targeted properties. However, in the search for new materials exhibiting properties and performance beyond that previously achieved, machine learning approaches are frequently limited by two major shortcomings. First, they are intrinsically interpolative. They are therefore better suited to the optimization of properties within the known range of accessible behavior than to the discovery of new materials with extremal behavior. Second, they require the availability of large datasets, which in some fields are not available and would be prohibitively expensive to produce. Here we describe a new strategy for combining genetic algorithms, neural networks and other machine learning tools, and molecular simulation to discover materials with extremal properties in the absence of pre-existing data. Predictions from progressively constructed machine learning tools are employed to bias the evolution of a genetic algorithm, with fitness evaluations performed via direct molecular dynamics simulation. We survey several materials design problems we have addressed with this framework. Specific examples for designing soft materials and 2D organic materials will be presented.

535265 Predicting Colloidal Crystals from Shapes Via Inverse Design and Machine Learning

Abstract:

A fundamental challenge in materials design is linking building block attributes to crystal structure. Addressing this challenge is particularly difficult for systems that exhibit emergent order, such as entropy-stabilized colloidal crystals. We combine recently developed techniques in inverse design with machine learning to construct a model that correctly classifies the crystals of more than ten thousand polyhedral shapes into 13 different structures with a predictive accuracy of 96% using only two geometric shape measures. With three measures, 98% accuracy is achieved. We test our model on previously reported colloidal crystal structures for 71 symmetric polyhedra and obtain 92% accuracy. Our findings (1) demonstrate that entropic colloidal crystals are controlled by surprisingly few parameters, (2) provide a quantitative model to predict these crystals solely from the geometry of their building blocks, and (3) suggest a prediction paradigm that easily generalizes to other self-assembled materials.

522234 Autonomous Crystal Structure Characterization with Neighborhood Graph Analysis

Abstract:

We have previously developed a template-free method for characterizing local crystalline structure called Neighborhood Graph Analysis. Unlike conventional methods, our technique uses nonlinear manifold learning to infer structural relationships directly from particle tracking data. While it provides exceptional flexibility and discriminating power, in the past our graph-based classifier was prohibitively slow to evaluate, making it impractical for everyday use. We present a modification to the original method which results in nearly identical results while speeding up the calculation by three orders of magnitude. As with the original technique, we are able to characterize particles near defects, grain boundaries, and interfaces of close-packed crystal structures. However, the increased speed of our new algorithm allows us to extend the analysis to more exotic structures such as ices, clathrates, perovskites, and pyrochlores. We can even handle superlattices composed of two or more species. We demonstrate how the resulting analysis can be used to enhance structural analysis whether the relevant structures are known or not.

537653 The Physical Analytics Pipeline - a Bayesian Optimization of the Hybrid Organic-Inorganic Perovskite Compositional Space

Abstract:

Hybrid Organic-Inorganic Perovskites (HOIPs) are an exciting class of photovoltaic materials due to their combination of high solar cell efficiency and the ability to be fabricated by a low-energy room temperature solution process. Perovskites are classified by their characteristic ABX3 structure, where A is an organic (or inorganic) cation (e.g., methyammonium, formamadinium or Cs), B is a metal cation (typically Pb or Sn), and X is a halide (where X= I, Br, Cl). Choice of solvent is also very important in the fabrication process and can be one of perhaps eight common solvent choices. With all these possible choices of ABX3, including mixing metal cations, mixing halides and use of binary solvent blends, the HOIP compositional space quickly becomes unwieldy, with at least 500,000 possible combinations. Experimental exploration of this space is Edisonian and limited by time and resources and a lack of underlying rational design guidelines. Even computational approaches are limited by the lack of available force fields for the multi-species HOIP systems, which restricts the calculations to expensive quantum mechanical ones. The Physical Analytics pipeLine (PAL), which we have developed, tackles this problem by bridging Gaussian Process Bayesian Optimization with computational methods to minimize the number of necessary calculations needed to locate the global maximum. As a proof of concept, PAL has been implemented for pure and mixed halide perovskites, and is benchmarked against the Sequential Model-based Algorithm Configuration (SMAC). PAL performs extremely well in comparison to SMAC, and is robust upon replication with different initial training sets. Further improvements to the PAL model have been made to take into account the ability to combine a variety of information sources, with different accuracy and cost, allowing us to use cheaper, less accurate calculations to aid in improving our predictions, whilst minimizing the total cost of optimization. This combination of exploration and exploitation allows for efficient location of broad parameter spaces where data may be sparse.

528859 A Computational Tool for Applying Optimization Under Uncertainty on Advanced Process Simulators

Abstract:

An optimization under uncertainty (OUU) tool implemented in the Framework for Optimization, Quantification of Uncertainty, and Surrogates (FOQUS) is presented. FOQUS (Miller et al., 2017) was developed as part of Carbon Capture Simulation Initiative (CCSI) and is a computational platform capable of interfacing with complex process models implemented in widely-used commercial process simulators. CCSI is a cooperation between national laboratories, industry, and universities with the goal of developing, demonstrating, and deploying computational tools to decrease the deployment time for new carbon capture technologies through modeling, optimization, and uncertainty quantification (Miller et al., 2014). OUU is a powerful tool for process design and optimization, which allows a process to be optimized for performance, while also minimizing risk given the significant uncertainties inherent in innovative technologies.

Accurately modeling complex chemical engineering processes (e.g., solid sorbent-based carbon capture plants) requires rigorous large-scale models. Optimizing these models with traditional mathematical optimization tools is challenging, especially when uncertainty is analyzed (Reddy et al., 2017). For instance, having uncertain variables with continuous distributions increases the size of the optimization problem, which in turn increases the computational burden. FOQUS can potentially reduce the time required for implementing OUU, thus making OUU more feasible, with its alternative approach of using process simulator models within an OUU framework.

The capabilities of FOQUS are demonstrated with the design and OUU of a solid sorbent-based carbon capture system for a coal-fired power plant, which was simulated in Aspen Custom Modeler (ACM). The uncertainties originate from material characterization (e.g., solid sorbent particle diameter and heat transfer coefficients) and from the operation of the power plant (e.g., the inlet flue gas flow rate). The reaction kinetic parameters for this system are well-studied, and a previous study found that the uncertainty of these parameters had negligible effect on the performance of the carbon capture plant (Eslick et al., 2015).

The objective of this study is to minimize the expected value of the cost of electricity (COE) while maintaining a minimum of 90% CO_2 capture. FOQUS uses a two-stage stochastic optimization framework, where the equipment design variables are the first-stage variables, and the operating variables are the second-stage variables. The parameter uncertainty distributions are represented by generating scenarios with different values of the uncertain parameters. The problem is solved as a nested optimization problem. In the first stage, a derivative-free optimization algorithm in FOQUS adjusts the values of the first-stage variables to be the same across all scenarios. Nested inside that optimization, the optimization algorithm in the process simulator is used to adjust the values of the second-stage variables. Second-stage variables are not required to be the same across all scenarios, which means that they provide recourse to uncertainties. Leveraging the optimization algorithm provided by the process simulator improves the performance of the solution method, and the second-stage optimization problems can be run in parallel, using the parallel simulation job queuing capabilities of FOQUS.

To highlight the importance of considering the uncertainties of variables prior to designing the plant, the expected value of the COE obtained from OUU is compared against the expected value of perfect information (EVPI) and the value of the stochastic solution (VSS) (Birge and Louveaux, 2011). The EVPI compares the expected value of the COE obtained from the case where the values of both the equipment design variables and the operating variables can be adjusted to respond to the perturbation in the value of the uncertain variable (i.e., the ideal but unrealistic situation) against the expected value of the COE obtained from executing OUU. A negative value of the EVPI suggests that the expected value of the COE generated from the ideal case is smaller compared to the expected value of the COE generated from performing OUU. On the other hand, the VSS compares the expected value of the COE obtained from executing OUU against the expected value of the COE obtained from the case where the carbon capture plant is built using the values of the equipment design variables that were optimized without any uncertainty considerations, and then when the uncertainties are introduced, only the values of the operating variables were optimized to minimize the expected value of the COE. A negative value of the VSS suggests that the expected value of the COE generated from performing OUU is smaller compared to the expected value of the COE generated from building the plant using the values of the equipment design variables generated after optimization without considering the uncertainty.

For the case where the value of the particle diameter is perturbed from -25% to +25% from the base case, the EVPI is -0.8 USD/MWh, and the VSS is -4.8 USD/MWh. Building the carbon capture plant using the values of the equipment design variables obtained from OUU instead of those obtained without performing OUU improves the flexibility of the plant's operation, thus reducing technical risk. These results show that FOQUS can easily facilitate OUU analyses in a simulation-optimization framework.

References

Birge, J.R., Louveaux, F., 2011. Introduction to Stochastic Programming, second ed. Springer Science+Business Media, LLC, New York, NY, USA.

Eslick, J.C., Tong, C., Ng, B., Lee, A., Dowling, A.W., Mebane, D.S., Chen, Y., Miller, D.C., 2015. Optimization under uncertainty with rigorous process models. 2015 AIChE Annual Meeting; 8–13 November 2015; Salt Lake City, UT, USA.

Miller, D.C., Agarwal, D., Bhattacharyya, D., Boverhof, J., Chen, Y., Eslick, J., Leek, J., Ma, J., Mahapatra, P., Ng, B., Sahinidis, N.V., Tong, C., Zitney, S.E., 2017. Innovative computational tools and models for the design, optimization and control of carbon capture processes, in: Papadopoulos, A.I., Seferlis, P. (Eds.), Process Systems and Materials for CO_2 Capture: Modelling, Design, Control and Integration. John Wiley & Sons Ltd, Chichester, UK, pp. 311–342.

Miller, D.C., Syamlal, M., Mebane, D.S., Storlie, C., Bhattacharyya, D., Sahinidis, N.V., Agarwal, D., Tong, C., Zitney, S.E., Sarkar, A., Sun, X., Sundaresan, S., Ryan, E., Engel, D., Dale, C., 2014. Carbon Capture Simulation Initiative: A case study in multiscale modeling and new challenges. Chemical and Biomolecular Engineering 5, 301–323.

Reddy, S.S., Sandeep, V., Jung, C.-M., 2017. Review of stochastic optimization methods for smart grid. Frontiers in Energy 11 (2), 197–209.

530323 A Novel Objective Reduction Method: Application to the Comparison of Risk Metrics

Abstract:

A novel objective reduction method: application to the comparison of risk metrics

Daniel Vázquez, Rubén Ruiz-Femenia, José A. Caballero

Keywords

Multi-objective Optimization, Inherent Safety, stochastic optimization, sample average approximation

Typically, stochastic programming models optimize the expected value of an indicator that asses the performance of the system, providing a design solution that behaves the best when considering all the scenarios simultaneously. However, this approach may provoke that the value of this indicator worsens after the uncertainty is revealed in a particular unfavorable scenario. As decision makers may have different attitude towards the risk associated with a design under uncertainty, it would be advisable to incorporate risk management techniques to identify the trade-off between risk and performance measures. This approach increases the complexity of the problem to be solved because not only the problem is transformed into a multi-objective optimization (MOO) problem –in which the expected performance measure and its associated risk must be simultaneously optimized– but also the simultaneous incorporation of uncertainty leads to a stochastic multi-objective optimization (SMOO) problem, which is inherently high-dimensional because it captures multiple realizations of random data[1].

Multi-objective optimization provides a set of Pareto optimal alternatives that feature the property that it is not possible to find another feasible solution which is better in one of the objectives without worsening at least one of the others. The main limitation of multi-objective optimization is that its computational burden exponentially grows with the number of objectives. Consequently, in many applications, it is impractical to compute the entire Pareto set. One possible way to overcome this limitation consists of omitting some of them, thereby reducing the associated complexity. However this dimensionality reduction should be applied trying to maintain the Pareto dominance structure, as Brockhoff and Zitzler [2] formally stated.

Following the dimensionality reduction approach, in this work we propose a modification of the objective reduction models 2 and 3 of the work by Vazquez et al.[3], in which a mixed-integer linear programming formulation, inspired in the Set Covering Problem, is presented. First, we apply the model 1 of the previous work in its original form to a case study of a petrochemical supply chain[4]. This model provides the minimum subset of objectives that preserve the dominance structure completely unaltered. Keeping intact the dominance structure of the Pareto frontier is often too strict for practical applications, especially if the minimum number of non-redundant objectives is still too large. For that reason, in models 2 and 3 the dominance structure is allowed to be modified when some objectives are removed, and its degree of the degeneration is quantified by an error (denoted as d-error). By relaxing this constraint of immutability, model 2 offers the minimum objective subset while keeping the d-error below some given threshold (this model belongs to the type of problems known as the d-Minimum Objective Subset, d-MOOS problems). Likewise, in the third model we fix the maximum number *k* of objectives to preserve, and the model finds the subset of at most *k* objectives with the minimum d-error (this model is also known as the Minimum Objective Subset of Size *k* with Minimum Error, *k*-EMOSS problem). In the original work, the d-error in models 2 and 3 are computed using the infinity norm, while in this work we propose the use of the norm 1 in order to increase the sensitivity of these models. In addition, we introduce a fourth model based on new metric denominated the non-redundancy ratio. This new concept is defined as the percentage of the Pareto solutions needed to maintain completely the dominance structure of the non-redundant subset of objectives of a given size.

On the other hand, in this work we consider the following metrics to control and manage risk: Downside Risk (DR_{Ω})[5], Financial Risk (FR_{Ω}) [6], Worst Case, Value-at-Risk (VaR) [7], Opportunity Value (OV) [8] and Conditional Value-at-Risk (CVaR)[9]. Figure 1 shows a geometric interpretation of these risk metrics. DR_{Ω} and FR_{Ω} require the definition of a target (denoted by Ω), whereas VaR, OV and CVaR are calculated for a probability target in the cumulative probability distribution function (usually 0.05). To deal with the high dimensionality of the SMOO problem formulated, we solved it using a decomposition strategy based on the Sample Average Approximation algorithm[10]. Under this approach, the stochastic problem (second stage) contains only one objective (the economic) and all the risk metrics are computed after solving this problem. For that reason, a Pareto filter is used to ensure that the solutions belong to the Pareto frontier.

The results highlight how the different risk metrics affect the dominance structure of the problem, and trade-off among them, with the additional key feature of measuring the error incurred after removing some objectives.

Figure 1. Geometric interpretation of the risk metrics.

References

(1) Cao, Y.; Fuentes-Cortes, L. F.; Chen, S.; Zavala, V. M., Scalable modeling and solution of stochastic multiobjective optimization problems. *Comput Chem Eng* **2017,** 99, 185-197.

(2) Brockhoff, D.; Zitzler, E., Are All Objectives Necessary? On Dimensionality Reduction in Evolutionary Multiobjective Optimization. In *Parallel Problem Solving from Nature - PPSN IX*, Springer Berlin / Heidelberg: 2006; Vol. 4193/2006, pp 533-542.

(3) Vázquez, D.; Fernández-Torres, M. J.; Ruiz-Femenia, R.; Jiménez, L.; Caballero, J. A., MILP method for objective reduction in multi-objective optimization. *Comput Chem Eng* **2018,** 108, 382-394.

(4) Guillén-Gosálbez, G.; Grossmann, I., Optimal design and planning of sustainable chemical supply chains under uncertainty. *AIChE J.* **2009,** 55, (1), 99-121.

(5) Eppen, G. D.; Martin, R. K.; Schrage, L., A Scenario Approach to Capacity Planning. *Operations Research* **1989,** 37, (4), 517-527.

(6) Barbaro, A.; Bagajewicz, M. J., Managing financial risk in planning under uncertainty. *AIChE J.* **2004,** 50, (5), 963-989.

(7) Jorion, P., *Value at risk : the new benchmark for managing financial risk.* 3rd ed.; McGraw-Hill: New York, 2007; p xvii, 602 p.

(8) Aseeri, A.; Bagajewicz, M. J., New measures and procedures to manage financial risk with applications to the planning of gas commercialization in Asia. *Comput. Chem. Eng.* **2004,** 28, (12), 2791-2821.

(9) Rockafellar, R. T.; Uryasev, S., Optimization of Conditional Value-at-Risk. *Journal of Risk* **2000,** 2, 21-41.

(10) Medina-González, S.; Pozo, C.; Corsano, G.; Guillén-Gósalbez, G.; Espuña, A., Using Pareto filters to support risk management in optimization under uncertainty: Application to the strategic planning of chemical supply chains. *Comput Chem Eng* **2017,** 98, 236-255.

536208 Stochastic Programming Framework for Electric Power Infrastructure Planning

Abstract:

Energy systems planning models can be used to study the impact of new technology developments, resource cost trends, and policy shifts on the projected generation mix to meet future demand. Such systems are highly stochastic due to the uncertainty in future fuel prices, load demand and renewable generation. However, because of the computational expense of integrating detailed operating decisions in the hourly (or sub-hourly) level with investment decisions over a few decades, and having a fairly good representation of the grid, most of the available planning tools are deterministic.

In this paper, we address the long-term planning of electric power infrastructures under uncertainty. We propose a multi-stage stochastic integer programming (MSIP) formulation that includes investment decisions on a yearly basis and operating decisions on an hourly basis. This model optimizes the generation expansion required to meet the projected electricity demand over the next few decades while considering detailed operational constraints (i.e., unit commitment), the variability and intermittency of renewable generation sources, and the power flow between regions. We consider scenarios of low, medium and high load demand in each stage.

The major challenge lies in the tractability of the model, as its deterministic version already has 1,000,000+ constraints and 400,000+ integer variables [1]. To be able to solve such a large-scale model, we decompose the problem using Stochastic Dual Dynamic Integer Programming (SDDiP) [2]. SDDiP is a stage-based decomposition that solves each node of the scenario tree separately. The algorithm solves the problem in a forward and backward fashion, in which the Forward Pass provides a feasible solution and upper bound to the expected value, and the Backward Pass projects the problem onto the subspace of the stage variables using Benders-like cuts. Additionally, to reduce solution time, we incorporate scenario sampling to the algorithm, solving the Forward and Backward Passes for a randomly selected subset of scenarios [2], and solve nodes within the same stage in parallel.

The proposed formulation and algorithm are applied to a case study in the region managed by the Electric Reliability Council of Texas (ERCOT), and we show that an initially intractable model can be solved in a reasonable amount of time.

[1] Lara, C. L., Mallapragada, D., Papageorgiou, D., Venkatesh, D., & Grossmann, I.E. (2017) "Electric Power Infrastructure Planning: Mixed-Integer Programming Model and Nested Decomposition Algorithm", *Submitted for publication.*

[2] Zou, J., Ahmed, S. & Sun, X.A., (2018). "Stochastic dual dynamic integer programming", Mathematical Programming. pp. 1-42., https://doi.org/10.1007/s10107-018-1249-5

524568 Strategic Planning of Supply Chains Considering Extreme Events: Novel Heuristic and Application to the Petrochemical Industry

Abstract:

Chemical supply chains today are a crucial component in the ongoing supply of large population centres. However, episodes of extreme weather in particular have, in recent years, revealed vulnerabilities in global supply networks to high-impact events. And with a possible increase in both frequency and intensity of these events due to climate change, supply chains are at risk of disruption now more than ever, with potentially dire economic, societal, and environmental consequences.

Applying stochastic programming to handle these uncertainties can quickly lead to prohibitively large models that cannot be solved in reasonable time. To remedy this, we propose a new methodology which combines the sample average approximation method with a selection heuristic for extreme event scenarios. Combined with multiobjective optimisation, this allows for the efficient analysis of the tradeoff between economic performance and disruption risk.

We demonstrate the effectiveness of this methodology in multiple case studies, achieving good solutions in reduced CPU time.

525269 Resilient Design and Operations of Chemical Process Systems Using Robust Optimization

Abstract:

A major goal of risk management is to avoid the occurrences of undesired events by implementing effective prevention and protection strategies [1]. However, many disruptive events suggest that not all unexpected events can be avoided. Disruptive events usually strike a process system and cause critical failures in vulnerable processes [2]. The undesired consequences of disruptive events highlight the need for enhancing the resilience of process systems. Although resilience may be interpreted by different terminologies in various contexts, a resilient system is always capable of absorbing a portion of the impacts from disruptive events and recovering to the original state rapidly. There are several research challenges to develop a general framework for resilience optimization. The first challenge is to propose a novel quantitative measure of resilience for process systems. The quantitative measure should be able to account for both performance degradation and system recovery. Additionally, resilience is an intrinsic property of a system, indicating that the quantitative measure should be independent of external systems or volatile markets. The second challenge is how to model resilience enhancement strategies, and how to integrate the resilience enhancement strategies and process models into an integrated systems analysis and optimization framework. Because the resilience of a process system is relevant to both safety and operability, the system performance under the worst-case realization of disruptive events is of paramount importance [3]. Adaptive robust optimization is an emerging method to handle uncertainty in sequential decision-making processes and to hedge against the worst-case realization of uncertainty. Although adaptive robust optimization could be an effective tool for addressing problems on resilient design and operations, there are research challenges on developing a two-stage adaptive robust optimization model for resilience optimization and on effective solution of the resulting optimization problem that cannot be tackled directly by any off-the-shelf optimization solvers due to its multilevel structure.

In this work, we address the resilient design and operations of process systems in response to disruption events. A novel quantitative measure of resilience is proposed as the ratio of the quantity of products manufactured with disruptive events to that without disruptive events. Five resilience enhancement strategies are then introduced, including selecting the most resilient technology/process alternatives, increasing the capacities of operating processes, employing parallel operating processes, building backup processes, and optimizing the operating levels after the occurrence of disruptive events. A general framework for resilience optimization is further proposed to incorporate the quantitative measure of resilience and the resilience enhancement strategies into process design and operations. In the first step of the proposed resilience optimization framework, a preliminary risk assessment is performed for a given system to identify the disruptive events that are worth considering in process design and operations. The numbers of failed processes for the identified disruptive events and the recovery time of each process are used as input parameters for resilient design and operations. In the second step, a multiobjective two-stage adaptive robust mixed-integer fractional programming model is formulated. There are two objective functions: the first objective function is to maximize the resilience under the worst-case realization of disruptive events, and the second objective function is to minimize the total capital cost. Both objective functions are independent of external processes and volatile markets, thus reflecting the intrinsic properties of the given system. The resulting optimization model has a three-level structure: the first level determines the optimal network configuration, equipment capacities, and capital costs; the second level determines the worst-case realization of disruptive events; the third level determines the optimal number of available processes and operating levels in each time period. To tackle the computational challenges stemming from the multilevel structure and the nonlinear objective function, a tailored global optimization method is proposed by integrating the inexact parametric algorithm and the column-and-constraint generation algorithm. The applicability of the proposed resilience optimization framework is illustrated through two applications on process design and planning of a chemical process network and superstructure optimization of shale gas processing and natural gas liquids recovery processes, respectively.

References

- Hosseini, K. Barker, and J. E. Ramirez-Marquez, "A review of definitions and measures of system resilience," Reliab. Eng. Syst. Safe., vol. 145, pp. 47-61, 2016.
- Bhamra, S. Dani, and K. Burnard, "Resilience: the concept, a literature review and future directions," Int. J. Prod. Res., vol. 49, pp. 5375-5393, 2011.
- T. T. Dinh, H. Pasman, X. D. Gao, and M. S. Mannan, "Resilience engineering of industrial processes: Principles and contributing factors," J. Loss Prevent. Proc., vol. 25, pp. 233-241, 2012.

525691 A Mixed-Integer Conic Programming Formulation for Computing the Flexibility Index Under Multivariate Gaussian Random Variables

Abstract:

Quantifying the flexibility of complex systems such as chemical plants and power distribution networks is drawing increased interest. Flexibility seeks to quantify the ability of a system to counteract disturbances and maintain feasible operation [2]. Over the past 35 years, Grossmann and co-workers have proposed several strategies to quantify flexibility [8, 9]. This can be done by solving the so-called flexibility index problem, which seeks to find the largest uncertainty set under which the system maintains feasible operation [3]. This problem is conceptually attractive, but it is computationally challenging to solve due to the presence of nested optimization problems [4]. Floudas and co-workers showed that, for hyperbox uncertainty sets and convex constraints, the flexibility index problem can be reformulated as a mixed-integer linear program (MILP). This approach is scalable, but hyperbox uncertainty sets do not capture correlation between random variables. As a result, hyperbox representations can provide highly conservative solutions. Motivated by this limitation, Straub and Grossmann proposed the stochastic flexibility index which represents the probability that the system will exhibit feasible operation [6, 7]. This index can be computed by using a quadrature scheme, but the approach suffers from scalability issues [5, 7].

In this work, we propose a scalable approach to compute the flexibility index for multivariate Gaussian variables. The Gaussian representation enables us to capture correlations. We show that the problem can be cast as a mixed-integer conic program (MICP) that seeks to find the largest radius of a hyperellipsoidal uncertainty set under which the system maintains feasible operation. We also prove that the radius can be directly related to the probability of feasible operation, which is precisely the definition of the stochastic flexibility index. Consequently, both indexes are equivalent in this case. The proposed approach can leverage the recent emergence of efficient MICP solvers [1].

References

1. Berthold, S. Heinz, and S. Vigerske. Extending a cip framework to solve miqcp s. In *Mixed integer nonlinear programming*, pages 427–444. Springer, 2012.
2. Deffuant and N. Gilbert. *Viability and resilience of complex systems: concepts, methods and case studies from ecology and society*. Springer Science & Business Media, 2011.
3. E. Grossmann, B. A. Calfa, and P. Garcia-Herreros. Evolution of concepts and models for quantifying resiliency and flexibility of chemical processes. *Computers & Chemical Engineering*, 70:22–34, 2014.
4. E. Grossmann, K. P. Halemane, and R. E. Swaney. Optimization strategies for flexible chemical processes. *Computers & Chemical Engineering*, 7(4):439–462, 1983.
5. Ma, V. Rokhlin, and S. Wandzura. Generalized gaussian quadrature rules for systems of arbitrary functions. *SIAM Journal on Numerical Analysis*, 33(3):971–996, 1996.
6. Pistikopoulos and T. Mazzuchi. A novel flexibility analysis approach for processes with stochastic parameters. *Computers & Chemical Engineering*, 14(9):991–1000, 1990.
7. A. Straub and I. E. Grossmann. Integrated stochastic metric of flexibility for systems with discrete state and continuous parameter uncertainties. *Computers & Chemical Engineering*, 14(9):967– 985, 1990.
8. E. Swaney and I. E. Grossmann. An index for operational flexibility in chemical process design. part i: Formulation and theory. *AIChE Journal*, 31(4):621–630, 1985.
9. E. Swaney and I. E. Grossmann. An index for operational flexibility in chemical process design. part ii: Computational algorithms. *AIChE Journal*, 31(4):631–641, 1985.

530733 Modeling for Reliability Optimization of System Design and Maintenance Based on Markov Chain Theory

Abstract:

Plant availability has always been a critical consideration for the design and operation of chemical processes as it represents the expected fraction of normal operating time, which impacts directly the ability of meeting customer demands. The failure of a key unit is likely to cause the shut-down of the entire plant. Effective ways to reduce the impact of equipment breakdown on the entire plant include installing back-up units and carrying out regular inspections and maintenance. Currently, discrete event simulation tools are used to evaluate the reliability/availability of selected alternatives to simulate the behavior of every asset in a plant using historical maintenance data and statistical models (Sharda and Bury, 2008). However, this approach does not systematically consider all the alternatives as it would be the case in an optimization approach.

A number of works have been reported to optimize the reliability of chemical plants. Pistikopoulos et al. (2001) and Goel et al. (2003b) formulate an MILP model for the selection of units with different reliability and the corresponding production and maintenance planning for a fixed system configuration. Terrazas-Moreno et al. (2010) formulate an MILP model using Markov chains to optimize the expected stochastic flexibility of an integrated production site by the selection of pre-specified alternative plants and the design of intermediate storage. Lin et al. (2012) model a simple utility system using Markov chain and carry out RAM (reliability, availability & maintainability) analysis iteratively to decide the optimal reliability design.

While mixed-integer optimization models have been proposed to address the reliability problem from various aspects, there are virtually no systematic modeling approaches proposed to make decisions regarding the number and selection of parallel units, as well as maintenance policy for the optimal profitability of reliable chemical processes. In response to this gap, this work extends our recent mixed-integer framework (Ye et al., in press) and introduces a systematic approach to model the stochastic process of system failures and repairs as a continuous-time Markov chain.

Two decision levels are considered to increase the availability of the system. The first decision level is to install parallel units for certain processing stages, such that when the primary unit fails, the other units can fill in its place in order to reduce system downtime. The second decision level is to carry out periodic inspections, and conditional maintenance if the inspection result indicates that the equipment will fail shortly. By that strategy the system can avoid a number of repairs, which are more costly than maintenances in terms of both time and money. Thus, there are two types of trade-offs in the model. The first type is the trade-off between the costs incurred and the revenue gained from increasing system availability. The second type is the budget balance between the two decision levels and among the processing stages. A non-convex MINLP model based on Markov chain assumption is proposed accordingly. It is worth mentioning that when inspections and conditional maintenance are not considered, the model reduces to an MILP, and a small superstructure with two stages is solved directly with global optimization solvers to show that improved system availability and net profit can be obtained compared to the case when disregarding maintenance, although at the expense of longer computing time.

However, the size of the non-convex MINLP model grows geometrically with the number of processing stages in the system. In addition, the combination of the two decision levels creates a large number of bilinear and multilinear terms. In order to overcome these computational difficulties and make the model applicable to practical problems, a decomposition scheme is proposed to reduce the model complexity. Moreover, scenario reduction, i. e., preprocess the scenarios and eliminate those with consistently low probabilities, is also applied to reduce the size of the model. The non-convex MINLP model with the specialized solution methods is applied to the two-stage problem and reaches global optimum in significantly shorter time than the global solvers. It is then used to solve larger examples that are not directly solvable and is able to obtained solutions in acceptable time.

References

Pistikopoulos, E. N., Vassiliadis, C. G., Arvela, J., and Papageorgiou, L. G. (2001). Interactions of maintenance and production planning for multipurpose process plants a system effectiveness approach. Industrial & engineering chemistry research, 40(14):3195-3207.

Goel, H. D., Grievink, J., and Weijnen, M. P. (2003b). Integrated optimal reliable design, production, and maintenance planning for multipurpose process plants. Computers & chemical engineering, 27(11):1543-1555.

Sharda, B. and Bury, S. J. (2008). A discrete event simulation model for reliability modeling of a chemical plant. In Proceedings of the 40th Conference on Winter Simulation, pages 1736-1740. Winter Simulation Conference.

Terrazas-Moreno, S., Grossmann, I. E., Wassick, J. M., and Bury, S. J. (2010). Optimal design of reliable integrated chemical production sites. Computers & Chemical Engineering, 34(12):1919-1936.

Lin, Z., Zheng, Z., Smith, R., and Yin, Q. (2012). Reliability issues in the design and optimization of process utility systems. Theoretical Foundations of Chemical Engineering, 46(6):747-754.

Ye, Y., Grossmann, I. E., Pinto, J. M., (in press). Mixed-integer nonlinear programming models for optimal design of reliable chemical plants. Computers & chemical engineering, doi: 10.1016/j.compchemeng.2017.08.013.

533520 Novel Method for the Integration of Flexibility and Stability in Design of Chemical Processes Under Parametric Uncertainties

Abstract:

Process optimization under parametric uncertainties is a challenging topic of practical importance in the Process Systems Engineering since uncertainties can always cause infeasibilities, detrimental production disturbances and extra capital or operating costs. Flexibility and stability are two crucial components of chemical process operability [1]. Flexibility guarantees feasible steady-state operation over a range of uncertain operating conditions, while stability represents the tolerance for slight perturbations and emphasizes the dynamic characteristics of a system [2]. In previous research, flexibility and stability have been studied separately, resulting in flexible regions containing unstable areas that are difficult to operate and control.

The major difficulty of integrating flexibility and stability lies in how to convert the Lyapunov stability conditions into specific constraints embedded within the flexibility analysis model. This paper proposes a novel method for integrating flexibility with stability, which incorporates stability constraints obtained by singularity theory based stability analysis method into the MINLP model to calculate flexibility index. The singularity theory based stability analysis method can characterize regions in parameter space over which different kinds of stability characteristics may exist.

As a result, a stable flexible region is obtained, which not only adapts to variations in uncertain parameters, but also ensures stable operation and process inherent safety. 1,3-propanediol fermentation system and methyl methacrylate polymerization process have been studied to demonstrate the effectiveness of the proposed methodology. The obtained results illustrate that the novel method for considering flexibility and stability simultaneously possesses distinct computational advantages over the eigenvalue optimization algorithm, so it has great potential for applications in large-scale systems.

References

[1] Floudas, C. A., GÃ¼mÃ¼ÅŸ, Z. H., & Ierapetritou, M. G. (2001). Global optimization in design under uncertainty: feasibility test and flexibility index problems. *Industrial & Engineering Chemistry Research*, *40*(20), 4267-4282.

[2] Jiang, H., Chen, B., Wang, H., Qiu, T., & Zhao, J. (2014). Novel method for considering process flexibility and stability simultaneously. *Industrial & Engineering Chemistry Research*, *53*(38), 14765-14775.

522027 Energy-Water Nexus Study for a Mushroom Farming Initiative in Nigeria

Abstract:

A water treatment and recycle system was designed for a mushroom farming initiative in Kaduna, Nigeria. The energy requirements of the proposed system were studied to determine the optimal system to minimize energy use for water treatment. Other considerations were ease of use and capital and operating costs.

The concept behind the cooperative is for 200 women farmers pool their resources to pay for water and energy costs. Electricity will be provided through solar panels. We were consulted to design a water treatment and recycle system for a demonstration of the concept. The concept will likely be replicated in remote areas without an abundance of water, so water recycle is essential. Reliable electricity will also likely be an issue so solar panels will be used.

The first step was to collect samples of water from the potential water source: the Kaduna River. Preliminary tests showed that the water was relatively clean but contained bacteria and some hardness. A treatment and recycle system was designed appropriate for the impurities in the water. Due to limited energy provided by the solar panels, energy requirements were studied for all appropriate treatments to minimize energy use of the final system. The final design and final energy use of the water system will be presented.

523808 Sustainable Optimal Strategic Planning for Shale Water Management

Abstract:

The development of new additives in the shale gas industry, which tolerate the use of high Total Dissolved Solids (TDS) base fluid, has allowed reusing the wastewater in the drilling of subsequent wells (U.S. Environmental Protection Agency, 2016). This practice is currently the most popular and cost-effective option for shale gas water management. Although it minimizes freshwater consumption, producers should take into consideration possible long-term issues and challenges. For instance, the TDS concentration will increase significantly, around 2-3 times, which can represent a cost barrier to reuse the water for fracturing operations. Besides, as the number of drilled wells decrease this practice becomes less attractive. Specifically, the volume of fracturing fluid required to fracture new wells may be less than the volume of water generated by producing wells in the area.

Currently, several works have been reported on the optimization of shale gas water management (Gao and You, 2015; Guerra et al., 2016; Yang et al., 2015). However, most of the works published in the literature consider that the water blending ratio is restricted or the return to pad operations are not allowed. Drouven and Grossmann (2017) assume that the water-blending ratio is unrestricted, they over-estimate the friction reducers expenses. The MILP model that they proposed does not account the salt concentration of impaired water. Moreover, they only distinguish between two types of water: impaired water and freshwater. Hence, the model cannot handle any water management option when drilling operation decrease.

In this work, we propose a mixed-integer non-linear programming (MINLP) model considering the TDS concentration of flowback and impaired water as a function of time and for different water treatment solutions. We estimate the friction reducers expenses as a function of TDS concentration to determine if the level of TDS in impaired water is an impediment to reusing it in fracturing operations. Moreover, the model distinguishes between four types of water: impaired water, flowback water, desalinated water and freshwater. The objective is to maximize the "sustainability profit" in order to obtain a compromise solution between economic, environmental and social aspects. Only a single objective function is necessary since all the indicators are expressed in monetary terms (Zore et al., 2017).

The water management system comprises wellpads, shale gas wells in each wellpad, centralized water treatment technologies (CWT), freshwater sources, fracturing crews, and disposals wells.

After hydraulic fracturing, a portion of water called flowback water returns to the wellhead. The flowback water is stored in fracturing tanks (FT) onsite before basic treatment (pre-treatment) in mobile units or transport to CWT facility or Class II disposal. Pre-treatment includes technologies to remove suspended solids, oil and grease, and bacteria, certain ions that can cause the scale to form on equipment and interfere with fracturing chemical additives. After pre-treatment, the water can be used as a fracturing fluid in the same wellpad or in the neighboring wellpad, or can be desalinated in the onsite TDS removal technologies. The flowback water reused for fracturing operations is called impaired water and is stored in impaired water tanks (IWT). Several desalination technologies can be selected such as multi-stage membrane distillation (MMD), multi-effect evaporation with mechanical vapor recompression (MEE-MVR) or forward-reverse osmosis (FO-RO) hybrid. We consider that the outflow brine salinity in the onsite treatment is close to salt saturation conditions to achieve ZLD operation. Cost and salinity levels restrict the type of desalination unit that can be used for TDS removal. The onsite desalinated water can also be used as a fracturing fluid in the same wellpad, transported to the next wellpad or discharge for other usages. The flowback water can also be transported and treated in CWT. Desalinated water from CWT can select the same routes as the desalinated water in onsite technologies. Freshwater is obtained from an uninterruptible fresh water source. Desalinated water and freshwater are stored in freshwater tanks (FWT) and/or water impoundment. Transportation of freshwater, impaired water, flowback water and desalinated water can only be through trucks. Storage tanks and mobile treatment are assumed to be leased.

The problem is to determine: (1) wellpad fracturing start date (fracturing schedule), (2) number of tanks leased at each time period, (3) number of trucks needed in each time period, (4) flowback destination, reuse (impaired water) , treatment (onsite or offsite) or disposal, and (5) quality of water used to fracture each well.

The optimization problem is formulated as an MINLP model that includes: assignment constraints, material balance in storage tanks, mixer and splitters, logic constraints and an objective function. Salt material balances are modeled using total flows and salt composition as variables (bilinear terms). An advantage of using this representation is that the variables involved in the bilinear terms are well bounded, allowing us to define tight under and over estimators.

Them objective function includes the economic profit, the eco-profit and the social-profit.

- *Economic profit* includes revenues from natural gas minus the sum of the following expenses: wastewater disposal cost, storage tank cost, freshwater cost, friction reducer cost, wastewater and freshwater transport cost and onsite and offsite treatment cost.
- *Eco-profit* distinguishes between eco-benefit (raw material and products that unburden the environment) and eco-cost (raw material al products that burden the environment). Both terms are calculated by using eco-cost coefficients (Delft University of Technology, 2017). In our problem, impaired water and desalinated water used to fracture a neighboring well exhibit unburdening effect on the environment. However, natural gas, freshwater withdrawal, disposal and transportation burden the environment.
- *Social profit* includes social security contributions paid for the employed people to fracture a well, plus the social transfer by hiring people, minus social cost. We only take into account the numbers of jobs on a fracturing crew and the time that they are working to fracture a specific well. Once the well is completed, the number of jobs generated by truck drivers or maintenance team are not considered.

The proposed model is applied to a case study in Marcellus Play with 3 wellpads and 20 wells, three years discretized at one week per time period, four interruptible sources of fresh water, three class II disposal wells, two CWT plants and one fracturing crew. The MINLP problem consists of 2808 binary variables, 14977 continues variables and 12010 constraints.

The solution of the model shows that the capabilities to reuse impaired water to fracture other wells is the best economic and environmental practice for shale gas water management. The model, which can be effectively solved with the proposed decomposition technique, reveals that the level of TDS in impaired water is not an obstacle to reusing it for fracturing purposes. Also, it has been shown that onsite desalination treatment can be cost-effective for operators once no more wells to fracture are available.

References:

Delft University of Technology, 2017. The Model of the Eco-costs/Value Ratio (EVR). URL http://www.ecocostsvalue.com (accessed 12.1.17).

Drouven, M.G., Grossmann, I.E., 2017. Optimization models for impaired water management in active shale gas development areas. J. Pet. Sci. Eng. 156, 983–995.
Gao, J., You, F., 2015. Shale Gas Supply Chain Design and Operations toward Better Economic and Life Cycle Environmental Performance: MINLP Model and Global Optimization Algorithm. ACS Sustain. Chem. Eng. 3, 1282–1291.
Guerra, O.J., Calderón, A.J., Papageorgiou, L.G., Siirola, J.J., Reklaitis, G. V., 2016. An optimization framework for the integration of water management and shale gas supply chain design. Comput. Chem. Eng. 92, 230–255.
U.S. EPA, 2016. Technical Development Document For Effluent Limitations Guidelines and Standars for the Oil and Gas Extraction Point Source Category. Washington, DC.
Yang, L., Grossmann, I.E., Mauter, M.S., Dilmore, R.M., 2015. Investment optimization model for freshwater acquisition and wastewater handling in shale gas production. AIChE J. 61, 1770–1782.
Zore, Ž., Čuček, L., Kravanja, Z., 2017. Syntheses of sustainable supply networks with a new composite criterion – Sustainability profit. Comput. Chem. Eng. 102, 139–155.

534130 A Multi-Objective Energy-Water Nexus Planning Model: A Case Study of the Power Systems in Texas Edwards Aquifer

Abstract:

Increasing power demands and water scarcity as well as environmental concerns give rise for the need of a planning strategy for power generating systems that takes into consideration the nexus between energy and water [1]. Power systems planning problems are complex and comprise of decisions related to: (i) the ability to add additional power plants with different types of cooling and power generating technologies, (ii) the ability to convert the cooling and or power generating technologies of existing power plants, (iii) uncertainties, and (iv) environmental and social impacts [2]. Given that the type of power generating and cooling technology of a power plant directly affects its water usage, the optimization of a power system is a multi-objective problem [3]. Therefore, a systematic methodology based on the Energy-Water Nexus (EW-N) is necessary for modeling and solving the optimization problem on power systems planning [4].

In this work, we develop a novel superstructure for planning strategies of power generating systems based on EW-N connections. The EW-N problem is formulated as a two-stage stochastic multi-objective mixed integer linear program [5]; with "here-and-now" decision variables such as additional power plants with specific power generation and cooling technologies and "wait-and-see" decision variables comprising of and not limited to the capacity and water utilization of power plants. The model determines the optimal conversion and expansion policies for the current infrastructure, based upon uncertain future water availability and power demands. The model is implemented into a case study regarding EW-N in the Edward Aquifer, aiming to provide policy makers with a systematic tool to analyze various scenarios and technology options. Moreover, the model is also flexible in that it can also be integrated into the Edwards Aquifer Groundwater and River System Simulation Model (EDSIMR) [6, 7], which is a regional model for the comprehensive decision-making and resources allocation regarding Food-Energy-Water Nexus in the Edwards Aquifer.

References:

[1] Diangelakis, N. A., & Pistikopoulos, E. N. (2017). A multi-scale energy systems engineering approach to residential combined heat and power systems. Computers & Chemical Engineering, 102, 128-138.

[2] Koltsaklis, N. E., Dagoumas, A. S., Kopanos, G. M., Pistikopoulos, E. N., & Georgiadis, M. C. (2014). A spatial multi-period long-term energy planning model: a case study of the Greek power system. Applied Energy, 115, 456-482.

[3] Mohtar, R. H., & Daher, B. (2016). Water-energy-food nexus framework for facilitating multi-stakeholder dialogue. *Water International*, *41*(5), 655-661.

[4] Garcia, D. J., & You, F. (2016). The water-energy-food nexus and process systems engineering: a new focus. *Computers & Chemical Engineering*, *91*, 49-67.

[5] Birge, J. R., & Louveaux, F. (2011). Introduction to stochastic programming. Springer Science & Business Media.

[6] Gillig, D., McCarl, B. A., & Boadu, F. (2001). An economic, hydrologic, and environmental assessment of water management alternative plans for the south central Texas region. Journal of Agricultural and Applied Economics, 33(1), 59-78.

[7] McCarl, B. A., Dillon, C. R., Keplinger, K. O., & Williams, R. L. (1999). Limiting pumping from the Edwards Aquifer: an economic investigation of proposals, water markets, and spring flow guarantees. Water Resources Research, 35(4), 1257-1268.

534167 Optimal Use of Thermal Membrane Distillation (TMD) for Treatment of Flowback Water

Abstract:

Optimal use of thermal membrane distillation (TMD) for treatment of flowback water

Kaiyu Cao[1], Priscille I. Etoughe[2], Rajib Mukherjee[2,3], Debalina Sengupta[3], Joseph S. Kwon[1,2,3*], Mahmud M. El-Halwagi[1,2,3]

1. *Artie McFerring Department of Chemical Engineering, Texas A&M Univrsity, College Station, TX*
2. *Texas A&M Energy Institute, Texas A&M University, College Station, TX*
3. *Gas and Fuels Research Center, Texas A&M Engineering Experiment Station, College Station, TX*

The flow rate of flowback water from hydraulic fracturing varies with time. Most of the water are produced within the first two weeks after hydraulic fracturing is completed. The concentration of total dissolved solids (TDS) in the flowback water also changes with time. For efficient recovery and reuse of flowback water, treatment with thermal membrane distillation (TMD) for the removal of TDS has been found to be the method of choice [1]. Even though flowback water is a continuous process, in this work, for modeling purpose a multi-period optimization problem has been formulated to account for the time-dependent variation of the flow rate and TDS concentration of flowback water, accounting for the time-; hereeach day is treated as a period. Then, we have integrated a TMD unit, considering the largest capacity of flowrate from a well in any given period. An upper limit on the concentration of TDS in the discharged (reject) water is considered to avoid precipitation. Utilizing the field data of flowback water from a given well for each period, the optimization formulation has been solved to determine the maximum period of time for optimal use of TMD, maximizing annualized profit from recovered water per period. Our result shows that it is not profitable to use water recovery beyond first few days, depending on the formation. The performance of the optimization formulation is measured in terms of percentage of recovery and is compared with cost associated with them.

References:

[1] Elsayed, N. A., Barrufet, M. A., Eljack, F. T., & El-Halwagi, M. M. (2015). Optimal design of thermal membrane distillation systems for the treatment of shale gas flowback water. *Int J Membr Sci Technol*, *2*, 1-9.

536214 The Energy-Water Nexus of Thermoelectric Power Generation and Its Impacts in the Muskingum River Watershed in Ohio

Abstract:

Thermoelectric power plants generate electricity and flue gas scrubbers mitigate emissions from the plants. However, at the same time the power plants withdraw large quantities of water from the watershed, which corresponds to 45% of the total 2010 US water use [1], and emissions from the scrubbers cause deposition of nutrients in the watershed. Therefore, increasing electricity generation could increase water stress and deteriorate water quality. Also, in the watershed where power plants are located, other activities, such as farming, interact with the power plants since these activities also require water as well as energy from the power plants and release nutrients to the watershed. To prevent shifting of environmental impacts across multiple flows [2], the Energy-Water nexus between different activities in the watershed needs to be understood in assessing the impacts of power plants.

Muskingum River Watershed (MRW) in Ohio ranks fourth as the most polluted watershed in the US [3], and two coal-fired power plants and three natural gas-fired power plants are located in the MRW. In this work, a holistic analysis approach is employed to evaluate the water-energy nexus in the MRW. Various technological and ecological systems are considered. Technological systems include mining, thermoelectric power generation, and CO_2 conversion activities. Ecological systems include land use and farming practice. Alternative scenarios for each component are analyzed to address the nexus. An agro-ecological model, which is developed by the Soil and Water Assessment Tool (SWAT), is used to assess water quality and quantity [4], while engineering models are used for the power plants and scrubbing processes.

Fossil fuel power plants not only require a huge amount of water, but also emit 28% of the 2016 US greenhouse gas emissions [5], 67% of the 2014 US SO_2 emissions, and 12% of the 2014 US NO_x emissions [6]. Therefore, ecosystem services, such as water quality regulation, air quality regulation, and climate change regulation, play an important role in mitigating nutrient runoffs and air emissions. To account for the role of ecosystems, Techno-Ecological Synergy framework is applied in analyzing the impacts of activities in the MRW [7,8]. In this analysis, the trade-offs between water quality, water quantity, net electricity generation, carbon footprint, air quality, and net present value are identified. Various scenarios are considered to suggest better solutions that are 'win-win' in terms of multiple objectives.

Considering the Energy-Water nexus in the analysis will avoid shifting of the environmental impacts across multiple flows because interactions between flows are identified and all flows are included as objectives [2]. This work could be applied to any watershed and could be extended to address the Food-Energy-Water nexus by including food productions from farming activities as another objective. Addressing the nexus could help businesses and policy makers in making decisions to improve the sustainability of the watershed.

References

[1] Maupin, M.A. et al. Estimated use of water in the United States in 2010. U.S. Geological Survey, Reston, Virginia (2014).

[2] Bakshi, B. R., Gutowski, T. G. & Sekulic, D. P. Claiming Sustainability: Requirements and Challenges. ACS Sustain. Chem. Eng. 6, 3632–3639 (2018).

[3] Inglis, J. & Dutzik, T. Wasting our waterways: Toxic industrial pollution and restoring the promise of the clean water act. Environment America Research & Policy Center, Boston, MA (2006).

[4] Khanal, S., Lal, R., Kharel, G. & Fulton, J. Identification and classification of critical soil and water conservation areas in the Muskingum River basin in Ohio. J. Soil Water Conserv. 73, 213–226 (2018).

[5] Inventory of U.S. greenhouse gas emissions and sinks: 1990-2016. U.S. Environmental Protection Agency, Washington, D.C. (2018).

[6] 2014 National Emissions Inventory (NEI). US Environmental Protection Agency. https://www.epa.gov/air-emissions-inventories. Accessed on Apr 13 2018.

[7] Bakshi, B. R., Ziv, G. & Lepech, M. D. Techno-ecological synergy: A framework for sustainable engineering. Environ. Sci. Technol. 49, 1752–1760 (2015).

[8] Gopalakrishnan, V., Bakshi, B. R. & Ziv, G. Assessing the capacity of local ecosystems to meet industrial demand for ecosystem services. AIChE J. 62, 3319-3333 (2016).

537140 Systematic Analysis and Optimization of Water-Energy Nexus

Abstract:

Securing scarce energy and water resources from the environment, while at the same time satisfying on-growing water and energy demands for end-use consumption is crucial [1]. The intermediate processing facilities, energy sources (acting also as water sinks) and water sources (acting as energy sinks), receive energy and water natural resources respectively, process, exchange and deliver them as usable electricity, fuels and water for consumption. It is important to quantify the performance and the efficiency of this water-energy nexus, and understand how it is affected by the selection of the energy and water sourcesâ€™ technologies, operating capacities and their connectivity.

To this end, we present a scalable, graph-theoretic representation of a nexus for the design and optimization of complex water-energy systems [2]. We define a nexus as a directed bipartite graph with water and energy flows and show that for specified external grid demands, the optimal nexus configuration with minimum water and energy generation is the one without any redundant subgraphs. We then introduce the water-energy (WEN) diagram that can systematically identify and eliminate those redundant subsystems. This leads to (i) minimum generation/extraction of water and energy resources from the environment, or (ii) maximum yield of water and energy to meet external demands. Our approach is simple to implement and results in optimal nexus configurations that also consider operational constraints, connectivity restrictions and water quality specifications. In the second part, we develop mathematical programming-based algebraic method that provides the same results but for more complicated and large-scale nexus design. This also takes into account the intensity factors of the sources, restrictions on source-sink matching, and quality specifications. The model is further extended to include capital and operating costs of the processing facilities, as well as geographic distances. The approach is demonstrated using case studies on water-energy nexus systems focusing on power generation, seawater desalination, groundwater and surface water at regional and national scales.

References:

[1] D. J. Garcia and F. You, "The water-energy-food nexus and process systems engineering: a new focus," Computers & Chemical Engineering, vol. 91, pp. 49-67, 2016.

[2] S. Tsolas, M. N. Karim, M. M. F. Hasan, " Optimization of Water-Energy Nexus: A Network Representation-based Graphical Approach," Applied Energy, *Submitted.*

538380 Thermo-Economic Optimization Based Comparison of Membrane Distillation Vs Mechanical Vapor Recompression for Shale Gas Produced Water Treatment

Abstract:

With advancements in horizontal drilling and hydraulic fracturing methods, unconventional shale gas extraction is rapidly developing. However, shale gas production is accompanied with a host of environmental challenges. One such critical challenge is the sustainable management of large quantities of high salinity wastewater generated in the process of hydraulic fracturing. While recycling may be a short-term strategy, long term solutions are needed for economical and environmentally conscious management of high salinity wastewaters. Desalination is one possible technological solution to treat the produced water and reduce the brine volume for final disposal while simultaneously producing fresh water for other beneficial uses. Membrane distillation (MD) system is a vapor pressure driven emerging membrane technology capable of handling hypersaline wastewater. Since MD can be operated at low temperature (30-90^0C), it has the potential to be integrated with low-grade waste heat sources. Another advantage of MD is the low capital investment of the process. However, low energy efficiency of MD results in high operating costs when waste heat is not available. Mechanical vapor recompression (MVR) is another desalination technology capable of treating hypersaline produced water. Unlike MD, MVR system is a well-established work driven desalination process, in which the heat recovery and energy efficiency is relatively high by utilizing the vapor compression cycle. In contrast with MD, MVR system requires considerable amount of high quality energy sources. Considering the competitive desalination efficiency of these two technologies, there is a need for a systematic way to comparison their energetic and economic performance. To date, very limited work exists on techno-economic optimization and evaluation of MD and MVR technologies for produced water treatment.

We present a mixed-integer non-linear programming (MINLP) formulation of a direct contact membrane (DCMD) process design for produced water treatment while minimizing the total treatment cost. The proposed model simultaneously optimizes the operating conditions (i.e. the temperature, pressures, and flowrate of all streams) and DCMD plant configuration (number of membrane stages, number of modules in each stage, and membranes geometry). Also, the optimization model, takes into account the possibility to improve system performance by including brine recirculation and heat integration. The results of the proposed optimization model are compared with our previous work on MVR optimization for application in produced water treatment. The optimization model is tested for a hypothetical desalination plant with 0.5 million gallons per day (MGD) capacity, in which produced water with 10% salinity is concentrated to 30% salinity of the exiting brine. The results are also validated through ASPEN simulations. Sensitivity analysis is also conducted to analyze the effect of input parameters on the system performance and economics. Tradeoffs between MD and MVR technologies for treatment of shale gas produced water will be discussed.

529522 Application of Adsorbate Solid Solution Theory to Design Novel Adsorbents for Arsenic Removal Using Computer-Aided Molecular Design

Abstract:

Application of Adsorbate Solid Solution Theory to Design Novel Adsorbents for Arsenic Removal Using CAMD

Rajat K. Doshi[1,2], Arti Ayodhyaprasad Rajput[3], Rajib Mukherjee[1], Suresh Gupta[3] , Urmila M. Diwekar[1*],

1. Vishwamitra Research Institute, Crystal Lake, IL, USA
2. Henry B. Plant High School, 2415 S. Himes Ave., Tampa, USA
3. Department of Chemical Engineering, Birla Institute of Technology & Science, Pilani, India

*Corresponding: Urmila@vri-custom.org

Arsenic is a carcinogenic contaminant that pollutes the groundwater, a consequence of poor arsenic disposal. Various techniques are used to remove arsenic, such as oxidation, coagulation-flocculation, ion exchange, and adsorption, among which adsorption is the most efficient method. Current arsenic separating agents have a limited adsorption capacity. The overall objective of this work is to develop a computational tool for the design of novel adsorbents for arsenic remediation using clay materials including beidellite, zeolite, and sepiolite that are cheap and readily available [1]. In the first part of this research, we use the Group Contribution Method (GCM) to predict thermodynamic properties and calculate the UNIFAC interaction parameters between arsenic and other functional groups that are selected from clay materials. In the second part of this research, we utilize a computer-aided molecular design (CAMD) framework that develops new adsorbent candidates with enhanced adsorption capacities based on the group interaction parameters generated in the first part. The efficient ant colony optimization (EACO) algorithm maximizes the adsorption capacity with certain structural possibilities, thermodynamic property correlations, and process conditions. It was found that the newly designed adsorbent has an order of magnitude higher removal capacity than the adsorbents' reported in the literature [2]. The adsorption capacity estimated for the designed adsorbent was simulated in groundwater conditions; however, arsenic contamination is not solely confined to groundwater as it is also highly prevalent in wastewater flows from coal-burning power plants. An expansion of our work will be carried out by conducting experiments on the designed adsorbents. The goal is to test the performance of the designed adsorbent in coal power plant's wastewater flows and understand how the different ionic environment affects the adsorption capacity. The maximum amount of arsenic that the designed adsorbent can adsorbed per unit mass of adsorbent from wastewater found in coal-burning power plants will be presented. Moreover, the novel adsorbent has also been synthetically prototyped to determine the actual adsorption capacity of the novel adsorbent in varying ionic environments and arsenic concentrations. Our initial results from conducting a set of experimentation has demonstrated an enhanced adsorption capacity of the synthetic adsorbent that is higher than any naturally occurring adsorbent.

References:

1. Bektaş, N., Aydın, S., & Öncel, M. S. (2011). The adsorption of arsenic ions using beidellite, zeolite, and sepiolite clays: a study of kinetic, equilibrium and thermodynamics. Separation Science and Technology, 46(6), 1005-1016.
2. Doshi, R. K., Mukherjee, R., & Diwekar, U. M. (2018). Application of Adsorbate Solid Solution Theory To Design Novel Adsorbents for Arsenic Removal Using CAMD. ACS Sustainable Chemistry & Engineering, 6(2), 2603-2611.

542620 The Impact of Shale Gas and Oil on the Chemical Industry

Abstract:

Lower alkanes and aromatics are very useful as both fuels and chemical raw materials. Occasionally supply disruptions constrained availability but also encouraged investigation of alternative feedstocks and chemistries. Recent new technologies for the production of unconventional gas and oil sources from tight shale formations significantly increased availability of both methane and condensate, radically altered process economics, resulted in the construction of much new intermediates capacity, and multiplied the ways traditional and alternative chemistries might be exploited. This presentation will outline historically how natural gas components became useful as chemical feedstocks, how government regulations occasionally led to perturbations in availability and price, how supply crises led to the development of alternative chemistries, catalysts, and ultimately the exploitation of shale resources. More specifically, will be discussed how excess coproducts, especially condensate, from increasing shale gas and oil production for energy are having significant unexpected impacts on the chemical process industries and opportunities for new process chemistry research.

541186 Sustainable Energy and Chemicals: Past, Present, and Future

Abstract:

The last 40 years has seen the U.S. energy situation vacillate from gasoline rationing and concerns over Peak Oil, to projected energy and feedstock prices that were noncompetitive in global markets for chemicals production, to the economic revolution of frack gas and light tight oil. The importance of low cost energy and feedstocks in driving economic opportunity will be discussed relative to this history, along with the importance of sustainability in decision making to gain stakeholder buy-in and acceptance. Strategies for managing career paths and competencies despite changing targets and high uncertainty will be addressed.

542618 Disruptions: What the Future May Hold

Abstract:

Disruptive forces are all around us. They change the way we communicate and work. They change the way we develop goods and services. They also change the types of products that are manufactured. Examples such as the way in which we produce and consume energy or the feedstocks we use to produce product have seen many changes over the course of the last few decades, and these disruptions are expected to continue. The chemical engineering discipline and the expectations for a chemical engineerâ€™s skills has also changed and adapted as a result of these disruptive forces. A discussion of the broad themes associated with these forces and the implications that they may have to chemical engineers in the future will be discussed.

542621 Geopolitical Factors Influencing the Evolution of the Chemical Industry

Abstract:

The basic chemicals and plastics industry is built on a simple principle: find a cheap feedstock and convert it into something more valuable. However, the right feedstock depends on place and time. Certain policies and actions of various governments have influenced the evolution and profitability of the chemical industry during the past 50 years, sometimes in surprising ways. For example, high oil prices and US government funding encouraged the development and commercialization of coal gasification technology during the late 1970s and early 1980s. This enabled China to develop a competitive basic chemical industry using coal as a feedstock during the last 25 years. Similarly, an abrupt increase in oil prices during the first decade of the 21st century encouraged the present shale gas revolution, causing a major shift in the economy of producing basic chemicals and polymers in the US. If history is our guide, then what governments choose to fund and regulate may be as important as market forces in determining the shape of our industry and profession in the future.

542622 Agility & Resilience: How to Maintain Career Competitiveness in the Changing Chemical Industry

Abstract:

The world around is changing. The shift to protectionism, global instability and price volatility of raw materials and feedstocks has the potential to wreak havoc on the chemical industry. With the revolution of shale gas in North America and the sudden drop in feedstock costs, the chemical sector in the US has reasserted its historic dominance. Learn how the role of a chemical engineer can contribute to industry growth and resilience, while adapting to allow for personal agility throughout the career life cycle.

535092 Leveraging Deep Learning for Pharmaceutical Discovery Lead Profiling

Abstract:

As part of small molecule drug discovery, potential lead compounds are studied to ensure that they have a low risk of adverse behavior. Cytochrome P450s (CYP) are a family of enzymes that are an important drug metabolism pathway; drug candidates that inhibit the activity of these enzymes can lead to dangerous drug-drug interactions. Once synthesized, candidate compounds are submitted to high throughput *in vitro* assays to screen for CYP inhibition. By prioritizing more promising options and measuring the effect of changes to the core structure, the ability to virtually screen compounds for CYP inhibition would save teams time and effort during drug design.

This work describes the development of a model that can predict CYP inhibition from chemical structure using in-house data from BMS. Such quantitative structure activity relationship (QSAR) models have been successfully developed from publicly available data sets, however, the limited amount of public CYP inhibition data hinders their practical utility. As part of an innovative multidisciplinary team at BMS, analytics experts from different functional areas contributed to exploring and analyzing the BMS CYP assay database. Using this much larger set of private data (>1M entries) combined with recent advancements in machine learning algorithms, a model suitable for virtual screening was developed. Of the many modeling approaches that were explored, >85% classification accuracy was achieved using a deep learning approach. Important caveats for handling real-world data as well as a comparison of different modeling methodologies will be presented.

528297 Learning to Design and Validate Small-Molecule Synthetic Routes from Historical Reaction Data

Abstract:

Advances in laboratory automation promise to decrease the manual effort of synthesis, but determining *how* to synthesize a compound currently requires time and effort investment from expert chemists. To achieve full autonomous chemical synthesis, one must have robust synthesis planning software that can propose fully-specified synthetic routes to target molecules.

In this talk, we will describe our recent efforts to develop such software. The overarching theme of our work is how to most effectively leverage historical reaction data to inform decision-making in small molecule pathway design.

The overall synthesis planning workflow contains a number of interconnected modules. We focus on two critical aspects of computer-aided synthesis planning and how machine learning and other data-driven techniques have enabled new approaches to both challenges. First, we discuss the problem of retrosynthetic planning (i.e., identification of suitable starting materials) and how the recursive expansion and search strategy are both conducive to machine learning approaches. Second, we discuss the challenge of *in silico* reaction validation, which can be addressed by solving the inverse problem of forward reaction prediction. We summarize neural network-based approaches we have taken to develop models that can anticipate the products of a chemical reaction after being trained on previously published reactions. Here we use the model for reaction validation, but its utility extends to prediction of side products and impurities. Finally, we describe how these techniques for retrosynthesis and forward prediction are integrated into an overall workflow that, for a given molecular target, predicts a rank ordered list of reaction paths that connect the target to purchasable starting materials via a series of plausible reaction steps. The integrated program offers additional features for excluding specific reactions or chemicals, e.g., for IP or toxicity concerns.

530970 Using Machine Learning to Recommend Reaction Conditions and Quantifying Similarity of Catalysts, Solvents, and Reagents

Abstract:

Using Machine Learning to Recommend Reaction Conditions and Quantifying Similarity of Catalysts, Solvents, and Reagents

Hanyu Gao, Thomas J. Struble, Connor W. Coley, William H. Green, Klavs F. Jensen

One essential element for synthetic planning is the reaction conditions. Properly designed reaction conditions can promote desired reactivity, and sometimes small changes in reaction context can lead to drastically different reaction outcomes. While other aspects of computer-aided synthetic planning has seen rapid development, including retrosynthesis[1] and evaluation of reaction outcomes, [2,3] the suggestion of reaction conditions is primarily considered a human task and relies heavily on chemists' knowledge and experience, and computer aided condition recommendation remains a challenging and under-explored problem. [4–6]

In this work we develop a neural-network based model to predict suitable reaction conditions for organic transformations. The model is trained on roughly 10 million examples from the Reaxys[7] database in order to predict the exact chemicals used as catalysts, solvents, reagents, and an appropriate temperature for the reaction. Prediction results are evaluated both quantitatively, using a variety of accuracy metrics, and qualitatively, using multiple sets of representative examples. The model is able to predict a closely similar context combination, based on strictly defined criteria, to recorded catalyst, solvent and reagent in the top-ten predictions with accuracy of 67.4%, with accuracies for individual elements reaching 80%~90%. Empirical evaluation shows that many of the failed predictions are potential alternatives to the true conditions, or due to data quality issues. Temperature is predicted to be within 20K deviation from the recorded temperature in 60%~70% of the test cases, with dependence on the correctness of chemical context prediction.

Besides making predictions, the model can also generate information that can be used to quantify the similarity of catalysts, solvents and reagents. We discovered that the weight matrix in the last layer for predicting catalysts/solvents/reagents can be effectively used to calculate a score that represents the functional similarity between catalysts/solvents/reagents. The identified near neighbors of common catalysts, solvents and reagents agree well with established knowledge. This work demonstrates the possibility of inferring functional similarity of chemicals purely from reaction data, without relying on property calculations.

The reaction condition suggestions and similarity information provided by this tool can be used to aid experimental design, improve the accuracy of in silico evaluation of reactivity, and pathway-level assessment and improvement of chemical synthesis processes.

Reference

(1) Segler, M. H. S.; Preuss, M.; Waller, M. P. Planning Chemical Syntheses with Deep Neural Networks and Symbolic AI. *Nature* **2018**, *555* (7698), 604.

(2) Jin, W.; Coley, C.; Barzilay, R.; Jaakkola, T. Predicting Organic Reaction Outcomes with Weisfeiler-Lehman Network. In *Advances in Neural Information Processing Systems*; 2017; pp 2604–2613.

(3) Coley, C. W.; Barzilay, R.; Jaakkola, T. S.; Green, W. H.; Jensen, K. F. Prediction of Organic Reaction Outcomes Using Machine Learning. *ACS Cent. Sci.* **2017**, *3* (5), 434–443.

(4) Marcou, G.; Aires de Sousa, J.; Latino, D. A. R. S.; de Luca, A.; Horvath, D.; Rietsch, V.; Varnek, A. Expert System for Predicting Reaction Conditions: The Michael Reaction Case. *J. Chem. Inf. Model.* **2015**, *55* (2), 239–250.

(5) Segler, M. H. S.; Waller, M. P. Modelling Chemical Reasoning to Predict and Invent Reactions. *Chem. Eur. J.* **2017**, *23* (25), 6118–6128.

(6) Lin, A. I.; Madzhidov, T. I.; Klimchuk, O.; Nugmanov, R. I.; Antipin, I. S.; Varnek, A. Automatized Assessment of Protective Group Reactivity: A Step toward Big Reaction Data Analysis. *J. Chem. Inf. Model.* **2016**, *56* (11), 2140–2148.

(7) Reaxys https://new.reaxys.com/.

537077 Applying Data Science Techniques to Solubility Data for Synthetic Compounds: An Expedited End-to-End Workflow from Data Collection to Crystallization Process Design

Abstract:

Crystallization is an important process often employed several times within synthetic routes for drug substances. The development of robust crystallization processes can be separated into two tasks: (1) the selection of a solvent-system, and (2) the design of a process within that solvent system that affords materials with desired chemical/physical purity and meets yield requirements. An end-to-end workflow relying on data science techniques was developed for capturing, visualizing, and interpreting solubility data to facilitate the consistent and rapid execution of these tasks.

This workflow starts with the collection of solubility data using standardized equipment sets and approaches into templated tables. These tables contextualize the solubility data by joining each measured value (concentration, temperature, composition, etc.) with relevant meta-data (solute purity, x-ray diffraction results, equipment, date, etc.). The contextualized solubility data is ingested within a database – providing a single source for all solubility data. A templated visualization then consumes the data from this source. Further it can be filtered within the visualization as necessary (e.g., limited to a specific solute and limited to only data collected for a specific lot of solute).

To facilitate solvent selection as the first task in crystallization process development, this visualization automatically applies a decision tree to collected solubility data to classify solvents with regards to crystallization as solvents that are likely: "good for a thermal process", "solvent within antisolvent driven crystallization", or "good antisolvents". Based on this classification the scientist working with the system can either begin the task of process design or apply a predictive solubility model that has been integrated within the visualization to determine other solvent systems that may be worth investigating. The application of this model uses simple R scripts with open source libraries (e.g., non-linear optimization packages) doing the "heavy-lifting".

Once a solvent system has been selected, the task of process design begins using an automated script to fit and select the best-model for solubility data across ranges of temperature/antisolvent ratios within that system. The contextualization of the solubility data allows for visual identification and rapid exclusion of outliers within the model fitting step. Once a solubility model has been selected and fit for a given system – a constrained optimization algorithm is applied to determine the process that affords the highest yield given user supplied constraints (which are modified to ensure the process meets desired chemical/physical purity needs). This initial crystallization process conditions are then attempted and refined as necessary.

This end-to-end workflow has resulted in significant time savings, and allowed for the setting of consistent expectations for initial designs across projects. Further, it is an early demonstration of the integration of modelling and data science techniques within process development.

534161 Optimizing Drying Profile of Polymeric Drug Products Using Machine Learning and First Principle Modeling

Abstract:

Drying is one of the critical steps for manufacturing of polymeric drug products. The key quality attributes for the dried product can be the residual solvent content, and the flowability. The flowability of the powders can be dependent on the glass transition temperature which in turn depends on the solvent content. Hence it requires special precautions to keep the temperature of the drug product during drying below a critical transition temperature (Tg). This work shows a case study where such type of issues can be tackled by employing a hybrid model combining machine learning and first principles to optimize the drying profile in minimal number of experimental runs.

Typically the drug product is dried in an agitated Nutsche filter type of a dryer. This drying requires special care to ensure that at all points of drying; the temperature of the product is below the critical value (Tg). While a simple heat transfer model (like in Figure 1, section A) can be used to link the jacket temperature to the product temperature, the presence of solvent and its evaporation needs additional considerations as detailed below. In the model building step, Model 1 was a Support Vector Machine (SVM) based model to predict the Tg for different residual solvent levels (based on experimental data). The model predicted with accuracy over 98%. Model 2 captures the mass transport of the solvent from the drug product, accounting for the temperature dependent mass transport of the solvent.

Here K_i and C_i are the diffusivity and concentration of solvent i, and T is the product temperature.   is calculated as a function of temperature of each solvent in the system, based on the regression of the experimental data.

Figure 1: Modeling approach for drying

To predict the optimal jacket temperature profile for drying to meet the product requirements, the heat transfer model, Model 1 and Model 2 are solved using an algorithm like the one shown in Figure 2. For a given temperature profile, and initial solvent levels, Model 2 predicted the time evolution of the solvent concentration. At each time, Model 1 calculated the Tg. The optimal input temperature profile was chosen such that the product temperature at any time was lower than the predicted Tg at that time.

Figure 2: Modeling algorithm for predicting the optimal temperature profile

The above model building approach of combining machine learning and first principle modeling can be extended to other unit operations as well where only part of the underlying physics is understood. In combination with a mixing type of a model, this approach can be applied to scale-up drying (Figure 1). The presentation shows specific example cases of such applications.

525894 A Reaction Database for Small Molecule Pharmaceutical Processes Integrated with Process Information: Future Developments

Abstract:

To resolve process engineering issues during the development of synthetic routes for new drugs at an early stage of process development, access to chemical information is needed, which can be provided through knowledge databases, experience, literature review and/or computer-aided tools [1,2]. The available data (knowledge) is used for similarity search, reaction data analysis, synthesis route planning, drug discovery-development and prediction of physicochemical properties [3]. A computer-aided tool for reaction synthesis based on efficient retrieval of chemical information that are organized in chemical reaction databases have been developed. The reaction data in the developed databases consist of individual reactions and chemical information for the associated molecules involved in the reactions.

At the current stage, the database covers 350 chemical and biochemical reaction systems with the data-information mentioned above. Extension of the database is an-going objective and involves several tasks. One current task is to develop and use algorithms for efficient collection of reaction data integrated with process information. This task is crucial and very demanding and requires tests to verify the accuracy of the data and check for their consistency, in the same way phase equilibria data are checked through established consistency tests. Another task is to create an efficient interface for data import-export integrated with tools for synthesis-design of pharmaceutical processes.

The overall objective is for the reaction database to cover a very wide range chemical and biochemical reaction systems and integrate it with generic model-based prediction tools that combines reaction pathways synthesis algorithms, solvent selection tools [5] consistency tests for reaction kinetics data, prediction of missing thermodynamic properties, process synthesis/design algorithms [6, 7], process analysis [6] and optimization tools [8]. Such an integrated computer aided tool will enhance the amount of data-information available, thereby reducing time and cost for problem solution. Also, the tool would provide useful insights for multi-objective optimization involving decision making related to screening of different alternatives for potential active ingredients candidates at the early stage of drug/process development.

This presentation will give a description of the developed reaction database where the objective is to collect-store data for reactions involved in the synthesis of small molecules in pharmaceutical processes together with a search engine to retrieve the data when required. For example, in studies involving the synthesis-design of reaction-separation schemes and/or the role of organic solvents in reaction performance improvement. The goal is to provide a data rich environment by making useful process information available during the early stage synthesis of pharmaceutical products. The database is structured based on a specially developed ontology in terms of reaction classification (reaction types); compounds participating in the reaction; need-use of organic solvents and their function; information for single step and multistep reactions; target products; reaction conditions and reaction kinetic data. Information related to reactor scale-up together with information on downstream separation and other relevant information for each reaction are also available in the database, including the references for the collected data. Additionally, the information available in the database allows the calculation of well-known "green" sustainability metrics [4]. The application of the database will be illustrated through the synthesis of ibuprofen, for which data on different reaction pathways are available whose performances are compared using "green" sustainability metrics.

[1] Caron, S.; Thomson, N.M. Pharmaceutical process chemistry: Evolution of a contemporary data-rich laboratory environment. J. Org. Chem. 2015, 80, 2943–2958.

[2] Gasteiger, J. Chemoinformatics: Achievements and challenges, a personal view. Molecules 2016, 21, 151.

[3] Warr, W.A. A short review of chemical reaction database systems, computer-aided synthesis design, reaction prediction and synthetic feasibility. Mol. Inform. 2014, 33, 469–476.

[4] Papadakis, E.; Anantpinijwatna, A.; Woodley, M.J.; Gani R. A reaction database for small molecule pharmaceutical processes integrated with process information. Processes. 2017, 5(4), 58.

[5] Gani, R.; Jiménez-González, C.; Ten Kate, A.; Jones, M.; Powell, L.; Atherton, J.H.; Cordiner, J.L.A modern approach to solvent selection product formulations. Chem. Eng. 2006, 113, 30-43

[6] Tula, A.K.; Eden, M.R.; Gani, R. Process synthesis, design and analysis using a process-group contribution method. Comput. Chem. Eng. 2015, 81, 245-259

[7] Emenike, V. N.; Schenkendorf, R.; Krewer, U. A systematic reactor design approach for the synthesis of active pharmaceutical ingredients. Eur. J. Pharm. Biopharm. 2018, 126, 75-88.

[8] Diab, S.A.; Jolliffe, H.G.; Gerogiorgis, D. I. Plantwide technoeconomic analysis and separation solvent selection for continuous pharmaceutical manufacturing: Ibuprofen, artemisinin, and diphenhydramine. Comput. Chem. Eng. 2018, 41, 85-120

526686 Multiscale CFD Model Parallelization with Application to PECVD of Thin Films

Abstract:

Accurate and efficient modeling of plasma-enhanced chemical vapor deposition (PECVD) has remained the goal of many research institutions due to far reaching applications to the manufacturing of silicon thin films for use in the photovoltaic and microelectronics industries [1],[2]. In the effort to improve product quality and to cut down on manufacturing costs due to the difficulties associated with in situ measurements and the fabrication of reaction chamber components, three-dimensional (3D) modeling which captures the physical geometry of common PECVD reactors has become invaluable [3]. Recently, Crose et al. proposed a novel multiscale computational fluid dynamics (CFD) model, which combined a macroscopic CFD domain with a microscopic surface domain. While this model was successful in capturing the behavior of the PECVD reactor with regard to the non-uniform deposition of *a*-Si:H films, and in allowing for improved reactor geometries to be explored, the computational costs of the model were non-trivial. Specifically, single batch simulations required days to complete rendering batch-to-batch studies (e.g., run-to-run control strategies) infeasible.

Motivated by the above considerations, an improved parallel computation strategy is applied which allows for the discretization of both the macroscopic CFD volume and the microscopic kMC algorithm across 64 cores of UCLA's Hoffman2 computational cluster. The outlined multiscale model is applied to the deposition of 300 nm thick *a*-Si:H films in two distinct PECVD reactor geometries. The results suggest a significant improvement to the uniformity of the thin-film product (~4% change in the film thickness uniformity) and the reduction in computational costs suggest future implementation of run-to-run control strategies similar to those demonstrated by Crose et al. [4] for two-dimensional systems.

[1] Collins, D., Strojwas, A., White, D., "A CFD Model for the PECVD of Silicon Nitride," IEEE Trans. Semicond. Manuf., 7, 176–183, 1994.

[2] Da Silva, A. and Morimoto, N., "Gas flow simulation in a PECVD reactor," In Proceedings of the 2002 International Conference on Computational Nanoscience and Nanotechnology, San Juan, Puerto Rico, 22–25, 2002.

[3] Crose, M., W. Zhang, A. Tran and P. D. Christofides, "Multiscale Three-Dimensional CFD Modeling of PECVD of Amorphous Silicon Thin Films," Comp. & Chem. Eng., 113, 184-195, 2018.

[4] Crose, M., JSI Kwon, A. Tran and P. D. Christofides, "Multiscale Modeling and Run-to-Run Control of PECVD of Thin Film Solar Cells," Renewable Energy, 100, 129-140, 2017.

538604 An Analysis of the Nonlinear Behavior of the Autothermal Reactor

Abstract:

The autothermal reactor presented by [1] shows the multiplicity of the equilibrium profiles for an axial nonisothermal tubular reactor described by Arrhenius type nonlinear models. The unusual behavior is not expected for a tubular reactor with no diffusion effects and linear kinetics. This behavior is similar to the exothermic continuous stirred tank reactor model [2], which shows multiple steady states, stable or unstable, and in practical applications, the unstable steady states may correspond to the operating points of interest. The present work proposes an extended study on the non-linear behavior of the autothermal reactor. Different types of configurations for the reactor (for instance, co-current flow heating jacket and no temperature dependence of the kinetic constant) are studied to compare with the original system.
In [3], an analysis of a simplified tubular reactor with diffusion and temperature dependence kinetics was able to show the conditions necessary for the existence of multiple equilibrium profiles. In this work, a similar analysis is made on a simplified autothermal reactor to find the conditions necessary for the multiplicity of steady-states.
To demonstrate the nonlinear behavior of the full original autothermal model, a numerical analysis is performed to investigate how the system behaves as the model parameters are changed. Different discretization techniques are used to show the influence on the nonlinear behavior obtained.

[1] Bonvin, D., Rinker, R.G., Mellichamp, D.A. (1980). Dynamic analysis and control of a tubular autothermal reactor at an unstable state. Chemical Engineering Science, 35 (3), 603-612.

[2] Vejtasa, S.A., Schmitz, R.A. (1970). An Experimental Study of Steady State Multiplicity and Stability in an Adiabatic Reactor. AIChE Journal, 16(3), 416-419.

[3] Dochain, Dennis., (2016). Analysis of the multiplicity of equilibrium profiles in tubular reactor models. IFAC-PapersOnLine 49-18, 903-908.

534962 Accurate and Efficient Discrete Finite Volume Approximations for Population Balances Incorporating Coagulation and Fragmentation

Abstract:

This work is focused on developing two finite volume schemes for approximating the 1D and 2D simultaneous coagulation-fragmentation population balance equations. The two finite volume schemes are different in a sense that one scheme is merely focused on conserving the total mass of the system whereas the other scheme preserves the total number of particles as well as conserves the total mass of the system. Both schemes rely on introducing weights in their formulations to retain different properties such as total mass and total number of particles in the system. The inherit issue of the existing finite volume schemes is that it required very fine grid for predicting the number density function accurately (Kumar et al. 2009). In addition to this, the sectional methods such as fixed pivot technique and cell average technique are computationally very expensive due to their complex formulations although highly accurate. Moreover, solving a system involving simultaneous coagulation-fragmentation mechanisms rather add values to the computational cost. In order to overcome these issues, two finite volume schemes are developed and compared with the cell average technique along with newly developed exact solutions. The comparison reveals that both new finite volume schemes approximate the number density functions and their corresponding different order moments with higher accuracy even on a coarse grid at a lesser computational cost. It is also shown that considering fine grid for the approximation of numerical results lead to equal accuracy for cell average method and finite volume methods, however, both finite volume schemes are still computationally less expensive. Additionally, it is shown that both schemes can be easily extended to solve higher dimensional problems and retains the numerical accuracy and efficiency.

Keywords: Particle, Fragmentation, Coagulation, Population Balances, Finite volume Scheme, Cell Average Technique.

536345 Comparison of Global, Stochastic Optimization Algorithms Using Toy Problems and Multi-Parameter Models to Kinetic Fermentation and Rheological Data

Abstract:

The evaluation of multiple parameters involved in nonlinear dynamic simulation models based on a minimization of the sum of the squares of the differences between the predictions and experimental data is often a non-trivial step. Classical least squares minimization methods are all local methods that only converge to a local minimum that is not necessarily the global one, the answer depending crucially of the initial guess. The evaluation of the sensitivity of the local minimization solution to the initial guess can be quite time-consuming, especially when the model predictions involve the solution of a set of ODEs the time-integration of which requires by itself a non-negligible computational time. In this work we present a stochastic method to evaluate that dependence effectively with an application to the evaluations of parameters in the dynamic simulation of a thixotropic concentrated suspension system using Large Amplitude Oscillatory Shear (LAOS) and other transient data, as well as kinetic data from a set of fermentation reactions. In addition, with this work we compare several other contemporary stochastic methods using several classic chemical engineering "toy problems" from literature [1,2].

A classical approach to fitting parameters has been the minimization of the sum of least squares, also known as the weighted residuals least-squares technique, using a local, gradient method. There are several variations of this basic strategy currently in use. Some examples are the general Newton's method, trust-region methods, line search methods, and Levenberg-Marquardt methods. Alternatively, stochastic methods have been developed in order to better explore the available parameter space in search of a global minimum and/ or alleviate the dependence of the solution on the initial guess. These methods include simultaneous perturbation stochastic approximation methods (SPSA), singular simulated annealing, particle swarm (PSO) and genetic algorithms (GA) [1,3].

The approach developed in the present work incorporates a hybrid parallel, simulated annealing, "Metropolis" –like algorithm, within a least squares fitting algorithm recently developed by Armstrong et al. [1]. The parallel simulated annealing is used to develop a good initial guess of the values of the parameters space using as an objective function the squares of the differences of the model predictions from oscillatory in time experimental data obtained over a period. The selection of the initial guess is followed by an application of a modified least squares local minimization procedure to determine accurately the "global" minimum. An important advantage of the proposed method is that all parameters needed to execute it are evaluated numerically based on few simulated annealing runs. A further advantage of the proposed algorithm is the potential speedup by executing its most time-consuming step, a series of simulated annealing simulations, in parallel where possible.

The *parallel* simulated annealing algorithm proposed in the present work attempts to surpass the shortcomings of previous stochastic algorithms by more systematically exploring the parameter space without the need of complicated adjustable parameters. The parallel simulated annealing algorithm employs features of the more traditional simulated annealing 'Monte-Carlo' energy minimization approach, with the pseudo-energy to be minimized is derived from the least squares error function constructed between the model predictions and available data. In contrast to the standard simulated annealing process, the parallel simulated annealing procedure that is proposed to be followed here involves utilization of a series of stochastic simulated annealing runs each one built around its own Boltzmann energy level. However, the parallel simulated annealing sequences are not completely independent. From time to time, there is the possibility offered for a parameter information exchange between 'nearest neighbor' Boltzmann energy levels. When this exchange is conducted appropriately it allows for a maximum efficiency search for the optimum parameter values in a wide parameter space, without requiring finely tuned parameters or a sensitive dependence to the initial guess [1, 4, 5, 6, 7].

[1] Armstrong et al. "An Adaptive Parallel Tempering Method for the Dynamic Data-Driven Parameter Estimation of nonlinear Models." AIChE J. (2016).

[2] Battles and Trefethen "An extention of MATLAB to continuous functions and operators." SIAM J Sci Com. (2005).

[3] Armstrong. PhD Thesis, University of Delaware (2015).

[4] Spall. "Adaptive Stochastic Approximation by the Simultaneous Perturbation (SPSA) Method." IEEE Transactions of Automatic Control (2000).

[5] Spall, J.C., Introduction to Stochastic Search Optimization. Wiley Interscience. 2003, Hoboken, New Jersey: John Wiley and Sons Inc.

[6] Earl, D.J., M.W. Deem, Parallel Tempering: Theory, applications and new perspectives. Phys. Chem. Chem, Phys., 2005. 7: p. 3910-3916.

[7] Amar, J.G., The Monte Carlo Method in Science and Engineering. Institution of Electrical Engineering Computer Science and American Institute of Physics, 2006. MAR/ APR p. 9 - 19.

537644 Multilevel Monte Carlo Applied for Efficient Estimation of Observables in Multiscale Stochastic Systems

Abstract:

The characteristic trait of stochastic systems is the noise in their observables. Insufficient knowledge about the variability in the observables limits the ability to design processes robust to the uncertainty. The simplest way to accurately quantify the observables is to collect many samples from the system of interest and conduct a statistical analysis of the observed variables. However, this brute-force approach is resource- and/or time-intensive and thus decreases the utility of the results obtained from it, especially in time-sensitive situations. Many methods of efficient quantification of noise in the observables have been developed, such as Latin Hypercube sampling, Power Series and Polynomial Chaos Expansions [1]. A more recent computational technique is Multilevel Monte Carlo (MLMC) sampling [2], where many coarse inexpensive samples are used to obtain a crude estimate and assess its variability, and progressively fewer samples of increasing accuracy (higher discretization "levels") are used to refine the approximation. MLMC has been developed in computational finance, and has since then been applied in many other fields.

In our previous work, we applied MLMC to uncertainty quantification in classic chemical engineering systems [3]. In this work, we extend and adapt MLMC to perform accurate estimation of observables in stochastic multiscale systems. We used as a case study a model of thin film formation by Chemical Vapour Deposition (CVD) [4], which was also previously used to quantify uncertainty and estimation of observables in multiscale systems [5],[6]. In this model, a kinetic Monte Carlo (kMC) solid-on-solid simulation represents the microscale formation of a thin film, while continuum mass, energy and momentum transfer equations are used to represent the macroscale supply of the chemical vapor to the substrate where the film forms. The non-closed-form microscale representation and the deterministic macroscale equations are coupled through a boundary condition and thus form a stochastic multiscale system.

Most of the studies that use MLMC, including our previous work [3], discretize the time domain of continuous equations and use finer discretization for higher levels of approximation accuracy. In the CVD system, the accuracy of approximation increases with larger lattice sizes of the kMC simulation. We established an empirical relationship that relates the kMC time step to the lattice size and used it to control the definition of higher accuracy levels in the MLMC framework. We demonstrate that MLMC can be used to accurately estimate the observables of this multiscale stochastic system in a fraction of the time required when conducting the conventional Monte Carlo sampling of the full multiscale simulation.

References

[1] Eldred, M.S., "Recent Advances in Non-Intrusive Polynomial Chaos and Stochastic Collocation Methods for Uncertainty Analysis and Design", In Proceedings of the 50th AIAA Structures, Structural Dynamics and Materials Conference, Palm Springs, California, 2009.

[2] Giles, M. "Multilevel Monte Carlo Path Simulation", Operations Research, 56, 3, 2008.

[3] Kimaev, G. and Ricardez-Sandoval, L.A., "Multilevel Monte Carlo Applied to Chemical Engineering Systems Subject to Uncertainty", AIChE Journal, 64, 5, 2018.

[4] Vlachos, D., "Multiscale Integration Hybrid Algorithms for Homogeneous–Heterogeneous Reactors", AIChE Journal, 43, 11, 1997.

[5] Rasoulian, S. and Ricardez-Sandoval, L.A. "Uncertainty Analysis and Robust Optimization of Multiscale Process Systems With Application to Epitaxial Thin Film Growth", Chemical Engineering Science, 116, 590, 2014.

[6] Rasoulian, S. and Ricardez-Sandoval, L.A. "Robust Multivariable Estimation and Control in an Epitaxial Thin Film Growth Process Under Uncertainty", Journal of Process Control, 34, 70, 2015.

519763 Tightening Mccormick Relaxations Via Reformulation of Intermediate Functions into Schema

Abstract:

Generalized McCormick relaxations [2] can present significant boons in global optimization applications. This framework allows for the relaxation and subsequent optimization of algorithmically-defined functions that naturally arise during flowsheet modeling. Relaxations of functions considered under this theory exhibit quadratic convergence, helping to reduce the degree of clustering in spatial branch-and-bound algorithms [1-4]. Moreover, this simulation-based feasible-path approach enables the solution of flowsheeting problems in a reduced-dimension decision space, in contrast to the higher-dimensionality "lifting" of the decision space required by state-of-the-art auxiliary variable methods (AVMs) [5-6].

McCormick relaxations are typically calculated by overloading common arithmetic operators [2,3] in object-oriented programming languages [7]. Most recent research in McCormick-related relaxations have focused on providing improvements to the properties of these calculations to yield tighter relaxations, faster. Mitsos et al. developed tighter multivariate relaxations [8]. Khan et al. extended this to develop differentiable analogs [9]. Recently, Najman et al. extended this standard framework; provided tighter relaxations by using affine relaxations to shrink the associated interval bounds [10]. This represents one of the first usages of a reformulation of the directed-acyclic-graphs (DAGs) to tighten McCormick relaxations.

Analysis of the DAG representation of functions has been used in many applications ranging from constraint-satisfaction problems to scheduling [11,12]. Recently, the usage of convex-transformable functions was explored within the BARON context resulting in a significant improvement to performance on standard benchmark libraries [13]. Khajavirad et al. identified a number of regular expressions in DAGs that allow for tighter relaxations when relaxed directly [14]. In this work, we discuss the potential of tightening McCormick relaxations via the reformulation of intermediate functions.

Rather than define a series of regular expressions, we define a series of schema based on various functional attributes (convexity information). A graph theoretic approach to identifying potential intermediate function reformulations is developed that allows for reformulation presented in terms of cut vertices [15-17]. At these potential points of reformulation, a series of general tests based on interval arithmetic are applied to determine the function schema. This approach makes use of the abstract-syntax tree (AST) feature of Julia [18] to dynamically build tighter convex and concave relaxations of expressions belonging to these schemata. A series of examples from both the COCONUT library and extant literature are presented [19].

[1] McCormick, G.P. Computability of Global Solutions to Factorable Nonconvex Programs: Part I – Convex Underestimating Problems. Mathematical Programming 10: 147-175, 1976.

[2] Scott, J.K., Stuber, M.D., and Barton, P.I. Generalized McCormick Relaxations. *J Global Optim*, 51:569-606, 2011.

[3] Mitsos, A., Chachuat, B., and Barton, P.I. McCormick-Based Relaxations of Algorithms. SIAM J. Optim. 20(2): 573-601.

[4] Bompadre A. and Mitsos A. Convergence rate of McCormick relaxations. *J Global Optim*, 52(1):1-28, 2012.

[5] Tawarmalani, M. and Sahinidis, N. V. A polyhedral branch-and-cut approach to global optimization. Mathematical Programming, 103(2), 225-249, 2005.

[6] Sahinidis, N. V., BARON 14.4.0: Global Optimization of Mixed-Integer Nonlinear Programs, User's manual, 2014.

[7] Chachuat, B. MC++ 2.0 [Computer Software]. (2017) Retrieved from https://omega-icl.github.io/mcpp.

[8] Tsoukalas, A. and Mitsos, A. Mutlivariate McCormick Relaxations. *J Global Optim*, 59:633-662, 2014.

[9] Khan, K.A., Watson, H.A., and Barton, P.I. Differentiable McCormick Relaxations. *J Global Optim*, 67(4):687-729, 2017.

[10] Najman, J. and Mitsos, A. Tighter McCormick Relaxations through Subgradient Propagation, arXiv preprint arXiv:1710.09188, 2017

[11] X. H. Vu, H. Schichl and D. Sam-Haroud. Using directed acyclic graphs to coordinate propagation and search for numerical constraint satisfaction problems. *16th IEEE International Conference on Tools with Artificial Intelligence*, pp. 72-81, 2004.

[12] Baskiyar, Sanjeev, and Rabab Abdel-Kader. "Energy aware DAG scheduling on heterogeneous systems." *Cluster Computing* 13.4 (2010): 373-383.

[13] Nohra, C. "Global Optimization of Nonconvex Problems with Convex-Transformable Intermediates" 2017 Fall AIChE Meeting, Minneapolis, MN.

[14] A. Khajavirad, J. J. Michalek, and N. V. Sahinidis. Relaxations of factorable functions with convex-transformable intermediates. Mathematical Programming, 144:107–140, 2014

[15] Merris, R. Graph Theory. John Wiley & Sons, Inc., Hoboken, NJ, USA. doi: 10.1002/9781118033043, 2000.

[16] Schmidt, Jens M. A Simple Test on 2-Vertex- and 2-Edge-Connectivity. *Information Processing Letters*, 113 (7): 241-244, 2013.

[17] Fourer, R. et al. Convexity and Concavity Detection in Computational Graphs: Tree Walks for Convexity Assessment. *INFORMS Journal on Computing* 22.1: 26-43, 2010.

[18] Bezanson, J. et al. Julia: A fresh approach to numerical computing. *SIAM Review*. 59, 65-98, 2017.

[19] O. Shcherbina, et al. Benchmarking Global Optimization and Constraint Satisfaction Codes, In *Global Optimization and Constraint Satisfaction*, Bliek, Ch., Jermann, Ch. and Neumaier, A. Springer, Berlin 2003, 212-222.

538462 Optfill: A Novel Optimization-Based Tool to Automate the Gapfilling of Genome-Scale Metabolic Models

Abstract:

OptFill: A Novel Optimization-Based Tool to Automate the Gapfilling of Genome-Scale Metabolic Models

Wheaton Schroeder and Rajib Saha

The University of Nebraska – Lincoln, Lincoln, NE

Computational modeling of metabolism is now an indispensable tool to drive the processes of understanding, discovering, and redesigning of biological systems. By defining the metabolic space, genome-scale metabolic (GSM) models can assess allowable cellular phenotypes and explore metabolic potential and restrictions under specific environmental and/or genetic conditions. GSM model curation processes typically involve gleaning information on gene annotations and reactions from major public databases such as KEGG, Uniprot, Metacyc and ModelSeed. However, inconsistencies across these databases and incomplete gene annotations leave gaps in any GSM models. One of the major tools called Gapfill (as well as its many offshoots) applies Mixed Integer Linear Programming (MILP)-based approach and utilizes the additional functionalities from closely related organisms or changing the direction of existing reactions in order to fill gaps in any GSMs. Although GapFill automates the model building processes, it fixes the gaps individually without any consideration of not creating thermodynamically infeasible cycles (TICs). Hence, Gapfill always makes redundant changes and increases the number of TICs in GSM models, which ultimately require further manual scrutiny.

In order to address these issues as well as increase the automation of GSM model reconstruction, introduced here is an improved method, namely OptFill, to fill gaps in GSMs. OptFill applies a multi-level Mixed Integer Non-Linear Programming (MINLP) optimization approach which addresses the fixes needed on a per-GSM model basis (as opposed to a per-metabolite/gap basis in Gapfill), which seeks to obtain three objectives: maximize the number of gaps fixed, minimize the number of reactions/functionalities added, and minimize the number of TICs created. As each imbedded optimization level is MINLP in form, Lagrangian duality is used to reduce the multi-level multi-objective formulation to a single-level single-objective formulation for the ease of solution. OptFill is currently being applied to the development of a GSM model of a poorly annotated black yeast strain *Exophiala dermatitidis*. Since approximately 4% of open reading frames *Exophiala dermatitidis* are annotated with enzyme classifications, the initial GSM reconstruction has a huge number of metabolic gaps. Thus, OptFill provides a distinct advantage over the traditional Gapfill approach in the extent of automation, speed of model development, and needed manual curation after filling the gaps and this is more evident for any non-model, poorly annotated, and under-studied organisms such as *Exophiala dermatitidis*.

532722 A Novel Homotopy Continuation Technique to Locate All Real Solutions of a Nonlinear System of Algebraic Equations

Abstract:

A Novel Homotopy Continuation Technique to Locate All Real Solutions of a Nonlinear System of Algebraic Equations

Saeed Khaleghi Rahimian[a,*], J.D. Seader[b]

[a] San Jose, CA 95129, USA

[b] Department of Chemical Engineering, University of Utah, Salt Lake City, UT 84112, USA

* Email: s.khaleghi.r@gmail.com (AIChE Member)

One of the most critical limitations of all homotopy continuation methods, in dealing with solving a system of nonlinear equations, is to track all the real solutions including those located on isolated branches called "Isolas" [1]. To tackle this difficulty, a new system of equations is developed where all real roots are also bifurcation points, through which new stemming branches connect all the solutions. The homotopy function is defined as: $H(x,t) = 1+g[f(x)]-t$, where $f(x)$ and t are the system of equations and homotopy parameter, respectively. Function $g[f(x)]$ requires two properties. First, it must only vanish where $f(x)=0$ (i.e. $g[f(x)]=0$, iff $f(x)=0$), thus $g[f(x)]=0$ has the same roots as $f(x)=0$. The derivative of $g(x)$ at roots also needs to be zero (i.e. $g'(0)=0$). The latter property makes the Jacobian matrix H_x singular at all real roots with dim $N\{H_x(0)\}=n$, where n is number of equations. Thus, we would have 2^{n-1} maximum unique bifurcated branches at all roots [2]. The initial point on the homotopy paths is found readily by solving a nonlinear system of equations with the degree of freedom equal to one. To show the robustness of the new algorithm, two smooth (i.e. twice differentiable) functions are used for $g(f)$ including square functions (i.e. $g(f)=f^2$) [3] and a combination of cosine and linear functions. The new approach with both g functions was able to find successfully all the reported roots for more than 20 test problems, which included a variety of algebraic and transcendental terms.

[1] Choi, S. H.; Book, N. L. Unreachable roots for global homotopy continuation methods. *AIChE J.* 1991, 37, 1093–109.

[2] Decker, D.W.; Keller, H.B. Multiple limit point bifurcation, *J. Math. Anal. Appl.*, 1980, 75, 417–430.

[3] Khaleghi Rahimian, S.; Jalali, F.; Seader, J. D.; White, R. E. A robust homotopy continuation method for seeking all real roots of unconstrained systems of nonlinear algebraic and transcendental equations. *Ind. Eng. Chem. Res.* 2011, 50, 8892–8900.

524384 The Industrial Implementation of Validated Dynamic Simulation and Optimisation Tools Towards Superior Beer Fermentation

Abstract:

Beer fermentation is an extremely complex biochemical phenomenon (Vanderhaegen et al., 2006), whose mathematical representation and parameter estimation are hitherto pursued only via reduced order models (Dochain, 2003), which consider only the most essential subset of important biochemical reactions via selective aggregation. Subsequent process optimisation studies have been successfully performed on the basis of these foregoing models, primarily addressing batch time minimisation or ethanol yield maximisation (de Andrés-Toro, 1998; Carrillo-Ureta, 2001), or both (Rodman and Gerogiorgis, 2016). A necessary assumption of process optimisation studies is that all parameter datasets published in the original fermentation modelling papers are accurate and consistent; even more so, the complexity (and deterioration) of yeast performance perplexes further industrial endeavours.

Our recent work has involved the comprehensive computation and visualisation of attainable envelopes for beer ethanol maximisation and batch duration minimisation, under explicit flavour-modifying compound concentration constraints, pursued via simulation-based optimisation (Rodman and Gerogiorgis, 2016), as well as by rigorous dynamic optimisation, via an NLP formulation employing orthogonal collocation on finite elements (Rodman and Gerogiorgis, 2017). More recently, we have also explored the problem of obtaining highest ethanol in minimum time, employing a multi-objective formulation via the Strawberry stochastic evolutionary algorithm, achieving the detailed visualisation of Pareto fronts (Rodman et al., 2018). Therein, a clear operational asymmetry is observed: while acceptable fermentation rapidly diminishes below a minimum batch duration threshold, the subsequent improvement in final ethanol concentration changes very little with increasing batch time, rendering long fermentations meaningless (and implying a "sweet spot", which is highly dependent upon several underlying biochemical characteristics).

The present paper illustrates the industrial implementation of these novel computational tools for optimising fermentation operations in a major multinational brewing corporation, under the auspices of a Royal Academy of Engineering (RAEng) Industrial Fellowship (2017-2018). By combining the published results of our foregoing studies with additional sensory (product), but even more so operational (plant management) constraints, we have arrived at superior temperature manipulation profiles for industrial fermentors, which achieve higher final ethanol concentration, reduced batch duration, and constraint satisfaction for key beer flavour modifiers (esters and diacetyls). In contrast to our previous studies, these optimal fermentation T(t) protocols also consider a minimum number of turning points, reflecting the prominent control room culture (and the operational preference) to avoid frequent adjustments - a fact well justified by the limited instrumentation and online data acquisition, which in turn is due to the limited accessibility, and the multiphase, inhomogeneous nature of the fermenting mixture.

REFERENCES

Vanderhaegen, B., et al., 2006. The chemistry of beer aging - a critical review. *Food Chem.* **95**(3): 357-381.

de Andrés-Toro, B., et al., 1998. A kinetic model for beer production under industrial operational conditions. *Math. Comput. Simulat.* **48**(1): 65-74.

Carrillo-Ureta, G., P. Roberts, and V. Becerra, 2001. Genetic algorithms for optimal control of beer fermentation. *Proc. IEEE Int. Symp. Intell. Control*, 391-396.

Dochain, D., 2003. State and parameter estimation in chemical and biochemical processes: a tutorial, *J. Proc. Control* **13**(8): 801-818.

Gee, D. A., Ramirez, W. F. 1988. Optimal temperature control for batch beer fermentation. *Biotechnol. Bioengineering*, **31**: 224-234.

Rodman, A. D., Gerogiorgis, D.I., 2016. Multi-objective process optimisation of beer fermentation via dynamic simulation, *Food Bioprod. Proc.* **100**(A): 255-274.

Rodman, A.D. and Gerogiorgis, D.I, 2017. Dynamic optimization of beer fermentation: sensitivity analysis of attainable performance vs. product flavor constraints. *Comput. Chem. Eng.* **106**(C): 582-595.

Rodman, A.D., Fraga, E.S. and Gerogiorgis, D.I., 2018. On the application of a nature-inspired stochastic evolutionary algorithm to constrained multi-objective beer fermentation optimisation. *Comput. Chem. Eng.* **108**: 448-459.

524592 From Academia to Industry: Optimization Models for Shale Gas Development

Abstract:

From Academia to Industry: Optimization Models for Shale Gas Development

Markus G. Drouven[1], Ignacio E. Grossmann[2]

[1]Optimization Engineering

EQT Corporation

Pittsburgh, PA 15222

[2]Department of Chemical Engineering

Carnegie Mellon University

Pittsburgh, PA 15213

[1]mdrouven@eqt.com, [2]grossmann@cmu.edu,

Abstract

For the past four years the Center for Advanced Process Decision-Making (CAPD) at Carnegie Mellon University (CMU) and EQT Corporation have been actively collaborating to develop optimization models for shale gas development. The CAPD is a research center at CMU that is engaged in Process Systems Engineering research for the process industries. The CAPD focuses on a number of different research areas including a) modeling and optimization algorithms, b) process synthesis and product design, c) enterprise-wide optimization and process operations, and d) molecular computing. EQT, on the other hand, is the leading natural gas producer in the United States with a strategic focus on shale gas development in the Marcellus Play. The company locates natural gas, develops acreage and transports produced gas to distribution systems.

In this presentation, we review how the shale gas development research project at CMU started and how it progressed over time. We show that a number of researchers associated with the CAPD made important contributions to this domain. In particular, Knudsen et al. (2014) developed a Lagrangean relaxation scheme for shut-in scheduling in shale-gas multi-pad systems. Thereafter, Cafaro et al. (2014) proposed a mixed-integer nonlinear programming (MINLP) model for the strategic planning of the shale gas supply chain. Next, Yang et al. (2014) developed a stochastic two-stage mixed-integer linear programming (MILP) scheduling models for shale plays water management. Most recently, Drouven et al. (2016) developed a multiperiod MINLP model for long-term, quality-sensitive shale gas development. Finally, Ondeck et al. (2018) developed a bi-criterion optimization approach for simultaneous planning and scheduling of shale gas production. In this talk we review the key findings of the aforementioned work, and we discuss their significance for the shale gas industry as a whole.

We also review how the research within the CAPD led to the formation of the Optimization Engineering team at EQT. We show how this team currently translates practical business and/or engineering problems within the company into mixed-integer programming models; and we demonstrate how these models are used to support faster and better decision-making across the organization.

Finally, we present a real-world case study to demonstrate how optimization models at EQT are currently being used to support safer and more economic shale gas development. The case study reveals how a rigorous optimization model for water management applied to an active shale gas development area can reduce the number of water hauling trucks on the road (a safety benefit), limit impaired water disposal volumes (an environmental benefit) and how it minimizes operational costs (an economic benefit).

References

[1] Knudsen BR, Grossmann IE, Foss B, Conn AR. Lagrangian relaxation based decomposition for well scheduling in shale-gas systems. *Computers & Chemical Engineering*. 2014; 63:234-249.

[2] Cafaro, D. C.; Grossmann, I. E. Strategic Planning, Design, and Development of the Shale Gas Supply Chain Network. *AIChE Journal*. 2014. doi: 10.1002/aic.14405.

[3] Yang L, Grossmann IE, Manno J. Optimization models for shale gas water management. *AIChE Journal*. 2014; 60 (10):3490-3501.

[4] Drouven, M. G.; Grossmann, I. E. Multi-Period Planning, Design and Strategic Models for Long-Term Quality-Sensitive Shale Gas Development. *AIChE Journal*. 2016;62(7): 2296-2323.

[5] Ondeck, A., M.G. Drouven, N. Blandino, I.E. Grossmann, "Multi-System Development Planning for Shale Gas Production," *Proceedings of the 13th International Symposium on Process Systems Engineering – PSE 2018*, July 1-5, 2018, San Diego, California, USA

524722 Reconciling Ranking Criteria with Fuzzy Sets for Effective Use in Project Portfolio Selection

Abstract:

Maintenance activities at process plants can be divided into "run-the-plant" and improvement categories. The first category is focused on the daily maintenance activities that ensure safe plant operation. The maintenance improvement category are larger projects aimed at further increasing operating reliability. Within a specified investment budget, it is a challenge to determine the right mix of these improvement projects that will meet performance and economic targets. One approach is portfolio optimization. Although optimization for Project Portfolio selection has been researched since the 1950's resulting in models with high degrees of mathematical sophistication (Salo et al 2011), there is a significant gap between the extents of application in the financial sector compared with project investment selection, for example, the selection of maintenance improvement projects at multiple process plants within a specified budget. A contributing factor to the lack of application is that the selection of maintenance improvement projects is complicated by scaling issues of consequences and decision criteria that are different than traditional monetary considerations like an expected rate of return. A recent paper discusses this complexity for portfolio selection of maintenance infrastructure improvement projects (Mild 2015). One common approach for multiple quantitative and qualitative criteria is to use 1,3,9 ranking matrices. It can be a challenge to harmonize these rankings especially when they are provided by a large cohort of SME's across disparate plants. Ayyub suggested that for system variables that are qualitative in nature, fuzzy set theory can be used in defining the potential states together with a suitable observation channel that yields a quantitative equivalent for each state (Ayyub 2001). In this work we demonstrate how we implemented this idea for the selection of an optimal portfolio of maintenance improvement projects from a list of project proposals from 1300 plants using fuzzy set theory based on the work done by Dubois and Prade (Dubois 1987), Carlsson et al. (Carlsson 2001), and Vercher et al.(Vercher 2007).

Salo, J. Keisler and A. Morton (eds.) **Advances in Portfolio Decision Analysis: Improved Methods for Resource Allocation**. 2011

Mild, Juuso Liesiö, Ahti Salo, **Selecting infrastructure maintenance projects with Robust Portfolio Modeling**, Decision Support Systems, Volume 77, 2015

Ayyub, Bilal M. **Elicitation of expert opinions for uncertainty and risks** 2001

Dubois and H. Prade. **The mean value of a fuzzy number**. Fuzzy Sets and Systems, 24:279-300, 1987.

Carrlson, Fuller, R. and Majlender, P. **A possibilistic approach to selecting portfolios with highest utility score**. TUCS Technical Report No. 355, August 2000

529287 Multi-Scale Modeling and Design Optimization of an Industrial Hydrogen Production Plant with High-Resolution PSA and Steam-Methane Reformer Models

Abstract:

Hydrogen, a widely used and important feedstock for many industrial chemical processes, is typically produced at high purity through steam-methane reforming (SMR) processes. While steam-methane reforming can be very energy-intensive and is subject to catalyst deactivation, it is still the most widespread hydrogen production process and accounts for nearly half of the world production of hydrogen [1]. Such processes have very high theoretical energy efficiencies, with some elaborately integrated plants approaching 95% energy efficiency (accounting for export of steam), but the majority operate well below their design efficiencies due to non-ideal behaviors unaccounted for during design and operational optimization [2].

An SMR-based plant closely follows the general â€œreaction-separation-recycleâ€ chemical process layout: furnaces with catalyst-packed tubes serve as the reforming reactors to produce syngas, from which hydrogen is subsequently separated by pressure swing adsorption (PSA). The carbon monoxide and unreacted methane are recycled to provide fuel for the furnace. Inclusion of more detail and relevant physical phenomena in the reaction and separation system models can help bridge the gap between optimal predicted and actual plant behavior [3]. However, there are many numerical and data-collection challenges related to simulating spatially distributed hydrogen reforming furnaces and PSA units. In our previous work [2], we established a detailed, physics-based SMR furnace model and showed that it could be seamlessly incorporated into a complete hydrogen plant model for optimization by using pseudo-transient models [4]. Using this extended model, the optimization of an industrial hydrogen-production plant improved the overall thermal efficiency by ~1.5%, which translated into a savings of $1.14 MM/year.

While a detailed reactor model was successfully employed in our previous work, the PSA unit was modeled as a component splitter with fixed recovery and product purity. This assumption clearly excludes the key relationship(s) between reactor furnace design and product separation efficiency. In the present work, we rely on a recently developed pseudo-transient approach for PSA modeling and optimization to seamlessly integrate a detailed, distributed (semi-continuous) PSA model with the other (continuous) units of the hydrogen plant. The full multi-scale process model with distributed reaction and separation unit operation models allows for hydrogen recovery and product purity to be optimized by altering the PSA parameters, and for the PSA design to be simultaneously specified and considered in the plant-wide efficiency.

[1] T da Silva Veras, TS Mozer, D da Costa Rubim Messeder dos Santos, A da Silva Cesar. Hydrogen: Trends, production, and characterization of the main process worldwide. *International Journal of Hydrogen Energy*, 42(4):2018-2033, 2017.

[2] A Kumar, TF Edgar, M Baldea. Multi-resolution model of an industrial hydrogen plant for plantwide operational optimization with non-uniform steam-methane reformer temperature field. *Comp. Chem. Eng.*, 107:271-283, 2017.

[3] RC Pattison, C Tsay, M Baldea. Pseudo-transient models for multiscale, multiresolution simulation and optimization of intensified reaction/separation/recycle processes. *Comp. Chem. Eng.*, 105:161-172, 2017.

[4] RC Pattison and M Baldea. Equation-oriented flowsheet simulation and optimization using pseudo-transient models. *AIChE Journal*, 60(12):4104-4123, 2014.

[5] C Tsay, RC Pattison, M Baldea. A pseudo-transient optimization framework for periodic processes: Pressure swing adsorption and simulated moving bed chromatography. *AIChE Journal*, 2017. Doi:10.1002/aic.15987

529957 Optimization of Circuitry Arrangements for Heat Exchangers

Abstract:

Heat exchanger performance is important in various heating and air-conditioning systems that are widely used in residential and commercial applications. Although these systems are manufactured in various shapes and configurations [1], a common configuration is that of the crossflow fin-and-tube type, in which a refrigerant flows through a set of pipes and moist air flows across a possibly enhanced surface on the other side of the pipe, where thermal energy is transferred between the air and the refrigerant.

Heat exchanger performance optimization can be achieved by optimizing a number of different metrics including maximization of heating or cooling capacity, size reduction, component material reduction, manufacturing cost reduction, reduction of pumping power, or a combination of these metrics. Systematic optimization of heat exchanger design has been a long-standing research topic [2–4]. It is a particularly challenging problem mainly for the following reasons: (i) there is a highly discontinuous and nonlinear relationship between the circuitry design and the heat exchangers performances, and (ii) the decision space is extremely large making exhaustive search algorithms insufficient for searching the entirety of the solution space.

In this paper, we propose a new formulation for the refrigerant circuitry design problem. We model this problem as a constrained binary optimization problem. We incorporate only realistic manufacturing constraints to the optimization problem without requiring extensive domain knowledge. As a result, the proposed approach can be applied to different types of heat exchangers. In order to evaluate the heat exchanger performance using different refrigerant circuitry designs, we use CoilDesigner [5], a steady-state simulation and design tool for air to refrigerant heat exchangers, which acts as a black-box since the exact relationship of the objective functions with the decision variables is not explicit. We apply derivative-free optimization algorithms [6] to optimize heat exchanger performance. Although the derivative-free optimization literature has recently been attracting significant attention, it currently lacks systematic comparisons between mixed-integer constrained derivative-free optimization algorithms on industrially-relevant problems. Another contribution of this work is the systematic comparison of four different mixed-integer constrained derivative-free optimization algorithms (MIDACO [7], NOMAD [8], TOMLAB/glcDirect [9], and TOMLAB/glcSolve [9]) and a box-bounded derivative-free optimization algorithm (CMAES [10]) that are applied to optimize heat exchanger circuitry using two different thermal efficiency criteria.

Computational results on 17 different circuitry architectures show that TOMLAB/glcDirect and TOMLAB/glcSolve have the best performance among all derivative-free optimization solvers. Timewise, TOMLAB/glcSolve is faster than TOMLAB/glcDirect on smaller, but TOMLAB/glcDirect is much faster on larger instances and on average. We also formulated the circuitry optimization problem as a constraint satisfaction problem using Choco solver in order to automate the procedure of finding all possible feasible circuitry designs and verify the results of the derivative-free methods for small heat exchangers. The results show that the proposed method provides optimal refrigerant circuitries satisfying realistic manufacturing constraints.

References cited

[1] G. F. Hewitt, G. L. Shires, and T. R., Bott. Process heat transfer, vol. 113. CRC press, Boca Raton, FL, 1994.

[2] D. H. Fax, and R. R. Mills. Generalized optimal heat exchanger design. ASME Transactions, 79:653–661, 1957.

[3] C. P. Hedderich, M. D. Kelleher, G. N. Vanderplaats. Design and optimization of air-cooled heat exchangers. Journal of Heat Transfer, 104:683–690, 1982.

[4] F. O. Jegede, and G. T. Polley. Optimum heat exchanger design: process design. Chemical Engineering Research & Design, 70:133–141, 1992.

[5] H. Jiang, V. Aute, and R. Radermacher. CoilDesigner: A general-purpose simulation and design tool for air-to-refrigerant heat exchangers. International Journal of Refrigeration, 29:601–610, 2006.

[6] L. M. Rios, and N. V. Sahinidis. Derivative-free optimization: a review of algorithms and comparison of software implementations. Journal of Global Optimization, 56:1247–1293, 2013.

[7] M. Schlüter, and M. Munetomo. MIDACO User Guide (Current as of 17 April, 2018). http://www.midaco-solver.com/, 2018.

[8] M. A. Abramson, C. Audet, G. Couture, J. E. Dennis Jr., and S. Le Digabel. The Nomad project (Current as of 17 April, 2018). http://www.gerad.ca/nomad/, 2018.

[9] K. Holmström, A. O. Göran, and M. M. Edvall. User's Guide for TOMLAB 7, (Current as of 17 April, 2018). http://tomopt.com, 2018.

[10] N. Hansen. The CMA Evolution Strategy: A Tutorial (Current as of 17 April, 2018). http://www.lri.fr/hansen/cmaesintro.html, 2018.

534376 An Integrated Data-Driven Modeling & Global Optimization Approach for Production Planning Under Uncertainty

Abstract:

The tight competition, environmental regulations, and lower profit margins are some of the factors that drive the petrochemical industry to make planning and scheduling operations more efficient over the last few decades. Planning and scheduling have been topics of special interest for many process systems engineers as constrained optimization problems and the refinery planning problem has received considerable attention since the introduction of linear programming in 1950s [1]. Complex production facilities such as petroleum refineries are highly interconnected and the objectives of most of the individual units are conflicting and optimizing a single unit's operation can end up the overall operation to be suboptimal if not infeasible [2], thus making a holistic approach both necessary and advantageous.

Traditional production planning approaches rely on LP principles with fixed-yield planning models, even though refinery operations such as distillation, processing, and pooling are highly nonlinear in nature. While there are commercially available highly detailed mathematical models for simulation of processing units, such detailed models cannot be used in enterprise-wide optimization-based approaches due to high computational expense. Recent studies showed that, data-driven modeling offers a promising way to obtain inexpensive nonlinear models, which can relate relevant inputs to relevant outputs to describe each individual process accurately, and therefore making a plant- or enterprise-wide optimization-based approach realizable [3,4]. The increase in computational power of global optimization solvers such as ANTIGONE and BARON can open the door to planning frameworks that are formulated as NLPs or MINLPs [5,6].

In this work, we are proposing a framework for integrated data-driven modeling and global optimization to solve production planning problems. In the first step of the work, we organize, analyze, and process the real plant data provided by Hyundai Oilbank's Daesan Refinery located in South Korea. Linear and nonlinear models of different complexities are trained, validated, and then compared to find the best models to describe the processes. In the second step of the work, we create a superstructure containing all possible connections and operating modes in the refinery. The resulting multi-period planning model is a non-convex nonlinear optimization model (NLP), which is solved to e-global optimality using ANTIGONE. Results of several case studies are provided to illustrate the efficiency of our proposed model and global optimization approach compared to the actual plant operation. Multi-period formulation is compared with single-period formulation, and the uncertainty in demands and prices are addressed by robust optimization.

1. CA Floudas, AM Niziolek, O Onel, LR Matthews, 2016, Multi-Scale Systems Engineering for Energy and the Environment: Challenges and Opportunities, AIChE Journal, 62 (3), pp. 602-623.
2. JM Pinto, M Joly, LFL Moro, 2000, Planning and scheduling models for refinery operations, Computers and Chemical Engineering, 24 (9-10), pp. 2259-2276.
3. F Boukouvala, CA Floudas, 2017, Argonaut: Algorithms for global optimization of constrained grey-box computational problems, Optimization Letters, 11(5), pp. 895-913.
4. J Lie, X Xiao, F Boukouvala, B Zhao, G Du, X Su, H Liu, CA Floudas, 2016, Data-Driven Mathematical Modeling and Global Optimization Framework for Entire Petrochemical Planning Operations, AIChE Journal, 62 (9), pp. 3020-3040.
5. R Misener, CA Floudas, 2014, ANTIGONE: Algorithms for coNTinuous Integer Global Optimization of Nonlinear Equations, Journal of Global Optimization, 59 (2–3), pp. 503–526.
6. M Tawarmalani, N Sahinidis, 2005, A polyhedral branch-and-cut approach to global optimization, Mathematical Programming, 103(2), pp. 225-249.

535475 Robust Multi-Period Vehicle Routing: Construction of Uncertainty Sets and Evaluation Via Rolling-Horizon Simulations

Abstract:

The Vehicle Routing Problem (VRP) is fundamental to transportation operations in industrial and commercial supply chains. It is concerned with the design of cost-optimal routes for a fleet of capacity-constrained vehicles (e.g., trucks) to serve a set of geographically distributed customers.

Traditional VRP variants are of an operational nature since they involve the design of vehicle routes over a single period (e.g., a work day). However, several transportation problems arising in practice are of a tactical nature since they involve the scheduling and routing of vehicles over multiple time periods (e.g., a week). This is particularly the case when customers may dynamically place service requests/orders on any given day, and each request specifies a set of future days (also called "day window") during which service can take place. At the end of each day, the distributor must make scheduling decisions consisting of assigning a visit day to each unfulfilled service request, along with the standard routing decisions, so as to minimize the long-run transportation costs. It is important to note that the computed schedule is implemented in a rolling horizon fashion: only routes of the first day are executed while new orders are received; unfulfilled requests at the end of the first day and new customer requests accumulated during the day constitute the new portfolio of orders to be considered for scheduling the following day. Such decision-making setups are typical in systems in which services are provided by appointment; relevant examples include scheduling of maintenance personnel [1, 2], food distribution and collection [3], and auto-carrier transportation [4]. We refer to this class of problems as Multi-Period VRPs.

Existing approaches for the Multi-Period VRP (e.g., see [3, 5]) essentially ignore the uncertainty in future customer orders, arising from the dynamic aspect of the problem. In other words, the models are agnostic to potential service requests that may be received in the future, even those that may be received within the current planning horizon. Consequently, the computed schedules lead to situations which can either be infeasible, leading to additional costs that must be incurred to recover feasible schedules (e.g., commissioning additional vehicles), or too expensive in terms of transportation costs. In contrast, we can generate better, risk-averse schedules by explicitly accounting for customer order uncertainty during the planning process. This is supported by the fact that companies and distributors often have large amounts of historical data, which can be used to obtain demand forecasts and provide information regarding calls for service in future time periods.

In a recent paper [6], we proposed a modeling and solution methodology that explicitly hedges against customer order uncertainty in Multi-Period VRP variants. The specific highlights of this paper were as follows. First, we modeled future customer orders as binary random variables, and encapsulated the set of probable future scenarios in an uncertainty set of finite (but possibly very large) cardinality. Second, we used this model of uncertainty in a robust optimization framework. The "true" model is multi-stage with

binary recourse decisions and, hence, is computationally intractable. Therefore, we developed two tractable two-stage surrogate models (each with binary recourse decisions) that provide an upper bound and a lower bound to the true multi-stage model, respectively. Notably, the upper bounding model is practically implementable and its solutions can be adopted by the practitioner for daily operations. Third, we developed an efficient integer programming formulation and branch-and-cut solution algorithm for the two-stage models. The computational experiments showed that (i) we can obtain robust schedules for 100-customer, 5-period problems in less than 2 hours, by paying an additional transportation cost of about 1.5% (on average), and, (ii) the approximation gap of our two-stage model against the "true" multi-stage model is less than 1% (on average).

In this work, we extend the above mentioned paper in three important directions. The contributions are enabled by a rolling-horizon simulation platform that we developed to mimic the day-to-day operations of distributors. Notably, the simulation platform allows us to evaluate the true long-run cost of any decision policy for the Multi-Period VRP, and thus, allows us to draw practically meaningful insights.

Our specific contributions are as follows. First, we use statistical models to construct uncertainty sets from historical data, and show how to update these sets in a rolling-horizon context. The size of our constructed uncertainty sets can be easily tuned to reflect distributors' risk tolerance. Second, we use our simulation platform to identify the ideal length of the planning horizon which appropriately trades-off computational tractability with cost savings. To elaborate, a longer horizon can yield higher cost savings, but is also more difficult to optimize; therefore, the potential benefits may not be realizable as the computed schedules may be suboptimal in practice. Third, we adopt the constructed uncertainty sets and optimal planning horizon in our simulations, and analyze the impact of multiple operating policies on distribution costs. Specifically, we analyze the impact of (i) day-window widths (e.g., next-day service vs. two-day service), (ii) fleet size and capacity, (iii) vehicle commissioning costs, and (iv) advance notice periods (i.e., number of days elapsed between the day an order was placed and the first feasible service day). In all, our contribution is a holistic decision-making framework that practitioners can utilize for tactical planning of their distribution operations.

References:

[1] N. Bostel, P. Dejax, P. Guez and F. Tricoire (2008). Multiperiod planning and routing on a rolling horizon for field force optimization logistics. In *Vehicle Routing: Latest Advances and New Challenges*, B. Golden, S. Raghavan, and E. Wasil (editors), 503—525, Springer US, Boston, MA.
[2] H. Tang, E. Miller-Hooks, and R. Tomastik (2007). Scheduling technicians for planned maintenance of geographically distributed equipment. *Transportation Research Part E: Logistics and Transportation Review*, 43(5):591—609.
[3] M. Wen, J.-F. Cordeau, G. Laporte, and J. Larsen (2010). The dynamic multi-period vehicle routing problem. *Computers & Operations Research*, 37:1615—1623.
[4] J.-F. Cordeau, M. Dell'Amico, S. Falavigna, and M. Iori (2015). A rolling-horizon algorithm for auto-carrier transportation. *Transportation Research Part B: Methodological*, 76:68—80.
[5] R. Baldacci, E. Bartolini, A. Mingozzi, and A. Valetta (2011). An exact algorithm for the period routing problem. *Operations Research*, 59(1):228—241.
[6] A. Subramanyam, F. Mufalli, J.M. Pinto, and C.E. Gounaris (2017). Robust multi-period vehicle routing under customer order uncertainty. Available at *Optimization Online* (2017/04/5947).

538525 Data-Driven Modelling and Optimization of Compressor Operations

Abstract:

Compressors are crucial pressurization equipment commonly used in applications such as air separation, natural gas (NG) processing and liquefaction, petrochemicals, liquefied natural gas (LNG) regasification, etc. They are energy-intensive and usually consume a significant portion of the overall energy expense of a plant.

Generally, multiple stages of compressors mounted on a single shaft (called trains) operate in different configurations such as series or parallel. These trains may operate at different flow capacities such as 25%, 50%, 75%, and 100%. To handle high loads and promote reliability, several trains of compressors exist in a process plant. Starting with the inlet gas conditions and desired pressure ratio, it is not obvious to assign optimal loads in the compressor network. Without a systematic guiding program, operators usually assign unnecessarily high loads to a compressor train and/or choose an inefficient combination of compressors that results in significantly higher power consumption. Therefore, it is important to optimize the compressor operations for varying loads and gas conditions. However, as a preceding step, we must accurately predict the performance of all individual compressors to account for the effects of their geometry, ageing, etc.

Some literature works optimized the compressor operations for specific applications such as LNG regasification. For instance, Shin et al.[1] and Jang et al.[2] distributed optimal loads to the boil-off gas (BOG) compressors in an LNG regasification plant to minimize their power consumption using a mixed integer linear problem (MILP) formulation. Most importantly, they did not consider stages in a compressor train and various modes of their operations. Besides, they used simplistic compressor models without accurately characterizing the performance of individual compressors.

In this work, we begin with the performance prediction of the reciprocating compressors using operational data. Noting the variation in compressor performance with the gas conditions, we develop correlations to estimate the adiabatic efficiency and constant volumetric efficiency loss. These developed correlations enable the prediction of accurate power consumption and outlet conditions for varying inlet gas conditions.

Then, given the inlet gas conditions and desired outlet pressure, we develop a nonlinear programming (NLP) model to find the configuration and capacities of the compressor trains that minimizes the overall power consumption. Our model results guide the plant operators to assign optimal loads to compressors in the network for varying gas conditions.

As a case study, we illustrate our procedure for the modeling and optimization of compressor operations on a real LNG regasification terminal. We develop correlations to estimate the parameters for the BOG compressors using vendor and operational data. Subsequently, using an accurate predictive model for the compressors, we optimize their operations and achieve significant savings in power consumption.

References:

1. Shin MW, Shin D, Choi SH, Yoon ES, Han C. Optimization of the Operation of Boil-Off Gas Compressors at a Liquified Natural Gas Gasification Plant. *Ind Eng Chem Res.* 2007/09/01 2007;46(20):6540-6545.
2. Jang N, Shin MW, Choi SH, Yoon ES. Dynamic simulation and optimization of the operation of boil-off gas compressors in a liquefied natural gas gasification plant. *Korean Journal of Chemical Engineering.* May 01 2011;28(5):1166-1171.

531802 Optimization-Based Predictive Control of Networked Process Systems with Discrete and Delayed Sensor-Controller Communication

Abstract:

A key problem in networked control systems is the design of resource-aware control methods that can systematically balance the inherent tradeoff between the desired closed-loop stability and performance requirements on the one hand and the constraints on the sensing, computation and communication resources of the networked devices on the other. It is well known that maximizing closed-loop system performance often requires frequent sensor-controller communication and increased levels of network resource utilization, whereas reducing communication costs favors keeping the sensing and communication levels to a minimum which can compromise the control system performance.

An approach that seeks to enforce closed-loop stability with reduced sensor-controller communication is model-based networked control [1]. The basic idea of this approach is to embed a predictive model of the plant within the control system to generate the control action when sensor-controller communication over the network is suspended and state measurements are unavailable to the controller, and to update the model state when communication is restored. Through a closed-loop stability analysis, an explicit characterization of the minimum allowable model update frequency needed to achieve closed-loop stability can be obtained. This approach has subsequently been extended and applied to address a wide range of problems, including control of large-scale process networks [2], distributed control of diffusion-reaction processes with spatially-distributed sensors and actuators [3], and fault-tolerant stabilization of networked process systems [4]. While these approaches help reduce network resource utilization without compromising closed-loop stability, a close examination of existing model-based networked control strategies reveals a general lack of explicit control performance optimization in the controller design framework. In the context of balancing the intrinsic tradeoff between control performance and network resource utilization, there has been an apparent increased emphasis on reducing network utilization, often at the expense of the achievable control system performance.

An effort to address this issue was made in [5], [6] where performance optimization considerations and control constraints were explicitly incorporated in the networked controller design methodology through the use of model predictive control formulations. The focus of these studies, however, has been on optimizing the closed-loop performance, with respect to the response speed and the control effort, without explicit characterization or consideration of the sensor-controller communication cost. The problem of reducing sensor-controller communication costs was addressed indirectly through the use of a separate event-triggered control approach.

In this work, we present a methodological framework for the design and implementation of optimization-based predictive control systems for networked processes with discrete and delayed sensor-controller communication over a resource-constrained network. The framework aims to optimize the control system performance, while simultaneously minimizing sensor-controller communication costs and overall network utilization. Initially, an auxiliary model-based networked controller with an embedded propagation unit that compensates for the communication delay is designed. The closed-loop stability region is explicitly characterized in terms of the controller gain, the communication delay and the maximum allowable model update period that ensures closed-loop stability under periodic sensor-controller communication. Based on this characterization, a finite-horizon optimization problem in which the cost function imposes explicit penalties on the model state, the control action and the model update frequency is formulated subject to the delay-dependent stability constraints. The optimization problem is solved in a receding horizon fashion to determine the optimal controller gain and optimal model update period needed at each update time. The receding horizon strategy results in time-varying controller gain and communication scheduling policies that allow the process to adapt more effectively to environmental uncertainties and perturbations. The implementation and efficacy of the developed optimization-based networked control approach are illustrated using a chemical process example.

References:

[1] L. A. Montestruque and P. J. Antsaklis, "On the model-based control of networked systems," *Automatica*, vol. 39, pp. 1837–1843, 2003.

[2] D. Xue and N. H. El-Farra, "Supervisory logic for control of networked process systems with event-based communication," in *Proceedings of American Control Conference*, Boston, MA, 2016, pp. 7135–7140.

[3] Z. Yao and N. H. El-Farra, "Model-based networked control of spatially distributed systems with measurement delays," in *Proceedings of American Control Conference*, Montreal, Canada, 2012, pp. 2990–2995.

[4] Z. Yao and N. H. El-Farra, "Data-driven actuator fault identification and accommodation in networked control of spatially-distributed systems," in *Proceedings of American Control Conference*, Portland, OR, 2014, pp. 1021–1026.

[5] Z. Yao and N. H. El-Farra, "Resource-aware model predictive control of spatially distributed processes using event-triggered communication," in *Proceedings of 52nd IEEE Conference on Decision and Control*, Florence, Italy, 2013, pp. 3726–3731.

[6] Y. Hu and N. H. El-Farra, "Adaptive quasi-decentralized MPC of networked process systems," in *Distributed Model Predictive Control Made Easy*, vol. 69, Maestre, J. M. and Rudy R. Negenborn (Eds.), Springer: Berlin, 2014, pp. 209–223.

532837 Stochastic-Tube MPC for Offset-Free Tracking in the Presence of Plant-Model Mismatch

Abstract:

Model predictive control (MPC) has been successfully applied for high-performance control of a wide range of multivariable constrained systems [1]. Receding-horizon implementation of MPC provides some degree of robustness to system uncertainties, which generally stem from structural plant/model mismatch, exogenous disturbances, and measurement noise. However, marginal robust performance may not be adequate in practice to prevent, for example, unstable process operation or violation of quality constraints for products with stringent quality requirements.

Tube-based MPC has emerged as a popular approach for MPC of linear systems with bounded uncertainty. The key advantage of tube-based MPC is that it can provide guaranteed constraint satisfaction through an offline tightening of the constraints, while its complexity is comparable to that of a nominal MPC problem. The concept of tubes, originally developed for set-based uncertainty descriptions [2], has also been extended to stochastic systems [3][4]. This work aims to address an open challenge in stochastic-tube MPC related to systematically handling mismatch between the process model and the true plant, which is particularly important in process systems applications. The key notion of this work is to separate the system uncertainty into two distinct sources of additive bounded state error [5][6]. The first source is a non-random uncertainty that represents mismatch between the linear model and the true plant (or other types of persistent disturbances that cannot be modeled with a probability distribution). The second source represents random variations due to either intrinsic stochastic variability in the system or exogenous disturbances. Recursive feasibility and stability of the proposed stochastic-tube MPC strategy is guaranteed by defining a straightforwardly constructible terminal invariant set that includes changes in the steady-state operating conditions/setpoints. This is an extension of the terminal invariant set for tracking, proposed in [7], to stochastic-tube MPC. Offset-free tracking is achieved with a disturbance estimator that can be tuned completely independently of the proposed controller. A new filter model for the deterministic disturbance that is based on a bound on the variation in the plant-model mismatch is also proposed, which allows for further reduction in the conservatism of the control performance.

The proposed stochastic-tube MPC is demonstrated on two simulation case studies to highlight its advantages over alternative tube-based MPC methods. First, a benchmark DC-DC converter problem is used to illustrate how chance constraints can be guaranteed in the presence of plant-model mismatch as well as persistent and random disturbances. In this case study, we also observe that the controller has a significantly enlarged domain of attraction since the terminal set is formulated in terms of a more general tracking problem (as opposed to regulation). The second case study is based on an industrial fluidized-bed catalytic cracking (FCC) unit wherein we demonstrate that setpoints, which are unreachable by standard tube-based MPC, can be tracked without offset using the proposed approach.

References

[1] M. Morari, J. H. Lee. Model predictive control: Past, present and future, Computers & Chemical Engineering, 23, 667–682, 1999.

[2] D. Q. Mayne, M. M. Seron, S. V. Rakovic, Robust MPC of constrained linear systems with bounded disturbances, Automatica, 41, 219–224, 2005.

[3] B. Kouvaritakis, M. Cannon, Model predictive control: Classical, robust and stochastic, Springer, 2015.

[4] M. Lorenzen, F. Dabbene, R. Tempo, F. Allgower, Constraint-tightening and stability in stochastic model predictive control, IEEE Transactions on Automatic Control, 62, 3165–3177, 2017.

[5] J. A. Paulson, L. Xie, A. Mesbah, Offset-free robust MPC of systems with mixed stochastic and deterministic uncertainty, IFAC-PapersOnline, 50, 3530—3535, 2017.

[6] J. A. Paulson, T. L. M. Santos, A. Mesbah, Mixed stochastic-deterministic tube MPC for offset-free tracking in the presence of plant-model mismatch, Journal of Process Control, Submitted, 2018.

[7] D. Limon, I. Alvarado, T. Alamo, E. F. Camacho, MPC for tracking piecewise constant references for constrained linear systems, Automatica, 44, 2382–2387, 2008.

533600 An Efficient Distributed Algorithm for Multistage Scenario Model Predictive Control Using Primal Decomposition

Abstract:

Model predictive control is a mature and widely used optimal control strategy in many applications. Research on model predictive control (MPC) under uncertainty has been steadily gaining more interest and several forms of robust and stochastic MPC formulations have been studied in the control literature, which optimizes a single open-loop control trajectory that optimize over all possible realizations of the uncertainty. This can lead to overly conservative solution. As noted by the authors in [1] a better strategy would be to optimize over different control trajectories rather than a single control trajectory. One such approach is the multistage scenario model predictive control, also widely known as feedback min-max MPC or robust multistage MPC.

In multistage scenario MPC, the uncertainty is sampled into finite number of discrete realizations and the future propagation of the uncertainty in the prediction horizon is represented via a scenario tree. By allowing the control trajectories to vary for each scenario, we explicitly take into account the notion of feedback. In order to capture the real time decision-making process accurately, the so-called non-anticipativity constraints must be enforced which says that the control input leading to any branching must be the same. Consequently, the control input at the first sample must be the same for all the scenarios, thus enabling closed-loop implementation [1].

However, the major challenge with multistage scenario MPC is that it leads to large optimization problems and hence computationally demanding. Since the different scenarios are all independent except for the non-anticipativity constraints, which couples the different scenarios, one can easily decompose the scenario MPC problem into smaller problems and use a master problem to co-ordinate the different subproblems. Scenario decomposition using dual decomposition was proposed in [2] and [3]. Dual decomposition (also known as Lagrangian decomposition) method solves the subproblems by relaxing the coupling constraints. A master problem then co-ordinates the individual subproblems iteratively. The previously relaxed non-anticipativity constraints are feasible only upon convergence. The authors in [3] indicate that such methods require a relatively large number of iterations between the master problem and the subproblem to converge, leading to challenges with practical implementation.

The risk of dual decomposition is then that the master problem may not converge within the required time. This leads to infeasibility of the non-anticipativity constraints, the implications of which are that the different subproblems may give different control inputs at the first sample time in the prediction horizon. This is not acceptable for real-time closed-loop implementation. Therefore, we propose an alternative approach to scenario decomposition using the primal decomposition approach, which ensures the non-anticipativity constraints are always feasible. This is because, in contrast to dual decomposition, primal decomposition produces a primal feasible solution with monotonically decreasing objective value at each iteration [4]. In addition, we also present a novel backtracking algorithm to suitably choose the step-size of the master problem update.

Since the different scenarios differs only in the uncertain parameters, the distributed scenario MPC problem can be recast as a parameteric NLP problem. By exploiting the NLP sensitivities, we only need to solve one subproblem as a full NLP. The remaining subproblems can be solved using the path-following predictor corrector QP algorithm that approximates the NLP [5]. This results in a computationally efficient formulation of the distributed scenario MPC framework. Simulation results using a CSTR case study show that the sensitivity-based distributed scenario MPC provides very good approximation of the fully centralized scenario MPC and the full NLP distributed scenario MPC.

References

1. Scokaert P.O, Mayne D.Q. Min-max feedback model predictive control for constrained linear systems. IEEE Transactions on Automatic control. 1998 Aug;43(8):1136-42.
2. Lucia, S. Subramanian, and S. Engell, "Non-conservative robust nonlinear model predictive control via scenario decomposition," in Control Applications (CCA), 2013 IEEE International Conference on. IEEE, 2013, pp. 586–591.
3. Marti, S. Lucia, D. Sarabia, R. Paulen, S. Engell, and C. de Prada, "Improving scenario decomposition algorithms for robust nonlin-ear model predictive control," Computers & Chemical Engineering, vol. 79, pp. 30–45, 2015.
4. P. Bertsekas, Nonlinear programming. Athena Scientific, 1999.
5. Kungurtsev and J. Jaschke, "A predictor-corrector path-following al-gorithm for dual-degenerate parametric optimization problems," SIAM Journal on Optimization, vol. 27, no. 1, pp. 538–564, 2017.

537071 Approximate Dynamic Programming Based Control of Hydraulic Fracturing Process to Achieve Uniform Proppant Concentration Level

Abstract:

Approximate dynamic programming based control of hydraulic fracturing process to achieve uniform proppant concentration level

Harwinder Singh Sidhu[1, 2], Prashanth Siddhamshetty [1, 2], Abhinav Narasingam[1, 2] ,

Joseph Sang-Il Kwon [1, 2]

[1] Artie McFerrin Department of Chemical Engineering, Texas A&M University, College Station, TX 77845 USA

[2] Texas A&M Energy Institute, Texas A&M University, College Station, TX 77845 USA

Petroleum and natural gas remain an important part of the global energy supply. In recent years, the extraction of underground resources such as shale gas and oil has been stimulated by hydraulic fracturing [1]. In hydraulic fracturing, achieving the desired fracture geometry, and uniform proppant concentration at the end of pumping are two primary objectives, because they are directly related to the overall efficiency of the process.

Over the last ten years, significant efforts have been made in the application of model predictive control (MPC) to the drilling processes to enhance pressure control flexibility, efficiency, and process safety [2, 3]. Recently, some work has been done in this direction with a focus on regulating the proppant concentration across the fracture [4, 5]. However, there are two remaining issues with MPC, which are both theoretical and practical in nature. First, the online computational burden can be significant to calculate the optimal control moves at each sampling time, when the underlying system is nonlinear and high-dimensional, which is usually the case in hydraulic fracturing. The second issue is its inability to consider the future interplay between uncertainty and estimation in the optimal control calculation. Alternatively, these issues can be addressed by the approximate dynamic programming (ADP) approach [6]. ADP is a model-based control technique, and can be employed to derive an improved control policy given some starting sub-optimal control policies (or alternatively, closed-loop identification data), while circumventing the â€˜curse-of-dimensionalityâ€™ of the traditional dynamic programming (DP) approach.

Motivated by the recent success of ADP in several applications [7], we present an ADP based control framework for the closed-loop operation of a hydraulic fracturing process. First, we will derive a high-fidelity model for hydraulic fracturing process based on first-principles. Second, a numerical scheme will be described to handle the high computational requirement caused by coupling of PDEs defined over a time-varying spatial domain. Third, a reduced-order model (ROM) will be developed by using these simulation results. Fourth, a nonlinear MPC theory is employed to obtain the starting sub-optimal control policies in ADP. Lastly, ADP approach will be utilized for the design of the feedback control system that provides a foundation for the online control of proppant bank height to achieve uniformity at the end of the treatment, which is directly related to the overall efficiency of the hydraulic fracturing process.

[1] Economides, M.J. and Martin, T., 2007. *Modern fracturing: Enhancing natural gas production* (pp. 978-1). Houston: ET Publishing.

[2] Breyholtz, Ø., Nygaard, G. and Nikolaou, M., 2010, June. Automatic control of managed pressure drilling. In *American Control Conference (ACC), 2010* (pp. 442-447). IEEE.

[3] Asgharzadeh Shishavan, R., Hubbell, C., Perez, H., Hedengren, J., Pixton, D.S. and Pink, A.P., 2014, October. Multivariate control for managed pressure drilling systems using high speed telemetry. In *SPE Annual Technical Conference and Exhibition*. Society of Petroleum Engineers.

[4] Gu, Q. and Hoo, K.A., 2015. Model-based closed-loop control of the hydraulic fracturing process. *Industrial & Engineering Chemistry Research*, *54*(5), pp.1585-1594.

[5] Siddhamshetty, P., Yang, S. and Kwon, J.S.I., 2017. Modeling of hydraulic fracturing and designing of online pumping schedules to achieve uniform proppant concentration in conventional oil reservoirs. *Computers & Chemical Engineering*.

[6] Lee, J.H. and Lee, J.M., 2006. Approximate dynamic programming based approach to process control and scheduling. *Computers & Chemical engineering*, *30*(10-12), pp.1603-1618.

[7] Tosukhowong, T. and Lee, J.H., 2009. Approximate dynamic programming based optimal control applied to an integrated plant with a reactor and a distillation column with recycle. *AIChE journal*, *55*(4), pp.919-930.

528250 Advanced-Step Multistage Nonlinear Model Predictive Control

Abstract:

Model predictive control (MPC) has been a mature technique used to control and optimization of chemical processes, especially in dealing with Multiple-Input-Multiple-Output (MIMO) and interactions between variables and constraints [1]. Its nonlinear counterpart, Nonlinear Model Predictive Control (NMPC), has received growing attentions recently due to its efficacy of representing the nonlinear dynamics for a wide state space. However, various sources of uncertainty have deleterious effects on the performance of NMPC and it becomes necessary to robustify against these uncertainties.

We distinguish the sources of uncertainty into two categories: the first type is due to uncertain model parameters and the second one is unmeasured disturbance. The first type uncertainty assumes to be realized within one sampling time despite staying unknown for the current step, while the second type uncertainty requires more than one sampling time to resolve, or will never be resolved, even given longer time. Different approaches should be used to deal with two types of uncertainty: a scenario tree structure [3] captures the evolution of type 1 uncertainty, and robust sensitivity-based method [5] needs to be applied to consider type 2 uncertainty.

In contrast, competing robust NMPC approaches suffer from the following disadvantages. Min-max NMPC [2] seeks to optimize the worst-case scenario and also satisfy constraints for all the cases, which is very conservative. Tube-based NMPC [4], an extension of tube-based MPC, incorporates an ancillary controller that enforces the trajectory of the uncertain system to stay inside a tube which centers around the nominal trajectory. However, tube-based NMPC fails to address the optimal control performance under the influence of uncertainty. Ideal multistage NMPC [3] models the uncertainty evolution as a multistage scenario tree, which takes future control actions as the recourse variables to counteract the effect from uncertainty. However, the inevitably increased optimization problem size presents a difficult task to implement online.

In this talk, we propose parallelizable advanced-step multistage NMPC (as-msNMPC) as a real-time implementable robust NMPC scheme that interactively addresses both sources of uncertainty. This framework precomputes a set of solutions based on predicted state information with advanced-step fashion [6], and updates the control action online with NLP sensitivity, which are two orders of magnitude faster than solving NLP from scratch. The as-msNMPC framework has been applied to a CSTR and a distillation case study.

References:

[1] Grne, L., & Pannek, J. (2013). Nonlinear model predictive control: theory and algorithms.

[2] Lazar, M., De La Peña, D. M., Heemels, W. P. M. H., & Alamo, T. (2006). Min-max nonlinear model predictive control with guaranteed input-tostate stability. In *17th Symposium on Mathematical Theory for Networks and Systems. Kyoto, Japan*.

[3] Lucia, S., Finkler, T., & Engell, S. (2013). Multi-stage nonlinear model predictive control applied to a semi-batch polymerization reactor under uncertainty. *Journal of Process Control*, *23*(9), 1306-1319.

[4] Mayne, D. Q., Kerrigan, E. C., Van Wyk, E. J., & Falugi, P. (2011). Tubeâ€based robust nonlinear model predictive control. *International Journal of Robust and Nonlinear Control*, *21*(11), 1341-1353.

[5] Pirnay, H., LÃ³pez-Negrete, R., & Biegler, L. T. (2012). Optimal sensitivity based on IPOPT. *Mathematical Programming Computation*, *4*(4), 307-331.

[6] Zavala, V. M., & Biegler, L. T. (2009). The advanced-step NMPC controller: Optimality, stability and robustness. *Automatica*, *45*(1), 86-93.

528111 Decomposition of Optimization Problems Using Community Detection and Its Application in Nonlinear Model Predictive Control

Abstract:

For the nonlinear model predictive control of large-scale and complex process systems, centralized optimization can be computationally challenging. ***Decomposition*** is an important class of techniques to deal with problems on large scales, where the entire system is partitioned into multiple non- or weakly interacting sub-problems which can be solved with significantly reduced complexity. For nonlinear model predictive control, the complexity can be reduced with

- A decomposition on the level of linear algebra (see e.g. [1]) – The matrices involved in the optimization algorithm can be large but sparse and structured (tri-diagonal). This special structure can be utilized to decompose and parallelize the matrix inversion operations.
- A decomposition on the level of plant flowsheet (see e.g. [2]) – By analyzing the interactions between the process variables, the plant can be decomposed into distributed subsystems. Distributed model predictive controllers are assigned to the corresponding subsystems and make control decisions in a coordinated way through communication.

In this work, we seek a decomposition of the optimal control problems directly ***on the level of optimization formulations***. This is inspired by the recent development of distributed optimization in machine learning (see e.g. [3]), which allows one to partition the monolithic optimization problem into smaller optimization problems.

Specifically, we represent optimization problems as ***bipartite networks***, whose nodes represent the optimization variables and constraints, and edges capture the variable-constraint interactions. ***Community detection*** algorithms in network science, which aim to partition the nodes into groups with the largest statistical difference between intra- and inter-group interactions, are employed to generate the most suitable decomposition of optimization problems. For the optimal control problem of a benchmark reactor-separator process, the decompositions are examined with two different optimization algorithms – alternating direction method of multipliers (ADMM) [3], and a decomposed version of IPOPT algorithm [5]. Significant improvement in the computational performance of nonlinear model predictive control is observed.

References:

[1] Wan, W., Eason, J. P., Nicholson, B., & Biegler, L. T. (2017). Parallel cyclic reduction decomposition for dynamic optimization problems. ***Comput. Chem. Eng.***, in press.

[2] Daoutidis, P., Tang, W., & Jogwar, S. S. (2017). Decomposing complex plants for distributed control: Perspectives from network theory. ***Comput. Chem. Eng.***, in press.

[3] Boyd, S., ***et al.*** (2011). Distributed optimization and statistical learning via the alternating direction method of multipliers. ***Found. Trends Mach. Learn.***, ***3***(1), 1-122.

[4] Fortunato, S., & Hric, D. (2016). Community detection in networks: A user guide. ***Phys. Rep.***, ***659***, 1-44.

[5] Wächter, A., & Biegler, L. T. (2006). On the implementation of an interior-point filter line-search algorithm for large-scale nonlinear programming. ***Math. Prog.***, ***106***(1), 25-57.

526409 Safeness Index-Based Economic Model Predictive Control of Stochastic Nonlinear Systems

Abstract:

Process operational safety plays an important role in designing control systems for chemical processes since failure to ensure process safety often leads to disastrous incidents potentially causing human and capital loss [1]. Recently, a new class of economic model predictive control systems (EMPC) was developed to simultaneously ensure process operational safety and economic optimality based on constraints expressed in terms of a function called Safeness-Index [2] which characterizes the relative safeness of different points in the state-space. At this point, the approach in [2] can handle sufficiently-small disturbances but its applicability to nonlinear chemical processes with stochastically-modeled disturbances of unbounded variation remains an open issue.

Motivated by this, in this work, we develop a process Safeness Index-based economic model predictive control method for a broad class of stochastic nonlinear systems with input constraints. A stochastic Lyapunov-based controller is first utilized to characterize a region of the state-space surrounding the origin, starting from which the origin is rendered asymptotically stable in probability [3]. Using this stability region characterization and a process Safeness Index function which characterizes the region in state-space in which it is safe to operate the process, an economic model predictive control system is then developed using Lyapunov-based constraints to ensure economic optimality, process operational safety and closed-loop stability in probability. A chemical process example is used to demonstrate the applicability and effectiveness of the proposed approach.

[1] Sanders R E. Chemical process safety: learning from case histories. Butterworth-Heinemann, 2015.

[2] Albalawi F, Durand H, Christofides P D. Process operational safety using model predictive control based on a process Safeness Index. Computers & Chemical Engineering, 2017, 104: 76-88.

[3] Wu Z, Zhang J, Zhang Z, et al. Economic Model Predictive Control of Stochastic Nonlinear Systems. AIChE Journal, 2018. In press

525751 Dual Offset Blocking Strategy for Computationally-Efficient Model Predictive Control

Abstract:

One of the most outstanding issues with model predictive control (MPC) is the on-line computational load required to solve the optimization problem within a sampling period, and move blocking is one of the most commonly used input parameterization methods to reduce the computational load by fixing the control inputs over arbitrary time intervals denoted as blocks [1,2]. However, it is difficult to guarantee the recursive feasibility of move blocked MPC because of the extra constraints imposed by move blocking. An offset blocking based on LQR solution with a time-varying blocking structure, denoted as moving window blocking (MWB), is suggested. This scheme has the advantage of stabilizing a system *a priori* [3]. Another offset blocking scheme retains previous control inputs as the base sequence. This scheme can handle a relatively large domain and is flexible in choosing the blocking structure without additional constraints [4,5].

However, the offset blocking schemes can leave room for optimality because they are only focused on the recursive feasibility. In particular, if the base sequence deviates from the feasible solution set or the optimal solution set, the closed-loop performance can be degraded. For example, the offset blocking with LQR solution using MWB may suffer from performance degradation if the LQR solution deviates from the feasible solution set. In addition, since the terminal constraint sets of these schemes are much more restrictive than that of regular MPC, the reduction in domain would be considerable.

In this study, we present a geometric interpretation of move blocking, and analyze the existing offset blocking schemes in terms of cost optimality based on the interpretation. Then, we present the dual offset blocking scheme that inherits only the advantages and overcome the disadvantages of the existing schemes. The proposed scheme parameterizes the input sequence in terms of deviations from linear interpolation between unconstrained infinite-horizon LQR solution and the retained solution from the previous sampling instant as described in Figure 1. By this, we can improve the cost optimality of the controller by adding a degree of freedom to the selection of the base sequence, allowing the controller to select a more desirable base sequence. Additionally, this scheme always guarantees the recursive feasibility by utilizing the feasibility of the previous solution only with the same terminal constraint of regular MPC, and improve stabilizing performance by utilizing the pre-stabilizing property of the LQR solution. Moreover, we apply the dual mode prediction paradigm [6] to this scheme by fixing the input solution as the unconstrained infinite horizon LQR solution when the state reaches the maximal positively invariant set subject to the LQR gain. We could further improve the optimality of the controller with this. A LQR example of a ball-plate system and other chemical process control examples are illustrated to show the efficacy of the suggested approach.

Figure 1. A schematic illustration of the dual offset blocking strategy.

REFERENCES

[1] Hara, N., & Kojima, A. (2012). Reduced order model predictive control for constrained discrete time linear systems. *International Journal of Robust and Nonlinear Control*, *22*(2), 144-169.

[2] Li, S. E., Jia, Z., Li, K., & Cheng, B. (2015). Fast online computation of a model predictive controller and its application to fuel economy–oriented adaptive cruise control. *IEEE Transactions on Intelligent Transportation Systems*, *16*(3), 1199-1209.

[3] Cagienard, R., Grieder, P., Kerrigan, E. C., & Morari, M. (2007). Move blocking strategies in receding horizon control. *Journal of Process Control*, *17*(6), 563-570.

[4] Ong, C. J., & Wang, Z. (2014). Reducing variables in Model Predictive Control of linear system with disturbances using singular value decomposition. *Systems & Control Letters*, *71*, 62-68.

[5] Shekhar, R. C., & Manzie, C. (2015). Optimal move blocking strategies for model predictive control. *Automatica*, *61*, 27-34.

[6] Kouvaritakis, B., & Cannon, M. (2016). *Model predictive control*. Springer, Switzerland.

526391 On Integration of Feedback Control and Safety Systems: Studying a High-Pressure Flash Drum Separator

Abstract:

The continued occurrence of incidents in the chemical process industries, despite efforts to prevent them is a testament to the need for continued work focused on enhancing process operational safety to protect human lives and the environment [1]. Traditional approaches to process safety like process design modifications neglect important aspects impacting process operational safety, such as multivariable interactions of process components and variables, limited control system authority due to limitations on the capacity of control actuators, and the manner in which the safety or relief system response may impact the effectiveness of the process control system. When a safety system identifies unsafe behaviors, the safety system automatically take actions, such as on/of-type valves and pressure relief valves, which drives the process state back to a safe operating region. Since these drastic actions change the dynamic of the process, research is conducted to investigate how the safety system influences control system functions.

In the present work, we focus on a high-pressure flash drum separator under proportional-integral (PI) control systems, where a pressure relief valve is installed to protect the drum from extraordinarily high pressure. Using a large-scale dynamic process simulator, we demonstrate that modifying the tuning parameters of one of these PI controllers based on the safety system being on or off leads to improved closed-loop performance compared to the case in which the tuning parameters of the PI controller remain the same regardless of the state of the safety system. In addition to PI control, we also investigate how model predictive control systems can be modified to handle safety system actions. We will present a comprehensive simulation study to evaluate the approach and demonstrate that the integration of safety and control system can significantly aid chemical process operational safety and performance.

[1] T. Marlin. Operability in process design: Achieving safe, profitable, and robust process operations. Ontario, Canada: McMaster University, 2012.

525251 Two Feedback Control Schemes for the Size and Shape of Needle-like Crystals Growing in Suspension

Abstract:

The purification and the solidification of products by batch crystallization from solution is an essential unit operation in the chemical and pharmaceutical industries. The particle size and shape distribution (PSSD) is considered to be one of the central attributes of the solid products of such a batch crystallization process, because it affects not only downstream operations, but also product quality [1].

Since feedback control-based approaches have the potential to provide robustness with respect to process disturbances and to mitigate batch-to-batch variations, they are considered to be promising for operating batch processes. Furthermore, online monitoring tools for the size and shape evolution of crystals growing in suspension have evolved significantly in the past few years [see, e.g., Ref. 2 and references therein], enabling the development of tailored feedback control schemes for such processes.

In this contribution, we present two different feedback controllers for the evolution of the average length and width of needle-like crystals during seeded, growth-dominated batch cooling crystallization, where both strategies rely on the capability to measure PSSDs online [2]. The first scheme is a path following controller (PFC) that operates without a kinetic model for crystal growth. The second scheme is a more complex nonlinear model predictive controller (NMPC) that relies on a multidimensional crystal growth rate model.

Both controllers were designed, and their performance was tested, using a custom process simulation framework that features a morphological population balance model based on a polyhedral model for the needle-like particles and a detailed model of the characteristics of the online PSSD monitoring device presented in Ref. [2]. Therefore, this framework takes into account the discrepancy between the faceted shape of real crystals and the simpler generic particle shape models commonly applied for monitoring and modeling purposes.

In terms of reaching target average sizes and shapes chosen to lie within the attainable region of a given seed population, both controllers performed similarly. The feedback mechanism in the NMPC scheme was able to compensate for the model-plant mismatch. Additionally, the PFC was able to robustly achieve final average dimensions with respect to variations in the crystal growth rates. Since acquiring multidimensional growth models requires considerable efforts [see, e.g., Ref. 3], we conclude that PFC is a more adequate feedback control scheme for the considered process.

References

[1] Davey, R.; Garside, J., *From Molecules to Crystallizers: An Introduction to Crystallization*. Oxford University Press: New York, NY, 2000.

[2] Rajagopalan, A. K.; Schneeberger, J.; Salvatori, F.; Bötschi, S.; Ochsenbein, D. R.; Oswald, M. R.; Pollefeys, M.; Mazzotti, M., A comprehensive shape analysis pipeline for stereoscopic measurements of particulate populations in suspension. *Powder Technol.* **2017**, *321*, 479-493.

[3] Eisenschmidt, H.; Voigt, A.; Sundmacher, K., Face-Specific Growth and Dissolution Kinetics of Potassium Dihydrogen Phosphate Crystals from Batch Crystallization Experiments. *Cryst. Growth Des.* **2015**, *15*, 219-227.

531599 Robust Fault Tolerant Control of Hydraulic Pipeline Systems

Abstract:

We study the application of the fault tolerant control for hydraulic pipeline systems described by the so-called water hammer equations [1]. By taking the benefits of backstepping transformations [2] and the superposition principle, novel observers are proposed to realize fault detection and estimation while fault detectability analysis is facilitated in backstepping coordinates. Moreover, fault detectability conditions are given and namely once a fault occurs the detection residual (the difference between the measured physical output and the estimated output) exceeds the predefined threshold. Upon a fault detection, the integration of novel plant state observer design and the design of novel adaptive tuning update law is proposed without consideration model uncertainties and external disturbances present [2]. Nevertheless, due to the coupling between the fault parameters and state/input, the conventional observer cannot undertake the tasks of fault estimations upon the fault detection. By decoupling the observation error system, this paper first develops a novel approach to solve the actuator fault detection and estimation simultaneously. In the proposed approach, only detection residual is needed. However, the observer error system cannot be transformed into the desired form for the solution of the sensor estimation problem. Later, the observer is modified based on superposition principle, and novel methods are proposed with available detection residual. The developed methods cannot only address both fault detection and estimation issues but also complete plant state estimation in presence of faults. In particular, bounded model uncertainties and disturbances are considered, and by applying proposed methods the state estimation error and fault parameter estimation error are ultimately bounded. Based on resulting fault detection and estimation, a novel robust output regulation scheme is proposed to complete output tracking issues. Finally, a representative example is given to verify theoretical results.

[1] Van Pham, T., Georges, D., & Besancon, G. (2014). Predictive control with guaranteed stability for water hammer equations. IEEE transactions on automatic control, 59(2), 465-470.

[2] Diagne, A., Diagne, M., Tang, S., & Krstic, M. (2017). Backstepping stabilization of the linearized Saint-Venantâ €“Exner model. Automatica, 76, 345-354.

[3] Xu, X., and Dubljevic, S., Adaptive fault detection and estimation for a class of PIDE systems based on boundary observers, Automatica, Under review, 2018.

532925 Learning-Based Nonlinear Model Predictive Control with Chance Constraints for Stochastic Systems

Abstract:

Design of model predictive controllers (MPC) faces an inherent trade-off between performance and robustness. In general, stability and robustness of MPC can be derived using approximate models, but MPC performance critically hinges on the model quality [1, 2]. In this work, we present a learning-based nonlinear MPC (NMPC) strategy with chance constraints for stochastic nonlinear systems. The learning-based NMPC strategy provides guarantees on stability and state chance constraint satisfaction, while the controller performance is enhanced by identifying more accurate models online without compromising the aforementioned guarantees. The key notion of the proposed strategy is that robustness guarantees can be decoupled from control performance by using two models of the system: the first is an approximate model derived offline with bounds on its uncertainty and is used for providing guarantees on stability and constraint satisfaction, and the second model is derived online using, for example, machine learning tools and is applied for computing the control cost function. Thus, the learning-based NMPC strategy minimizes a control cost subject to learned dynamics, while robustness is ensured by verifying that the designed control inputs preserve the stability and chance constraint satisfaction of the initial approximate model [3, 4].

To this end, we present a new constraint tightening approach by extending the concept of stochastic tubes [5] to a nonlinear setting. The constraint tightening approach is used for ensuring stability, recursive feasibility, and constraint satisfaction properties of the learning-based NMPC with individual chance constraints [6]. In this approach, state constraints are tightened recursively by constructing a sequence of sets that bound the one-step ahead disturbance propagation of the nominal model predictions. The sets are computed from an initial constraint set, which is obtained from the empirical cumulative distribution of the system uncertainty and is subsequently tightened with an appropriate backoff parameter to account for the individual chance constraints.

The proposed learning-based NMPC strategy has an online computational complexity comparable to that of nominal MPC. We demonstrate the performance of the learning-based NMPC on two nonlinear systems, including a benchmark DC-DC converter case study, where Gaussian process modeling [7] is used for online learning of the nonlinear system dynamics based on closed-loop data. Simulation results indicate that the learning-based NMPC leads to an enlarged domain of attraction and improved control performance due to, respectively, effective handling of chance constraints and computing the control cost based on learned dynamics.

References

[1] D. Mayne, M. Seron and S. Rakovic, "Robust model predictive control of constrained linear systems with bounded disturbance.," *Automatica,* 2005.

[2] A. Mesbah, "Stochastic model predictive control: An overview and perspectives for future research," *IEEE Control Systems,* 2016.

[3] A. Aswani, H. Gonzalez, S. Shankar Sastry and C. Tomlin, "Provably safe and robust learning-based model predictive control," *Automatica,* 2013.

[4] D. Limon, J. Calliess and J. Maciejowski, "Learning-based Nonlinear Model Predictive Control," in *IFAC,* 2017.

[5] B. Kouvaritakis and M. Cannon, "Model predictive control: Classical, robust and stochastic.," *Springer,* 2015.

[6] T. L. Santos, A. D. Bonzanini and A. Mesbah, "A Constraint-Tightening Approach to Nonlinear Model Predictive Control with Chance Constraints for Stochastic Systems," in *Proceedings of IEEE Conference on Decision and Control,* Submitted, 2018.

[7] C. E. Rasmussen and K. I. Williams, Gaussian Processes for Machine Learning, Cambridge, Massachusetts; London, England: MIT Press, 2006.

532982 Control Under Uncertainty in Automated Drug Delivery

Abstract:

Biological processes are complex dynamical systems with significant uncertainties and exogenous disturbances that render their regulation and control a challenging problem. Model predictive control (MPC) formulations, often used in biomedical applications, particularly for controlled drug delivery, are effective in rejecting disturbances and mitigating their effects through receding horizon control. MPC is commonly used in artificial pancreas (AP) systems to automatically compute the required amount of insulin dose to administer in patients living with the chronic disease of type 1 diabetes to regulate their glucose concentrations [1]. Despite recent advancements, the development and commercialization of an AP is challenging because of the complex, nonlinear, and only approximately known biochemical and physiological kinetics and dynamics of glucoseâ€"insulin metabolism [2].

High-fidelity predictive models are not readily available for metabolic processes due to the significant variability in human physiology and obscure information on the time and amount of carbohydrate consumption and exercise or physical activity levels. A fully automated AP system eliminating the need for patients to interact with the system and manually enter user inputs for meal and exercise announcements represents a substantial step towards achieving better insulin delivery systems [3]. Recognizing and attenuating the challenging and uncertain disturbances such as meals and exercise that affect the glycemic homeostasis is therefore necessary [4]. Concerning the automation of AP systems, several studies have incorporated unannounced meals through the estimation of time-varying parameters or analysis of glucose trends. Furthermore, additional physiological variables related to physical activity are also considered to automatically accommodate exercise [5]. Despite these efforts, automatically handling of unannounced meals and exercise in adaptive and personalized glycemic models for predictive control in AP systems is not sufficiently studied, while the future progression of the meal and exercise effects is not elucidated because their temporal evolution is uncertain.

Motivated by the above considerations, a stochastic MPC formulation is proposed for handling the uncertain effects of meals and exercise on the future glycemic predictions. The proposed MPC utilizes adaptive models to accurately characterize the time-varying glycemic dynamics. To address the uncertainty in the future projections of the meals and exercise effects, the future evolution of the uncertain variables is modelled using uncertainty sets from which realizations of the uncertain variables are considered in a stochastic control formulation. The optimal insulin infusion rates are then computed with respect to the possible realizations concerning the uncertain variables. The efficacy of the proposed stochastic MPC for handling uncertainty in meals and exercise for glucose control is demonstrated using simulation case studies.

[1] Gondhalekar, R., Dassau, E., and Doyle III, F. J. (2016). Periodic zone-MPC with asymmetric costs for outpatient-ready safety of an artificial pancreas to treat type 1 diabetes. Automatica, 71, 237-246.
[2] Messori, M., Incremona, G. P., Cobelli, C., & Magni, L. (2018). Individualized model predictive control for the artificial pancreas: In silico evaluation of closed-loop glucose control. IEEE Control Syst., 38(1), 86-104.
[3] Turksoy, K., Hajizadeh, I., Samadi, S., Feng, J., Sevil, M., Park, M., Quinn, L., Littlejohn, E., and Cinar, A. (2017).

Real-time insulin bolusing for unannounced meals with artificial pancreas. Control Eng. Pract., 59, 159-164.
[4] Bequette, B. W., Cameron, F., Baysal, N., Howsmon, D. P., Buckingham, B. A., Maahs, D. M., and Levy, C. J. (2016). Algorithms for a Single Hormone Closed-Loop Artificial Pancreas: Challenges Pertinent to Chemical Process Operations and Control. Processes, 4(4), 39.
[5] Turksoy, K., Quinn, L., Littlejohn, E., and Cinar, A. (2014). Multivariable adaptive identification and control for artificial pancreas systems. IEEE Trans. Biomed. Eng., 61(3), 883-891.

533982 Predictive Control with Model Performance Monitoring and Re-Identification

Abstract:

The operation of chemical plants faces numerous challenges such as inherent nonlinearity, complex variable interactions and process constraints. The most common control method that can handle these challenges is model predictive control (MPC) [1, 2]. In several industrial applications of MPC, a linear model is used, in part due to the simplicity of developing linear models and in part due to the computational ease with using linear models. In order to handle the resultant plant-model mismatch, MPC with re-identification is proposed.

There exist some results on MPC with re-identification (IMPC) where model validity is accounted for by requiring excitation constraints to ensure that the model parameters remain identifiable [3]. In this approach, identification is performed at every time step. Furthermore, the approach requires finding the right trade-off between the inevitable performance deterioration (due to excitation conditions) and the possibility of loss of model validity.

But, in these approaches, the original training data is not retained in the new model identification [4], and these methods are designed to address situations where the system is changed significantly and previous data are not at all representative of the plant in question. In situations where plant model mismatch arises due to change in operating condition (with the possibility of reverting back to the nominal plant operation), it becomes useful to merge old and new plant data in the re-identification step.

In this work we address the problem of handling plant-model mismatch by designing a subspace identification based MPC framework that includes model monitoring and closed-loop identification components.

Key Words: Model Predictive Control, Subspace Identification Methods, Re-Identification

[1] Nagy, Z. K., & Braatz, R. D. (2003). Robust nonlinear model predictive control of batch processes. *AIChE Journal*, *49*(7), 1776-1786.

[2] Schäfer, J., & Cinar, A. (2004). Multivariable MPC system performance assessment, monitoring, and diagnosis. *Journal of process control*, *14*(2), 113-129.

[3] Genceli, H., & Nikolaou, M. (1996). New approach to constrained predictive control with simultaneous model identification. *AIChE journal*, *42*(10), 2857-2868.

[4] Kheradmandi, M., & Mhaskar, P. (2018). Model predictive control with closed-loop re-identification. *Computers & Chemical Engineering*, *109*, 249-260.

534373 Advanced Biomimetic Control Approach Integrated with Multi-Agent Optimization for Nonlinear Chemical Processes

Abstract:

A Biologically-Inspired Optimal Control Strategy, denoted as BIO-CS, has been recently developed. This strategy has been implemented on various advanced energy systems applications such as the Integrated Gasification Combined Cycle (IGCC), the Hybrid Performance (HYPER) system and a fermentation process (Mirlekar et al., 2017; 2017a; Lima et al., 2016; Li et al., 2016). In such applications, BIO-CS has shown to have features for handling process model nonlinearities as well as flexibility for employing different optimal control solvers and termination criteria. Also, in the past, multi-agent-based optimization techniques that mimic ant-colony-based behavior with improved efficiency have been studied under the name of Efficient Ant Colony Optimization (EACO) for molecular design and solvent selection case studies (Gebreslassie & Diwekar, 2015). The abilities of heuristic-based methods such as EACO, Efficient Genetic Algorithm (EGA) and Efficient Simulated Annealing (ESA) were also explored to develop homogenous Multi-agent Optimization (MAO) techniques by establishing communication protocols between the algorithm procedures and the global information sharing environment (Gebreslassie & Diwekar, 2017). However, the combination of biomimetic control strategies and agent-based optimization methods for nonlinear systems has not yet been addressed in an integrated fashion. To bridge this gap, in this work, BIO-CS is integrated with MAO to design a novel framework that leads to optimal process operations. The developed framework yields optimal setpoints or a trajectory of setpoints for a nonlinear, multivariable system considering an overall process objective by employing MAO. This system is then optimally controlled by BIO-CS to achieve such desired output setpoints.

The applicability of the proposed method is demonstrated using a nonlinear, multivariable fermentation process model (Li et al., 2016) for bioethanol production. In this multivariable system, finding the optimal production rate (or profitability) as well as performing the simultaneous control of product concentration and temperature of the fermentor at their setpoints are critical for optimal performance. The proposed framework is implemented on the fermentation process to address all these challenges. Specifically, implementation scenarios of setpoint tracking and plant-model mismatch are considered. The results of the developed framework are compared to a gradient-based Sequential Quadratic Programming (SQP) technique and a classical proportional-integral (PI) controller in terms of optimization and control studies, respectively. As an additional control development, BIO-CS is cast as a Model Predictive Controller (MPC) and the resulting formulation (BIO-CS as MPC) is compared to the agent-based BIO-CS approach in terms of computational time, time to reach steady state and tracking performance. The implementation results of the proposed framework will be discussed to show its potential for simultaneous optimization and control of advanced energy systems.

References:

1. G. Mirlekar, S. Li, and F. V. Lima. Design and implementation of a Biologically-Inspired Optimal Control Strategy (BIO-CS) for chemical process control. *Ind. Eng. Chem. Res.*, 56: 6468â€"6479, 2017.
2. G. Mirlekar, P. Pezzini, M. Bryden, D. Tucker, and F. V. Lima. A Biologically-Inspired Optimal Control Strategy (BIO-CS) for hybrid energy systems. *In Proceedings of the American Control Conference*, 4821-4826, 2017a.
3. F. V. Lima, S. Li, G. Mirlekar, L. N Sridhar, and G. J. Ruiz-Mercado. Modeling and advanced control for sustainable process systems. In Sustainability in the design, synthesis and analysis of chemical engineering processes, G. Ruiz-Mercado and H. Cabezas (eds.), *Elsevier*, 2016.
4. S. Li, G. Mirlekar, and F. V. Lima. Development of chemical process design and control for sustainability. *Processes*, 4(3):23, 2016.
5. B. H. Gebreslassie and U. M. Diwekar. Efficient ant colony optimization for computer aided molecular design: Case study solvent selection problem. *Computers and Chemical Engineering*, 78:1-9, 2015.
6. B. H. Gebreslassie and U. M. Diwekar. Homogenous multiagent optimization for process systems engineering problems with a case study of computer aided molecular design. *Chemical Engineering Science*, 159:194-206, 2017.

536170 Nonlinear Optimal Control Structure Design

Abstract:

Nonlinear Optimal Control Structure Design

Temitayo Bankole, Debangsu Bhattacharyya

Department of Chemical and Biomedical Engineering, West Virginia University, Morgantown, WV 26506, USA

Economically optimal operation of the process plants at the face of disturbances without violating process and environmental constraints is becoming increasingly important. To achieve this goal, not only controller design, but optimal design of the control structure is absolutely critical.

While there are numerous works on controller design, research in the area of optimal control structure design is still nascent. Traditionally, heuristics and process knowledge have been employed for control structure design. More recently, methodical approaches have been proposed (Halvorsen et al., 2003) based on economics. Optimal selection of the primary controlled variables (CVs) by considering plant economics as well as controllability in presence of closed-loop dynamics and time delay has been recently proposed by some of the co-authors of this work (Jones et al., 2014a, 2014b).

The methods proposed so far for optimal CV selection is based on linear models. Since these linear models are developed by linearizing the nonlinear process model under design condition, their accuracy is highly limited especially for highly nonlinear plants under off-design conditions. Thus the economic loss function generated by using such a local approximation can have high inaccuracy under off-design condition. Furthermore, controllability as well as closed-loop dynamics and time delay can significantly differ under off-design conditions. The current approach in this area is to use the nonlinear model for a posteriori analysis of top few optimal CV sets obtained using the linear analysis and select the 'best' set that shows a good trade-off between the economic and control performance (Jones et al., 2014a, 2014b). Since only top few CV sets are extracted for this analysis due to computational expense, it may be possible that the 'best' set might remain unexamined. This can lead to a suboptimal solution. However, evaluation of large number of CV sets using the nonlinear model is not computationally tractable.

With this incentive, a nonlinear measure of the economic loss function and controllability is developed in this work and used for CV selection problem. In a process plant of reasonable scale, there can be trillions of candidate CV sets. Therefore, it is intractable to use the full nonlinear model for the large-scale combinatorial optimization problem that is solved for CV selection. A novel approach is developed in this work by making use of stochastic sampling methods so that the economic and control performance of the candidate sets can be accurately evaluated under off-design conditions. This approach embodies novel features of an intelligent system, especially for systems with high nonlinearity which cannot be approximated by linear models at the desired operating range. The method is applied to a large-scale process model of an integrated gasification combined cycle process.

References

Halvorsen, I. J., Skogestad, S., Morud, J. C., & Alstad, V. (2003). Optimal selection of controlled variables. Industrial & Engineering Chemistry Research, 42, 3273-3284.

Jones, D., Bhattacharyya, D., Turton, R., & Zitney. (2014a). Plant-wide control system design: Primary controlled variable selection. Computers & Chemical Engineering, 220-234.

Jones, D., Bhattacharyya, D., Turton, R., & Zitney. (2014b). Plant-wide control system design: Secondary controlled variable selection. Computers & Chemical Engineering, 71, 253-262.

541565 25 by 25: Chemical Engineering in the Next 25 Years

Abstract:

In 2008, the centennial year for AIChE, we asked 25 chemical engineers for their visions of the profession from around the US and the world: Nine faculty, eight from industry, and eight post-docs and graduate students. We posed four questions related to each individual's industrial sector or area of research, asking for extrapolation, new impacts on existing sectors, new sectors, and an optional, more open-ended comment on the future of the profession as a whole.

In this 110th year of AIChE, we have revisited these questions with many of the same people. Ten years has seen many developments in our profession, in society, and in our thought-leaders' experiences. Here we report on their perspectives, comparing them to the 2008 visions.

541542 The Future of Chemical Engineering Itself

Abstract:

Chemical engineeringâ€™s future is bright because it is vital to so much of society and industry -- and its future is complicated by its rapidly expanding universe of activities. It must represent depth in core skills centered on the chemical sciences including biology and materials. At the same time, it must be transdisciplinary; it must exemplify the need and value of reaching across disciplines in borderless ways.

One way is to create and disseminate a compelling vision and strategy for chemical engineering, research, and education for the future. AIChE and the National Academies are working together to that end. In 1988, the National Academies published "Frontiers in Chemical Engineering: Research Needs and Opportunities.â€ Its committee of authors outlined a wide-ranging roadmap to turn promising research opportunities into reality, while calling for researchers to embrace new frontiers.

Since 1988, that has happened -- along with many other transformative changes like the Web, fracking, personalized medicine, and social media. A new nationwide effort, aided by international contributions, will address both the identity of chemical engineering and its future directions. Its goal is to ensure that the people in chemical engineering are primed to meet society's needs and rapid changes.

544708 Accelerating Innovation through Academic-Industrial Partnerships

Abstract:

In 2011, The Dow Chemical Company launched the University Partnership Initiative (UPI) to sponsor collaborative research at leading U.S. Universities. This 10-year commitment aims to address fundamental questions in industrial chemistry and chemical engineering with alignment to Dowâ€™s next-generation process technology and product development efforts. To date, this portfolio of projects has generated an impressive array of intellectual property, publications, and numerous examples of the collaborative developments implemented at Dow. This presentation will introduce UPI program and highlight several contributions from our collaborators and leaders in the chemical engineering profession.

540559 Maximizing Uptime, Efficiency, and Safety of Industrial Operations through Early Risk Detection

Abstract:

Achieving zero incidents while maximizing process uptime and efficiency is a key vision for operating companies. An important building block towards this vision is establishment of a proactive risk mitigation culture, supported by effective systems and a strong management. Studies have shown that most of the incidents and unexpected process failures can be avoided if the operating teams get timely information about developing risks and take preventive actions early on. While advances have been made in process monitoring and automation in the past decades, there still remains significant technological and behavioral gaps that prevent plant operations from proactively mitigating process risks.

A modern industrial plant monitors thousands of parameters, generating upwards of 50-100 million data points every day. Thanks to recent breakthroughs in machine learning and artificial intelligence approaches, today there are autonomous systems that can sift through this data and point out meaningful and timely insight. This can help operating teams ascertain process issues that are hidden in the data, long before process variables reach critical levels.

In this presentation, a new autonomous system, Dynamic Risk Analyzer^{TM} (DRA), will be introduced that points out early indicators of process issues using proprietary machine learning. The risk indicators enable operations team to respond and make required process changes, days and sometimes weeks ahead of any alarm, avoiding late stage (expensive) fixes. To realize these objectives fully, a proactive management workflow must also be in place. The presentation will discuss new initiatives and workflows developed by plant operations to implement a proactive culture. Real-life case studies from plants using DRA technology will be presented on how this culture change resulted in increased uptime, efficiency and safety of plant operations.

531309 Gaussian Processes for Hybridizing Analytical & Data-Driven Decision-Making

Abstract:

Surrogate models are widely appreciated in process systems engineering [1]. The typical setting focuses on expensive-to-evaluate, possibly uncertain functions. Examples include: modular process simulators [2], integrated gasification combined cycle processes [3], a carbon capture absorber [4], and many other applications, e.g. [5-8]. Resources are typically limited, so effective decision making requires data-efficient learning.

The data science and statistical machine learning communities typically focus on models learned solely from observed data. But chemical engineering applications may also require explicit, parametric models, e.g. modeling known process constraints, operations constraints, and cost objectives [9]. So, work has integrated semi-algebraic functions with those learned from data [10] or developed semi-physical modeling techniques [11, 12].

This presentation surveys the state-of-the-art in hybridizing analytical and data-driven decision making [13]. We consider three probabilistic modeling applications to these hybrid situations:

Design of experiments for model discrimination [14]. We bridge the gap between classical, analytical methods [15] and Monte Carlo-based approaches [16]. Classical methods may have difficulty managing non-analytical model functions and data-driven Monte Carlo approaches come at a high computational cost. We replace the original, parametric models with probabilistic, non-parametric Gaussian process surrogates learned from model evaluations. The surrogates are flexible regression tools that extend classical analytical results to non-analytical models, while providing us with model prediction confidence bounds and avoiding the computational complexity of Monte-Carlo approaches.

Multi-objective optimization [17, 18]. We make novel extensions to Bayesian multi-objective optimization in the case of one analytical objective function and one black-box, i.e. simulation-based, objective function. The resulting method has been applied to a bone neotissue application [19] and a more general test suite.

Scheduling plant operations under uncertainty. For processes with equipment degradation, we use Gaussian processes to approximate large-scale, mixed-integer optimization problems.

We close by offering a broad outlook on applying probabilistic surrogate models to chemical engineering. Statistical machine learning has recently attracted significant interest in process systems engineering [20]. Here we show that state-of-the-art research in Gaussian processes [21, 22] and probabilistic modeling more generally [23] can have a big impact on chemical engineering.

References

[1] Bhosekar A, Ierapetritou M. Advances in surrogate based modeling, feasibility analysis and and optimization: A

review. *Comput Chem Eng*. 108: 250-267, **2018**.

[2] Caballero JA, Grossmann IE. An algorithm for the use of surrogate models in modular flowsheet optimization. *AIChE J*. 54(10):2633-2650, **2008**.

[3] Lang Y, Zitney SE, Biegler LT. Optimization of IGCC processes with reduced order CFD models. *Comput Chem Eng*. 35(9):1705-1717, **2011**.

[4] Cozad A, Sahinidis NV, Miller DC. Learning surrogate models for simulation-based optim-ization. *AIChE J*. 60(6):2211-2227, **2017**.

[5] Boukouvala F, Ierapetritou M. Surrogate-based optimization of expensive flowsheet modeling for continuous pharmaceutical manufacturing. *J Pharm Innov*. 8: 131, **2013**.

[6] Soepyan FB, Cremaschi S, Sarica C, and others. Estimation of percentiles using the Kriging method for uncertainty propagation. *Comput Chem Eng*. 93:143-59, **2016**.

[7] Shokry A, Ardakani MH, Escudero G, Graells M, Espuña A. Dynamic Kriging-based fault detection and diagnosis approach for nonlinear noisy dynamic processes. *Comput Chem Eng*. 106:758-76, **2017**.

[8] Tran AP, Georgakis C. On the estimation of high-dimensional surrogate models of steady-state of plant-wide processes characteristics. *Comput Chem Eng*. DOI 10.1016/j.comp-chemeng.2018.02.014, **2018**.

[9] Boukouvala F, Hasan MF, Floudas CA. Global optimization of general constrained grey-box models: new method and its application to constrained PDEs for pressure swing adsorption. *J Glob Optim*. 67(1-2): 3-42, **2017**.

[10] Boukouvala F, Floudas CA. ARGONAUT: AlgoRithms for Global Optimization of coNstrAined grey-box compUTational problems. *Optim Lett*. 11(5): 895-913, **2017**.

[11] Pearson RK, Pottmann M. Gray-box identification of block-oriented nonlinear models. *J Process Contr*. 10(4):301-315, **2000**.

[12] Cozad A, Sahinidis NV, Miller DC. A combined first-principles and data-driven approach to model building. *Comput Chem Eng*. 73: 116-27, **2015**.

[13] Boukouvala F, Misener R, Floudas CA. Global optimization advances in mixed-integer nonlinear programming, MINLP, and constrained derivative-free optimization, CDFO. *Eur J Oper Res*. 252: 701-727, **2016**.

[14] Olofsson S, Deisenroth MP, Misener R. Design of Experiments for Model Discrimination Hybridising Analytical and Data-Driven Approaches. *arXiv:1802.04170*. **2018**.

[15] Buzzi-Ferraris G, Forzatti P, Emig G, Hofmann H. Sequential experimental design for model discrimination in the case of multiple responses. *Chem Eng Sci*. 39(1): 81-85, **1984**.

[16] Vanlier J, Tiemann CA, Hilbers PAJ, van Riel NAW. Optimal experiment design for model selection in biochemical networks. *BMC Syst Biol*, 8(20), **2014**.

[17] Olofsson S, Mehrian M, Geris L, Calandra R, Deisenroth MP, Misener R. Bayesian Multi-Objective Optimisation of Neotissue Growth in a Perfusion Bioreactor Set-Up. *Computer Aided Chemical Engineering*. 40: 2155-2160, **2017**.

[18] Beykal B, Boukouvala F, Floudas CA, Pistikopoulos EN. Optimal design of energy systems using constrained grey-box multi-objective optimization. *Comput Chem Eng*. DOI 10.1016/j.compchemeng.2018.02.017, **2018**.

[19] Mehrian M, Guyot Y, Papantoniou I, Olofsson S, Sonnaert M, Misener R, Geris L. Maximizing neotissue growth kinetics in a perfusion bioreactor: An *in silico* strategy using model reduction and Bayesian optimization. *Biotechnol Bioeng*. 115(3):617-29, **2018**.

[20] Lee JH, Shin J, Realff MJ. Machine learning: Overview of the recent progresses and implications for the process systems engineering field. *Comput Chem Eng*. DOI 10.1016/j.compchemeng.2017.10.008, **2017**.

[21] Damianou A, Lawrence N. Deep Gaussian processes. In *Artificial Intelligence and Statistics*. 207-215, **2013**.

[22] Deisenroth MP, Ng JW. Distributed Gaussian processes. In *32nd International Conference on International Conference on Machine Learning*. 37: 1481-1490, **2015**.

[23] Ghahramani Z. Bayesian non-parametrics and the probabilistic approach to modelling. *Phil Trans R Soc A*. 371(1984):20110553, **2013**.

532317 Designing Difficult-to-Cyberattack Process Control Systems

Abstract:

Cybersecurity of chemical process control systems is critical because attacks on these processes can harm physical equipment as well as people [1]. Cyberattacks can take many forms, including sending false sensor measurements to operators in an attempt to make them think that the process is operating normally when it is not [2] or manipulating the control actions applied to the process [3]. In [4], it is noted that physical attacks on plants may be prevented by making the plants less attractive to attack by reducing the consequences of an attack by making a plant more inherently safe (e.g., using less hazardous materials); a similar concept should apply in the context of cybersecurity.

Motivated by the above considerations, we develop an implementation strategy for model predictive control (MPC) that seeks to make the control system more difficult to cyberattack, when the specific type of attack considered is one in which the attacker provides false state measurements to the MPC. We develop an implementation strategy in which a variety of control designs with stability-based constraints (e.g., constraints based on Lyapunov and Control Lyapunov-Barrier functions) are developed and one is randomly selected at every sampling time. Each controller on its own can guarantee closed-loop stability and recursive feasibility for initial states within a well-characterized region of state-space, and the implementation strategy ensures that the only controllers available at a given sampling time to be selected between are those which would be feasible, such that the implemented control actions guarantee closed-loop stability even when randomly selected. The goal of the random selection is to make it difficult for a cyberattacker to determine which incorrect value of the measured state to provide to the controller at a given sampling time to achieve a certain goal since the controller which will be used with that measurement, and therefore the input to be applied when the state measurement is provided, is difficult to discern *a priori*. A chemical process example demonstrates the use of this randomized MPC implementation strategy.

[1] A. A. Cárdenas, S. Amin, Z.-S. Lin, Y.-L. Huang, C.-Y. Huang and S. Sastry, "Attacks against process control systems: Risk assessment, detection, and response," In *Proceedings of the ACM Asia Conference on Computer & Communications Security*, Hong Kong, China, 2011.

[2] O. Linda, M. Manic and M. McQueen, "Improving control system cyber-state awareness using known secure sensor measurements," In *Proceedings of the International Conference on Critical Information Infrastructures Security*, Lillehammer, Norway, 2012.

[3] A. Roisch, H. Voos, Y. Li and M. Darouach, "A model predictive approach for cyber-attack detection and mitigation in control systems," In *Proceedings of the IEEE Conference on Decision and Control*, pages 6621-6626, Florence, Italy, 2013.

[4] S. Bajpai and J. P. Gupta, "Site security for chemical process industries," *Journal of Loss Prevention in the Process Industries*, 18, 301-309, 2005.

526412 Cyber Security of Model Predictive Control Systems for Chemical Processes

Abstract:

Recently, the security of process control systems has become crucially important since control systems are vulnerable to cyber-attacks, which is a series of computer actions to compromise the security of control systems (e.g., integrity, stability and safety) [1]. Among cyber-attacks, targeted attacks are severe threats for control systems because of their specific designs with the aim of damaging the control actions applied to a chemical process (for example, Stuxnet worms aims to modify the data sent to a Programmable Logic Controller) [2]. Additionally, targeted attacks are usually stealthy and difficult to detect using classical detection methods since they are designed based on some known information of control systems (perhaps they make explicit use of the process state measurement). Therefore, designing an advanced detection system [3] and a suitable model predictive control system for nonlinear processes in the presence of targeted cyber attacks is an important open issue.

Motivated by this, in this work, we develop a model predictive control system for nonlinear systems [4] subject to targeted cyber attacks. Specifically, a cyber attack that aims to damage close-loop stability via a sensor tamper is first considered and applied to the closed-loop process, and it is demonstrated that if not suitable action is taken by the model predictive controller the closed-loop system loses its stability. Subsequently, an anomaly-based detection method is developed using a priori knowledge of control system, and a suitable model predictive control method is developed to reduce the impact of cyber attacks and re-stabilize the closed-loop system in finite time. A chemical process example is used to demonstrate the applicability of the proposed approach.

[1] Ye N, Zhang Y, Borror C M. Robustness of the Markov-chain model for cyber-attack detection. IEEE Transactions on Reliability, 2004, 53: 116-123.

[2] Cárdenas A A, Amin S, Lin Z S, et al. Attacks against process control systems: risk assessment, detection, and response. Proceedings of the 6th ACM symposium on information, computer and communications security. ACM, 2011: 355-366.

[3] Ozay M, Esnaola I, Vural F T Y, et al. Machine learning methods for attack detection in the smart grid. IEEE Transactions on Neural Networks and Learning Systems, 2016, 27: 1773-1786.

[4] Mhaskar P, El-Farra N H, Christofides P D. Stabilization of nonlinear systems with state and control constraints using Lyapunov-based predictive control. Systems & Control Letters, 2006, 55: 650-659.

542728 Keynote Talk: Technology Development for Cybersecure Fossil Power Generation

Abstract:

Distributed sensor networks, secure sensor data transmission, and automated controls are key elements to better power plant flexibility, performance, and reliability. The prevalence of wireless sensors is likely to increase dramatically, supported by Internet of Things (IOT) infrastructure, and driven by a need to use data analytics approaches for condition-based monitoring and fault detection to reduce forced outages, extend asset life, and maximize plant productivity. Additionally, many utilities are leveraging a centralized monitoring and diagnostic approach, either in-house or through a vendor service, aggregating data streams associated with many power plants requiring remote access to plant data. These technologies need to be developed and integrated in a manner that doesnâ€™t create additional system vulnerabilities â€“ particularly related to cybersecurity. The National Energy Technology Laboratory is sponsoring research on behalf of the US Department of Energyâ€™s Office of Fossil Energy to identify and mitigate cybersecurity vulnerabilities related to fossil-based power generation. A review of preliminary cybersecurity vulnerabilities will be presented along with an update on pertinent cybersecurity R&D underway.

543156 Keynote Talk: Cybersecurity Realities in Industrial Control System Environments

Abstract:

Cybersecurity, particularly in the Industrial Control System (ICS) discipline, is hardly a rare topic these days. There are endless discussions of IT/OT convergence, cybersecurity frameworks, IIoT, the growth of Advanced Persistent Threats, and myriad similar topics. However, there are essential differences between the traditional IT approach to cybersecurity and the paradigm that exists in an ICS environment that create unique challenges and provide likewise unique opportunities. This keynote will highlight some of the less obvious aspects of defending these environments and engage the audience to think, perhaps, a little differently. The talk will cover the organizational capability, governance, and philosophical distinctions encountered in real-world cybersecurity deployments and some of the lessons learned by major Oil, Gas, and Chemical producers. Additionally, this session will consider some of the disruptive technologies that have emerged as a result of the rapid expansion of the ICS cybersecurity market. This particular landscape is in rapid transition and it is important to understand some of the esoteric approaches necessary for successful cybersecurity development early in the journey.

530253 A Metaheuristic Approach to Best Subset Selection for the Development of Regression-Based Surrogate Models

Abstract:

Learning algebraic models from data is becoming an increasingly popular method to model complex chemical engineering systems. This trend is driven by the difficulty of constructing and solving physics-based models, increases in available system and experimental data, and improved strategies for regression. In this work, we focus on predicting the value of a single output variable using a linear combination of features. These features include the regressor variables as well as nonlinear transformations of the regressors.

Well-known strategies for linear regression include Forward/Backward Selection, Ordinary-Least Squares, and Best Subset Selection (BSS). BSS seeks to select coefficient values for the features as to minimize the least-squares error between the actual and predicted values of the output while also penalizing the inclusion of features in the objective. This task can be posed as a mixed-integer optimization (MIO) problem and leads to simple models with good accuracy. However, as the amount of data points and the number of potential features grows, the resulting combinatorial explosion makes the MIO problem intractable to solve in a reasonable amount of time.

In this paper, we present a metaheuristic-based algorithm for solving the Best-Subset-Selection problem. Our strategy includes a stochastic global search and a deterministic local search. We begin by defining neighborhoods in the combinatorial space whose sizes are based-on the Hamming distance from a current point. Within this neighborhood, we randomly select a new point in the combinatorial space and optimize the continuous variable. Then, we perform a local search in the neighborhood of the current point. The local search is done for small neighborhoods via exhaustive enumeration of combinatorial solutions and application of well-known updates for the coefficient vector or by using a commercial solver for large neighborhoods. If a better solution is found, the local search is re-centered at the current best point. If no new solution is found, the neighborhood size is increased until the maximum neighborhood size is reached and the algorithm moves back into the global search phase. We demonstrate our approach on a wide range of test data and show that our approach yields accurate models in a tractable manner.

525545 Data-Driven Stochastic Robust Optimization for Process Operation Under Uncertainty

Abstract:

Optimization under uncertainty has attracted tremendous attention from both academia and industry (Grossmann et al., 2016). Uncertain parameters, if not being accounted for, could render the solution of an optimization problem suboptimal or even infeasible. To this end, a plethora of mathematical programming techniques have been proposed. These techniques have their respective strengths and weaknesses, which lead to different application scopes. Stochastic programming focuses on the expected performance of a solution by leveraging the scenarios of uncertainty realization and their probability distribution. However, this approach requires accurate information on the probability distribution, and the resulting optimization problem could become computationally challenging. Robust optimization provides an alternative paradigm that does not require accurate knowledge on probability distributions. The state-of-the-art approaches for optimization under uncertainty leverage the synergy of different optimization methods to inherit their corresponding strengths and complement respective weaknesses (Yue and You, 2016).

Labeled multi-class data are ubiquitous in a variety of areas and disciplines. For example, process data are labeled with the operating modes of chemical plants (Afzal et al., 2017). Due to the massive amount of available uncertainty data and dramatic progress in big data analytics, data-driven optimization under uncertainty emerges as a promising paradigm (Shang et al., 2017). Most existing data-driven optimization methods are restricted to unlabeled uncertainty data. Discarding labels and applying the existing methods to labeled uncertainty data ignores information embedded in labels, possibly compromising results and leading to sub-optimal solutions. Consequently, new modeling frameworks and computational algorithms are required for handling labeled multi-class uncertainty data.

This paper proposes a data-driven stochastic robust optimization (DDSRO) framework that systematically and automatically handles labeled multi-class uncertainty data. In large datasets, uncertainty data are often attached with labels to indicate their data classes. For the labeled multi-class uncertainty data, we propose a data-driven uncertainty modeling process that includes two parts, i.e. probability distribution estimation for different data classes, and uncertainty set construction. The probability distribution for data classes is learned from labeled uncertainty data through maximum likelihood estimation for the multinoulli distribution. To capture the nature of uncertainty data with different labels, a group of Dirichlet process mixture models is employed to model uncertainty with a variational inference algorithm (Ning and You, 2017). These two pieces of uncertainty information are incorporated into the DDSRO framework through a bi-level optimization structure. Two-stage stochastic programming is nested in the outer problem to optimize the expected objective over categorical data classes; robust optimization is nested as the inner problem to hedge against high-dimensional uncertainty and ensure computational tractability. To demonstrate effectiveness of the proposed method, process network planning application (You and Grossmann, 2011) is presented. References

M. Afzal, W. Tan, T. Chen, 2017, Process Monitoring for Multimodal Processes With Mode-Reachability

Constraints. IEEE Transactions on Industrial Electronics, 64, 4325-4335.

I. Grossmann, R. Apap, B. Calfa, P. García-Herreros, Q. Zhang, 2016, Recent advances in mathematical programming techniques for the optimization of process systems under uncertainty, Computers & Chemical Engineering, 91, 3-14.

C. Ning, F. You, 2017, Data driven adaptive nested robust optimization: General modeling framework and efficient computational algorithm for decision making under uncertainty. AIChE Journal, 63, 3790-3817.

C. Shang, X. Huang, F. You, 2017, Data-driven robust optimization based on kernel learning. Computers & Chemical Engineering, 106, 464-479.

F. You, I. Grossmann, 2011, Stochastic inventory management for tactical process planning under uncertainties: MINLP models and algorithms. AIChE Journal, 57, 1250-1277.

D. Yue, F. You, 2016, Optimal supply chain design and operations under multi scale uncertainties: Nested stochastic robust optimization modeling framework and solution algorithm. AIChE Journal, 62, 3041-3055.

530400 Uncertainty Quantification and Stochastic Programming Strategies for Energy Market Participation

Abstract:

During the past few years, announcements of several utility-scale energy storage projects have made the headlines, promising to improve the grid reliability, ease renewable integration, curtail emissions, and reduce overall systems cost [1]. The financial soundness of these investments has yet to be proven because energy storage economics heavily depends on the surrounding electricity market dynamics, energy policies, and long-term performance (e.g., degradation, wear-and-tear).

A standard technique for economic assessment is to formulate market participation as an optimization problem and calculate the maximum possible revenue in hindsight, i.e., with perfect information [2]. Here physical limitations and market rules are modeled as constraints. Our recent analysis of California market data revealed over 60% of revenue opportunities come from the fastest market layers (i.e., real-time market, ancillary services) for batteries, concentrated solar thermal power, and industrial systems [3-5]. The finding that energy arbitrage in the day-ahead market is the least profitable participation strategy is consistent with other studies [2,6,7].

An important limitation of these studies is the assumption of perfect information, while in reality, market participants must bid under price and weather uncertainties [8,9]. We develop an integrated approach for uncertainty quantification and stochastic optimization of energy systems biding into complex energy markets. First, Gaussian Process (GP) statistical models [10] are trained using historical data and used to generate probabilistic forecasts for market prices. These forecasts are input parameters for a two-stage stochastic model predictive control framework to compute optimal market participation policies. Preliminary benchmarks indicate the proposed approach, implemented in a receding horizon framework, is capable of capturing over 90% of theoretical market revenues, compared to only 75% with simple backcasting. We also explore different strategies to integrate probabilistic forecasts including sampling and Smolyak quadrature rules (sparse grids) [11-13]. Finally, we discuss how the proposed framework enables comparison of different energy storage technologies based on the ability to cope with exogenous (market, weather, etc.) uncertainty.

References:

[1] Cardwell and Krauss (2017). A Big Test for Big Batteries. *New York Times.* https://www.nytimes.com/2017/01/14/business/energy-environment/california-big-batteries-as-power-plants.html

[2] Walawalkar, Apt, and Mancini (2007). Economics of electric energy storage for energy arbitrage and regulation in New York. *Energy Policy* 4, pg. 2558-2568.

[3] Dowling, Kumar, Zavala (2016). A Multi-Scale Optimization Framework for Electricity Market Participation. *Applied Energy* 190, pg. 147-164.

[4] Dowling, Zheng, and Zavala (2018). A decomposition algorithm for simultaneous scheduling and control of CSP systems. *AIChE Journal*, in press.

[5] Dowling and Zavala (2017). Economic opportunities for industrial systems from frequency regulation markets. *Computers & Chemical Engineering*, in press.

[6] Fares and Webber (2014). A flexible model for economic operational management of grid battery energy storage. *Energy* 78, pg. 768-776.

[7] Lizarraga-Garcia, Ghobeity, Totten, and Mitsos (2013). Optimal operation of a solar-thermal power plant with energy storage and electricity buy-back from grid. *Energy* 5, pg. 61-70.

[8] Dominguez, Baringo, and Conejo (2012). Optimal offering strategies for a concentrating solar power plant. *Applied Energy* 98, pg. 316-325.

[9] Kumar, Wenzel, Ellis, ElBsat, Drees, and Zavala (2018). A Stochastic Model Predictive Control Framework for Stationary Battery Systems. *IEEE Transactions on Power Systems*, in press.

[10] Bishop (2006). Pattern Recognition and Machine Learning. Springer.

[11] Chen, Mehrotra, and Papp (2015). Scenario generation for stochastic optimization problems via the sparse grid method. *Computational Optimization and Applications* 62, pg. 669-692.

[12] LaÃnez-Aguirre, Mockus, and Reklaitis (2015). A stochastic programming approach for the Bayesian experimental design of nonlinear systems. *Computers & Chemical Engineering* 72, pg 312â€"324.

[13] Renteria, Cao, Dowling, and Zavala (2017). Optimal PID controller tuning using stochastic programming techniques, *AIChE Journal*, in press.

538865 Predicting Future Production for Unconventional Resources: A Data-Driven Approach

Abstract:

Hydraulic fracturing for horizontal wells is essential for oil and gas production from unconventional low-permeability reservoirs. A key challenge is to predict the effect of basic fracture design and construction variables with anticipated production and ultimate recovery. Decisions on such variables would optimize a related criterion, e.g. production, recovery, or NPV. Ideally, it would be desirable to understand the effect of several decision variables as well as reservoir-related parameters on the chosen objective. Fundamental equations could be used that describe phenomena related to hydraulic fracturing and subsequent production from a fractured reservoir. This is currently the prevailing approach in industry. While effective, this approach is cumbersome. As a simpler alternative, one could capitalize on the significant body of data on production from unconventional fractured wells. An opportunity exists to explore how such data can be used to develop methodologies that efficiently suggest optimal fracturing designs through the use of data-driven predictive models. Such models would capture the combined effect of model input variables, namely fracturing-job variables and reservoir properties, on the chosen objective. In a previous presentation, we explored how models can be developed that predict future production based on the initial production. This presentation focuses on a more ambitious objective, namely predicting future production, well before production commences. This would be useful, as it could help guide well placement and fracture design, both important tasks. To be able to build such a data-driven model, it is a key requirement that historical pressure data should be available along with production data. The purpose of this presentation is to explore how to build a model in that case.

Standard multivariate modeling techniques are used to build a model connecting various input variables (such reservoir properties, wellbore geometry, fracture construction process, production pressure, and others) with production.

The data-driven models developed would be useful for design purposes and what-if analyses. They can also provide insight into the variables that have strong impact on fractured horizontal well performance. Therefore, the proposed models are envisioned to offer initial decision support for well placement and fracturing design. In addition, they may will also be able to indicate parameters that should be improved/modified to get desired results in existing wells.

528072 Distributed Approximate Dynamic Programming (dADP) for Data-Driven Optimal Control of Nonlinear Systems

Abstract:

Data-driven techniques have been developed in the field of chemical engineering for multiple purposes, including process monitoring and control. Data-driven control offers an alternative to model-based control strategies that does not rely on an accurate control-oriented process model. Among the wide spectrum of methods of data-driven control [1], approximate dynamic programming (ADP) is a type of algorithms which has been successfully used for discrete-state systems in artificial intelligence applications, and discussed for process control (see e.g. [2]).

ADP is a data-driven optimal control strategy, where historical datasets are exploited to train the control policy and value function iteratively towards the optimal solution which satisfies the Bellmanâ€™s principle of optimality. For systems with continuous (infinite) states, the optimality principle assumes a specific form of the Hamilton-Jacobi-Bellman (HJB) equations. For nonlinear input-affine systems, the HJB equations can be transformed such that the model functions can be substituted with some data information, so that by choosing suitable basis functions for the optimal control policy and the value function, the policy and value iterations can be approximately solved as a regression problem [3].

In this work, we propose a novel approach different from that of [3], which directly formulates the HJB equations as a nonlinear regression problem, so that the approximate control policy and value function can be directly obtained. The framework is also extended to the cases where input constraints are present. This formulation is suitable for solving ADP in a big-data setting where a centralized optimization exploiting all the data in the regression procedures is infeasible. Specifically, we employ the alternating direction of multipliers (ADMM) [4], which is the most widely used distributed optimization algorithm, as well as its accelerated version [5] to regress the parameters of the optimal control policy and value function. We call the resulting framework *distributed adaptive dynamic programming* (dADP) as it adaptively updates the parameters to approach the optimum throughout the distributed optimization iterations, and we will illustrate this method in a chemical reactor example.

References

[1] Hou, Z. S., & Wang, Z. (2013). From model-based control to data-driven control: Survey, classification and perspective. *Inf. Sci.*, *235*, 3-35.

[2] Lee, J. H., & Wong, W. (2010). Approximate dynamic programming approach for process control. *J. Process Control*, *20*(9), 1038-1048.

[3] Luo, B., *et al.* (2014). Data-based approximate policy iteration for affine nonlinear continuous-time optimal control design. *Automatica*, *50*(12), 3281-3290.

[4] Boyd, S., *et al.* (2011). Distributed optimization and statistical learning via the alternating direction method of multipliers. *Found. Trends Mach. Learn.*, *3*(1), 1-122.

[5] Goldstein, T., *et al.* (2014). Fast alternating direction optimization methods. *SIAM J. Imaging Sci.*, *7*(3), 1588-1623.

535235 Data-Driven Evolution Equation Reconstruction for Parameter-Dependent Nonlinear Dynamical Systems

Abstract:

When studying observations of chemical reaction dynamics, closed form equations based on a putative mechanism may not be available. Yet when sufficient data from experimental observations can be obtained, even without knowing what exactly the physical meaning of the parameter settings or recorded variables are, data-driven methods can be used to construct minimal (and in a sense, robust) realizations of the system. The approach attempts, in a sense, to circumvent physical understanding, by building intrinsic "information geometries" of the observed data, and thus enabling prediction without physical/chemical knowledge. Here we use such an approach to obtain evolution equations for a data-driven realization of the original system – in effect, allowing prediction based on the informed interrogation of the agnostically organized observation database. We illustrate the approach on observations of (a) the normal form for the cusp singularity, (b) a cusp singularity for the nonisothermal CSTR, and (c) a random invertible transformation of the nonisothermal CSTR, showing that one can predict even when the observables are not "simply explainable" physical quantities. We discuss current limitations and possible extensions of the procedure.

534609 Integrated Data-Driven Monitoring & Explicit Fault-Aware Control of Chemical Processes: An Adaptive Approach for Smart Operation

Abstract:

The emergence of Industrie 4.0, and Smart Manufacturing initiatives along with the advances in computational power has significantly valorized process data for decision-making [1]. Today, industrial "Big Data" plays a significant role in process monitoring, and accordingly correction/prevention of process failures during operation. Process faults/failures arise when disturbances, that change process dynamics, occur in a chemical process and cannot be reversed by the existing controlling scheme. The common practice to handle this undesired condition in industry is to adopt different maintenance strategies including corrective actions. Corrective maintenance strategies aim rapid reaction and recovery of the process from faulty operation. One of the most common corrective actions is re-tuning of the existing controllers within the system while the process continues under faulty condition [2]. However, the major pitfall of this approach is the dead time spent under faulty operation during traditional controller re-tuning which may possibly harm the end-product quality, and/or cause process delays to achieve certain product quality.

In order to minimize the process delay between control action changes to correct any undesired changes during process operation, we propose an alternative approach that integrates a novel machine learning based process monitoring framework with multi-parametric Model Predictive Controller (mp-MPC) design. Central to the process monitoring framework are novel theoretical and algorithmic developments in SVM-based feature selection [3-5] which encapsulates highly nonlinear relationships between features, thus improves fault detection model accuracy. On the other hand, multi-parametric Model Predictive Controller (mp-MPC) design via the Parametric Optimization and Control (PAROC) framework [6] enables having the offline map of explicit fault-aware control strategies. Here, the control strategies are affine functions of the system states and the magnitude of the detected fault. By integrating a novel and accurate data-driven monitoring framework and explicit fault-aware control technology, we aim to produce an adaptive approach for smart operation where rapid switches between *a priori* mapped control action strategies are enabled by continuous monitoring information of the chemical processes. The results will be presented through the penicillin production benchmark model [7].

References

[1] Edgar, T. F., & Pistikopoulos, E. N. (2017). Smart manufacturing and energy systems. Computers & Chemical Engineering.

[2] Huang, B., & Shah, S. L. (2012). Performance assessment of control loops: theory and applications. Springer Science & Business Media.

[3] Onel, M., Kieslich, C. A., Guzman, Y. A., Floudas, C. A., & Pistikopoulos, E. N. (2018). Big Data Approach to Batch Process Monitoring: Simultaneous Fault Detection and Diagnosis Using Nonlinear Support Vector Machine-based Feature Selection. Computers & Chemical Engineering.

[4] Kieslich, C. A., Tamamis, P., Guzman, Y. A., Onel, M., & Floudas, C. A. (2016). Highly Accurate Structure-Based Prediction of HIV-1 Coreceptor Usage Suggests Intermolecular Interactions Driving Tropism. PloS one, 11(2), e0148974.

[5] Guzman, Y. A. (2016). Theoretical advances in robust optimization, feature selection, and biomarker discovery (Doctoral dissertation, Princeton University).

[6] Pistikopoulos, E. N., Diangelakis, N. A., Oberdieck, R., Papathanasiou, M. M., Nascu, I., & Sun, M. (2015). PAROC—An integrated framework and software platform for the optimisation and advanced model-based control of process systems. Chemical Engineering Science, 136, 115-138.

[7] Birol, G., Ündey, C., & Cinar, A. (2002). A modular simulation package for fed-batch fermentation: penicillin production. Computers & Chemical Engineering, 26(11), 1553-1565.

539355 Why Plant Operations Are Unstable after All the Design and How Data Science Can Help

Abstract:

Modern chemical plant operations involve a great deal of energy and material integration to be efficient. However, even though the plants were designed with sound principles, minor deviations or uncertainties in real-world operations can break the assumptions and lead to drastically undesirable operation behaviors, resulting in waste of energy, waste of raw material, loss of productivity, and compromised quality.

In this talk we demonstrate with many industrial examples how plant operations are oscillatory persistently without an effective way to identify the causes. We then present the use of data analytics and machine learning methods for troubleshooting and root cause diagnosis. Plant troubleshooting is different from process monitoring where a normal batch of data is assumed available or given. Plant troubleshooting deals with data that contain existing problems to be diagnosed. Dynamic component analysis and causality analysis of time series data are proposed for feature detection and root cause analysis. Data visualization is made possible by using the dynamic dimension reduction property of the methods. Industrial applications are used to demonstrate the effectiveness of the proposed methods.

543554 Gas Capture by Ionic liquids

Abstract:

With the rapid development of the global industrialization process, a large number of gases, such as CO_2, SO_2, NH_3 and H_2S, have been produced. Among of them, CO_2 and SO_2 have caused air pollution problems, resulting in a serious harm to the environment and human health. On the other hand, CO_2 and SO_2 are also important source for synthesizing chemicals. Recently, ionic liquids (ILs) have been regarded as one of the most promising alternative solvents for capturing CO_2 or SO_2 due to their special physicochemical properties, such as negligible vapor pressure and designable character. For example, it was found that N-butylpyridinium thiocyanate ([C_4Py][SCN]) can absorb 0.841 gSO_2/g IL at 293 K and 0.10 MPa[1].

In this work, the mechanism of CO_2/SO_2 absorption in different kinds of ILs were systemically studied by molecular simulations. The microscopic structure, interaction energy and dynamic properties of the mixtures of ILs and CO_2/SO_2 were investigated. The results of interaction energy and fractional free volume of ILs showed that unlike non-polar gas CO_2, the interaction energy of cation-anion and free volume do not play a dominant role in SO_2 absorption. The solvation of SO_2 in ILs is controlled by enthalpy. Both the cations and anions play an important role in SO_2 solubility. Based on the above studies, a serial of ILs were synthesized for capturing CO_2/SO_2. Moreover, the energy and economic analysis of whole process of CO_2 capture with ILs were simulated. It was found that the IL process saved about 15% regeneration heat energy consumption and 11.5% total capture cost compared to the MEA process.

Acknowledgment

This work was supported by the National Natural Scientific Fund of China 51674234.

Reference

1. J. Zeng, H. S. Gao, X. C. Zhang, H. F. Dong, X. P. Zhang, S. J. Zhang, *Chem. Eng. J*, 2014, 251, 248-256.

533203 Thermoeconomic Optimization of Reverse Brayton Cycle Based Cryocooler for HTS Power Transmission Cable

Abstract:

In the era of the global sustainable energy crisis, an efficient power transmission system is required to meet the increasing power demand. After the discovery of superconductivity in 1911, many materials with zero resistance at a temperature below 25 K, known as Low-Temperature Superconductor (LTS), were discovered and a possible alternative for efficient power transmission. However, the cost of these materials and cooling system to maintain them in the superconducting state was very high. Thus, it is practically not feasible to use such materials for the transmission system. In 1986, the discovery of High Temperature Superconductors (HTS), which can be maintained in the superconducting state using liquid nitrogen, made the practical and economically feasible for use HTS power cable. These cables have to be maintained in a temperature range of 65 to 80 K with a cooling load of 2 to 5 kW for 1 km long transmission cable. Reverse Brayton cycle based Cryocooler (RBC) is the most suitable cooling system because of a wide range of cooling capacity at different temperature level. However, the capital and operational costs of RBC is high compared to the cryogenic roadmap (Gouge, 2004). The current cost of refrigeration for HTS devices is five to ten times of cryogenic roadmap ($25 per watt of refrigeration at 65-80 K). Hence, a thermoeconomic optimisation of RBC has been carried out to enhance the performance of the system from the thermodynamic and economic point of view.

Fig. 1 Reverse Brayton Cycle based cryocooler for HTS power transmission cable.

In the present study, a reverse Brayton cryocooler consists of a compressor, main heat exchanger, turboexpander and load heat exchange as shown in Fig.1, has been investigated and optimised for 10 kW cooling load at 65 K. A commercial process simulator Aspen HYSYS V8.6 has been used for the study. Helium, more efficient than neon (Dhillon, 2017) has been used as working fluid for the cryocooler. The thermodynamic and economic models have been developed for analysing of the cryocooler. It is observed that the exergy efficiency of the system has maxima at 1.9 pressure ratio whereas cost per hour has manima at 2.3 pressure ratio as shown in Fig. 2. The inlet pressure and pressure ratio of the compressor have been varied to optimise the system process parameter. The effect of design parameters of equipment on the thermoeconomic performance of the system has been studied. The results obtained for RBC are presented in the present work and may be helpful for engineers and scientist to reduce the cost of RBC cryocooler without compromising the performance of the system.

Fig. 2 Variation of cost per hour and exergy efficiency of RBC with different pressure ratios

1. Gouge, M. Cryogenic cooling technology for HTS electric machinery, 2004. IEEE Power Engineering Society General Meeting, 1-3, DOI: 10.1109/PES.2004.1373239.

2. Dhillon, A. K., Dutta R. and Ghosh P. Exergetic analysis of reverse Brayton cryocooler for different cooling loads at 65 k for HTS cables, 2017. 14th Cryogenics 2017 IIR International Conference, DOI: 10.18462/iir.cryo.2017.027

530055 Technology Readiness Levels As a Framework for Techno-Economic Assessment

Abstract:

Techno-economic assessment (TEA) is an important basis for decisions about research, development and deployment (RD&D). Methods applied in TEA depend heavily on the availability of data which is closely linked to the development progress. Technology readiness analysis and especially its rating scheme, technology readiness levels (TRL), are popular concepts for rating the maturity of RD&D according to available data. TRLs can therefore serve as a framework that facilitates the choice of methods along RD&D by sorting them according to their required input data.

TRL rating is widely adopted in the evaluation of RD&D projects by government (*e.g.* policy makers) [1,2], industry and academia. However, as the TRL concept was developed for space exploration [3], a domain that integrates a variety of disciplines from mechanical, electrical and chemical engineering to aviation and medicine, TRL guidelines and criteria remain unspecific to technologies, which can lead to misinterpretation and overly optimistic evaluation. A variety of TRL definitions evolved over the years, for example for biomanufacturing [4] or for the defense sector [5], defining specific TRL criteria for their respective domains. The need for a specification of TRL analysis for the chemical industry was identified in a report of the European association of research and technology organizations (EARTO) [6].

In this contribution, we summarize existing TRL concepts and propose adaptions and extensions to criteria for TRL rating in the chemical industry: Firstly, TRLs are characterized by the description of the project phase, its tangible deliverables and other qualitative criteria. Secondly, for a more objective and comprehensible TRL, specific indicators representing known information of various aspects of chemical engineering such as reaction engineering or process development are proposed. In addition, quantitative indicators such as the production capacity are often used and therefore discussed for different types of processes in this contribution.

Furthermore, we outline how methods of (capital) cost estimation and profitability calculation can be sorted by TRL in order to give guidance for sound TEA.

We gratefully acknowledge the financial support by the European Institute of Innovation and Technology (EIT) and the German Federal Ministry of Education and Research (BMBF).

[1] European Commission. EN HORIZON 2020 WORK PROGRAMME 2016 â€" 2017, General Annexes (European Commission Decision C (2017) 2468 of 24 April 2017), Annex G, Technology Readiness Levels (TRL); 2017.

[2] BMBF, Definition des Technologischen Reifegrades [â€¦] accessed on 08.02.2017 from https://www.ptj.de/lw_resource/datapool/_items/item_5502/definition_des_technologischen_reifegrades.pdf

[3] S. R. Sadin, F. Povinelli, R. Rosen, contribution at 39thIAF Congress, Bangalore, India, Oct. 8-15, 1988

[4] NSF Engineering Research Center for Biorenewable Chemicals (CBiRC). Technology Readiness Levels https://www.cbirc.iastate.edu/industry/technology-readiness-levels/ (accessed Oct 31, 2017).

[5] U.S. DEPARTMENT OF DEFENCE. Technology Readiness Levels in the Department of Defense (DoD) https://www.army.mil/e2/c/downloads/404585.pdf (accessed Oct 31, 2017)

[6] EARTO, The TRL Scale as a Research & Innovation Policy Tool, EARTO Recommendations, 2014

536017 Techno-Economic Modeling and Optimization of Catalytic Reactive Distillation for Bio-Oil Upgradation

Abstract:

The most significant benefit of pyrolysis bio-fuel is its availability from virtually every waste biological materials, which provides a double edge in sustainable waste management in addition to meeting our energy requirements. However, unfavorable characteristics of crude bio-oils received from pyrolysis process compared to petroleum fuels means that it must undergo upgradation before using in combustion engines. Reactive distillation (RD) is emerging as a powerful technology for these chemical reactive separation processes. RD, which has tremendous advantages over traditional reaction and separation unit operations including handling of unfavorable chemical and phase interactions, enhancing conversion and selectivity, circumventing azeotropic mixtures and reducing capital and energy costs, has been exploited for the reactive separation of various equilibrium controlled chemical reaction systems. Pyrolysis bio-oil upgradation is usually carried out by transesterification and esterification reactions with an alcohol and RD is suited pefectly to these types of equilibrium controlled reactions.

Here, equilibrium RD process simulation and economic evaluation was carried out for the esterification of n-butanol with complex mixtures of several fatty acids, phenolics and water (represent pyrolysis crude bio-oils) using Aspen PLUS process simulator and Aspen Process Economic Analyzer, respectively at both atmospheric and high pressure (10 bar). Complete design and optimization was carried out using evolutionary techniques from simplest system (ordinary distillation) to the most rigorous one (complex RD) with introduction of progressive complexities. RD simulation was performed using UNIQUAC as base-property method in Aspen RADFRAC module by minimizing Gibbs free energy. Prior to distillation design, esterification reaction kinetics and binary/ternary interactions were analyzed using Aspen RGibbs reactor module and Aspen Property PLUS, respectively. Simple distillation column was designed and optimized to obtain the initial trial parameters to be used in RD simulation, which was initially optimized only for acetic acid-water (75% acetic acid-25% water) mixture before RD design for complex fatty acid mixtures based on techno-economic feasibility and optimization. The reaction of phenol with fatty acids was incorporated and phenol was found to react with formic acid entirely, which increases the reaction conversion 5-10% due to its favorable effects. The conversions for the esterification reactions were found to be 88-99% for various simulated bio-oils with phenol composition 5-20% for butanol:acid (B:A) ratio of 2.5, and water was almost completely separated from the ester products which should increase the fuel quality. Higher water percentage in bio-oil feed was found to reduce acid conversion and ester separation. A reflux ratio of 0.95, distillate to feed ratio 0.505 and total 18 stages (3 rectifying, 9 reactive and 4 stripping) with n-butanol and bio-oil feed stage of 4 and 13, respectively, was found to be optimum at 1 bar based on reaction conversion, ester separation, heat duties and capital and operating costs. Effects of B:A feeding ratio and water percent in bio-oil is also analyzed and B:A = 2.25 was found to be optimum for 25% water in bio-oil. These works and optimized parameters can serve as a design platform for pyrolysis bio-oil upgradation to transportation fuels.

Reference:

1. Khan, M.A., Adewuyi, Y.G. High pressure reactive distillation simulation and optimization for the esterification of pyrolysis bio-oil. Process Engineering Journal, 1 (2017) 73-85.
2. Khan, M.A., Adewuyi, Y.G. Simulation of reactive distillation for esterification of pyrolysis bio-oil. The Proceeding of AIChE Annual Meeting, 2013, San Francisco, California, USA.

536658 Superstructure Optimization of Membrane-Based Carbon Capture Systems

Abstract:

This work presents the optimal design, configuration and operating conditions of a membrane-based post-combustion CO_2 capture process using computationally identified mixed matrix membranes. A new mixed integer nonlinear model is presented, which includes discrete and continuous decisions to minimize the cost of electricity (COE). The simultaneous optimization of process configuration, membrane materials, and operating conditions ensures rigorous exploration of the entire feasible design space.

Membrane systems often consist of multiple stages of membranes and compressors with intercooling. Flue gas source plays a very important role, since the CO_2 concentration highly affects the final plant configuration. While a simple multi-stage configuration may be optimal for concentrated CO
integrated with the boiler is often more effective for coal fired power plants.[1] Given the importance of finding the right configuration, two approaches have been used to optimize these systems. The first approach uses simulation-based optimization that leverages commercial process simulators such as Aspen Custom Modeler or gPROMS to develop rigorous models[2,3,4]. Previous work has used the Framework for Optimization, Quantification of Uncertainty and Surrogates (FOQUS)[2] to optimize the design and operating conditions of specific process configurations. In these approaches, the number of potential configurations that can be analyzed is limited since each must be set up individually.[3,4]

The second approach focuses on superstructure-based mathematical optimization of multi-stage membrane systems, which often requires simplified models for the equipment.[5,6] The advantage of the second approach is that the configuration as well as the design and operating conditions may be optimized simultaneously, resulting in more cost-effective plants.

Our work leverages efforts at NETL for developing new materials for membrane-based post-combustion capture processes.[2,3,4,7] An advanced superstructure-based process optimization model is developed using rigorous mathematical models for compressors, pumps, mixers, and heat exchangers and reduced order models for liquefier columns and property calculations. The current work makes use of results from a recent study[4], which analyzed a large number of hypothetical mixed matrix membrane materials for carbon capture. This work addresses the aforementioned limitations by analyzing multiple potential advanced process configurations and the selection of the membrane material from a wide set of potential options while optimizing the plant operation.

[1] Merkel T. C., Lin H., Wei X., Baker R. (2010). Power plant post-combustion carbon dioxide capture: An opportunity for membranes. Journal of Membrane Science. 359, 126-139.

[2] Miller D. C., Agarwal D., Bhattacharyya D., Boverhof J., Cheah Y. W., Chen Y., Eslick J., Leek J., Ma J., Mahapatra P., Ng B., Sahinidis N. V., Tong C., Zitney S. E. (2016). Innovative computational tools and models for the design, optimization and control of carbon capture processes, Computer Aided Chemical Engineering, 38, 2391-2396.

[3] Morinelly J. E. M., David C., 2012, Post-Combustion Gas Permeation Carbon Capture System Models, AIChE Annual Meeting, Pittsburgh, PA, October 28-November 2, 2012.

[4] Budhathoki S., Ajayi O., Steckel J.A., Wilmer C.E. (2018). High-throughput computational prediction of the cost of carbon capture using mixed matrix membranes. In preparation.

[5] Hasan M. M., Baliban R. C., Elia J. A., Floudas C. A. (2012). Modeling, Simulation, and Optimization of Postcombustion CO_2 Capture for Variable Feed Concentration and Flow Rate. 1. Chemical Absorption and Membrane Processes. Industrial and Engineering Chemistry Research. 51, 15642-15664.

[6] Arias A. M., Mussati M. C., Mores P. L., Scenna N. J. (2016). Optimization of multi-stage membrane systems for CO_2 capture from flue gas. International Journal of Greenhouse Gas Control. 53, 371-390.

[7] Zamarripa M. A., Ajayi O., Matuszewski M., Miller D. C., 2017, Superstructure-based optimization of membrane-based carbon capture systems. AIChE Annual Meeting, Minneapolis, MN, October 28-November 2, 2017.

543630 Making Hydrogen from Water with a Protein Organized Electrode: Ultra-High Utilization of Noble Metal in Proton Exchange Membrane Electrolysis for Capital Cost Reduction

Abstract:

Renewable hydrogen could play a critical role in low-carbon energy systems and in fulfilling intermittent energy needs. Proton exchange membrane (PEM) electrolysis technology for hydrogen production is safe, simple, and highly efficient. PEM electrolysis has ability to be integrated in a variety of energy systems and infrastructures, such as transportation, renewable electricity, and biogas upgrading. It also has the ability to address seasonal-variation in long timescales. But, one of the major challenges remains the high cost of noble-metal electrodes in the membrane electrode assembly (MEA). Conventional catalysts like platinum and iridium are required to facilitate the hydrogen evolution reaction (HER) and the oxygen evolution reaction (OER) at electrodes for superior stability in long-term. The contribution of catalyst cost in the MEA becomes a more significant fraction of the capital cost when the production is scaled up.

To reduce the capital cost, we have developed protein-integrated electrodes which are designed to allow lower amounts of noble metal catalyst while maintaining high electrolysis performance. The purpose of our designed protein, elastin-like polypeptides (ELP), is to organize ionomer networks and create efficient pathways between the ionomer and catalyst sites, thereby significantly improving catalyst utilization. A preliminary cost analysis indicates that even at current prices for engineered proteins and using reasonable ratios of protein to catalyst particles (~6.5 metal:protein by mass), a 10X reduction in noble metal could be achieved for the same cost. Considering this technology is expected to lower loadings by more than 10X, and that the future cost to produce these proteins will decrease while the cost of noble metals will increase, the technology has great economic potential. For example, if the cost of protein manufacturing drops by a reasonable 25%, and a 20X reduction in metal loading is achieved, the total electrode cost would be 60% lower. ELP is suitable as electrode organizing molecules for its flexibility in design, modular format, and manufacturability. Our results using a quartz crystal microbalance and a transmission electron microscope show a robust binding of ELP to platinum catalyst and ionomers which affects the resulting ionomer-catalyst structures. This system provides an innovative, multidisciplinary solution for improving catalyst utilization in membrane electrode assemblies.

522132 Design Optimization and Economic Evaluation of LNG/NGL Integrated Processes Under Lean Feed Composition

Abstract:

The economic evaluation of the integrated liquefaction and natural gas liquids (NGL) recovery process for the lean feed is required to meet recent increasing demand for lean gas reservoirs. This paper performs economic analysis for the conventional co-production of liquefied natural gas (LNG) and NGL process compared to the proposed simplified integrated process with consideration of the LNG higher heating value (HHV) specification under the lean feed condition. The dual mixed refrigerant (DMR) cycle was selected for liquefaction part due to high efficiency reported by previous studies. The schemes were built by Aspen HYSYS commercial software and the object function was optimized with a genetic algorithm (GA). The results show that the simplified integrated process has economic advantages compared to the conventional integrated scheme under the lean feed composition, which shows different results in terms of previous studies conducted by normal or rich feed condition. Additionally, the capital cost can be reduced remarkably by simplifying NGL recovery section with almost no loss of liquefaction efficiency in the proposed simplified integrated process.

534601 A General Framework for the Evaluation of Direct Nonoxidative Methane Conversion Strategies

Abstract:

Natural gas is a versatile and clean chemical feedstock. The majority of natural gas is currently used for electricity generation, heating and cooking, and it is underutilized as a feedstock for the production of liquid transportation fuels and chemicals, including synthetic materials and plastics. The development of natural gas conversion technologies has recently attracted significant attention due to the increase of natural gas supply in the United States and low natural gas prices relative to crude oil. The conversion of natural gas into chemicals presents a promising mean of utilizing an abundant resource, while achieving energy security and mitigating pollutant emissions. Yet, the direct conversion of methane to olefins and higher hydrocarbons is still at the basic research level and it is unclear which, and to what extent, these technologies must be improved to develop a commercial process.

In this work, we study single-step natural gas conversion technologies that directly convert methane to olefins and higher hydrocarbons. Despite the relative simplicity of these technologies, the development of processes based on these approaches remains challenging. Accordingly, we utilize process synthesis and modeling to assess the economic feasibility of direct nonoxidative methane conversion strategies. We first develop an integrated strategy, which is a significant undertaking because any strategy employing a nonoxidative catalytic system must include compression and separation sections for product recovery and purification as well as recovery and recycle of unreacted methane, along with refrigeration, power generation, and utility sections. We then develop a flexible approach that allows for the systematic evaluation of various technology alternatives and for the identification of the key technology gaps that must be overcome. The key variables we considered include conversion temperature, methane one-pass conversion, hydrocarbon products selectivities, and natural gas feed rate. We also develop surrogate unit models and a database that allows to easily calculate key flows. The results of our analyses (see Figure 1) demonstrate that an economically feasible direct methane conversion process is contingent upon fundamental research advances in the area of catalytic conversion to increase methane conversion to hydrocarbon products (e.g., coke formation less than 20% and a minimum conversion to products of 25%). Upon this development, further efforts can be devoted to improve ethylene selectivity as well as reduce catalyst cost and overall capital costs.

Figure 1. Impact of Key Process Variables on Process Economics. Impact of methane one-pass conversion and (a) conversion temperature on NPV, (b) hydrocarbon products selectivity/coke selectivity on NPV, (c) ethylene selectivity in hydrocarbon products on NPV, and (d) ethylene selectivity in hydrocarbon products on MESP.

References

Huang, K.; Miller, J.B.; Huber, G.W.; Dumesic, J.A.; Maravelias, C.T.; 2018. A General Framework for Evaluation of Direct Nonoxidative Methane Conversion Strategies. *Joule* 2, 349-365.

536721 Upstream Process Optimization to Reduce the Cost of Air Pollution Control

Abstract:

Air Pollution Control (APC) systems are often thought of as solely end-of-pipe solutions, especially when new APC systems are installed on existing production processes. Frequently, design modifications to the upstream process can significantly reduce the capital and operating costs of the APC system required, but many companies neglect to include a comprehensive process design review as part of an APC projectâ€™s scope.

This case study will illustrate the project execution steps followed for a VOC abatement project on a complex batch production process involving multiple production lines where relatively low-cost modifications to the existing upstream process resulted in significant capital and operating cost savings associated with the new APC design. Several design alternatives were modeled to evaluate their impact on the process operations and APC system design.

The time taken to holistically consider the impact that changes to the existing process would have on the process design basis of the APC system ultimately lead to the implementation of specific measures that resulted in a pronounced reduction in the size of the APC system required and identified special features to include in the design to decidedly reduce the utility consumption of the system.

537434 Nonsmooth Simulation of Dry and Vaporless Tray Distillation Columns

Abstract:

Rigorous simulation of tray distillation columns requires the solution of the system of MESH (Mass balance, Equilibrium, Summation and Energy balance) equations, which assume liquid-vapor equilibrium exists at the conditions of each stage. However, as one or more input parameters such as the reflux ratio are varied, vapor and liquid streams can disappear within the column, which in turn allows for the remaining vapor/liquid phases to become superheated/subcooled. In that case, the "E" and "S" equations can no longer be satisfied simultaneously, leading to failure of any MESH-based simulation tool such as Aspen Plus' RadFrac model. This is especially concerning for flowsheet simulation with recycle streams and for optimization, since model convergence is required under multiple input conditions.

Alternative formulations can be proposed to create a single model that is valid both in the 2-phase and 1-phase regimes, with the correct equations being enforced automatically without prior knowledge of the effective regime for a given set of inputs. However, these strategies introduce nonsmooth function behavior and require special mathematical tools for model simulation.

To the best of our knowledge, dry/vaporless distillation simulation has only been directly addressed in the literature in [1], [2]. In [1], slack variables are introduced and the original equation-solving task is transformed into an optimization problem, solved with a penalty-based iterative linear programming strategy. In [2], the KKT conditions for this optimization problem, with the corresponding complementarity constraints, are solved iteratively using smoothing approximations. However, all of these approaches introduce artificial variables, increase problem complexity and require the solution of a sequence of equation systems or of an optimization problem. Additionally, simulation results in the 1-phase regime were presented only for a couple of isolated input conditions.

Instead, we propose the reformulation of the MESH model by introducing explicitly nonsmooth equations, which allows for simulation in the 2-phase and 1-phase regimes by direct equation-solving and doesn't introduce any extra variables. As similarly done in [3] for the simpler problem of single-stage flash calculations, automatic lexicographic directional differentiation [4] can be employed to obtain exact generalized derivative elements for the nonsmooth MESH equation system, which can in turn be solved with a suitable method such as the semismooth Newton or the linear programming Newton methods.

In this work, we will present simulation results for dry/vaporless tray distillation columns using novel nonsmooth modeling approaches. The results include detailed input parameter sensitivity analysis and a new type of bifurcation behavior which hasn't been reported in the distillation literature so far. Additionally, we will present nonsmooth strategies to avoid infeasible input specification automatically and create a distillation simulation model that is robust to failure associated with the disappearance of liquid and/or vapor phases.

References

[1] L. G. Bullard and L. T. Biegler. Iterated linear programming strategies for non-smooth simulation: a penalty based method for vapor-liquid equilibrium applications. Computers & Chemical Engineering, 7(1):95-109, 1993.

[2] V. Gopal and L. T. Biegler. Smoothing Methods for Complementarity Problems in Process Engineering. AIChE Journal, 45(7):1535-1547, 1999.

[3] H. A. J. Watson, M. Vikse, T. Gundersen, and P. I. Barton. Reliable flash calculations: Part 1. Nonsmooth inside-out algorithms. Industrial & Engineering Chemistry Research, 56(4):960-973, 2017.

[4] K. A. Khan and P. I. Barton. A vector forward mode of automatic differentiation for generalized derivative evaluation. Optimization Methods and Software, 30(6):1185-1212, 2015.

535821 Analysis of Chemical Process System Analysis with Entropy Generation

Abstract:

Systematic analyses of chemical process systems with the two steady-state balance equations from the First and Second Laws of Thermodynamics are used to provide insights about both energy efficiency and property models, using entropy generation as the indicator. Such evaluations of overall processes and of subsections provide reliable comparisons for energy requirements of process alternatives, indicate the impacts of potential changes, and check property model and calculational consistencies. Especially useful are the constraints that entropy generation must be positive for all sections, and that the sums of section values for heat rejection and entropy generation must equal those from an overall analysis. Examination has been made of three different post-combustion carbon-capture processes with the same flue-gas feed and products. Entropy generation allows comparison of their overall energy requirements, suggests outcomes for improving efficiency, indicates inefficiencies of process sectors, and reveals inconsistencies in absorption solvent property models for MEA and NH_3. Extensions to publications[1,2] will be shown.

References:

1. J.P. O'Connell, "Chemical process systems analysis using thermodynamic balance equations with entropy generation", Comp. & Chem. Eng. 2017, **107**, 3-15.
2. J.P. O'Connell, "Chemical process systems analysis using thermodynamic balance equations with entropy generation. Revaluation and extension", Comp. & Chem. Eng. 2018, **111**, 37-42.

532195 Effective Generalized Disjunctive Programming (GDP) Models for Modular Plant Design

Abstract:

Evolving demands and increasing competition in the chemical processing industries drive a need for the most effective process design, at the right time. Engineers must identify the optimal configurations among many flowsheet alternatives in order to arrive at the desired design. One of the major approaches to synthesis is superstructure optimization, which involves postulating a superstructure representation of the alternatives, formulating a mathematical program to capture the decision logic, and applying a solution algorithm to determine the optimal configuration and operating conditions (Chen and Grossmann, 2017). Recently, there has been increased interest in modular designs due to their potential for improving safety, reliability, flexibility, and time-to-market (Baldea et al., 2017). While modularity on the product side has been addressed in depth (Rogers and Bottaci, 1997), new model extensions and methodologies for superstructure optimization must be developed to address modular process design challenges.

A hallmark of modular design is an emphasis shift from economies of scale towards economies of mass production, also referred to as "numbering-up" (Lier et al., 2015). This requires standardization of modules, a concept that has been explored in terms of aggregation of process units (Kampczyk et al., 2004) and cluster analysis to identify heat exchanger modules (Eilermann et al., 2017). This standardization implies that continuous design variables in the optimization formulation will now take values from some discrete set. While the introduction of discrete variables adds combinatorial complexity, it can also induce linearity in synthesis models, making them easier to solve (Voudouris and Grossmann, 1992).

In this work, motivated by Stephanopoulos and Westerberg (1980), we present a modification of the stage-wise heat exchanger network synthesis model from Yee and Grossmann (1990) that allows for incorporation of modular exchangers, selected from among a set of standard sizes. The modular design problem is formulated as a Generalized Disjunctive Programming (GDP) model Using this demonstration model, we show that discretization of the exchanger area allows for linear reformulations of the basic heat exchanger design equation, which is usually bilinear. Reformulation strengthens linear relaxations of the overall design problem, therefore aiding in the convergence of optimization solvers. Therefore, the consideration of modular process units and resulting discretization in design variables can even simplify solution of synthesis problems. We express the GDP model using the Pyomo optimization modeling library (Hart et al., 2017) in Python. Pyomo enables programmatic access to modify and manipulate modeling objects such as variables and constraints, allowing for automatic model interrogation and reformulation (Chen et al., 2018). Using Pyomo, we develop a plugin to automatically detect induced linearity from discrete modular design choices and reformulate the model to exploit the special structure.

Another central point of modular design is improved time-to-market due to expedited design, construction, commissioning, and startup of modular facilities. Several authors have examined the net present value (NPV) economic impact of switching to a modular design, including analysis on the sensitivity to the number of modules employed (Brodhagen et al., 2012; Lier and Grünewald, 2011; Mothes, 2015). We extend these studies using a model adapted from (Menezes et al., 2015) to examine the flowsheet design and construction scheduling of a process as a simultaneous GDP optimization problem, and demonstrate its value in balancing the trade-off between investment costs, operating costs, and time-to-market.

In this work, we demonstrate the value of effective GDP models both in capturing the logic and addressing the mathematics of modular chemical plant design. Through various case studies, we show how GDP enables investigation of key tenants of modular design.

References

Baldea, M., Edgar, T.F., Stanley, B.L., Kiss, A.A., 2017. Modular manufacturing processes: Status, challenges, and opportunities. *AIChE J.* 63, 4262–4272.

Brodhagen, A., Grünewald, M., Kleiner, M., Lier, S., 2012. Erhöhung der Wirtschaftlichkeit durch beschleunigte Produkt- und Prozessentwicklung mit Hilfe modularer und skalierbarer Apparate. *Chemie Ing. Tech.* 84, 624–632.

Chen, Q., Grossmann, I.E., 2017. Recent Developments and Challenges in Optimization-Based Process Synthesis. *Annu. Rev. Chem. Biomol. Eng.* 8, 249–283.

Chen, Q., Johnson, E.S., Siirola, J.D., Grossmann, I.E., 2018. Pyomo.GDP: Disjunctive Models in Python, in: Eden, M.R., Ierapetritou, M.G., Towler, G.P. (Eds.), *Proceedings of the 13th International Symposium on Process Systems Engineering.* Elsevier B.V., San Diego.

Eilermann, M., Post, C., Schwarz, D., Leufke, S., Schembecker, G., Bramsiepe, C., 2017. Generation of an equipment module database for heat exchangers by cluster analysis of industrial applications. *Chem. Eng. Sci.* 167, 278–287.

Hart, W.E., Laird, C.D., Watson, J.-P., Woodruff, D.L., Hackebeil, G.A., Nicholson, B.L., Siirola, J.D., 2017. *Pyomo — Optimization Modeling in Python*, 2nd ed, Springer Optimization and Its Applications. Springer International Publishing, Cham.

Kampczyk, B., Hicking, B., Schmidt-Traub, H., 2004. More Efficient Plant Design by Modularization? *Chem. Eng. Technol.* 27, 481–484.

Lier, S., Grünewald, M., 2011. Net Present Value Analysis of Modular Chemical Production Plants. *Chem. Eng. Technol.* 34, 809–816.

Lier, S., Wörsdörfer, D., Grünewald, M., 2015. Wandlungsfähige Produktionskonzepte: Flexibel, Mobil, Dezentral, Modular, Beschleunigt. *Chemie Ing. Tech.* 87, 1147–1158.

Menezes, B.C., Kelly, J.D., Grossmann, I.E., Vazacopoulos, A., 2015. Generalized capital investment planning of oil-refineries using MILP and sequence-dependent setups. *Comput. Chem. Eng.* 80, 140–154.

Mothes, H., 2015. No-Regret-Lösungen - Modulare Produktionskonzepte für komplexe, unsichere Zeiten. *Chemie Ing. Tech.* 87, 1159–1172.

Rogers, G.G., Bottaci, L., 1997. Modular production systems: a new manufacturing paradigm. *J. Intell. Manuf.* 8, 147–156.

Stephanopoulos, G., Westerberg, A.W., 1980. Modular Design of Heat Exchanger Networks. *Chem. Eng. Commun.* 4, 119–126.

Voudouris, V.T., Grossmann, I.E., 1992. Mixed-integer Linear Programming Reformulations for Batch Process Design with Discrete Equipment Sizes. *Ind. Eng. Chem. Res.* 31, 1315–1325.

Yee, T.F., Grossmann, I.E., 1990. Simultaneous optimization models for heat integration—II. Heat exchanger network synthesis. *Comput. Chem. Eng.* 14, 1165–1184.

538285 ProCAFD: A Computer-Aided Tool for Sustainable Process Synthesis, Design, Analysis and Improvement

Abstract:

The chemical process industry is always looking for innovative designs, which are both efficient and sustainable. To generate a truly optimal process flowsheet a systematic method to identify the types of tasks-operations that need to be performed, the corresponding design of the operation-equipment, their configuration, mass-energy flows, environmental impacts, etc., need to be employed. That is, the synthesis problem (processing route generation), the design problem (design of the individual operations), the analysis problem (simulation and evaluation of alternatives) and the innovation-intensification problem (more sustainable designs) need to be considered within the same method. In this case, development and use of a computer aided tool that combines the best features of knowledge-based and optimization-based approaches into one method is an attractive option.

This work focuses on the development of a computer aided software tool, capable of enumerating all the feasible alternatives within the entire search space defined by a process synthesis problem to determine the most sustainable process using a hybrid approach. In the synthesis problem a superstructure of all feasible alternatives is created using a technique similar to computer aided molecular design (CAMD) for molecular design. According to this technique [1], chemical process flowsheets are synthesized in the same way as atoms or groups of atoms are combined to form molecules in CAMD techniques. The building blocks in the process flowsheet synthesis problem are called process-groups, which represent a single or set of unit operations that are selected by employing a thermodynamic insights-based method. These building blocks are then combined using connectivity rules to generate a superstructure of feasible process flowsheet alternatives, from which the optimal process configuration is determined in two ways: a) Group contribution based flowsheet property models that enumerates and evaluates all possible alternatives (generate and test approach); and b) Superstructure based optimization (direct solution approach) In the first approach, flowsheets are generated and tested by estimating performance indicators of the generated flowsheets through the contributions of the associated process-groups and flowsheet property models (energy consumption, carbon foot-print, , product recovery, product purity etc). In this way, the entire list of feasible process flowsheets are quickly generated, screened and selected. In the second approach, a superstructure based optimization problem is solved by first formulating the synthesis problem as a mixed integer linear or nonlinear programming problem and solved through a suitable solver (GAMS) [2, 3]. Here, the software interface, Super-O helps to set up the optimization problem by creating an input file for GAMS, which is then solved to find the optimal processing route. In design and analysis problems, additional design details not covered by the synthesis problem are added to the process tasks-operations so that simulations with rigorous process models can be performed. The simulation results serve as input for equipment sizing and utilities requirement calculations, which in turn serve as input for process analysis (economic, sustainability factors, safety, LCA factors, etc.). The analysis problem also includes hazards analysis through PHASuite [4], which involves identification, assessment and mitigation of process hazards. This comprehensive process analysis identifies process hotspots, where the corresponding task-operation within an analyzed process could be improved, that is, helps to define targets for improvement. Through the solution of the innovation-intensification problem, any processing route that has been analyzed and targets for improvement defined, is furthered improved by applying various synthesis-intensification methods [5] to determine process alternatives that match the targets for improvement. Any match of the design targets therefore corresponds to a more sustainable process alternative.

The developed methods and associated computer-aided tools have been integrated into ProCAFD â€“ a tool for process synthesis, design, analysis and intensification. ProCAFD has been tested on various case studies involving new as well as retrofit chemical and biochemical process synthesis-design-intensification. The paper will present results from a selection of the developed case studies.

References:

[1] Tula, A.K., Eden, M. R., and Gani, R. (2015). Process synthesis, design and analysis using a process-group contribution method. Computers and Chemical Engineering, 81, 245-259.

[2] Quaglia, A., Sarup, B., Sin, G., and Gani, R. (2012). Integrated Business and Engineering Framework for Synthesis and Design of Enterprise-Wide Processing Networks. Computers and Chemical Engineering, 38, 213-223.

[3] Bertran, M-O., Frauzem, R., Sanchez-Arcilla, A. S., Zhang, L., Woodley, J., and Gani, R. (2017). A generic methodology for processing route synthesis and design based on superstructure optimization. Computers and Chemical Engineering, 106, 892-910

[4] Zhao, C., Venkatasubramanian, V. (2005). Learning in intelligent systems for process safety analysis. Computer Aided Chemical Engineering, 20, 1123-1128.

[5] Babi, D. K., Holtbruegge, J., Lutze, P., Gorak, A., Woodley, J. M., Gani, R. (2015), Sustainable process synthesisâ €“intensification, Computers and Chemical Engineering, 81, 218-244

537856 Systematic Process Design and Innovation Using Building Blocks

Abstract:

Discovery of novel chemical processes often requires creativity and "out-of-the-box" thinking as innovative solutions may lie outside the boundaries of the prevailing design paradigm. An example is process intensification that aims to outperform the existing processes with drastic improvement in performance metrics such as size, energy efficiency, cost, operability and safety [1-2]. However, identification of such intensified solutions at the conceptual design stage is a challenging task as there can be myriad of candidate process configurations. While optimization-based process synthesis approaches provide methodical tools for process design, they request pre-postulated superstructures with fixed connectivity and equipment types. This limits their scope for the discovery of unconventional design solutions.

To this end, building block-based process synthesis [3-6] offers significant advantages over classic superstructure-based process synthesis, as it relies on fundamental phenomena and tasks that constitute most chemical processes. In this work, first we will describe how a single block-based representation of these phenomena and tasks in a two-dimensional grid can be formulated, modeled using an MINLP and applied to solve wide ranges of process synthesis, integration and intensification problems. We will also discuss how this new representation can be used to generate novel process configurations without postulating the unit operations *a priori*. Each block in the grid can be assigned with a reaction operation and separation operations are represented via two neighboring blocks separated by a separation boundary. When linear split models are translated into block-variable (e.g. composition) dependent non-linear split models, rigorous separation models (e.g. tray-by-tray models) can be formulated within the same functional form. This enables systematic identification of the synergy between different operations (e.g. reaction and separation). We will demonstrate the benefits of the proposed approach via several problems and show that systematic process innovation and novel process discovery can be made possible via utilization of the building block-based superstructure.

[1] Stankiewicz, A. I., Moulijn, J. A. (2000). Process Intensification: Transforming Chemical Engineering. *Chemical Engineering Progress*, 1, 22–34.

[2] Lutze, P., Gani, R., Woodley, J. M. (2010). Process intensification: a perspective on process synthesis. *Chemical Engineering Processing: Process Intensification*, 49 (6), 547–558.

[3] Demirel, S. E., Li, J., and Hasan, M. M. F., (2017). Systematic Process Intensification using Building Blocks, *Computers and Chemical Engineering*, 105, 2-38.

[4] Li, J.; Demirel, S.E.; Hasan, M.M.F. Process Synthesis using Block Superstructure with Automated Flowsheet Generation and Optimization. *AIChE Journal*, 2018, under review.

[5] Li J., Demirel S.E., Hasan M.M.F. Process Integration using Block Superstructure. *Industrial & Engineering Chemistry Research*, 2018, 57: 4377–4398.

[6] Li J., Demirel S.E., Hasan M.M.F. Fuel Gas Network Synthesis Using Block Superstructure. *Processes*. 2018, 6(3): 23.

530233 Surrogate-Based Optimization for Biocatalytic Manufacturing of Diabetes Drug

Abstract:

Along with the rapid economic development and improved quality of life, the world diabetes prevalence has climbed to over 8% in recent years [1], and even more than 10% of population suffers from diabetes in countries such as the United States and so on. Type 2 diabetes accounts for the majority of patients [2]. Dipeptidyl peptidase IV (DPP-4) inhibitors have shown potent effect for treating type 2 diabetes, including sitagliptin. The mechanism of action of DPP-4 inhibitors is inhibiting the DPP-4 enzyme to reduce the rate of degradation of glucagon-like peptide-1, a hormone whose effect is increasing insulin and decreasing glucagon secretion [3]. Sitagliptin has advantages over other active pharmaceutical ingredients such as: instead of intravenous injection, it is orally active and patients do not suffer from hypoglycemia, etc. [4] In addition, sitagliptin has the highest market share of 14.8% in diabetes market in 2016, i.e., widely used by a majority of patients [5].

This paper demonstrates two case studies, maximizing product concentration and productivity, respectively, to show the application of surrogate-based optimization for sitagliptin manufacturing as this method has not been fully explored in the pharmaceutical industry. The advantages of this method include its open-source nature and being capable of approximating the global optimum of high-dimension objective function in a short time. A model of continuous manufacturing using plug flow reactors for the biocatalytic synthesis was proposed, based on the synthesis of sitagliptin using novel transaminase developed by Savile et al., that claims 10-13% higher yield, 53% higher productivity and 19% reduction in total waste [6]. The reaction of the biocatalytic transformation is described as below:

Prositagliptin ketone + *i*-$PrNH_2$ -> Sitagliptin + Acetone

The solutions of optimization were determined through pySOT package in Python 3.6 [7]. The sensitivity analysis was also conducted to identify critical process parameters (CPPs). The reaction kinetics constant was identified by assuming a first-order reaction due to the excess of *i*-$PrNH_2$. Also according to the study of Savile et al., we assumed feeding time to be proportional to the concentration of impurity formed, and the concentration of enzymes proportional to the overall catalyzing efficiency [6]. The results show that in case 1 of maximizing product concentration, the flow is extremely low and the influence of the concentration of enzymes is insignificant; while in case 2 of maximizing productivity, the flow rate is relatively high but still below the upper bound due to the formation of impurity. The results of sensitivity analysis show that in both case 1 and 2, the most influential CPP is the concentration of reactant. Most importantly, this method shows its ability to obtain the global optimum within a short computational time, and its application in pharmaceutical industry.

The present study also proposes several feasible downstream pathways including 1) the denaturation of enzymes, filtration, extraction and chromatography in series [6]; 2) differential extraction with membrane and chromatography in series [6]; and 3) ion-exchange adsorption of product, hydrophobic adsorption of enzyme, wash and chromatography in series [8][9]. However, it is similar to purify the transaminases as biocatalysts as to purify therapeutic proteins, which includes filtration, cell lysis, chromatography, etc. This may disobey the spirit of green chemistry. Therefore, we propose a manufacturing process in which the immobilized cell is adopted as biocatalyst instead of the enzyme, followed by differential extraction and chromatography steps. The techno-economic analysis is conducted to support such design decision.

In the future work, optimization could be conducted based on the whole process, from reaction to crystallization. Most importantly, not only the process optimization, but an economic optimization can also be performed with more information of the costs and revenues to obtain the maximized profits. Furthermore, life cycle assessment can be conducted to evaluate the environmental impact of each manufacturing manner. In the end, the optimal solution can be determined based on the trade-off between the economic and environmental performance.

Reference:

[1] World Health Organization, "Global Report on Diabetes," p. 27, 2016. Retrieved from < http://apps.who.int/iris/bitstream/10665/204871/1/9789241565257_eng. pdf >

[2] World Bank database, Diabetes prevalence refers to the percentage of people ages 20-79 who have type 1 or type 2 diabetes. Retrieved from < https://data.worldbank.org /indicator/SH.STA.DIAB.ZS?view=chart >

[3] N. A. Thornberry and B. Gallwitz, "Mechanism of action of inhibitors of dipeptidyl-peptidase-4 (DPP-4)," *Best Pract. Res. Clin. Endocrinol. Metab.*, vol. 23, no. 4, pp. 479–486, Aug. 2009.

[4] E. J. Verspohl, "Novel therapeutics for type 2 diabetes: Incretin hormone mimetics (glucagon-like peptide-1 receptor agonists) and dipeptidyl peptidase-4 inhibitors," *Pharmacol. Ther.*, vol. 124, no. 1, pp. 113–138, Oct. 2009.

[5] Evaluate Pharma, "World Preview 2017, Outlook to 2022," *Evaluate Pharma*, no. 10th Edition. pp. 1–41, 2017.

[6] C. K. Savile *et al.*, "Biocatalytic Asymmetric Synthesis of Chiral Amines from Ketones Applied to Sitagliptin Manufacture," *Science (80-.).*, vol. 329, no. 5989, pp. 305–309, Jul. 2010.

[7] http://pysot.readthedocs.io/en/latest/index.html

[8] M. D. Truppo, J. D. Rozzell, and N. J. Turner, "Efficient Production of Enantiomerically Pure Chiral Amine at Conc 50 g/L Using Transaminase," *Org. Process Res. Dev.*, vol. 14, no. 1, pp. 234–237, 2010.

[9] M. D. Truppo, H. Strotman, and G. Hughes, "Development of an Immobilized Transaminase Capable of Operating in Organic Solvent," *ChemCatChem*, vol. 4, no. 8, pp. 1071–1074, 2012.

COMPARISON OF FIRST PRINCIPLES AND PARAMETRIC MODELS FOR THE DESIGN AND OPERATION OF A SPRAY DRYER FOR WHEY PROTEIN PRODUCTION

ANIBAL JOSÉ BARRIOS QUANT* 1, JOHN TELOTTE 2, JOSE ANDRES PEREZ MENDOZA 1.

1, Chemical engineering department, Universidad del Atlántico, Km 7, Puerto Colombia, Atlántico, Colombia

2, Department of Chemical and Biomedical Engineering, FAMU-FSU College of Engineering, Florida State University, 2525 Pottsdamer St, Tallahassee, FL, USA

*Corresponding author:abarriosquant@fsu.edu; Tel: +57 3053058185

Abstract

In this work we analyze four models for spray dryer design for a whey protein process. The models include both first principles approaches and empirical correlations. For a large set of given set of design specifications (inlet and outlet moisture contents, inlet air conditions and residence time) the physical size of the spray drying system was determined from all correlations and then the designs were compared to both industry heuristics for size ratios and actual dryer performance (where possible) to assess the validity of the models. Specific consideration was given to the correlation between the systems that met design heuristics and those that produced usable product. An economic analysis was developed to compare the prices that could be obtained designing a spray dryer based on the Masters restrictions. Finally a simulation on ANSYS CFD represented the flowfield of the spray dryer was developed.

KEYWORDS: Spray Dryer, design, sizing, dimensional analysis, critical variables, whey protein.

INTRODUCTION

Whey proteins with improved flavour and increased functionality are obtained with new technologies, as spray drying. While choosing a particular whey product, it is essential to match its function to the characteristics we want to achieve. For example, high protein concentrates or isolates are usually produced in spray dryers due with this process is possible to guarantee high quality powder [1,2].

Spray-drying is a rapid, continuous, cost-effective, reproducible and scalable process for the production of dry powders from a fluid material by atomization through an atomizer into a hot drying gas medium, usually air [3]. The design of spray dryers has remained largely empirical; it is still mainly based on the extensive experience and the vast body of operating data which the manufacturers have acquired over the years. To design spray dryers usually designers develop assumptions based on variables that allow an easier design process for spray dryers [4]. The focus of this study is to compare four different design models and their responses over common variables adjusted to evaluate the sizing of a spray dryer for whey protein production.

COMPARATIVE MODELS

The numerical equations who represented a phenomenon is better called design model, and is commonly used as a representation of spray dryers as well as digital simulations of spray drying. In this study the method was applicable to represent the particle movement inside the spray chamber and the dimensional structure of the equipment.

The dimensional structure of the spray dryer was explained by the models of Mujumdar [7], Suryanarayana [8], while the behavior of the drops inside the spray and the atomization process were showed by the models of Gluckert [9] and Masters [10].

The comparative analysis of these models was focused on study each models and evaluate their responses over the same operational conditions. The equations below show the structure of the models represented only by the diameter, length and volume. The complete structure of each model can be found in the supplementary information.

Mujumdar Model [7]

$$D = \sqrt{4\,G / 3600\,\pi\, j\, u}$$

$$L = 4V / \pi D^2 \qquad (1)$$

$$V = t\,F_2 / H\ \rho_{whey}$$

Suryanarayana Model [8]

$$D_{Spray} = \sqrt{4\,A_c\ / \pi}$$
$$L = 0.6\,D_{spray} \qquad (2)$$
$$V = G_s\,V_{avg}\ t_d$$

Gluckert Model [9]

$$D = 2\sqrt{(t_T\,2.24\,V_{res}\,(b\,(D_r\ / 2\,)^{12} + (D_r / 4)}$$
$$L = 1.5\,D_{spray} \qquad (3)$$
$$V_t = G_s\,V_{avg}\ t_d$$

Masters Model [10]

$$Dvs = 14000\ Q\ 10.24 / n\ h\ 0.12\ N\ d\ 0.6$$
$$L = 1.5\,D_{spray} \qquad (4)$$
$$V = 1/4\ \pi\,D_i\ 2\ h_{spray}$$

The comparison of the four models was evaluated by the variation of key variables of the model in range adjusted by the study based on avoidance of whey protein denaturation. The key variables were selected according to their influence over the four models. The values assigned to the key variables were reported in the literature and are shown in Table 1.

Table 1. Variables and values

Independent Variables (Factors)	Values	
X1= Humidity content (kg/kg db)	0.60	0.90
X2= Inlet air temperature (°C)	85	180
X3= Outlet air temperature (°C)	80	120
X4= Outlet whey humidity (kg/ kg dry)	0.05	0.10
X5= Air flow (m3/h)	123	288

Key variables

The key variables were organized randomly and set to obtain thirty- two responses. The analysis was evaluated by the effect of the variables on the responses obtained in the models and the relation between the size and the different values that were applied to the models.

The *treatments* were organized by considering the main factor of protein denaturation and this terminology was used to explain the thirty-two possible variables used on this study to determine the value of the responses (spray diameter, length and volume) in each particular model.

Economic analysis

The economic analysis was carried out in this study to determine the cost of building and operating a spray dryer over a specific range of operational conditions. Then, a comparative analysis was made to find out which model was the most economically efficient.

The cost estimation for the spray dryer was calculated following the economic considerations of Mujumdar [7]. This determined the total cost of a dryer taking into consideration the operation costs and the equipment costs, based on the quantity on the material and the dimensions of the diameter and the length.

The dryer unit was determined by the cost of wet product ($ / kg wet base) with Equation 5.

$$\begin{aligned} Cp &= CTtop\,Fs\,(1 + Xs) \\ Ct &= eCeq + Cop \\ e &= i\,(1+i)N\,/\,(1+i)N - 1 \\ Ceq &= D\,An'D + \alpha nZZn'z \qquad (5) \\ Cop &= hPCetop + Z\,CzTop \\ hp &= qN'D\,(M + W') \\ W &= M\ (\,2D^2/\,4) + \pi DL)\,dx \end{aligned}$$

Size analysis and CFD simulation

A CFD simulation was carried out to investigate the fundamental flow behaviour in the spray dryers and their modeling. The simulation is performed using a mixed Eulerian/Lagrangian approach, in which the single single phase Reynolds Averaged Navier Stokes (RANS) equations are solved to determine the flowfield. This work used the responses given by the models to employ them as boundary conditions to simulate the spray dimensions and the CFD. The results of this simulations are shown in Figure 2.

DISCUSSION AND RESULTS

Depending on the modeling and boundary conditions of each author, variations in the dimensional parameters (diameter, length and volume) were obtained in the design models. When the models of Mujumdar [7], Suryanarayana [8], Gluckert [9] and Masters [10] were analyzed using the residence time and air velocity levels (high and low) as variables, only four responses of the possible thirty-two were obtained for the models of Mujumdar, Suryanarayana, and Gluckert [7-9] and one of thirty-two for Masters [10].

Model responses

The responses of the models were calculated by the combination of the five variables selected for this study in the two chosen levels (Figure 1). This procedure showed thirty-two responses for each dimensional variable (diameter, length and volume) for every model. The average responses of each model were compared to describe the behavior with the same operational conditions.

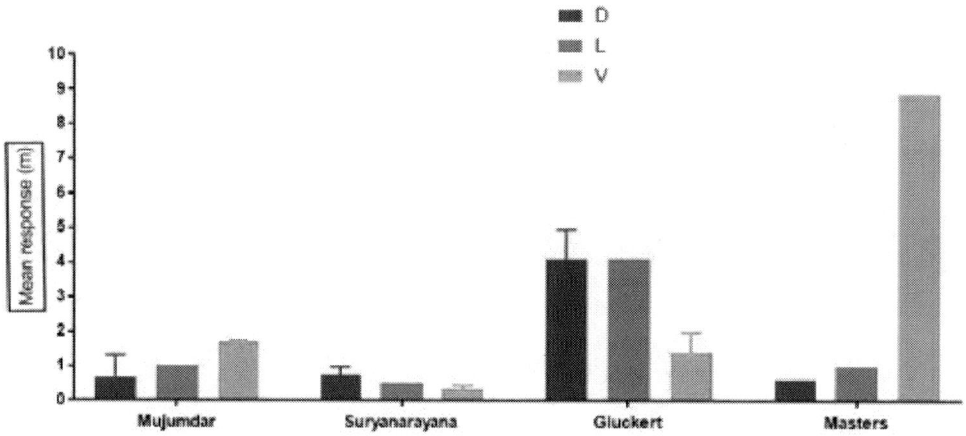

Figure 1. Model responses.
* "T" on top of the bars means average value conquered

The models of Mujumdar, Suryanarayana, Masters and Gluckert did not consider the denaturation conditions of whey protein for the designing of a spray dryer.

Models-comparison

The models comparison was developed with the analysis of the four models on two categories: sizing and modeling and cost estimation, this conditions were adjusted to analyze the scalability of the spray dryer of each model in terms of size of the chamber and cost estimation.

The sizing and modeling of the spray dryers were carried out in AutoCAD Inventor 2019, in this program was simulated the structure of the spray dryers according to the responses obtained on the variables analysis, the results demonstrated that exist appreciable differences on the sizes of the sprays dryers by each model.

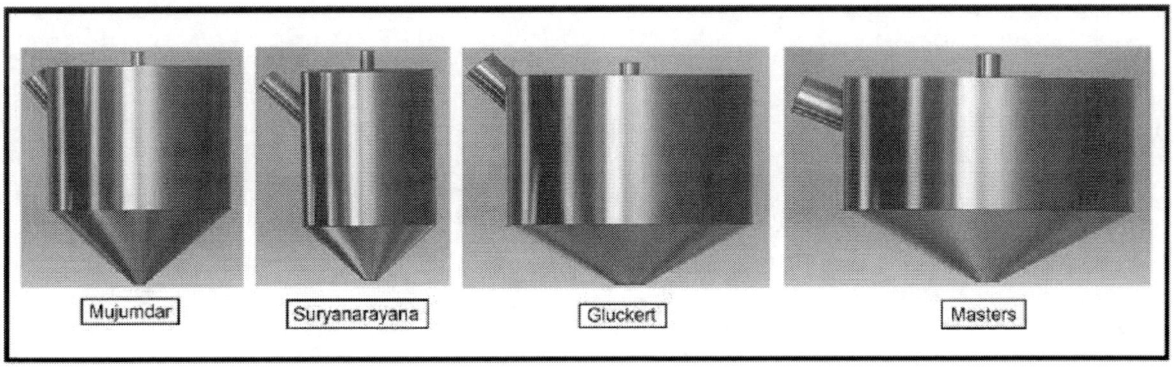

Figure 2. Size analysis

The cost estimation responses were calculated over a set of initial conditions for the 4 models expressed on the Table 2.

The model prices of the spray dryers varied from $6,417 USD to $14,803,215 USD. The lowest-priced option was the Suryanarayana [8] model with dimensional responses (diameter, length and volume) of 0.454 m, 0.313 m and 0.193 m3 respectively, and the highest-priced model was Gluckert [9] with dimensional responses (diameter, length and volume) of 5.301 m, 7.952 m and 0.912 m3 respectively.

Table 2. Cost estimation

Model	**Responses (U.S. Dollars)**
Mujumdar	$18,264 - $205,755
Suryanarayana	$6,417 - $91,540
Gluckert	$2,890,141 - $14,803,215
Masters	$27,711

CFD results

The results showed on the Figure 3 represent the flow patterns inside the spray chamber, the conditions proposed to carry out the simulation were the dimensions obtained on the responses of this work and the values of the temperature and the whey flow were adjusted according the conditions of non-denaturation presented on Table 1.

In this simulation (Figure 3) was possible to observe the changes that suffered the whey liquid particles inside the spray dryer, the particles traveled since the atomizer until the outlet hole, also is appreciable to see the the mass and energy transfer, as well as the the temperature profile in the middle of the spray dryer, represented by the mixture of colors in the center of the Figure 3.

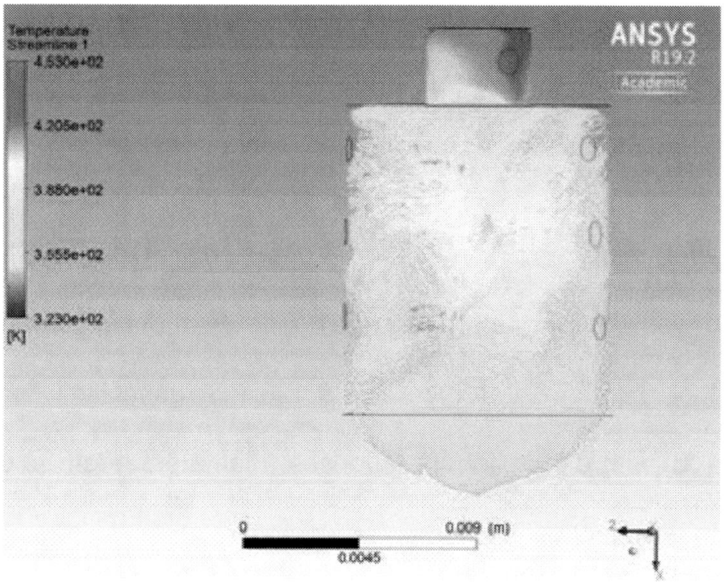

Figure 3. ANSYS CFD simulation

CONCLUSIONS

The models analyzed in this study presented different behaviors when evaluated over the same operational conditions. This is because the creator of each model used different mathematical correlations to describe the same phenomenon. This resulted in different responses in the spray dryer dimensional variables, (diameter, length and volume), that were analyzed by the simulations on ANSYS CFD who showed the behavior of the drop responses inside the spray chamber. Also in this work was studied the economic conditions for design a spray dryer to offer a guidance that allowed to analyze the cost estimation to develop the equipment. The analysis was carried with the key variables and responses adjusted in this research, obtaining a guidance to select a specific model design according the budget.

Acknowledgments

The authors gratefully acknowledge the Universidad del Atlántico in Barranquilla, Colombia for all the support and knowledge delivered. Thank you to Savannah Cayson for editing the writing of this article.

REFERENCES

[1] C. V. Morr, E. Y. W. Ha, Whey protein concentrates and isolates: Processing and functional properties. *Critical Reviews in Food Science and Nutrition.* **29** (2009) 431-476

[2] P. Schuck, R. Jeantet, B. Bhandari, X. D. Chen, Í. T Perrone, A. Fernandes de Carvalho, Recent advances in spray drying relevant to the dairy industry: A comprehensive critical review. *Drying Technol.* **34** (2016) 1773-1790

[3] M. Mezhericher, A. Levy, I. Borde, Spray drying modelling based on advanced droplet drying kinetics. *Chemical Engineering and Processing: Process Intensification.* **49** (2010) 1205-1213

[4] M. Mezhericher, A. Levy, I. Borde, Multi-Scale Multiphase Modeling of Transport Phenomena in Spray-Drying Processes. *Drying Technol.* **33** (2015) 2-23

[5] V. S. Birchal, L. Huang, A. S. Mujumdar, M. S. Passos, Spray Dryers: Modeling and Simulation. *Drying Technol.* **3** (2006) 275-279

[6] H. M. Lisboa, M. E. Duarte, M. E. Cavalcanti-Mata, Modeling of food drying processes in industrial spray dryers. *Food and Bioprod process.* **107** (2018) 49-60

[7] A.S. Mujumdar, *Handbook of industrial Drying.* (2014), 182-200

[8] A. Suryanarayana, *Mass Transfer Operations.* (2002), 571- 578

[9] F.A Gluckert, A theorical correlation of spray-dryer performance. *American Institute of Chemical Engineers*. **22** (1962) 713-724

[10] K. Masters, *Spray Drying Handbook*. (1979), 115-120

[11] M. Tunick, A. Thomas-Gahring, D. Van Hekken, S. Iandola, M. Singh, P. Qi, D. Ukuku, S. Mukhopadhyay, C. Onwulata, P. Tomasula, Physical and chemical changes in whey protein concentrate stored at elevated temperature and humidity. *Journal of Dairy Science*. **99** (2016) 2372-2383

[12] M.A. Haque, P. Aldred, J. Chen, C. Barrow, B. Adhikari, Comparative study of denaturation of whey protein isolate (WPI) in convective air drying and isothermal heat treatment processes. *Food chemistry*. **141** (2013) 702-711

536291 Molecular Thermodynamic Model for Aqueous Na^+–K^+–Mg^{2+}–Ca^{2+}–Cl^-/SO_4^{2-} Quinary Electrolyte Systems

Abstract:

Managing produced water is one of the major challenges in oil and gas production [1]. Produced water can be as much as seven times saline compared to sea water systems, and desalination of this water depends on novel material and process innovations. We present a comprehensive thermodynamic model aimed to support heat and mass balance calculations and predict mineral scale deposition of the saline water. The six major ions that control salinity of the water are: Na^+, K^+, Mg^{2+}, Ca^{2+}, Cl^-, and SO_4^{2-}. Current investigation focuses on establishing a thermodynamic model for the Na^+–K^+–Mg^{2+}–Ca^{2+}–Cl^- and the Na^+–K^+–Mg^{2+}–Ca^{2+}–SO_4^{2-} aqueous quinary electrolyte systems. To model the systems, we adopt electrolyte Non-Random Two Liquid theory (eNRTL) for electrolyte solutions [2] which describes concentration dependency of the solution nonideality with two binary interaction parameters for each water-electrolyte and electrolyte-electrolyte pair. Temperature dependency of each binary parameters is further correlated with a Gibbs-Helmholtz type expression with three temperature coefficients representing Gibbs energy, enthalpy, and heat capacity contributions. With the binary parameters and thermodynamic constants of the precipitating salts regressed from phase equilibria, calorimetric and salt solubility data, the model reliably calculates various thermodynamic properties i.e., salt solubility of the quaternary subsystems as well as quinary systems with temperatures from 273.15 to 473.15 K. The model is being extended to cover additional ions such as Sr^{2+}, Ba^{2+}, and HCO_3^- without losing inherent consistency and it should become a very useful tool for chemical engineering calculations.

Reference:

[1] A. V. García, K. Thomsen and E. H. Stenby, "Prediction of Mineral Scale Formation in Geothermal and Oilfield Operations Using the Extended UNIQUAC Model: Part I. Sulfate Scaling Minerals," Geothermics. 2005, 34(1): 61-97.

[2] Y. Song, C.-C. Chen, "Symmetric Electrolyte Nonrandom Two-Liquid Activity Coefficient Model," Ind. Eng. Chem. Res. 2009, 48:7788-7797.

530194 Numerical Investigation of Heat Transfer in Fixed-Bed Reactors Filled with Complex Particle Shapes Using CFD

Abstract:

Catalytic fixed-bed reactors with a low tube-to-particle diameter ratio are widely used in the chemical and process industry. They are characterized by a heterogenous packing morphology that lead to a strong interplay between local flow effects and the heat and mass transfer within the system. For that reason, simplified pseudo-homogenous models lead to an erroneous numerical description of that reactor type which is why more sophisticated numerical methods are needed.

Particle-resolved computational fluid dynamics (CFD) is a first-principle modeling approach that has been proven to be a reliable and predictive tool to analyze the flow, temperature and species field within fixed-beds in a spatially resolved manner as recently reviewed by *Jurtz et al.* This allows the method to be used as a design tool for new particle shapes to optimize the trade-off between a high active catalytic surface, a low pressure drop and a good heat transfer characteristic.

Within the scope of this presentation a fully automated workflow for the simulation of fluid flow and heat transfer in fixed-beds filled with cylindrical particles, Raschig rings and multi-hole cylinders will be presented. The different particle shapes will be compared regarding their thermal performance and advanced post-processing methods will be used to understand the complex interplay between particle orientation, convective heat transfer within the inner particle voids and the overall thermal performance. It will be shown how effective transport parameters can be extracted from the simulation results that afterwards can be used in simplified numerical models.

Literature

Jurtz, N., Kraume, M. and Wehinger, G. (2018). *Advances in fixed-bed reactor modeling using particle-resolved computational fluid dynamics (CFD).* Reviews in Chemical Engineering. Published: 2018/02/02. Available at: https://doi.org/10.1515/revce-2017-0059 [Epub ahead of print]

534211 Optimal Front-End Crude Scheduling for Refinery with Consideration of Proactive Unit Maintenance

Abstract:

The front-end crude scheduling (FECS) has been well studied and convincingly proven very effective for the scheduling of crude supply process [1,2]. It can be used to tackle crude movement problems from crude importing by vessels, unloading through berths, temporarily storing at port-side storage tanks, transferring via long-distance pipeline, blending at refinery-side charging tanks, to continuously feeding to refineries. Through FECS, optimal crude transfer schedules in terms of profitability and operability can be determined. In previous studies, most works were developed based on the assumption that all processing units were under normal working status without consideration of any possible abnormal performance. In reality, however, the performance of every processing unit will be inevitably decaying along with the processing time and load. Thus, regular maintenance operations have to be scheduled in advance to restore unit performing efficiency and more importantly to prevent potential unit failures [3,4]. Meanwhile, the unit maintenance schedule needs to be proactively integrated with the crude movement schedule to ensure the real operability as well as the minimization of the overall cost. Some research works did deal with maintenance issues by identifying post remedial strategies for unit failures [5,6]. However, the reactive response will unavoidably and stiffly disrupt the original schedule, resulting in higher operating expense and even infeasible operations. Therefore, proactive scheduling of regular unit maintenance is a promising opportunity to further improve and significantly benefit the crude supply process.

In this study, a novel continuous-time FECS model combined with proactive scheduling of unit maintenance has been developed. It involves a large-scale mixed-integer nonlinear programming problem, covering all crude movements from shipping all the way to the refinery processing. Through this model, the optimal schedule for both crude supply and unit maintenance operations will be simultaneously identified for the minimal operating cost. It can not only determine when and how to isolate the designated unit(s) from the process system for the maintenance preparation (i.e., how to discharge the leftover inventory in units), but also identify the best opportunity for the maintenance-completed unit(s) to rejoin the system for continuous production during the scheduling time horizon. In addition, real process requirements, including tank inventory, transfer flowrate, blending specifications are all taken into account to guarantee the obtained schedule is practical. The efficacy and performance of the proposed model has been tested and demonstrated through various unit(s) and maintenance scenarios.

References

[1] Zhang, S., & Xu, Q., (2015) Refinery continuous-time crude scheduling with consideration of long-distance pipeline transportation. Computer and Chemical Engineering, 75, 74-94.

[2] Li, J., Misener, R., & Floudas, C., (2012) Continuous-time modeling and global optimization approach for scheduling of crude oil operations. AIChE Journal, 58, 205-226.

[3] Amaran, S., Sahinidis, N., Sharda, B., Morrison, M., Bury, S., Miller, S., & Wassick, J., (2015) Long-term turnaround planning for integrated chemical sites. Computer and Chemical Engineering, 72, 145-158.

[4] Vieira, M., Varela, T., Povoa, A., (2017) Production and maintenance planning optimization in biopharmaceutical processed under performance decay using a continuous-time formulation: A multi-objective approach. Computer and Chemical Engineering, 107, 111-139.

[5] Zhang, S., Xu, Q., (2014) Reactive scheduling of short-term crude oil operations under uncertainties. Industrial & Engineering Chemistry Research, 53, 12502-12518.

[6] Zhang, S., Wang, S., & Xu, Q., (2015) A new reactive scheduling approach for short-term crude oil operations under tank malfunction. Industrial & Engineering Chemistry Research, 54, 12438-12454.

534540 Thermodynamic Modeling of CO_2 Absorption in Aqueous Amino Acid Salt Solutions with Symmetric Electrolyte NRTL Model

Abstract:

Development and large-scale deployment of renewable energy is the need of the hour, but it is still desirable to mitigate the increased CO_2 emissions as we make a transition from fossil fuel-based energy sources to renewable energy sources. In this perspective, many separation technologies have been designed for the large-scale removal of CO_2 from industrial sources. The most developed technology for CO_2 removal is the solvent based absorption using monoethanolamine (MEA)[1]. However, alkanolamines suffer from volatility, oxidative degradation and toxicity making them less suitable for solvent-based absorption[2,4]. Recently, aqueous solutions of amino acid salts (AAS) have gained considerable attention as alternative solvents for carbon capture because the inherent ionic nature of the system makes it non-volatile[2] which is advantageous at the stripper conditions. Also, their higher oxidative stability is favorable for oxygen-rich flue gas streams[4]. They also have comparable absorption kinetics to alkanolamines[2]. Thus, it is desirable to develop an extensive thermodynamic model of AAS-H_2O-CO_2 system to support process development and optimization. However, scarcity of extensive experimental data on thermodynamic aspects of the binary (AAS-H_2O) and ternary (AAS-H_2O-CO_2) systems[2] as well as the dissociation of salt to zwitterion, anion and cation[3] add to the complexity of modeling the liquid phase chemical equilibrium. Limited work has been reported for the experimental determination of CO_2 solubility in aqueous solutions of potassium salts of Glycine[4] (KGly), Sarcosine[2] (KSar), Proline[5] (KPro) and Taurine[6] (KTau). Prior models using extended UNIQUAC framework[2] and Deshmukh-Mather mehtodology[4] have been developed for potassium salts of sarcosine and glycine respectively.

In this study, we develop a symmetric electrolyte NRTL model to represent liquid phase non-idealities in conjunction with Redlich-Kwong equation of state to model vapor phase fugacity. Solubility data and pH data is used to identify eNRTL binary interaction parameters for one binary system (KSar-H_2O) and three ternary (KSar-H_2O-CO_2), (KGly-H_2O-CO_2), (KPro-H_2O-CO_2) systems. The models accurately represent the VLE behavior and liquid phase speciation in the three AAS-H_2O-CO_2 systems, covering a temperature range of 313-353 K and CO_2 loading of 0.2-1. The models can be further advanced upon the availability of more extensive experimental data on the absorption enthalpy for the three AAS-H_2O-CO_2 systems.

References:

1. Zhang, H.Que, C.-C.-Chen, Fluid Phase Equilibria 311 (2011) 67-75
2. E.Aronu, E.T.Hessen, T H.-Warberg, K.A.Hoff, H.F.Svendsen, Chemical Engineering Science 66 (2011) 2191-2198
3. -C.-Chen, Y.Zhu, Biotechnology Progress, Vol. 5 No. 3 (1989) 111-118
4. F.Portugal, J.M.Sousa, F.D.Magalhaes, A.Mendes, Chemical Engineering Science 64 (2009) 1993-2002
5. E.Majchrowicz, D.W.F.Brilman, Chemical Engineering Science 72 (2012) 35-44
6. S.Kumar, J.A.Hogendoorn, P.H.M.Feron and G.F.Veersteg, J.Chem.Eng.Data 46 (2001) 1357-1361.

525846 Plantwide Process Design with Automatic Column Optimization, Sequencing and Stacking Using a Rigorous Process Simulator

Abstract:

Column sequencing refers to the problem of determining the optimal column sequence of distillation columns for multicomponent separation. For example, for the separation of three components, ABC, there are two possible sequences: the direct split (A/BC followed by B/C) and the indirect split (AB/C followed by B/C). The number of possible sequences grows rapidly as the number of components to be separated grows. The column sequencing problem has been widely studied in the literature [1–11], although most researchers use shortcut equations (e.g. the Fenske-Underwood-Gilliland column design method) for the column design in order to reduce the computational load.

Column stacking refers to the process of adjusting the operating pressure of one or more columns so that the reboiler of one or more columns can be powered by condenser of another column. Depending on the temperature difference between the top and bottom of the columns and the temperatures of the utilities, three or more columns can be stacked on top of each other. This type of heat integration can significantly reduce energy consumption. Column stacking has also been widely studied in the literature [12–16], although again mostly using shortcut models.

Finally, it is well known that for many problems of practical interest, there is a tradeoff between the design of the reactor or reactor network and the cost of separation [17]. Specifically, for many problems of practical interest, selectivity decreases as per-pass conversion increases, so there is a tradeoff between selectivity losses at high conversion and separation cost (for recycle of unconsumed reactant) at low per-pass conversion. Thus in order to determine the optimal plant-wide process design, it is necessary to consider the design of the reactor or reactor network and the separation system simultaneously. For greater accuracy, it would be desirable to use a rigorous process simulator rather than shortcut column design equations.

In the present contribution, these problems are approached using a rigorous process simulator (Aspen Plus) coupled with a high-level procedural programming language (Matlab). The two programs are connected using a client-server architecture, in which Matlab is the client and Aspen Plus is the server. A process flowsheet consisting of a single distillation column is constructed in Aspen Plus, and this flowsheet is used to model each column in every sequence by changing the feed conditions. The reflux rate and boilup rate are manipulated (varied) in order to achieve desired specifications on the top and bottom product distributions. The optimal number of trays and feed tray location are determined by optimization using simulated annealing programmed in Matlab.

The code determines the optimal design of each column in every sequence and adds up the cost to determine the optimal column sequence. Once the optimal sequence is identified, the code automatically implements column stacking by adjusting the operating pressures of the columns within the limits of feasibility (given temperatures for the utilities and a minimum temperature driving force). Finally, this entire process (automatic column design, sequencing and stacking) is coupled with a model of a chemical reactor (in a separate Aspen flowsheet) to determine the optimal plantwide process design by adjusting reactor network design parameters (temperature and residence time) to find the optimal tradeoff between selectivity losses and separation costs using the optimized separation system design in each case.

References

[1] Nishida N.; G. Stephanopoulous; A. W. Westerberg. A review of process synthesis. AIChE Journal 1981, 27, 321-351.

[2] Tedder D. W; Rudd D. F. Parametric Studies in Industrial Distillation Part I. Design Comparisons. AIChE Journal 1978, 24, 303-315.

[3] Stephanopoulos G; Westerberg A. W. Studies in Process Synthesis–II. Evolutionary Synthesis of Optimal Process Flowsheets. Chemical Engineering Science 1976, 31, 195-204.

[4] Malone M. F.; K. Glinos; F. E. Marquez; J. M. Douglas. Simple analytical criterion for the sequencing of distillation columns. AIChE Journal 1985, 31, 686-689.

[5] P. B. Gadkari; R. Govind. Analytical Screening Criterion for Sequencing of Distillation Columns. Computer & Chemical Engineering 1988, 12, 1199-1213.

[6] M. Bezzina; L. Pibouleau; S. Domenech. A Statistical Approach to the Synthesis of Separation Sequences. Computer & Chemical Engineering 1989, 13, 651-660.

[7] Floquet, P.; L. Pibouleau.; S. Domenech. Separation Sequence Synthesis: How to Use Simulated Annealing Procedure? Computer. Chem. Eng 1994, 18, 1141-1148.

[8] An, W-Z.; Yuan X-G. A Simulated Annealing-Based Approach to the Optimal Synthesis of Heat-Integrated Distillation Sequences. Computers and Chemical Engineering 2009, 33, 199-212.

[9] F. Wang; Y. Luo; X. Yuan; A Formulation Methodology for Multicomponent Distillation Sequences Based on Stochastic Optimization. Chinese Journal of Chemical Engineering 2016, 24, 1229-1235.

[10] Boozarjomehry R. B.; A. Pourahmadi Laleh; W. Y. Svrcek. Automatic Design of Conventional Distillation Column Sequence by Genetic Algorithm. The Canadian Journal of Chemical Engineering 2009, 87, 477-492.

[11] Boozarjomehry R. B.; A. Pourahmadi Laleh; W. Y. Svrcek. Evolutionary Design of Optimum Distillation Column Sequence. The Canadian Journal of Chemical Engineering 2012, 90, 956-972.

[12] Andrecovich, M. J.; A. W. Westerberg. A Simple Synthesis Method Based on Utility Bounding for Heat-Integrated Distillation Sequences. AIChE Journal. 1985(b); 31: 363-375.

[13] Andrecovich, M. J.; A. W. Westerberg. An MILP Formulation for Heat-Integrated Distillation Sequence Synthesis. AIChE Journal 1985(a), 31, 1461-1474.

[14] Andrecovich, M. J.; A. W. Westerberg. Utility Bounds for Nonconstant QΔT for Heat-Integrated Distillation Sequence Synthesis. AIChE Journal 1985(c), 31, 1475-1479.

[15] C. A. Floudas; G. E. Paules IV. A Mixed-Integer Nonlinear Programming Formulation for the Synthesis of Heat-Integrated Distillation Sequences. Computers & Chemical Engineering 1988, 12, 531-546.

[16] Jaroslav, J.; Radim Ptacnlk. Synthesis of heat integrated rectification systems. Comput. Chem. Engng 1988, 12, 427-432.

[17] Douglas, J. M. Conceptual design of chemical processes. New York, McGraw-Hill, 1988.

521797 A Sigmoidal Approximation for Chance-Constrained Nonlinear Programs

Abstract:

Chance constraints (CCs) are powerful decision-making paradigm that enable decision makers to directly control risk. A CC (also known as a probabilistic constraint) imposes a requirement on the probability that a constraint is satisfied. Mathematically, this a CC can be expressed as $P(g(x,W)\leq 0)\geq \alpha$, where x is the decision variable and W is a random parameter vector. Unfortunately, nonlinear programs (NLPs) with CCs cannot be handled by standard optimization solvers and specialized reformulations and/or approximations are needed.

The most straightforward and general reformulation of a CC introduces binary variables to count the number of realizations under which the constraint $g(x,w)\leq 0$ is satisfied/violated, giving rise to a large-scale MINLP [1]. This reformulation exploits the fact that that the CC can be expressed as the expectation of the indicator function.

A popular conservative approximation for a CC requires that the constraint is satisfied over all possible realizations w of the random variable W (i.e., $g(x,w)\leq 0$ is satisfied with probability one) [2]. This scenario-based approximation can be easily handled using structured NLP solvers but can give very conservative solutions.

A CC can also be approximated in a conservative manner by using a conditional value-at-risk (CVaR) constraint [3]. This approach has been shown to be equivalent to an outer-approximation of the indicator function. The CVaR formulation can also be handled by structured NLP solvers but also yields very conservative solutions. The authors in [4] propose a difference of convex functions (DC) approximation for the indicator function and they show that the approximation can be made asymptotically exact. Unfortunately, this formulation requires specialized solution algorithms that are not guaranteed to work in a general nonconvex NLP setting.

In this talk, we propose a new approximation for CCs based on a risk metric that we call the sigmoidal value at risk (SigVaR). This metric is constructed by using a smooth outer-approximation of the indicator function. We prove that this approximation is conservative but that the level of conservatism can be made arbitrarily small for limiting parameter values. The SigVaR approximation brings computational benefits over exact mixed-integer and difference of convex functions reformulations because, as with CVaR and the scenario-based approximation, it can be handled using structured NLP solvers. We establish formal connections between SigVaR and CVaR that allow us to identify suitable parameter values for the approximation. We present small- and large-scale numerical studies including a wind turbine and a flare system optimization study [5,6] to demonstrate that SigVaR dramatically reduces conservatism of existing approaches.

[1] J. Luedtke, S. Ahmed, and G. L. Nemhauser, An integer programming approach for linear programs with probabilistic constraints, Mathematical Programming, 122 (2010), pp. 247-272.

[2] G. Calafiore and M. C. Campi, Uncertain convex programs: randomized solutions and confidence levels, Mathematical Programming, 102 (2005), pp. 25-46.

[3] A. Nemirovski and A. Shapiro, Scenario approximations of chance constraints, in Probabilistic and randomized methods for design under uncertainty, Springer, 2006, pp. 3-47.

[4] L. J. Hong, Y. Yang, and L. Zhang, Sequential convex approximations to joint chance constrained programs: Amontecarloapproach, Operations Research, 59 (2011), pp. 617-630.

[5] Y. Cao, F. D'Amato, and V. M. Zavala, Stochastic optimization formulations for wind turbine power maximization and extreme load mitigation, Under Review, 2017.

[6] J. Tovar-Facio, Y. Cao, J.M. Ponce-Ortega, and V. M. Zavala. Scalable Solution Strategies for Chance-Constrained Nonlinear Programs. Under Review, 2018.

536055 A Global Optimization Algorithm for Nonconvex Chance-Constrained Programs with Continuous Random Variables

Abstract:

A new algorithm is presented for globally solving nonconvex stochastic optimization problems subject to general chance constraints. Stochastic optimization is a powerful framework for determining optimal design and operational decisions under uncertainty, and is extensively used in manufacturing, chemical processing, and power distribution, to name only a few. In this framework, chance constraints refer to constraints that are required to hold with a specified (typically high) probability, but not with certainty. Chance constraints are often essential for obtaining sensible optimization results for problems with large uncertainties, specifically in cases where it would be impossible or inordinately costly to ensure that a constraint could be satisfied in every possible realization of uncertainty, no matter how unlikely (e.g., committing enough power generation units to satisfy demand under every possible realization of random demands, renewable power outputs, contingency scenarios, etc.) However, effective methods for rigorously solving chance constrained programs are currently limited to special cases where equivalent deterministic constraints can be derived, such as problems with independent linear chance constraints affected by random variables with simple (e.g., Gaussian) distributions. A number of other approaches are available for deriving and solving deterministic restrictions of the original problem by replacing the chance constraints with more conservative deterministic constraints, often derived from probabilistic inequalities such as the Cantelli-Chebyshev inequality. However, such constraints are often significantly more restrictive than the original chance constraints, and no general-purpose approaches are available for refining these approximations. Finally, chance constraints are commonly approximated using sample-based methods, but this often requires many samples to achieve high accuracy. Moreover, without further approximations, sample-based approximations of chance constraints are discontinuous with respect to the optimization variables, which can greatly hinder efficient optimization.

In this talk, we propose a new branch-and-bound (B&B) algorithm for solving nonconvex chance-constrained stochastic optimization problems to epsilon-global optimality. Specifically, we consider chance constraints that may be nonconvex with respect to both the decision variables and the random variables, and may depend on random variables described by a very general class of probability density functions. We assume that the objective function and any other constraints are written as expected-values of other potentially nonconvex functions over continuous random variables. However, we do not consider robust constraints or recourse decisions in our formulation. The core of our B&B algorithm is a novel method for generating tractable convex and concave relaxations of nonconvex chance constraints by an extension of the Jenson-McCormick relaxation technique for expected-value functions presented in Shao and Scott, *AIChE J.*, 2018 (10.1002/aic.16064). In contrast to other relaxations and restrictions of chance constraints in common use, our relaxations can be made to converge to the original constraints by partitioning the uncertainty space, which enables finite convergence of the overall B&B algorithm. Moreover, our relaxations provide rigorous, deterministic bounds on the original chance constraints (as well as the objective and expected-value constraints) which enables regions of the decision space to be fathomed with certainty, even when using a course partition of the uncertainty space. Thus, the uncertainty partition can be refined dynamically as the B&B search proceeds, leading to significant computational savings relative to sample-based approaches, which must use a fixed number of samples throughout the search. The advantages of our new B&B approach will be demonstrated through case studies in chemical reactor design and renewable energy systems.

524155 On Solving Nonconvex Two-Stage Stochastic Programs with Generalized Benders Decomposition

Abstract:

Stochastic programming is one of the major mathematical frameworks in the area of decision-making under uncertainty. A special case is two-stage stochastic programming in which stage-one decisions correspond to "here and-now" decisions that must be taken before the uncertainties are revealed; stage-two decisions correspond to "wait and see" decisions or recourse decisions that can be adjusted in the face of the realization of the uncertainties.. Solving the deterministic equivalent of the two-stage stochastic programs directly can be prohibitive when the number of scenarios is large because the computational time generally grows exponentially with the number of scenarios. Generalized Benders decomposition [1] can be applied to problems with continuous variables and convex constraints in the second stage. However, it cannot be applied directly to the problems with nonconvex second stage constraints.

In the area of nonconvex nonlinear stochastic programming, Li et al [2] has proposed a NGBD algorithm for problems with pure binary first stage variables. Cao and Zavala [3] propose a scalable algorithm that can globally optimize nonconvex two-stage stochastic programs. Ogbe and Li [4] propose a joint decomposition algorithm for general mixed-integer nonlinear nonconvex stochastic programs.

In this work, we further explore algorithms based on generalized Benders decomposition for solving two-stage stochastic programs with nonconvex constraints and mixed-integer variables in both first and second stage decisions. The validity and tightness of different types of cuts including Benders cuts, strengthened Benders cuts, and Lagrangean cuts are studied. Moreover, different convexification schemes for nonconvex problems are tested under this framework. In order for the algorithm to have finite convergence, a spatial branch and bound search is performed on the first stage decisions.

The algorithm is tested with several applications in process systems engineering, that includes the generalized pooling problem under uncertainty. The decomposition algorithm is benchmarked against the full-space solution with state-of-art global solvers, such as BARON, ANTIGONE.

[1] Arthur M. Geoffrion "Generalized benders decomposition. Journal of Optimization Theory and Applications 237-260, 1972.

[2] Xiang Li, Asgeir Tomasgard, and Paul I Barton. Nonconvex generalized benders decompo-sition for stochastic separable mixed-integer nonlinear programs. Journal of Optimization Theory and Applications, 151(3):425, 2011.

[3] Yankai Cao and Victor M Zavala. A scalable global optimization algorithm for stochastic nonlinear programs. Under Review, 2017.

[4] Emmanuel Ogbe and Xiang Li. A joint decomposition method for global optimization of multiscenario nonconvex mixed-integer nonlinear programs. arXiv preprint arXiv:1802.07342, 2018.

529193 New Developments in Flexibility Analysis in the Framework of Design Space Definition

Abstract:

The importance of the Design Space (DS) definition lies in the assurance of quality, a key goal of Quality by Design, and in the broadening from an acceptable operating set point, to a collection of tolerable operating regions. Working within the DS is considered acceptable and not additional actions are needed (e.g., notification). Operating outside the DS represents a deviation and would normally initiate an investigation. Defining properly the limits of the design space is vital to provide manufacturing flexibility and quality assurance. Even though process DS has been only recently formally introduced in the pharmaceutical industry, it is not a completely new concept. Identifying design and operating constraints as well as the necessary control scheme for a process are concepts and ideas that have been extensively studied in other industries and literature (Boukouvala ***et al.***, 2017). The use of concepts introduced by the process system literature, such as process feasibility and flexibility, have proven to be useful in the DS identification (García-Muñoz ***et al.***, 2015).

Flexibility addresses the guaranteed feasibility of operation of a plant over a range of conditions, with the ultimate goal being on how to design a process for guaranteed feasible operation over a specified parameter range (Grossmann ***et al.***, 2014). There are three main problems to address: the flexibility test problem, flexibility index calculation and the design under uncertainty with flexibility constraints. First, the flexibility test problem (Halemane and Grossmann, 1983) for a given design consists in determining whether by proper adjustment of the controls variables the process constraints hold for any realization of uncertain parameters. To indicate how much flexibility can actually be achieved in a given design (Swaney and Grossmann, 1985), the flexibility index is defined as the largest scaled factor that multiplies the deviation of the uncertain set for which the inequalities hold. As for design optimization problem, the selection of the design variables is involved so as to minimize cost and either satisfy the flexibility test, or maximize the flexibility index, where the latter problem gives rise to a multi-objective optimization problem.

However, these formulations are based implicitly on the assumption that control variables can compensate any variation in the uncertain parameter set and that during operation stage uncertain parameters can be determined with enough precision. Ostrovsky ***et al.*** (2004) and Rooney and Biegler (2004) extended the analysis by taking into account the level of parametric uncertainty in the mathematical models at the operation stage, by grouping the uncertain parameters, Θ, into two types: unmeasured and measured. The flexibility constraint was then extended to account for unmeasured process variability, . The flexibility constraint was re-defined as a multi-level optimization problem.

In this work, we propose new formulations for the extended flexibility analysis. The basic idea relies on recursively reformulating the inner optimization problems by the Karush-Kuhn-Tucker conditions, and with a mixed-integer representation of the complementarity conditions to solve the resulting multilevel optimization problem. Three types of models with uncertain parameters are tackled and solved with the proposed strategy: 1) linear programming model, 2) nonlinear programming model with monotonic variation of

unmeasured uncertain, and finally, 3) nonlinear programming model. Similar mixed-integer formulations to the traditional flexibility analysis are obtained for the first two cases, whereas a novel nonconvex MINLP is obtained for the latter case. Specifically, for the nonlinear programming model, we study a thermal de-protection reaction performed in a CSTR followed by an evaporation step. Numerical results are compared to the traditional flexibility analysis (Grossmann and Floudas, 1987), leading to larger values of constraint violation and lower values of flexibility indices due to the fact that the recourse action can only compensate for measured uncertain parameter variations.

Keywords

Flexibility Analysis, Quality by Design, Design Space

References

Boukouvala, F., Muzzio, F. J. and Ierapetritou, M. G. (2017) Methods and Tools for Design Space Identification in Pharmaceutical Development, in Comprehensive Quality by Design for Pharmaceutical Product Development and Manufacture (eds G. V. Reklaitis, C. Seymour and S. García-Munoz), John Wiley & Sons, Inc., Hoboken, NJ, USA.

García-Muñoz, S., Luciani, C.V., Vaidyaraman, S., Seibert, K.D. (2015). Definition of Design Space Using Mechanistic Models and Geometric Projections of Probability Maps. Organic Process Research & Development, 19, 1012–1023.

Grossmann, I.E. and Floudas, C. (1987). Active constraint strategy for flexibility analysis in chemical processes. Comput. chem. Engng. 11(6) 675-693.

Grossmann, I. E., Calfa, B. A. & Garcia-Herreros, P. (2014). Evolution of concepts and models for quantifying resiliency and flexibility of chemical processes. Computers & Chemical Engineering, 70, 22-34.

Halemane, K. P. & Grossmann, I. E. (1983). Optimal Process Design under Uncertainty. AIChE Journal, 29, 425.

Ostrovsky, G. M., Achenie, L.E.K., Datskov, I., Volin, YU. M. (2004). Uncertainty at both design and operation stages. Chem. Eng. Comm., 191:105-124.

Rooney, W. C. & Biegler, L. T. (2004). Optimal Process Design with Model Parameter Uncertainty and Process Variability. AIChE Journal, 49(2), 438-449.

Swaney, R. E. & Grossmann, I. E. (1985). An Index for Operational Flexibility in Chemical Process Design. Part 1 and Part 2 Formulation and Theory. AIChE Journal, 31, 621.

525541 Adaptive Robust Optimization Under Uncertainty with Regret

Abstract:

In recent years, decision making under uncertainty has received tremendous attention from both academia and industry [1]. Uncertainty could render the deterministic solution suboptimal or even infeasible. To this end, extensive research efforts have been made towards the development of systematic methodologies for handling uncertainty. Robust optimization has attracted considerable interest from the process systems engineering community due to its good balance between solution quality and computational tractability [2]. By introducing recourse actions, adaptive robust optimization (ARO) captures the sequential nature of decision-making processes, and is typically less conservative than static robust optimization [3, 4].

Although conventional ARO has a number of attractive features, it does not account for an important evaluation metric, known as regret, in decision-making theory [5]. In practice, the solution of an optimization problem under the presence of uncertainty is usually evaluated *ex post* by comparing its performance with that of the perfect information solution. The objective value of the perfect information solution is the best possible value that could be achieved by assuming that the decision maker had known the exact uncertainty realization *a priori*. While the perfect information solution is a utopia in the sequential decision making under uncertainty, this solution serves as an ideal benchmark for the quality of the optimized solution. The minimax regret criterion holds the promise to significantly improve the aforementioned evaluation performance, since it aims for the least deviation from the ideal benchmark.

In this work, we propose a general bi-criterion ARO (BCARO) framework that effectively accounts for the conventional robustness and minimax regret criteria. By minimizing the worst-case regret over all uncertainty realizations within an uncertainty set, the proposed framework pushes the performance of the resulting solution towards the utopia one. In this way, the evaluation performance against the ideal benchmark is significantly improved. Since the worst-case uncertainty realization is not uniquely defined for stochastic multiobjective optimization, different concepts of robust efficiency have been proposed. In this work, we employ the concept of point-based robust efficiency due to its clear interpretation and wide applications [6]. The proposed BCARO framework generates a set of Pareto-optimal solutions to reveal the systematic trade-offs between the conventional robustness and minimax regret criteria. This framework is general enough to be applied to various types of uncertainty sets, such as the polyhedral and data-driven uncertainty sets [7]. The resulting optimization problem is formulated as a bi-criterion multi-level mixed-integer programming problem, which cannot be solved directly by any off-the-shelf optimization solvers. Therefore, we further propose decomposition based algorithms for its efficient solution. To illustrate the applicability of the proposed framework, we present two applications on process network planning [8], and batch production scheduling [9]. In the process network planning problem, the proposed approach reduces the worst-case regret significantly by 52.3%, while decreasing the worst-case net present value (NPV) by merely 3.2%. In the batch production scheduling, the solution determined by the minimax regret criterion has the lowest worst-case regret of $13.6, and decreases the worst-case profit by only 0.3% compared with solution under the conventional robustness criterion.

References

[1] I. E. Grossmann, R. M. Apap, B. A. Calfa, P. García-Herreros, and Q. Zhang, "Recent advances in mathematical programming techniques for the optimization of process systems under uncertainty," *Computers & Chemical Engineering*, vol. 91, pp. 3-14, 2016.

[2] D. Bertsimas, D. B. Brown, and C. Caramanis, "Theory and applications of robust optimization," *SIAM Review*, vol. 53, pp. 464-501, 2011.

[3] A. Ben-Tal, A. Goryashko, E. Guslitzer, and A. Nemirovski, "Adjustable robust solutions of uncertain linear programs," *Mathematical Programming*, vol. 99, pp. 351-376, 2004.

[4] C. Ning and F. You, "A data-driven multistage adaptive robust optimization framework for planning and scheduling under uncertainty," *AIChE Journal*, vol. 63, pp. 4343â€“4369, 2017.

[5] D. E. Bell, "Regret in decision making under uncertainty," *Operations Research*, vol. 30, pp. 961-981, 1982.

[6] J. Ide and A. SchÃ¶bel, "Robustness for uncertain multi-objective optimization: A survey and analysis of different concepts," *OR Spectrum*, vol. 38, pp. 235-271, 2016.

[7] C. Ning and F. You, "Data-driven adaptive nested robust optimization: General modeling framework and efficient computational algorithm for decision making under uncertainty," *AIChE Journal*, vol. 63, pp. 3790-3817, 2017.

[8] D. Yue and F. You, "Planning and scheduling of flexible process networks under uncertainty with stochastic inventory: Minlp models and algorithm," *AIChE Journal*, vol. 59, pp. 1511-1532, 2013.

[9] H. Shi and F. You, "A computational framework and solution algorithms for two-stage adaptive robust scheduling of batch manufacturing processes under uncertainty," *AIChE Journal*, vol. 62, pp. 687-703, 2016.

536827 Hybrid Decision Rules in Multistage Adaptive Optimization

Abstract:

Multistage adaptive optimization is an important technique for addressing dynamic decision making problem under uncertainty. It has been employed in numerous process systems engineering applications, including planning, scheduling and supply chain management. The optimal solution of a multistage adaptive optimization problem can be cast as a set of decision policies/functions of the uncertainty. While the presence of the uncertainty induces a set of semi-infinite constraints that are computationally intractable, various approximation methods have been used in the past.

The first method approximates the decision policy through a discrete set of scenario dependent decision values. For instance, scenario-based multistage stochastic programming has been extensively used in different areas, where uncertainty is characterized with a finite number of scenarios using a scenario tree. The stochastic problem can then be reformulated into a deterministic one with the scenarios: the objective function is calculated as the expectation and the constraints are enforced based on scenarios. However, this method suffers from the curse of dimensionality: the problem size can quickly increase as the dimension of the uncertainty increases. The second method relies on decision rules, which has received increasing attention in recent years for solving such problems. In this type of method, the adaptive decisions are approximated as a function of the observed uncertainty to reduce the model complexity. The simplest type of decision rule, a linear decision rule (LDR), assumes an affine relationship between the decision and uncertain parameters. Using an LDR, strong duality of convex optimization can often be utilized to construct the tractable counterpart of the stochastic problem. The significance of the approach was demonstrated by Ben-Tal et al. [1].

While the LDR method can lead to significant computational advantage compared to the scenario method, its solution quality is limited by its simplicity. To address this limitation, several variants of nonlinear decision rules have been proposed to improve the approximation quality, but with increased computational load (e.g. [2]). The decision rule based method is governed by two competing objectives: tractability and approximation quality. A piecewise linear decision rule (PLDR) can be regarded as a special type of nonlinear decision rule. It elevates the approximation quality by having multiple slopes (i.e. decision variables) while inheriting the tractability features of an LDR due to its linear nature [3]. In this work, we investigate the hybrid strategy of using LDRs and PLDRs for solving multistage adaptive optimization. Several guiding strategies were proposed towards reaching a good balance between computational complexity and solution quality.

First, we investigate how best to lift an uncertainty set while using PLDR in multistage adaptive optimization. We demonstrate that defining a hybrid decision rule where the uncertainties in the near future are modelled with more details than later stages is attractive. The modelling detail of the uncertainty is reflected by the number of breakpoints used for lifting; the more breakpoints used to lift the uncertainty set, the more detailed is the uncertainty represented. This is similar to a construct used in scenario-based stochastic programming in which a scenario tree contains many branches per node (i.e. captures more detail) in early stages and relatively few branches per node (i.e. captures less detail) in later stages can be attractive [4]. The proposed method is applied to a multistage stochastic transportation problem and the trade-off between solution quality deterioration and computational load reduction is promising. Second, we study the dependence on the history of observed uncertainty in the decision rule. Ideally, decision rules are defined as a function of the full history of observed uncertainty. Nonetheless, solving a multistage adaptive optimization problem with full history LDR can be computationally challenging [5]. In this work, we exploit the correlation in the uncertainty to restrict the history dependence to reduce the computational load. We show empirically via the multistage stochastic newsvendor problem a relationship between the correlation lag in the uncertainty set and the minimum history lag in the decision rule required for a satisfactory trade-off between deterioration in solution quality and reduction in the computational burden.

References

[1] A. Ben-Tal, A. Goryashko, E. Guslitzer, and A. Nemirovski. Adjustable robust solutions of uncertain linear programs. Mathematical Programming, 99(2):351–376, 2004.

[2] D. Bertsimas, D. A. Iancu, and P. A. Parrilo. A hierarchy of near-optimal policies for multistage adaptive optimization. IEEE Transactions on Automatic Control, 56(12): 2809–2824, 2011.

[3] A. Georghiou, W. Wiesemann, and D. Kuhn. Generalized decision rule approximations for stochastic programming via liftings. Mathematical Programming, 152(1-2):301–338, 2015.

[4] A. N. Arslan and D. J. Papageorgiou. Bulk ship fleet renewal and deployment under uncertainty: A multi-stage stochastic programming approach. Transportation Research Part E: Logistics and Transportation Review, 97:69–96, 2017.

[5] M. Ang, Y. F. Lim, and M. Sim. Robust storage assignment in unit-load warehouses. Management Science, 58(11):2114–2130, 2012.

531460 An Algorithmic Cutting Plane Method for Solving Robust Optimization Problems with Endogenous Uncertainty

Abstract:

Robust Optimization (RO) is one of the major systematic approaches for mitigating effects of parameter uncertainty in mathematical optimization models [1]. The fundamental characteristic of RO is that it seeks the best possible solution that remains feasible for every possible realization of uncertainty within a postulated uncertainty set. The selection of the uncertainty set itself is of critical importance, since it can affect both the optimality of the solution, as well as the level of risk against which it is hedged.

Traditionally in the RO literature, uncertainty sets are constructed under the assumption that the decision maker's strategy cannot have an impact on the realization of uncertainty ("*exogenous uncertainty*"). However, there are well documented cases from the Stochastic Programming literature where the realization of uncertainty and/or its underlying distributional support can be affected by the decisions of the policy makers ("*endogenous uncertainty*") [2], with examples including oilfield production planning [3,4], R&D portfolio optimization [5], capacity expansion [6], clinical trial planning [7] and network interdiction [8], among others.

Recently, Lappas & Gounaris [9] proposed the extension of the traditional polyhedral uncertainty sets to decision-dependent uncertainty sets (DDUS), and showcased the flexibility of the introduced modelling capabilities in terms of capturing the effects of endogenous uncertainty in a variety of relevant problems, while maintaining the favorable computational tractability characteristics of RO. The solution approach utilized was based on a reformulation of the original problem to a monolithic robust counterpart. While the presented methodology is sufficient for small sized problems and continuous valued uncertain parameters, it can be limiting if the reformulated model is of size prohibitive to be loaded in memory, or when the uncertainty set involves uncertain parameters of discrete nature. The limitations of the reformulation approach become even more pronounced when a decision rule scheme is desired in the context of multi-stage decision-making contexts [10].

To alleviate the above limitations of the reformulation approach, we propose a novel algorithmic cutting plane method for solving robust optimization problems with endogenous uncertainty. In particular, we develop a branch & bound algorithm that gradually enforces the robustness of the solution by hedging against violating scenarios that are relevant to the optimal decision strategy. A comprehensive computational study across various problems illustrates the computational advantages of the proposed algorithm over the old reformulation approach, while the impact of various algorithmic enhancements is also investigated. Finally, the flexibility of the new algorithmic approach is presented with problems involving discrete uncertain parameters and elaborate decision-rule schemes.

References:

[1] A. Ben-Tal, L. El Ghaoui, and A. Nemirovski, *Robust optimization*. Princeton University Press, 2009.

[2] T. W. Jonsbråten, R. J. B. Wets, and D. L. Woodruff, "A class of stochastic programs with decision dependent random elements," *Annals of Operations Research*, vol. 82, pp. 83–106, 1998.

[3] V. Gupta and I. E. Grossmann, "A new decomposition algorithm for multistage stochastic programs with endogenous uncertainties," *Computers & Chemical Engineering*, vol. 62, pp. 62–79, 2014.

[4] R. M. Apap and I. E. Grossmann, "Models and computational strategies for multistage stochastic programming under endogenous and exogenous uncertainties," *Computers & Chemical Engineering*, vol. 103(8), pp.233-274, 2017.

[5] S. Solak, J. P. B. Clarke, E. L. Johnson, and E. R. Barnes, "Optimization of R&D project portfolios under endogenous uncertainty," *European Journal of Operational Research.*, vol. 207, no. 1, pp. 420–433, 2010.

[6] V. Goel and I. E. Grossmann, "A class of stochastic programs with decision dependent uncertainty" *Math. Program.*, vol, 108 (2–3) , pp. 355-394, 2006.

[7] M. Colvin and C. T. Maravelias, "A stochastic programming approach for clinical trial planning in new drug development." *Computers & Chemical Engineering,* vol. 32(11), pp. 2626-2642, 2008.

[8] S. Peeta, F. S. Salman, D. Gunnec, and K. Viswanath, "Pre-disaster investment decisions for strengthening a highway network.", *Computers & Operations Research*, vol. 37(10), pp. 1708-1719, 2010.

[9] N.H. Lappas and C. E. Gounaris, "Robust Optimization for decision-making under endogenous uncertainty.", *Computers & Chemical Engineering,* vol. 111(3), pp. 252-266, 2018.

[10] D. Bertsimas, and A. Georghiou, "Design of near optimal decision rules in multistage adaptive mixed-integer optimization.", *Operations Research*, vol. 63(3), pp. 610-627, 2015.

536209 Robust Explicit Optimization and Control within the Paroc Framework

Abstract:

The presence of uncertainty in process systems is one of the key reasons for deviation from set operation policies, resulting in suboptimal or even infeasible operation. As these uncertainties realize themselves on different time scales such as on a control, scheduling or design level, an integrated, comprehensive approach to consider uncertainty is required. For that reason, the PARametric Optimization and Control Framework (PAROC) which decomposes this challenging problem into a series of steps and computes explicit solutions was developed [1-4]. Robust optimization aims to immunize against uncertainty by guaranteeing constraint satisfaction [5]. In robust rolling horizon optimization of discrete-time systems, such as MPC, the derivation of robust optimization policies is mainly characterized by the challenge associated with multiplicative uncertainty [6-9]. The concept of explicit/multiparametric robust rolling horizon optimization aims to provide the explicit form of the closed-loop robust rolling horizon policy known "in the present" that can guarantee feasibility "in the future" [10-12].

In this work, we consider the development of closed-loop explicit robust rolling horizon solutions that correspond to MPC and scheduling problems and their applicability within the PAROC framework, i.e. their utilization in design, control, scheduling problems and their interactions. We consider box-constrained model uncertainty on linear state-space problems upon which we (i) formulate the robust counterpart in a single-step problem formulation, (ii) apply linear transformations to acquire the feasible space as a function of the initial state value realizations and the degrees of freedom and (iii) apply multi-parametric (mixed-integer) programming to acquire the explicit solution of the problem. With this approach, we show (i) how the need for a dynamic programming approach can be alleviated while (ii) guaranteeing the robust nature of the final solution and (iii) we extend the approach to the hybrid case where both integer and continuous variables can be considered via [13]. We present the developments through examples of control, scheduling and design optimization problems.

References

[1] Pistikopoulos, E.N., Diangelakis, N.A., Oberdieck, R., Papathanasiou, M.M., Nascu, I., Sun, M.. PAROC - An integrated framework and software platform for the optimisation and advanced model-based control of process systems (2015) Chemical Engineering Science, 136, pp. 115-138.

[2] Burnak, B., Katz, J., Diangelakis, N.A., Pistikopoulos, E.N., Simultaneous Process Scheduling and Control: A Multiparametric Programming-Based Approach (2018) Industrial and Engineering Chemistry Research, 57 (11), pp. 3963-3976.

[3] Diangelakis, N.A., Burnak, B., Katz, J., Pistikopoulos, E.N., Process design and control optimization: A simultaneous approach by multi-parametric programming (2017) AIChE Journal, 63 (11), pp. 4827-4846

[4] Diangelakis, N. A., Burnak, B., Pistikopoulos, E. N., A multi-parametric programming approach for the simultaneous process scheduling and control - Application to a domestic cogeneration unit (2017) Foundations of Computer Aided Process Operations / Chemical Process Control.

[5] Ben-Tal, A., Nemirovski, A., Robust convex optimization (1998) Mathematics of Operations Research, 23 (4), pp. 769-805.

[6] Mayne, D. Q.; Rawlings, J. B.; Rao, C. V.; Scokaert, P. O. M. (2000) Constrained model predictive control: Stability and optimality. Automatica, 36(6), 789 – 814.

[7] Kerrigan, E. C.; Maciejowski, J. M. (2004) Feedback min-max model predictive control using a single linear program: robust stability and the explicit solution. International Journal of Robust and Nonlinear Control, 14(4), 395 – 413.

[8] Wan, Z.; Kothare, M. V. (2003) An efficient off-line formulation of robust model predictive control using linear matrix inequalities. Automatica, 39(5), 837 – 846.

[9] Bemporad, A.; Borrelli, F.; Morari, M. (2003) Min-max control of constrained uncertain discrete-time linear systems. IEEE Transactions on Automatic Control, 48(9), 1600 – 1606.

[10] Pistikopoulos, E.N., Kouramas, K.I., Faísca, N.P., Robust Multi-Parametric Model-Based Control, (2009) Computer Aided Chemical Engineering, 26, pp. 297-302.

[11] Kouramas, K.I., Panos, C., Faísca, N.P., Pistikopoulos, E.N., An algorithm for robust explicit/multi-parametric model predictive control (2013) Automatica, 49 (2), pp. 381-389.

[12] Kouramas, K.I., Sakizlis, V., Pistikopoulos, E.N., Robust Parametric Model-Based Control (2014) Process Systems Engineering, 2-7, pp. 49-76.

[13] Oberdieck, R., Pistikopoulos, E.N., Explicit hybrid model-predictive control: The exact solution (2015) Automatica, 58, pp. 152-159.

527346 Estimating Uncertain Atmospheric Aerosol Dynamics with an Input Observer

Abstract:

Atmospheric models simulate mass balances of gaseous pollutants and particulate matter – or atmospheric aerosols – in the atmosphere. Typical methods to solve atmospheric inverse problems minimize the model error based on a simplified version of the model [1]. However, these methods may have issues estimating nonlinear aerosol dynamics. In this paper we propose a passivity-based input observer (PBIO) that estimates parameters based on nonlinear conservation laws that can be represented exactly. The method used here is similar to inventory control in which the model states and measurements are projected to a set of inventories. Inventories are positive, additive, and continuously-differentiable variables that follow general conservation laws (like mass or energy).

The PBIO estimates a set of parameters by forcing the model bias, in terms of the inventory variables, to zero. We formulate a quadratic storage function in terms of the model bias. Additionally, the storage function rate of change is constrained to a negative quadratic in terms of the model bias. The parameter update law is derived using these two constraints, leading to a proportional feedback and nonlinear feedforward decoupling on the model bias to update the parameters. This approach reduces the closed-loop system's sensitivity to model mismatch, which can be tuned by adjusting the feedback gain.

Atmospheric models include significant model mismatch due to uncertainty in model inputs (emission fields, meteorology, land type, etc.), model parameters (reaction rate constants, empirical parameters, etc.), and sometimes unknown atmospheric physics and chemistry. The Intergovernmental Panel on Climate Change has characterized how this model mismatch propagates to uncertainty in aerosol's radiative effects in the atmosphere and in clouds [2]. Predicted concentrations of cloud condensation nuclei (CCN), or particles that activate to form cloud droplets, is one of the significant drivers of uncertainty in recent climate change. The particle number size distribution and composition are necessary to predict CCN concentrations. Therefore, in this work, we focus on improving predictions of the number size distribution by estimating uncertain aerosol dynamics based on ground-site measurements.

We choose to estimate scaling factors for the rates of key uncertain processes in the model so that the system is input-affine and is easily inverted for the unknown scaling factors. Key uncertain aerosol processes that we consider are primary organic aerosol emissions, nucleation of new particles from condensable vapors, and condensation of volatile organic compounds (VOCs). Primary aerosol emissions are difficult to monitor and calculate due to the variability in emission fluxes and intensity levels within each type of source. The chemistry that drives nucleation of particles from molecular clusters of compounds possibly including sulfuric acid, water, amines, and organic compounds is still an active area of research. There are large uncertainties in the identity of VOCs and the yield of VOCs partitioning to the particle phase.

Here, we present the design of a PBIO to estimate aerosol dynamics and preliminary applications to a zero-dimensional or box model. We use a zero-dimensional version of the TwO-Moment Aerosol Sectional (TOMAS) microphysical code, which models size-resolved aerosols with the discretized general dynamic equation. We generate synthetic measurements by running TOMAS with a set of "true" rates for primary emissions, nucleation, and SOA production. Then, we start a TOMAS simulation with a set of process rates biased from their "true" value. We quantify the PBIO performance in terms of the bias in the estimated inventory variables, process rates, and the full number size distributions. The methodology developed here will facilitate future work that estimates primary aerosol emissions, nucleation, and VOC condensation rates in a global model based on a measurement network.

[1] Rao, K. S. (2007). Source estimation methods for atmospheric dispersion. *Atmospheric Environment, 41,* 6964–6973. https://doi.org/10.1016/j.atmosenv.2007.04.064

[2] U. D. Cubasch, D. Chen, M. Facchini, D. Frame, N. Mahowald, and J. G. Winther. *Climate Change 2013: The Physical Science Basis. Contribution of Working Group I to the Fifth Assessment Report of the Intergovernmental Panel on Climate Change* [Stocker, T.F., D. Qin, G.-K. Plattner, M. Tignor, S.K. Allen, J. Boschung, A. Nauels, Y. Xia, V. Bex and P.M. Midgley (eds.)]. Cambridge University Press, 2013.

530416 Adaptive Model Predictive Control with Recursive Subspace Identification

Abstract:

Adaptive model predictive control with recursive subspace identification

Iman Hajizadeh, Mudassir Rashid, Ali Cinar

Department of Chemical and Biological Engineering, Illinois Institute of Technology, Chicago, IL, USA
(Corresponding author: cinar@iit.edu)

Control of chemical processes with time-varying parameters and nonlinear dynamics is challenging. The presence of stochastic disturbances, random measurement noises and unknown time-varying delays can make the problem more complex. A time-invariant model cannot describe the dynamic behaviour of these processes accurately and an offline-tuned controller cannot perform satisfactorily when the process is subjected to major disturbances[1,2].

In this work, we propose a data-driven recursive subspace identification approach and an adaptive model predictive control (AMPC) with a comprehensive controller performance assessment system (CPAS). To obtain an accurate and reliable time-varying model of the process, we extended the optimized version of the recursive predictor-based subspace identification method to better handle stochastic disturbances, measurement noises, and variable delays[3â€"5]. This is done by incorporation of constraints on the fidelity and accuracy of the identified models, correctness of the sign of the input-to-output gains, and the integration of heuristics to ensure stability of the recursively identified models. Then, an AMPC is designed using this recursively updated model to handle constraints on the manipulated variables and outputs as well as any safety limits of the process[6]. To make the controller more efficient for mitigating the effects of unknown disturbances, the controller parameters are modified using information from a feature extraction method that can detect significant changes based on the rate and shape of variations in the outputs in real-time. Furthermore, to monitor the performance of the AMPC and retune the controller parameters, we also propose a detailed CPAS to evaluate the closed-loop performance on-line based on predefined performance indices in different time windows.

We have tested our proposed techniques for an artificial pancreas (AP) system where the model of glucose-insulin dynamics in people with type 1 diabetes is the underlying system. This process has all the aforementioned challenges in the modelling and controller design. The nonlinearities and time-varying changes of blood glucose dynamics, the occurrence of nonstationary disturbances, time-varying delays on measurements and insulin infusion, and noisy data from sensors provide challenges for the AP. Furthermore, an AP system is challenged by several factors such as meals, exercise, sleep and stress (MESS) that may have significant effects on glucose dynamics in the body[7].

The FDA approved UVa/Padova metabolic simulator is used to design and test closed-loop algorithms for the AP[8]. This simulator has a set of virtual patients with type 1 diabetes on which to run the in-silico experiments. The results based on new developed techniques show an excellent performance in increasing the average percentage of time spent in the target range (Blood Glucose Concentration [BGC] â^ [70, 180] mg/dL) by 12 % and reducing the mean and standard deviation of BGC by 10 and 5 mg/dL, respectively. The average maximum value of BGC was reduced by 17 mg/dL without any significant change in the average minimum value of BGC (a decrease of 0.5 mg/dL) and the injected insulin (manipulated variable) (an increase of 0.5 Unit of insulin per day). These outcomes demonstrate a significant improvement in the AP control system and their potential to be used in developing a fully automated AP that can function without any manual information and accommodate major disturbances to the BGC, such as MESS.

References

1. SchÃ¤fer J, Cinar A. Multivariable MPC system performance assessment, monitoring, and diagnosis. *J Process Control*. 2004;14(2):113-129.
2. Ljung L, Söderström T. Theory and Practice of Recursive Identification. *Electr Eng*. 1983;4:541.
3. Houtzager I, van Wingerden J-W, Verhaegen M. Recursive predictor-based subspace identification with application to the real-time closed-loop tracking of flutter. *IEEE Trans Control Syst Technol*. 2012;20(4):934-949.
4. Hajizadeh I, Rashid M, Cinar A. Ensuring Stability and Fidelity of Recursively Identified Control-Relevant Models. In: *The 18th IFAC Symposium on System Identification*. Stockholm, Sweden; 2018 (Accepted).
5. Hajizadeh I, Rashid M, Turksoy K, et al. Multivariable recursive subspace identification with application to artificial pancreas systems. *IFAC-PapersOnLine*. 2017;50(1):886-891.
6. Rashid M, Hajizadeh I, Cinar A. Plasma Insulin Cognizant Predictive Control for Artificial Pancreas. In: *The American Control Conference*. Milwaukee, WI; 2018 (Accepted).
7. Cinar A, Turksoy K. *Advances in Artificial Pancreas Systems: Adaptive and Multivariable Predictive Control*. Springer; 2018.
8. Dalla Man C, Micheletto F, Lv D, Breton M, Kovatchev B, Cobelli C. The UVa/Padova type 1 diabetes simulator: new features. *J Diabetes Sci Technol*. 2014;8(1):26-34.

532502 Robust Model Predictive Control with Decomposed Disturbance Subsets for Less Conservative Control

Abstract:

Robust Model Predictive Control with Decomposed Disturbance Subsets for Less Conservative Control

Tae Hoon Oh and Jong Min Lee*

School of Chemical and Biological Engineering, Seoul National University,

1 Gwanak-ro, Gwanak-gu, Seoul, 08826, Korea

Robust Model Predictive Control (RMPC) has been developed over two decades in order to handle the uncertainty associated with model prediction. The uncertainties originate from model-plant mismatch or unknown disturbances. Most of the existing formulations of RMPC minimize the worst-case cost, and its resulting control actions are excessively conservative and significantly affected by the knowledge of the uncertainties. Although several approaches [1,2] are proposed to reduce the conservatism, it is difficult to apply these methods to real processes due to the lack of systematic consideration of the probability of having the worst-case scenario [3]. Conservative control actions, which are often far from optimal ones, are attributable to the worst-case scenario. The probability of having such a worst-case is negligible because it is a single point in a disturbance set. Therefore, it is much more reasonable to design a controller that represents the worst-case as a set so that the probability of the predicted scenario is nontrivial.

This study proposes a computationally tractable linear RMPC formulation with additive disturbances. The proposed formulation involves a parameter that trade-offs control performance against the probability of the worst-case scenario. In order to consider the predicted disturbance as a set rather than a point, the original disturbance set is decomposed into several finite convex subsets at each time step within the prediction horizon of N. Collecting N-1 decomposed subsets gives the sequence of disturbance subsets where each sequence is considered as the candidate for the worst-case scenario and the maximization is taken over those finite sequences. To ensure that the cardinality of the sequences is finite even for a large N, minimum response probability is proposed so that the probability of each sequence should be higher than this threshold value. Then the cost for each sequence is evaluated using the parameterized tube MPC with the sequence of decomposed disturbance subsets [4].

The minimum response probability serves as a tuning parameter to adjust the conservatism of control. If the minimum response probability is 1, the formulation is identical with the original PTMPC approach. On the other hand, if the probability approaches 0, the pessimistic control action is taken which is similar to the original closed-loop worst-case RMPC [5]. Besides the minimum response probability, there are additional possibilities to extend the problem formulation. First, the way of decomposing the disturbance set into given number of subsets is not unique. This is crucial because the control actions are affected by the shape of decomposed subsets. The ideal method would be partitioning the disturbance set in the order of increasing cost, but the resulting subsets are not convex. In addition, it is well known that partitioning an arbitrary compact set into several convex subsets is an NP-hard problem [6]. In this study, the finite identical set approximates partitioning of the arbitrary set, although more precise decomposition could be done depending on the shape of disturbance set. The distribution of the disturbance probability is not unique because the minimum response probability constraints the probability of the sequence of N subsets only. Since it is practical to give more weights on the risk of near future, one suggestion is decomposing the disturbance set into smaller subsets as the time step is closer to the current time. The recursive feasibility and stability are analyzed with the general form of problem formulation. The simulation results suggest that this method could be utilized to adjust the control performance.

References

[1] Stoorvogel, A.A., Weeren, A.J.T.M., "The discrete-time Riccati equation related to the control problem" *IEEE Transactions on Automatic Control* 39(3) (1994): 686-691.

[2] Gabrel, V., Murat, C., Thiele, A., "Recent advances in robust optimization: an overview" *European journal of operational research* 235 (3) (2014): 471-483.

[3] Saltik, M.B., Ozkan, L. Ludlage, J.H., Weiland, S., Van den Hof, P.M., "An outlook on robust model predictive control algorithms: Reflections on performance and computational aspects." *Journal of Process Control* 61 (2018): 77-102.

[4] Rakovic, S. V., Kouvaritakis, B., Cannon, M., Panos, C., Findeisen R. "Parameterized tube model predictive control." *IEEE Transactions on Automatic Control* 57(11) (2012): 2746-2761.

[5] Scokaert, P.O.M., Mayne D.Q. "Min-max feedback model predictive control for constrained linear systems." *IEEE Transactions on Automatic Control* 43(8) (1998): 1136-1142.

[6] Bernard, C. "Convex partitions of polyhedral: a lower bound and worst-case optimal algorithm." *SIAM Journal on Computing* 13(3) (1984): 488-507.

532878 A Nonlinear Programming Framework for Estimating Spatial Coupling and Seasonal Transmission Parameters in Disease Transmission

Abstract:

Highly infectious pathogens can cause localized outbreaks that are followed by regional extinction as susceptible hosts are exhausted. Spatial transmission of these pathogens can once again spark new outbreaks. Study of spatial transmission parameters is essential to predicting and mitigating these outbreaks, as well as quantifying disease persistence within a large population.

We address the issue of estimating disease transmission parameters using a flexible, scalable modeling framework. Spatial coupling parameters are a measure of the level of mixing of individuals between regional subpopulations [1]. Seasonal transmission parameters describe the impact of seasonal variation due to effects like degree of interaction [2]. This work focuses on estimating both spatial coupling and seasonal transmission parameters with a nonlinear programming approach for a network of subpopulations.

A framework is presented for efficient estimation of city-to-city spatial transmission rates by inferring transport information from localized disease case data using a statistical, hazard-based SIR model [3]. The estimation is demonstrated using records of measles outbreaks in 954 cities across England and Wales between 1944 and 1964. First, a stochastic model is constructed to predict spatio-temporal disease dynamics and accurately match existing datasets. A statistical hazard-based approach focusing on disease fade-out periods provides the basis for the estimation. Then, the model is extended to simultaneously estimate the city-to-city spatial transmission parameters, as well as seasonal transmission parameters. The proposed approach for this large-scale estimation accurately reproduces existing parameter estimates from [1,4], is readily scalable to larger problem sizes, and substantially reduces solution times.

References:

[1] Bjørnstad, O. N. and Grenfell, B. T. (2008). Hazards, spatial transmission and timing of outbreaks in epidemic metapopulations. *Environmental and Ecological Statistics*, 15:265–277.

[2] Word, D. P., Cummings, D. a. T., Burke, D. S., Iamsirithaworn, S., and Laird, C. D. (2012). A nonlinear programming approach for estimation of transmission parameters in childhood infectious disease using a continuous time model. *Journal of the Royal Society, Interface / the Royal Society*, 9:1983–97.

[3] Grenfell, B. T. (2000). Time series modelling of childhood diseases: a dynamical systems approach. *Appl. Statist.*

[4] Xia, Y., Bjørnstad, O. N., and Grenfell, B. T. (2004). Measles metapopulation dynamics: a gravity model for epidemiological coupling and dynamics. *The American naturalist*, 164(2):267–281.

533000 Active Fault Diagnosis for Stochastic Linear Systems: Design Criteria and Implementation Issues

Abstract:

Reliable diagnosis of faults in engineering systems is increasingly challenging because of their growing complexity and stricter requirements on performance, safety, and availability in operating such systems. The majority of available methods for fault detection and diagnosis passively relies on input-output data, which may not be sufficiently informative for timely and reliable diagnosis [1]. Two of the primary reasons the data may fail to reveal that a fault has occurred is the presence of feedback controllers and system uncertainty [2].

Methods for *active fault diagnosis*, or AFD, improve diagnosis time and confidence by increasing the amount of diagnostically relevant information in the input-output data. AFD involves acting on the system through injecting input signals that are specifically designed for investigating whether potential faults have occurred. When uncertainties are described using probability distributions, the input signals generally increase some statistical distance between the predicted outputs associated with a set of models for the faults. The larger the statistical distance, the higher the chance of reconciling the input-output data with the model that corresponds to the current fault status, which in turn results high-confidence diagnosis. However, input design for AFD is a challenging problem even in the case of linear systems, generally relying on solving nonconvex optimization problems; see, e.g., [3,4,5,6].

We here present results on a recently developed method for AFD for linear stochastic systems. The method involves minimizing a bound on the probability of misdiagnosis [7], subject to a set of models for the potential faults, bounds on the input, and probabilistic constraints on the state. Our discussion is focused on the tightness of the bound and numerical challenges in its minimization, and we demonstrate the performance of the proposed approach on a benchmark problem. Preliminary results suggest that locally optimal solutions can be determined efficiently and result in fast and reliable diagnosis. We further compare the method to other similar approaches, including [4] and [6], and discuss challenges of state chance constraints, receding-horizon implementation, and global optimization.

References

[1] M. Blanke, M. Kinnaert, J. Lunze, and M. Staroswiecki, Diagnosis and Fault-Tolerant Control. Berlin, Germany: Springer, 2nd ed., 2006.

[2] S.L. Campbell and R. Nikoukhah, Auxiliary Signal Design for Failure Detection. Princeton, NJ: Princeton University Press, 2004.

[3] R. Nikoukhah, "Guaranteed active failure detection and isolation for linear dynamical systems," Automatica, vol. 34, no. 11, pp. 1345–1358, 1998.

[4] L. Blackmore and B.C. Williams, "Finite horizon control design for optimal discrimination between several models," in Proceedings of the IEEE Conference on Decision and Control, (San Diego, CA), pp. 1147–1152, 2006.

[5] J.K. Scott, R. Findeisen, R.D. Braatz, and D.M. Raimondo, "Input design for guaranteed fault diagnosis using zonotopes," Automatica, vol. 50, no. 6, pp. 1580–1589, 2014.

[6] J.A. Paulson, T.A.N. Heirung, R.D. Braatz, and A. Mesbah, "Closed-loop active fault diagnosis for stochastic linear systems," in Proceedings of the American Control Conference, (Milwaukee, WI), 2018.

[7] T. Kailath, "The divergence and Bhattacharyya distance measures in signal selection," IEEE Transactions on Communication Technology, vol. 15, no. 1, pp. 52–60, 1967.

533078 Stochastic Multiscale Model-Based Predictive Control Via Polynomial Chaos Theory: Manufacturing of Thin Films for Pharmaceutical Applications

Abstract:

This presentation describes an integrated approach for end-to-end process optimization from mathematical modeling through to the control that explicitly addresses uncertainty throughout. The approach, which is based on polynomial chaos theory [1, 2], is demonstrated through its full implementation to a continuous pharmaceutical manufacturing process in which drying is used to produce a thin film that enables precisely controllable drug delivery [3]. A multi-scale model is constructed that consists of a two-dimensional differential algebraic equation (DAE) model for the polymer-solvent in the film [4, 5], a three-dimensional computational fluid dynamics (CFD) model for considering air flow dynamics, and a data-driven model for foil/support temperature dynamics. The multi-scale model is expanded to a stochastic model using non-intrusive polynomial chaos expansion (PCE). Since the multiscale model is too computationally costly to be used for parameter estimation, a PCE-based surrogate model [6,7] is constructed prior to estimation. The experiments are designed by D-optimal design method using the Fisher information matrix, and parameter estimation is carried out by using maximum-likelihood method with selected optimal parameter subset by Sobol indices [8]. The surrogate model is employed in a stochastic model predictive control (SMPC) algorithm for startup through quasi-steady operations. All of the algorithms are implemented on an experimental continuous thin film dryer process located at the Massachusetts Institute of Technology [3]. In the SMPC, the PCE method is combined with the input-output-based dynamic matrix control (DMC) to generate a quadratic program with feasible online computational time. The startup control was handled by extending the SMPC algorithm in [9] to apply to time-varying systems. Experimental results showed that the proposed method achieved the startup control, which aims to heat and cast simultaneously as well as reducing off-spec product. Furthermore, improved robustness with respect to the parametric uncertainty compared to the conventional DMC is demonstrated in a rigorous, non-conservative, and highly accurate uncertainty analysis applied to the full-order multiscale simulation model involving coupled nonlinear partial differential algebraic equations. Overall, this study proposes and demonstrates an approach to the self-consistent and systematic design of a control system for nonlinear distributed parameter systems in which the information is seamlessly passed from on task to the next, the effects of uncertainty are explicitly addressed in each task, no unrealistic assumptions are used at any stage of the analysis or synthesis, and the approach is computationally appropriate.

[1] R. G. Ghanem and P. D. Spanos, Spectral stochastic finite-element formulation for reliability analysis. *Journal of Engineering Mechanics*, 1991. **117**(10):2351-2372.

[2] D. Xiu and G. E. Karniadakis, The Wiener-Askey polynomial chaos for stochastic differential equations. *SIAM Journal on Scientific Computing*, 2002. **24**(2):619-644.

[3] B. L. Trout, T. A. Hatton, E. Chang, J. M. B. Evans, S. Mascia, W. Kim, R. R. Slaughter, Y. Du, H. H.

Dhamankar, and K. M. Forward, *Layer Processing for Pharmaceuticals*. International Patent PCT/US2012/035464, 2012.

[4] A. Mesbah, A. N. Ford Versypt, X. Zhu, and R. D. Braatz, Nonlinear model-Based control of thin-film drying for continuous pharmaceutical manufacturing. *Industrial & Engineering Chemistry Research*, 2014. **53**(18):7447-7460.

[5] G. D. Verros, Application of non-equilibrium thermodynamics and computer aided analysis to the estimation of diffusion coefficients in polymer solutions: The solvent evaporation method. *Journal of Membrane Science*, 2009. **328**(1):31-57.

[6] R. Schobi, B. Sudret, and J. Wiart, Polynomial-chaos-based Kriging. *International Journal for Uncertainty Quantification*, 2015. **5**(2):171-193.

[7] C. V. Mai, M. D. Spiridonakos, E. N. Chatzi, and B. Sudret, Surrogate modeling for stochastic dynamical systems by combining nonlinear autoregressive with exogenous input models and polynomial chaos expansions. *International Journal for Uncertainty Quantification*, 2016. **6**(4):313-339.

[8] A. Saltelli, M. Ratto, T. Andres, F. Campolongo, J. Cariboni, D. Gatelli, M. Saisana, and S. Tarantola, *Global Sensitivity Analysis. The Primer*, John Wiley & Sons, 2008.

[9] J. A. Paulson, A. Mesbah, S. Streif, R. Findeisen, Fast stochastic model predictive control of high-dimensional systems. in *Proceedings of the 53rd IEEE Conference on Decision and Control*, pp. 2802-2809, 2014.

534224 An Improved Set-Based State Estimation Method for Fault Detection and Diagnosis in Highly Nonlinear and Uncertain Chemical Processes

Abstract:

Due to the level complexity, integration, and automation in modern chemical processes, abnormal events such as equipment malfunctions and failures pose a serious threat to safe and profitable operation. Therefore, there is an urgent need for algorithms that can achieve early and accurate fault detection and diagnosis (FDD), which can effectively mitigate the safety risks associated with abnormal operations, as well as the associated economic losses caused by off-spec production, maintenance, and downtime. Classical data-based FDD methods have proven to be effective in many industrial chemical processes. However, the lack of high quality historical data that is relevant to current operating conditions (e.g., under a novel fault or abnormal disturbance) often leads to false alarms, missed faults, and misdiagnoses. This is a significant limitation, especially for systems with large uncertainties. A promising alternative is to exploit first principles process models, which are available at least at the level of individual process units and subsystems in many applications of interest. In the model-based approach, faults are detected and diagnosed by comparing the set of process outputs that are consistent with the model (under all relevant uncertainties) to the outputs observed from the real process. This talk specifically considers the set-based FDD approach, where this comparison is done using rigorous enclosures of the possible model outputs computed through a set-based state estimation method. The set-based approach is advantageous because it eliminates the possibility of false alarms and can even guarantee the detection of certain faults using active excitations. However, existing set-based estimation methods are either too computationally demanding for online FDD or produce bounds that are too conservative for effective FDD, especially for highly nonlinear and uncertain systems.

In this talk, we will present an improved set-based state estimation method for nonlinear discrete-time systems with large uncertainties. This method extends a novel reachability analysis algorithm based on discrete-time differential inequalities that we presented in last year's AIChE meeting. Our estimation method is performed recursively in two steps. First, the prediction step computes an enclosure of the possible model outputs under uncertainty over one discrete time step. Next, the correction step uses the processes measurements to update this enclosure by eliminating regions that are not consistent with the measurements. In contrast to existing set-based estimation methods, our prediction step makes use of our previously developed discrete-time differential inequalities method, which uses very efficient interval computations, but is effective at mitigating some key sources of conservatism typically associated with such computations in discrete-time systems. This approach also enables the use of redundant model equations to tighten the prediction bounds, which is not possible using existing approaches and has been shown to lead to much tighter enclosures for many representative reaction and separation models. Finally, we show that this reachability approach can be modified in the context of set-based state estimation to exploit past process measurement in a novel way even in the prediction step, leading to further improvements in bound accuracy. Numerical examples will be provided demonstrating the effectiveness of this approach for quickly and accurately detecting and diagnosing faults. Moreover, the advantages of our approach will be highlighted through comparisons with existing set-based FDD methods and more conventional data-based approaches.

536717 Robust Model Based Control Via Closed-Loop Reference Trajectory Optimization

Abstract:

Model predictive control (MPC) relies on a dynamic model for prediction of future process behavior in order to calculate the control input. No model is perfect, and depending on the level of uncertainty, poor control performance can result if the uncertainty is not accounted for in the control calculation, particularly with regard to process constraint violations. This has led to the development of several robust MPC strategies over the past three decades, that can be broadly classified into open-loop and closed-loop strategies. In the former, the uncertainty propagation in the predicted response does not take future input action into account, whereas some form of feedback action is considered in the latter.

Lee and Yu (1997) present robust MPC algorithms that minimize the worst-case cost (min-max) based on either open-loop or and closed-loop prediction, the latter determined via a dynamic programming approach. Kothare et al. (1996) present a robust MPC algorithm that minimizes an upper bound on a worst-case performance objective, and utilizes a state feedback control law at each prediction step. Sakizlis et al. (2004) present a multi-parametric programming formulation of robust MPC, where they consider additive time-varying uncertainty. In a subsequent contribution, Pistikopoulos et al. (2009) consider uncertainty in the state-space matrices in a multi-parametric robust MPC formulation. Multi-scenario based robust MPC formulations have been proposed by several workers. Huang and Biegler (2009) present a robust nonlinear MPC formulation, in which the predicted response is propagated for a discrete number of uncertainty realizations in an open-loop manner. Lucia et al. (2013) present a multi-stage nonlinear MPC strategy in which the predicted future states and inputs evolve in response to uncertainty realizations through a scenario tree, and in Lucia et al. (2014), compare the multi-stage scenario-based approach to an open-loop prediction strategy and one that utilizes an affine control policy. A robust MPC strategy for supply chain operation proposed by Mastragostino et al. (2014) follows a two-stage scenario-based approach.

In previous work, the idea of rigorous closed-loop prediction has been applied in a dynamic real-time optimization (DRTO) formulation that computes set-point trajectories for an underlying MPC system based on the prediction of the future plant response under the action of constrained MPC (Jamaludin and Swartz, 2017a,b). This closed-loop DRTO formulation results in a multilevel optimization problem due to the MPC optimization subproblems along the DRTO prediction horizon, and is solved by reformulating the inner subproblems by algebraic constraints corresponding to their Karush-Kuhn-Tucker (KKT) optimality conditions, resulting in a single mathematical program with complementarity constraints (MPCC).

In the present study, a robust control strategy is proposed that follows a two-layer control structure in which set-point trajectories are computed at the upper layer and passed to an underlying plant MPC. This scheme is similar in structure to that of reference management or command governors (Bemporad et al., 1997; Bemporad and Mosca, 1998; Sugie and Yamamoto, 2001). However, in reference management, the primal control system is typically an unconstrained, linear control system, with the reference signal manipulated in order to handle constraints on the closed-loop response. The proposed scheme, on the other hand, utilizes constrained MPC as the core controller, with set-point trajectories calculated based on the predicted closed-loop response under uncertainty. The reference optimization layer considers multiple scenario realizations of the plant that covers the range of uncertain parameters. The predicted closed-loop response for each plant scenario is generated through a sequence of control calculations using the nominal MPC model, and a plant response calculation, with a separate closed-loop trajectory generated for each uncertain plant scenario. Only one reference trajectory is generated by an optimization problem and shared by all nominal MPC optimization subproblems along the closed-loop prediction horizon. Two case studies based on a linear transfer function with uncertain plant gain and a nonlinear evaporator process with an uncertain plant parameter are conducted, and demonstrate that the proposed control structure can avoid constraint violations and provide faster tracking capabilities despite different real plant parameters or sudden changes in such parameters along the simulation.

References

Bemporad, A., Casavola, A., Mosca, E. (1997). Nonlinear control of constrained linear systems via predictive reference management. IEEE Trans. Auto. Control 42(3), 340-349.

Bemporad, A. and Mosca, E. (1998). Fulfilling hard constraints in uncertain linear systems by reference managing. Automatica, 33(4), 451-461.

Huang, R. and Biegler, L.T. (2009). Robust nonlinear model predictive controller design based on multi-scenario formulation. Proc. American Control Conference, St. Louis, MO, 2341-2342.

Jamaludin, M.Z. and Swartz, C.L.E. (2017a). Dynamic Real-time Optimization with Closed-loop Prediction. AIChE J., 63, 3896 â€“ 3911.

Jamaludin, M.Z. and Swartz, C.L.E. (2017b). Approximation of Closed-loop Prediction for Dynamic Real-time Optimization Calculations," Comput. Chem. Eng., 103, 23-38.

Kothare, M.V., Balakrishnan, V. and Morari, M. (1996). Robust constrained model predictive control using linear matrix inequalities. Automatica, 32(10), 1361â€"1379.

Lee, J.H. and Yu, Z. (1997). Worst-case formulations of model predictive control for systems with bounded parameters. Automatica, 33(5), 763-781.

Lucia, S., Finkler, T., Engell, S. (2013). Multi-stage nonlinear model predictive control applied to a semi-batch polymerization reactor under uncertainty. Journal of Process Control 23, 1306â€"1319.

Lucia, S., Andersson, J.A.E., Brandt, H., Diehl, M. and Engell, S. (2014). Handling uncertainty in economic nonlinear predictive control: A comparative case study. J. Process Control, 24, 1247-1259.

Mastragostino, R., Patel, S. and Swartz, C.L.E. (2014). Robust decision making for hybrid process supply chain systems via model predictive control. Comp. Chem. Eng, 62, 37-55.

Pistikopoulos, E.N., Faisca, N.P., Kouramas, K.I. and Panos, C. (2009). Explicit robust model predictive control. In Proceedings of the International Symposium on Advanced Control of Chemical Processes, 243-248.

Sakizlis, V., Kakalis, N.M.P., Dua, V. Perkins, J.D. and Pistikopoulos, E.N. (2004). Design of robust model-based controllers via parametric programming. Automatica, 40(2), 189-201

Sugie, T. and Yamamoto, H. (2001). Reference management for closed loop systems with state and control constraints. Proc. ACC, 1426-1431.

534745 Identifying Nucleation and Growth Kernels of Crystallization

Abstract:

Crystallization, a separation process, is simply attachment and detachment of molecules across time scales. Minor changes in the operating conditions have proven to change the outcome of crystallization and controlling the process of crystallization remains a challenge. Nucleation and crystal growth are the two processes of the crystallization and are modeled using the thermodynamic approaches. Mechanism of both processes is the same, but the time scales of the two processes are different. The challenge in modeling the process of crystallization is to derive nucleation rate of crystal structure and growth rates of energetically stable faces on the crystal. The theories used to find these rates rely on calculating interfacial free energies which is often difficult and lacks generality. With the help of a novel approach based molecular simulations, we were able to find the nucleation and growth rates. Once the rates are known, it is necessary to connect the operating conditions to outcome of crystallization. A set of operating conditions which lead to a specific outcome of crystallization will be called as kernel of crystallization. To effectively capture the time scales and the mechanism of crystallization without lack of generality, we propose the use of Population Balance Modeling (PBM). PBM finds the rate of change of internal coordinate and for crystallization, the internal coordinates are nucleation and growth rates. The rates obtained from different operating conditions can be connected to the outcomes of crystallization and hence PBM successfully finds the kernel of crystallization. Knowledge of these kernels can successfully control the process of crystallization and also shows potential to solve the problem of polymorphism.

538036 A Novel Mode of Supersaturation Feedback Control: Semi-Batch Cooling Crystallization By Feeding Flow Rate Profiles

Abstract:

Concentration Feedback Control (CFC) approaches, based on ATR-FTIR, ATR-UV/Vis or Raman spectroscopy coupled with chemometrics, have been widely applied in cooling and antisolvent crystallization systems at laboratory as well as industrial scales, by which the improved product quality have been extensively demonstrated [1]. In the feedback loop, traditionally the supersaturation is controlled to an optimal set point by manipulating the solution or cooling jacket temperature. However, the traditional batch cooling crystallization is limited by the temperature dependency and has limited applicability in some specific, but often encountering situations like crystallization of heat sensitive materials or polymorphic crystallization systems.

In this work, we propose a novel CFC implementation of semi-batch mode (feed hot saturated solution into lower temperature saturated solution), which can achieve the feedback control of supersaturation by manipulating the feeding flow rate rather than the temperature. In such, the crystallization temperature can be fixed throughout the process and thus eliminate the temperature dependence of the system. The semi-batch mode complements the classical mode in CFC implementation to some crystallization systems, such as enantiotropic polymorphic system, heat sensitive system, temperature dependent reversible reaction-crystallization and also could be designed as a start-up strategy for continuous MSMPR crystallizers [2].

Using the proposed operational method for enantiotropic polymorphic system, in the feeding step the process temperature is fixed or varied in a narrow range. This is advantageous over the cooling crystallization of much more stable polymorphic form, because the polymorph stability order is a function of temperature for enantiotropic polymorphic system. For instance, the transition temperature of α and β form of para-amino benzoic acid (pABA) in pure ethanol [3], is 13.8 (β form is stable under 13.8). Therefore, it is difficult to obtain pure β form with CFC in batch cooling mode because the operating temperature range was limited by this transition point. Whereas the crystallization temperature is fixed in the proposed CFC with semi-batch mode, which can be flexible for crystallization of easily design for each polymorphic form. In this context, the semi-batch operation to generate different polymorphic forms combines the advantages of batch and continuous crystallization: applicable to small scale production and stably produces one or the other pure polymorphic forms.

We modelled, simulated and evaluated three supersaturation control procedures for cooling crystallization processes: batch crystallization (T-SSC), semi-batch crystallization (F-SSC) and combined cooling and semi-batch crystallization (TF-SSC). For this, the monovariate population balance equation was used, involving nucleation and growth, solved by the method of moments and extended with the solute mass balance equation [4]. In this work, the comparison of these three scenarios will be presented through numerical simulations, which enables the identification of optimal operation modes of cooling crystallizers for various types of crystallization systems. Proof of concept experimental demonstrations for cooling and semi-bath crystallization will be also presented to support the numerical results followed by the implementation of T-SSC and F-SSC for enantiotropic polymorphic system (o-ABA) to achieve the efficient crystallization of both polymorphic forms.

References:

[1] Nagy Z K, Braatz R D. Advances and new directions in crystallization control[J]. *Annual Review of Chemical & Biomolecular Engineering, 2012, 3(5):55.*

[2] Su Q, Nagy Z K, Rielly C D. Pharmaceutical crystallisation processes from batch to continuous operation using MSMPR stages: Modelling, design, and control[J]. *Chemical Engineering & Processing Process Intensification, 2015, 89:41-53.*

[3] Hao H, Barrett M, Hu Y, et al. The Use of in Situ Tools To Monitor the Enantiotropic Transformation of p-Aminobenzoic Acid Polymorphs[J]. *Organic Process Research & Development, 2012, 16(16):35-41*

[4] Poehlein G W, Wenzel L A. Theory of particulate processes-analysis and techniques of continuous crystallization: By Alan D. Randolph and Maurice A. Larson, Academic Press, New York, *Journal of Colloid & Interface Science, 1972, 40(1):130-130.*

524824 Parameter Range Optimization and Modeling for a Reactive Crystallization Process

Abstract:

Chemical purity control is an essential component of quality risk management in active pharmaceutical ingredient (API) manufacturing. Impurity control typically relies on optimization of reaction parameters to minimize impurity formation, with downstream unit operations such as extractions and crystallizations serving as additional means of purity upgrade. In cases where crystallization occurs during the reaction, process performance can be more difficult to predict as the nucleation and uncontrolled crystallization of the product introduces stochastic variability with respect to impurity entrapment in the crystal lattice. This presentation details an industrial case study focused on process analysis and design of a robust impurity control strategy for an API step. In this step, due to a large solubility difference between API and penultimate starting material, high supersaturation conditions are achieved in the reaction, resulting in API self-nucleation and crystallization. Variability in nucleation and crystallization dynamics observed during process parameter space development resulted in variable levels of starting material entrapped in the isolated product. Extensive experimental studies were conducted to determine factors that influenced residual starting material impurity levels in isolated API, including reagent charge rate, solvent composition, and mixing conditions. A model was produced to assess process robustness over wide parameter space and quantify risks associated with operation on edges of or outside of the recommended ranges.

524971 *Enhanced Mass Transfer Process Control Via Hollow Fiber Membrane Assisted Antisolvent Crystallization*

Abstract:

Enhanced Mass Transfer Process Control via Hollow Fiber Membrane Assisted Antisolvent Crystallization

Xiaobin Jiang[a*], Linghan Tuo[a], Xuehua Ruan[b], Wu Xiao[a], Gaohong He[a,b*]

a State Key Laboratory of Fine Chemicals, Engineering Laboratory for Petrochemical Energy-efficient Separation Technology of Liaoning Province, School of Chemical Engineering, Dalian University of Technology, Dalian, Liaoning 116024, China

b School of Petroleum and Chemical Engineering, State Key Laboratory of Fine Chemicals, Dalian University of Technology at Panjin, Panjin 124221, China

Corresponding author: xbjiang@dlut.edu.cn; hgaohong@dlut.edu.cn

Abstract

Antisolvent crystallization plays an important role in various fields as an effective separation[1-4] and crystal manufacture technology[5-10]. Due to the operation under ambient temperature and atmospheric pressure, antisolvent crystallization is environment friendly and low energy consumption [11,12].

In this work, a novel hollow fiber membrane assisted antisolvent crystallization (MAAC) was proposed to enhance the antisolvent crystallization process control. PES membrane module was introduced as the key device for antisolvent mass transfer and solution mixing. An antisolvent liquid film formed on the membrane surface was renewed by the crystallization solution. The liquid film can additionally prevent the membrane from contacting with crystallization solution. By controlling the shell side flow velocity and the antisolvent mass transfer process, the antisolvent permeate rate can achieve sensitive, stable and accurate control during long term utilization.

Benefited from the nano and sub-nano size pores on the membrane surface and high packing density of the membrane module, MAAC greatly improved the micromixing between antisolvent and crystallization solution and reducing the local supersaturation degree. The calculated interfacial mass transfer rate of MAAC was approximately 1/50 of the one of the conventional antisolvent crystallization. On the basis of this effective and accurate antisolvent mass transfer control, MAAC provided the erythritol crystal products with better morphology, narrower CSD and smaller C.V. compared to the products manufactured by conventional antisolvent crystallization under the same antisolvent adding rate and operation duration.

Additionally, due to the the intrinsic feature of antifouling of the proposed MAAC operation, the permeate rate of MAAC could keep a stable value in a long running and the permeate performance of the membrane module was repeatable after more than twenty four times utilization. With well-controlled nucleation and crystal growth in the membrane module reducing the required retention time and the size of the tank for suspension liquid in MAAC, the production capacity *P.C.* (kg products¡¤m^{-3}¡¤h^{-1}) can be also enhanced via MAAC.

Figure. 1 Comparison of size distributions of aspect ratio (A), crystal morphology (B) and interfacial mass transfer rate (C) between erythritol produced by MAAC and conventional antisolvent crystallization after 1 hour operation.

Acknowledgements

We acknowledge financial contribution from National Natural Science Foundation of China (Grant No. 21527812, 21676043, U1663223, 21606035), Changjiang Scholars Program (T2012049), the Fundamental Research Funds for the Central Universities (DUT16TD19, DUT17ZD203) and Education Department of the Liaoning Province of China (No. LT2015007).

References

1. Biradha K, Su CY, Vittal JJ. Recent Developments in Crystal Engineering. *Crystal Growth & Design.* 2017;11(11):875-886.

2. Cisternas LA, V¨¢squez CM, Swaney RE. On the design of crystallization-based separation processes: Review and extension. *Aiche Journal.* 2010;52(5):1754-1769.

3. Simon L, Allan MS. Continuous antisolvent plug-flow crystallization of a fast growing API. Paper presented at: International Symposium on Industrial Crystallization - Isic2011.

4. Ulrich J, Frohberg P. Problems, potentials and future of industrial crystallization. *Frontiers of Chemical Science and Engineering.* 2013;7(1):1-8.

5. Lu H, Wang J, Wang T, Wang N, Bao Y, Hao H. Crystallization techniques in wastewater treatment: An overview of applications. *Chemosphere.* 4// 2017;173:474-484.

6. Park M-W, Yeo S-D. Antisolvent crystallization of carbamazepine from organic solutions. *Chemical Engineering Research and Design.* 2012;90(12):2202-2208.

7. Zhang C-t, Wang H-r, Wang Y-l. Internally generated seeding policy in anti-solvent crystallization of ceftriaxone sodium. *Chemical Engineering and Processing: Process Intensification.* 2010;49(4):396-401.

8. Aditya NP, Yang H, Kim S, Ko S. Fabrication of amorphous curcumin nanosuspensions using ¦Â-lactoglobulin to enhance solubility, stability, and bioavailability. *Colloids and Surfaces B: Biointerfaces.* 2015;127:114-121.

9. Zijlema TG, Geertman RM, Witkamp GJ, Rosmalen GMV, Graauw JD. Antisolvent Crystallization as an Alternative to Evaporative Crystallization for the Production of Sodium Chloride. *Industrial & Engineering Chemistry Research.* 2000;39(5):1330-1337.

10. † GXZ, Mitsuko Fujiwara, Yi Woo, ¡ì Xing, et al. Direct Design of Pharmaceutical Antisolvent Crystallization through Concentration Control. *Crystal Growth & Design.* 2006;6(4):892-898.

11. Lindenberg C, Krättli M, Cornel J, Mazzotti M, Brozio J. Design and Optimization of a Combined Cooling/Antisolvent Crystallization Process. *Crystal Growth & Design.* 2009;9(2):1124-1136.

12. Nowee SM, Abbas A, Romagnoli JA. Antisolvent crystallization: Model identification, experimental validation and dynamic simulation. *Chemical Engineering Science.* 2008;63(22):5457-5467.

525202 Experimental Implementation of a Model-Free Feedback Controller for the Size and Shape of Needle-like Crystals Growing in Suspension

Abstract:

Particle size and shape have a major impact on downstream processes in a crystallization process. The size and shape influences the physical characteristics of the final product obtained at the end of a crystallization step. Feedback control-based approaches have been proposed to control the size and shape of a population of crystals. Application of feedback control in a process can be particularly powerful as they have the ability to guide a population towards a desired target despite the presence of un-modeled phenomena and process disturbances. The prerequisite for the development and implementation of such feedback controllers are the availability of online monitoring and characterization tools for both the solid and liquid phase.

Studies aimed at experimental application of feedback control for size and shape control are scarce in literature [1]. This could be partly attributed to the unavailability of tools that could characterize the size and shape of a population of crystals. Recent advancements in imaging based tools have nevertheless addressed the above-mentioned issue of solid phase characterization [2]. Independent simulation studies performed using model-based and model-free feedback controllers in a growth-dominated seeded batch cooling crystallization process have led us to conclude that a model-free controller can robustly reach final target dimensions with a comparatively low level of complexity. The model-free approach requires only knowledge of the thermodynamics eliminating the need to develop multi-dimensional kinetic models for the system-of-interest [3].

In this work, a feedback controller was implemented in a lab scale batch crystallizer to guide two different seed populations of β L-glutamic acid in water towards desired target dimensions under growth-dominated conditions. The size and shape of the population of crystals was characterized using the online monitoring tool, μ-DISCO [2]. The experimental campaign was a two-step process, involving namely constant supersaturation controller (CSC) and model-free path following controller (PFC) experiments. The CSC experiments were performed at four different constant supersaturation levels to serve two purposes. First, to verify the monotonic behavior of the ratio of the rate of change of the average particle dimensions in the population with temperature, which is essential for the PFC. Second, to estimate an attainable region in the particle size and shape space where the controller could act on the population of crystals.

Upon the completion of the first step, PFC experiments were performed on both population of crystals, to reach two different target dimensions that falls within the attainable region. Repeatability experiments for all the PFC experiments were also performed. The controller was able to successfully reach target dimensions with only the thermodynamic knowledge of the system despite the controller acting on different batches of seed crystals, and operating conditions. The robustness provided by the controller used and experimental verification of the same leads us to conclude that model-free approach towards size and shape control could be a promising alternative to model-based control strategies.

References

[1] Eisenschmidt, H.; Bajcinca, N.; Sundmacher, K. Optimal Control of Crystal Shapes in Batch Crystallization Experiments by Growth-Dissolution Cycles. *Cryst. Growth Des.* **2016**, *16* (6), 3297–3306.

[2] Rajagopalan, A. K.; Schneeberger, J.; Salvatori, F.; Bötschi, S.; Ochsenbein, D. R.; Oswald, M. R.; Pollefeys, M.; Mazzotti, M. A Comprehensive Shape Analysis Pipeline for Stereoscopic Measurements of Particulate Populations in Suspension. *Powder Technol.* **2017**, *321*, 479–493.

[3] Ochsenbein, D. R.; Schorsch, S.; Vetter, T.; Mazzotti, M.; Morari, M. Growth Rate Estimation of β L-Glutamic Acid from Online Measurements of Multidimensional Particle Size Distributions and Concentration. *Ind. Eng. Chem. Res.* **2014**, *53* (22), 9136–9148.

537109 Morphology Based Adsorption Kinetics of a Selectively Modulated Metal-Organic Framework

Abstract:

Metal organic frameworks (MOFs) have gained interest in recent decades due to their high porosity, tunable pore aperture, and easily synthesized crystalline structures for applications such as catalysis, separations, sensing, gas storage, medicine, and energy. MOFs are composed of metal ions or oxo-metallic secondary building units (SBUs) coordinated with organic linkers to form a porous, open framework. The rational selection of these components in addition to modulating species and structure directing agents allows for the control of pore geometry and size at the surface of MOF crystals. By tuning the crystal size, morphology, and the dominant surface pore as determined by the crystal orientation, it is possible to tune the selectivity of a MOF to enhance the adsorption and separation of a given species of gas molecule. Previous results have shown that a copper based MOF with anisotropic pore distribution in the unit cell, $[Cu_2(ndc)_2(dabco)]_n$, can show controlled crystalline morphology through the addition of modulators during the synthesis. In this work, we leverage the modulation of crystallization growth processes to synthesize MOF surfaces with selective pore sizes and orientations. We demonstrate how changing the morphology of MOFs can affect the selectivity and adsorption kinetics of select species. The relationship between particle morphology, surface pore size and the adsorbed species is discussed. Additionally, tuning MOF morphology and orientation for higher selectivity and separations is discussed.

525827 Effects of Scale-up on the Mechanism and Kinetics of Crystal Nucleation

Abstract:

Crystal nucleation is an important step in solution crystallization of pharmaceutical compounds as it determines the crystal shape, polymorphic form, chiral form and the crystal size distribution of the product. Understanding this fundamental phase separation of molecules from the liquid to the solid phase is essential to exhibit process control. Nucleation is a stochastic process for unseeded crystallization processes involving low supersaturation levels or for small crystallizer volumes. Stirred small volume solution methods can be utilized to measure the stochastic nature of nucleation and to estimate nucleation kinetics. This way, probability distributions of induction times can be obtained from isothermal induction time experiments which can be used in conjunction with the classical nucleation theory (CNT) to estimate the nucleation rate and other important nucleation parameters. However, it remains unclear how scale-up affects the nucleation mechanism and nucleation kinetics in solution crystallization processes as the isothermal induction time method is currently limited to small scale experiments.

In this work, the effects of scale-up on the mechanism and kinetics of crystal nucleation will be presented. Four different crystallization setups were investigated, ranging from small magnetically-stirred 10 mL solutions to overhead-stirred solutions of 680 mL. The onset of crystallization was characterized using a Mettler Toledo Focussed Beam Reflectance Measurement (FBRM) probe. The crystallizer was programmed to start a new experiment after the FBRM probe detected crystallization. This way, many induction times were recorded in an automated fashion. The nucleation kinetics were estimated from the probability distributions of the induction times.

Nucleation kinetics were estimated for paracetamol in 2-propanol in the different crystallization platforms and at different supersaturation ratios. The nucleation rate was an order of magnitude faster in the magnetically-stirred crystallizer as compared to the crystallizers involving overhead stirring. The thermodynamic part of nucleation did not significantly change the nucleation rate whereas the kinetic nucleation parameter was found to be the rate-determining process when the crystallization process was scaled-up. In particular, the shear rate was rationalized to be the part of the kinetic parameter that changes most significantly when the crystallization process was scaled-up. The shear rate increases with volume and this effect becomes stable after the experiment has sufficiently increased in volume. In addition to these kinetic results, experiments involving the chiral sodium chlorate showed that the single nucleus mechanism is the underlying nucleation mechanism in all four tested crystallization setups. On the other hand it was found that cooling crystallization experiments were driven by a multi-nucleus mechanism. The presence of the single nucleus mechanism provides support for the use of probability distributions of induction times to estimate nucleation kinetics across all four tested crystallization setups.

In conclusion, a robust methodology was used to automatically record induction times using an FBRM probe. This approach is easily transferable and allows to capture induction times in different crystallizer types and volumes. The results presented herein on paracetamol and sodium chlorate contribute to a better understanding of the scale-up effects on the mechanism and kinetics of crystal nucleation of fine chemicals.

530421 Accelerating High-Throughput Experimentation with Templated Statistical Analysis

Abstract:

Automated high throughput experimentation for process development allows researchers to rapidly evaluate the effects of process conditions on the performance of synthetic steps. At BMS, an established workflow is used to conduct highly automated experimental designs. The final step of the workflow involves merging the automated reaction conditions with sample chromatograms (HPLC, GC, or LC-MS reports) and analyzing the data using statistical software. Often this is a bespoke, manual process that may be challenging to reproduce when the need arises in the future.

This presentation reports the development of a statistical analysis template that automates much of the repetitive, manual work of data analysis. Through a web interface, statistical analysis reports are reproducibly created from the raw sample data and shared with the project team. The work of merging data files, empirical model building, plotting trends in the data, response surfaces, and suggested future experiments are performed automatically with a minimal amount of user input. In addition to standardized data products, project teams have the flexibility to customize the analysis to address specific hypotheses. This template has been successfully applied at BMS to both automated and manually executed experimental designs.

536403 Process Fingerprinting Tools in the Development of an Alkylation Reaction

Abstract:

The final chemical transformation in the synthesis of doravirine (an NNRTI drug intended for the treatment of HIV) is an alkylation reaction with several key process variables and interactions that had historically produced high sensitivities in reaction performance. In this work, high-throughput experimentation and data-intensive PAT tools are applied to characterize and optimize this reaction for greatly improved performance. With the use of automated reactors and sampling devices, large data sets are produced to measure reaction kinetics, build models, and define ideal operating parameters to reduce impurity generation and process variability. In-situ FTIR, FBRM, and PVM are also applied to characterize the multi-phase reaction mixture by measuring solubility and addressing scale-up effects. This case study highlights the ability of these tools to enable rapid, data-driven decision making in various phases of chemical process development.

537623 Reactor Design for Continuous Synthesis of Pharmaceutical Intermediates: Correlation of Computational Fluid Dynamics and Design of Experiment Approaches

Abstract:

Reactor design principals as applied to the continuous production of pharmaceutical intermediates are important due to the increasing need to control processes and reduce costs. The Bill and Melinda Gates Foundation is dedicated to the reduction of costs of critical AIDS treatments for the developing world. Investigations into the cost reduction of the integrase inhibitor, dolutegravir, have encouraged the exploration of continuous reactor design and optimization in the conversion of keto-ester 1 into intermediate enamine-ester **3**. Thermodynamic and kinetic studies of the reaction have been used as inputs into comprehensive computational fluid dynamic studies of the reaction in order to model mixing efficiency and extent of reaction. In a parallel approach, design of experiment was used in order to study reactor design factors using conversion as an output. The result of both approaches will be discussed along with attempts to develop a predictive correlation for the production of this important intermediate in the dolutegravir synthesis.

528173 Optimal Design of Experiments for Building Fundamental Models of Pharmaceutical Production Processes

Abstract:

In many chemical and pharmacological systems, fundamental models play a significant role in design and optimization of industrial processes. When building a fundamental model, knowledge about chemical, physical and biological laws is often used to develop a model or propose rival models that can describe the underlying process behaviour. Usually, these models contain adjustable parameters that need to be estimated using available experimental data. Often available experimental data are not sufficiently informative for model building and validation, so modeler developers must perform additional experiments to obtain data to achieve improved parameter estimates and model predictions. Experimental runs may be expensive and time-consuming, especially if the new operating conditions lie outside of the routine operating range of the system. Therefore, experiments should be designed carefully to save time and effort while obtaining sufficient additional information.

Different experimental design strategies can be applied to obtain improved parameter estimates and model predictions Factorial and fractional-factorial experimental designs are commonly used because they are easy to implement and interpret. However, these designs are somewhat inflexible because they do not account for prior experimental data, nor for any structural information contained in the fundamental model equations. To overcome these difficulties, sequential model-based optimal design techniques can be used to select appropriate experimental settings that will lead to accurate parameter estimates and model predictions

Objective functions for sequential design may be A-optimal (aimed at reducing variance of parameter estimates), D-optimal (aimed at reducing the size of the joint confidence region for the parameters) or V-optimal (aimed at reducing uncertainty in model predictions at key operating conditions). An important computational step when performing traditional A-,D- and V-optimal design calculations is inversion of the Fisher information matrix (FIM). Unfortunately, in practical sequential design situations for fundamental models with numerous parameters, the FIM may be non-invertible. The main objective of the current study is to investigate the effectiveness of the several approaches for addressing the problem of noninvertible FIM when designing new sequential model-based experiments for pharmaceutical production processes. Proposed approaches for dealing with this problem include: i) a â€œleave outâ€ approach where problematic parameters are identified and left out of the design problem (LO approach), ii) a pseudo-inverse approach where the Moore Penrose pseudo inverse is used (PI approach) and iii) an approach where modelers select new experiments based on their prior knowledge without using a formal experimental design technique (MS approach). For the MS approach, the new experimental settings are chosen randomly to be either the lower or the upper bounds for the corresponding decision variables.

This study focuses on sequential A-optimal design because the A-optimality criterion is relatively easy for modelers to interpret and is the most readily usable design criterion for situations where the FIM is noninvertible. Two case studies are considered. The first case study is a simple linear example model with 7 parameters. In this example, old experimental results are available that result in perfectly correlated parameters, which makes the FIM noninvertible. Additional sequential experiments are designed to obtain parameter estimates with the lowest total variance. The second case study involves a Michael addition reaction used by a pharmaceutical company. Old experimental data are available at settings that result in a noninvertible FIM. New A-optimal sequential experiments are designed to select the recipes for one or two additional batch experiments that will lead to the best possible parameter estimates. The investigations are carried out using Monte Carlo simulations. Simulated datasets with random noise are generated for both case studies. The effectiveness of the LO, PI and MS approaches are compared using an A-optimal criterion.

The results for the two case studies show that the LO approach is better on average for achieving parameter estimates and model prediction that are closer to their true values. In addition, the results illustrate that, even with a badly conditioned or non-invertible FIM, it is advantageous to design sequential experiments using the LO approach rather than arbitrarily choosing new experimental settings at the bounds of the operating range. Designing more than one new experiment improves the quality of the parameter estimates. Also, if the number of sequential experiments is large enough, the FIM can become invertible so that the LO and PI approaches give the same results.

525884 Model-Based Design of Experiments for Pharmaceutical Reaction Development

Abstract:

Optimal experimental design has long been a strategy for maximizing knowledge capture while minimizing experimental cost in many modeling applications for diverse fields such as geology, physics, biology, and economics. Where physical models are available to describe data, model-based design of experiments provides a framework for efficient experimentation to develop and refine model systems and their parameters. Pharmaceutical process development is no exception to the need for efficient experiments with limited material, resources, and time. In this work, model-based design of optimal experiments has been applied to systems of synthetic organic chemical reactions relevant to pharmaceutical process chemistry. Model discrimination techniques are used to discern potential mechanisms with targeted experimentation. Parameter identifiability and estimability analysis is applied to justify model structure and suggest optimal analytical methods. Parameter precision is maximized by predictions of sensitivity within traditional and novel objective functions applied to parallel experimental design. These model-based optimization strategies are combined in a workflow that provides resource-efficient reaction model building in short development timelines.

534959 Defining the Optimal Operating Window for Pharmaceutical Reactions Using the Dynamic Response Surface Methodology for All Measured Species

Abstract:

For pharmaceutical reactions, it is of interest to identify the end-of-reaction conditions that lead to a final mixture in which the product is above a certain benchmark concentration and the concentrations of the by-products or intermediates are below target values. For robust manufacturing processes, the end of reaction conditions must be stable to ensure that impurity growth does not occur during the in-process control sample procedure or during other events that may require extended cycle times. Numerically, this robustness can be defined as the conditions that maximize the time window where the batch concentrations remain within the product quality constraints. This is achieved here by utilizing the Dynamic Response Surface Methodology (DRSM) Model[1,2] a recently proposed data-driven method that models time-resolved outputs.

Here, we apply a constrained variation of the DRSM-2[2] model in a simulated pharmaceutical reaction network of eight reactions involving ten species, as well as experimental data from the two collaborating pharmaceutical companies. The DRSM models for the time evolution of the measured species concentrations under different operating conditions play a crucial role in the solution of the above optimization problem. The optimal operating widows is initially estimated from the normal predictions of the DRSM model. This can be updated by accounting for the prediction interval of the DRSM. We examine an additional robustness issue related to the sensitivity of the calculated operating window with respect to a possible variation of the operating conditions from the calculated optimal ones.

References:

1. Klebanov N, Georgakis C. Dynamic Response Surface Models: A Data-Driven Approach for the Analysis of Time-Varying Process Outputs. *Industrial & Engineering Chemistry Research.* 2016;55(14):4022-4034.
2. Wang ZY, Georgakis C. New Dynamic Response Surface Methodology for Modeling Nonlinear Processes over Semi-infinite Time Horizons. *Industrial & Engineering Chemistry Research.* 2017;56(38):10770-10782.

534317 Sensitivity Analysis and Identification of Feasible Region of a Wet Granulation Continuous Pharmaceutical Manufacturing Process

Abstract:

Driven by the advances in continuous pharmaceutical manufacturing when compared to batch processes, extensive research is focused on developing predictive models for continuous unit operations. While this is leading to detailed and more accurate models, it is accompanied by an increase in model complexity and computational expense. This is further amplified when several such models are integrated to develop a flowsheet model that simulates the dynamic behavior of a plant scale manufacturing process. The limitations are particularly exacerbated for integrated models simulating the wet granulation solid oral dosage manufacturing route as different approaches are used to model unit operations in this route. For example, feeders and blenders are modeled using semi-empirical approaches where as units involving particle size changes such as granulation and milling are simulated using population balance models[1]. Discrete element method models are developed to understand and capture particle scale interactions that would render more accuracy to the model. In order to fully realize the potential of the developed models through applications such as sensitivity analysis and feasibility analysis, there is a need to develop and implement strategies to overcome the computational limitations.

Sensitivity analysis is the study of how uncertainty in a model output can be assigned to different sources of uncertainty in the model input[2]. Sources of uncertainty exist through potential variability in raw material quality, operating conditions and production rate demands. As flowsheet models have several input and output variables, it is important to identify the variables that have a significant effect on the properties of interest. This allows prioritization of the input factors and simplification of the high dimensional problem that aids in further applications such as feasibility analysis. Feasibility analysis is used to identify the feasible region of a process that is characterized by the range of input variables within which the process meets equipment, quality and production constraints[3]. Traditional methods to identify feasible regions incur large sampling costs which is prohibitive for a wet granulation flowsheet model. In addition, some constraints are not available in closed form expression. Alternatively, they may be available but expensive to evaluate[4].

Current work utilizes surrogate based feasibility analysis method[5] that is based on building surrogate models used to approximate the function of interest. In this work, a reduced order model is built for feasibility function, which characterizes the maximum constraint violation. The surrogate model accuracy is improved through selection of new samples that are identified using an adaptive sampling strategy. The strategy uses a modified Expected improvement function to direct search towards feasible region boundaries and unexplored regions[6]. The developed approach is utilized to identify feasible regions for an integrated wet granulation line that simulates a plant scale operation.

References

1. Boukouvala F, Chaudhury A, Sen M, et al. Computer-Aided Flowsheet Simulation of a Pharmaceutical Tablet Manufacturing Process Incorporating Wet Granulation. *J Pharm Innov.* 2013;8(1):11-27.
2. Saltelli A, Annoni P, Azzini I, Campolongo F, Ratto M, Tarantola S. Variance based sensitivity analysis of model output. Design and estimator for the total sensitivity index. *Computer Physics Communications.* 2010;181(2):259-270.
3. Wang Z, Escotet-Espinoza MS, Ierapetritou M. Process analysis and optimization of continuous pharmaceutical manufacturing using flowsheet models. *Computers & Chemical Engineering.*
4. Rogers A, Ierapetritou M. Feasibility and flexibility analysis of black-box processes Part 1: Surrogate-based feasibility analysis. *Chem Eng Sci.* 2015;137:986-1004.
5. Wang Z, Ierapetritou M. A novel feasibility analysis method for blackâ€box processes using a radial basis function adaptive sampling approach. *Aiche J.* 2017;63(2):532-550.
6. Boukouvala F, Ierapetritou MG. Derivativeâ€free optimization for expensive constrained problems using a novel expected improvement objective function. *Aiche J.* 2014;60(7):2462-2474.

533385 Dynamic Modeling of Phase Changes in Liquid-Liquid Mixtures

Abstract:

Dynamic simulation of chemical processes is important for optimization and control of transient process behavior, e.g., batch processes, startup and shut-down. Most chemical processes contain separation of mixtures. These unit operations are commonly described using a phase equilibrium assumption, e.g., vapor-liquid equilibrium (VLE) for distillation, liquid-liquid equilibrium (LLE) for extraction, or vapor-liquid-liquid equilibrium (VLLE) for heteroazeotropic distillation. Fundamentally, phase equilibrium at given pressure and temperature is found at the global minimum of the Gibbs free energy of the system [1, 2, 3]. In dynamic simulation, the number of phase varies due to vanishing and (re)appearing phases. This poses a major challenge for simulation. The different approaches for phase equilibrium calculation can be classified according to the most popular modes for chemical process simulation as sequential-modular (SM) and equation-oriented (EO) [4]. In the SM approach, the correct phase regime and the corresponding phase compositions are determined in an iterative procedure by repeatedly solving flash equations with a different phase number until a physical solution determines the actual state of the system. A major challenge in coupling these sequential-modular phase calculations to dynamic equation-oriented simulation is the handling of vanishing and (re)appearing phases and their composition. For dynamic simulation, we ideally aim at continuous and differentiable formulations. Thus, assigning arbitrary value for mole fractions of non-existing phases is not desirable because it leads to discontinuities at phase transition. In addition, dynamic simulation is more efficient if derivatives are used that, however, are not directly available from SM flash calculations. Using finite difference may lead to inaccuracy and round-off errors [5]. Within the past decades, different works contributed to the development of EO formulations starting with the formulation of Gibbs free energy minimization and derivation of first-order (necessary but not sufficient) optimality conditions thereof [6]. They then further relax the (necessary but not sufficient) equilibrium condition if a phase vanishes. This allows tracking a local solution with continuous and differentiable equations. The resulting algebraic system of equation enables dynamic simulation in all phase regimes of most VLE systems. We examine the application of this approach to LLE problems as proposed in literature [6, 7, 8] and show limitations of the formulation. The limitations arise because the equilibrium condition for each component is relaxed by a single auxiliary variable and there is no desired solution in certain areas of the single-phase regime. In addition, we propose a hybrid continuous formulation to overcome these limitations for dynamic simulation of phase changes in LLE systems and demonstrate this in an illustrative case study.

References:

[1] Baker LE, Pierce AC, Luks KD. Gibbs Energy Analysis of Phase Equilibria. Society of Petroleum Engineers Journal. 1982;22(05):731-742.

[2] Michelsen ML. The isothermal ash problem. Part I. Stability. Fluid Phase Equilibria. 1982;9(1):1-19.

[3] Mitsos A, Barton PI. A dual extremum principle in thermodynamics. AIChE Journal. 2007;53(8):2131-2147.

[4] Biegler LT, Grossmann IE, Westerberg AW. Systematic methods of chemical process design. Upper Saddle River, New Jersey: Prentice Hall. 1997.

[5] Kamath RS, Biegler LT, Grossmann IE. An equation-oriented approach for handling thermodynamics based on cubic equation of state in process optimization. Computers & Chemical Engineering. 2010;34(12):2085-2096.

[6] Gopal V, Biegler LT. Smoothing methods for complementarity problems in process engineering. AIChE Journal. 1999;45(7):1535-1547.

[7] Biegler LT. Nonlinear programming: Concepts, algorithms, and applications to chemical processes. MOS-SIAM series on optimization. Philadelphia: Society for Industrial and Applied Mathematics and Mathematical Optimization Society. 2010.

[8] Sahlodin AM, Watson HAJ, Barton PI. Nonsmooth model for dynamic simulation of phase changes. AIChE Journal. 2016;62(9):3334â€“3351.

534009 From Graphical to Optimization-Based Distillation Column Design: A Mccabe-Thiele-Inspired Math Program

Abstract:

The design and optimization of distillation columns have drawn considerable attention in the chemical engineering literature for more than hundred years. In general, there are three types of design methods: (1) graphical methods such as the McCabe-Thiele diagram [1]; (2) short-cut methods such as the Underwood equation [2]; and (3) rigorous tray-by-tray methods [3, 4]. While the graphical methods are straightforward and intuitive, they are not suitable for math-programming-based models. On the other hand, equation-based shortcut methods can be easily implemented in an optimization model, but they are based on various assumptions and simplifications, the most important being the approximation of the vapor-liquid equilibrium (VLE) behavior. Finally, the rigorous methods rely on minimum assumptions and can be formulated as mathematic programming models, but the energy balances rely on detailed calculations of thermodynamic properties (e.g. enthalpy), which can be computationally demanding. Not surprisingly, model accuracy comes at a steep price: complexity and computational intractability.

While graphical methods have some important limitations, such as limited degrees of freedom, they also have two important advantages: (1) they are intuitive and offer significant insights, and (2) â€œcomplex thermodynamicsâ€ can be readily represented graphically. As a result, they are still used nowadays for preliminary design. The natural question then becomes: can we develop flexible math-programming-based approaches than maintain the advantages of graphical methods? To the best of our knowledge this seemingly straightforward question has not been addressed. Accordingly, the goal of this work is to address it.

First, we propose a mixed integer nonlinear programming (MINLP) model for distillation column design based on the concepts and equations underpinning the McCabe-Thiele method as well as ideas used in rigorous methods. Specifically, the model involves tray-by-tray concentration calculations (as in rigorous models), which are based however on material balances written around different sections of the distillation column (as in the McCabe-Thiee method). Furthermore, similar to the McCabe-Thiele method, the concentration relationship between vapor and liquid is described by a vapor-liquid equilibrium (VLE) line, which can be approximated using piecewise linear functions or polynomial functions. Going beyond the graphical methods, the model employs binary variables to determine the optimal number of trays and optimal feed locations.

The proposed model is extended to account for multicomponent mixtures and complex columns (e.g., multiple feeds and multiple side streams). In that respect, it is a generalization of the McCabe-Thiele method and its representation as a math program. We also show that, in some cases, major assumptions (e.g., equimolar overflow) can be relaxed.

The decision variables of the model include not only column design specifications (e.g. number of trays and diameter) and operating conditions (e.g. reflux ratio and heat duty), but also flowrate and composition of streams entering (e.g., feed) and leaving (e.g., top and bottom) the column. Therefore, the proposed model can be readily used as a sub-model for optimization-based process synthesis. Finally, the applicability of our model is demonstrated through a cryogenic air separation example.

Reference

[1] W. L. McCabe and E. Thiele, "Graphical design of fractionating columns," *Industrial & Engineering Chemistry*, vol. 17, pp. 605-611, 1925.

[2] A. Underwood, "Fractional distillation of multicomponent mixtures," *Industrial & Engineering Chemistry*, vol. 41, pp. 2844-2847, 1949.

[3] J. Viswanathan and I. E. Grossmann, "Optimal feed locations and number of trays for distillation columns with multiple feeds," *Industrial & engineering chemistry research*, vol. 32, pp. 2942-2949, 1993.

[4] H. Yeomans and I. E. Grossmann, "Optimal design of complex distillation columns using rigorous tray-by-tray disjunctive programming models," *Industrial & engineering chemistry research*, vol. 39, pp. 4326-4335, 2000.

528581 An MINLP Formulation for the Optimization of Heat-Pump Assisted Distillation Configurations

Abstract:

Distillation is the predominant choice of unit operation for the separation of a multicomponent mixture into its constituent components. It is well-known that the number of distillation configurations available increases combinatorially with increase in number of components. All the configurations differ in terms of energy requirement. This gives rise to two questions: For a given system, (a) what is the optimal configuration? (b) What are the optimal operating conditions?

Here, we propose a method that allows the screening of distillation configurations based on the minimum heat pump work required for the separation, which is equal to the net exergy added to the system. Exergy of a stream is its maximum work potential when it is brought to a reference state along a reversible path. In addition to the overall energy requirement, the proposed screening criterion also accounts for temperature levels of condensers and reboilers. This criterion is especially useful for cryogenic distillations, which are work-driven, like air and Natual Gas Liquid (NGL) separations.

Agrawal and Herron (1997) described a method to compute the exergy input solely in terms of relative volatilities and composition of various streams. Using this approach, we formulate a novel Mixed-Integer Nonlinear Program (MINLP) for the identification of exergetically attractive distillation configurations. However, the model, in its natural form, has a large number of nonlinear and non-convex terms. This is exacerbated by the fact that the model is numerically sensitive, which leads to errors in computation of lower bounds. We address these challenges by employing a simple, yet powerful, variable elimination process. This process significantly reduces the number of non-convexities by collapsing the non-convex equilibrium equations onto a single equation. The new equation has several interesting properties that can be exploited to improve the convergence of branch-and-bound method. Furthermore, the reduced model is free from numerical sensitivities. Finally, with the net heat pump work as the objective, the MINLP is solved to global optimality with BARON [2]. The utility of the formulation will be illustrated with a few case studies.

1. Agrawal, R., & Herron, D. M. (1997). Optimal thermodynamic feed conditions for distillation of ideal binary mixtures. AIChE journal, 43(11), 2984-2996.
2. Tawarmalani, M., & Sahinidis, N. V. (2005). A polyhedral branch-and-cut approach to global optimization. Mathematical Programming, 103(2), 225-249.

523814 New Isotherm Model for S-Shaped Isotherm Data to be Used in Process Modeling and Its Model Reduction with Machine Learning Techniques

Abstract:

Title: New Isotherm Model for S-shaped Isotherm Data to be Used in Process Modeling and its Model Reduction with Machine Learning Techniques

Seongbin Ga, Sangwon Lee, Jihan Kim*, Jay H. Lee*

Email: jayhlee@kaist.ac.kr

This work suggests a new isotherm model for S-shaped isotherm data with an offset. This model enable the process simulation to express many adsorption isotherm phenomena with offset such as H_2O adsorption. Although molecular simulations can predict this type of pure component isotherm data with the consideration of the molecular interactions, an ordinary isotherm model covering the S-shaped isotherm has been absent. In particular, the process level simulation requires the ordinary isotherm model representing the sorbent property. The absence of the model hinders the process level evaluation of many adsorbents. Furthermore, the isotherm information provided by its experiments or its molecular simulations cannot be directly used in the process simulation because the information is, in many cases, about the pure component rather than about the mixture gas. For this, the spreading pressure for the suggested isotherm model is also derived by integral of the uptake-pressure ratio in order to use ideal adsorption solution theory (IAST), which is a method to describe the mixture isotherm behaviors from the pure isotherm models. The computation cost reduction from this derived analytical equation is shown in the comparison with that of a numerical integral method. During this computation, the implicit structure of IAST equations leads to the increased computation cost as well as divergence issues. To solve this problem, a simple machine learning method is used, and the neural network model with an explicit set of equations is found. The performance of this model is also tested.

529988 Heat Pump Assisted Configurations for Amine Based Natural Gas Sweetening Units

Abstract:

Natural gas, taken from gas wells, is a mixture of several hydrocarbons namely methane, ethane and slightly heavier hydrocarbons like propane and butane; amongst which, methane is the chief and the predominant component which constitutes almost 80-90 % of natural gas. Natural gas, from gas wells, also has impurities in the form of CO_2, H_2S, water, mercaptans etc. These impurities have to be removed for further usage of natural gas such as for pipeline distribution, liquefaction and/or fractionation. The acid gases, a mixture of CO_2 and H_2S, are very critical impurities and have to be removed as they can cause corrosion to the equipments used in the processing and distribution of natural gas. Even for natural gas distribution, the pipeline limits specified in United States are $CO_2 < 2$ mol% and $H_2S < 4$ ppm [1]. The pivotal and foremost step in the natural gas processing is the acid gas removal which removes acid gases from the natural gas. This process is also called natural gas sweetening. Among the different acid gas removal processes, amine based absorption is the most commonly used process because of its simplicity, flexibility in handling large quantity of feeds and easy retrofit ability [2]. A simplified flowsheet of an amine based natural gas sweetening system contains an absorber and a solvent regenerator (distillation column). The feed natural gas contacts with the lean amine based solvent in the absorber. Here an exothermic chemical reaction takes place between the solvent and the gas stream; the resultant sweet gas exits out from the same. The rich solvent (now loaded with acid gas) is then stripped from the acid gases in a regenerator. Large amounts of energy (reverse of the chemical reaction, thus endothermic), however, is required for the regeneration of the solvent. This makes this process highly energy intensive (3-4 MJ/kg of CO_2), which happens to possibly be a major drawback of this process [2, 3].

For reducing the energy consumption in the amine based natural gas sweetening systems, different studies have been carried which includes development of new and mixed amine solvents, optimizing the process and structural configuration of the system etc. [4-6]. An interesting development in this regard is the usage of heat pump in the acid gas removal system. The heat pump is for the solvent regeneration unit. The concept in a heat pump assisted scheme is to effectively heat integrate the condenser (rectification section) and reboiler (stripping section) thermal energy to reduce the heat to be supplied to the reboiler. Since the temperature of the rectification section is lesser than the stripping section, a mechanical means (compressor) is used where heat is moved from the heat source at a lower temperature to a heat sink at a higher temperature. This kind of a heat pump is called a mechanical vapor recompression (MVR). An interesting way to increase the energy efficiency of MVR is to utilize the available heat within the MVR to further heat other process streams in the process. This is called self-heat recuperation (SHR); the combination of MVR and SHR has been extensively used in many distillation based systems [7-10].

The objective of this work is to develop different process configurations of the solvent regeneration unit within an amine based natural gas sweetening system utilizing the concepts of MVR and SHR; and, to study the energy consumption of the same with respect to the conventional acid gas removal system. The process flowsheets were developed and simulation using Aspen Hysys V8.4 and the feed considered was 10 MMscfd of natural gas (82.5% methane, 10% ethane, 5% CO_2 and 2.5% H_2S; 824.7 psia and 37.8^{0}C). 45 wt.% methyl diethanolamine (MDEA) was used as the amine solvent. Four different and new configurations were developed and the results were analyzed. It was observed that our new configurations showed at least 15-20% savings in the energy consumption in comparison to the conventional process for the same feed and almost the same acid gas recovery. Consequently, there is also reduction in the CO_2 emissions compared to the conventional acid gas removal units. Future work will seek to optimize these configurations within a simulation-optimization framework with respect to the cost of the process.

References:

1. Mokhatab S., Poe W.A., Speight J.G. Handbook of Natural Gas Transmission and Processing, Gulf Professional Publishing. 2012.
2. Song C, Liu Q, Ji N, Deng S, Zhao J, Kitamura Y. Natural gas purification by heat pump assisted MEA absorption process. Applied Energy. 2017; 204:353-61.
3. Niu MW, Rangaiah GP. Retrofitting amine absorption process for natural gas sweetening via hybridization with membrane separation. International Journal of Greenhouse Gas Control. 2014; 29:221-30.
4. Muhammad A, GadelHak Y. Simulation based improvement techniques for acid gases sweetening by chemical absorption: A review. International Journal of Greenhouse Gas Control. 2015; 37:481-91.
5. El Hadri N, Quang DV, Goetheer EL, Zahra MR. Aqueous amine solution characterization for post-combustion CO2 capture process. Applied Energy. 2017; 185:1433-49.
6. Wang M, Joel AS, Ramshaw C, Eimer D, Musa NM. Process intensification for post-combustion CO2 capture with chemical absorption: A critical review. Applied energy. 2015; 158:275-91.
7. Long NV, Lee M. Novel acid gas removal process based on self-heat recuperation technology. International Journal of Greenhouse Gas Control. 2017; 64:34-42.
8. Gao H, Zhou L, Liang Z, Idem RO, Fu K, Sema T, Tontiwachwuthikul P. Comparative studies of heat duty and total equivalent work of a new heat pump distillation with split flow process, conventional split flow process, and conventional baseline process for CO2 capture using monoethanolamine. International journal of greenhouse gas control. 2014; 24:87-97.
9. Christopher CC, Dutta A, Farooq S, Karimi IA. Process Synthesis and Optimization of Propylene/Propane Separation Using Vapor Recompression and Self-Heat Recuperation. Industrial & Engineering Chemistry Research. 2017; 56:14557-64.
10. Kazemi A, Mehrabani-Zeinabad A, Beheshti M. Recently developed heat pump assisted distillation configurations: A comparative study. Applied Energy. 2018; 211:1261-81.

533030 Conceptual Design and Exergy Analysis of the Cryogenic Energy Storage System Integrated with LNG Cold Utilization

Abstract:

According to the Outlook for Energy (ExxonMobil, 2016), liquefied natural gas (LNG) trade is expected to almost triple in the next three decades.[1] Natural gas is transported in liquid state for long-distance because the volume of LNG is approximately 600 times smaller than that of the same mass of gas phase natural gas.[2] LNG must be re-gasified before it is transported to end users.[3] Generally, the cold exergy of LNG is wasted to seawater during the regasification process.[4] Thus, recovering the cold exergy of LNG is one of the most important issue in LNG regasification step.[5]

This study focuses on the development of an efficient cryogenic energy storage (CES) process using the cold exergy from liquefied natural gas (LNG) regasification. While LNG has low internal energy, it has high exergy because of its cryogenic characteristics, and much of this exergy is wasted in the regasification process.[6] Thus, this work focuses on the recovery of LNG cold exergy to store cryogenic energy using air as a working fluid. The power generation process by LNG direct expansion requires pumping work to increase the pressure, heat for the LNG vaporization, and the power is generated by an expansion step. The available energy forms produced by LNG regasification process are cold and shaft work. On the other hand, a CES process using air as the working fluid requires shaft work to compress the air and cold to liquefy it. Therefore, the supply and demand of the LNG regasification power generation process and the CES process are well aligned. In this study, we discuss the integration of shaft work and cold to LNG regasification power generation process with direct expansion and a CES process using air. The cold of LNG is transferred in two forms: cold transfer by heat exchange to liquefy air, and shaft work transfer by direct expansion of LNG to compress the air. Thermodynamic analysis of the proposed combined CES and LNG regasification process is carried out in three exergy flow steps: the LNG regasification step, the air liquefaction step, and the air expansion step. Based on exergy analysis, the exergy efficiency is calculated to be 54.4, 94.2, and 61.1 %, for the LNG regasification, air liquefaction, and liquid air expansion steps, and 32.6 % for the overall proposed process. The net amount of work produced by CES system is 67.44 kJ/kg-LNG with 93.48 kJ/kg-LNG of storage work and 160.92 kJ/kg-LNG of work produced by air release process. In addition, the proposed system has an advantage which system can storage and release the energy simultaneously. Therefore, daily produced energy per 1 kg of LNG by CES system is more than double compare to the most recent contributions with storage and release time divided systems. This study not only proposes an efficient energy storage process that can generate power flexibly but also highlights further possibilities for performance enhancement by thermodynamic analysis.

REFERENCES

[1] ExxonMobil, "The Outlook for Energy: A view to 2040," (2016).

[2] **Lee, I.,** *et al.*, "Design and optimization of a pure refrigerant cycle for natural gas liquefaction with subcooling.," *Ind. Eng. Chem. Res.,* **53** (25), 10397-403 (2014).

[3] **Lee, I.,** *et al.*, "Decision Making on Liquefaction Ratio for Minimizing Specific Energy in a LNG Pilot Plant," *Ind. Eng. Chem. Res.,* **54** (51), 12920-12927 (2015).

[4] **Szargut. J., and Szczygiel. I.,** "Utilization of the cryogenic exergy of liquid natural gas (LNG) for the production of electricity," *Energy* **34** (7), 827-837 (2009).

[5] **Gómez. M.R.,** *et al.*, "Review of thermal cycles exploiting the exergy of liquefied natural gas in the regasification process," *Renew. Sust. Energy. Rev.,* **38**, 781-795 (2014).

[6] **Lee, I.,** *et al.*, "Conceptual design and exergy analysis of combined cryogenic energy storage and LNG regasification processes: Cold and power integration," *Energy,* **140**, 106-115 (2017).

533756 Global Deterministic Surrogate-Based Process Design

Abstract:

Surrogate-based process optimization is increasingly popular and is believed to gain more importance as digitalization of industry further progresses because more data becomes available. In the previous literature, there have been many efforts to combine surrogate modeling with (global) optimization. For instance, Gaussian processes (GPs) have been used in the field of Bayesian optimization for optimization of expensive-to-evaluate black-box functions (e.g., [1,2]). In adaptive approaches, GPs emulate black-box objectives, e.g., measurements of a chemical experiment [3], in every iteration to assist a following query point selection. ALAMO is another adaptive sampling approach [4-6] that aims the development of simple surrogate models in the light of small data sets. Artificial neural networks (ANNs) are one of the most commonly used machine-learning tools and are well-established in both academia and industry [7]. In contrast to standard GPs, ANNs can have multiple outputs and can handle large training sets (i.e., GP training requires N by N matrix inversions, where N is the number of training points).

Various works on ANN-based process optimization are available (e.g. [7-10]). In these works, ANNs learn unit operations or sub-process from simulations or experimental data. Then, the ANNs are aggregated to entire (hybrid) process models, which are finally optimized. In the previous literature, ANN-based process optimization has mostly been limited to local or stochastic global solution approaches (e.g., genetic algorithm). These methods can fail to identify global optima and cannot give a guarantee for global optimality. Recent advances in global deterministic optimization with ANNs embedded allow to overcome this limitation [11].

The emerging research field of global deterministic ANN-based process optimization develops new potentials but raises research questions, which are discussed using an organic Rankine cycle process optimization as an illustrative case study. Herein, the complex Helmholtz thermodynamics of the ethanol working fluid is learned via ANNs to eliminate implicit functions and auxiliary variables from the optimization problem. The results show that the solution time of the problem is reduced significantly using the proposed approach. Further, the presentation discusses the influence of the ANN size on its prediction accuracy and the efficiency of global optimization, i.e., tightness of its relaxations.

[1] Bradford, E., Schweidtmann, A.M., Lapkin, A.A.: Efficient multiobjective optimization employing gaussian processes, spectral sampling and a genetic algorithm. Global Optimization Journal (2018).

[2] Forrester, A.I., Keane, A.J.: Recent advances in surrogate-based optimization. Progress in Aerospace Sciences 45(1-3), 50-79 (2009).

[3] Schweidtmann, A.M., Holmes, N., Bradford, E., Bourne, R., Lapkin, A.: Multi-objective self-optimization of continuous flow reactors. In preparation (2018).

[4] Wilson, Z.T., Sahinidis, N.V.: The ALAMO approach to machine learning. Computers & Chemical Engineering 106, 785-795 (2017).

[5] Cozad, A., Sahinidis, N.V., Miller, D.C.: A combined first-principles and data-driven approach to model building. Computers & Chemical Engineering 73, 116-127 (2015).

[6] Cozad, A., Sahinidis, N.V., Miller, D.C.: Learning surrogate models for simulation-based optimization. AIChE Journal 60(6), 2211-2227 (2014).

[7] Meireles, M., Almeida, P., Simoes, M.G.: A comprehensive review for industrial applicability of artificial neural networks. IEEE Transactions on Industrial Electronics 50(3), 585-601 (2003).

[8] Nascimento, C.A.O., Giudici, R., Guardani, R.: Neural network based approach for optimization of industrial chemical processes. Computers & Chemical Engineering 24(9-10), 2303-2314 (2000).

[9] Henao, C.A., Maravelias, C.T.: Surrogate-based superstructure optimization framework. AIChE Journal 57(5), 1216-1232 (2011).

[10] Smith, J.D., Neto, A.A., Cremaschi, S., Crunkleton, D.W.: CFD-based optimization of a flooded bed algae bioreactor. Industrial & Engineering Chemistry Research 52(22), 7181-7188 (2013).

[11] Schweidtmann, A.M., Mitsos, A.: Global Deterministic Optimization with ANNs Embedded. Preprint available at arXiv:1801.07114 (2018).

533563 Scheduling with Preemption

Abstract:

The most well-known scheduling formulations from the Process Systems Engineering (PSE) community do not account for the possibility of processing tasks to be interrupted. Yet, preemption is often encountered in daily practice, with the ISA-95 standard, which can act as a data-exchange platform for production scheduling [1], including planned break periods (e.g., due to preventive maintenance, weekends), number of stops allowed per task, minimum duration of each partial task, and possible penalties for interrupting a task.

Models allowing for preemption exist in the context of project scheduling. In [2], an activity with discrete duration *d*can be interrupted up to *d*times, with the time(s) of interruption being determined by a genetic algorithm. In this paper, we consider a more constrained form of preemption, where tasks can be interrupted at a given point in time, provided that they continue immediately after the break. Long tasks may be interrupted multiple times.

We will show how to handle preemption in discrete- and continuous-time formulations. On the one hand, we focus on the Resource-Task Network (RTN) discrete-time formulation [3], which is known to be very tight and better at handling events such as time-dependent electricity pricing [4]. The required changes occur at the level of the structural parameters and excess resource balance constraints, with the model variables remaining the same (it is straightforward to extend a State-Task Network based discrete-time formulation, such as the one in [5-6], to handle preemption in a similar way). On the other hand, we use Generalized Disjunctive Programming (GDP) [7-9] and a compact convex hull reformulation to derive the complex timing constraints of the continuous-time formulation. These are incorporated into a model for multistage batch plants that relies on the general precedence concept [10-11].

The performance of the proposed formulations is evaluated on an industrial problem featuring 24 orders and 4 stages (96 tasks) [4] considering the objective of makespan minimization. The results show a better performance for the discrete-time formulation (DT), which is able to solve the problem to optimality in less than one hour, whereas the continuous-time model (CT) can only handle 14 orders in the same computational time. This, despite DT having a worse relaxation and generating MILPs that are one order of magnitude larger in size. Preemption allows to reduce the makespan without bringing too much complexity into the model. In fact, CT can actually be solved faster compared to the non-preemptive case, probably due to the lower integrality gap.

Future work will deal with single stage batch plants with parallel units, to see if the ranking is altered when using a tighter continuous-time formulation, one with multiple time grids [9] and without super big-M constraints.

Acknowledgments: Financial support from Fundação para a Ciência e Tecnologia (FCT) through projects IF/00781/2013 andUID/MAT/04561/2013.

References:

[1] I. Harjunkoski, R. Bauer, 2014, Sharing Data for Production Scheduling using the ISA-95 Standard, Frontiers in Energy Research, doi: 10.3389/fenrg.2014.00044.

[2] V.V. Peteghem, M. Vanhoucke, 2010, A Genetic Algorithm for the Preemptive and Non-Preemptive Multi-mode Resource-Constrained Project Scheduling Problem, European Journal of Operational Research, 201,409-418.

[3] C.C. Pantelides, 1994, Unified Frameworks for the Optimal Process Planning and Scheduling, In Proceedings of the Second Conference on Foundations on Computer Aided Operations, Cache Publications, New York, 253.

[4] P.M. Castro, L. Sun, I. Harjunkoski, 2013, Resourceâ€"Task Network Formulations for Industrial Demand Side Management of a Steel Plant. Ind. Eng. Chem. Res. 52, 13046.

[5] E. Kondili, C.C. Pantelides, R.W.H. Sargent, 1993, A General Algorithm for Short-term Scheduling of Batch Operations â€" I. MILP Formulation. Comp. Chem. Eng. 2, 211.

[6] H. Lee, C.T. Maravelias, 2017, Discrete-time Mixed-integer Programming Models for Short-term Scheduling in Multipurpose Environments. Comp. Chem. Eng. 107, 171-183.

[7] E. Balas, 1979, Disjunctive programming, Annals of Discrete Mathematics 5, 3-51.

[8] R. Raman, I.E. Grossmann, 1994. Modeling and computational techniques for logic based integer programming. Comput. Chem. Eng. 18, 563-578.

[9] P.M. Castro, I. E. Grossmann, 2012, Generalized disjunctive programming as a systematic modeling framework to derive scheduling formulations. Ind. Eng. Chem. Res. 51, 5781-5792.

[10] I. Harjunkoski, I.E. Grossmann, 2002, Decomposition Techniques for Multistage Scheduling Problems using Mixed-Integer and Constraint Programming Methods. Comput. Chem. Eng.26, 1533.

[11] C. Méndez, J. Cerdá, 2002, An efficient MILP continuous-time formulation for short-term scheduling of multiproduct continuous facilities. Comput. Chem. Eng. 2002, 26, 687.

525315 Robust Planning and Scheduling for Processes with Equipment Degradation

Abstract:

Process planning and scheduling frequently assumes perfect equipment availability and performance. In reality, equipment degradation can cause deteriorating process performance or equipment failure if maintenance is not carried out frequently enough. Process performance and equipment availability are therefore dependent on the selected maintenance strategy. Furthermore, the rate at which equipment health degrades may be partially random and affected by the selected operating strategy. Process scheduling and planning are both known to be hard problems. Integrating them and including effects of maintenance and uncertain equipment degradation is therefore challenging.

Multiple authors have addressed planning and scheduling with equipment degradation by proposing integrated process and maintenance models which explicitly incorporate unit degradation, e.g. [1–4]. Most of these works assume deterministic relationships between degradation, process performance, operating variables, and/or time. In contrast, the field of Condition-based maintenance (CBM) has attracted significant attention in recent years by using data-informed, stochastic degradation models to infer equipment health, albeit mainly at unit level [5]. We argue that properly accounting for equipment degradation requires new integrated process and maintenance planning and scheduling models which exploit the more sophisticated data-driven stochastic degradation models developed in CBM.

To this end, we show how Lévy type models [6], a class of stochastic processes commonly used for degradation modelling in CBM, can be incorporated into a MILP scheduling and/or planning model using the Lappas and Gounaris [7] adjustable robust optimization approach. Robust optimization has been applied to scheduling and planning by multiple authors, e.g. [7-9], but it has not been applied, to the best of our knowledge, to uncertainty in equipment degradation. We account for effects of the operating strategy on degradation by allowing the Lévy models parameters to depend on a set of discrete operating modes. As demonstrated by Li and Li [10], selecting an appropriate uncertainty set size in robust optimization is challenging and may be treated as its own optimization problem. We describe the uncertainty set size through a single parameter which we optimize by solving the robust MILP model repeatedly. Since this model can be computationally expensive, we propose using Bayesian optimization, which is known to work well for low dimensional problems with expensive to evaluate objective functions. Bayesian optimisation can also manage the noise introduced when some of the MILP sub-solves cannot be solved to optimality in a reasonable amount of time.

We furthermore propose a cheap way of estimating probabilities of equipment failure. We generate data regarding the relative frequency of occurrence of operating modes by solving a short-term scheduling model repeatedly. This data can be used to construct a Markov chain from which a large number of long-term schedules can be generated cheaply. These schedules are not necessarily feasible in the original problem but can be used to obtain a good estimate of equipment failure probabilities.

We apply our framework to an integrated planning and scheduling model recently proposed by Biondi et al. [3] with explicit treatment of unit degradation for the state-task-network (STN) originally developed by Kondili et al. [12]. We demonstrate that robust optimization is capable of trading of equipment availability and cost of maintenance for a number of STN instances and we show that Bayesian optimization can be used to optimize the uncertainty set size in a computationally efficient way.

[1] N. I. Zulkafli and G. M. Kopanos, "Integrated condition-based planning of production and utility systems under uncertainty," *J. Clean. Prod.*, vol. 167, pp. 776–805, 2017.

[2] S. Liu, A. Yahia, and L. G. Papageorgiou, "Optimal Production and Maintenance Planning of Biopharmaceutical Manufacturing under Performance Decay," *Ind. Eng. Chem. Res.*, vol. 53, no. 44, pp. 17075–17091, 2014.

[3] M. Biondi, G. Sand, and I. Harjunkoski, "Optimization of multipurpose process plant operations: A multi-time-scale maintenance and production scheduling approach," *Comput. Chem. Eng.*, vol. 99, pp. 325–339, 2017.

[4] C. G. Vassiliadis and E. N. Pistikopoulos, "Maintenance scheduling and process optimization under uncertainty," *Comput. Chem. Eng.*, vol. 25, no. 2–3, pp. 217–236, 2001.

[5] S. Alaswad and Y. Xiang, "A review on condition-based maintenance optimization models for stochastically deteriorating system," *Reliab. Eng. Syst. Saf.*, vol. 157, pp. 54–63, 2017.

[6] D. Applebaum, "Lévy processes-from probability to finance and quantum groups," *Not. Am. Math. Soc.*, vol. 51, no. 11, pp. 1336–1347, 2004.

[7] N. H. Lappas and C. E. Gounaris, "Multi-stage adjustable robust optimization for process scheduling under uncertainty," *AIChE J.*, vol. 62, no. 5, pp. 1646–1667, 2016.

[8] Z. Li, R. Ding, and C. a Floudas, "A Comparative Theoretical and Computational Study on Robust Counterpart Optimization: I. Robust Linear Optimization and Robust Mixed Integer Linear Optimization," *Ind. Eng. Chem. Res.*, vol. 50, no. 18, pp. 10567–10603, 2011.

[9] C. Ning and F. You, "Data-driven robust MILP model for scheduling of multipurpose batch processes under uncertainty," *2016 IEEE 55th Conf. Decis. Control.*, 2016.

[10] Z. Li and Z. Li, "Optimal robust optimization approximation for chance constrained optimization problem," *Comput. Chem. Eng.*, vol. 74, pp. 89–99, 2015.

[11] D. R. Jones, M. Schonlau, and W. J. Welch, "Efficient Global Optimization of Expensive Black-Box Functions," *J. Glob. Optim.*, vol. 13, pp. 455–492, 1998.

[12] E. Kondili, C. C. Pantelides, and R. W. H. Sargent, "A general algorithm for short-term scheduling of batch operations—I. MILP formulation," *Comput. Chem. Eng.*, vol. 17, no. 2, pp. 211–227, 1993.

537894 A Simultaneous Process Scheduling and Personnel Allocation Framework for Industrial-Scale Multipurpose Facilities

Abstract:

Process scheduling aims to allocate tasks to resources at specific time intervals with the aim to meet an objective, e.g. minimization of operating costs, turnaround time or maximization of profits. Often, the processing times of the resources (e.g. machines) are either specified *a priori* or estimated from the amount of material to be processed in the machines [1]. However, there are industrial applications where the processing time also depend on the skills of the technical staff operating the machine. This condition is often found in manufacturing facilities that involve human intervention, e.g. weighting of samples before processing or start-up of machines. A more skilled (senior) operator may perform an activity in processing times that are shorter than those required by a junior (inexperienced) operator thus impacting operations management and plant productivity. Consequently, process scheduling and personal allocation decisions must be simultaneously considered, i.e. assign operators to resources while performing the optimal allocation of tasks to resources. A naive approach to address this problem consists in solving both problems sequentially. While this approach is simple and easy to implement, it may lead to sub-optimal or even infeasible solutions [2]. Therefore, integrated approaches have been proposed in the literature to address this problem [2], [3], [4]; however, industrial-scale applications aiming at solving personnel allocation and process scheduling are rather limited [5], [6].

The aim of this study is to present a mathematical formulation that integrates process scheduling and personnel allocation decisions for job shop large-scale plants. This formulation is based on network flows and comprises a large number of variables and constraints, which requires large computational efforts. To circumvent this issue, we propose in this work an alternative solution method with optimal guarantees that evaluates solutions faster for some groups of instances. The performance of the proposed formulation and methods was tested using instances based on an actual industrial-scale facility previously introduced in the literature [7], [8]. Optimal integrated schedules were obtained using a rolling horizon framework and compared to the case where personnel allocation decisions are not considered together within the process scheduling. The results show that significant improvements can be achieved when using the proposed integrated approach.

References

[1] Ierapetritou M., Floudas C. Effective continuous-time formulation for short-term scheduling. 1. Multipurpose batch processes. Industrial & Engineering Chemistry Research. 1998; 37(11): 4341-4359.

[2] Daniels R.L., Mazzola J.B. Flow shop scheduling with resource flexibility. Operations Research. 1994;42:504–522.

[3] Artigues C., Gendreau M., Rousseau L., Vergnaud A. Solving an integrated employee timetabling and job-shop scheduling problem via hybrid branch-and-bound. Computers & Operations Research. 2009;36:2330–2340.

[4] Guyon O., Lemaire P., Pinson E., Rivreau D. Solving an integrated job-shop problem with human resource constraints. Annals of Operations Research. 2014;23:147–171.

[5] Schulze M., Zimmermann J. Staff and machine shift scheduling in a German potash mine. Journal of Scheduling. 2017;20:1–22.

[6] Ahmadi-Javid A., Hooshangi-Tabrizi P. Integrating employee timetabling with scheduling of machines and transporters in a job-shop environment: A mathematical formulation and an anarchic society optimization algorithm. Computers & Operations Research. 2017;84:73–91.

[7] Lagzi S., Fukasawa R., Ricardez-Sandoval L. A multitasking continuous time formulation for short-term scheduling of operations in multipurpose plants. Computers & Chemical Engineering. 2017;97:135–146.

[8] Lagzi S., Lee D.Y., Fukasawa R., Ricardez-Sandoval L. A computational study of continuous and discrete time formulations for a class of short-term scheduling problems for multipurpose plants. Industrial & Engineering Chemistry Research. 2017;56(31):8940–8953.

536883 A Novel Mathematical Model for Short-Term and Medium-Term Scheduling of Multipurpose Batch Plants

Abstract:

Short-term scheduling of multipurpose batch plants has gained a lot of attention during the past three decades [1-3]. Different mathematical models such as discrete-time models [4-5] and continuous-time models including sequence-based models [8-9], process slot-based models [6], unit-slot models [7], global event-based models [10-11], and unit-specific event-based models [12-15] have been developed. The scheduling horizon is represented using time points, batches, slots, or event points. The computational efficiency and solution optimality of all these models largely depend on the number of time points (batches, slots, or event points) that are required to generate the optimal solution. In other words, reducing the required number of time points (batches, slots or event points) could lead to substantial decrease in computational expense. The advantages of the unit-specific event-based modelling approach have been well established in the literature [13-14], which often requires fewer number of event points than discrete-time, process slot-based, and global event-based modelling approaches. However, it seems that the proposed unit-specific event-based models are still computationally inefficient especially for large-scale industrial problems since a high number of event points are still required. Furthermore, a consumption task is always sequenced with any production task in different units related to the same state regardless of whether this consuming task consumes materials either from a specific production task or from storage tank [15-16]. Although this unconditional sequencing has been mitigated in recent two models [15-16], they may either lead to large model sizes which require excessive computational effort to generate the optimal solution or even sub-optimality and infeasibility.

In this work, we develop a novel unit-specific event-based formulation for short-term scheduling of multipurpose batch plants, where production and consumption tasks related to the same state are allowed to take place at the same event, which could significantly reduce the number of event points that are required to generate optimal solution. The definition of recycling tasks from Li and Floudas [14] and recycling states is employed to avoid infeasibility arising from allowing recycling tasks to take place at the same event points with their related consumption tasks with the same recycling states. To further reduce the number of event points that are required, we introduce a new binary variable to explicitly identify the specific production task or storage that provide materials to a consumption task. We sequence a consumption task only with the related specific production task that provides materials to this consumption task, not all related production tasks. Additionally, when a task consumes materials that already exist in storage, we do not need to sequence it with any production task. To illustrate the capability of our proposed model, we solve a number of examples that are well established in the literature. The computational results demonstrate that the proposed model significantly reduces the number of event points that are required and decreases the model size. As a result, our proposed model requires less computational time to generate the optimal solution compared to the existing models. We then incorporate our proposed model into the rolling horizon framework [17] to solve medium-term scheduling of large-scale industrial batch plants. The comparison results demonstrate that better profit could be generated with our proposed model compared to that of Janak et al. [17].

References

[1] Floudas, C. A.; Lin, X. Continuous-time versus discrete-time approaches for scheduling of chemical processes: A review. *Computers & Chemical Engineering,* 2004, 28, 2109–2129.

[2] Méndez, C. A.; Cerdá, J.; Grossmann, I. E.; Harjunkoski, I.; Fahl, M. State-of-the-art review of optimization methods for short-term scheduling of batch processes. *Computers & Chemical Engineering,* 2006, 30, 913–946.

[3] Harjunkoski I.; Maravelias, C. T.; Bongers, P.; Castro, P. M.; Engell, S; Grossmann, I. E.; Hooker, J.; Mandez, C.; Sand, G.; and Wassick, J. Scope for industrial applications of production scheduling models and solution methods. *Computers & Chemical Engineering*, 62:161 – 193, 2014.

[4] Kondili E.; Pantelides C. C.; Sargent R. W. H. A general algorithm for short-term scheduling of batch operations-I MILP formulation, *Computers & Chemical Engineering,* 1993;17(2);211-227

[5] Lee H.; Maravelias C. Discrete-time mixed-integer programming models for short-term scheduling in multipurpose environments. *Computers & Chemical Engineering* 2017, 107, 171-183

[6] Sundaramoorthy A., Karimi I. A. A simpler better slot-based continuous-time formulation for short-term scheduling in multipurpose batch plants, *Chemical* engineering science, 2005;60(10);2679-2702

[7] Susarla, N.; Li, J.; Karimi, I. A. A novel approach for short-term scheduling of multipurpose batch plants. *AIChE J.* 2010, 56, 1859-1879.

[8] Mendez, C. A.; Henning, G. P.; Cerda, J. An MILP continuoustime approach to short-term scheduling of resource-constrained multistage flowshop batch facilities. *Computers & Chemical Engineering*, 2001, 25, 701–711.

[9] Hui, C. W.; Gupta, A. A novel MILP formulation for short-term scheduling of multistage multi-product batch plants. *Computers & Chemical Engineering*, 2000, 24, 1611–1617.

[10] Castro P., Barbosa-Póvoa A. P. F. D., Matos H. An improved RTN continuous-time formulation for the short-term scheduling of multipurpose batch plants, *Industrial and Engineering Chemistry research,* 2001;40(9);2059-2068

[11] Maravelias, C. T.; Grossmann, I. E. New general continuous-time state-task network formulation for short-term scheduling of multipurpose batch plants. *Industrial and Engineering Chemistry research,* 2003, 42, 3056–3074

[12] Ierapetritou M., G. Floudas C., A. Effective continuous-time formulation for short-term scheduling. 1. Multipurpose batch processes, *Industrial & Engineering Chemistry,* 1998;37(11);4341-4359

[13] Shaik M. A., Floudas C. A. Novel Unified Modeling Approach for Short-Term Scheduling, *Industrial & Engineering Chemistry Research,* 2009;48(6);2947-2964

[14] Li, J.; Floudas, C. A. Optimal event point determination for short-term scheduling of multipurpose batch plants via unit-specific event-based continuous-time approaches, *Industrial & Engineering Chemistry Research*, 2010, 49(16), 7446-7469.

[15] Vooradi R; Shaik M. A. Rigorous unit-specific event based model for short term scheduling of batch plants using conditional sequencing and unit-wait times. *Industrial and Engineering research,* 2013, 52(36), 12950-12792

[16] Seid R.; Majozi T. A robust mathematical formulation for multipurpose batch plants. *Chemical Engineering Science,* 2012, 68(1), 36-53

[17] Janak S. L.; Floudas C. A.; Kallarth J.; Vormbrock N. Production scheduling of a large-scale industrial batch plant. I. Short-term and medium-term scheduling. *Industrial and Engineering Chemistry Research*, 2006, 45(25), 8234-8252

528598 An Algorithm Combining the Strengths of Discrete- and Continuous-Time Scheduling Models

Abstract:

While a wide variety of approaches have been proposed for chemical production scheduling, mathematical-programming-based methods are the most popular (Harjunkoski *et al.*, 2014). One of the most important attributes of the modeling approaches adopted in mathematical programming formulations is the representation of time. Mainly two different time approaches have been adopted: discrete and continuous. While the two representations have some distinct advantages, they are also limited in some different ways (Sundaramoorthy and Maravelias, 2011). Specifically, discrete-time models may provide inaccurate solutions if the discretization step is not chosen appropriately. Continuous-time models, on the other hand, are slower than their discrete-time counterparts for large-scale instances and have limited modeling flexibility.

In order to overcome this issue, a framework that allows to harness the advantages of both modeling approaches while overcoming their limitations was recently proposed (Lee and Maravelias, 2017). The framework consists of three stages: (1) obtain an approximate solution using a discrete-time MIP model with a relatively large discretization step, (2) map the solution obtained in the first stage onto multiple unit-specific and material-specific continuous-time grids, and (3) obtain an accurate solution containing the number of batches found in the first stage, by solving a continuous-time LP model. The framework is shown to offer the modeling flexibility and computational efficiency of discrete-time models, while retaining the accuracy of continuous-time models.

The goal of this work is to discuss how to enhance the three-stage algorithm. Specifically, after briefly reviewing the overall framework, we describe how to select some key user-defined parameters, namely, the discretization step and the relaxation of the scheduling horizon, both used in the 1st stage model. We show that the selection of these two parameters is nontrivial but, nonetheless, very important because the choice of these parameters impacts both the computational performance and solution quality. In particular, a larger discretization that well approximates time-related data leads to significantly smaller models as well as accurate solutions, while appropriately chosen horizon relaxation compensates for solutions that may be cut off due to discretization errors.

Specifically, we provide a way to systematically predict the best parameters, and show that it is applicable to a wide range of scheduling instances. To achieve this, we introduce several error schemes that account for various aspects, including importance of tasks, busyness of units, etc. By evaluating these errors, the best set of parameters can be determined. Through several case studies, we show that seemingly obvious parameter choices may lead to very low quality solutions, while highly counterintuitive choices result in surprisingly good solutions. We show that our method accurately identifies the best set of parameters, which in turn enhances the performance of the framework. We present computational results for a large library of instances.

Finally, we note that the proposed methods are applicable not only to the 3-stage discrete-continuous algorithm but also to any discrete-time scheduling model. In particular, we provide a method for the selection of the â€œoptimalâ€ fixed discretization step, which can in fact be rather different from what would have been chosen using previously proposed heuristic guidelines. Thus, our methods address one of the major limitations of the discrete-time models, namely, the selection of the discretization step when *coarse* steps are necessary for computational tractability.

Reference

1. Harjunkoski, I., Maravelias, C. T., Bongers, P., Castro, P. M., Engell, S., Grossmann, I. E., Hooker, J., Mendez, C., Sand, G. and Wassick, J. (2014) Scope for industrial applications of production scheduling models and solution methods. *Computers & Chemical Engineering*, **62**, 161-193.
2. Sundaramoorthy, A. and Maravelias, C. T. (2011) Computational Study of Network-Based Mixed-Integer Programming Approaches for Chemical Production Scheduling. *Industrial & Engineering Chemistry Research*, **50**, 5023-5040.
3. Lee, H. and Maravelias, C. T. (2017) Combining the advantages of discrete- and continuous-time scheduling models: Part 1. Framework and mathematical formulations, *Computers & Chemical Engineering* (in press)

528580 Batch Scheduling with Quality-Based Changeovers

Abstract:

Batch scheduling is a frequent and complex operation performed in all process industries. The combinatorial nature of the problem difficults finding non intuitive solutions to capture additional value. Because of the high frequency of these operations, even small improvements can quickly translate into millions of dollars in savings. The planning horizon spans from two days to a few weeks. In the case of discrete-time models, the horizon is divided in hourly periods. The resulting models are difficult to solve, thereby motivating intensive research in this area[1]. One of the features that adds more complexity to the problem is the presence of sequence-dependent changeovers (SDC)[2], in which the length of the cleaning task performed between two successive processing tasks depends on the sequence in which the processing tasks are performed. For example, if in a given reactor, task B is performed after task A, then a cleaning of 2 hours might be required between the tasks. However, if task A is performed after task B, maybe the cleaning time could be longer, not required, or the transition from B to A might be even forbidden. The cleaning is frequently required to prevent cross contamination between products.

In many applications, like the production of chemicals of the same product family, the cleaning can be avoided by making sure enough batches of the second product are performed in a row, and any impurities from the previous product are diluted enough such that the resulting product meet quality standards. We refer to this type of transitions as quality-based changeovers (QBC). This kind of transitions have not been addressed in the Process Systems Engineering (PSE) literature. The required models need to consider the option of performing either a quality-based changeover or a traditional cleaning changeover. We implement and test quality-based changeovers in three of the major general purpose scheduling frameworks in PSE, namely State-Task-Network (STN)[3], Resource-Task-Network (RTN)[4], and Unit-Operation-Port-State-Superstructure (UOPSS)[5]. Implementing in the three frameworks allows to compare their computational efficiency, and main features, including extensibility to accommodate novel features such as quality based changeovers. The implementations are tested in a small process network, and in an industrial-sized network. For the second case, a richer objective function is proposed, including minimization of priority-dependent backlog, changeover time, and deviation from a previously determined plan.

The results obtained give valuable insights on the three frameworks. Some highlights of the comparisons performed are:

QBC can seamlessly integrated in the STN and UOPSS frameworks by extension of their SDC constraints. On the other hand, RTN requires the addition of many auxiliary tasks and resources.

QBC reduces computational burden to the problem by tightening the SDC constraints.

Since few SDC constraints are active at the optimum, they can be effectively handled with use of lazy constraints.

UOPSS requires the addition of less SDC constraints, which makes it more efficient for large problems.

The modeling extensions allow to better represent industrial problems, while the insights obtained yield guidelines for model selection. These two aspects combined represent an important contribution to the adoption optimization of scheduling operations.

References

[1] Harjunkoski, I., Maravelias, C.T., Bongers, P., Castro, P.M., Engell, S., Grossmann, I.E., Hooker, J., Méndez, C., Sand, G. and Wassick, J., 2014. Scope for industrial applications of production scheduling models and solution methods. *Computers & Chemical Engineering*, *62*, pp.161-193.

[2] Kim, S.C. and Bobrowski, P.M., 1994. Impact of sequence-dependent setup time on job shop scheduling performance. *The International Journal of Production Research*, *32*(7), pp.1503-1520.

[3] Kondili, E., Pantelides, C.C. and Sargent, R.W.H., 1993. A general algorithm for short-term scheduling of batch operationsâ€"I. MILP formulation. *Computers & Chemical Engineering*, *17*(2), pp.211-227.

[4] Pantelides, C.C., 1994, July. Unified frameworks for optimal process planning and scheduling. In *Proceedings on the second conference on foundations of computer aided operations* (pp. 253-274). New York: Cache Publications.

[5] Zyngier, D. and Kelly, J.D., 2012, June. UOPSS: a new paradigm for modeling production planning & scheduling systems. In *Symposium on Computer Aided Process Engineering* (Vol. 17, p. 20).

531370 A Computational Comparison of New Models for the Multi-Mode Resource Constrained Project Scheduling Problem with Optional Activities

Abstract:

Project scheduling is a fundamental component of project management. However, the often large number of activities to be performed over long horizons, in conjunction with complicating constraints and sometimes conflicting objectives can pose serious challenges to obtaining optimal project scheduling solutions [1]. As a result, the multiple variants of the resource constrained project scheduling problem (RCPSP) have been a very active topic of research with a wide variety of proposed solution approaches that involve mathematical optimization, constraint programming, propositional satisfiability algorithms, and heuristic based methods [2].

The original RCPSP problem involves a set of activities that are represented via a directed graph. These activities have to be executed exactly once throughout the given horizon, while not exceeding the available resources and while not violating the prescribed precedence between them. The typical objective of this problem is to minimize the total makespan. An extension of the RCPSP is the Multi-mode RCPSP (MMRCPSP) [3], where activities are allowed to be executed in one out of multiple available modes, where each mode can have a different set of characteristics (such as duration, resource consumption, cost of execution).

While both of the aforementioned variants have illustrated proven modelling flexibility for a wide range of problems, they do rely on the fact that the directed graph of activities is fixed and is not part of the decision making. This is an assumption that can be rather limiting in environments where a decision may lead to different sets of tasks that have to be executed, or in cases where one may select to execute an optional activity that can lead to higher profit in the future.

To that end, we propose an extension to the original MMRCPSP to allow for the execution of optional activities with the use of propositional logic. To solve the MMRCPSP with optional activities to optimality, we will present a set of new Mixed Integer Linear Optimization models, and we will further extend existing MMRCPSP models [4,5,6] to handle optional tasks. Since the above models address generalizations to the RCPSP and plain MMRCPSP variants, we will conduct a comprehensive computational study across the models to compare their performance using both established RCPSP & MMRCPSP benchmarks from the PSPLIB [7], as well as synthetic benchmarks with optional activities. Finally, preprocessing and fine-tuning approaches will be presented that can further improve the computational performance of the newly introduced models.

References:

[1] J. Weglarz, "On certain models of resource allocation problems.", *Kybernetes*, vol. 9(1), pp. 61-66, 1980.

[2] C. Schwindt and J. Zimmermann, Handbook on Project Management and Scheduling Vol. 1., *Springer International Publishing*, 2015.

[3] F. B. Talbot, "Resource-constrained project scheduling with time-resource tradeoffs: The nonpreemptive case.", *Management science*, vol. 28(10), pp. 1197-1210, 1982.

[4] O. Koné, C. Artigues, P. Lopez, and M. Mongeau, "Event-based MILP models for resource-constrained project scheduling problems.", *Computers & Operations Research*, vol. 38(1), pp. 3-13, 2011.

[5] T. S. Kyriakidis, G. M. Kopanos, and M. C. Georgiadis, "MILP formulations for single-and multi-mode resource-constrained project scheduling problems.", *Computers & Chemical Engineering*, vol. 36, pp. 369-385, 2012.

[6] R. K. Chakrabortty, R. A. Sarker, and D. L. Essam, "Event based approaches for solving multi-mode resource constraints project scheduling problem.", *Proceedings of IFIP International Conference on Computer Information Systems and Industrial Management*, Springer, pp. 375-386, 2014.

[7] R. Kolisch and A. Sprecher, "PSPLIB-a project scheduling problem library: OR software-ORSEP operations research software exchange program.", *European Journal of Operational Research*, vol. 96(1), pp. 205-216, 1997.

530169 Scheduling and Analytics â€“ Towards Better Planning

Abstract:

There is a lot of hype ongoing on big data analytics (Qin, 2014) and machine learning. Among others, AIChE conference has also arranged topical conferences on the topic, and for a good reason. Most companies collect continuously data from sensors that is stored for a certain time but never actually used, unless there is a need for post analytics as part of trouble shooting (Yidan et al., 2016). One attempt to create true value from the data is to use it proactively to improve the quality and actuality of planning. Nevertheless, often a schedule that is based on statistical average data is outdated already by the time it gets sent to the plant floor and due to the hierarchical planning structures it is very difficult to quickly adapt a schedule to changing conditions. This is a challenge that has also been looked into in integration of scheduling and control studies (Touretzky et al., 2017). The presented project SINGPRO will merge Big Data platforms, machine learning and data analytics methods with process planning and scheduling optimization. The goal is to create online, reactive and anticipative tools for more sustainable and efficient operation. The currently employed classical mathematical optimization models (Harjunkoski et al., 2014) are often limited by fixed parameter sets, which are commonly updated off-line and represent only statistical averages. Such parameters could be estimated much more precisely in an on-line fashion using Big Data technologies. By creating collaboration interfaces between scheduling optimization, big data analytics and machine learning, the process related decision-making loop will become much more agile, self-aware and flexible.

With sophisticated data analytics methods, one can embed to the overall key performance indicators (KPI) also all information about the process, e.g., tracking abnormal situations (anomaly detection), individual process equipment performance degradations (predictive maintenance), anticipated process timings (prediction of process behavior) and scenario simulation (e.g., artificial intelligence AI planning). Such an approach will help to select the best production strategies in order to maintain, e.g., production and energy efficiency as well as sustainability in rapidly changing market situations through data-driven self-adaptive scheduling models. The topic of data-driven models have already been investigated in other domains (Van der Aalst et al., 2004) and tools become available for the process industry (Wilson and Sahinidis, 2017). It can be expected that Industrial Internet of Things (IIoT) provides the needed seamless connectivity, cloud computing infrastructure and service-based business models to realize this vision.

References

Qin, S. J. (2014), Process data analytics in the era of big data. AIChE J., 60, 3092-3100. doi:10.1002/aic.14523

Yidan Shu, Liang Ming, Feifan Cheng, Zhanpeng Zhang, Jinsong Zhao (2016), Abnormal situation management: Challenges and opportunities in the big data era. Computers & Chemical Engineering, 91m 104-113. doi:10.1016/j.compchemeng.2016.04.011

Touretzky, C. R., Harjunkoski, I., & Baldea, M. (2017). Dynamic models and fault diagnosis-based triggers for closed-loop scheduling. AIChE Journal, 63(6), 1959-1973. doi:10.1002/aic.15564

Harjunkoski, I., Maravelias, C. T., Bongers, P., Castro, P. M., Engell, S., Grossmann, I. E., . . . Wassick, J. (2014). Scope for industrial applications of production scheduling models and solution methods. Computers and Chemical Engineering, 62, 161-193. doi:10.1016/j.compchemeng.2013.12.001

Van der Aalst, W., T. Weijters, L. Maruster, (2004). Workflow mining: Discovering process models from event logs", *Knowledge and Data Engineering IEEE Transactions on*, vol. 16, no. 9, 1128-1142

Wilson, Z. T., & Sahinidis, N. V. (2017). The ALAMO approach to machine learning. Computers and Chemical Engineering, 106, 785-795. doi:10.1016/j.compchemeng.2017.02.010

534610 Designing Stochastic Model Predictive Control Based Neural Interface to Restore Communication between Brain Regions

Abstract:

Cerebellar (ischemic) stroke results in the loss of brain functions, which ultimately leads to severe disability in stroke survivors. An abundance of evidence shows a cerebellar stroke damages the postsynaptic dendrites of neurons, even in the absence of neuronal loss, and the synapses through which neurons communicate. In this work, we theoretically explore a possibility of whether an optimal controller can be used as an interface between two brain regions to restore the lost communication due to the loss of synapses. Particularly, we consider two interconnected networks (Network 1 and Network 2), each consists of 100 spiking neurons, as the representation of two distinct brain regions. Neurons in each network are synaptically connected within the network with probability 0.1 (sparse connectivity). Similarly, neurons in Network 1 are synaptically connected to neurons in Network 2 and vice-versa with probability 0.1. Neurons in both networks receive Poisson inputs which represent the inputs from other cortical areas of the brain. We first simulate the whole network and record the firing activity (timing of action potentials) of neurons in Network 1 before any loss in the number of synapses between Network 1 and Network 2. We then randomly remove a fraction of synapses (~30%) from Network 1 to Network 2 which results in altered dynamics and a poor communication between these two networks. Our objective is to restore this lost communication by developing a neural (brain-brain) interface where an optimal controller uses the firing activity of neurons from Network 1 to stimulate a fraction of neurons in Network 2. We particularly consider closed-loop optogenetic stimulation of neurons in Network 2 because of its capability in activating neurons selectively within a population. To design our controller, we first stimulate the neurons in Network 2 using a variety of stimulation patterns and record the firing activity (the timing of action potentials) of neurons in Networks 1 and 2. We then use these input-output data to develop a point-process generalized linear model (GLM) of neurons in Network 1. Using this model, we formulate an optimal control problem within the framework of stochastic model predictive control to restore the firing patterns of neurons in Network 1. We solve our control problem numerically and demonstrate the capability of the designed controller in restoring the lost communication between these two networks.

525047 Predictive Strategies for Control of Indoor Air Quality

Abstract:

Americans spend more than 90% of their time indoors. Air inside our homes can contain high concentrations of particulate matter (PM) of different sizes and chemical pollutants such as ozone (O_3), formaldehyde (HCHO) and volatile organic compounds (VOCs) [1,2]. Hence, controlling these contaminants of concern (COCs) is an important factor for human health. The American Society of Heating, Refrigerating, and Air-Conditioning Engineers (ASHARE) prescribes fixed ventilation rates for the operation of heating, ventilation and air conditioning (HVAC) systems, which is suboptimal in terms of pollution control and energy consumption. Moreover, different pollutants react differently to a change in manipulated variables, in this case the ventilation rate. For example, an increase in the ventilation rate might be favorable by reducing PM but could increase the concentration of ozone in the room. The aforementioned issues make indoor air quality control a primary target for model-based analysis, control and optimization. The primary objective of this work is to identify the operating conditions of the HVAC system that minimize the energy usage while keeping the concentration of COCs within acceptable limits.

We develop a first-principles model that relates the concentration of PM, HCHO and O_3 in a room to the ventilation rate and energy consumption of the HVAC system. Different pollutants have different penetration factors, deposition velocities and indoor emission rates. Therefore, the concentration of each type of pollutant is described by a separate differential equation with the above mentioned quantities as parameters determined from experiments in the published literature. The model captures real-world situations such as pollutants entering the room from outside and pollutants generated in the room due to human activities like cooking.

Next, we use the model to identify optimal operational settings. We solve two problems: (1) identify "steady-state" ventilation rate for idealized periodic variations in pollutants in the room, and (2) develop and deploy model predictive control as an online optimization strategy to calculate optimal ventilation rate at each time instant based on instantaneous measured fluctuations in indoor and outdoor conditions. The objective function of the optimization problem is the energy consumption of the system and the pollutant concentrations (either instantaneous or time-averaged) are expressed as constraints. The proposed optimization strategies show considerable reduction in energy usage compared to the existing standards, without jeopardizing indoor air quality or posing health concerns.

REFERENCES

[1] Boor, B.E., Siegel, J.A., Novoselac, A., "Monolayer and multilayer particle deposits on hard surfaces: literature review and implications for particle resuspension in the indoor environment." *Aerosol Science and Technology*, 47(8): 831-847 (2013)

[2] Morawska, L., Afshari, A., Bae, G.N., Buonanno, G., Chao, C.Y.H., Hanninen, O., Hofmann, W., Isaxon, C., Jayaratne, E.R., Pasanen, P., Salthammer, T., "Indoor aerosols: from personal exposure to risk assessment." Indoor Air, 23(6): 462-487 (2013)

532907 An Optimization-Based Feedback Control Strategy for Spatially-Uniform Dose Delivery Using Atmospheric Pressure Plasma Jets

Abstract:

Atmospheric pressure plasma jets (APPJs) are increasingly used as alternative or complementary treatments for cancer, wound healing, and surface sterilization [1]. APPJs are capable of locally generating and delivering electric fields, chemical species, and thermal effects, making them a unique device for plasma medicine. A critical challenge in application of APPJs arises from ensuring safe and reproducible delivery of a cumulative effect of plasma, i.e., plasma dose, in a spatially uniform manner across the target substrate. The sharp spatial gradients observed in temperature, species concentrations, and electric field in APPJs can significantly complicate uniform dose delivery [2]. Furthermore, internal variability [3] as well as the effect of external conditions (such as varying properties of substrate [4], or changes in ambient conditions) can drastically affect the APPJ operation. Considering that medical therapies require stringent safety considerations, the need for in-situ diagnostics and feedback control is becoming increasingly critical for high-performance (safe, reproducible, and therapeutically effective) operation of plasma medical devices [5].

In this work, we investigate the use of a model-based control strategy for spatially-uniform delivery of the thermal effects of an APPJ on an inert substrate via real-time control experiments. APPJs typically have an effective treatment area of few mm^2 because of their small dimensions. As practical applications in plasma medicine commonly require treatment of larger areas, the jet must often be translated over the treatment area. We first show that proportional-integral-derivative (PID) control cannot achieve uniform dose delivery due to the integrating (i.e., non-retractable) and spatially distributed nature of the dose delivery problem. We then present a hierarchical model-based feedback control strategy where an optimization-based supervisory controller, implemented online in a shrinking-horizon manner, determines the optimal spatial delivery of thermal effects in terms of a temperature setpoint profile, which is fed to a lower-level PID controller. It is shown that the hierarchical control strategy enables effective planning of spatial dose delivery in real time while the PID controller enables fast disturbance rejection. We demonstrate that this control strategy significantly improves the spatial uniformity of the thermal dose delivered to the target relative to the case of PID control only.

[1] T. von Woedtke, H.-R. Metelmann, and K.-D. Weltmann, "Clinical Plasma Medicine: State and Perspectives of in Vivo Application of Cold Atmospheric Plasma," *Contrib. to Plasma Phys.*, vol. 54, no. 2, pp. 104–117, Feb. 2014.

[2] J. J. Liu and M. G. Kong, "Sub-60 °C atmospheric helium–water plasma jets: modes, electron heating and downstream reaction chemistry," *J. Phys. D. Appl. Phys.*, vol. 44, no. 34, p. 345203, 2011.

[3] J. Shin and L. L. Raja, "Run-to-run variations, asymmetric pulses, and long time-scale transient phenomena in dielectric-barrier atmospheric pressure glow discharges," *J. Phys. D. Appl. Phys.*, vol. 40, no. 10, pp. 3145–3154, 2007.

[4] S. A. Norberg, E. Johnsen, and M. J. Kushner, "Helium atmospheric pressure plasma jets touching dielectric and metal surfaces," *J. Appl. Phys.*, vol. 118, no. 1, pp. 1–13, 2015.

[5] D. Gidon, D. B. Graves, and A. Mesbah, "Effective dose delivery in atmospheric pressure plasma jets for plasma medicine: A model predictive control approach," *Plasma Sources Sci. Technol.*, vol. 26, no. 8, pp. 85005–85019, 2017.

527202 Dynamic Real-Time Optimization of a Gas-Phase Polymerization Reactor

Abstract:

Continuous polyolefin manufacture is widely implemented, due to its high production rate and strong economic advantages compared with batch polymerizations. A plant produces polymers with different properties by traversing through a "wheel" of different product grades. When the grades do not have overlapping property specifications, the transitions between grades can take hours to complete and generate a large amount of "off-specification" polymers sold at substantially lower price than "on-specification" grades. The development of optimal grade transition strategies to reduce this "off-spec" offers a way to improve the line profitability / capacity without CAPEX.

The goal of this work is to develop a dynamic real-time optimization (DRTO) for transitions in a gas-phase propylene-ethylene copolymerization reactor. The optimization requires a mathematical model to accurately predict process dynamic behavior (e.g. monomer concentrations, reactor temperatures) and polymer product properties (e.g. melt flow, polymer composition). Therefore, a detailed process model is developed first. In some processes, a fluidized-bed reactor is used to carry out polymerization reactions by gas-phase monomers/ comonomers and solid catalysts. The reactor is represented by a two-phase model, consisting of an emulsion phase and a bubble phase. In order to prevent runaway reactions due to the highly exothermic reactions, the inlet gases are cooled down to a low temperature or even partially condensed before entering the reactor in industrial operations. However, this condensed mode operation is hardly considered in most fluidized-bed polymerization studies. In this study, we address the condensation of inlet gas flow by constructing a surrogate vapor-liquid equilibrium model, derived from Benedict-Webb-Rubin (BWR) equation of state, to reduce computational complexity.

With a concrete process model, the DRTO is implemented for the online optimal grade transition. Although the process is operated continuously, the grade transition can be treated as a semi-batch process, requiring satisfaction of the end-point quality constraints defined by the desired polymer grade. Consequently, a framework to combine a shrinking horizon nonlinear model predictive control (sh-NMPC) and an expanding horizon least square estimation (eh-LSE) [1] can be implemented. At each time step, eh-LSE estimates states and parameters based on measurements obtained in the past. The updated model is then implemented in the sh-NMPC to calculate optimal controls in future steps to reduce the transition time. A multi-stage transition formulation [2] is applied to directly optimize economic objective function augmented with regularization terms. With this online estimation-optimization framework, the optimal control scheme is able to reduce the transition time and better preserve path and end-point constraints under parameter uncertainties.

References

[1] Jung, T. Y., Nie, Y., Lee, J. H., & Biegler, L. T. (2015). Model-based on-line optimization framework for semi-batch polymerization reactors. *IFAC-PapersOnLine*, *48*(8), 164-169..

[2] Shi, J. , Biegler, L. T. and Hamdan, I. (2016). Optimization of grade transitions in polyethylene solution polymerization processes. AIChE J., 62: 1126-1142.

534667 Model Predictive Control of a Nonisothermal and Nonisobaric Membrane Reactor for Water-Gas Shift Reaction Applications

Abstract:

Modern hydrogen production units are tasked with producing the most hydrogen possible while dealing with flow variations caused by changing power demands. Classical methods for hydrogen production by the water-gas shift reaction are governed by equilibrium limitations that take effect at high temperatures and high concentrations of H_2. The implementation of a membrane reactor with temperature control enables the hydrogen concentration and temperature to reach an equilibrium at a higher concentration of H_2 [1]. Another challenge that is prevalent in this process is the cyclical input flowrate of the syngas produced from changing upstream reforming process conditions [2]. These cyclical changes in syngas production can cause drastic temperature changes within the reactor. These challenges can be addressed by the implementation of advanced controllers that can cope with dynamic changes associated with different conditions, such as temperature oscillations and mitigation of hot spots [1].

In this presentation, both linear and nonlinear model predictive control (MPC) methods are implemented on a designed water-gas shift membrane reactor model. The implementation aim is to maximize the production of hydrogen by considering the temperature control performed by manipulating the flow rates of the coolant entering the cooling jacket at different reactor zones. The control strategies considered for this application are: Dynamic Matrix Control (DMC), Quadratic DMC (QDMC), Nonlinear MPC (NMPC), and a Biomimetic-based controller cast as MPC (BIO-CS as MPC) [3]. The coolant usage is constrained by the use of quadratic programing (QP) or sequential quadratic programing (SQP) formulations, depending on the employed MPC type, to match industrial standards. To mimic industrial cycling conditions, the flow rate of the syngas is changed around ± 10% from its operating steady state [2]. The MPC results that will be discussed show an increase in the production of hydrogen due the temperature control under cycling process conditions.

References:

1. Georgis, D., Lima, F. V., Almansoori, A., and Daoutidis, P. (2014). Thermal Management of a Water-Gas Shift Membrane Reactor for High-Purity Hydrogen Production and Carbon Capture. *Industrial & Engineering Chemistry Research*, 7461-7469.
2. Kyriakides, A.-S., Seferlis, P., Voutetakis, S., and Papadopoulou, S. (2016). Model Predictive Control for Hydrogen Production in a Membrane Methane Steam Reforming Reactor. *Chemical Engineering Transactions*, 991-996.
3. Mirlekar, G., Li, S., and Lima, F. V. (2017). Design and Implementation of a Biologically-inspired Optimal Control Strategy (BIO-CS) for Chemical Process Control. *Industrial & Engineering Chemistry Research*, 6468-6479.

536936 Operator-Triggered Advisory System for Electric Arc Furnace Process Optimization

Abstract:

Electric arc furnaces (EAFs) are widely used in the steel industry worldwide for production of steel by melting scrap steel collected from different sources. Approximately 25% of the world's steel is produced using EAFs [1]. EAFs are huge consumers of electricity and approximately 400 kilowatt-hours is used per tonne of steel produced [2]. EAFs are currently operated with a low level of process automation and control, and decisions related to input amounts and timings are determined by the operators. Due to large fluctuations during the process, harsh operating conditions and the limited availability of measurements, real-time process optimization is very challenging [3]. The desired product quality is achieved by altering the chemistry of scrap by addition of fluxes. EAF operating decisions are taken by the operators based on what has worked in the past. Although operator's experience is important for safe process operation, complex interacting relationships are difficult to take into account. Given the high operating cost due to the level of electricity consumption, considerable savings may accrue through the development and deployment of model-based optimization strategies to support EAF operators' decision making. A real-time advisory system which computes economically optimal input trajectories based on a first-principles model is proposed in this work.

Optimization and control strategies developed in the past focused only on a sub-system of the EAF, such as electrode movement [4], rather than the entire process dynamics. It is only recently, Swartz and MacRosty studied dynamic optimization based on a detailed first principles EAF model [5]. However, a single optimization at the start of the process is not sufficient to counter various uncertainties such as initial condition discrepancy, process disturbances, plant-model mismatch and measurement noise. The effect of these uncertainties can be tackled by input trajectory manipulation during the batch process at various triggered time points [6]. The re-optimizations need the current knowledge of the plant states for accurate initialization. In this work, we introduce an operator-triggered advisory system for EAF operation that uses a multi-tiered optimization strategy coupled with multi-rate Moving Horizon Estimation (MHE) to provide decision support to the plant operators. MHE is particularly attractive due to its direct handling of nonlinear system model, easy incorporation of constraints and straightforward handling of multi-rate measurements [7, 8]. MHE runs in parallel with the plant so that the advisory system is aware of the current state of the process. The advisory system uses a multi-tiered initialization scheme in combination with state-of-the-art NLP solvers to obtain fast solutions to the optimal control problems at each sampling time. The state estimates are subsequently provided to multi-tiered optimizer to calculate the optimal input sequence to maximize the economics. However, the optimizer is called upon only when the operator triggers the advisory system. The tiered-optimization coupled with MHE has a novel mechanism to handle potential infeasibilities due to end-point constraints.

The real-time economics-based dynamic optimization problem with end-point constraints is handled through multi-tiered optimization. The multi-tiered strategy incorporates three sequential dynamic optimization tiers where the last 2 tiers come into effect if the previous tier fails to give a solution. In the first tier, *Direct optimization*, we attempt to solve the original end-point constrained dynamic optimization problem. If we detect an infeasible problem or a maximum number of specified optimization iterations is reached then we proceed to the second tier. In tier 2, *Feasibility restoration through horizon extension*, we reformulate the original dynamic optimization by extending the time horizon in integral steps to find the minimum integral extension required to yield a feasible solution. If feasibility cannot be achieved, then we move to the third tier. In the third tier, *End-point relaxation,* we remove the end-point constraint and instead minimize the state variable deviation from the upper and lower bounds at the extended end-point. Further, we propose a Sub-tier: *Extended implementation* which gets activated only at the end-time if the estimated state is not within the specified upper and lower bounds.

The advisory system is applied to a multi-part EAF case study to demonstrate its performance. The energy required to melt the scarp steel is provided in form of electrical and chemical energy. A detailed first-principles DAE model is employed that takes the complex physical phenomena into account [9, 3]. The model considers four zones for the furnace (gas, slag-metal, slag-metal, solid scrap) with material movement between the zones limited by mass transfer. Chemical equilibrium in the slag-metal and gas zones is modeled by embedding equations for Gibbs free energy minimization. The DAE model is discretized and a simultaneous approach employed for solving the dynamic optimization problems. The case studies demonstrate a significant economic benefit if the advisory system is used. We show that the advisory system is capable of providing decision support to the operators in real-time without significant delays. As a future work, we intend to investigate alternate objective function formulations where environmental effects such as emissions are also considered.

References

[1] "Steel Statistical Yearbook of World Steel Association," [Online]. Available: 1. Steel Statistical Yearbook of World Steel Association, http://www.worldsteel.org/steel- by-topic/statistics/steel-statistical-yearbook-.html. 2016.

[2] R. J. Fruehan, The Making, Shaping, and Treating of Steel: Ironmaking volume, vol. 2, AISE Steel Foundation, 1999.

[3] R. D. M. MacRosty and C. L. E. Swartz, "Dynamic modeling of an industrial electric arc furnace," *Ind.Eng.Chem.Res.,* vol. 44, pp. 8067-8083, 2005.

[4] R. Nadira and P. Usoro, "Self-adjusting model algorithmic control of a three-phase electric arc furnace," *Journal of dynamic systems, measurement, and control,* vol. 110, no. 4, pp. 361-366, 1988.

[5] R. D. M. MacRosty and C. L. E. Swartz, "Dynamic optimization of electric arc furnace operation," *AIChE journal,* vol. 53, no. 3, pp. 640-653, 2007.

[6] J. Flores-Cerrillo and J. F. MacGregor, "Control of batch product quality by trajectory manipulation using latent variable models," *Journal of Process Control,* vol. 14, no. 5, pp. 539-553, 2004.

[7] F. Allgöwer, T. A. Badgwell, J. S. Qin, J. B. Rawlings and S. J. Wright, "Nonlinear predictive control and moving horizon estimation—an introductory overview," *Advances in control,* pp. 391-449, 1999.

[8] R. López-Negrete and L. T. Biegler, "A moving horizon estimator for processes with multi-rate measurements: A nonlinear programming sensitivity approach," *Journal of Process Control,* vol. 22, no.

4, pp. 677-688, 2012.

[9] S. Shyamal and C. L. E. Swartz, "Real-Time Energy Management for Electric Arc Furnace Operation," in *Journal of Process Control*, 2018 (in press).

538717 Geosteering Using Image Logs and an Intelligent Bottomhole Assembly

Abstract:

Geo-steering refers to making decisions about directing a horizontal drilling system towards a desired trajectory. Geo-steering decisions are made at the surface by a team of engineers and geologists to direct (control) the bit in the appropriate direction within the reservoir. Logging tools in the bottom hole assembly (BHA) sense oil pockets at the bit, and the BHA moves in a reasonable direction, to drill a well that will drain the maximum possible sweet spots within the reservoir, after production start. Assessing this direction is done by identifying the floor and the ceiling of the reservoir and staying within these limits. This approach has two issues on which improvements can be envisioned: (a) there is considerable dead-time (due to very limited bandwidth of measuring/logging-while-drilling (M/LWD) tools) which reduces achievable control performance, and (b) considerable manpower is required.

The proposed approach attempts to address both of these issue by using an automation system at the BHA, with two capabilities: (a) Process images in real time, to detect floor and ceiling of the reservoir, and (b) perform rapid model predictive control (MPC) calculations, to steer the bit in the right direction.

The proposed approach would eliminate dead-time from the overall geo-steering control loop, thus resulting in tighter control. This would reduce diversion of the bit outside the reservoir pocket, resulting in a well with higher contact area with the reservoir. Benefits for the drilling operation would also result, as the bit would wear less, as there would be less need for demanding excursions from the ideal path, and MTBF for each component of the tool would reduce.

This presentation shows simulations of how this overall architecture would work. Specifically, edge detection and other image processing algorithms are used to detect reservoir floor and ceiling, and MPC is used to minimize both distance from the desired trajectory and overall mechanical energy. Issues on numerical efficiency, computer chip requirements, and overall feasibility are discussed.

536902 Intelligent Pressure Control Method for Managed Pressure Drilling

Abstract:

Objective/Scope

Tradition drilling technique using manual choke poses problems with the regularity of pressure management and higher unproductive drilling time. Managed Pressure Drilling (MPD) with automated choke controller has become increasingly popular as it provides precise control over Bottom Hole Pressure (BHP) during various drilling activities. The main challenges with automated choke control are the ability of the system to work effectively in highly dynamic operating conditions and measurement of BHP, which can be viewed as an unmeasured state. While measuring BHP using downhole sensors, the measurements are communicated with slow mud pulse telemetry. In addition, several uncertain factors such as drill pipe movement and reservoir influx lead to uncertainty. Invariably, measurement is not available when circulation is low or during pipe connection. Even models used for predicting BHP are unreliable due to uncertainties in friction loss and fluid density change.

This study presents a robust and accurate intelligent choke controller. The system uses a simple hydraulic model for the estimation of the BHP. Real-time tuning of the hydraulic model is performed using an Unscented Kalman Filter (UKF) optimization technique. Once the sensor measurements are available, UKF makes an estimate of real-time density and friction factor and makes an update to the hydraulic model. The control of the BHP is performed by using L1 adaptive control algorithm. It is a robust control method suitable for a highly non-linear system like drilling. The main components of the L1 adaptive algorithm are a state predictor, an adaptive law, and a low pass filter. The low pass filter prevents high-frequency noise from entering the controller. The adaptation law allows for the faster adaptation of the system, which in turn allows for the faster convergence of the bottom hole pressure. The L1 adaptive architecture permits the adaptive gain to be relatively high. In fact, the only limitation to the adaptive gain is determined by the hardware. This would allow for faster adaptation without losing the robustness.

Method, Procedure, Process

Governing equations were derived for the drilling system using the mass and momentum balance of the fluid flow. These equations are used for the design of estimation and control methods. L1 adaptive algorithm consists of a state predictor, an adaptive law, and an estimator. For the state predictor design, the equations were rearranged such that it contains bottom hole pressure as a state variable and three uncertain system parameters. The adaptive law makes prediction of these uncertain parameters. The estimator of L1 adaptive control law consists of the hydraulic model which makes prediction of the bottom hole pressure. Furthermore, the estimation performance is improved by real-time tuning of the method by using Unscented Kalman Filter (UKF). A process noise covariance matrix and measurement noise covariance matrix need to be designed for UKF. Measurement noise depends upon the characteristics of the sensor. Determining process noise matrix is more cumbersome. Inverse quadratic method was used for determining a relationship between elements of process noise matrix. In addition, a tuning parameter is introduced to scale the process noise matrix in relation to measurement noise matrix. The performance of UKF depends upon this scaling parameter.

Complete design of the system is implemented as blocks in Simulink. The plant was designed to simulate drilling condition. The simulation of the system was done to study its performance in various drilling scenarios like normal circulation condition, pipe connection, time delay, and in the presence of noise in the system.

Results

An innovative control method was designed for real-time estimation and control utilizing a combination of L1 adaptive algorithm and UKF. The system performance was studied with operating parameters of a drilling operation in Marcellus Shale play. UKF was observed to be effective for estimation of uncertain system variables of the drilling system. The estimation error for UKF was obtained to be less than 5 % with convergence time of 10 seconds. The simulation of combined L1 adaptive algorithm and UKF showed a stable performance for highly transient input parameters. The performance of the system was studied with changing set point of 420 bar, 460 bar, and 480 bar. The system was able to maintain a new set point within 5 seconds. In addition, the steady state error was observed to be less than 1 %. For the simulation of pipe connection, the pump flow rate was ramped down from 0.01 m3/sec to 0 m3/sec. A small amount of overshoot of about 10 bar was observed during the transient phase. For the simulation with time delay added to the sensors, stable output response was obtained for a delay of up to 3 seconds.

527229 Addressing Uncertainty in Large-Scale Bioconversion Product and Process Networks with Two-Stage Adaptive Robust Optimization

Abstract:

Addressing Uncertainty in Large-Scale Bioconversion Product and Process Networks with Two-Stage Adaptive Robust Optimization

Daniel J. Garcia[a], Jian Gong[b],Fengqi You[b]*

1. Department of Chemical and Biological Engineering, Northwestern University, Evanston, IL, 60626
2. Department of Chemical and Biomolecular Engineering, Cornell University, Ithaca, New York, 14853

*Email: Fengqi.you@cornell.edu

Submitted for consideration in session 10A01 Design and Operations Under Uncertainty

While enthusiasm towards production of biofuels and bioproducts has fallen somewhat out of favor in recent years, these products could be a renewable alternative to petroleum fuels and chemicals in the future. There are many questions on which biofuels to produce, which feedstocks to use, and if the net life cycle CO_2 emissions of biofuels even are lower than their petroleum counterparts.[1] Biofuel production is fraught with a host of uncertainties. If biofuels or bioproducts are to be economically and environmentally feasible, then identifying optimal production pathways from feedstock to fuels and chemicals is crucial.[2-5] However, previous works that aim to optimize biofuel production both for cost and environmental impacts typically treat the input parameters as deterministic rather than uncertain. In reality, there are many parameters of the bioconversion processes that are uncertain. For example, feedstock price and product demand often vary significantly over time. If bioconversion pathways are optimized without considering these uncertainties, then it is possible that the identified designs might be infeasible for certain realizations of the uncertainties.[6] This uncertainty and others throughout the bioconversion process must be considered if optimal bioconversion pathways are to be identified.[7]

Thus, in this work, we construct a two-stage adaptive robust mixed integer nonlinear programming (MINLP) problem to identify optimal bioconversion pathways from a large-scale bioconversion product and process network of alternatives. Feedstock price and final product demand are considered uncertain and are modeled with uncertainty sets. The network we consider has 194 different technologies and 139 feedstocks, intermediates, byproducts, and final products.[2] We choose an adaptive robust approach rather than a conventional robust optimization approach to identify less conservative solutions that could be more practical.[8-10] Stochastic programming could be another modeling alternative, but this method requires knowledge of the probability distribution of values the uncertain parameter can take, which is often difficult to obtain.[11,12] After formulating a two-stage adaptive robust MINLP model, we propose a solution method that combines a column-and-constraint generation algorithm and a branch-and-refine algorithm. We compare results from the two-stage adaptive robust model, a deterministic model with no consideration of uncertainty, and results from a conventional static robust version of the model. Decisions to be made in the model include technology selection, capacity sizing of each chosen technology, operating level of each technology, CAPEX and OPEX, and quantities of feedstocks to purchase and products to produce. Nominal values for biomass feedstock prices and product demands were taken from previous works, and readily available statistical reports or official government reports were used to determine the ratio of the largest deviation of the uncertain parameters from their nominal values.[1,2,13,14]

Adaptive robust optimization provides optimal solutions for each uncertainty level of each uncertain parameter. The minimum total annualized cost of the deterministic version of the model (uncertainty budgets of 0 for all uncertain parameters) is the lowest at \$17.9M/y, and the most conservative two-stage adaptive robust optimal solution has the highest cost at \$22.5M/y. As the uncertainty budgets decrease, the overall annualized cost decreases, but the risk of being unable to meet demand at the lowest possible cost increases. Soybeans and softwood are used in all solutions in varying amounts to produce ethanol, gasoline, biodiesel, and polyhydroxybutyrate (PHB). In the adaptive robust solutions, more softwood is purchased and more gasoline is produced via gasification, syngas reforming, methanol synthesis, and the methanol to gasoline process than in the deterministic solution. Increasing the feedstock price or product demand uncertainty budgets increase the total annualized cost in all cases. However, if the feedstock price uncertainty budget is higher than 2, the uncertainty budget of product demand does not influence the minimum annualized cost much. Therefore, the final solutions are more sensitive to product demand uncertainty than feedstock price uncertainty, an important result for the biofuels industry. The range of results provided from the two-stage adaptive robust approach in this work provide decision-makers with more information, allowing for more tailored decisions.

References

1. Garcia D, You F. Network-Based Life Cycle Optimization of the Net Atmospheric CO2-eq Ratio (NACR) of Fuels and Chemicals Production from Biomass. *ACS Sustainable Chemistry & Engineering.* 2015/08/03 2015;3(8):1732-1744.
2. Garcia D, You F. Multiobjective optimization of product and process networks: General modeling framework, efficient global optimization algorithm, and case studies on bioconversion. *AIChE Journal.* 2015;61(2):530-554.
3. Garcia D, You F. Life Cycle Network Modeling Framework and Solution Algorithms for Systems Analysis and Optimization of the Water-Energy Nexus. *Processes.* 2015;3(3):514-539.
4. Rizwan M, Lee JH, Gani R. Optimal processing pathway for the production of biodiesel from microalgal biomass: A superstructure based approach. *Computers & Chemical Engineering.* 11/11/ 2013;58(0):305-314.
5. SantibanÌ*f*ez-Aguilar JE, GonzaÌlez-Campos JB, Ponce-Ortega JMa, Serna-GonzaÌlez M, El-Halwagi MM. Optimal Planning of a Biomass Conversion System Considering Economic and Environmental Aspects. *Industrial & Engineering Chemistry Research.* 2011/07/20 2011;50(14):8558-8570.
6. Sahinidis NV. Optimization under uncertainty: state-of-the-art and opportunities. *Computers & Chemical Engineering.* 2004;28(6-7):971-983.
7. Gong J, You F. Sustainable design and synthesis of energy systems. *Current Opinion in Chemical Engineering.* 2015;10:77-86.
8. AtamtÃ¼rk A, Zhang M. Two-stage robust network flow and design under demand uncertainty. *Operations Research.* 2007;55(4):662-673.
9. Bertsimas D, Iancu DA, Parrilo PA. A hierarchy of near-optimal policies for multistage adaptive optimization. *IEEE Transactions on Automatic Control.* 2011;56(12):2809-2824.
10. Bertsimas D, Litvinov E, Sun XA, Zhao J, Zheng T. Adaptive robust optimization for the security constrained unit commitment problem. *IEEE Transactions on Power Systems.* 2013;28(1):52-63.
11. Pistikopoulos E, Ierapetritou M. Novel approach for optimal process design under uncertainty. *Computers & Chemical Engineering.* 1995;19(10):1089-1110.
12. Gupta V, Grossmann IE. Multistage stochastic programming approach for offshore oilfield infrastructure planning under production sharing agreements and endogenous uncertainties. *Journal of Petroleum Science and Engineering.* 2014;124:180-197.
13. EIA. Petroleum & Other Liquids - Prices, Sales Volumes & Stocks by State. In: DOE, ed2015.

14. ycharts. Corn Farm Price Received. 2015; https://ycharts.com/indicators/corn_price.

534309 A Flexible Design Framework for Process Systems with Demand Response Objectives

Abstract:

The fluctuation of electricity prices due to the changing power demand and generation profiles during a specific period is already playing a significant role in industrial energy management strategies, and is likely to continue with the increasing power grid penetration of renewable resources. The challenge of mitigating the impact of volatile electricity price profiles evoked the concept of demand response (DR), where electricity consumption is adjusted depending on its prices or events triggered by grid operators during a specific period. While higher electricity prices encourage consumers to lower their power demand, low prices offer an incentive to increase it. As a result, fluctuating electricity price profiles are smoothed benefiting both the energy suppliers and consumers.

Demand response has been studied extensively in the context of chemical process systems [1]; however, the primary focus of current research efforts is on the process scheduling and control aspects of the problem; which are typically addressed by increasing or decreasing production rates during the day or by shifting production load among specific units. In our previous work [2], we developed a proactive operational reconfiguration strategy for heat exchanger networks (HENs) operating under varying electricity prices and subject to intermittent renewable energy resource availability to satisfy DR objectives. However, the design of DR-enabled process networks has received comparatively little attention, and has mainly focused on air separation units [3], where variable capacity operations are accounted for during the process design step. The design of a process network, allowing varied operating levels as well as enabling optimal process flowsheet reconfiguration under different DR scenarios is a novel concept. Intuitively, reconfiguration suggests the purchase of more (redundant) process equipment which generally results in a higher capital investment. Whether the economic benefits from DR operation can compensate for the higher investment and thereafter reduce the overall cost remains an open-research question.

In this work, a classical pump network problem [4] is used as a motivating example to illustrate the flexible design concept utilizing the reconfiguration. We first reformulate the pump network problem to incorporate different electricity prices and production demand as uncertain parameters, design decision and reconfiguration decision. The design problem is then cast as a stochastic mixed-integer nonlinear programming problem, with mixed-integer recourse variables. Li et al. [5] have first developed decomposition algorithm to solve two-stage mixed-integer convex nonlinear stochastic programs with mixed-integer variables in both the first and second stages; however, as the resulting pump network model is nonconvex, a heuristic based progressive hedging algorithm is used to solve the problem. We demonstrate the reconfiguration of the pump network to satisfy different DR-related scenarios, showing the potential benefits of a reconfigurable design under demand response.

[1] Harjunkoski,I., Scholtz, E., and Feng, X., 2014. Industry meets the smart grid. Chemical. Engineering Progress., 110, pp.45–50.

[2] Wang, X., El-Farra, N.H. & Palazoglu, A., 2015. Proactive Reconfiguration of Heat-Exchanger Supernetworks. Industrial & Engineering Chemistry Research, 54(37), pp.9178–9190.

[3] Pattison, R.C. & Baldea, M., 2014. Optimal Design of Air Separation Plants with Variable Electricity Pricing. Computers & Chemical Engineering, 34, pp.393–398.

[4] Westerlund, T., Pettersson, F. & Grossmann, I.E., 1994. Optimization of pump configurations as a MINLP problem. Computers & Chemical Engineering, 18(9), pp.845–858.

[5] Li, C. & Grossmann, I.E., 2018. An improved L-shaped method for two-stage convex 0-1 mixed integer nonlinear stochastic programs. Computers & Chemical Engineering, 112, pp.165–179.

534948 Accounting for Uncertainty Via Scheduling Informed Optimal Design: A Renewable Ammonia Case Study

Abstract:

The optimal design of systems with renewable energy generation requires taking into account decision making on the time scale of scheduling due to the time-varying nature of renewable energy availability. As a result, such optimization problems tend to be large as operating states at each time point over at least one year need to be tracked. Because of the innate size of the deterministic problem, it is extremely difficult to include nonlinear relations for capital cost and other phenomena, or further take into account uncertainty in parameters such as renewable energy availability using stochastic or robust programming, as such formulations of the problem quickly become intractable. However, uncertainty can play an important role in deciding on an optimal design, since a deterministic design may not include enough operating flexibility to manage uncertainty. As an example of this, consider a renewable energy powered chemical plant designed to run in the electricity market structure presented in [1], where day-ahead power exchange commitments must be made. Market parameters govern how much commitments in consecutive time periods can vary and how close a realized exchange must be to its commitment value to avoid paying a penalty, a scenario referred to as a "commitment violation." Designing a plant in such a market structure necessitates considering operation under uncertainty, otherwise one would never design for the possibility of commitment violations due to imperfect forecasts.

In this work, we consider the case study of finding the optimal design of a wind-powered ammonia generation system operating in the aforementioned market structure. To keep the problem tractable, we propose a method where the optimization problems for scheduling and design are decoupled. To do so, we develop a receding horizon formulation for scheduling a system with a wind turbine which services a renewable ammonia plant as well as other uncertain but uncontrollable loads. These problems make hourly decisions about the operating state of the system to minimize operating cost based on uncertain forecasts of weather and power load over the next 48 hours. We then simulate many yearlong scheduling problems for wind-powered ammonia plants of various designs to obtain annual operating costs. From this data, we use a method similar to that described in [2] to identify a small number of key design parameters on which operating cost depends and develop simple correlations relating operating cost to these parameters. We repeat this analysis for different values of the electricity market parameters and obtain different sets of correlations for tight, loose, and no market parameters.

The correlations found by scheduling the renewable ammonia plant are then embedded in a design optimization problem attempting to find unit sizes which can meet a given power and ammonia demand at minimal net present cost. Because scheduling is embedded in the correlations for operating cost, the design optimization problem no longer needs to include variables for operating states, making the resulting optimization problem much smaller and allowing more accurate nonlinear capital cost functions to be included. Our results show how the optimal design of the ammonia system can change based on the market parameters, as a tight market structure where one cannot deviate much from their power exchange commitment requires a design that can be operated more flexibly than a loose market structure. We also compare our results with those obtained from attempting to design without taking into account uncertainty, and show that the deterministic optimal design is not as economical in the electricity market structure.

[1] Zachar, M. and Daoutidis, P. Microgrid/macrogrid energy exchange: A novel market structure and stochastic scheduling. IEEE Trans. Smart Grid (8), 2017, pp. 178-189.
[2] Allman, A. and Daoutidis, P. Optimal scheduling for wind-powered ammonia generation: E ects of key design parameters. Chem. Eng. Res. Des., 2017, in press.

536224 Design and Optimization of Multifunctional Processes for Utilizing Unconventional and Distributed Feedstocks

Abstract:

Unconventional resources and distributed feedstocks such as shale gas and associated natural gas are forecasted to meet majority of future energy demands [1]. For tapping into these unconventional resources in remote locations, multifunctional chemical processes with hybrid reaction-separation phenomena, as in sorption enhanced reaction processes (SERP), can be effectively deployed for reaction, conversion and storage purposes. In SERP technology, *in-situ* removal of reaction byproducts from the reaction gas mixture, by a solid adsorbent, enhances production of the desired product. The advantages of SERP concept include compactness, lower energy consumption and equipment costs, better process performance, higher reaction conversions, and enhanced product quality. In this work, we develop a high-fidelity generalized reaction-adsorption modeling and simulation (GRAMS) platform for accurately capturing reaction and adsorption dynamics in SERP systems packed with solid catalysts and porous adsorbents [2]. GRAMS is based on a one-dimensional, pseudo-homogeneous, non-isothermal, non-adiabatic, and non-isobaric NAPDE-based model. The model predictions have been extensively validated with experimental and industrial data for conventional steam methane reforming (SMR), methanol synthesis, sorption-enhanced SMR (SE-SMR), sorption-enhanced water gas shift reaction (SE-WGSR), and pressure swing adsorption (PSA). GRAMS is designed to obtain novel multifunctional processes by optimally combining several catalysts and adsorbents in a single intensified column.

To this end, we further develop a systematic design and synthesis framework by coupling GRAMS with an in-house simulation-based constrained grey-box optimizer [3][4]. The formulated framework simultaneously optimizes SERP cycle configuration, column design specifications and process operating conditions. The major modeling contributions include a generalized boundary-condition formulation for representing different operation modes in an SERP cycle, and the use of continuous pressure variables for selecting discrete operation modes and flow directions. The framework has been successfully demonstrated for optimal operation of three SERP systems, namely SE-SMR, SE-WGSR and sorption-enhanced methanol (SE-MeOH). For periodic SE-SMR, we design a process that produces 95% pure hydrogen from natural gas with 35% higher productivity and 10.86% lower cost in comparison to existing small-scale, distributed systems. In case of SE-MeOH, a novel process is obtained where methanol is synthesized from syngas with single-pass carbon conversions >80% and a production capacity of 109500 tons per year, with significant savings in energy consumption and process economics.

References

[1] M. Guarnone, F. Rossi, E. Negri, C. Grassi, D. Genazzi, and R. Zennaro, "An unconventional mindset for shale gas surface facilities," *J. Nat. Gas Sci. Eng.*, vol. 6, pp. 14–23, 2012.

[2] A. Arora, S. S. Iyer, and M. M. F. Hasan, "GRAMS: A General Framework Describing Adsorption, Reaction and Sorption-Enhanced Reaction Processes," *Submitted*, 2018.

[3] A. Arora, I. Bajaj, S. S. Iyer, and M. M. F. Hasan, "Optimal Synthesis of Periodic Sorption Enhanced Reaction Processes with Application to Hydrogen Production," *Comput. Chem. Eng.*, vol. 115, pp. 89–111, 2018.

[4] I. Bajaj, S. S. Iyer, and M. M. F. Hasan, "A Trust Region-based Two Phase Algorithm for Constrained Black-box and Grey-box Optimization with Infeasible Initial Point," *Comput. Chem. Eng.*, 2017.

538020 Simultaneous Process Synthesis and Heat Integration Using Building Block Superstructure

Abstract:

Efficient energy utilization is a paramount factor for the profitability and sustainability of a chemical plant as cost of the end-product and environmental impact of the process is highly dependent on the energy consumption of the process. Hence, the degree of heat integration within the process is an important decision that has to be handled together with the synthesis and design of the process. Currently, there are several methods available in the literature for the simultaneous process synthesis and heat integration [2-6]. These methods are mainly demonstrated for the classic superstructure representations which are based on pre-specified equipment types and priori postulated flowsheet connectivity.

Building-block based superstructure [7], on the other hand, can yield non-intuitive flowsheets while considering different mass integration alternatives without *a priori* postulation of the processing steps and the connectivity between them. This representation is based on a two-dimensional grid structure where several different phenomena, e.g. reaction, V-L equilibrium, G-L equilibrium, etc., can be represented either as a 'single block' or via multiple neighboring blocks sharing a common boundary. This provides a generic representation method for novel intensified equipment design and synthesis. Recently, this method has been also demonstrated for equipment-based representation along with detailed design formulations [8]. Furthermore, the method has been applied for different process integration problems including mass integration, heat integration and property integration [9-10]. In this work, we will show that the building block representation method can be also used to perform simultaneous flowsheet synthesis and heat integration. The formulation is based on a single mixed integer nonlinear programming (MINLP) model. As the flowsheet connectivity and equipment types are not specified *a priori*, the heat integration formulation is based on a generic formulation that can handle streams with unknown identity (i.e. hot/cold). Owing to its fundamental outlook toward process synthesis, building-block superstructure with heat integration enables to design, synthesize and discover new processes including refrigeration cycles without specification of the processing routes beforehand. With this, proposed method becomes a powerful tool for the synthesis of more sustainable and energy efficient processes for a variety of applications which will be demonstrated by several case studies.

References:

[1] Chen, Q. and Grossman, I.E., 2017. Recent Developments and Challenges in Optimization-Based Process Synthesis. *Annual Review of Chemical and Biomolecular Engineering, 8*(1).

[2] Papoulias, S.A. and Grossmann, I.E., 1983. A structural optimization approach in process synthesis—III: total processing systems. *Computers & chemical engineering, 7*(6), pp.723-734.

[3] Duran, M.A. and Grossmann, I.E., 1986. Simultaneous optimization and heat integration of chemical processes. AIChE Journal, 32(1), pp.123-138.

[4] Yee, T.F. and Grossmann, I.E., 1990. Simultaneous optimization models for heat integration—III. Process and heat exchanger network optimization. *Computers & Chemical Engineering, 14*(10), pp.1165-1184.

[5] Grossmann, I.E., Yeomans, H. and Kravanja, Z., 1998. A rigorous disjunctive optimization model for simultaneous flowsheet optimization and heat integration. *Computers & chemical engineering, 22*, pp.S157-S164.

[6] Kong, L., Avadiappan, V., Huang, K. and Maravelias, C.T., 2017. Simultaneous chemical process synthesis and heat integration with unclassified hot/cold process streams. *Computers & Chemical Engineering, 101*, pp.210-225.

[7] Demirel, S. E., Li, J., and Hasan, M. M. F., (2017). Systematic Process Intensification using Building Blocks, *Computers and Chemical Engineering*, 105, 2-38.

[8] Li, J.; Demirel, S.E.; Hasan, M.M.F. Process Synthesis using Block Superstructure with Automated Flowsheet Generation and Optimization. *AIChE Journal*, 2018, under review.

[9] Li J., Demirel S.E., Hasan M.M.F. Process Integration using Block Superstructure. *Industrial & Engineering Chemistry Research*, 2018, 57: 4377–4398.

[10] Li J., Demirel S.E., Hasan M.M.F. Fuel Gas Network Synthesis Using Block Superstructure. *Processes*. 2018, 6(3): 23.

526057 Green Operation of an Air Separation Unit Using an Efficient MILP Optimal Scheduling Framework

Abstract:

Greenhouse gas emissions in the United States increased by 7% from 1990 to 2014, and carbon dioxide accounted for 82% of greenhouse gas (GHG) emissions in 2015 [1,2]. Electricity transmission and production accounts for 29% of the greenhouse gas emissions, and industry accounts for 21% [1]. In this context, our work analyzes the possibility of simultaneously lowering the industrial and electricity sector greenhouse gas (particularly, carbon dioxide) emissions by changing the electricity demand profile of electricity-intensive industrial manufacturing plants.

In this context, we define green operation as a strategy for modulating plant production rates in a manner that favors electricity consumption at times when the contribution of renewable sources to the energy generation mix is maximized. Specifically, we postulate a general production scheduling framework that utilizes information from the power grid (and the instantaneous contribution of e.g. wind and solar generation) and demand data to, i) raise production levels when renewable generation is high, generating product in excess of demand, ii) store the excess product, and iii) lower production rates when non-renewable (i.e., fossil fuel-based) power generation increases. In the latter case, stored products are used to meet demand. In order to account for significant daily fluctuations in the rate of power generation by renewables, the scheduling framework relies on dynamic models of the manufacturing process.

We illustrate this framework by performing optimal production scheduling calculations for greener operation of an air separation unit (ASU) as a prototypical electricity-intensive chemical process. ASUs are typically large plants, with dominant time constants in the order of hours. As a consequence, production rate changes are scheduled over hourly intervals, under a set of previously derived [3,4] dynamic constraints. The overall scheduling problem is posed as a Mixed Integer Linear Program (MILP).

An extensive computational case study reveals that green operation can significantly reduce CO2 emissions without jeopardizing the production process. The cost of green operation is compared with conventional demand-response scheduling (which is typical in the interaction of ASUs with the power grid). Our study reveals new potential strategies and electricity pricing schemes that can be used to incentivize greener operation by the chemical industry.

[1] EPA: Overview of Greenhouse Gases. (2018). Retrieved October 4, 2018, from https://www.epa.gov/ghgemissions/overview-greenhouse-gases

[2] IPCC (Intergovernmental Panel on Climate Change). 2013. Climate change 2013: The physical science basis. Working Group I contribution to the IPCC Fifth Assessment Report. Cambridge, United Kingdom: Cambridge University Press.www.ipcc.ch/report/ar5/wg1.

[3] Kelley, M. T., Pattison, R. C., Baldick, R., & Baldea, M. An Effcient MILP Modeling Framework for Demand Response Operation of Air Separation Units. Applied Energy, Accepted.

[4] Pattison, R. C., Touretzky, C. R., Johansson, T., Harjunkoski, I., & Baldea, M. (2016). Optimal Process Operations in Fast-Changing Electricity Markets: Framework for Scheduling with Low-Order Dynamic Models and an Air Separation Application. Industrial & Engineering Chemistry Research, 55(16), 4562â€"4584. https://doi.org/10.1021/acs.iecr.5b03499

526147 Graph Representation and Decomposition of Diffusion-Convection-Reaction Processes for Distributed Control

Abstract:

Chemical plants are large networks of process systems with varied dynamics, broadly classified as lumped and distributed parameter systems (LPS's and DPS's). The control of such networks following a fully centralized approach is impractical whereas decentralized control is generally ineffective. Distributed control, a promising middle-ground between these approaches, overcomes these limitations [1] as it entails the decomposition of the chemical network into sub-networks controlled using intercommunicating local controllers. In order to identify the decomposition of a network for optimal control performance, recently, network theory based algorithms have been developed that identify communities in a network with high intra-community interactions and low inter-community interactions by maximizing modularity [2,3]. These algorithms are based on the equation graphs of the process networks. The equation graphs, which consist of nodes representing variables and directed edges representing the interactions among the variables, are currently limited to networks of LPS's [3] and convection-reaction systems modeled by first-order hyperbolic partial differential equations (PDEs) [4].

However, most DPS's, in reality, are diffusion-convection-reaction systems, whose dynamics are captured by parabolic PDEs (PPDEs). In this work, we propose an equation graph representation of the variables of PPDE systems. The manipulated inputs in a PPDE system are of three types - velocity manipulation, distributed manipulation, and boundary manipulation. The outputs of the PPDE system are of three types - boundary outputs (e.g., stream composition or temperature at the exit of a column), distributed outputs (e.g., concentration or temperature profiles in the column), and spatially varying time-dependent outputs (e.g., hot spot temperature in a column). For the equation graph representation, we incorporate the boundary manipulations in the PPDEs using a Dirac delta function, and define modified output variables as functions of the weighted averages over the spatial domain of the state variables, where the weight is a piece-wise smooth function of the spatial coordinate that attains values close to 1 where the original outputs are located (e.g., at the boundary, at a finite number of locations along the spatial coordinate, or at the hot-spot location), and a small non-zero value at the location of the boundary manipulations (if any). The spatial averaging converts the distributed output variables into vectors with elements that are functions of time alone. The weight function ensures that the values of the modified outputs are close to those of the original outputs while collocating the manipulated variables with the modified outputs. The collocation ensures that the strength of the interaction between a manipulated variable and a modified output can be captured in the form of an explicit dependence of the time-derivatives of the modified output on the manipulated variable. In the equation graph, we represent each input variable, distributed state variable, and the modified output variable with a node. An edge connects an input or a state node to another state node if the former variable appears explicitly in the PPDE corresponding to the latter variable's dynamics. An edge connects a state node to an output node if they are algebraically related. We show that in this graph, the length of an input-output path, i.e., the smallest number of edges joining the two variables, is directly related to the strength of the structural interaction between the variables as captured by the lowest order time-derivative of the output that explicitly depends on the input variable. This is analogous to the equation graphs of LPS's [5] and hyperbolic PDE systems. Thus, this representation allows us to capture on a common equation graph all the variables and interconnections of a generic chemical network to determine the optimal network decomposition for distributed control.

As a case study, we consider a typical reaction-separation plant which comprises four tubular reactors with interstage cooling using three heat exchangers, and two flash columns, with material recycle and heat integration. A first-order, reversible, exothermic (in the forward direction) reaction occurs in the tubular reactors, which have significant convection and diffusion phenomena, and thus are described by quasi-linear PPDEs. The heat exchangers and flash columns are modeled as LPS's. This model is used for implementing model predictive control (MPC). The network decomposition algorithm identifies three sub-networks (communities) in the plant's optimal decomposition, while the edges across the three communities on the equation graph represent the communication among the local controllers of each sub-network. We demonstrate the optimality of the decomposition by comparing the control performance of distributed MPC implemented on the optimal decomposition to the MPC based on other architectures such as the sub-optimal decompositions determined from the algorithm and the conventional centralized and decentralized architectures.

References

[1] R. Scattolini. Architectures for distributed and hierarchical model predictive control-a review. *Journal of Process Control*, 19(5):723-731, 2009.

[2] D. B. Pourkargar, A. Almansoori, and P. Daoutidis. Impact of decomposition on distributed model predictive control: A process network case study. *Industrial & Engineering Chemistry Research*, 56(34):9606-9616, 2017.

[3] S. S. Jogwar and P. Daoutidis. Community-based synthesis of distributed control architectures for integrated process networks. *Chemical Engineering Science*, 172:434-443, 2017.

[4] M. Moharir, L. Kang, P. Daoutidis and A. Almansoori. Graph representation and decomposition of ode/hyperbolic pde systems. *Computers & Chemical Engineering*, 106:532-543, 2017.

[5] P. Daoutidis and C. Kravaris. Structural evaluation of control configurations for multivariable nonlinear processes. *Chemical Engineering Science*, 47(5):1091-1107, 1992.

527143 Machine Learning-Based Model Predictive Control of Distributed Chemical Processes

Abstract:

Many important commodity and specialty chemicals are produced from highly exothermic processes in packed-bed tubular reactors, such as phthalic anhydride synthesis, ammonia synthesis and light olefin synthesis, at chemical manufacturing industry for several decades. In today's saturated markets where continuous improvement of process efficiency to reduce the production costs is necessary to keep chemical manufacturing plants from being obsoleted by global competition, packed-bed modeling and control have become an active research area for both academia and chemical manufacturing industry. Open literature reports suggest that the intrinsic characteristics of packed-bed tubular reactors with exothermic processes, i.e., nonlinearity, nonstationary (due to reactor fouling and catalyst deactivation), nonlinear distributed dynamics and limited on-line measurements with large delays, pose substantial difficulties to process control [1, 2, 3], and model-based control algorithms developed from more high-fidelity models yield better performance in practice [4]. Hence, the need for high-fidelity computationally efficient models to simulate the dynamic behavior of packed-bed tubular reactors controlled variables (e.g., the hot-spot temperature and product composition) as a function of the packed-bed tubular reactors manipulated inputs (e.g., the jacket temperature profile) becomes apparent. Traditional mechanistic models and computational fluid dynamics (CFD) models, although often yield adequate descriptions of the process dynamics, would not be suitable for designing model-based controllers because solving these models repeatedly in real-time as part of a model predictive control (MPC) optimization problem that seeks to reduce the hot-spot temperature and maximize the product composition may require significant computational time. Data-based modeling is an appealing alternative as data-driven models are computationally inexpensive and can have reasonable accuracy.

Motivated by the above considerations, the present work introduces a statistical-based machine learning algorithm that generates a computationally efficient Bayesian recurrent neural network (RNN) to simulate the dynamics of the hot-spot temperature and product composition at the exit of the packed-bed reactor as a function of the jacket temperature profile and the feed conditions. Next, a CFD model for the commercial-scale packed-bed tubular reactor used in phthalic anhydride synthesis is developed [1, 2, 3, 5] and is used to create a sufficiently large database, from which the Bayesian RNN is derived. Then, the Bayesian RNN is integrated within an MPC framework to create an MPC that optimizes the jacket temperature profile to maximize the product composition and to prevent the hot-spot temperature from exceeding a critical temperature value. Subsequently, the present work outlines a novel procedure that creates the communication pathway between a CFD solver and a nonlinear programming (NLP) solver so that CFD model for the packed-bed tubular reactor can be used to simulate the process dynamics of the physical reactor in the closed-loop (under machine-learning MPC) system. Finally, the closed-loop control system is subjected to various feed disturbances to demonstrate the effectiveness of the data-based control scheme proposed in this work.

[1] Hua, X., Jutan, A., 2000. Nonlinear inferential cascade control of exothermic fixed-bed reactors. AIChE journal 46, 980-996.

[2] Chen, C.Y., Sun, C.C., 1991. Adaptive inferential control of packed-bed reactors. Chemical engineering science 46 , 1041-1054.

[3] Wu, W., Huang, M.Y., 2003. Nonlinear inferential control for an exothermic packed-bed reactor: Geometric approaches. Chemical engineering science 58, 2023-2034.

[4] Logist, F., Smets, I.Y. and Van Impe, J.F., 2008. Derivation of generic optimal reference temperature profiles for steady-state exothermic jacketed tubular reactors. Journal of Process Control 18, 92-104.

[5] De Klerk, A., 2003. Voidage variation in packed beds at small column to particle diameter ratio. AIChE journal 49, 2022-2029.

531473 Optimization-Based Sensor/Actuator Scheduling and Control of Sampled-Data Distributed Processes

Abstract:

Despite the extensive body of research work on control of spatially-distributed systems, the problem of designing sampled-data feedback controllers for spatially-distributed processes has received relatively limited attention (e.g., [2], [3],[7] for some notable exceptions). This is an important problem given the prevalent use of digital sensors and controllers in industrial control systems. To date, the overwhelming majority of studies on the analysis and design of sampled-data control systems have focused primarily on lumped parameter systems modeled by ordinary differential equations (e.g., see [1],[4],[5],[6],[8],[9]).

An effort to address the sampled-data control problem for spatially-distributed systems was undertaken in a series of earlier works (e.g., [10],[11]). The main idea was to include an approximate finite-dimensional model of the infinite-dimensional system in the controller to provide estimates of the dominant slow states that compensate for the unavailability of output measurements between sampling times, and to update the model states using the actual measurements at each sampling time. A key objective of this model-based control approach was to guarantee closed-loop stability with minimal sampling frequency. This is an appealing goal in situations where technological constraints on the measurement sensing techniques make it difficult or costly to collect measurements frequently. To this end, an exact characterization of the maximum allowable sampling period required for stabilization was obtained in the earlier studies and found to depend on the spatial placement of the measurement sensors and control actuators. An implication of this finding is that one could use this characterization to identify the set of feasible sensor and actuator locations that can achieve stabilization with minimal sampling frequencies.

While this is important from a sampling cost savings standpoint, the resulting actuator and sensor placement does not consider the impact of discrete measurement sampling on closed-loop performance. It is well known, for example, that faster sampling rates are typically required to enhance the control system performance. Ultimately, there is a need to resolve the inherent conflict that arises between the tight restrictions on sampling rates which are required to maintain the control system performance on the one hand, and the usual demand for limited sampling frequency which is desired to minimize measurement costs.

Motivated by these considerations, we present in this work an optimization-based methodology for the placement and scheduling of measurement sensors and control actuators in spatially-distributed processes with low-order dynamics and discretely-sampled output measurements. Initially, a sampled-data observer-based controller, with an inter-sample model predictor, is designed based on an approximate finite-dimensional system that captures the infinite-dimensional system's dominant dynamics. An explicit characterization of the interdependence between the stabilizing locations of the sensors and actuators and the maximum allowable sampling period is obtained. Based on this characterization, a constrained finite-horizon optimization problem is formulated to obtain the sensor and actuator locations, together the corresponding sampling period, that optimally balance the trade-off between the control performance requirements on

the one hand, and the demand for reduced sampling, on the other. The objective function penalizes both the control performance cost, expressed in terms of the response speed and the control effort, and the sampling cost, expressed in terms of the sampling frequency. The optimization problem is solved in a receding horizon fashion, leading to a dynamic policy that varies the sensor and actuator spatial placement, together with the sampling period, over time. The developed methodology is illustrated through an application to a simulated diffusion-reaction process example.

References:

[1] Chen, T. and Francis, B. (1995). *Optimal Sampled-Data Control Systems*. Springer, London; New York.

[2] Cheng, M., Radisavljevic, V., Chang, C., Lin, C., and Su, W. (2009). A sampled-data singularly perturbed boundary control for a heat conduction system with non-collocated observation. *IEEE Trans. Automat. Contr.*, **54**, 1305-1310.

[3] Fridman, E. and Blighovsky, A. (2010). Sampled-data stabilization of a class of parabolic systems. In *Proceedings of the 19th International Symposium on Mathematical Theory of Networks and Systems*. Budapest, Hungary.

[4] Fujioka, H. (2009). A discrete-time approach to stability analysis of systems with aperiodic sample-and-hold devices. *IEEE Trans. Automat. Contr.*, **54**, 2440-2445.

[5] Hu, L., Bai, T., Shi, P., and Wu, Z. (2007). Sampled-data control of networked linear control systems. *Automatica*, **43**, 903-911.

[6] Hu, Y. and El-Farra, N.H. (2011). Control, monitoring and reconfiguration of sampled-data hybrid process systems with actuator faults. In *Proceedings of 50th IEEE Conference on Decision and Control*, 736-3741. Orlando, FL.

[7] Logemann, H., Rebarber, R., and Townley, S. (2005). Generalized sampled-data stabilization of well-posed linear infinite-dimensional systems. *SIAM J. Cont. Optim.*, **44**, 1345-1369.

[8] Naghshtabrizi, P., Hespanha, J., and Teel, A. (2008). Exponential stability of impulsive systems with application to uncertain sampled-data systems. *Syst. Contr. Lett.*, **57**, 378-385.

[9] Nesic, D. and Grune, L. (2005). Lyapunov-based continuous-time nonlinear controller redesign for sampled-data implementation. *Automatica*, **41**, 1143-1156.

[10] Yao, Z. and El-Farra, N.H. (2011). Robust fault detection and reconfigurable control of uncertain sampled-data distributed processes. In *Proceedings of 50th IEEE Conference on Decision and Control*, 4925-4930. Orlando, FL.

[11] Yao, Z. and El-Farra, N.H. (2012). A predictor-corrector approach for multi-rate sampled-data control of spatially distributed systems. In *Proceedings of 51st IEEE Conference on Decision and Control*, 2908-2913. Maui, HI.

531070 Adaptive Model Reduction for Dissipative PDE Systems with Strong Convective Phenomena

Abstract:

Many industrial chemical processes exhibit spatial variation due to diffusion and convection. Examples include plasma enhanced, chemical vapor/physical vapor deposition and tubular reactors. Controlling these processes is of significant safety and economic importance. Since these processes are infinite dimensional in nature because of the spatial variation, the observation and control questions become nontrivial. These systems are called distributed parameter system (DPS) and most of them can be mathematically described by dissipative partial differential equations (PDEs). Since the dynamic behavior of systems described by dissipative PDEs can be approximated by finite dimensional systems, low order models can be obtained via method of weighted residuals (MWR). However, as convection becomes increasingly more dominant compared to diffusion, the accuracy of the reduced order model of a certain order decreases since the part of the system energy that is captured by the modes of the dominant model and the ones of the neglected model become closer. In the extreme case in which convection is dominant and diffusion is negligible, this approach is no longer applicable, since **all** the modes in WMR are needed to capture the system energy.

To deal with the accuracy issue of such model order reduction methods for processes with significant convection behavior, we propose to use discrete adaptive proper orthogonal decomposition (DAPOD) to recursively update the reduced order model (ROM) when new observations become available. We first use MWR and proper orthogonal decomposition (POD) to construct a reduced order model offline. When new snapshots become available online, the ROM is updated by DAPOD. DAPOD can mitigate the impact of historic observations compared with POD and has lower computational cost than applying POD iteratively.

Improving the accuracy of ROM for systems with strong convective phenomena can also improve the process controllability. Actuation in PDEs can be divided into 3 categories: source terms/interior control, boundary control and coefficients control. Most of the studies focus on source terms and boundary control, while control via coefficients has been rarely discussed. Solving the accuracy issue makes control by manipulating the convection coefficient feasible.

In this work, the performance of DAPOD in systems with strong convective phenomena is illustrated through a tubular reactor example. First, a ROM of a system with time-varying velocity is constructed via DAPOD to demonstrate the improvement in the accuracy of ROM. Then velocity is actuated to control the system. It's also demonstrated that APOD can improve the accuracy of the ROM when unexpected disturbances occur.

REFERENCES

[1] Amit Varshney, Sivakumar Pitchaiah, and Antonios Armaou. Feedback control of dissipative PDE systems using adaptive model reduction. AIChE journal, 55(4), 2009.

[2] Manda Yang and Antonios Armaou. Revisiting APOD accuracy for nonlinear control of transport reaction

processes: A spatially discrete approach. Chemical Engineering Science, 181:146â€“158, 2018.

537000 Equation-Free Multiparametric Model Predictive Control for Dissipative PDEs

Abstract:

Constrained model predictive control (MPC) is a powerful control strategy where on-line optimization and a receding horizon are employed to compute the optimum trajectory for the manipulated variables [1]. The use of MPC for distributed parameter systems, consisting of partial differential equations (PDEs) is computationally expensive [2]. The dissipative nature of such systems, expressed as a separation of scales in the eigenvalues of the linearized models [3], has been exploited for model order reduction (MOR) in control of (usually parabolic) PDEs [4], projecting the infinite dimensional states onto a small finite subspace. Nevertheless, the real-time control of PDEs requires the knowledge of the first principle equations that model the system's dynamics. To tackle this problem equation-free methodologies have been proposed [5]–[7] that solves the non-linear dynamic optimization problem; although the computational burden can significantly be dropped with the use of the reduced order gradients, they usually require a considerable amount of cpu-time, which can make them unsuitable for real-time control applications. The computation of the on-line control action can be enhanced with the use of multiparametric (mp) optimization [8] that computes an off-line map of the control law as a function of the states. Therefore, the computational effort is redistributed off-line, overcoming the on-line computational burden of the classic MPC. Nevertheless, despite the fact that the majority of computations are performed off-line, the large number of variables of the physical system may produce an intractable computational approach making the use of model reduction necessary. Rivotti et al. [9] combine non-linear model order reduction, based on balancing of empirical gramians to reduce the size of the system with mp approximate non-linear MPC, which can still be prohibitive for complex large-scale systems, modelled by a large set of PDEs.

In this work, an equation-free multiparametric MPC is proposed where there is no use of equations and only an available black-box simulator (such as COMSOL [10]) is employed, performing input-output tasks. An equation-free non-linear dynamic optimization has been developed based on its static counterpart [11] utilizing an equation-free model reduction. The mp-optimization takes advantage of the reduced gradients produced by the non-linear dynamic optimizer. The multiparametric approach aims to compute an off-line map that approximates the reduced non-linear problem efficiently. First, an initial solution is computed for the non-linear optimization problem employing the matrix-free Arnoldi iterations [12] for constructing a model reduction orthonormal basis without the use of full gradients. As a result a reduced order piecewise affine (PWA) model is produced as a good approximation of the model. Then, the mp-optimization is solved computing an initial set of critical regions (CR). Afterward, the CRs are refined using the non-linear dynamic optimization in order to approximate the non-linear problem within an arbitrary tolerance. The results show that only a small number of CR regions are necessary in order to sufficiently approximate the problem. The CRs are constructed in a form of a search tree, due to the refinement algorithm, which accelerates the on-line search. In order to demonstrate the effectiveness of this algorithm, the aforementioned methodology has been applied to a chemical engineering application modelled by COMSOL.

[1] Camacho and Bordons, *Model Predictive Control*, vol. 7, no. 11. 2007.

[2] T. F. Edgar, W. J. Campbell, and C. Bode, "Model-based control in microelectronics manufacturing," *Proc. 38th IEEE Conf. Decis. Control (Cat. No.99CH36304)*, vol. 4, no. December, pp. 85–91, 1999.

[3] W. Xie, I. Bonis, and C. Theodoropoulos, "Data-driven model reduction-based non-linear MPC for large-scale distributed parameter systems," *J. Process Control*, vol. 35, pp. 50–58, Nov. 2015.

[4] A. Varshney, S. Pitchaiah, and A. Armaou, "Feedback control of dissipative PDE systems using adaptive model reduction," *AIChE J.*, vol. 55, no. 4, pp. 906–918, 2009.

[5] C. Theodoropoulos, Y.-H. Qian, and I. G. Kevrekidis, "'Coarse' stability and bifurcation analysis using time-steppers: A reaction-diffusion example," *Proc. Natl. Acad. Sci.*, vol. 97, no. 18, pp. 9840–9843, Aug. 2000.

[6] C. Theodoropoulos and E. Luna-ortiz, "A Reduced Input / Output Dynamic Optimisation Method," in *Model Reduction and Coarse-Graining Approaches for Multiscale Phenomena*, 2006, pp. 535–560.

[7] A. Armaou, I. G. Kevrekidis, and C. Theodoropoulos, "Equation-free gaptooth-based controller design for distributed complex/multiscale processes," *Comput. Chem. Eng.*, vol. 29, no. 4, pp. 731–740, Mar. 2005.

[8] E. N. Pistikopoulos, N. A. Diangelakis, R. Oberdieck, M. M. Papathanasiou, I. Nascu, and M. Sun, "PAROC—An integrated framework and software platform for the optimisation and advanced model-based control of process systems," *Chem. Eng. Sci.*, vol. 136, pp. 115–138, Nov. 2015.

[9] P. Rivotti, R. S. C. Lambert, and E. N. Pistikopoulos, "Combined model approximation techniques and multiparametric programming for explicit non-linear model predictive control," *Comput. Chem. Eng.*, vol. 42, pp. 277–287, Jul. 2012.

[10] "COMSOL Multiphysics® v. 5.2. www.comsol.com. COMSOL AB, Stockholm, Sweden." .

[11] I. Bonis and C. Theodoropoulos, "Model reduction-based optimization using large-scale steady-state simulators," *Chem. Eng. Sci.*, vol. 69, no. 1, pp. 69–80, 2012.

[12] Y. Saad, "Numerical methods for large eigenvalue problems," *Algorithms Archit. Adv. Sci. Comput.*, p. 346 p., 1992.

528346 Product Uniformity in PECVD Systems: Applying Run-to-Run Control to a Multiscale Three-Dimensional CFD Model

Abstract:

Uniform deposition of thin film layers remains a challenge in silicon processing industries due to the lack of in situ measurements and poor distribution of deposition species across the substrate surface [1],[2]. In particular, for plasma-enhanced chemical vapor deposition (PECVD) of amorphous silicon (a-Si:H), growth rate non-uniformities greater than 20% across the surface of the wafer are common [3], which may lead to inefficient solar cells and microelectronics of poor quality. Process operators typically avoid this issue by preconditioning the reaction chamber and manually adjusting deposition conditions such that thickness non-uniformities may be reduced. Last year Crose et al. [3] developed a run-to-run-based, control scheme which has been demonstrated to reduce the offset in product thickness from 5% to less than 1% within ten batches of operation [3]; likewise, in recent years three dimensional computational fluid dynamics (CFD) models of PECVD reaction chambers have demonstrated exceptional fidelity to industrially used systems [4],[5]. Motivated by this, in this work, we investigate the application of the newly developed run-to-run control strategy to the simulated deposition of amorphous silicon (a-Si:H) thin films using a transient, three-dimensional CFD model as the process simulator. The multiscale three-dimensional CFD model recently developed by Crose et al. [5] has been shown to accurately capture the complex behavior of the PECVD reactor and to reproduce experimentally observed non-uniformities with respect to the film thickness and porosity, and will be used throughout this work. A computationally efficient parallel processing scheme allows for the application of the R2R algorithm to 20 serial batch simulations which are shown to reduce the product offset from the 300 nm thickness set-point.

[1] A. Armaou, P.D. Christofides, "Plasma enhanced chemical vapor deposition: modeling and control," Chem. Eng. Sci., 54, 3305-3314, 1999.

[2] O. Gabriel, S. Kirner, M. Klick, B. Stannowski, R. Schlatmann, "Plasma monitoring and PECVD process control in thin film silicon-based solar cell manufacturing," EPJ Photovolt., 55202, 1-9, 2014.

[3] Crose, M., Kwon, JSI., Tran, A. and Christofides, P.D., "Multiscale modeling and run-to-run control of PECVD of thin film solar cells," Renewable Energy, 100, 129-140, 2017.

[4] Crose, M., Tran, A. and Christofides, P.D., "Multiscale computational fluid dynamics: methodology and application to PECVD of thin film solar cells," Coatings, 7 (2), 22, 23 pages, 2017.

[5] Crose, M., W. Zhang, A. Tran and P. D. Christofides, "Multiscale
Three-Dimensional CFD Modeling of PECVD of Amorphous Silicon Thin Films," Comp. & Chem. Eng., 113, 184-195, 2018.

530732 Process Monitoring and Leakage Diagnosis for Distributed Pipeline System Based on Discrete Observer and Moving Horizon Estimation

Abstract:

Distributed pipeline systems are increasingly constructed onshore and offshore for oil, gas and water transportation, which makes cheaper, safer and better product delivery available to local suppliers and consumers [1]. However, such benefits heavily rely on security working condition of the distributed pipeline systems, and any singularity (such as leakage, break and bend) may lead to catastrophe including casualties, property damage, resource loss, and trigger stringent environmental issues. Considering that, this work focuses on distributed pipeline system working condition monitoring and operating state estimation by using Luenberger observer design, moving horizon estimation (MHE) [2] and extreme learning machine (ELM) [3].

In terms of so-called water hammer equation based modelling, a linearized first-order hyperbolic partial differential equations (PDEs) based model is derived around equilibrium points of velocity, pressure and density by applying the linearization method. Given that Euler discretization method cannot guarantee the spatial discretization accuracy, Cayley-Tustin discretization scheme (Crank-Nicolson integration scheme) is applied in this work so that the energy and quadratic invariants are preserved without any model reduction or spatial approximation of distributed parameter system [4-5]. Given that a discretized-time distributed pipeline model is a infinite-dimensional system, a discrete Luenberger observer [6] is designed for pipeline process monitoring and ultimately the-state-of-art moving horizon estimation is utilized for multi-input and multi-output (MIMO) infinite-dimensional pipeline system state estimation under coloured noise and physical constraints. In particular, various leakage scenarios are generated by adjusting the boundary condition of upstream velocity and downstream pressure and ultimately extreme learning machine algorithm is utilized for leak detection, localization and size estimation intelligently. Finally, a series of simulations are presented based on a long-range infinite dimensional pipeline system and the proposed method is tested and demonstrated.

[1] Billmann L., Isermann R., 1987. Leak detection methods for pipelines. Automatica. 23 (3): 381-385.

[2] Rawlings J. B. 2009. Model Predictive Control: Theory and Design. Mathematics in Science and Engineering. Madison, WI: Nob Hill Publishing, LLC. p. 576. ISBN 978-0-9759377-0-9.

[3] Huang G., Zhu Q., Siew, C. K. 2006. Extreme learning machine: theory and applications. Neurocomputing. 70 (1): 489â€“501.

[4] Xu Q., Dubljevic S., 2017. Linear model predictive control for transport-reaction processes. AIChE Journal 63 (7): 2644-2659.

[5] Havu V., Malinen J., 2007. The cayley transform as a time discretization scheme. Numerical Functional Analysis and Optimization 28 (7-8): 825-851.

[6] Luenberger D., 1971. An Introduction to Observers, IEEE Transactions on Automatic Control, 16 (6): 596-602.

535193 Dynamic Modeling and Control of a Natural Gas Combined Cycle (NGCC) Power Plant with a Damage Model

Abstract:

As the penetration of intermittent renewable energy sources into the electric grid keeps increasing, natural gas combined cycle (NGCC) plants are expected to follow their load much more frequently than in the past. While cycling operation helps to improve the viability of a power plant, it can lead to its lifetime consumption due to creep, fatigue and corrosion damages in the long term [1]. With this motivation, a dynamic model of a NGCC plant coupled with a lifetime estimation model is developed. Advanced model-based controllers are developed to optimize the transient operation under different scenarios while maintaining acceptable component life.

A high-fidelity dynamic model of a NGCC plant is developed with rigorous equipment level models to capture the impact of load-following on the equipment health and life. A thermo-hydraulic model of the heat recovery steam generator (HRSG) is developed and used to obtain the optimal design of various sections of the HRSG including the number of passes, rows of tubes and tube numbers, dimensions of the tubes, and fin geometry. A stage-by-stage model of the steam turbine is developed to estimate the performance of the triple-pressure steam turbine with multiple steam addition and extraction points under nominal and off-design operations.

The high pressure superheater, intermediate pressure reheater and high pressure steam drum are the most stressed components during load variations. Both creep and thermo-mechanical fatigue damage models are developed for the superheater/reheater. A fatigue damage model is developed for the steam drum. The creep assessment is conducted based on the R5 British procedure and the fatigue lifetime calculation is computed according to the EN 13345 Standard and UNI EN 12952[2-4]. The Rainflow cycle counting method is used to determine stress ranges and cycles. According to the overall stress amplitude, the lifetime consumption of each cycle is estimated.

A model predictive controller (MPC) is developed for fast load following while keeping the stress within admissible bounds. Our study shows that a model based approach can help to accomplish fast ramp rates while maintaining acceptable component life.

[1] Benato, A., et al. "LTE: A procedure to predict power plants dynamic behaviour and components lifetime reduction during transient operation." Applied energy 162 (2016): 880-891.

[2] British Energy. R5 Assessment procedure for the high temperature response of structures, Issue 3. British Energy; Gloucester, UK; 2003.

[3] European Committee for Standardization. EN 13345 part 3 unfired pressure vessels, clause 17; simplified assessment of fatigue life, and clause 18; detailed assessment of fatigue life; 2002.

[4] UNI â€" Ente Nazionale di Normazione. UNI EN 12952-5:2011. Water-tube Boilers Standards; 2011.

531887 A Reactive Distillation Optimization Model for Low- and High-Temperature Fischer–Tropsch Syntheses for Multiphase Product Recovery

Abstract:

Liquid fuels and chemicals are predominantly produced from crude oil. The abundance of shale gas has attracted attention for alternatives to be sought since natural or shale gas can be converted to the same liquid fuels and chemicals through integrated gas-to-liquid (GTL) processes. Biomass and carbonaceous wastes as well can also be converted to liquid fuels and chemicals through biomass-to-liquids and waste-to-liquids processes respectively. However, natural/shale gas is the most hydrogen-rich carbon source and can provide the most carbon-efficient route to fuels.

GTL processes require high capital cost but the main detractor of widespread adoption is investment risk linked to feed and products price volatilities. The decision to build a gas-to-liquid plant has to be based on the perceived future prices of oil and gas; and the prices of specialty chemicals produced from downstream processes. Therefore, robust technology that can offer product flexibility in response price volatilities is desirable to mitigate the investment risk.

Process intensification where syngas conversion in multiple reaction zones and product separation are combined in one unit can lead to smaller overall plants with lower capital and running costs. Consequently, we have developed an equation-oriented framework for optimal synthesis of integrated reactive distillation (RD) systems for FT processes in GAMS in which the phase equilibrium is given by a cubic equation of state, the reaction rate expressions are expressed in terms of fugacities, and the product selectivity of catalysts are based on the Anderson–Schulz–Flory distribution and experimental data. Comparing the RD for FT synthesis against conventional slurry reactors, provided results show that RD has a potential edge in industrial processes.

To explore the potential of producing a predominantly wax product on a low-temperature FT process, we are expanding the FT-RD optimization model to optimize tray numbers, feed tray location and catalyst loading on trays. We are also implementing the FT-RD optimization model within the PyOMO optimization framework. To produce a high gasoline fraction, light olefins, and some oxygenates (mainly alcohols) on a high-temperature FT process, we are developing a more detailed kinetic model to account for the various products.

536606 Design and Intensification of Sorption-Enhanced Reaction Processes for Methanol Production

Abstract:

Methanol is a viable alternate fuel to depleting oil and gas reserves, and is one of the most advantageous chemicals due to clean burning, CO_2 utilization and as a raw material for synthesizing hydrocarbons [1]. Since the commercialization of methanol production in 1923, the process has undergone significant improvements [2]. However, even after improved process efficiency, the current methanol production process from syngas is equilibrium-limited and is one of the largest energy consumers in chemical industry [3]. The process inefficiencies include low single-pass conversions, low methanol yield and selectivity, high energy consumption and equipment cost, and catalyst deactivation [4].

The objective of this work is to counter currently-existing process inefficiencies by enhancing single-pass reaction conversions and preventing catalyst deactivation. To this end, we apply the sorption enhanced reaction process (SERP) concept for designing a periodic sorption-enhanced methanol synthesis (SE-MeOH) process from syngas; *in-situ* removal of water byproduct favorably shifts the equilibrium towards enhanced methanol synthesis in accordance with the Le Chatelier's principle. For capturing the dynamics of the SE-MeOH process, we use the generalized adsorption-reaction modeling and simulation (GRAMS) platform. GRAMS is based on a one-dimensional, pseudo-homogeneous, non-isothermal, non-adiabatic, and non-isobaric NAPDE-based model, and has been extensively validated with experimental data for several SERP case studies [5]. GRAMS is coupled with an in-house simulation-based constrained grey-box optimizer for optimizing SE-MeOH process cycle configuration, design parameters and operating conditions [6][7]. We obtain single-pass carbon conversions as high as 80% in the Lurgi-type Methanol reactor via SE-MeOH with production capacity exceeding 109500 tons/yr, which is the typical capacity of an industrial plant [8]. The developed SE-MeOH process has smaller carbon footprint, enhanced product quality, and smaller reactor, condenser and recompressor size leading to significant savings in energy consumption and process economics. This presentation also covers the integration of the developed SE-MeOH process with a steam methane reforming reactor unit for converting natural gas to methanol via syngas route.

References

[1] G.A. Olah, Beyond oil and gas: the methanol economy, Angew. Chemie Int. Ed. 44 (2005) 2636–2639.

[2] A. Cybulski, Liquid-phase methanol synthesis: catalysts, mechanism, kinetics, chemical equilibria, vapor-liquid equilibria, and modeling—A Review, Catal. Rev. Eng. 36 (1994) 557–615.

[3] A. V Kruglov, Methanol synthesis in a simulated countercurrent moving-bed adsorptive catalytic reactor, Chem. Eng. Sci. 49 (1994) 4699–4716.

[4] M. Kuczynski, M.H. Oyevaar, R.T. Pieters, K.R. Westerterp, Methanol synthesis in a countercurrent gas—solid—solid trickle flow reactor. An experimental study, Chem. Eng. Sci. 42 (1987) 1887–1898.

[5] A. Arora, S.S. Iyer, M.M.F. Hasan, GRAMS: A General Framework Describing Adsorption, Reaction and Sorption-Enhanced Reaction Processes, Submitted. (2018).

[6] A. Arora, I. Bajaj, S.S. Iyer, M.M.F. Hasan, Optimal Synthesis of Periodic Sorption Enhanced Reaction Processes with Application to Hydrogen Production, Comput. Chem. Eng. 115 (2018) 89–111.

[7] I. Bajaj, S.S. Iyer, M.M.F. Hasan, A Trust Region-based Two Phase Algorithm for Constrained Black-box and Grey-box Optimization with Infeasible Initial Point, Comput. Chem. Eng. (2017).

[8] N. Rezaie, A. Jahanmiri, B. Moghtaderi, M.R. Rahimpour, A comparison of homogeneous and heterogeneous dynamic models for industrial methanol reactors in the presence of catalyst deactivation, Chem. Eng. Process. Process Intensif. 44 (2005) 911–921.

525385 Comparison of Adsorptive Separation Performance of a Hollow Fiber Bed and a Packed Bed: A Modeling Study Using Zeolite 13X and a Mixture of Propylene and Propane

Abstract:

Since its discovery by Carl Wilhelm Scheele in 1773, adsorption has become one of the commonly used industrial techniques for fluid (gas/liquid) separation, mostly fueled by the invention of tunable synthetic zeolites in the 1940s and the development of cyclic schemes which allowed for product recovery and adsorbent regeneration. [1] Traditionally, adsorbent crystals are formed into pellets (with or without a binder) and then packed into beds. However, pressure drops are usually incurred due to the random and tortuous nature of the packing void space which influences the cost of pressurizing the feed gas and the shape of the adsorption front. Heat transfer is inefficient for operating a TSA cycle and there is scope for improvement of fluid to adsorbent mass transfer coefficient. [2-4]

To overcome these difficulties, several structured configurations including, monoliths, laminates and foams, have been developed. Detailed reviews and comparisons of these configurations are available in literature [2,3,5,6]. More recently, a novel hollow fiber-based solid sorbent system has been proposed and experimentally validated[4,7-9] which provides an additional degree of control over the heat transfer efficiency as described here. The hollow fiber configuration relies on solid sorbents embedded in a porous polymeric hollow fiber matrix. Several identical fibers are assembled inside a module which resembles a shell and tube heat exchanger. The bore of the hollow fiber has an impermeable lumen layer which allows the flow of a cooling/heating medium to mitigate the temperature of operation of the process despite heat of adsorption/desorption. [4]

Compared to the traditional packed bed, the configuration of the hollow fiber bed offers several advantages: (1) the structured packing has a significantly lower pressure drop (2) the diffusion length for the fluid to the interior of the hollow fibers is significantly lower than the spherical packed bed (3) the bore allows for a heating/cooling medium to efficiently transfer heat to/from the adsorbent phase. However, in terms of adsorbent density per unit bed volume, the hollow fiber bed has a clear disadvantage.

From the previous discussion it is clear that while the use of a hollow fiber adsorbent bed has several advantages, it also has at least one significant drawback. It has been hypothesized that the advantages will outweigh the disadvantages of using a hollow fiber bed, but testing and understanding the drivers of the relative performance is needed. The aim of this computational study is therefore, to perform a detailed comparison of the properties and separation performance of the novel hollow fiber bed configuration with that of the traditionally used packed bed. To enable unbiased comparison, an equivalent model for the packed bed and the hollow fiber bed was developed. A preliminary parametric comparison of properties such as heat and mass transfer co-efficients, adsorbent density, and pressure drop was performed on the basis of these models. Optimized five step single bed cycles for the separation of propylene and propane on zeolite 13X were also studied. For the same recovery and purity, the hollow fiber bed was found to have a productivity that was five times higher than the packed bed. The hollow fiber bed also showed higher productivity, when parameters such as the desorption pressure and the ratio of the purge to feed velocity were varied in both beds.

1. DÄ…browski, A. Adsorptionâ€"from theory to practice. *Advances in colloid and interface science* **2001**, *93*, 135-224.
2. Rezaei, F.; Mosca, A.; Webley, P.; Hedlund, J.; Xiao, P. Comparison of traditional and structured adsorbents for CO2 separation by vacuum-swing adsorption. *Industrial & Engineering Chemistry Research* **2010**, *49*, 4832-4841.
3. Rezaei, F.; Webley, P. Optimum structured adsorbents for gas separation processes. *Chemical Engineering Science* **2009**, *64*, 5182-5191.
4. Lively, R. P.; Chance, R. R.; Kelley, B.; Deckman, H. W.; Drese, J. H.; Jones, C. W.; Koros, W. J. Hollow fiber adsorbents for CO2 removal from flue gas. *Industrial & Engineering Chemistry Research* **2009**, *48*, 7314-7324.
5. Rezaei, F.; Grahn, M. Thermal management of structured adsorbents in CO2 capture processes. *Industrial & Engineering Chemistry Research* **2012**, *51*, 4025-4034.
6. Akhtar, F.; Andersson, L.; Ogunwumi, S.; Hedin, N.; Bergström, L. Structuring adsorbents and catalysts by processing of porous powders. *Journal of the European Ceramic Society* **2014**, *34*, 1643-1666.
7. Lively, R.; Chance, R. R.; Koros, W. J.; Deckman, H. W.; Kelley, B. T.: Sorbent fiber compositions and methods of temperature swing adsorption. Google Patents, 2008.
8. Fan, Y.; Labreche, Y.; Lively, R. P.; Jones, C. W.; Koros, W. J. Dynamic CO2 adsorption performance of internally cooled silicaâ€supported poly (ethylenimine) hollow fiber sorbents. *AIChE Journal* **2014**, *60*, 3878-3887.
9. DeWitt, S. J.; Sinha, A.; Kalyanaraman, J.; Zhang, F.; Realff, M. J.; Lively, R. P. Critical Comparison of Structured Contactors for Adsorption-Based Gas Separations. *Annual review of chemical and biomolecular engineering* **2018**.

526769 Towards a Systematic Process Intensification Framework for Advanced Distillation Systems

Abstract:

Towards a Systematic Process Intensification Framework for Advanced Distillation Systems

Yuhe Tian[a,b,c], M. Sam Mannan[a,b,c], and Efstratios N. Pistikopoulos[a,b]

[a] Artie McFerrin Department of Chemical Engineering, Texas A&M University, College Station, TX

[b] Texas A&M Energy Institute, Texas A&M University, College Station, TX

[c] Mary Kay O'Connor Process Safety Center, Department of Chemical Engineering, Texas A&M University, College Station, TX

Process Intensification (PI) offers the potential to drastically reduce the energy consumption and processing cost by utilizing multifunctional phenomena at different time and spatial scales. Distillation, the most widely utilized but highly energy-intensive separation technology, accounts for over 40% of the energy used in the chemical process industry and over 95% of that in separation industry [1]. Various advanced distillation technologies (e.g., heat-integrated distillation, reactive distillation, extractive distillation, and membrane distillation) have been developed based on PI principles to maximize synergy effects and can achieve around 30% energy savings [2]. Despite recent advances [3-6], systematic synthesis methods and software prototypes to explore phenomena-based process optimization and intensification for advanced distillation design are still lacking.

To address this challenge, in this work, we propose a systematic process intensification and synthesis framework based on recent extensions of the phenomenological Generalized Modular Representation Framework [7-9]. Herein, distillation systems are represented as aggregated multifunctional mass/heat exchange modules based on which intensification possibilities can be revealed without a pre-postulation of equipment or flowsheet configurations. Process driving forces, described by Gibbs free energy changes, are optimized to enhance fundamental transport and reaction performances. The derived intensified designs can then be analyzed by detailed steady-state simulation and design optimization. The capabilities of the proposed framework to deliver advanced distillation systems with improved energy efficiency are demonstrated with three case studies: (i) heat-integrated distillation, (ii) extractive distillation, and (iii) reactive distillation. Extensions of the framework to include process safety and dynamic operability assessment will also be discussed [10].

Reference

1. Kiss, A. A. (2014). Distillation technology–still young and full of breakthrough opportunities. *Journal of Chemical Technology and Biotechnology*, *89*(4), 479-498.

2. Olujić, Ž., Jödecke, M., Shilkin, A., Schuch, G., & Kaibel, B. (2009). Equipment improvement trends in distillation. *Chemical Engineering and Processing: Process Intensification*, *48*(6), 1089-1104.

3. Tian, Y., Demirel, S. E., Hasan, M. M. F., & Pistikopoulos, E. N. An Overview of Process Systems Engineering Approaches for Process Intensification: State of the Art. *Chemical Engineering and Processing: Process Intensification*, Submitted.

4. Tula, A. K., Babi, D. K., Bottlaender, J., Eden, M. R., & Gani, R. (2017). A computer-aided software-tool for sustainable process synthesis-intensification. *Computers & Chemical Engineering*, *105*, 74-95.

5. Pichardo, P., & Manousiouthakis, V. I. (2017). Infinite DimEnsionAl State-space as a systematic process intensification tool: Energetic intensification of hydrogen production. *Chemical Engineering Research and Design*, *120*, 372-395.

6. Demirel, S. E., Li, J., & Hasan, M. F. (2017). Systematic process intensification using building blocks. *Computers & Chemical Engineering*, *105*, 2-38.

7. Papalexandri, K. P., & Pistikopoulos, E. N. (1996). Generalized modular representation framework for process synthesis. *AIChE Journal*, *42*(4), 1010-1032.

8. Ismail, S. R., Proios, P. and Pistikopoulos, E. N. 2001. Modular synthesis framework for combined separation/reaction systems. *AIChE Journal*, *47*(3), 629-649.

9. Proios, P., Goula, N. F., & Pistikopoulos, E. N. (2005). Generalized modular framework for the synthesis of heat integrated distillation column sequences. *Chemical engineering science*, *60*(17), 4678-4701.

10. SYNOPSIS – Synthesis of Operable Process Intensification Systems. Awarded project [DOE RAPID Manufacturing Institute].

535419 An MILP-Based Operability Approach for Process Design, Intensification and Modularity of Nonlinear and High-Dimensional Energy Systems

Abstract:

Operability emerged as a tool for the selection of the necessary inputs to achieve specified desired outputs in a given process [1]. In a previously developed mixed-integer linear programming (MILP)-based operability approach [2], the operability concept was extended to attain process intensification towards obtaining modular designs for energy systems. While a computational time improvement was achieved, the developed framework is still restricted to computational geometry techniques applied in two-dimensional spaces. The extension of this approach to systems with input and output spaces of higher dimensions consists of a challenge that still remains to be addressed. High dimensions give rise to new concerns such as volume of simulated data and overall increase in algorithm computational complexity, potentially impacting the computational cost. In this work, an improved MILP-based operability framework is derived with the utilization of computational geometry techniques suitable for higher-dimensional and nonlinear systems. This presentation will be focused on the discussion of the feasibility of the new approach to produce intensified modular designs for energy systems applications.

Process operability has been successfully applied to high-dimensional and nonlinear systems by the utilization of nonlinear programming (NLP) tools [3]. Potential tractability issues associated with the high computational expense of nonlinear algorithms have been tackled with the employment of parallel computing [4]. In the proposed framework, the addressed high-dimensional and nonlinear system is decomposed into several linearized subsystems. These subsystems are obtained using spatial discretization techniques that produce a set of convex polytopes in each n-dimensional space. Then, an MILP optimization problem is formulated based on spatial search and barycentric interpolation. For this application, the goal of the optimization problem is to minimize physical size, satisfying all process constraints and intensification targets. The outcome of this optimization problem is a modular design of high efficiency, associated with the most intensified operating point. As MILP tools have higher computational efficiency than NLP tools, a single processor will be considered for the proposed framework calculations, in contrast to the parallel workers present in the NLP approach.

The developed framework will be tested and compared to past NLP approaches for two case-studies: a membrane reactor for the direct methane aromatization conversion to hydrogen and benzene and a natural gas combined cycle (NGCC) process. These systems are characterized by dimensionalities as high as 5x5 and 8x8 dimensions, respectively [4]. The trade-off between accuracy and computational cost will be discussed for the validation of the proposed framework when compared to NLP approaches for high-dimensional system extrapolations.

References

1. Vinson D. R. and Georgakis C. "A new measure of process output controllability". J. Proc. Cont., 10(2-3), 185-194 (2000).
2. Gazzaneo V., Carrasco J.C. and Lima F. V. "An MILP-based operability approach for process intensification and design of modular energy systems." Accepted for publication in Proceedings of Process Systems Engineering (PSE), July (2018).
3. Carrasco J.C. and Lima F. V. "Operability-based approach for process design, intensification, and control: application to high-dimensional and nonlinear membrane reactors". In Proceedings of the FOCAPO/CPC (2017).
4. Carrasco J.C. and Lima F. V. "Bilevel and parallel programming-based operability approaches for process intensification and modularity." Accepted for publication in AIChE Journal. DOI10.1002/aic.16113 (2018).

525666 Compact, Nonsmooth Operators for Single-Component Mass and Water Integration

Abstract:

Focus in modern plant retrofit, expansion, and design has shifted towards the use of intensification techniques to implement processes that are both sustainable and economically competitive. By calculating the minimum resource usage required for a process with optimal reuse, process integration methods can be used to identify targets where there is significant scope for intensification and motivate the design of novel technologies, such as multistream heat exchangers for optimal heat recovery. Therefore, process integration can be used both to intensify existing plants where a fundamental redesign of the process is infeasible and also to ensure that proposed intensified processes operate optimally. To further decrease resource use, these integration methods have expanded beyond integrating individual processes to sharing resources between plants in the form of eco-industrial parks. However, as process integration is applied to increasingly complex problems, improved numerical methods are required to efficiently consider resource usage and constraints.

Current methods for solving process integration problems fall into two general categories: pinch analysis or superstructure approaches. Pinch analysis methods are physically intuitive and relatively easy to implement; however, current techniques are often heuristic and limited to a small class of problems. Even algorithmic pinch point approaches are typically mathematical programming problems with numbers of constraints that increase significantly with the number of sources and sinks. In the context of fixed-flow rate water allocation problems, these approaches are also unable to solve for concentration process variables and cannot be used for simultaneous process integration and optimization. On the other hand, superstructure techniques are adaptable to many different integration problems. However, these formulations are mixed-integer programming problems that often involve nonconvex functions, are difficult to solve, and whose complexity increases rapidly with problem size.

To address these limitations, we have developed compact, nonsmooth operators for both single-contaminant mass and water integration problems. These operators extend the work of Watson *et al.*, who simulated multistream heat exchangers using a nonsmooth system of equations adapted from a mathematical programming pinch analysis approach proposed by Duran and Grossmann [1,2]. Our formulation consists of systems of only two equations: an overall resource balance for the process and a single nonsmooth equation describing the resource balance below potential pinch points. Automatic lexicographic-directional differentiation is used to calculate exact generalized derivative elements for the nonsmooth equation system [3]. These elements are used in a suitable method, such as Semismooth or Linear Programming Newton, to solve the system without requiring smoothing approximations or additional variables.

We will present these novel operators and their solution methods and demonstrate how they can be used both to solve for optimal resource targets and for process parameters given resource constraints, including source and sink concentrations in water allocation problems. In addition, we will show other unique benefits of this approach,

including the ability to simulate both pinch and threshold problems using a single model. Overall, our results will highlight how these nonsmooth operators significantly reduce problem complexity compared to optimization approaches and can provide computationally tractable solutions to large-scale interplant integration problems.

References

[1] H .A. J. Watson, K. A. Khan, and P.I. Barton. Multistream Heat Exchanger Modeling and Design. AIChE Journal, 61(10):3390-3403, 2015.

[2] M.A. Duran and I.E. Grossmann. Simultaneous Optimization and Heat Integration of Chemical Processes. AIChE Journal, 32(1):123-138, 1986.

[3] K. A. Khan and P. I. Barton. A vector forward mode of automatic differentiation for generalized derivative evaluation. Optimization Methods and Software, 30(6):1185-1212, 2015.

535724 Optimal Mass Exchanger Network Synthesis Using a 2-Step Hybrid Algorithm Including Packed Column Design

Abstract:

A method for the synthesis of mass exchanger networks (MENs) is presented that focuses on including detailed packed column design considerations at the network optimization level to deliver designs that are optimal and more realistic. Most approaches to simultaneous optimization of MENs have been based upon a stage-wise superstructure with simplified representations of the individual mass exchangers. Even with these simplified representations, the MINLP is still non-convex and computationally difficult to solve with modern deterministic solvers. Additionally, it is not possible to guarantee that when a solution is found that it is globally optimal.

The method developed in this study, uses a version of this aforementioned stage-wise MINLP formulation in the first step. Once the solution to this is found, a secondary nonlinear programming (NLP) step, based on orthogonal collocation on finite elements, is performed to find the optimal column dimensions, packing sizes, fluid velocities, and pressure drops, based on the flows and stream matches from the network optimization. Once the new individual column designs are found, the MINLP is then updated with a series of correction factors based on the solution of the detailed designs. This procedure repeats between the MINLP and NLP sub-optimization until the differences between the solutions converge or until a maximum number of iterations is reached.

The methodology, presented in detail in the work of Short, et al. (2018), is moved from the GAMS environment into PYOMO, allowing for powerful new ways to enhance the algorithm. The previous methodologyâ€™s correction factor determination is improved upon in order to speed up and increase the chances of convergence. Additionally, through the application of different MINLP solvers, larger examples that were previously not able to be solved are included.

Reference: Short, M., Isafiade, A.J., Biegler, L.T.,Kravanja, Z., 2018, Synthesis of mass exchanger networks in a two-step hybrid optimization strategy, Chemical Engineering Science, 178, 118 - 115.

529420 Mathematical Optimization of Sustainable Water Desalination Processes Using Directional Solvent Extraction

Abstract:

Access to clean fresh water is an ever-growing concern for modern society as it is critical to ensure human health, to protect threatened ecosystems, and to promote economic growth and prosperity. Modern seawater desalination technologies remain energy intensive; they required three to four times the theoretical minimum energy for separation [1]. Further research is required to both reduce the energy intensity of seawater desalination and design new systems that are driven by renewable solar thermal and waste process heat.

Directional solvent extraction (DSE) is a promising new technology for solar desalination that can effectively utilize low temperature (40 °C to 80 °C) heat. DSE exploits special solubilities of certain solvents (typically saturated fatty acids) to extract fresh water from saline sources at elevated temperature and release the water when the solution is cooled. This approach has several unique features: (1) water can dissolve in the solvent and the solubility increases with temperature; (2) the solvent is virtually insoluble in water; (3) the solvent does not dissolve salts; and 4) no membrane is required. Previous work includes initial concept demonstration as a batch process [2,3], molecular simulation to understand solvent performance [4], and heat integration for a single-stage continuous process [5]. The predicted energy requirement for DSE, without process intensification or optimization, is >160 kWh/m^3 of fresh water.

This presentation explores new process intensification opportunities for DSE. Most notably, previous work has focused on single-stage configurations [5]. Preliminary results suggest a counter-current configuration with a moderate number of equilibrium stages can reduce the energy requirement by up to 70%. Moreover, a mathematical optimization framework is proposed to optimize the DSE process and systematically explore trade-offs between product quality, energy requirements, and capital costs. This is done using an equation-oriented approach that simultaneously optimizes process operating conditions (such as temperatures, flow rates, compositions), design parameters (such as equipment sizes), and heat recovery opportunities [6,7]. Thus, the simultaneous approach can exploit tens to hundreds more degrees of freedom than conventional targeting methods (i.e., as employed in [5]), which consider optimal heat integration with fixed flowrates and temperatures. This capability is critically to fully optimize the DSE process and inform ratio molecular design of novel thermoresponsive solvents.

References:

[1] M. Elimelech, W. A. Philip (2011). The Future of Seawater Desalination: Energy, Technology, and the Environment. *Science* 333 (6043), pp. 712-717.

[2] A. Bajpayee, T. Luo, A. Muto, G. Chen (2011). Very low temperature membrane-free desalination by directional solvent extraction. *Energy Environ. Sci.* 4, pp. 1672â€"1675.

[3] D. Rish, S. Luo, B. Kurtz, T. Luo (2014). Exceptional ion rejection ability of directional solvent for non-membrane desalination. *Appl. Phys. Lett.* 104 (2), p. 024102.

[4] T. Luo, A. Bajpayee, G. Chen (2011). Directional solvent for membrane-free water desalinationâ€"A molecular level study. *J Appl. Physics.*, 110 (5), p. 054905.

[5] S. Alotaibi, O. M. Ibrahim, S. Luo, T. Luo (2017). Modeling of a continuous water desalination process using directional solvent extraction. *Desalination* 420, pp. 114-124.

[6] A.W. Dowling, L. T. Biegler (2015). A Framework for Efficient Large Scale Equation-Oriented Flowsheet Optimization. *Comp. Chem. Eng.* 72, pp. 3-20.

[7] A. W. Dowling, L. T. Biegler (2013). Optimization-based process synthesis for sustainable power generation. *Chem. Eng. Trans.* 35, pp. 1-12.

523960 System Decomposition for Distributed Multivariate Statistical Process Monitoring

Abstract:

Monitoring of complex large-scale chemical plants to ensure safe and cost effective operation is a challenging task. Multivariate Statistical Process Monitoring (MSPM) methods like Principal Component Analysis are popular in the chemical industry because they are easy to implement and do not require a priori knowledge of the process mathematical model. In MSPM methods, the magnitude of deviation of the measured variables from their mean values is combined to calculate a test statistic(s) which is compared with a threshold value to determine if a fault is affecting the process. MSPM methods underperform in detecting those faults that affect only a small subset of a large process system's measured variables because most measured variables do not contribute significantly to the test statistic when such faults occur. Implementing MSPM methods in a distributed configuration, wherein the MSPM method is applied to each subsystem and the monitoring results of the subsystems are combined using a consensus strategy, could in principle address this problem. Implementing MSPM methods in a distributed configuration also makes the monitoring system more fault tolerant. It may also be necessary to implement a MSPM method in a distributed configuration when the layout of a chemical plant is such that the communication of all the plant's measurements to a central location is infeasible.

The system decomposition (i.e. the partitioning of a system's measured variables into subsystems) has a significant impact on the performance of a distributed MSPM (DMSPM) method. The DMSPM method should ideally be implemented using a decomposition for which its performance, in monitoring a set of faults, is optimal so that the faults can be detected with greater speed and accuracy. The optimal decomposition for a DMSPM method depends on the set of faults that tend to affect the system. In an optimal decomposition of the system, the different measured variables of a subsystem tend to be affected by similar faults thus ensuring that most measured variables in the subsystem contribute significantly to the subsystem test statistic when a fault occurs thereby making the statistical hypothesis tests of the DMSPM method more sensitive to the faults. The optimal decomposition for a DMSPM method also depends on various other factors such as:

1. The number of subsystems that the system is partitioned into.
2. The MSPM method that is implemented in a distributed configuration.
3. Consensus between the subsystems.

Therefore, finding the optimal decomposition for a DMSPM method is a difficult task. An effective strategy to find the optimal decomposition would be to use simulation optimization wherein the performance of the DMSPM method is simulated for a set of candidate decompositions and the decomposition with the best performance is considered optimal.

In this work, we propose and present a novel simulation optimization method, called the Performance Driven Agglomerative Clustering (PDAC) method, which finds a near optimal system decomposition for a DMSPM method. The PDAC method uses the greedy search of Ward's agglomerative clustering algorithm to generate a set of candidate decompositions (the decision variables). Normal operation and faulty data, input by the user, is then used to simulate the performance of the DMSPM method for a candidate decomposition and calculate its missed detection rate (MDR) which is the objective function. The MDRs of the candidate decompositions having the same number of subsystems are compared. After applying the agglomerative clustering procedure, a near optimal decomposition is generated for every possible value of the number of subsystems. The number of subsystems can range from one to the number of measured variables in the system. A fine tuning procedure is used to slightly modify some of the decompositions output by the agglomerative clustering procedure and further reduce their MDRs. The monitoring performance of the decompositions output by the agglomerative clustering and fine tuning procedures is then compared to find the optimal number of subsystems and hence the optimal system decomposition for the DMSPM method. The PDAC method is a completely data-driven system decomposition method and can be automated. The PDAC method can also, in principle, be applied to most DMSPM methods since it only requires simulation of the DMSPM method using process data. To illustrate its effectiveness, PDAC is used to find the decomposition of the benchmark Tennessee Eastman Process case study for which the monitoring performance, using a distributed Principal Component Analysis based monitoring scheme, is optimal.

525556 Transfer Entropy-Based Dynamic Causal Maps Generation for Fault Diagnosis in Systems with Cycles

Abstract:

On-line fault detection and diagnosis in process industries has always been a challenge due to the large size and complexity of the plant resulting in enormous amount of potential interactions between variables. Causality represented in the form of a directed graph (digraph) is one of the most widely used tools for root cause analysis, once the presence of a fault is identified [1]. In conventional approaches of graph-based diagnosis, a digraph representing the actual desired process is developed and by tracing the path of variables/nodes showing deviation from desired behavior, root cause is identified. The performance of this approach depends on the accuracy of the causal map developed. Bauer et al. [2] proposed an algorithm using transfer entropy to detect direct causality and construct a causal map. This technique works well for systems that can be represented as directed acyclic graphs. In a chemical plant, however, feedback loops and recycling streams (e.g., refluxes) are common. Such cyclic effects make the problem of direct causality detection and root cause analysis harder.

In the proposed work, we use transfer entropy as a measure of causality between variables, as it is capable of capturing non-linear interactions between them. A weighted adjacency matrix derived from transfer entropy (based on process data) between each pair of variables is used to construct a weighted causal map. Onset of fault is accompanied by change in transfer entropy, which leads to change in the weighted causal graph identified. Tracking this change in the dynamic causal map serves as a useful metric for identification of faults, instead of relying on accurate estimation of the true map itself. In this work, we report on the performance of this approach on various simulated case studies, including the Tennessee Eastman process. The technique is found to satisfactorily locate the fault, either as a faulty cycle or a faulty node.

References

[1] V. Venkatasubramanian, R. Rengaswamy, and S. N. Kavuri, "A review of process fault detection and diagnosis part II: Qualitative models and search strategies," *Comput. Chem. Eng.*, vol. 27, no. 3, pp. 313–326, 2003.

[2] M. Bauer, J. W. Cox, M. H. Caveness, J. J. Downs, and N. F. Thornhill, "Finding the Direction of Disturbance Propagation in a Chemical Process Using Transfer Entropy," *IEEE Trans. Control Syst. Technol.*, vol. 15, no. 1, pp. 12–21, 2007.

526663 Efficient Process Monitoring and Causality Analysis of Processes Via the Integrated Use of the Graphical Lasso and Markov Random Fields Modeling

Abstract:

Process monitoring is an important aspect of safe operation of process plants. Various methods exist which monitor the process using data-driven methods, but they all have certain limitations. Conventional methods, such as PCA and PLS, are effective in monitoring large systems by analyzing the reduced system space, but assumes the variables are linearly correlated, resulting in inaccurate results for faults occurring in small regions. Also, since the individual characteristics of the variables are mitigated, it is difficult to analyze and isolate the source of the fault. Recent methods have tried resolve these problems by developing new methods for process monitoring. An example of these methods are the use of variant auto-encoders and SVMs, since these methods can deal with nonlinear and non-Gaussian variables, and show more accurate monitoring performances. However, they are still limited in terms of fault isolation and propagation path analysis, due to the difficulty in causality analysis. The use of Bayesian networks is another exemplary effort to enable fault detection and isolation, and a few studies evaluate causality analysis performances of Bayesian network monitoring. But a critical limitation in using Bayesian networks is that the user should be readily aware of the causal relationship between variables, and that cyclic relationships cannot be modeled.

In this study, a novel monitoring method for accurately detecting the faults and analyzing the cause of the faults is proposed. Named the Glasso-MRF monitoring framework, this method integrates the use of the graphical lasso algorithm (G-lasso) and the Markov random field (MRF) modeling framework to divide the monitored variables into relevant groups and then detect the faults separately. Graphical lasso uses the lasso constraint on the coefficient of variables, making the inverse covariance matrix of variables to be of sparse form. The use of G-Lasso downsizes the process into groups that are highly correlated, enabling the fault propagation path to be identified rather simply, and relieving the computational complexity of the MRF-based monitoring. MRF modeling can extensively model the variable relationships including cyclic ones. Also, since the probability of MRF can be calculated using factor graphs, the individual contribution of each variable to the fault can be calculated, enabling effective fault isolation. The inference of MRFs are usually complex due to the existence of partition functions, but by down-sizing the system using G-lasso, this problem is resolved as well. The proposed method was applied to the well-known Tennessee Eastman process to test out its performance. Surprisingly, the detection rate was higher than any other state-of-the-art monitoring methods, including auto-encoders and Bayesian networks, showing more than 96% detection rates for all of the 21 faults encoded in the Tennessee Eastman process. Also, the fault propagation path and causality analysis results proved to be reliable, with respect to the related results in previous studies. These results prove that, the proposed methodology can effectively detect the fault as well as identify their causes, and show its propagation throughout the process, without any a priori knowledge of the process variables.

531348 Transfer Learning Method for Chemical Plant Fault Diagnosis

Abstract:

Modern chemical and petrochemical engineering plants are working within complex and large variety of operating and performance conditions. In particular, many flammable and explosive chemicals are involved in these complex reactions and/or are stored in plant areas and harsh operational/working conditions might be required, such as high temperature, high pressure, acid/alkaline environment and etc [1]. Hence, any component singularity or failure may lead to severe catastrophe, property damage, and environmental contamination. On the other hand, multi-reactions and multi-equipment cascading systems often cause the uncertainty, nonlinearity and multi-working modes (changeable working steady states) of chemical plants, which poses a great challenge of plant condition monitoring and fault detection, alarming and diagnosis.

Hence, considering the complex and varying chemical process, well-trained traditional fault detection and diagnosis models for one working condition may tend to lapse under another working condition. Given that, more effective and efficient online monitoring and fault diagnosis techniques need to be proposed and applied. This work introduces state-of-the-art transfer learning (TL) methods into the area of chemical plant condition monitoring and fault intelligent diagnosis.

Compared to traditional machine learning (ML) methods, transfer learning theories specialize in cross-domain learning, when the testing data are following different data distributions [2-3], which provides the natural benefits for varying condition fault diagnosis. In particular, stemming from well-formed transfer component analysis algorithm [3], more representative and adaptive cross-domain features will be extracted, selected and fused for singularity detection and fault recognition by combining classical classifiers such as support vector machine (SVM), extreme learning machine (ELM) application, etc [4-5].

In order to eliminate differences brought by changeability of working conditions, several similarity evaluation criteria such as Euclidean distance and Kullback-Leibler divergence (KL) are utilized for addressing the nonlinearities and uncertainties masked in the monitoring data and bridging the data distribution gaps of difference working conditions. Finally, Tennessee Eastman process (TEP) dataset is utilized for fault detection and diagnosis to test the performance of the proposed method. In particular, reactor cooling water inlet temperature change and valve sticking faults are detected and diagnosed by the proposed enhanced transfer component analysis technique, and other conventional feature selection and fusion methods such as principle component analysis (PCA) and kernel principle component analysis (KPCA) are used as baseline methods to verify the superior performance of the presented approach.

[1] Frank P. M. 1990. Fault diagnosis in dynamic systems using analytical and knowledge-based redundancy: A survey and some new results. Automatica, 26(3), 459-474.

[2] Pan S. J., Tsang I. W., Kwok J. T. and et al. 2011. Domain adaptation via transfer component analysis, Neural Networks IEEE Transactions, 22(2):199-210.

[3] Pan S. J., Yang Q. 2010. A survey on transfer learning, IEEE Transactions on Knowledge and Data Engineering, 22(10): 1345-1359.

[4] Wang J., Xie J., Zhao R. and et al. 2016. A New Probabilistic Kernel Factor Analysis for Multisensory Data Fusion: Application to Tool Condition Monitoring. 65(1): 2527-2537.

[5] Xie J., Zhang L., Duan L. and et al. 2016. On cross-domain feature fusion in gearbox fault diagnosis under various operating conditions based on Transfer Component Analysis. 2016 IEEE International Conference on Prognostics and Health Management (ICPHM). 1-6.

537200 Multi-Class Classification of Process Faults Using Nonlinear Support Vector Machine Based Feature Selection Algorithm

Abstract:

The advances in information, and sensing technology has paved the way for large amount of process data collection in real time, initiating the â€œ*Big Data*â€ era in industrial decision making processes [1]. While the modern process industry aims for smarter, safer, and more efficient operation, more operating variables are integrated under closed loop control resulting in increased process structure complexity, which obfuscates process control [2]. Thus, the industrial data outbreak immensely facilitates process monitoring by use of machine learning techniques. One of the most popular machine learning techniques is Support Vector Machines (SVMs) [3] which allows the use of high dimensional feature sets for learning problems such as classification and regression. Yet reducing the dimensionality of the feature space in data-driven modeling, known as dimensionality reduction and feature selection, is still a key task in improving model accuracy [4] as well as decreasing a priori data collection, which in turn yields enhanced efficiency in chemical processes.

In this work, we perform multi-class classification using nonlinear Support Vector Machine based feature selection algorithm for fault identification and diagnosis. Previously, we have presented a novel data-driven framework for simultaneous fault detection and diagnosis for chemical processes [5-6] that uses nonlinear Support Vector Machines (SVM) based feature selection algorithm [7]. The framework produces two-class fault-specific SVM models to detect known faults individually. In order to improve the decision making, we propose to extend the current framework by performing multiclass classification of process faults and normal operation. Here, the idea is to create a decision-tree based classification scheme where the process data is initially classified as either normal or faulty, and then examined with a series of trained models, where each defining a region of data space that separates one class from all others (one-versus-all) or from a single other class (one-versus-one) for fault detection. By incorporating multi-class SVM models, we aim to minimize the number of inspections to identify the process fault, thus enhance the current fault detection and diagnosis framework. We present results for the Tennessee Eastman process as a case study and compare our approach to existing approaches for fault detection and diagnosis.

References

[1] Reis, M. S., Gins, G. (2017). Industrial Process Monitoring in the Big Data/Industry 4.0 Era: From Detection, to Diagnosis, to Prognosis. Processes, 5(3), 35.

[2] Chiang, L. H., Russell, E. L., & Braatz, R. D. (2000). Fault detection and diagnosis in industrial systems. Springer Science & Business Media.

[3] Cortes, C., Vapnik, V. (1995). Support-vector networks. Machine Learning, 20(3), 273-297.

[4] Onel, M., Beykal, B., Wang, M., Grimm, F. A., Zhou, L., Wright, F. A., Phillips, T. D., Rusyn, I., Pistikopoulos, E.

N. Optimal Chemical Grouping and Sorbent Material Design by Data Analysis, Modeling and Dimensionality Reduction Techniques. 28th European Symposium on Computer-Aided Process Engineering (ESCAPE-28); Elsevier, 2018; p Accepted manuscript.

[5] Onel, M., Kieslich, C. A., Guzman, Y. A., Floudas, C. A., & Pistikopoulos, E. N. (2018). Big Data Approach to Batch Process Monitoring: Simultaneous Fault Detection and Diagnosis Using Nonlinear Support Vector Machine-based Feature Selection. Computers & Chemical Engineering.

[6] Onel, M.; Kieslich, C. A.; Guzman, Y. A.; Floudas, C. A.; Pistikopoulos, E. N. Simultaneous Fault Detection and Diagnosis of Continuous Processes via Nonlinear Support Vector Machine-based Feature Selection. 13th International Symposium on Process Systems Engineering (PSE 2018); 2018; p Accepted manuscript.

[7] Guzman, Y. A. (2016). Theoretical advances in robust optimization, feature selection, and biomarker discovery (Doctoral dissertation, Princeton University).

527861 In Situ FT-IR Quantitative Analysis of Amine Concentration and CO_2 Loading Amount in Mixture Solvent Using Deep Neural Network

Abstract:

Title : In situ FT-IR Quantitative Analysis of Amine Concentration and CO_2 Loading Amount in Mixture Solvent using Artificial Neural Network

Yo Sung Yoon, Jay H. Lee*

Email: jayhlee@kaist.ac.kr

This work presents the quantitative analysis of amine based CO_2 capture process for reclaimer and make up. The process has been widely investigated with the first generation amine solvent, but in the long term operation, many challenging problems regarding stability originated from amine degradation and corrosion have been raised. These problems may cause severe solvent loss, foaming, and fouling. Therefore, it is required to develop online monitoring method to cope with those unfavorable phenomena. As a breakthrough, this study suggests the in-situ FT-IR quantitative analysis model of the amine concentration and CO_2 loading amount in MEA, MDEA, AMP mixture, which are the well-known commercialized absorbents in the CO_2 capture process, by employing reduced order artificial neural network.

The input data of neural network model is FT-IR data of calibration sets and the concentrations of MEA, MDEA, and AMP as well as CO_2 loading amount are set as supervision of corresponding data. A total of 60 calibration set is composed of all FT-IR data, recorded in the range of 790cm^{-1} to 4000cm^{-1} and divided by 6660 intervals, resulting in the input matrix of [60X6660]. In the preprocessing of data, principal component analysis(PCA) is used for model order reduction due to highly redundant FT-IR raw data. This study involves 7 principal components in which their cumulative variance is above 99% of the total variance. Then, the size of input matrix becomes [60X7] with minor loss of data information. The artificial neural network with two hidden layers is applied in regression with these reduced order input data. As a result, the coefficients of determination of the model over 0.999 are achieved and validation results with test sets fall within approximately 5% relative error. The leave-one-out cross validation is also applied.

537182 Deep Learning for Pyrolysis Reactor Monitoring: From Thermal Imaging Towards Smart Monitoring System

Abstract:

The pyrolysis reactor is a crucial component of chemical industry, which is used to crack heavier hydrocarbons to lower molecular weight hydrocarbons in a fired furnace. Condition monitoring of pyrolysis reactors is of paramount importance for manufacturing effectively and safely. While in practice, such monitoring is always challenging, particularly the operation condition inside the reactor, due to the high temperature (over 800) in the cracking regions where normal sensors can hardly be implemented. Infrared thermography is one of the practical solutions; it provides temperature information inside the fired reactor, where detailed operation conditions such as overheating, tube coking and tube deformation can be observed clearly. To take fully advantage of such information, an automatic monitoring system is built using deep learning techniques, which helps plant operators make correct decisions about the reactor operating.

Recent years, data-driven approaches using machine learning algorithms and statistics have been widely applied in feature learning. As the most advanced branch in machine learning, deep learning (DL) methods bring huge breakthroughs in many areas particularly for image analysis. In our work, based on the infrared thermography, we opted a deep convolutional neural network (DCNN) to build a monitoring system for the surface temperature and shape changing of pyrolysis tubes inside the fired furnace. To recognize the tube regions from the bulk background, a 50-layer ResNet[1] was used to segment tube areas from the raw images. In order to transfer the image classification model into a segmentation model, the strategy of fully convolutional network (FCN)[2] was applied in the ResNet model that tube areas are predicted by the upsampling of multiple-layer feature maps in the ResNet. After the segmentation of detailed tube regions, detailed surface temperature and area of the tube can be identified from the raw photos. Therefore, a monitoring system can be constructed where process operators can easily observe the overheated areas and shape changes through such system. Besides the monitoring efforts by operators, an automatic monitoring system can be further composed by the introduction of multivariate statistics. In this work, a PCA-based monitoring model[3] is introduced to detect faulty conditions from the information provided by six infrared cameras. The monitoring system successfully identified the overheated spots on the tube, as well as abnormal vibration of one infrared camera due to installation problem. This work provides contribution of combining state-of-the-art DL approaches into industrial monitoring systems.

[1] K. He, X. Zhang, S. Ren, and J. Sun, "Deep Residual Learning for Image Recognition," in *2016 IEEE Conference on Computer Vision and Pattern Recognition (CVPR)*, 2016, pp. 770–778.

[2] J. Long, E. Shelhamer, and T. Darrell, "Fully convolutional networks for semantic segmentation," *Proc. IEEE Comput. Soc. Conf. Comput. Vis. Pattern Recognit.*, vol. 07-12-June-2015, pp. 3431–3440, 2015.

[3] L. H. Chiang, E. Russell, and R. D. Braatz, "Fault detection and diagnosis in industrial systems," *Time*, vol. 12, no. 10–11, pp. 285–294, 2001.

531004 Model-Based Stochastic Fault Detection and Diagnosis of Lithium-Ion Batteries

Abstract:

Lithium-ion battery (Li-ion) is becoming the dominant energy storage solution in hybrid electric and electric vehicles, due to its higher energy density and longer life cycle [1]. For these applications, the battery must perform reliably and pose no safety threats. However, the performance of the Li-ion battery can be affected by abnormal thermal behaviors, which are defined as faults. For example, as previously reported, thermal runaway can cause fire, and temperature imbalances in battery cells can affect the aging behavior of battery [2, 3]. It is essential to develop reliable thermal management system to accurately predict and monitor thermal behaviors of Li-ion battery [4]. Using the first-principle models of Li-ion battery, this work presents a stochastic fault detection and diagnosis (FDD) algorithm to identify two particular faults in the Li-ion battery cell, i.e., thermal runaway fault and convection fault.

Models of Li-ion battery are typically derived from the underlying physical phenomena. However, to make model useful and tractable, it is common to make assumptions and simplification during the model development, which consequently may introduce mismatch between the model and the battery under study. An iterative real-time optimization is developed in this work, which can correct the model by incorporating the error between measurements and model prediction of temperatures.

The performance of the FDD algorithm can be affected by uncertainty, which may originate from intrinsic time varying phenomena or may results from model calibration with noisy data [5]. To evaluate the effect of uncertainty on FDD, a generalized polynomial chaos (gPC) expansion is used to firstly approximate uncertainty in thermal dynamics and then propagate such an uncertainty onto model predictions that can be used for FDD. As compared to sampling-based uncertainty analysis techniques such as the Monte Carlo (MC) simulations, the gPC-based approximation of uncertainty in thermal dynamics can significantly reduce the computational complexity and improve the accuracy of FDD [6].

Additionally, most of the available FDD algorithms in the literature use fault signatures identified from a pattern recognition method or residual calculated from an observer to detect the occurrence of faults [7, 8]. These techniques provide little information about the probability that a fault has occurred. Since faults occurring in the battery may be of stochastic nature and since the uncertainty in thermal dynamic behaviors is considered in this work, the use of fault signature or residual for FDD may not be effective. A joint confidence region (JCRs)-based optimization is developed for stochastic fault detection, which can provide a probabilistic description of the occurrence of faults.

The model correction and stochastic fault detection algorithms are applied to a two-state Li-ion battery system with stochastic thermal faults that can affect the core and surface temperatures of battery. The efficiency of the proposed FDD algorithm is quantified in terms of fault detection rate for both individual and simultaneous faults in Li-ion battery.

References

[1] J. M. Tarascon and M. Armand, "Issues and challenges facing rechargeable lithium battries," *Nature,* vol. 414, no. 6861, pp. 359-367, 2001.

[2] S. Dey, Z. A. Biron, S. Tatipamula, N. Das, S. Mohon, B. Ayalew and P. Pisu, "Model-based real-time thermal fault diagnosis of Lithium-ion batteries," *Control Engineering Practice,* vol. 56, pp. 37-48, 2016.

[3] Q. Wang, P. Ping, X. Zhao, G. Chu, J. Sun and C. Chen, "Thermal runaway caused fire and explosion of lithium ion battery," Journal of Power Sources, vol. 208, pp. 210-224, 2012.

[4] M. Charkhgard and M. Farrokhi, "State-of-Charge Estimation for Lithium-Ion Batteries Using Neural Networks and EKF," IEEE Transactions on Industrial Electronics, vol. 57, no. 12, pp. 4178-4187, 2010.

[5] Y. Du, H. Budman and T. Duever, "Fault detection and diagnosis with parametric uncertainty using generalized polynomial chaos," Computers and Chemical Engineering, vol. 76, pp. 63-75, 2015.

[6] D. Xiu, Numerical Methods for Stochastic Computation: A Spectral Method Approach, Princeton, New Jersey, USA: Princeton University Press, 2010.

[7] C. Chien, C. Hsu and P. Chen, "Semiconductor fault detection and classification for yield enhancement and manufacturing intelligence," Flex. Serv. and Manuf. J., vol. 25, no. 3, pp. 367-388, 2013.

[8] M. C. Keir and A. G. Alleyne, "Dynamic model-based fault detection and diagnosis residual considerations for vapor compression systems," in American Control Conference, 2006, Minneapolis, 2006.

536052 Implementation of Model-Predictive Safety Systems to Detect Predictively Operation Hazards in Non-Minimum-Phase Processes

Abstract:

In 2016 [1] we introduced a method of model-predictive safety (MPS) system design, and in 2017 [2] min-max optimization methods that can be used offline to calculate systematically (a) the most aggressive control action that minimizes each process-constraint index when uncertain model parameters take their nominal values, and (b) the most aggressive control action that minimizes each process-constraint index when uncertain model parameters take their worst-case values.

An MPS system [1] generates alarm signals that are predictive and systematically account for process nonlinearities and interactions, while typical existing functional safety systems generate reactive, non-interacting alarm signal(s) when a process variable exceeds a threshold, For the first time, the MPS system design method proposed a systematic utilization of the dynamic process models to generate predictive alarm signals (alerts) for the detection of present and future operation hazards (OHs) in real time. An MPS system uses a process model to project in real-time the process operability status and to generate alarm signal(s) indicating the presence of a present or future OH with reasonable accuracy; it generates predictive alarm signals that alert the process personnel to imminent and potential future OHs before the actual OHs occur. No control system is (will be) able to assure safe process operation in the presence of a current (future) OH in a process.

In this paper, the min-max optimization methods presented in Ref. [2] are used for a non-minimum-phase (NMP) chemical reactor to calculate offline the two most aggressive control actions that minimize each operability-constraint index. Because this particular NMP reactor shows inverse temperature response, a successful implementation of the MPS system on this reactor requires the use of a large prediction horizon for the MPS system. The performance of the MPS system implemented with difference prediction horizons will be shown using numerical simulations.

References

[1] Mohseni Ahooyi, T., J.E. Arbogast, W.D. Seider, U.G. Oktem, and M. Soroush, "Model-Predictive Safety System for Proactive Detection of Operation Hazards," AIChE J., 62, 2024-2042 (2016).

[2] Soroush, M., J.E. Arbogast, and W.D. Seider, â€œModel-Predictive Safety System for Predictive Detection of Operation Hazards: Off-Line Calculation of Most Aggressive Control Actions and Worst-Case Uncertainties,â€ CAST Division 10 Plenary Session at the 2017 AIChE Annual Meeting, Minneapolis, MN, November (2017)

535954 Towards Global Optimization on Low-Dimensional Surrogates Via Manifold Learning

Abstract:

Global optimization typically requires explicit forms of the objective function and constraints to provide valid lower and upper bounds on the solution. Thus, it is a challenge to perform global optimization for problems where the objective function or constraints are not known in explicit form, but are available only through sampling of a simulation, a black-box or a grey-box [1-3]. In these cases, global optimization can only be performed on surrogate models constructed on the data, and not on the original problem. These problems become increasingly difficult to solve as dimensionality increases; retrieving enough samples to develop meaningful surrogate models can become difficult, and efficient means for solving such problems will be required for many applications in the future.

In this work, we seek to conduct optimization (global as well as local) of high dimensional black-box problems for which manifold learning may be applied to effectively reduce the problem dimensionality. Our core assumption is the existence of a low-dimensional manifold in search (parameter) space that is approached quickly by an "inner" optimization algorithm (e.g. a steepest descent method) applied to the objective function. On the manifold, local optimization is then assumed to converge much slower. The approach we discuss does not require information about the manifold beyond its existence. It has already been shown that identifying the underlying manifolds of the objective function, and taking steps on them, can quickly speed up traditional local optimization algorithms [4]. From a global optimization perspective, identifying low-dimensional underlying manifolds for non-convex objective functions in high-dimensional search space has the potential to significantly decrease the required sampling and surrogate modeling. That is, rather than spending the majority of the time working in a high-dimensional space, the problem can be addressed based on its representative low-dimensional manifold. Building a surrogate model at this coarser level may allow the use of global optimization solvers, such as ANTIGONE [5], as soon as an explicit form of the interpolant of the objective function on the manifold is constructed.

The approach presented begins with the identification of the lower dimensional manifold through brief spurts of local optimization algorithms starting at sample points in the high dimensional space. As we assume that the manifold is strongly attracting for, say, steepest descent methods, we only need to sample the space of fibers over the manifold, and not the full, high-dimensional space. After fast convergence towards the manifold, Diffusion Maps are then applied to the points close to the manifold, identifying a ("nonlinear") parametrization [6]. Then, a surrogate model may be built on this low-dimensional manifold, providing the ability to utilize standard global optimization algorithms, and the optimal solutions may then be lifted back from the diffusion map coordinates to the original high-dimensional forms. We apply this algorithm to multiple examples to explore its efficiency and effectiveness for this class of problems.

1. Boukouvala, F., & Floudas, C. A. (2017). ARGONAUT: AlgoRithms for Global Optimization of coNstrAined grey-box compUTational problems. *Optimization Letters*, *11*(5), 895-913.
2. Rios, L. M., & Sahinidis, N. V. (2013). Derivative-free optimization: a review of algorithms and comparison of software implementations. *Journal of Global Optimization*, *56*(3), 1247-1293.
3. Boukouvala, F., Misener, R., & Floudas, C. A. (2016). Global optimization advances in mixed-integer nonlinear programming, MINLP, and constrained derivative-free optimization, CDFO. *European Journal of Operational Research*, *252*(3), 701-727.
4. Pozharskiy, D. et al. Manifold Learning for Coarse-Grained Optimization, in preparation.
5. Misener, R., & Floudas, C. A. (2014). ANTIGONE: algorithms for continuous/integer global optimization of nonlinear equations. *Journal of Global Optimization*, *59*(2-3), 503-526.
6. Coifman, R. R., & Lafon, S. (2006). Diffusion maps. *Applied and computational harmonic analysis*, *21*(1), 5-30.

534057 On the Derivation of Piecewise Linear Continuous Approximating Functions

Abstract:

Continuous piecewise linear functions have been widely used in fitting data points or approximating nonlinear functions. While many papers have focused on how to model continuous piecewise linear functions and how to incorporate them into mixed-integer programing (MIP) models, there has been limited research on how to obtain the piecewise linear functions. For example, how many segments are needed, given an error tolerance, to approximate data; or given a fixed number of segments, where to place the break points so that the fitting error is minimized. In this work, we consider the problem of fitting one-dimensional (i.e., one independent variable) data set with continuous piecewise linear functions with variable number/location of break points. While some papers have considered optimizing the number and/or location of break points [1-4], none has addressed the same problem as in this work. For example, the approach by Toriello and Vielma [1] deals only with convex piecewise linear functions; the MINLP models in Rebennack and Kallrath [2, 3] are for approximating functions, instead of data points; and the approach by Yang et al. [4] leads to discontinuous piecewise linear approximation.

In this paper, we present two mixed-integer models. The first one (M1) minimizes the fitting error given a fixed number of break points, while the second one (M2) finds the minimum number of break points to satisfy an error tolerance for every data point. In both models, binary variables and mixed-integer constraints are introduced to assign data points to the correct interval/segment of the piecewise linear functions. Most importantly, unlike known approaches that enforce continuity via nonlinear constraints, we first derive a proposition that allow us to enforce continuity through an equivalent condition, which involves comparing the approximations of two special data points with piecewise linear functions from two adjacent intervals. Then, we present linear constraints to enforce that the obtained approximation satisfies this condition. Thus, we propose a mixed-integer linear fitting model, rather than mixed integer nonlinear model, which is significantly faster than all previously proposed approaches.

Further, we show how the proposed models can be extended to approximate continuous functions. In particular, we show how to find the minimum number of linear segments required such that error between the piecewise linear approximation and the original function is within a tolerance. This involves solving M2 iteratively, with the data set coming from evaluating the original function at discrete locations.

The models are tested with a series of real-world examples using data that come from experiments. The models are also applied to find piecewise linear approximations of some benchmark univariate functions with different error tolerances. Solution quality and computational performance are compared against known approaches.

Reference

[1] A. Toriello and J. P. Vielma, "Fitting piecewise linear continuous functions," *European Journal of Operational Research,* vol. 219, pp. 86-95, May 16 2012.

[2] S. Rebennack and J. Kallrath, "Continuous Piecewise Linear Delta-Approximations for Univariate Functions: Computing Minimal Breakpoint Systems," *Journal of Optimization Theory and Applications,* vol. 167, pp. 617-643, Nov 2015.

[3] S. Rebennack and J. Kallrath, "Continuous Piecewise Linear Delta-Approximations for Bivariate and Multivariate Functions," *Journal of Optimization Theory and Applications,* vol. 167, pp. 102-117, Oct 2015.

[4] L. J. Yang, S. S. Liu, S. Tsoka, and L. G. Papageorgiou, "Mathematical programming for piecewise linear regression analysis," *Expert Systems with Applications,* vol. 44, pp. 156-167, Feb 2016.

532437 Surrogate-Based Optimization Framework in Process Systems Engineering

Abstract:

Simulations are widely used in engineering and scientific disciplines to describe realistic and complex systems and interactions. Computational advances over the last decades have enabled the use of simulations for process analysis and optimization leading to best economic, environmental friendly and safe solutions. However, there are multiple challenges in the objective of optimizing computationally expensive computer simulations.

In the pursuit of more realistic descriptions of reality, simulations may become computationally expensive leading to interesting decision-making problems where information is limited. Surrogate modeling allows us to build lower fidelity models that approximate the original simulation utilizing limited data. Surrogate modeling has been an active area of research over past two decades and number of techniques exist in the literature [1]. Furthermore, surrogate models can be more efficiently used if one makes use of sequential sampling to gather information.

This work discusses a framework for surrogate-based optimization. First, surrogate models are built from an initial data set obtained using design of experiments techniques. New samples are then generated using adaptive sampling schemes that balance between finding a better optimal solution and reducing the uncertainty in the surrogate model. Surrogate models are iteratively updated using these samples until desired accuracy is achieved or a stopping criteria is met. Specifically, this work focuses on Kriging and radial basis function surrogate models. Adaptive sampling strategies used in this work make use of an infill criteria based on expected improvement function. Finally, this framework is applied to solve problems in three different application domains: integrated scheduling and control, pharmaceutical feasibility analysis, and supply chain optimization. For integrated scheduling and control, adaptive sampling is used to minimize uncertainty in surrogate models that represent the control-level simulation [2]. For feasibility analysis problems, a modified version of expected improvement function is used to understand feasible region boundary [3]. In supply chain simulation optimization, the problem of finding optimal warehouse capacities that minimize the total cost of the supply chain where the supply chain network consists of more than one enterprise is considered[4]. With the help of Kriging surrogate model, adaptive sampling is used to guide the global search in the proposed optimization framework.

References

[1] A. Bhosekar and M. Ierapetritou, "Advances in surrogate based modeling, feasibility analysis, and optimization: A review," *Comput. Chem. Eng.*, vol. 108, pp. 250–267, 2018.

[2] L. S. Dias and M. G. Ierapetritou, "Integration of scheduling and control under uncertainties: Review and challenges," *Chem. Eng. Res. Des.*, vol. 116, pp. 98–113, 2016.

[3] Z. Wang, M. S. Escotet-Espinoza, R. Singh, and M. Ierapetritou, "Surrogate-based Optimization for Pharmaceutical Manufacturing Processes," in *27th European Symposium on Computer Aided Process Engineering*, vol. 40, A. Espuña, M. Graells, and L. Puigjaner, Eds. Elsevier, 2017, pp. 2797–2802.

[4] N. Sahay and M. Ierapetritou, "Multienterprise supply chain: Simulation and optimization," *AIChE J.*, vol. 62, no. 9, pp. 3392–3403, 2016.

537826 Spatial Branch-and-Bound Optimization Using Surrogate Approximations

Abstract:

Data-driven modeling and optimization using surrogate approximations has drawn growing attention for the optimization of complex, multiscale chemical process systems [1]. Specifically, the involvement of multidisciplinary design objectives, computationally expensive simulations and costly physical experiments often hinder the attempts to find an accurate analytical model for these systems, which can be used for traditional derivative-based optimization. In these cases, surrogate modeling or metamodeling is often used to create analytical and cheap-to-evaluate models for design space exploration, model approximation, and global optimization [2-4]. Despite the widely available surrogate techniques and surrogate-based optimization algorithms, many methodologies do not guarantee optimality and are prone to the inconsistency caused by the selection of different surrogate models. Specifically, the challenge of data-driven DFO is to develop approaches that are capable to provide consistent convergence to good solutions, despite the lack of abundant data, and the uncertainty of not knowing the best approximation model for the underlying phenomena.

Common approahces in data-driven optimization iteratively update and optimize the surrogate models with adaptive sampling strategies aiming to speed up the search for better solutions. In this work, we propose new ideas for data-driven optimization, which aim to utilize the uncertainty and variability that unavoidably exists, to search the space efficiently. Specifically, several approaches are proposed that take advantage of statistical information from commonly used sampling strategies (i.e., space filling designs and sparse grids) and best performing surrogate models (i.e., support vector regression, kriging, and orthogonal polynomials), to derive over and under estimators that are used within a custom-based spatial branch-and-bound framework. Three specific ideas are explored and compared in this work. First, a purely machine-learning based approach, utilizes soft margin loss of support vector regression to approximate the over and under estimators and gradually tighten the search space until convergence. The second approach is based on the concept of quantifying the uncertainty of the fitted surrogate models to derive the upper/lower bounds using robust optimization. Finally, the third approach uses sparse grids, orthogonal polynomials and approximation error bounds to bound the surrogate predictions. Using the above three ideas, we develop a branch-and-bound algorithm, and heuristics that are based on data-driven analysis. Finally, we compare this spatial branch-and-bound framework with existing data-driven methods and provide results on the computational efficiency, sampling reuqirments and convergence on a general class of benchmark problems.

References:

- Chiang, L., B. Lu, and I. Castillo, *Big Data Analytics in Chemical Engineering.* Annual Review of Chemical and Biomolecular Engineering, 2017. **8**(1): p. 63-85.
- Wang, G.G. and S. Shan, *Review of Metamodeling Techniques in Support of Engineering Design Optimization.* Journal of Mechanical Design, 2006. **129**(4): p. 370-380.
- Bhosekar, A. and M. Ierapetritou, *Advances in surrogate based modeling, feasibility analysis, and optimization: A review.* Computers & Chemical Engineering, 2018. **108**: p. 250-267.
- Forrester, A.I.J. and A.J. Keane, *Recent advances in surrogate-based optimization.* Progress in Aerospace Sciences, 2009. **45**(1): p. 50-79.
- Bertsimas, D., V. Gupta, and N. Kallus, *Data-driven robust optimization.* Mathematical Programming, 2018. **167**(2): p. 235-292.
- Bungartz, H.-J. and M. Griebel, *Sparse grids.* Acta Numerica, 2004. **13**: p. 147-269.

533682 Reduced Space Formulation for Global Optimization with Artificial Neural Networks Embedded

Abstract:

Artificial neural networks (ANNs) are known to be universal approximators [1]. They are utilized in numerous applications for data-driven black-box modeling [2]. In addition, they are likely to gain further importance as digitalization of industry further progresses. In many industrial applications, it is desired to incorporate one or several ANNs into an optimization framework for determining optimal operation strategies or optimal process design [2]. This means that it is desired to embed ANNs in optimization problems. In the previous literature (e.g., [3]), optimization problems with ANNs embedded have mostly been solved by local or stochastic global approaches, e.g., genetic algorithms. However, these methods can fail to identify global optima and give no guarantee of optimality.

Global deterministic optimization of problems with ANNs embedded was done by Smith et al (2013) who used BARON to optimize a flooded bed algae bioreactor that was modeled by an ANN with one hidden layer and three neurons [4]. However, ANNs often comprise numerous neurons in several hidden layers. In these cases, the consideration of ANNs as equality constraints leads to large-scale nonlinear optimization problems. Furthermore, the activation function that is used in each neuron is often a non-linear function, e.g., the hyperbolic tangent function. Currently, the hyperbolic tangent function is not directly available in many global solvers and when represented as summation of terms, its relaxations are relatively weak.

We present a method for deterministic global optimization of optimization problems with ANNs embedded [5], based on recognizing the structure and utilizing it. More specifically, the presented method propagates convex and concave relaxations without any extra variables through ANNs by means of McCormick-based relaxations of algorithms [6,7]. This approach reduces the dimensionality of the optimization problem drastically. In addition, the convex and concave envelopes of nonlinear activation functions are implemented in the presented framework tightening the overall ANN relaxations. The resulting optimization problem is solved using our in-house global deterministic solver MAiNGO [8,9].

The computational performance of the proposed method is illustrated and compared to the solver BARON on four case studies: a test function, a fermentation process, a compressor plant, and a chemical process optimization. Moreover, the scaling of the solution approach with the ANN size is discussed. The results show that the dimensionality of the reduced-space optimization problem is drastically reduced compared to the full-space formulation. Further, the proposed method performs favorable on the case studies compared to BARON, i.e., it shows a speedup.

[1] Hornik, K., Stinchcombe, M., White, H.: Multilayer feedforward networks are universal approximators. Neural Networks 2(5), 359-366 (1989).

[2] Meireles, M., Almeida, P., Simoes, M.G.: A comprehensive review for industrial applicability of artificial neural networks. IEEE Transactions on Industrial Electronics 50(3), 585-601 (2003).

[3] Nascimento, C.A.O., Giudici, R., Guardani, R.: Neural network based approach for optimization of industrial chemical processes. Computers & Chemical Engineering 24(9-10), 2303-2314 (2000).

[4] Smith, J.D., Neto, A.A., Cremaschi, S., Crunkleton, D.W.: CFD-based optimization of a flooded bed algae bioreactor. Industrial & Engineering Chemistry Research 52(22), 7181-7188 (2013).

[5] Schweidtmann, A.M., Mitsos, A.: Global Deterministic Optimization with ANNs Embedded. Preprint available at arXiv:1801.07114 (2018).

[6] Mitsos, A., Chachuat, B., Barton, P.I.: McCormick-based relaxations of algorithms. SIAM Journal on Optimization 20(2), 573-601 (2009).

[7] McCormick, G.P.: Computability of global solutions to factorable nonconvex programs: Part I - convex underestimating problems. Mathematical Programming 10(1), 147-175 (1976).

[8] Bongartz, D., Najman, J., Scholl, S., Mitsos, A.: MAiNGO: McCormick based Algorithm for mixed integer Nonlinear Global Optimization. Technical report (2018).

[9] Bongartz, D., Mitsos, A.: Deterministic global optimization of process flowsheets in a reduced space using McCormick relaxations. Journal of Global Optimization 20(9), 419 (2017).

537745 Optimization of Data-Dependent Mixed-Integer Nonlinear Problems

Abstract:

Many engineering problems involve complex computer simulations to describe phenomena that cannot be explicitly captured by physics-based algebraic equations. In order to optimize such systems, one needs to rely on black/grey-box, or derivative-free optimization methods [1, 2]. Due to the increased interest in optimization with data, several exciting developments can be found in the literature, ranging from methods which are purely sample-driven, to methods that employ a combination of sampling, fitting of surrogate models and derivative-based optimization [2]. A lot of the developments in the area of derivative-free optimization focus on nonlinear constrained optimization problems, in which all of the variables are continuous, while the objective function and at least a subset of the constraints rely on a black-box simulation. However, many chemical process systems problems, such as flowsheet superstructure optimization (synthesis), contain discrete or binary variables [3]. While many algorithms exist to solve equation-based MINLPs, solving a black-box-MINLP (bb-MINLP) is very challenging since integrality constraints cannot be relaxed due to lack of algebraic equations and the inability to collect simulation samples at non-integral values.

Existing bb-MINLP solvers can be divided into two broad categories. The first category is based on local trust-region search, in which the pattern or mesh-guided search is used for optimization of both continuous and discrete variables [4, 5]. The second approach employs the concept of surrogate functions for optimization. In this case, the surrogate models are constructed for the objective and constraints as a function of continuous and discrete variables by assuming continuity of all variables, while after a solution is obtained for in the continuous space, the discrete variables are fixed to the nearest discrete value [6, 7]. Despite the above recent advances, current methods have only been tested on relatively low dimensional problems. In this work, we will present several approaches for solving bb-MINLP problems by employing various sampling strategies, surrogate models, decomposition strategies, machine-learning driven heuristics, and high-performance computing. Results will be presented for a set of benchmarking problems as well as process synthesis optimization case studies.

References:

1. Amaran, S., et al., *Simulation optimization: a review of algorithms and applications.* Annals of Operations Research, 2016. **240**(1): p. 351-380.
2. Boukouvala, F., R. Misener, and C.A. Floudas, *Global optimization advances in Mixed-Integer Nonlinear Programming, MINLP, and Constrained Derivative-Free Optimization, CDFO.* European Journal of Operational Research, 2016. **252**(3): p. 701-727.
3. Trespalacios, F. and E. Grossmann Ignacio *Review of Mixedâ€Integer Nonlinear and Generalized Disjunctive Programming Methods.* Chemie Ingenieur Technik, 2014. **86**(7): p. 991-1012.
4. Abramson, M.A., et al., *Mesh adaptive direct search algorithms for mixed variable optimization.* Optimization Letters, 2008. **3**(1): p. 35.
5. Liuzzi, G., S. Lucidi, and F. Rinaldi, *Derivative-Free Methods for Mixed-Integer Constrained Optimization Problems.* Journal of Optimization Theory and Applications, 2015. **164**(3): p. 933-965.
6. Müller, J., *MISO: mixed-integer surrogate optimization framework.* Optimization and Engineering, 2016. **17**(1): p. 177-203.
7. Davis, E. and M. Ierapetritou, *A kriging based method for the solution of mixed-integer nonlinear programs containing black-box functions.* Journal of Global Optimization, 2009. **43**(2): p. 191-205.

538907 On Discretization Based Global Optimization for Mixed-Integer Bilinear Programs

Abstract:

Mixed-integer bilinear programs (MIBLP) arise from many engineering problems [1]. A typical example is the multiperiod gasoline blending problem, where the bilinear terms are used to describe the contents of liquid flows after blending and each integer variable is used to indicate whether a blender is to receive or deliver flows in a certain time period. Due to the nonconvexity of the bilinear terms and the integer variables, solving MIBLP to global optimality is difficult, especially when the problem size is large (e.g., when the number of time periods is large).

Recently, discretization based global optimization methods have received a lot of attention for solving MIBLPs (e.g., [2][3]). The basic idea is to discretize one of the variables in each bilinear term and represent thediscretized variables withbinary variables, then reformulate the 'discretized problem' into a mixed-integer linear program (MILP) via exact linearization [4]. The MILP problem obtained from discretization is a restriction of the original MIBLP problem, and a slightly modified MILP is a relaxation of the original MIBLP.When the resolution of discretization is increased, the restriction MILP and the relaxation MILP both approach the original MIBLP, and eventually converge to a global optimal solution(in a finite number of iterations).

In this presentation, we first discuss alternative ways to discretization of continuous variables in MIBLPs. All existing discretization approaches employ a radix-based framework, where the integer variable that approximates the continuous variable is expressed through a numerical system with a predefined radix or base. In the process systems engineering community, the 10-based framework is often used for discretization (and the resulting method is called multiparamatric disaggregation technique [2]); while in the operations research community, the 2-based framework is often used [5]. We start our discussion with a general R-based framework, and point out the conditions under which the continuous relaxation of the 'discretized bilinear term' is a convex hull relaxation. This result can help people choose an appropriate base for discretization. Then we propose an enhanced R-based framework, which guarantees convex hull relaxation of the 'discretized bilinear term' but with the expense of additional constraints.

Second, we discuss a new iterative global optimization procedure for efficient solution of MIBLPs. The basic idea is to refine the discretization according to the solution of the previously solved MILP relaxation. For example, if the solution of a continuous variable in the MILP relaxation is at the middle of two adjacent discretization points of the variable, then the resolution of discretization is doubled in the next iteration. However, with a fixed radix, the adaptive discretization strategy may result in unnecessarily many binary variables. To overcome this difficulty, we propose a hybrid discretization framework that allows a numerical system with multiple bases, so that in each iteration a different base may be selected according to the desired resolution of discretization. We also prove that the hybrid framework guarantees convex hull relaxation of the 'discretized bilinear term'.

We verify our theoretical analysis results via a set of MIBLP problems from the literature. The simulation results will also demonstrate the computational advantages of the proposed global optimization method over a state-of-the-art global optimization solver.

References

[1]Castro, P. M., 2016. Normalized multiparametric disaggregation: an efficient relaxation for mixed-integer bilinear problems. Journal of GlobalOptimization 64 (4), 765-784.

[2]Kolodziej, S., Castro, P. M., Grossmann, I. E., 2013a. Global optimization ofbilinear programs with a multiparametric disaggregation technique. Journalof Global Optimization 57 (4), 1039-1063.

[3]Kolodziej, S. P., Grossmann, I. E., Furman, K. C., Sawaya, N. W., 2013b. Adiscretization-based approach for the optimization of the multiperiod blendscheduling problem. Computers & Chemical Engineering 53, 122-142.

[4]Bemporad,A., Morari, M, 1999. Control of systems integrating logic, dynamics, and constraints. Automatica 35, 407-427.

[5]Gupte, A., Ahmed, S., Cheon, M. S., Dey, S., 2013. Solving mixed integerbilinear problems using MILPformulations. SIAM Journal on Optimization23 (2), 721-744.

527215 Quadratic Cut Decomposition Method for Convex Mixed-Integer Nonlinear Programs

Abstract:

Quadratic cut decomposition method for convex mixed-integer nonlinear programs

David E. Bernal*, Lijie Su**, Lixin Tang**, Ignacio E. Grossmann*

*Chemical Engineering Department, Carnegie Mellon University, Pittsburgh PA.

** Institute of Industrial and Systems Engineering, Northeastern University, Shenyang, P. R. China

In this work, we address the incorporation of 2nd order derivative information to the Outer-Approximation (OA) method to solve convex mixed-integer nonlinear programming (MINLP) problems. An MINLP problem arises when we represent an optimization problem using nonlinear algebraic equations involving continuous and discrete variables. A method to solve this problem is OA [1], [2], where the original problem is decomposed into two subproblems. The first subproblem, a master problem, considers the discrete nature of the integer-constrained variables and approximates the nonlinear functions with linear equations through a Mixed-Integer Linear Program (MILP). The second subproblem optimizes the nonlinear objective and constraints for fixed discrete variables giving rise to a Nonlinear Program (NLP). The linearization point of the first subproblem is provided by the solution of the second one, while the values for the discrete variables fixed in the second subproblem are provided by the solution of the first one. The solution of the master problem yields a lower bound (minimization) while the NLP subproblem provides an upper bound for the objective of the original MINLP. Solving these problems iteratively results in the reduction in the difference between these bounds. The difference between the bound given by the linear approximated subproblem and the optimal solution of the MINLP relies heavily on how well the linearizations approximate the nonlinear equations of the original problem.

Several previous contributions have proposed using quadratic approximations for MINLP decomposition methods. Grossmann[3] and Fletcher and Leyffer[2] discussed including 2nd order derivative information in the master problem of the OA method. Specifically, considering the Hessian of the Lagrange function in the objective of the master problem gives rise to an Mixed –Integer Quadratic Problem (MIQP) problem. Also, Ding-Mei and Sargent[4] showed that the quadratic approximation helps to define NLP subproblems that have better initial guesses of the continuous variables from the MILP.

In this work, we propose a new approach for using quadratic accurately. In particular, we propose to use a scaled quadratic approximation of the nonlinear functions instead of a linear approximation and incorporate it into the OA method. This scaled quadratic approximation uses 2nd order derivative information and therefore approximates the nonlinear functions more accurately than the linearization, yielding stronger lower bounds while ensuring the validity of the relaxation. The scaled quadratic cut is proved to be a

stricter and tighter underestimation for convex nonlinear functions than classical supporting hyperplanes, which results in the improvement of the OA solution method. The resulting master problem becomes a Mixed-Integer Quadratically Constrained Program (MIQCP).

Having a quadratic approximation of the nonlinear constraints allows the master problem to provide a closer approximation of the continuous feasible region, and therefore, improve the lower bound provided by the relaxation[3]. Recently, MILP solvers like CPLEX[5] and Gurobi[6] have expanded their capabilities to efficiently solve MIQCP problems[7]. The crucial factor of MIQCP solvability is the Hessian matrix, the second-order partial derivatives matrix. Convex MIQCP problems, which have a positive semi-definite Hessian matrix, can be solved to optimality as one special convex MINLP with state-of-the-art MILP solvers. These developments allow us to consider quadratic objectives and constraints in the master problem. The Hessian of the nonlinear functions is available thanks to the NLP solvers, and therefore can be included in the problem efficiently.

The implementation of this method in the Mixed-Integer Nonlinear Decomposition Toolbox in Pyomo MindtPy is presented together with numerical results of benchmark MINLP problems demonstrate the effectiveness of the proposed MINLP solution methods with scaled quadratic cuts compared to the traditional OA method.

[1] M. A. Duran and I. E. Grossmann, "An outer-approximation algorithm for a class of mixed-integer nonlinear programs," *Math. Program.*, vol. 36, no. 3, pp. 307–339, 1986.

[2] R. Fletcher and S. Leyffer, "Solving mixed integer nonlinear programs by outer approximation," *Math. Program.*, vol. 66, no. 1–3, pp. 327–349, 1994.

[3] I. E. Grossmann, "Review of nonlinear mixed-integer and disjunctive programming techniques," *Optim. Eng.*, vol. 3, no. 3, pp. 227–252, 2002.

[4] Ding-Mei and R. W. H. Sargent, "A combined SQP and branch and bound algorithm for MINLP optimization," London, 1992.

[5] IBM Corp. and IBM, "V12.6: User's Manual for CPLEX," *Int. Bus. Mach. Corp.*, vol. 12, no. 1, p. 481, 2009.

[6] I. Gurobi Optimization, "Gurobi Optimizer Reference Manual," *Www.Gurobi.Com*, p. 572, 2016.

[7] C. Bliek, P. Bonami, and A. Lodi, "Solving Mixed-Integer Quadratic Programming problems with IBM-CPLEX : a progress report," *Proc. Twenty-Sixth RAMP Symp.*, no. M, pp. 171–180, 2014.

533015 Machine Learning for Automated Meal Detection in Glucose Control Systems for People with Diabetes

Abstract:

Feedforward disturbance announcements enable attenuation of the effects of disturbances on the controlled variable in closed-loop control. Meals are major disturbances to the glucose homeostasis and automated detection of meal consumption and carbohydrate estimation of the consumed meal are critical for fully automated artificial pancreas control systems [1]. In this study, a machine learning approach that integrates trend detection based on qualitative analysis, and meal estimation based on a fuzzy logic system is used to develop a meal detection and estimation module.

A qualitative analysis technique translates the variations in the time-series glucose concentration measurements to a set of seven qualitative variables (QV) [2]. The QVs are associated with different patterns of glucose change inferred by signs of the first and second derivatives. The meal is detected when the pattern of glucose variation among the seven QVs best matches the QV associated with an accelerating increase (i.e., both derivatives are positive). A fuzzy logic approach is employed to quantify the similarity between the glucose variations and the seven QVs. Positive, zero and negative fuzzy sets are defined for both the first and second glucose derivatives where the degrees of membership for the glucose derivatives belonging to these fuzzy sets are directly used for computing the similarities with the QVs [3].

Following meal detection, a fuzzy logic-based system quantifies the meal size, or more specifically the absorbed carbohydrates in the bloodstream. The development of the algorithm using fuzzy logic is motivated by the approximate reasoning ability and low computation requirements of fuzzy systems since the module will be used in computational environments such as smart phones that have limited computational power. The two inputs of fuzzy estimator are the ratio of measured glucose rise and the estimated effective insulin present in the bloodstream. The estimation algorithm determines a distributed approximation of the carbohydrates (meal size), followed by the calculation of the amount of insulin infusion (manipulated variable) to compensate for the unannounced meal. The learning component of the meal detection and carbohydrate estimation module captures the intra-and inter patient variability. Two parameters of the algorithm are adaptively modified depending on the risk function defined based on glucose deviation from the control target range.

The algorithm is employed for both in silico subjects and clinical experiments. For in silico study, the performance of glucose control using the proposed automated meal detection module is compared with a meal announced approach [3]. Using clinical experiments, we evaluate the accuracy of the meal detection and meal size estimation module over 117 consumed meals/snacks. Sensitivity, the percentage of correctly detected meals, is 93.5% for meals and 68.0% for snacks. Detection time defined as the time elapsed from the start of the meal to the detection is 34.8 22.8 min which is a reasonable due to physiological delay and gradual dynamic of CHO absorption into bloodstream. The estimation errors, the actual minus estimated CHO, approximately follow a Gaussian distribution centered around zero (mean: -1.7, standard deviation 28.1 g).

[1] A. Cinar, "Multivariable adaptive artificial pancreas system in type 1 diabetes". Curr Diab Rep 17:88, 2017.

[2] B. R. Bakshi and G. Stephanopoulos, "Representation of process trends-III. multiscale extraction of trends from process data," Comput. Chem. Eng., vol. 18, no. 4, pp. 267-302, 1994.

[3] S. Samadi, K. Turksoy, I. Hajizadeh, et al.: "Meal detection and carbohydrate estimation using continuous glucose sensor data". IEEE J Biomed Heal Informatics vol. 21, no. 3, pp. 619-627, 2017.

539719 Source Analysis of Process Variability in Multi-Step Bio-Process Manufacturing

Abstract:

Commercial bio-processes are comprised of multiple-step batch operations. However, its productivity is often affected by abnormal variations that lead to undesirable results. Though multivariate statistical methods have been in use in recent years to monitor and detect faults for bio-processes, it is still difficulty to identify the causes of multiple-step process variability.

In this paper, we analyze cell culture manufacturing variability in multiple-step processes using machine learning techniques, with the objective to identify the causes of high lactate production in recombinant cells. The proposed method is composed of three parts: 1) a unsupervised learning step that finds the most separable subspace for high / low lactate batches by principal component analysis (PCA); 2) a clustering step that classifies different causes in the space found in Step 1; and 3) a fault diagnosis step that applies robust linear discriminant analysis (LDA) to normal data and each class of fault data and characterizes faults based on robust LDA contribution analysis. The proposed method is demonstrated on a multi-stage batch process from Genentech, showing its ability to categorize and diagnose the sources of multi-step process viability.

529584 Multi-Rate Hard and Soft Sensors Fusion for Monitoring Chemical Processes

Abstract:

Product quality variables of chemical processes, such as product compositions and purities, are usually measured off-line using analytical techniques in a laboratory. The sampling frequencies are often too low for process monitoring and control purposes. To obtain more frequent estimations of product quality variables, software sensors (also known as soft sensors or inferential sensors), including partial least squares (PLS) models [1, 2] and other data-driven mathematical models [3, 4], have been widely applied over past three decades [5]. These data-driven soft sensors predict the infrequent measured product quality variables using frequently sampled inputs, such as temperature, pressure, and flow rate. To address the time-varying characteristics of chemical processes, caused by factors like fouling, catalyst aging and operating condition changes, several online adaptive soft sensor algorithms have been proposed, including the Kalman filter based method [6] and the means and variance update approach [7].

The accuracy of individual sensor can be easily impaired by varieties of factors, such as instrument malfunctioning, operator mistakes and the inherent measurement error. To improve the monitoring accuracy and reliability for chemical processes, the sensor fusion methodology has been proposed to combine the estimations from lab analyzer and soft sensor using Kalman filter [8, 9]. The lab measurements serve as references and are used to correct the soft sensor estimation when they are available. With the improving capability of sampling instruments, the frequent online measurements of product quality variables by hardware sensors (or online analyzers) become increasingly available. Yet the sensor fusion approaches using online analyzer as well as lab analyzer and soft sensor have not been well discussed.

Here we present a multi-rate sensor fusion scheme based on maximum-likelihood approach [10]. It was examined in a chemical process at Dow. The product impurity was estimated by fusing the measurements from a lab analyzer, an online analyzer and a PLS soft sensor with different sampling rates. The soft sensor was developed using mean and variance update algorithm to track the time-varying process dynamics. The online analyzer measurements were filtered to reduce their variability. These two sensor measurements were fused together with the lab measurements using maximum-likelihood approach. It has shown that the sensor fusion approach improves the process monitoring reliability, quantified by the rates of correctly identified impurity alarm, comparing to the case of using an individual sensor.

References

1. Wold, S., M. Sjöström, and L. Eriksson, PLS-regression: a basic tool of chemometrics. Chemometrics and intelligent laboratory systems, 2001. 58(2): p. 109-130.
2. Chiang, L.H., et al., Diagnosis of multiple and unknown faults using the causal map and multivariate statistics. Journal of Process Control, 2015. 28: p. 27-39.
3. Bishop, C.M., Neural networks for pattern recognition. 1995: Oxford university press.
4. Jang, J.-S.R., C.-T. Sun, and E. Mizutani, Neuro-fuzzy and soft computing: a computational approach to learning and machine intelligence. 1997.
5. Kadlec, P., B. Gabrys, and S. Strandt, Data-driven Soft Sensors in the process industry. Computers & Chemical Engineering, 2009. 33(4): p. 795-814.
6. Chen, K., et al., Soft Sensor Model Maintenance: A Case Study in Industrial Processesâˆ—âˆ—The authors would like to acknowledge the support from the DOW chemical company and the natural sciences and engineering research council of Canada (NSERC). IFAC-PapersOnLine, 2015. 48(8): p. 427-432.
7. Lu, B. and L. Chiang, Semi-supervised online soft sensor maintenance experiences in the chemical industry. Journal of Process Control, 2017.
8. Xie, L., et al. Kalman filtering approach to multirate information fusion for soft sensor development. in Information Fusion (FUSION), 2012 15th International Conference on. 2012. IEEE.
9. Fatehi, A. and B. Huang, Kalman filtering approach to multi-rate information fusion in the presence of irregular sampling rate and variable measurement delay. Journal of Process Control, 2017. 53: p. 15-25.
10. Ernst, M.O. and M.S. Banks, Humans integrate visual and haptic information in a statistically optimal fashion. Nature, 2002. 415(6870): p. 429-433.

534226 A New Big Data Benchmark Problem: Fluid Catalytic Cracker Under Model Predictive Control

Abstract:

The chemical industry has been for decades a major generator of large-scale data sets. Yet, the means for analyzing these data have started to come of age fairly recently. Their application in transforming the "four Vs" of data (volume, velocity, veracity, variety) to a fifth and most important "V" – Value – is still lagging. In effect, many process operators feel that they are "drowning in data," being overwhelmed by the sheer amount of data and the challenge of extracting information from them [1].

This work stems from the authors' assumption that advances in big-data applications in the chemical industry would benefit considerably from the availability of a set of benchmark problems. We base this assumption on the tremendous impact that the definition of a set of "standardized" control problems, such as the Tennessee Eastman challenge process [2] and the Pensim simulator [3] have had on the development of plant-wide control techniques. Further, we note that the simulation models of these problems have been "borrowed" by the data analysis community and turned into machines for generating data sets used to develop and test new algorithms.

While these benchmark problems typically provide sufficient dynamic complexity and the opportunity to simulate multiple types of process faults, we note that they feature rather simple control approaches (in its standard form, the Tennessee Eastman plant is an open-loop system). As such, a key advance of the present work is to propose a benchmark model together with an advanced control system of the type used in industry, i.e., MPC. We believe that this development will provide a much richer environment for developing new data-driven approaches, such as plant-model mismatch detection, controller performance monitoring, etc.

Our focus is on modeling a Fluid Catalytic Cracking process (FCC). The starting point is the Model IV industrial cracker described by McFarlane et al.[4] and the computational implementation of this model by Ali [5]. The original system consists of a regenerator, preheater, reactor and air blowers. To make the model more realistic, the following extensions are made in this work. First, we add a fractionator (modeled as a steady state flash unit) to separate the cracking product. Further, we model the reactor (riser) as a dynamic unit and incorporate a complete model of the cracking reaction system, that captures the temperature dependence of the reaction rates and of the product distribution. Further, we model the dynamics of the most important actuators and implement a regulatory control system based on those found in practice. Subsequently, we carry out a system identification effort to build a state-space model of the entire system, and formulate a state-space model predictive controller. Several scenarios are tested in closed-loop, including setpoint tracking, disturbance rejection and the ability to deal with plant model mismatch.

The model and the controller are implemented in MATLAB (www.mathworks.com) and are available upon request and free of charge to researchers interested in using them for research in the big data realm.

References

[1] Venkatasubramanian, Venkat. "DROWNING IN DATA: Informatics and modeling challenges in a data rich networked world." *AIChE J.* 55, no. 1 (2009): 2-8.

[2] Downs, James J., and Ernest F. Vogel. "A plant-wide industrial process control problem." *Computers & chemical engineering*17, no. 3 (1993): 245-255.

[3] Birol, Gülnur, Cenk Ündey, and Ali Cinar. "A modular simulation package for fed-batch fermentation: penicillin production." *Computers & Chemical Engineering* 26, no. 11 (2002): 1553-1565.

[4] McFarlane, Randy C., Ralph C. Reineman, James F. Bartee, and Christos Georgakis. "Dynamic simulator for a model IV fluid catalytic cracking unit." *Computers & chemical engineering* 17, no. 3 (1993): 275-300.

[5] E. Ali, http://faculty.ksu.edu.sa/Emad.Ali/Pages/SimulinkModule.aspx3. Last accessed November 21, 2017

537148 Computation-Driven Mechanistic Understanding of the Cellular Cost and Regulation of Melanin Production

Abstract:

Computation-Driven Mechanistic Understanding of the Cellular Cost and Regulation of Melanin Production

Wheaton Schroeder, Jyothi Kumar, Rajib Saha and Steve Harris

The University of Nebraska – Lincoln, Lincoln, NE

Exophiala dermatitidis, a melanized, extremophilic black yeast, is of interest primarily for its ability to synthesize melanin as a stress protectant. In addition, as an emerging fungal pathogen, melanin deposition in the cell wall hides this organism from the host's immune system, often leading to high mortality rates. Without melanin deposited in its cell walls, *E. dermatitidis*' pathogenic efficacy is significantly decreased. *E. dermatitidis*' genome is small, supporting the organism's dual purpose use: first, as a model system for melanocytes to explore treatments for melanin-related human skin disorders such as albinism, and second, as a model pathogenic organism to study how infections can be combated by disrupting melanin production, not only for *E. dermatitidis*, but also for other melanized fungal pathogens such as *Cryptococcus neoformans*. At present, relatively little is known about the regulation and kinetics of melanin production in *E. dermatitidis*, and in particular, whether synthesis of melanin imposes a severe metabolic cost on the fungus. Computational modeling, a promising approach that has not yet been systematically applied to extremophilic fungi or to human fungal pathogens, could fill this knowledge gap.

Computational modeling of metabolism is now an indispensable tool to drive the processes of understanding and redesigning of biological systems. Although Flux Balance Analysis (FBA) is the primary tool used for this purpose, it has significant limitations due to the lack of reaction kinetics, chemical species concentration, and metabolic regulation. In contrast, kinetic models of metabolism (kMMs) provide not only a more accurate method for designing novel biological systems but also for the characterization of reaction kinetics, metabolite concentration, and metabolic regulation in these systems. Hence, the development of a kinetic model framework will create a greater understanding of *E. dermatitidis* as a pathogen and as a melanin-producing model organism. However, the multi-omics data required for kinetic model construction is prohibitive to their development and widespread use. This is particularly true for understudied organisms, such as *E. dermatitidis* whose genome is less than 50% annotated and for which no current metabolic network model exists. To begin to address these issues, a metabolic network model was created for this organism, using its annotated genome and the metabolic network models from species in the related genus *Aspergillus*. The metabolic gaps in this model were then filled using reactions from *Saccharomyces cerevisiae* and from the KEGG database, using a novel optimization-based gap filling tool named OptFill. The completed metabolic network model includes approximately 2000 reactions and 2400 metabolites. FBA was then performed using 36 different nutrition scenarios that vary the carbon source and limiting micronutrient.

The kMM of *E. dermatitidis* was developed by applying an optimization-based method for hypothesizing kinetic mechanisms and regulation pioneered by this laboratory, named **K**inetic **OPT**imization using **I**nteger **C**onditions (**KOPTIC**). This method circumvents the omics data requirement and semi-automates kMM construction using reaction rates and concentration data derived from the metabolic network model to return plausible kinetic mechanisms. Using the FBA data, KOPTIC predicted the kinetic mechanisms and metabolic regulation for most reactions. These predictions form the kMM for *E. dermatitidis*, which can now be used to explore the important questions about the biosynthesis, regulation, and metabolic cost of melanin production. Furthermore, understanding melanin regulation and kinetics can help establish treatment options for melanin-related human skin disorders such as albinism, as well as for infections caused by *E. dermatitidis* and other melanized pathogens.

534680 Using Data Variety for Modeling and Control of Batch Processes

Abstract:

In this age of big data and Internet of Things (IOT), massive amount of data is generated by various sensing devices. Big data are generally characterized by 3 Vs namely, volume, data variety (temperatures and pressures along with infrared spectroscopy, acoustic, or video images) and velocity [1]. In addition to these, sometimes additional parameters - veracity and value, are included to characterize the challenges associated with big data. Several of these challenges (such as those related to databases) are best addressed within the fields of computer science and business analytics applications. An aspect of big data that falls within the realm of process control problems is that of handling data variety and veracity.

While not necessarily identified in the contributions as 'big data solutions', several researchers have developed feedback control schemes that utilize traditional (temperatures etc.) as well as distributed (in the form of particle size distribution measurements) [2] either as part of a first-principles based approach or in purely data driven models [3-6]. The present work aims to address the outstanding big data modeling and control problem (particularly in the context of batch operations), as detailed below.

The process model (used for feedback control) is generally understood to be one of two kinds- so called first principles models, where the structure of the model is given by governing equations (appropriate balances, laws of thermodynamics), or data driven model, where the structure of the model is chosen based on simplicity of model development (albeit cognizant of the ability to predict process dynamics). Both models, however have parameters that need to be identified/determined, and this can only be done via measurements. In this sense, the availability of more data (or variety of data) generally increases the ability to capture the process dynamics better and thus presents an opportunity for advanced control implementation.

A popular advanced model-based control approach, model predictive control (MPC), has increasingly been studied for the control of batch processes (see, e.g., MPC formulations that utilize first principles models [7-10] and for data-driven approaches [11-15]). The presence of variety of data (henceforth referred to as data variety), however, poses a challenge with first-principles model-based control. In this approach, the model complexity increases to explain data variety. For instance, in case of a particulate process, while the temperature and concentration measurements can be described using ordinary differential equations, prediction of particle size distribution requires population balance equations. Although results exist in the literature where first-principles based approach have been successfully used for control, use of alternative, data driven approaches remains attractive due to the abundance of data and the associated simplicity of model development.

This motivates the development of data driven modeling and control approaches to handle the emerging big data opportunity. Over past several decades, data-analytics in process systems engineering have gained prominence. Largely, methods have been developed and applications demonstrated for traditional data [16-

19] with limited results illustrating robust archiving (volume) aspect of the big data in batch processes [20]. One well known approach to data driven batch quality modeling and control is to use latent variable methods. Some of the contributions in this area have dealt with data variety issue directly [3-5] or in conjunction with first-principles model [6]. The simplest implementation of these methods uses so called batch-wise unfolding where each row represents a complete batch duration. However, this unfolding method introduces problems dealing with batches of unequal duration. To address these problems, two methods have been introduced. One method uses a monotonically increasing variable to align batches. The second, more advanced method, uses a technique developed for alignment of sound in voice recognition called dynamic time warping (DTW [20]. While these approaches are useful tools for batch monitoring, they are not as helpful for control because it is unclear how to map future variable trajectories back to the real time domain.

In summary, while possible for monitoring, these models cannot be directly utilized to vary batch durations to achieve improved product quality in a feedback control framework that uses big data. To account for these limitations, for traditional data, a multi-model approach was proposed in [12]. These models were based on the 'current measurements' of the process instead of the 'time'. In more recent work, a subspace-based modeling and control approach for batch quality control was proposed, comprising a state space dynamic model and a quality model, that also did not require batches to be of the same duration for building the model [14]. While these recent results are based on building models using traditional data, they present an opportunity to be adapted to handle the big data problem.

Motivated by these considerations, this work presents the use of data variety for modeling and control of batch particulate process using a two-tier subspace identification based approach. Two formulations are presented: a) minimizing the volume of fines in the final product by leveraging the variety of measurements and b) control of shape of the particle size distribution in the product.

[1] S. J. Qin, Process data analytics in the era of big data, AIChE Journal 60 (9) 360 (2014) 3092–3100.

[2] A. Kumar, M. Baldea, T. F. Edgar, Real-time optimization of an industrial steam methane reformer under distributed sensing, Control Engineering Practice 54 (2016) 140–153.

[3] J. Flores-Cerrillo, J. F. MacGregor, Control of particle size distributions in emulsion semibatch polymerization using mid-course correction policies, Industrial and Engineering Chemistry Research 41 (7) (2002) 1805-1814.

[4] J. Flores-Cerrillo, J. F. MacGregor, Control of batch product quality by trajectory manipulation using latent variable models, Journal of Process Control 14 (5) (2004) 539–553.

[5] M. Park, M. Dokucu, Regulation of the emulsion particle size distribution to an optimal trajectory using partial least squares model-based predictive control, Industrial & engineering (2004) 7227–7237.

[6] M.-J. Park, F. J. Doyle, Nonlinear model order reduction and control of particle size distribution in a semibatch vinyl acetate/butyl acrylate emulsion copolymerization reactor, Korean Journal of Chemical Engineering 21 (1) (2004) 168–176.

[7] D. Shi, N. H. El-Farra, M. Li, P. Mhaskar, P. D. Christofides, Predictive control of particle size distribution in particulate processes, Chemical Engineering Science 61 (1) (2006) 268–281.

[8] D. Bonvin, Optimal operation of batch reactors- a personal view, Journal of Process Control 8 (5-6) (1998) 355–368.

[9] P. D. Christofides, M. Li, L. M¨adler, Control of particulate processes: Recent results and future challenges, Powder Technology 175 (1) (2007) 1–7.

[10] P. D. Christofides, N. El-Farra, M. Li, P. Mhaskar, Model-based control of particulate processes, Chemical Engineering Science 63 (5) (2008) 1156–1172.

[11] S. Aumi, P. Mhaskar, Integrating Data-Based Modeling and Nonlinear Control Tools For Batch Process Control, Aiche Journal 58 (2012) 2105–2119.

[12] S. Aumi, B. Corbett, P. Mhaskar, T. Clarke-Pringle, Data-based modeling and control of Nylon-6, 6 batch polymerization, IEEE Transactions on Control Systems Technology 21 (1) (2013) 94–106.

[13] B. Corbett, B. Macdonald, P. Mhaskar, Model Predictive Quality Control of Polymethyl Methacrylate, IEEE Transactions on Control Systems Technology 23 (2) (2013) 3948–3953.

[14] B. Corbett, P. Mhaskar, Subspace identification for data-driven modeling and quality control of batch processes, AIChE Journal (2016) 62 (2016) 1581-1601.

[15] A. Garg, P. Mhaskar, Subspace Identification-Based Modeling and Control of Batch Particulate Processes, Industrial & Engineering Chemistry Research 56 (26) (2017) 7491–7502.

[16] R. J. Treasure, U. Kruger, J. E. Cooper, Dynamic multivariate statistical process control using subspace identification, Journal of Process Control 14 (3) (2004) 279–292.

[17] T. Kourti, Multivariate dynamic data modeling for analysis and statistical process control of batch processes, start-ups and grade transitions, Journal of Chemometrics 17 (1) (2003) 93–109.

[18] J. Flores-Cerrillo, J. F. MacGregor, Latent variable MPC for trajectory tracking in batch processes, Journal of Process Control 15 (6) (2005) 651–663.

[19] M. Golshan, J. F. MacGregor, M.-J. Bruwer, P. Mhaskar, Latent Variable Model

Predictive Control (LV-MPC) for trajectory tracking in batch processes, Journal of Process Control 20 (4) (2010) 538–550.

[20] T. Kourti, Abnormal situation detection, three-way data and projection methods; robust data archiving and modeling for industrial applications, Annual Reviews in Control 27 II (2003) 131–139.

534910 Optimal Input and Sensor Selection for System Health Assessment

Abstract:

Fault diagnosis is an area of significant research activity with a plethora of methods, as a result of the complexity of modern industrial engineered systems. Modern cyber-physical systems are in need for advanced fault detection and isolation (FDI) test design methods that are accurate, robust against uncertainty, capable of reducing maintenance costs, and can, thus, improve system performance and reliability. Modelling techniques such as unknown input observers can mitigate the implications of uncertainty in fault detection capability and false alarm rates. The latter can be more effective if reconfiguration of system inputs is feasible for the purpose of fault diagnosis in lieu of environmental and system uncertainty. In the same context, sensor design and selection can also improve the precision of fault diagnosis, and methods for singling out informative sensors in diagnosis have been extensively studied.[1] However, efforts that consider active test design and sensor selection simultaneously for the purpose of fault diagnosis are scarse.[2] In this presentation, we discuss an integrated approach to active FDI with sensor selection. In the proposed framework, test design optimization using system inputs and sensors is first performed. Then, an evaluation of the FDI test accuracy across a range of decision thresholds via classification techniques and receiver operating characteristics provides optimal FDI tests in terms of both design and execution.

The selection of active FDI test designs and sensors is formulated as a constrained optimization problem. Constraints of this problem include the allowable operating space, sensor availability and accuracy. This optimization is achieved using models that represent the anticipated steady-state or transient system performance within sufficient accuracy. It is assumed that each fault targeted for diagnosis can be characterized as a model parameter. System uncertainty caused by environmental variability or modelling error is also considered as parametric variance in the model. As shown in Palmer et al.,[3] the rate of successful fault diagnosis can be improved by maximizing the sensitivity of the outputs with respect to faults, and minimizing the correlation between faults and uncertain parameters. In this communication, we present a method for calculating optimal test designs using output sensitivities of the outputs for a fixed (limited) number of discrete sampling points and locations, subject to system faults and uncertainty. The sensitivities of all the feasible measurements with respect to all the anticipated faults and uncertain inputs or conditions are compiled into a variation of the Fisher Information Matrix (FIM). Sensor selection is incorporated into the FIM using a vector of binary elements.

Two optimal test design formulations are considered for active FDI: optimal designs for tests containing only steady-state information, and information obtained from steady-state and transient responses. We present the optimal test designs generated in a simulated case study that satisfy Ds-optimality. Ds-optimal criterion, also known as subset D-optimal criterion, was selected for its ability to selectively reduce the volume of the joint confidence region between parameters of interest (e.g. system faults). The resulting data generated from implementing FDI with the optimal test design are then used to train and classify the faults in the system. Principal Component Analysis (PCA) followed by *k*-Nearest Neighbor (*k*-NN) classification are performed to determine the condition of the system. The *k*-NN algorithm is preferred in fault diagnosis for its capability of multi-class classification and its simplicity.[4] However, in systems with large datasets, *k*-NN has been known to over-fit the data resulting in precision issues. Thus, PCA is implemented to reduce the dimensionality of the test data and identify the most informative system features. After PCA, Monte Carlo simulations are implemented at anticipated fault and fault-free scenarios to train the *k*-NN algorithm for each fault scenario. The effectiveness of the selected FDI test and sensor set is then evaluated by testing the system at various conditions to determine the frequency of correct classification in each scenario. The correct classification rate (CCR) and overall classification accuracy obtained using the selected FDI test designs are then compared to determine the success of the method. Probabilistic *k*-NN is also performed for a comprehensive evaluation of the trade-off between selectivity and sensitivity of the FDI test. A decision threshold is used to determine the maximum allowed probability for a system to be considered fault-free based on any given set of test data. Finally, receiver operating characteristic (ROC) curves are generated for comparing correct classification and false positives, using *k*-NN with a decision threshold based on the allowable proportion of neighbors.

The proposed method is demonstrated on the air-handling system of a diesel engine. This diesel engine subsystem was simulated using a model reported by Wahlström and Eriksson[5], and it is subject to multiple actuator faults and leaks in various components of the system. Four faults are considered: actuator fault in the exhaust gas recirculator (EGR); actuator fault in the variable geometry turbine (VGT); leak in the inlet manifold; and leak in the exhaust manifold. Including the fault-free system state, the FDI has to discriminate between each system state and determine the system health status. Application of the proposed method shows significant improvement in correct fault classification as a result of the test design optimization performed simultaneously with sensor selection. The classification accuracy for the Ds-optimal test design with sensor selection is shown to be superior than that with the nominal sensor selection set.

Acknowledgment

This work was sponsored by the UTC Institute for Advanced Systems Engineering (UTC-IASE) of the University of Connecticut and the United Technologies Corporation. Any opinions expressed herein are those of the authors and do not represent those of the sponsor.

References

1. Najjar, N., Gupta, S., Hare, J., Kandil, S. & Walthall, R. Optimal Sensor Selection and Fusion for Heat Exchanger Fouling Diagnosis in Aerospace Systems. *IEEE Sens. J.* **16,** 4866–4881 (2016).
2. Patan, M. & Ucinski, D. Sensor scheduling with selection of input experimental conditions for identification of distributed systems. in *2010 15th International Conference on Methods and Models in Automation and Robotics* 148–153 (IEEE, 2010). doi:10.1109/MMAR.2010.5587245
3. Palmer, K. A., Hale, W. T. & Bollas, G. M. Active Fault Identification by Optimization of Test Designs. *IEEE Trans. Control Syst. Technol.* **In review,** (2018).
4. He, Q. P. & Wang, J. Principal component based k-nearest-neighbor rule for semiconductor process fault detection. in *2008 American Control Conference* 1606–1611 (IEEE, 2008). doi:10.1109/ACC.2008.4586721
5. Wahlström, J. & Eriksson, L. Modelling diesel engines with a variable-geometry turbocharger and exhaust gas recirculation by optimization of model parameters for capturing non-linear system dynamics. *Proc. Inst. Mech. Eng. Part D J. Automob. Eng.* **225,** 960–986 (2011).

535069 Performance Characterization and Fault-Tolerant Control of Multi-Rate Sampled-Data Process Systems with Unknown Measurement Delays

Abstract:

A key capability of modern-day process control systems is that of fault handling. Faults in process elements such as measurement sensors and control actuators if unaccounted for can result in process instabilities and more commonly loss of product quality. Naturally the importance of this problem has led to a vast and growing body of research on the topic of fault-tolerant control (e.g., see [1],[2],[3]). Conventional methods of addressing this problem include control formulations with a controller driving the process to a desirable operating point through use of continuous state measurements. However, with real-time operation practices the availability of state measurements is governed by the sampling rate of the measurement sensor. Sampled-data systems are found in industrial processes in places where continuous measurements of system data prove untenable due to physical or logistical constraints.

Fault-tolerant control of sampled-data systems has been the focus of prior works, where stability-based reconfiguration [4] and accommodation [5] approaches were developed to address the problem of control actuator fault handling in processes with known measurement delays. These approaches where extended in [6] to include fault estimation capabilities through use of an optimization-based scheme to aid in identifying the fault size and location, and facilitate the implementation of the fault accommodation logic. A framework for the integration of data-based fault detection and performance-based fault accommodation was subsequently developed in [7] and applied to a distributed energy system.

A key underlying consideration of prior works is that all process states are sampled at the same rate. In many practical settings, limitations on the sensing and processing capabilities of different sensors may result in different sampling rates. The use of multi-rate sampling can also be triggered by the relative importance of the measurements collected, where a fast sampling rate is applied to sensors placed at critical locations in the process, while reduced sampling rates are applied to less critical sensors to optimize energy resource consumption and reduce cost. A more robust framework for fault-tolerant control should therefore allow for the possibility of sampling each state at a different rate. Furthermore, the assumption that the measurement delays are fixed and known a priori needs to be re-examined as the controller would not realistically have access to the exact size of the delay which can also be time varying in general.

The objective of this contribution is to address the problem of performance-based actuator fault accommodation in multi-rate sampled-data process systems with unknown measurement delays. A key feature of this contribution is the explicit integration of online delay and fault estimation into the performance-based fault accommodation framework that would allow for the measurement delay and fault magnitudes to be estimated and used as parameters in the fault accommodation strategy. Based on the estimated delay and fault sizes, the potential fault accommodation measure can then be evaluated with reference to the baseline to determine if more drastic measures are needed or if the fault can be sufficiently mitigated through minor adjustments in the controller design parameters. The approach is implemented

through leveraging the available sampled-data from the process to estimate the severity of the fault and the magnitude of the measurement delay through use of a data-based moving-horizon parameter estimation scheme. The accommodation scheme will then determine an appropriate response while using the estimated fault and delay values to meet performance baselines needed by the process and maintaining closed-loop stability. This will be achieved by characterizing the closed-loop stability region and the chosen performance metric of the system as functions of the fault size, the delay size, the frequencies of measurement sampling and other controller design parameters. These characterizations, which evolve in time as the estimated delay and fault values vary, provide the basis for selecting the appropriate accommodation strategy that accounts explicitly for both the faults and the delays in the process. The implementation of the developed fault-tolerant control approach will be demonstrated using a chemical process example.

References:

[1] Simani, S., Fantuzzi, C., and Patton, R. (2003). *Model-based Fault Diagnosis in Dynamic Systems Using Identification Techniques*. Springer, London.

[2] Blanke, M., Kinnaert, M., Lunze, J., and Staroswiecki, M. (2003). *Diagnosis and Fault-Tolerant Control*. Springer, Berlin-Heidelberg.

[3] Mhaskar, P., Liu, J., and Christofides, P.D. (2013). *Fault-Tolerant Process Control: Methods and Applications*. Springer-Verlag, London.

[4] Sun, Y. and El-Farra, N.H. (2011). Model-based fault detection and fault-tolerant control of process systems with sampled and delayed measurements. In *Proceedings of 18th IFAC World Congress*, 2749-2754. Milan, Italy.

[5] Napasindayao, T. and El-Farra, N.H. (2013). Fault detection and accommodation in particulate processes with sampled and delayed measurements. *Ind. & Eng. Chem. Res.*, 52(35), 12490-12499.

[6] Napasindayao, T. and El-Farra, N.H. (2015). Model-based fault-tolerant control of uncertain particulate processes: Integrating fault detection, estimation and accommodation. In *Proceedings of 9th IFAC Symposium on Advanced Control of Chemical Processes*, 872-877. Whistler, Canada.

[7] Allen, J.T. and El-Farra, N.H. (2017). A model-based framework for fault estimation and accommodation applied to distributed energy resources. *Renewable Energy*, 100, 35-43.

538702 Real Time Constrained Model Predictive Control for the Continuous Manufacturing of Pharmaceuticals

Abstract:

The cost of healthcare in the US is the fastest growing component of the inflation index. One method aimed at reducing these growing costs is in the development of processes for continuous pharmaceutical manufacturing. The continuous production of pharmaceuticals offers reduced costs, simplified and streamlined scale-up, and flexibility compared to batch process alternatives.

This research presents an approach to control system design for the continuous production of pharmaceuticals. First, a dynamic model of the continuous drug manufacturing process is developed that leads to a system of nonlinear, differential algebraic equations. This model is then linearized around typical process operating conditions. The linearized model is subsequently discretized and is used to develop a real time online model predictive control system that ensures the integrity of the pharmaceutical production process and helps satisfy the stringent FDA requirements typically required for safe drug production. Specifically, a constrained infinite time linear quadratic regulator control strategy is employed that can satisfy hard constraints in real time. For example, if there is variability in the quality of raw materials, the controller can ensure, in real time, that FDA requirements on drug purity are met.

The proposed approach is demonstrated with a case study based on a continuous Ibuprofen manufacturing process, as developed for steady-state conditions by Jolliffe & Gerogiorgis (2014), and references therein. In this work, a dynamic model of this process is developed, linearized, and controlled as described above.

534547 Model Predictive Control of a Rotational Molding Process

Abstract:

Rotational Molding, also known as rotomolding or rotational casting, is a batch process used for the processing of plastics through distinct heating and cooling phases [11]. In this process a mold filled with a powdered charge is slowly rotated in a heated oven, resulting in the softened material being dispersed and sticking to the walls of the mold. In order to maintain an even wall thickness, the mold is continuously rotated at all times both during the heating and cooling phase. One of the control challenges is to obtain consistent high-quality products from one batch to another while avoiding unfinished parts with incomplete curing or degradation due to prolonged overheating.

Traditional approach for control of rotational molding processes has been to use an open loop, recipe-based policy. In this approach, a predefined input trajectory is applied to each batch [1]. The approach assumes that the desired product quality can be obtained by repeating historically successful input profiles. Although these approaches are easy to implement and do not require a model for the process, they are incapable of rejecting disturbances that affect the process. This has motivated the use of feedback control strategies.

A classical closed loop control approach that could be used in the rotomolding process is trajectory tracking. In this approach, a predefined set-point trajectory for a measured process variable, such as internal mold air temperature, is tracked. PI controllers are often used for implementing trajectory tracking control. The challenge with trajectory tracking approach is that even with perfect tracking, it may not result in desired quality, as the relationship between the measured/tracked variable and final quality may change significantly with changes in the process conditions across batches.

These challenges are addressed by control strategies that are cognizant of the causal relationship, not only between the manipulated and the (online) measured variables, but also between the manipulated and the final quality variables. A popular model-based control approach, model predictive control (MPC), has increasingly been studied for the control of batch processes [2-5]. The reasons for the popularity of these control schemes are twofold: first, the feedback controller can counter the model uncertainties that are associated with model simplifications and measurement errors; the other being their ability to handle the ever-present input/output constraints. In implementing these formulations, where possible, good first principles models are preferred due to the ability to predict process behavior beyond the data set used and estimate model parameters. However, first principles models are difficult to develop, as it is often impractical to estimate the parameters through experiments, particularly because of the associated cost. Furthermore, these first principles models pose computational challenges for direct use in predictive control formulation, for instance, in case of distributed parameter systems such as batch particulate processes.

These issues have motivated the use of simpler, often linear, models derived from past batch data. A variety of approaches for the development of data-driven models have been proposed. One excellent approach is partial least squares (PLS), which models the process in a projected latent space [6]. These models are essentially time-varying linear models, linearized around mean past trajectories, and therefore require the batches to be the same length, or to recognize an appropriate alignment variable. To account for these limitations, a multi-model approach was proposed in [5]. The multiple models were based on the 'current measurements' of the process instead of the 'time'. These developments were followed by contributions in the area of integration of these data-driven models with the advanced control formulations [4-5]. More recently, a subspace identification based batch control approach was proposed in [7] where a LTI state-space model of the batch process is estimated. Subspace identification methods, in general, are a class of model identification methods that is non-iterative in nature and models a process as a state-space linear time invariant system, identifying the system matrices along with the order of the system. One of the major advantages with this approach is that it facilitates the accommodation of variable duration batches without the need for alignment of batch lengths as demonstrated through different applications in [7-10].

Motivated by these considerations, this work presents a subspace identification based state-space modeling and control approach for rotational molding process to achieve desired quality specifications. Contributions from this developed system represent the progression of this industry towards advanced manufacturing practices, with improved reliability between batches.

[1] D. Bonvin, B. Srinivasan, D. Hunkeler, Control and optimization of batch processes, IEEE Control Systems Magazine 26 (6) (2006) 34-45.

[2] D. Shi, N. H. El-Farra, M. Li, P. Mhaskar, P. D. Christofides, Predictive control of particle size distribution in particulate processes, Chemical Engineering Science 61 (1) (2006) 268-281.

[3] D. Bonvin, Optimal operation of batch reactors- a personal view, Journal of Process Control 8 (5-6) (1998) 355-368.

[4] S. Aumi, P. Mhaskar, Integrating Data-Based Modeling and Nonlinear Control Tools For Batch Process Control, AIChE Journal 58 (2012) 2105-2119.

[5] S. Aumi, B. Corbett, P. Mhaskar, and T. Clarke-Pringle. Data-based modeling and control of nylon-6, 6 batch polymerization. Control Systems Technology, IEEE Transactions on, 21(1):94-106, Jan 2013.

[6] J. Flores-Cerrillo, J. F. MacGregor, Control of particle size distributions in emulsion semibatch polymerization using mid-course correction policies, Industrial and Engineering Chemistry Research 41 (7) (2002) 1805-1814.

[7] B. Corbett, P. Mhaskar, Subspace identification for data-driven modelling and quality control of batch processes, AIChE Journal 62 (2016) 1581-1601.

[8] Abhinav Garg, P. Mhaskar (2017), Subspace Identification Based Modeling and Control of Batch Particulate Processes, Industrial and Engineering Chemistry Research, 56 (26) 7491-7502.

[9] Abhinav Garg, Brandon Corbett, Prashant Mhaskar, Gangshi Hu, Jesus Flores-Cerrillo (2017), High Fidelity Model Development and Subspace Identification of a Hydrogen Plant Startup Dynamics, Computers and Chemical Engineering, 106 183-190.

[10] Rashid, M. M., Mhaskar, P., & Swartz, C. L. (2017). Handling multiâ€rate and missing data in variable duration economic model predictive control of batch processes, AIChE Journal, 63(7), 2705-2718.

[11] Felipe P.C. Gomes, Abhinav Garg, P. Mhaskar and Michael R. Thompson, Quality monitoring of rotational molded parts using a nondestructive technique, ANTEC (2018), Accepted.

534349 A Multiparametric Programming Based Approach to Integrate Design, Scheduling, and Control of a Batch Process

Abstract:

Integrating short term (control), middle term (schedule), and long term (design) decisions has been shown to improve the operability and profitability of batch processes [1]. The interactions between these distinct time-scale decisions allow for opportunities to (i) replace fixed recipe-based schedules with detailed model-based schedules, (ii) build design dependent schedule aware controllers, and (iii) reduce capital costs by circumventing overdesigned processing units. While the efficiency of simultaneous decision making in process systems is greatly acknowledged, the methodology/procedure to achieve an effective integration is still an open question due to the order of magnitude differences in the time scales of the constituent problems, as well as their respective objectives. The most recent advances within the last decade [2-7] depict a promising basis for further improvement.

In this work, we present a unified framework to simultaneously determine the optimal codependent short term and middle term operating strategies of a batch process for a range of market conditions, while minimizing the capital costs. A single high fidelity model is utilized to develop (i) multiparametric rolling horizon optimization (mpRHO) strategies as a function of process states and time variant market conditions, and (ii) multiparametric Model Predictive Control (mpMPC) for effective set-point tracking, following a similar methodology described in Burnak *et al* [8]. The design of the process is accounted for as an additional unknown bounded parameter in the offline operational strategies. The offline nature of mpRHO and mpMPC strategies allows for integration in the design optimization of the process, i.e. minimizing the capital costs. The framework is showcased on (i) a single non-isothermal batch reactor, and (ii) a batch process comprising a preprocessing unit, a reactor, and a post-processing unit.

1. Pistikopoulos, E. N., Diangelakis, N. A., Towards the integration of process design, control and scheduling: Are we getting closer?, Computers and Chemical Engineering (2015), 91, pp. 85-92.
2. Zhuge, J., Ierapetritou, M. G., Integration of scheduling and control for batch processes using multi-parametric model predictive control, AIChE J. (2014), 60 (9), pp. 3169-3183.
3. Baldea, M., Harjunkoski, I., Integrated production scheduling and process control: A systematic review, Computers and Chemical Engineering (2014), 71, pp. 377-390.
4. Dias, L. S., Ierapetritou, M. G., Integration of scheduling and control under uncertainties: Review and challenges, Chemical Engineering Research and Design (2016), 116, pp. 98-113.
5. Patil, B.P., Maia, E., Ricardez-Sandoval, L.A., Integration of scheduling, design, and control of multiproduct chemical processes under uncertainty, AIChE Journal (2015), 61, pp. 2456-2470.
6. Dias, L. S., Pattison, R. C., Tsay, C., Baldea, M., Ierapetritou, M. G., A simulation-based optimization framework for integrating scheduling and model predictive control, and its application to air separation units, Computers and Chemical Engineering (2018), pp. 139-151.
7. Rossi, F., Casas-Orozco, D., Reklaitis, G., Manenti, F., Buzzi-Ferraris, G., A computational framework for integrating campaign scheduling, dynamic optimization and optimal control in multi-unit batch processes, Computers and Chemical Engineering (2017), 107, pp. 184-220.
8. Burnak, B., Katz, J., Diangelakis, N. A., Pistikopoulos, E. N. Simultaneous process Scheduling and control: A multiparametric programming based approach, Industrial & Engineering Chemistry Research (2018), 57 (11), 3963-3976.

524393 Multi-Objective Dynamic Optimisation of a Fed-Batch Nosiheptide Reactor

Abstract:

Thiopeptide antibiotics are highly potent and structurally complex; this growing class of secondary metabolites are formed by ribosomal peptide biosynthesis. Nosiheptide, a sulfur-containing peptide antibiotic obtained through fermentation, can be regarded as the structural prototype of the class e thiopeptide antibiotics. It exerts exceptional antibiotic activity in vitro and in a mouse model against critical Gram-positive pathogens such as MRSA, VRE, or *Clostridium difficile*. Shown non-toxic at high dosages it has previously been used for applications in farm animals (Benazet at al., 1980). Recently the first total synthesis of nosiheptide was reported, utilizing double macrocyclization of a fully functionalized linear precursor (Wojtas et al., 2016). Given low indusial yields, strong motivation exists to dynamically optimise the process for improved product yield while reducing production time and cost. Similar multi-objective optimisation problems are common in the chemical industry (Rodman, Fraga and Gerogiorgis, 2017), which can be solved with both stochastic and deterministic techniques.

The nosiheptide fermentation process is an extremely complex biochemical reaction, with the complete reaction mechanism not strictly understood. A simplified mechanistic fermentation process model has been proposed, considering only the most essential interactions in the biochemical network (Wojtas et al., 2016). The model considers the temperature and pH dependence of biomass growth, in addition to substrate consumption and product production, with the 19 model parameters determined from the model authors experimental campaigns.

To circumvent mass transfer inhibition at excessive substrate concentrations in the broth the nosiheptide reactor is operated in a fed-batch mode, where the reactor is only partially filled initially, with substrate supplemented over time. Herein the multi-objective dynamic optimisation problem can be explicitly formulated: our two minimisation criteria are batch time and inverse yield with four decision (control) profiles to compute towards optimising the system: both the dynamic feed flow rate and the dissolved oxygen concentration must be determined in addition to the reactor temperature and pH profile subject to operability constraints.

This contribution employs (epsilon constrained ***or*** variable weighted sum) handling of the bi-objectives problem, allowing the Pareto front trade-off between the competing objectives to be visualised. A direct method for dynamic optimisation (simultaneous strategy) has been performed in each case to compute the four optimal control trajectories. Orthogonal polynomials on finite elements are used to approximate the control and state trajectories allowing the continuous problem to be converted to NLP form (Biegler, Cervantes, and Wächter, 2007). The resultant large scale NLP problem is been solved for each instance using IPOT (Wächter and Biegler, 2006), and global optimality is ensured with a multi-start search via Latin Hypercube Sampling of the input space for initialisation.

The algorithm employed demonstrates that appropriate reactor operation and control can significantly improve product yield. Additionally, explicit visualization of the objective trade-offs provides valuable insight to decision makers towards critical economic process decisions.

REFERENCES

Benazet, F., Cartier, M., Florent, J., et al., 1980. Nosiheptide, a sulfurcontaining peptide antibiotic isolated from Streptomyces actuosus 40037. *Cell. Mol. Life Sci.* **36**(4): 414â"416.

Biegler, L. T., Cervantes, A. M. and WÃ¤chter, A., 2002. Advances in simultaneous strategies for dynamic process optimization. *Chemical Engineering Science* **57**: 575-593.

Biegler, L.T., 2007. An overview of simultaneous strategies for dynamic optimization. *Chemical Engineering and Processing: Process Intensification* **46**(11): 1043-1053.

Niu, D., Jia, M., Wang, F. and He, D., 2013. Optimization of nosiheptide fed-batch fermentation process based on hybrid model. *Industrial & Engineering Chemistry Research* **52**(9): 3373-3380.

Rodman, A.D. and Gerogiorgis, D.I, 2017. Dynamic optimization of beer fermentation: sensitivity analysis of attainable performance vs. product flavor constraints. *Computers and Chemical Engineering* **106**(C): 582-595.

Rodman, A.D., Fraga, E.S. and Gerogiorgis, D.I., 2018. On the application of a nature-inspired stochastic evolutionary algorithm to constrained multi-objective beer fermentation optimisation. *Computers and Chemical Engineering* **108**: 448-459.

Wojtas, K.P., Riedrich, M., Lu, J.Y., Winter, P., Winkler, T., Walter, S. and Arndt, H.D., 2016. Total Synthesis of Nosiheptide. *Angewandte Chemie International Edition*, **55**(33): 9772-9776.

534565 Big Data and the Chemical Industry: From Sensor to Value

Abstract:

With the Internet of Things, smart and wireless sensors, advanced control systems, wireless communications and smart devices, the amount of data collected and stored has grown exponentially. These bring big opportunities to the chemical industry where Big Data can be used to enhance manufacturing & operational excellence, forecast pricing, improve formulations, and improve marketing and support innovation.

We present the challenges and opportunities of Big Data in SABIC and a case study illustrating the use of BD technology to move from data, to knowledge, to value in support of Innovation.

536696 Big Data and Iot: A Demonstration Testbed of Multi-Stage Centrifugal Pumping System

Abstract:

Big data and IoT: a demonstration testbed of multi-stage centrifugal pumping system

Devarshi Shah[1], Q. Peter He[1], and Jin Wang[1]

[1]Department of Chemical Engineering, Auburn University, Auburn, AL, 36849, USA

Abstract:

A new industrial revolution is happening to make current manufacturing systems smarter, safer, and more efficient. With the emergence of the Industrial Internet of Things (IoT) and ever advancing computing power and expansion of wireless networking technologies, a new generation of networked, information-based technologies, data analytics, and predictive modeling are providing unprecedented embedded computing capabilities as well as access to previously unimagined potential uses of data and information. This revolution is known as smart manufacturing [1] or Industry 4.0.

Although many general frameworks have been proposed for IoT enabled systems for industrial application [1]-[3]. There is limited literature on demonstrations or testbeds of such systems. In addition, there is a lack of systematic study on the characteristics of IoT sensors and data analytics challenges associated with IoT sensor data. This study is an attempt to help fill this gap by exploring the characteristics of IoT vibration sensors and show how IoT sensors and big data analytics can be used to develop real time monitoring frameworks.

In this work, we will introduce the design of an IoT testbed using multi-stage centrifugal pumping system equipped with non-invasive IoT vibration sensors as shown in Fig. 1. We will discuss the system architecture and protocols designed for data collection, transmission, and storage as shown in Fig. 2. We will also discuss the characteristics of IoT sensors and data analytics challenges associated with data collected from these sensors. The main focus of this work is the development of data-driven predictive models based on the vibration signals to infer flowrate inside the pipe and rpm of the pump motor, and compare with experimental measurements. Altogether, this study serves as a demonstration of how IoT sensors and big data analytics can be integrated and utilized for real-time process monitoring.

Fig. 1 Actual setup of the testbed with five IoT vibration sensors attached to the pumping system at five different locations

References:

[1] J. Davis, T. Edgar, J. Porter, J. Bernaden, and M. Sarli, "Smart manufacturing, manufacturing intelligence and demand-dynamic performance," *Comput. Chem. Eng.*, vol. 47, pp. 145–156, Dec. 2012.

[2] G. Fortino and P. Trunfio, Eds., *Internet of Things Based on Smart Objects*. Cham: Springer International Publishing, 2014.

[3] H. S. Kang, J. Y. Lee, S. Choi, H. Kim, J. H. Park, J. Y. Son, B. H. Kim, and S. Do Noh, "Smart manufacturing: Past research, present findings, and future directions," *Int. J. Precis. Eng. Manuf. Technol.*, vol. 3, no. 1, pp. 111–128, Jan. 2016.

536078 Pressure Swing Adsorption Cycle Synthesis Utilizing Artificial Neural Networks As Surrogate Models

Abstract:

With fossil fuels expected to be a significant portion of the worldâ€™s energy mix for the near future, it is important to minimize CO_2 emissions from current power plants through carbon capture and sequestration (CCS).[1] In post-combustion CCS, CO_2 is separated from the power plant flue gas emissions, containing mainly of N_2 and CO_2. Of the technologies currently available for CCS, pressure swing adsorption (PSA) is one of the more promising due to low energy requirements and short cycle times compared to other adsorption based technologies. However, one challenge that exists in using PSA for CCS is designing the cycle to match newly developed materials for this application. Although the steps in all PSA cycles can be classified into six different possibilities (pressurization, feed, depressurization, light reflux, heavy reflux and pressure equilibration), the arrangement of the steps and interactions between steps lead to hundreds of potential different combinations. The objective of this work is to develop a new approach that is capable of synthesizing PSA cycles to capture the CO_2 from flue gas at the required purities and recoveries while minimizing energy requirements and maximizing adsorbent productivities.

In this work, we present a new framework for synthesizing the PSA cycle with the lowest predicted CO_2 capture costs. In this framework, we train artificial neural networks (ANNs) using Bayesian regularization methods as surrogate models for the various PSA steps. The ANNs are trained on simulation data collected from our PSA model consisting of a system of partial differential algebraic equations incorporating mass and energy balances, pressure drop across the column, competitive multi-site Langmuir isotherms and the linear driving force model.[2] With the ANN surrogate models, we propose a mixed integer nonlinear programming (MINLP) model to determine the ideal ordering and duration of steps in order to minimize the energy requirements and maximize the adsorbent productivity. We evaluate this model with several adsorbents, including Ni-MOF-74, UTSA-16 and zeolite 13X to compare the adsorbents under optimized cycle conditions.

1. MIT. *The Future of Coal*; 2007. http://web.mit.edu/coal/.
2. Leperi KT, Snurr RQ, You F. Optimization of Two-Stage Pressure/Vacuum Swing Adsorption with Variable Dehydration Level for Postcombustion Carbon Capture. *Ind. Eng. Chem. Res.* 2016;55:3338-3350.

537187 Continuous Control of a Polymerization System with Deep Reinforcement Learning

Abstract:

Improving and deriving noble control laws for polymerization reaction has been one of the uppermost tasks in the chemical industry. There have been extensive studies in control industries from proportional controls to various forms of model predictive controls, which have been benefiting in current chemical processes and plants[1]. Recent breakthroughs in deep learning area have started to inspire the development of Artificial Intelligence(AI)-based controllers in chemical reaction control area thanks to its known success in wide-range applications from gaming to robotics control[2]. These applications utilize deep reinforcement learning (Deep RL), of which the ultimate goal is to enable computers to make human-like policies based on AI's exploration of the environment[3]. However, some challenges are still faced for applying Deep RL in controlling chemical reactions, since the inputs of chemical reactors are usually high-dimensional, and the system dynamics are often sensitive and has considerable time-delay effect[2].

Previous studies have been done in controlling a free-radical polymerization process by following real-time measurement of weight-average molar mass[1], where the specific molar mass distribution can be achieved by following a trajectory of weight-average molar mass with respect to time[1]. In this work, we developed a deep learning based controller for a free radical poly-acrylamide polymerization system using Deep Deterministic Policy Gradient (DDPG). The DDPG utilizes actor-critic structure that is able to predict actions of infinite dimensions[4]. The controller calculates the control action a_t, which consists of the monomer flow rate F_m and initiator flow rate F_i to adjust at each time step t, and the system response of the action is recorded as a state s_t. The network is trained to maximize the cumulative reward r that accounts for the distance between current output and target output for each iteration[5]. The network is trained on an established kinetic model of the polymerization reaction, and the controller gradually learns the policy through exploration of the system. Then convergence is achieved when the average cumulative reward reaches a desired threshold. In our experiment, the controller successfully has learned the control policy to follow the target trajectory of the weight-average molar mass.

Overall the smart controller has shown robust control over a range of operating conditions, which indicates the deep reinforcement learning based approach's capability in controlling a nonlinear dynamic semi-batch system.

Reference

[1] N. Ghadipasha, W. Zhu, J. A. Romagnoli, T. Mcafee, T. Zekoski, and W. F. Reed, "Online Optimal Feedback Control of Polymerization Reactors : Application to Polymerization of Acrylamide – Water – Potassium Persulfate (KPS) System," 2017.

[2] S. P. K. Spielberg, R. B. Gopaluni, and P. D. Loewen, "Deep Reinforcement Learning Approaches for Process Control", 2017

[3] V. Mnih, D. Silver, and M. Riedmiller, "Playing Atari with Deep Reinforcement Learning," pp. 1–9.

[4] T. P. Lillicrap *et al.*, "Continuous learning control with deep reinforcement learning," , 2016.

[5] D. Silver, G. Lever, D. Technologies, G. U. Y. Lever, and U. C. L. Ac, "Deterministic Policy Gradient Algorithms."

526418 On Integration of Feedback Control and Safety Systems for Networked Chemical Processes

Abstract:

Safety is critical in the chemical process industries due to the severe consequences for both lives and property when safety is not maintained [1]. Despite many efforts to develop, characterize, and standardize effective safe process/plant design and operation procedures, accidents continue to occur causing significant human and capital losses [2]. Innovative operating strategies are required to eliminate unsafe process operation which may lead to catastrophic events. Recent advances in chemical process safety utilizing a system engineering perspective have sought to incorporate safety considerations within model predictive control (MPC) in the context of a single processing unit [3].

Motivated by the above considerations, we propose a distributed MPC design for chemical process networks that accounts for the activation of the safety system and changes the process model when the safety system is predicted to be activated at some time instant along the prediction horizon. An example of a continuous stirred-tank reactor (CSTR) coupled with a flash drum separator is considered to illustrate that model predictive control with varying models can control a process when the safety system is activated due to significant feed disturbances that lead to thermal runaway in the CSTR. This multi-unit chemical process is simulated in Aspen Plus Dynamics and a multi-variable distributed MPC is implemented to control the entire chemical process network as well as design and size the safety system. Disturbance and device failures are introduced to evaluate the distributed MPC performance for the cases in which it does and does not account for the activation of the safety system.

[1] The 100 largest losses 1974-2015: Large property damage losses in the hydrocarbon industry. Technical report, Marsh & McLennan Companies Inc., 2016.

[2] Center for Chemical Process Safety. Guidelines for Hazard Evaluation Procedures. John Wiley & Sons, Inc., New Jersey, 2008.

[3] Zhang, Z., Z. Wu, H. Durand, F. Albalawi and P. D. Christofides, "On Integration of Feedback Control and Safety Systems: Analyzing Two Chemical Process Applications," Chem. Eng. Res. & Des., 132, 616-626, 2018.

536070 Manifold Learning for Measurements across Several Sensors: Alternating Diffusion, Data Fusion and Constructing Nonlinear Observers for Complex Chemical Reaction Networks

Abstract:

We study the problem of reconciling measurements from different observations of simultaneously occurring uncoupled phenomena â€“ e.g. several simultaneous but noninteracting reaction networks occurring in an isothermal stirred tank (or, for that matter, in a cell). One needs to first understand which sets of variables ``belong togetherâ€, and should therefore constitute an irreducible building block. Novel data mining/ manifold learning techniques (in particular, Alternating Diffusion [1]) make this possible.
Once the several constituent building blocks (different reaction networks) are identified, it is possible to use data mining (diffusion maps) to fuse the information from different partial sensors of the same building block. It is also possible to exploit semi-supervised learning or even AI techniques (like deep neural networks or deep Gaussian processes) to reconstruct, from partial observations, the full state, including quantities that are difficult to measure. We implement and demonstrate these tools on ensembles of concentration measurements from several simultaneously occurring biochemical reactions.
[1] R. R. Lederman, R. Talmon, Learning the geometry of common latent variables using alternating-diffusion, Applied and Computational Harmonic Analysis, 44 (3) (2018), pp. 509-536, https://doi.org/10.1016/j.acha.2015.09.002.

535805 Emergence of Structural Features in Complex Networks Via Adaptation

Abstract:

Chemical and biological systems can often be represented in the form of networks which show interactions between the various variables involved. Understanding the complexity of these systems is crucial for any systems engineering application. One such measure to understand network complexity is entropy. Venkatasubramanian et al. [2006] have suggested entropy maximization as a general design principle for networked systems. More recently, Stevens [2016] observed that biological systems seem to follow a maximum entropic distribution. From a goal-driven setting, a system's objective is survival, which consists of a short-term component and a long-term component. Venkatasubramanian et al. [2004] suggested that the short-term and the long-term components can be represented using the efficiency of the networked system (to reach a goal) and robustness of the system to external disturbances (such as, node-deletion), respectively. In this work, we explore these design possibilities for various chemical and biological systems. We also study the correlation between the network topology exhibited by these systems and their robustness/efficiency metrics. This motivates us towards an information theory-based measure of complexity of networked systems. How this information theoretic measure relates to the systems' robustness and efficiency, is studied as well.

References

Charles F Stevens. A statistical property of fly odor responses is conserved across odors. Proceedings of the National Academy of Sciences, 113(24):6737–6742, 2016.

Venkat Venkatasubramanian, Santhoji Katare, Priyan R Patkar, and Fang-ping Mu. Spontaneous emergence of complex optimal networks through evolutionary adaptation. Computers & chemical engineering, 28(9): 1789–1798, 2004.

Venkat Venkatasubramanian, Dimitris N Politis, and Priyan R Patkar. Entropy maximization as a holistic design principle for complex optimal networks. AIChE journal, 52(3):1004–1009, 2006.

537393 Modeling the Stochastic Dynamics of Gene Regulatory Networks Using Probabilistic Boolean Networks

Abstract:

Gene regulatory networks (GRN) control nearly every aspect of a cellâ€™s behavior. There are numerous fields that can benefit from increased knowledge of the underlying mechanisms of gene regulation or the ability to manipulate behavior of GRNs. These networks are comprised of many layers of complexity and interwoven connections, making it a challenge to compute or simulate their dynamics. Current modeling methods leave the end user with a catch-22 of trying to choose between computational cost and system resolution.

Modeling by simulating or solving the Chemical Master Equation (CME) has the benefit of describing the system in high molecular detail and accounting for stochastic molecular processes. However, the high computational cost and large number of unknown parameters that come along with treating every molecule can overwhelm even state of the art computing clusters. It is possible to use the Boolean Network (BN) framework to model larger GRNs but this can result in an oversimplification of both the gene states and their interactions.

This work discusses using a Probabilistic Boolean Network (PBN) approach to approximating a CME model of GRN dynamics. Using the networkâ€™s eigenvalues and proposed global metrics to define the dynamics of the GRN, this method combines stochastic modeling and BNs to reduce computational cost while maintaining model accuracy.

536763 Koptic: A Novel Approach for *in silico* Prediction of Enzyme Kinetics and Regulation

Abstract:

KOPTIC: A novel approach for *in silico* prediction of enzyme kinetics and regulation

Wheaton Schroeder and Rajib Saha

The University of Nebraska – Lincoln, Lincoln, NE

Computational modeling of metabolism is now an indispensable tool to drive the processes of understanding and redesigning of biological systems. These tools enable the design of engineering interventions directed to the overproduction of a specific bioproduct or improvement of plant performance, particularly in the presence of feedback regulation. Although Flux Balance Analysis (FBA) is the primary tool used for this purpose, it has significant limitations due to the lack of reaction kinetics, chemical species concentration, and metabolic regulation. In contrast, kinetic models of metabolism (kMMs) provide not only a more accurate method for designing novel biological systems but also for the characterization of reaction kinetics, metabolite concentration, and metabolic regulation in these systems; however, the multi-omics data required for their construction is prohibitive to their development and widespread use. Here, we introduce **K**inetic **OPT**imization using **I**nteger **C**onditions (**KOPTIC**), which can circumvent the omics data requirement and semi-automate kMM construction by using reaction rates and concentration data derived from a metabolic network model to return plausible kinetic mechanisms through an optimization-based approach.

Arabidopsis thaliana (hereafter *Arabidopsis*) was chosen to verify KOPTIC. *Arabidopsis* is a model system for modern plant science due to its small genome, short lifecycle, and ease of genetic manipulation. Its prominent role in omics and plant science research makes it an ideal organism for the creation and integration of emerging omics tools, including computational modeling of metabolism. Indeed, several metabolic network models for *A. thaliana* already exist, and many regulatory interactions are well studied and documented. To verify KOPTIC, a four-tissue (leaf, root, seed, and stem) metabolic network model, consisting of the major primary carbon metabolism pathways (1015 reactions, 901 metabolites, 508 genes), was reconstructed for *Arabidopsis*. FBA was performed at 71 time-points in the of *Arabidopsis* lifecycle to simulate 50 days of growth and 20 days of plant senescence and death. KOPTIC was applied to the FBA data and predicted 3577 regulatory interactions involving metabolic, allosteric, and transcriptional regulatory mechanisms (with more than 30 verified by existing literature) and with a median fit error of 13.44%. This research showcases how an optimization-based approach can be used to create meaningful hypotheses of reaction kinetics, as well as discover new biosynthetic pathways, regulations, and genes. Additionally, KOPTIC serves as a novel tool to increase mechanistic understanding of plant-scale regulatory mechanisms and answer biologically relevant questions related to plant-scale metabolism, kinetics, and regulation.

536095 Lung Immunodynamics during pH1N1 Influenza Virus Infection

Abstract:

Lung Immunodynamics during pH1N1 Influenza Virus Infection

Emily E. Ackerman[1], Ericka Mochan[1,2], Jason E. Shoemaker[1,3,4]

[1]Dept of Chemical & Petroleum Engineering, University of Pittsburgh, [2]Dept of Mathematics, Carlow University,[3] Dept of Computation and Systems Biology,[4]McGowan Institute

Deadly influenza infections are often associated with distinct immune system dynamics. Distinct dynamics of key immune cells and inflammation-associated signaling pathways were observed in mouse and macaque studies of infection with a reconstructed strain of the 1918 pandemic influenza virus, with human-isolated H5N1 virus and with the H1N1 virus responsible for the 2009 pandemic (pH1N1). Moreover, studies in mice depleted of select immune cell groups have demonstrated improved survival to select infections. Together, all of these observations suggest that immunomodulatory treatments may help protect high risk groups during pandemics. But it is also clear that complete ablation of the immune response leads to uncontrolled virus replication and severe outcomes. Treatments must be optimized to promote healthy immune response profiles which minimize dangerous secondary processes, e.g. inflammation.

Our group has developed a method to maximally exploit available mouse host response data to develop an ensemble of models that characterize immune response dynamics in animals infected with pH1N1 virus. Previous models of influenza induced host responses focus primarily on lung cytokines and immune cell counts, but do not use gene expression data to inform model development. Here, we used an iterative, coexpression analysis approach to infer the log fold changes of key effector signaling molecules (e.g. IFNα, IFNβ, IRF3). The coexpression analysis allows us to dynamically characterize a larger number of molecules and improve time-course resolution for other model components, including changes in immune cell counts. After data normalization, we performed a series of parameterization training exercises to address several questions on influenza virus induced immune responses.

We used ensembles of parametrized models to determine if the kinetics of the immune response are conserved in mild and severe infections. We collected lung gene expression, cytokine and cell count data from animals infected with a mild, seasonal H1N1 virus or the more severe pH1N1 virus. We start with an initial ordinary differential equation model of the immune system that describes the chemical responses of epithelial cells, B cells, T cells, macrophages and several key cytokines (14 states in all). Using a Markov Chain Monte Carlo (MCMC) parameterization approach, we trained the model to the data derived from H1N1 infected animals. MCMC produces ensembles of parameterized models all of which may suitably characterize the data. We selected the top 1000 models which had the best fit and the least correlated parameters. We then challenged each model to correctly predict the immunodynamics observed in pH1N1-infected animals. To do so, we performed another MCMC parameterization but only allowed parameters associated with virus replication to vary. None of the models were capable of predicting the immunodynamics of pH1N1-infected animals. This suggests that either the kinetics driving immunodynamics are unique to each infection or that MCMC has simply not found the optimal solution. We then performed two more computational experiments: we applied the MCMC formulation to the pH1N1 data and then to the combined H1N1/pH1N1 data. After 10^8 parameter iterations, solutions were found which fit the pH1N1 data but the combined data never satisfactorily converged (based on the normalized sum squared errors and visual confirmation). With the caveat that the results are dependent on the model architecture, we concluded that pH1N1 and H1N1 induce unique host response kinetics. A direct comparison of the parameterizations between H1N1- and pH1N1-infected animals suggest that pH1N1 induces an unexpectedly strong interferon response. Further system architectures will be considered in the future and the modeling will be expanding to include data from more severe infections (e.g. H5N1 and H7N9 viruses).

534299 Understanding the Basal Ganglia Dynamic Transition from the Healthy to the Parkinsonian State

Abstract:

Most of the existing basal ganglia (BG) computational models have focused on providing an understanding of the network pathology associated with the Parkinsonâ€™s Disease (PD) by analyzing the healthy and PD discrete states. However, these models do not describe the dynamic transition between the two states as a function of the loss of dopamine neurons [1] [2]. An understanding of the BG dynamic transition between these two states is important for the design of future experiments and development of novel therapeutically effective treatments of PD. To better understand this dynamic transition, we developed a computational model of the basal ganglia that describes the transition from the healthy state to PD state as a function of the depletion of dopamine neurons.

A marker of the PD state is the emergence of beta-oscillation in subthalamic nucleus (STN) of the BG and is believed to be the cause of motor dysfunctions associated with PD [3]. One hypothesis is that the oscillations arise from an imbalance in the direct and indirect pathways that originate in the striatum of the BG [1]. Due to the high density of dopaminergic projection to the striatum, the loss of dopamine neuron reduces the dopamine concentration to pathological levels. The disruption of the striatum homeostasis creates an imbalance in the activity of neurons in the direct and indirect pathway, which are the D1 and D2 medium spiny neurons (MSNs), respectively [1]. Recent experimental data suggests that the asymmetric activity of the D1 and D2 medium spiny neurons (MSNs) may be due to a combination of factors, such as, a difference in the dopamine binding concentration of D1 and D2 receptors [4], the excess release of acetyl choline [3], and the different muscarinic (acetyl choline) receptors on the MSNs [5]. Based on these experimental results, we expanded the existing BG multi-compartment model developed by Kumaravelu *et al.* by including detailed chemical-level description of dopamine and acetyl choline dynamics in the striatum [2] as well as the dopamine neurons of the substantia nigra (SN) and their projections to the striatum [6]. To simulate the progression of PD, neurons in the SN were sequentially removed from the BG network.

A core result of this presentation is the emergence of beta-oscillation from the neurochemical imbalance in the striatum as the dopamine neurons die. Our computational model of BG shows the removal of dopamine neurons produce a neurochemical imbalance in the striatum that disrupts the overall homeostasis of the BG. The neurochemical imbalance leads to an asymmetric activation of the direct and indirect pathways and drives the BG network from the healthy state to the PD state.

References

[1] Humphries, M., Obeso, J. and Dreyer, J.K., 2018. Insights into Parkinson's disease from computational models of the basal ganglia. *bioRxiv*, p.260992.

[2] Kumaravelu, K., Brocker, D.T. and Grill, W.M., 2016. A biophysical model of the cortex-basal ganglia-thalamus network in the 6-OHDA lesioned rat model of Parkinsonâ€™s disease. *Journal of computational neuroscience*, *40*(2), pp.207-229.

[3] Kondabolu, K., Roberts, E.A., Bucklin, M., McCarthy, M.M., Kopell, N. and Han, X., 2016. Striatal cholinergic interneurons generate beta and gamma oscillations in the corticostriatal circuit and produce motor deficits. *Proceedings of the National Academy of Sciences*, *113*(22), pp.E3159-E3168.

[4] Shen, W., Flajolet, M., Greengard, P. and Surmeier, D.J., 2008. Dichotomous dopaminergic control of striatal synaptic plasticity. *Science*, *321*(5890), pp.848-851.

[5] Ztaou, S., Maurice, N., Camon, J., Guiraudie-Capraz, G., Kerkerian-Le Goff, L., Beurrier, C., Liberge, M. and Amalric, M., 2016. Involvement of striatal cholinergic interneurons and M1 and M4 muscarinic receptors in motor symptoms of Parkinson's disease. *Journal of Neuroscience*, *36*(35), pp.9161-9172.

[6] Yu, N. and Canavier, C.C., 2015. A mathematical model of a midbrain dopamine neuron identifies two slow variables likely responsible for bursts evoked by SK channel antagonists and terminated by depolarization block. *The Journal of Mathematical Neuroscience (JMN)*, *5*(1), p.5.

536572 Dynamic Transcriptomic Profiling of *Scheffersomyces Stipitis* Reveals Key Information of Its Gene Regulatory Network at Genome-Scale

Abstract:

Dynamic Transcriptomic Profiling of *Scheffersomyces stipitis* Reveals Key Information of Its Gene Regulatory Network at Genome-Scale

Matthew Hilliard[1], Thomas Jeffries[2,3], Q. Peter He[1] and Jin Wang[1]

[1] Department of Chemical Engineering, Auburn University, AL 36849

[2] Xylome, Madison, WI 53719

[3] Department of Bacteriology, University of Wisconsin at Madison, WI 53706

The rapid development of "omics" technologies offers unprecedented opportunities to help understand cellular metabolism at genome-scale. Among different "omics" data, RNAseq based transcriptome profiling has been used to enhance the understanding of the genome-scale response of the organism to different stimuli[1,2]. While these studies have provided insightful findings, they are most often limited to studying two steady-state conditions. From a control perspective, as cellular metabolism is a highly complex dynamic system, we hypothesize that the transient response could offer significantly more information on the system dynamics, especially the internal control mechanism.

In this work, we use *Scheffersomyces stipitis* as the model system to validate our hypothesis. *Scheffersomyces stipitis* is an important yeast species in the field of biorenewables due to its desired capacity for utilizing xylose[3], and it has been recognized that redox balance plays an important role in xylose fermentation[4]. However, there has not been any systems level understanding on how the shift in redox balance contributes to the overall metabolic shift in *S. stipitis* to cope with reduced oxygen uptake. To gain better understanding on how the cellular metabolism is regulated in response to reduced oxygen supply, we designed experiments to obtain its transcriptomic profiles during the dynamic transition from aerobic growth to oxygen-limited fermentation.

Compared to the analysis of RNAseq data among different steady-states, analysis of dynamic transcriptomic data that are time-dependent presents many new challenges. For example, while independent-sample is a basic assumption in many traditional analysis methods, the dynamic transient samples are not independent from each other. In this work, we will discuss how we address some of these challenges through adapting tools developed for dynamic systems in control engineering and present our findings. Specifically, compared to the initial (aerobic) and final (oxygen-limited) steady states, the transient states revealed significantly larger changes. Detailed examination of individual genes also confirmed this. Indeed, among all genes that showed differential expression during the transition, more than half of them only show differential expression during the transient states, and returned completely back to their initial steady state when the transition ended. In addition, when we used an improved genome-scale metabolic network model (GEM) developed by our group to integrate the transcriptomic data, the fluxes predicted by the GEM show 45% agreement with steady state transcription levels in terms of up or down regulations, however, they show 65% agreement when compared with initial transient expression levels. Finally, by integrating principal component analysis (PCA) with hierarchical clustering techniques, we were able to identify clusters of genes that showed similar dynamic responses to the perturbation, and to determine which metabolic pathways are affected the most by the change of the oxygen condition.

Our results confirm that the transient states indeed offer significantly important information that the steady-states cannot, which can be crucial in understanding the regulation of cellular metabolism.

(a)	(b)

Fig. 1 Dynamic transcriptomic profiles offer significantly more information than those of steady states. ***(a)*** 2D projection of global gene TPM data from all four trials. The circles and stars denote the aerobic and oxygen-limited steady-states, respectively, which are very close to each other; while the triangles denote the transition states, which show significantly larger range of changes. This is confirmed by a specific example of glutamate dehydrogenase (GDH) ***(b)*** where steady-states-only values (dashed lines) present a somewhat misleading picture (with mixed up-and down-regulations) compared to the full picture when the complete transition paths are revealed (solid lines).

[1] Ma, J., Zhang, B.-B., Wang, F., Sun, M.-M., Shen, W.-J., Lv, C., Gao, Z., Chen, G.-X. RNA-Seq Analysis of Differentially Expressed Genes in Rice under Photooxidation. *Russ J Plant Physl+*, (2017).

[2] Rong-Mullins, X., Ayers, M.C., Summers, M., Gallagher, J.E.G., Transcriptional Profiling of *Saccharomyces cerevisiae* Reveals the Impact of Variation of a Single Transcription Factor on Differential Gene Expression in 4NQO, Fermentable, and Nonfermentable Carbon Sources. *G3-Genes Genom Genet*, (2018).

[3] Jeffries, T.W., Grigoriev, I.V., Grimwood, J., Laplaza, J.M., Aerts, A., Salamov, A., Richardson, P.M., Genome sequence of the lignocellulose-bioconverting and xylose-fermenting yeast *Pichia stipitis*. *Nat Biotechnol*, (2007).

[4] Cho, J.-Y., Jeffries, T.W., Pichia Stipitis Genes for Alcohol Dehydrogenase with Fermentative and Respiratory Functions. *Appl Environ Microb*, (1998).

534887 The Impact of Microgrids on the Grid: Integration of Distributed Renewables Via Network-Constrained Affine Robust Unit Commitment

Abstract:

Distributed generation (DG) of electrical power is a promising alternative to the current paradigm of centralized generation followed by unidirectional distribution to the end user. It has the potential to reduce transmission congestion and losses while also providing resiliency to infrastructure disruptions. Additionally, local sources of renewable energy, such as wind and solar, can be incorporated to reduce emissions. Microgrids are an attractive approach to the technological realization of DG because they can use locally controllable resources such as energy storage systems (ESS) and fast-responding microturbines to mitigate the inherent intermittency and stochasticity of renewables. However, the current high cost of ESS technology means that standalone renewable DG is not yet economically viable. Thus, grid connection is required if widespread adoption of renewable DG is to be achieved and this results in increasing uncertainty and temporal variability of the residual load (demand minus local generation) which must be served by conventional generation.

In order to operate with grid-connected renewables in a way that is both reliable and economically optimal, the grid controlling entity (e.g. independent system operator) needs to determine optimal day-ahead unit commitments (UC) in the presence of uncertainty. A common approach is two- or multi-stage robust UC [1]. When complete recourse is allowed for wait-and-see decisions, network-constrained problems are computationally intensive without the use of approximations. However, accounting for these network constraints is important for capturing the coupled effects of renewable uncertainty and variability.

In view of this, we propose in this work a two-stage, network-constrained multi-objective unit commitment model based on affine robust optimization. The objectives are to minimize grid operating cost and emissions; the relative weighting of these is determined a-priori by a user-decided parameter. We consider uncertainty in residual load forecasts, that is, the difference between conventional load and local renewable generation. Restricting the recourse decisions to be affine with respect to this uncertainty allows for computationally efficient and heuristic-free solution of the model.

In addition to our UC model, we propose a grid configuration which corresponds to a truly distributed infrastructure. We include renewables-only microgrids at every bus in the transmission network, a departure from the common approach of considering larger capacity wind and/or solar farms at only one or a small number (compared to the total number of buses) of locations. The design of such a distributed system allows the controllable nature of microgrids to be fully exploited because local resources can be used to shape the residual load and dampen its uncertainty (see [2] for an example of a market structure which promotes this type of residual load regulation).

We solve the unit proposed commitment model to determine optimal operating cost and emissions for a grid represented by a version of the IEEE One Area Reliability Test System which has been modified to include (i) connection to an electricity market and (ii) microgrids at each bus which make up 15% to 25% of annual load through

solar generation. We find that incorporating distributed generation which tracks solar availability results in higher cost and emissions than a grid operating without renewables, except at low penetrations. We also investigate limiting the temporal variability and uncertainty of the residual load in a manner inspired by the energy exchange regulations proposed in [2]. This type of regulation provides increasingly superior economic and environmental performance compared to unregulated distributed generation as penetration of renewables increases, as well as lower operating costs and similar emissions to a present-day grid at all levels of renewable incorporation.

[1] M. Tahanan, W. V. Ackooij, A. Frangioni and F. Lacalandra, "Larges-scale unit commitment under uncertainty," *4OR*, vol. 13, no. 2, pp. 115-171, 2015.

[2] M. Zachar and P. Daoutidis, "Microgrid/macrogrid energy exchange: A novel market structure and stochastic scheduling," *IEEE Transactions on Smart Grid*, vol. 8, no. 1, pp. 178-189, 2017.

536588 Modeling and Optimization of Supercritical Pulverized Coal Power Plants Under Part Load Operation

Abstract:

Despite advances in commercial simulation packages for developing process models, the ability to use such models for optimizing complex plant flowsheets remains challenging. The Institute for the Design of Advanced Energy Systems (IDAES) is developing an open-source, advanced process system engineering framework [1] built on Pyomo [2] - a Python based algebraic modeling language. The primary objective is to leverage the capabilities of a high-level programming language to not only optimize and improve the efficiencies of existing power plants but also to accelerate the development of the advanced energy systems of the future.

The increase of renewable energy sources for power generation presents significant challenges for existing coal fired power plants that were primarily designed to provide constant, baseload power. In particular, these plants much often operate at part load due to the requirement of the evolving grid. Thus, it is important to consider how to best operate these plants under part load to ensure high efficiency and environmental compliance.

In this work, an equation-oriented plant wide model of a supercritical pulverized coal (SCPC) power plant is developed. The IDAES framework and its tools are leveraged to construct, initialize, and solve the resulting large-scale optimization problem. Due to the significant mass and energy integration in SCPC plants, we show that the entire plant must be considered when evaluating the impacts of several intricately-related operational and design decisions. For example, optimization considering only the boiler can lead to suboptimal conditions from the perspective of the entire plant.

Models for the different unit operations are developed by considering the tradeoff between accuracy and robustness to make them amenable for flowsheet optimization under varying operating scenarios. For example, a hybrid model that considers radiation effects in 3D and mass change in 1D for the fire side of the boiler is more preferable to a detailed CFD model for simulation studies [3]. However, the simplified first principles model still poses significant challenges for optimization studies. On the other hand, a reduced-order model can be utilized but the resulting inaccuracy risks leading the optimizer to invalid solutions. To this end, a trust region methodology [4] is applied to develop an adaptive surrogate model whose validity is checked at every iteration in the neighborhood of values of the decision variables. This model is coupled with a first-principles model of the water side inside the boiler tubes. This hybrid approach greatly reduces the computational expense with minor loss in the accuracy.

Models of other equipment items include the steam turbine that can handle choked flow conditions and steam condensation, feed water heaters and condenser with the ability to estimate the tube wall temperatures, and a selective catalytic reduction reactor. A smooth flash formulation is used that enables reliable vapor liquid equilibrium calculations over a broad range of operation that spans both subcritical and supercritical conditions [5]. It is coupled to an external function call that implements IAPWS steam

properties [6], returns exact 1st and 2nd derivatives, and is embedded directly into the optimization problems via the AMPL Solver Library. The plant-wide model is used to optimize the operating conditions under part load operation subject to critical operating and materials constraints so that the plant efficiency under these off-design conditions can be maximized without damaging the plant hardware.

REFERENCES:

[1] D. C. Miller *et al.*, "Next generation multi-scale process systems engineering framework," in *Proceedings of the 13th International Symposium on Process Systems Engineering, San Diego, California, USA*, 2018.

[2] W. E. Hart *et al.*, *Pyomo - Optimization Modeling in Python*, Second. Springer, 2017.

[3] J. Ma, J. P. Eason, A. W. Dowling, L. T. Biegler, and D. C. Miller, "Development of a first-principles hybrid boiler model for oxy-combustion power generation system," *Int. J. Greenh. Gas Control*, vol. 46, pp. 136–157, 2016.

[4] J. P. Eason and L. T. Biegler, "Reduced model trust region methods for embedding complex simulations in optimization problems," *Comput. Aided Chem. Eng.*, vol. 37, no. June, pp. 773–778, 2015.

[5] A. P. Burgard *et al.*, "A smooth, square flash formulation for equation-oriented flowsheet optimization," in *Proceedings of the 13th International Symposium on Process Systems Engineering, San Diego, California, USA*, 2018.

[6] IAPWS, "International Association for the Properties of Water and Steam, IAPWS R6-95(2016)," 2016.

537231 Cost-Effectiveness of Grid Energy Storage Technologies in Current and Future U.S. Power Systems

Abstract:

The integration of high and ultra-high shares of variable renewable energies (VREs)–wind and solar power– rises some technical challenges that need to be solved to maintain the reliability and cost-effectiveness of power systems. For example, VREs are weather-dependent and therefore their power generation is uncertain and exhibit variable diurnal and seasonal patterns. These properties cause more frequent and/or steeper net demand fluctuations that require the enhancement of power system flexibility [1,2]. Energy storage devices could play a pivotal role in the transition towards low-carbon and flexible power systems. For instance, these devices can provide a variety of services including energy arbitrage, transmission and distribution congestion relief and investment deferral, demand shifting and peak reduction, spinning and non-spinning reserves, and seasonal energy storage [3]. However, the assessment of the integration of VREs and energy storage technologies using model-based system analysis poses a challenge for the modeling of power system. For instance, the modeling of VREs and storage systems requires the accommodation of short-term dynamics – wind and solar generation patterns as well as evolution of storage levels– using a chronological hourly resolution in power planning models, which result in large computational needs [4,5]. This study focuses on the techno-economic assessment of grid-energy storage technologies from energy storage owner (ESO) and system operator (SO) perspectives using price-taker and production cost models, respectively. Different modeling approaches are evaluated in terms of both accuracy and computational cost. Finally, niche markets for grid energy storage in United States are identified and impacts on electricity prices and power system operations are quantified.

References

[1] IRENA. Planning for the renewable future: Long-term modelling and tools to expand variable renewable power in emerging economies. 2017.

[2] Kroposki B, Johnson B, Zhang Y, Gevorgian V, Denholm P, Hodge B, et al. Achieving a 100% Renewable Grid: Operating Electric Power Systems with Extremely High Levels of Variable Renewable Energy. IEEE Power Energy Mag 2017;15:61–73. doi:10.1109/MPE.2016.2637122.

[3] IEA. Technology Roadmap: Energy Storage. OECD Publishing; 2014. doi:10.1787/9789264211872-en.

[4] Wogrin S, Galbally D, Reneses J. Optimizing Storage Operations in Medium- and Long-Term Power System Models. IEEE Trans POWER Syst 2016;31:3129–38. doi:10.1109/TPWRS.2015.2471099.

[5] Xu B, Wang Y, Dvorkin Y, Fernandez-Blanco R, Silva-Monroy CA, Watson J-P, et al. Scalable Planning

for Energy Storage in Energy and Reserve Markets. IEEE Trans Power Syst 2017;32:4515–27. doi:10.1109/TPWRS.2017.2682790.

538031 New Operating Strategy for a Combined Cycle Gas Turbine Power Plant

Abstract:

Gas turbines are widely used in the power industry due to their high thermal efficiency, super flexibility, short delivery time, and excellent regulation capacity. The high-temperature exhaust from a gas turbine is normally utilized to produce extra power in a steam cycle that comprises a heat recovery steam generator (HRSG) and several steam turbines. This serial coupling of a gas turbine followed by a steam cycle is called a combined cycle gas turbine (CCGT) power plant.

CCGT plants run under part-load conditions in most of their lifetime. Several factors are responsible for this part-load operation. First, the power demand is hardly steady and rarely equals the design capacity. Second, a power plant is required to maintain spinning reserves (surplus capacity) to meet unforeseen peaks in demands. Third, a power plant may often be overdesigned to buffer against demand uncertainties. Hence, the study of CCGT plants under part-load conditions is of practical importance. Under part-load conditions, the plant operation drifts away from its design conditions and thus the plant thermal efficiency decreases. This wastes non-renewable fossil fuels and increases CO_2 emissions. Therefore, a proper operating strategy is needed for the part-load operation of a CCGT plant.

The operating strategies for gas turbine and power plants were studied in the literature. Kim and Hwang[1] investigated the part-load operation of recuperative gas turbines. They studied several operating strategies such as fuel flow control (FFC), variable speed control (VSC), and inlet guide vane control (IGVC) for the single-shaft configuration, and FFC and variable area nozzle control (VANC) for the two-shaft configuration. They found VSC and VANC gave better performance. Kim[2] analyzed the part-load performance of a combined cycle with different gas turbine design parameters under FFC and IGVC. They showed that the combined cycle exhibited superior performance under both operating strategies when the gas turbine pressure ratio and temperature were higher. Haglind[3,4] investigated the effects of variable geometry on the part-load performance of gas turbines and combined cycles, and found the use of IGVs and VANs improved their performance. Jimenez-Espadafor Auilar et al.[5] and Cheng et al.[6] studied several operating strategies for a combined heat and power (CHP) plant, and showed IGVC held the best regulation capacity and energy saving potential. Variny and Mieka[7] found that preheating the condensate and changing condensing pressure regulation strategy reduced the fuel consumption based on online monitoring data. Liu and Karimi[8] obtained an optimal operating strategy using a simulation-based optimization method to maximize the plant thermal efficiency at any part-load. Although these studies investigated different operating strategies, new strategy is needed to improve the plant part-load performance.

In this work, a new operating strategy called EGR-IGVC is proposed to improve the plant part-load performance. The new strategy manipulates exhaust gas recycle (EGR) ratio and inlet guide vane (IGV) sequentially to control gas turbine inlet and exhaust temperatures. It allows the CCGT plant to set its own acceptable gas turbine temperature limits at any time and thus maximize the plant performance. The plant performance under EGR-IGVC is evaluated and compared with that under the conventional inlet guide vane control (IGVC). The results show that EGR-IGVC improves the plant part-load performance. Specifically, the plant efficiency is increased by as much as 1.2% (actual) while the CO_2 emissions are reduced by as much as 13.8 kg MW^{-1} h^{-1}. Furthermore, EGR-IGVC enables the effective low-temperature heat utilization and alleviates the NO_x emissions from the power plant. Therefore, EGR-IGVC promises to a better strategy for the power plant part-load operation.

References

1. Kim T, Hwang S. Part load performance analysis of recuperated gas turbines considering engine configuration and operation strategy. *Energy*. 2006;31(2-3):260-277.
2. Kim TS. Comparative analysis on the part load performance of combined cycle plants considering design performance and power control strategy. *Energy*. 2004;29(1):71-85.
3. Haglind F. Variable geometry gas turbines for improving the part-load performance of marine combined cycles â €" Gas turbine performance. *Energy*. 2010;35(2):562-570.
4. Haglind F. Variable geometry gas turbines for improving the part-load performance of marine combined cycles â €" Combined cycle performance. *Applied Thermal Engineering*. 2011;31(4):467-476.
5. JimÃ©nez-Espadafor Aguilar F, Quintero RR, Trujillo EC, GarcÃa MT. Analysis of regulation methods of a combined heat and power plant based on gas turbines. *Energy*. 2014;72:574-589.
6. Yang C, Huang Z, Ma X. Comparative study on off-design characteristics of CHP based on GTCC under alternative operating strategy for gas turbine. *Energy*. 2018;145:823-838.
7. Variny M, Mierka O. Improvement of part load efficiency of a combined cycle power plant provisioning ancillary services. *Applied Energy*. 2009;86(6):888-894.
8. Liu Z, Karimi IA. Simulation and operation of a combined cycle gas turbine power plant for part-load opeartion. *Chemical Engineering and Research Design*. 2017;https://doi.org/10.1016/j.cherd.2017.12.009.

538684 Flexible Carbon Capture Exploiting Dynamic Changes in Electricity Price

Abstract:

While power plants produce electricity from fossil fuels, they are responsible for more than 30% of the global CO_2 emissions due to human activity. One way to prevent CO_2 emission and reduce its harmful contribution to global warming and climate change is to capture the CO_2 from the power plant flue gas using chemical separation. However, current capture processes are highly energy-intensive and need almost one-third of the electricity generated by the power plant itself. This is one of reasons why the carbon capture and storage (CCS) technology has not been commercially successful, albeit showing great potential for environmental benefits. In this work, we develop a framework for smarter design and operation of a flexible carbon capture process that changes its production capacity depending on the changes in electricity price. Given a forecast of electricity price over time, the operation of the capture process is scheduled to capture more CO_2 when the electricity price is low, and capture less CO_2 when the electricity demand and prices are high. This way, the power plant sustains a greater profit while reducing its CO_2 emission to the atmosphere. The overall optimization problem is formulated as a mixed integer linear program (MILP) to make decisions and schedules for the dynamically operated capture plant under changing electricity price and demand forecasts. The model is further extended to explore the possibility of a distributed capture system with multiple smaller capacities which provide more options for flexible operation under electricity price changes.

521753 Stability-Preserving Economic Optimization of Microgrids

Abstract:

Emerging decentralized energy resources are organized in the form of microgrids, which need to be coordinated with other microgrids and with the power grid to maintain stable power supply conditions [1]. The control of microgrids is challenging due their extremely fast dynamics (milliseconds), stochastic disturbances that cover a wide frequency spectrum (e.g., power loads), and due to component heterogeneity (e.g., each generator has a different frequency and power conversion system) [2].

The main control objective in a microgrid is to maintain voltages and frequencies at desired levels given steady-state set-points provided by a high-level energy management system. The energy management system is responsible for transacting and transporting power within the microgrid, with other microgrids, and with the power grid. This layer computes an economic-optimal steady-state by solving an AC power flow problem. The steady-state is usually updated every 5-10 min [3]. Due to the extremely fast dynamics of microgrids, the use of optimization-based control techniques such as model predictive control (MPC) for voltage and frequency stabilization is technically challenging. As a result, stabilization of microgrids usually relies on high-frequency low-level controllers such as droop control [2,4,5]. Droop controllers are highly robust controllers that can reliably stabilize the system in the face of high-frequency disturbances. This is because droop-controlled (closed-loop) microgrid dynamics have the key property that they form a Hamiltonian (dissipative) system around a stable equilibrium (steady-state) point [5]. Unfortunately, computing stable steady-states using AC power flow formulations is not straightforward [6]. In particular, the AC power flow layer might provide an economic-optimal point that is not a strict minimum of the Hamiltonian function (and therefore it is not a stable steady-state).

In this work, we propose modified AC power flow formulations to compute economic-optimal and stable equilibrium points for droop-controlled microgrids. We first consider a bilevel formulation in which we optimize an economic objective subject to the minimization of the Hamiltonian function. This formulation has the key advantage that the dynamics of the system do not need to be taken into consideration (it is entirely a steady-state problem). On the other hand, nonconvex bilevel problems are difficult to solve in general, because the inner problem needs to satisfy second order conditions. Motivated by this, we propose a stochastic optimal control formulation that seeks to stabilize the microgrid under a range of random perturbations of the economic-optimal equilibrium point [7,8]. This approach requires capturing the dynamics of the system and solving high-dimensional nonlinear programs (NLPs) but the NLPs are structured and can be handled in a scalable manner with existing tools [9,10]. We demonstrate the developments using a (343-bus) microgrid system, designed by Hart, Lasseter, and Jahns [11]. The proposed framework can be applied to various applications where finding economically-optimal and stable points for controllers is necessary.

References:

[1] Lasseter, R.H. Microgrids, in Power Engineering Society Winter Meeting, 2002. IEEE, vol. 1. IEEE, pp. 305–308, 2002.

[2] Dörfler, F., Simpson-Porco J.W., and Bullo, F. Breaking the hierarchy: Distributed control and economic optimality in microgrids. IEEE Transactions on Control of Network Systems 3, no. 3, pp. 241-253, 2016.

[3] Kundur, P., Balu, N.J., and Lauby, M.G. Power system stability and control. McGraw-hill New York, vol. 7, 1994.

[4] Simpson-Porco, J. W., Dörfler, F., and Bullo, F. Synchronization and power sharing for droop-controlled inverters in islanded microgrids, Automatica, vol. 49, no. 9, pp. 2603–2611, 2013.

[5] Schi er, J., Ortega, R., Astolfi, A., Raisch, J., and Sezi, T. Conditions for stability of droop-controlled inverter-based microgrids, Automatica, vol. 50, no. 10, pp. 2457–2469, 2014.

[6] Frank, S., Steponavice, I., and Rebennack, S. Optimal power flow: a bibliographic survey, Energy Systems, vol. 3, no. 3, pp. 221–258, 2012.

[7] Biegler, L.T. Nonlinear programming: concepts, algorithms, and applications to chemical processes. Vol. 10. Siam, 2010.

[8] Nemirovski, A. and Shapiro, A. Scenario approximations of chance constraints. In Probabilistic and randomized methods for design under uncertainty, pages 3–47. Springer, 2006.

[9] Chiang, N., Petra, C.G., Zavala, V.M. Structured Nonconvex Optimization of Large-Scale Energy Systems Using PIPS-NLP. 18th IEEE Power Systems Computations Conference, 2014.

[10] Shin, S., Hart, P., Jahns, T., and Zavala, V.M. A Hierarchical Optimization Architecture for Large-Scale Power Networks, Under Review, 2017.

[11] Hart, P.J., Lasseter, R., and Jahns, T. Reduced-order harmonic modeling and analysis of droop-controlled distributed generation networks, in Power Electronics for Distributed Generation Systems (PEDG), 2016 IEEE 7th International Symposium on. IEEE, pp. 1–9, 2016.

525801 Optimal Design of Aging Systems: A-Frame Coolers Design Under Fouling

Abstract:

Everyday operation causes fatigue to materials and dirt deposits build on the surface of equipment. As a result, these units suffer losses in performance that affect the overall yield of the unit and of the facility. Beyond the lost in performance, failures may occur if the units operate for longer periods. Eventually, some of the units or sections must be replaced or repaired. It is estimated that \$20,000 -\$30,000 of losses per hour can occur in refineries due to failures due to lack of proper maintenance (Tan and Kramer, 1997). Industry schedules maintenance stops to evaluate, mitigate, clean and repair sections and entire units. There is a large number of methods in the literature to evaluate maintenance strategies (Dekker, 1996). However, the most common optimization is minimum cost subject to average uptime or downtime (Wang, 2002) and mitigation strategies are developed based on the performance decay or aging process (Beaurepaire et al., 2012). To the best of our knowledge, simultaneous design and operation of the units considering performance decay is not addressed, although for instance the design of structures considers aging (Dekker, 1996).

Concentrated solar power (CSP) plants are typically located in regions with high solar incidence but limited water availability (Martín and Martín, 2013). The large cooling needs, as any other thermal facility, represent the strongest link between water consumption and power production (DOE, 2018). A frames are a special type of air coolers that allows decoupling the water –energy link in power plants. However, these systems consume up to 10% of the power generated in the facility to power the fans. The design of such units has typically been carried out based on rules (Krogen, 2004). Furthermore, A-frame units are located in open air and fouling, in particular, particle fouling is an important issue for their operation and performance. Fouling does not only affect the heat transfer coefficient, but deposit also block the cross sectional area, generating an additional pressure drop across the system (Pu et al., 2009; Sarfraz and Bach, 2016).

In this work we present a general methodology for the simultaneous optimal design and operation of units or processes whose yield is affected by aging or performance decays. We propose a parametric programming design procedure to determine the design and the cleaning/maintenance schedule. We use it for the design of A-frames under fouling conditions.

A detail equation based model for the units developed including mass and energy balances, design equations for the geometry and pipes layout, , the fan power curves, heat transfer coefficients, etc. and the aging process characterized. A two-stage procedure is proposed. First, the unit is optimally designed for the worst case scenario, just before maintenance. In a second stage, a multiperiod problem is solved for the optimal the operation of the unit over time including cleaning costs. The methodology is applied to A-frame dry cooling systems under fouling conditions, where fouling affects the pressure drop and the global heat transfer coefficient. Due to the sigmoidal deposition profile and assuming cleaning costs, the optimal cycle time is 8 yr. This design allows reducing the energy required to around 4% of the energy produced by the concentrated solar power plant. It is a promising result that can be affected by plant layout and ground availability.

References

Dekker, R. 1996 Application of maintenance optimization models: a review and analysis. Rel. Eng. Sys. Saf. 51 229-240
Beaurepaire, P., Valdebenito, M.A., Schueller, G.., Jensen, H.A. 2012. Reliability –based optimization of maintenance scheduling of mechanical components under fatigue. Compt. Methods. Appl. Mech Eng. 1(221-222), 24-40

Kröger, D.G., 2004. Air-cooled heat exchangers and cooling towers: thermal-flow performance evaluation and design. Volume II. Pennwell: Oklahoma, Tulsa.

Tan J.S.,Kramer M.A., 1997. A General Framework For Preventive Maintenance Optimization In Chemical Process Operations. Computers and Chemical Engineering, 21(12), 1451-1469.

Pu, H., Ding, G.-l., Ma, X.-K., Hu, H.-T., & Gao, Y.F., 2009. Effects of biofouling on air-side heat transfer and pressure drop for finned tube heat exchangers. International Journal of Refrigeration, 32(5), 1032-1040.

Sarfraz, O., Bach, C., 2016. A Literature Review On Heat Exchanger Air Side Fouling In Heating, Ventilation And Air conditioning (HVAC) Applications International Refrigeration and Air Conditioning Conference. Paper 1663.

529622 A Data-Driven Optimization Framework for Selection and Operation of Energy Storage Systems

Abstract:

Energy storage systems (ESS) have been widely used to provide benefit to the centralized power system through time shifting to improve the reliability and quality of power supply in the past (IEC, 2011). Along with peopleâ€™s rising awareness of fossil fuel depletion, climate change and pollution, there is a booming trend towards more research, development and utilization of renewable energy as well as the distributed energy systems based on renewable sources. However, due to their intermittent and fluctuating nature, the availability and amount of electricity generated from these sources are difficult to be matched with the demand of the electricity consumers. Energy storage systems therefore become essential to achieve demand-supply matching, reliability and quality control for the more complicated modern energy systems.

This work will answer two fundamental questions related to ESS: 1) how to choose a suitable energy storage technology for a specific application facing the large variety of technical features that could be considered as well as the diverse and abundant technologies that can be chosen; 2) how to effectively operate energy storage systems to maximize the payoff utility from decentralized demand side management (DSM), meanwhile benefiting the electrical grid.

First, a novel data-driven modeling and decision-making framework for the selection of energy storage systems is proposed to facilitate the decision-making process (Li *et al.*, 2018). The major criteria evaluated include the technical suitability, the economic feasibility, and the environmental impact. A machine learning method, framed as a supervised classification problem, is used to predict the technical suitability of the technology for the application based on the rated power and discharge duration features of the technologies. The up-to-date DOE Global Energy Storage Database is used as the training dataset for the modelling process of this data-driven approach. Furthermore, the economic feasibility of ESS is evaluated using the levelized cost of electricity (LCOE) and cost benefit analysis (CBA), while the environmental impact through life cycle assessment (LCA) method. With the established framework, a multi-objective optimization is employed to maximize technical suitability, maximize profits, and minimize environmental impact by selecting the optimal energy storage technology for a specific application.

Second, a game theoretic approach for DSM model incorporating energy storage components is developed. The proposed model is able to not only reduce the Peak-to-Average ratio to benefit the electrical grid, but also smoothen the dips in load profile caused by supply constraints. In a case study a variety of residential demand types are evaluated within the micro-grid to maximize the payoff of both individual users and the whole system. Moreover, the emerging blockchain technologies are introduced to guarantee the seamless and secure implementation of the proposed scheme, illustrating how a decentralized DSM approach could be implemented in practice.

Through the case study, a variety of energy storage technologies and applications including both traditional energy systems and emerging distributed energy systems are examined using the proposed data-driven decision-making framework. It shows a good prediction and can get easily updated with new input data. With such satisfactory ESS, the demonstrative micro-grid is able to maintain the supply and demand balance, reduce utility bills of consumers and reduce stress on grid in daily operation.

References

IEC (2011) *Electrical Energy Storage - White Paper*, *International Electrotechnical Commission*. doi: 10.1002/bse.3280020501.

Li, L. *et al.* (2018) â€˜A Multi-Objective Optimization Approach for Selection of Energy Storage Systemsâ€™, *Computers & Chemical Engineering*. doi: 10.1016/j.compchemeng.2018.04.014.

526408 Safe Economic Model Predictive Control of Nonlinear Systems

Abstract:

Economic model predictive control (EMPC) is a feedback control method which allows addressing process control tasks integrated with dynamic economic optimization of the process [1]. EMPC may continuously operate a chemical process in a time-varying fashion (off steady-state) to dynamically optimize process economic performance beyond what can be achieved via steady-state operation, and it also incorporates constraints to guarantee closed-loop stability within a well-characterized region of the operating state-space. On the other hand, process operational safety is of significant importance in the chemical process industries because unsafe process operation often results in very negative consequences [2]. Recently, a feedback control approach attempting to address stabilization and operational safety via Control Lyapunov-Barrier Functions (CLBFs) [3] has been investigated in the sense that the closed-loop state converges to the steady-state from a well-characterized set of initial conditions while avoiding an unsafe region the in state-space [4]. However, economic optimality is not addressed within this approach.

Motivated by the above, this work focuses on the design of a new class of economic model predictive control (EMPC) systems for nonlinear systems that address simultaneously the tasks of economic optimality, safety and closed-loop stability. Specifically, the EMPC is developed by incorporating an economics-based cost function and Control Lyapunov-Barrier Function (CLBF)-based constraints that ensure that the closed-loop state does not enter unsafe sets and remains within a well-characterized set in the system state-space while optimizing process economics. The new class of CLBF-EMPC systems is demonstrated using a nonlinear chemical process example.

[1] Ellis M, Durand H, Christofides P D. A tutorial review of economic model predictive control methods. Journal of Process Control. 2014, 24:1156-1178.

[2] Leveson NG, Stephanopoulos G. A system-theoretic, control-inspired view and approach to process safety. AIChE Journal. 2014, 60:2-14.

[3] Romdlony M Z, Jayawardhana B. Stabilization with guaranteed safety using control Lyapunovâ€"barrier function. Automatica, 2016, 66: 39-47.

[4] Wu Z, Albalawi F, Zhang Z, Zhang J, Durand H, and Christofides P D. Control Lyapunov-Barrier Function-Based Model Predictive Control of Nonlinear Systems. Proceedings of the American Control Conference, 2018, in press, Milwaukee, Wisconsin.

526778 Improving Flexibility and Energy Efficiency of a Post-Combustion CO_2 Capture Process Using Economic Model Predictive Control

Abstract:

Rising concerns for climate change has resulted in strict environmental regulations on emission of CO_2 into the atmosphere by emission sources such as fossil-fueled power plants. To reduce CO_2 emissions from power plants, electricity companies have diversified their generation sources. However, fossil fuels will remain an integral energy generation source as they are more reliable compared to the renewable energy sources. Therefore, to reduce emissions, carbon capture and storage (CCS) must be implemented.

Among the various options for CO_2 capture, Post-combustion CO_2 Capture (PCC) using reactive solvents is considered the most matured and viable option as it can easily be retrofitted to existing power plants. However, a major setback to the realization of this technology is its high regeneration heat requirement. Studies have shown that PCC attached to a power plant reduces the power plant's efficiency from about 40 % to 30 %. An overview of a PCC plant attached to a power generation plant is shown in the figure below.

As can be seen, steam from the power is sent to the PCC plant for CO_2 regeneration. This diversification as well as changing electricity demand could hinder effective economic operation of a PCC plant attached to the power plant to reduce CO_2 emissions. Again, unavailability of steam to the PCC plant during the peak periods could hinder process operation. To tackle this problem, efficient control algorithms are necessary. Many different control strategies have been applied to the post-combustion capture process in literature. The proportional-integral (PI) algorithm as well as tracking Model Predictive Control (MPC) has been applied to the process. However, economic MPC has not been applied to the PCC process, hence, the need for this to be investigated.

In this work, both tracking and economic model predictive controllers (MPC) were applied to a PCC plant and their economic performances compared under different scenarios namely: no emission limit, emission limit and cap-and-trade system. These scenarios were formulated in the optimization problem as soft constraints on the CO_2 emission. First, the tracking MPC was well designed and open-loop tests conducted to ensure the economic performance is as close as possible to that of economic MPC. Second, an appropriate setpoint update strategy from real-time optimization (RTO) was determined. Finally, the economic performances of the two controllers were compared for time-varying process operation. The results show that economic MPC was able to determine the optimal control policy depending on the scenario without further tuning while tracking MPC needed retuning in some instances. Again, the operation of the PCC plant under EMPC resulted in reduced heat duty. Therefore, economic MPC can potentially improve the operation of the post-combustion CO_2 capture process.

532246 Towards on-Line Development of Physically-Based Models for Model-Based Control Design

Abstract:

Many industrial model predictive controllers utilize linear empirical process models for making state predictions [1]. These have provided adequate state predictions around process steady-states; however, next-generation control designs like economic model predictive control (EMPC) [2] may not enforce steady-state operation, and therefore utilizing linear empirical models which may not be applicable in large regions of state-space may restrict the power of EMPC for optimizing profits. Furthermore, it may not be obvious how to design an EMPC without a more physically-meaningful process model, as a general economics-based objective function is required to be available and also important constraints (e.g., those related to safety when the process state is allowed to access broad regions of state-space) may be based on the process physics. Techniques for getting models from data (e.g., [3]) generally require some up-front guess of the types of terms which will appear in the model so that their coefficients can be selected and identified. Therefore, a critical step toward obtaining physically-based models for enabling practical development of next-generation model-based controllers is developing techniques for selecting the form of equations which describe the process physics so that the parameters can be fit in such equations utilizing available techniques.

Inspired by traditional experimentation techniques, which develop methods for obtaining very targeted data in an experimental setting, we seek to exploit the flexibility of the structure of EMPC and the closed-loop stability properties of EMPC designs with certain stability constraints (e.g., Lyapunov-based stability constraints [4]) to develop controllers with hard and soft constraints related to data which it is desired to gather from an on-line process for short periods of time. We explore formulations in which the controller objective function is augmented with terms that can be used in some sampling periods and not in others and which penalize the deviations of the closed-loop state from desired values to seek to obtain data with those desired values for short periods of time. We also explore formulations that enforce the data-gathering functionality through hard constraints on the state with and without feasibility guarantees and discuss some of the benefits and difficulties associated with utilizing such data-gathering EMPC's in a distributed architecture. An implementation strategy is developed that allows an initial EMPC design with a quadratic objective function and linear empirical model to be used before a more physics-based process model is developed from the data obtained during operation under this initial EMPC augmented with the data-gathering functionality. A benchmark continuous stirred tank reactor chemical process example demonstrates the developments.

[1] J. Qin and T. Badgwell, "A survey of industrial model predictive control technology," *Control Engineering Practice*, 11, 733-764, 2003.

[2] M. Ellis, H. Durand and P. D. Christofides, "A tutorial review of economic model predictive control methods," *Journal of Process Control*, 24, 1156-1178, 2014.

[3] S. L. Brunton, J. L. Proctor and J. N. Kutz, "Discovering governing equations from data by sparse identification of nonlinear dynamical systems," *Proceedings of the National Academy of Sciences*, 113, 3932-3937, 2016.

[4] M. Heidarinejad, J. Liu and P. D. Christofides, "Economic model predictive control of nonlinear process systems using Lyapunov techniques," *AIChE Journal*, 58, 855-870, 2012.

534054 Data Driven Economic Model Predictive Control for Unstable Systems

Abstract:

Control systems designed to manage chemical process operations often faces numerous challenges such as inherent nonlinearity, process constraints and uncertainty. Model predictive control (MPC) is a well-established control method that can handle these challenges [1, 2]. In MPC, the control action is computed by solving an open-loop optimal control problem at each sampling instance over a time horizon, subject to the model that captures the dynamic response of the plant, and constraints.

With increasing recognition (and ability) of MPC designs to focus on economic objectives, the notion of Economic MPC (EMPC) was developed for linear and nonlinear systems [3, 4], and several important issues (such as input rate-of-change constraint and uncertainty) addressed.

Most of the existing MPC formulations, economic or otherwise, have been illustrated using first principles models. With growing availability of data, there exists the possibility of enhancing MPC implementation for situations where a first principles model may not be available, and simple `step-test', transfer-function based model identification approaches may not suffice. One of the widely utilized approaches in the general direction of model identification are, subspace-based system identification methods which have been adapted for the purpose of model identification, where state-space model from measured data are identified using projection methods [5]. To handle the resultant plant model mismatch with data-driven model based approaches, monitoring of the model validity becomes especially important.

In a direction, the focus more explicitly on the model behavior, in a recent result [6] an adaptive data-driven MPC was proposed to evaluate model prediction performance and trigger model identification in case of poor model prediction. Data driven MPC or EMPC approaches, that utilize appropriate modeling techniques to identify data from closed-loop tests to handle operation around nominally unstable equilibrium points remain unaddressed [4].

Motivated by the above considerations, in this work we address the problem of data driven model based predictive control at an unstable equilibrium point. In order to identify a model around an unstable equilibrium point, the system is perturbed under closed-loop operation. Having identified a model, a Lyapunov-based MPC is designed to achieve local and practical stability. The Lyapunov-based design is then used as the basis for a data driven Lyapunov-based EMPC design to achieve economical goals while ensuring boundedness.

Key Words: Model Predictive Control, Subspace Identification Methods, Lyapunov-Based MPC, Unstable Equilibrium Point

[1] Mayne, D. Q., Rawlings, J. B., Rao, C. V., & Scokaert, P. O. (2000). Constrained model predictive control: Stability and optimality. *Automatica*, *36*(6), 789-814.

[2] Garcia, C. E., Prett, D. M., & Morari, M. (1989). Model predictive control: theory and practiceâ€"a survey. *Automatica*, *25*(3), 335-348.

[3] Alanqar, A., Ellis, M., & Christofides, P. D. (2015). Economic model predictive control of nonlinear process systems using empirical models. *AIChE Journal*, *61*(3), 816-830.

[4] Kheradmandi, M., & Mhaskar, P. (2018). Data Driven Economic Model Predictive Control. *Mathematics*, *6*(4), 51.

[5] Qin, S. J. (2006). An overview of subspace identification. *Computers & chemical engineering*, *30*(10-12), 1502-1513.

[6] Kheradmandi, M., & Mhaskar, P. (2018). Model predictive control with closed-loop re-identification. *Computers & Chemical Engineering*, *109*, 249-260.

537170 Incorporation of Sustainability and Economic Considerations in Process Control of Hydraulic Fracturing in Unconventional Reservoirs

Abstract:

Typically, the term shale oil refers to natural oil trapped in rock of low porosity and ultra-low permeability. What has made the recovery of shale oil and gas economically viable is the extensive use of hydraulic fracturing. Research on the relationship between the distribution of propping agent, called proppant, and well performance indicates that uniformity of proppant bank height and suspended proppant concentration across the fracture at the end of pumping determines the productivity of produced wells. Therefore, in our previous work [2], we calculated the optimal flow rate of fracturing fluids, the optimal number of horizontal wells and fractures per well, the length of the fracture, and the drainage area aspect ratio that maximize the overall productivity of the well-fracture system. It is important to note that the obtained pumping schedule did not consider the environmental and economic impacts of the post-fracking process such as treatment and reuse of flowback and produced water from fractured wells; therefore, from the sustainability point of view, utilizing energy and water resources just enough to stimulate the formation was not of our interest.

Motivated by this consideration, in this work, we propose a new control framework that has integrated sustainability considerations into the post-fracturing process. In this regard, a dynamic model is developed to describe the flow rate and the concentration of total dissolved solids (TDS) in flowback water from fractured wells, provided that the amount of water injected into the fracture well is known. Then, a thermal membrane distillation (TMD) unit is considered for the removal of TDS. A multiobjective optimization problem is formulated to determine the amount of water to be injected by considering the entire superstructure that consists of hydraulic fracturing, storage, transportation, and water treatment, maximizing annualized profit from recovered water per period while minimizing the water footprint of the process through Pareto optimal solution. The obtained water amount will be included in a model-based pumping schedule design technique to compute the optimal pumping schedule during a hydraulic fracturing process that will regulate the distribution of proppant bank height and suspended proppant concentration [2]; this flow may not be sufficient to achieve the desired fracture length, which will lower the productivity of the produced well, however, it will maximize the annualized profit of the entire superstructure as well as reduce environmental impact.

References:

[1] Siddhamshetty, P., Kwon, J.S., Liu, S., Valkó, P. (2017), Feedback control of proppant bank heights during hydraulic fracturing for enhanced productivity in shale formations. AIChE J., 64, 138-147.

[2]Elsayed, N. A., Barrufet, M. A., Eljack, F. T., El-Halwagi, M. M. (2015). Optimal design of thermal membrane distillation systems for the treatment of shale gas flowback water. Int. J. Membr. Sci. Technol., 2, 1-9.

529901 Multi-Objective Optimization of the Energy System in an Iron and Steel Plant Considering the Economic Cost and Life Cycle Environmental Impact

Abstract:

The iron and steel industry is one of the largest energy consumers and emission sources releasing large amount of CO_2 and some air pollutants such SO_2 and NO_x [1]. As reported, the energy consumption from the iron and steel industry could account for 18% of total industry final energy consumption, whilst CO_2 emissions could account for about 7% of the total CO_2 emissions in the world [2]. Such large energy consumption and CO_2 emissions mainly comes from the inefficient operation of its energy system involving the distribution of byproduct gases, steam and electricity among production units, boilers, turbines, combined heat and power (CHP) units, and waste heat and energy recovery units.

Many optimization models have been successfully developed for production processes in the iron and steel industry without consideration of its energy system, such as sintering process [3], blast furnace iron-making process [4], and steelmaking and continuous casting process [5]. Early work with respect to the energy system optimization in an iron and steel plant mainly focused on optimization of the byproduct gases distribution in gasholders and boilers [6-7] without simultaneous optimization of byproduct gases, steam and electricity distribution. Recently, Zeng et al. [8] developed a multi-period mixed-integer linear programming model for simultaneous optimization of byproduct gases, steam and electricity distribution, incorporating many significant operational features such as fuel selection, gasholder level control, ramp rate variation, mixing of different byproduct gases, piecewise constant generation rates of byproduct gases, and piecewise constant demand profiles of byproduct gases, steam and electricity. However, they mainly achieved the minimization of the economic cost without investigating the environmental impact of the operational performance. The inclusion of environmental impact makes the optimization problem more challenging, since the environmental and economic objectives are usually contradictive.

In this presentation, we use our previous work [8] as basis to develop a multi-period multi-objective planning model for optimization of the integrated energy system in an iron and steel plant. The objective is to minimize the economic cost and environmental impact simultaneously. The environmental performance is represented by the environmental impact (EI) potential rather than by emission release rates. The EI potential considers several environmental impacts caused by both the generation of the imported electricity, and the direct emissions from the operating process of the energy system, which are evaluated using life cycle assessment (LCA) method. To solve the proposed multi-objective optimization model, the augmented Îµ-constraint method is adopted to generate the Pareto frontier. A fuzzy preference selection method is used to determine the best compromising solution in which a fuzzy set is used to define the membership function of objectives and the non-dominated solution with the maximum normalized membership function value is generated. A real industrial example is investigated to demonstrate the capability of the proposed model and solution method. The computational results show that the total operating cost could be reduced by 6% and the environment impact could be reduced by 30% using our proposed model and solution approach in comparison with those from the actual operation.

Keywords: multi-objective optimization, environmental impact, Life cycle assessment, integrated energy system, iron and steel

References

[1] Jin P, Jiang Z, Bao C, Hao S, Zhang X. The energy consumption and carbon emission of the integrated steel mill with oxygen blast furnace. Resources Conservation & Recycling 2017; 117: 58-65

[2] International Energy Agency (IEA). Energy balance flows, http://www.iea.org/Sankey/index.html; 2017 [accessed 12 January 2017].

[3] Liu C, Xie Z, Sun F, Chen L. Optimization for sintering proportioning based on energy value. Applied Thermal Engineering 2016; 103: 1087-1094.

[4] Liu X, Chen L, Feng H, Qin X, Sun F. Constructal design of a blast furnace iron-making process based on multi-objective optimization. Energy 2016; 109: 137-151.

[5] Li J, Xiao X, Tang Q, Floudas C. Production Scheduling of a Large-Scale Steelmaking Continuous Casting Process via Unit-Specific Event-Based Continuous-Time Models: Short-Term and Medium-Term Scheduling. Industrial & Engineering Chemistry Research 2012; 51(21): 7300-7319.

[6] de Oliveira Junior VB, Pena JGC, Salles JLF. An improved plant-wide multiperiod optimization model of a byproduct gas supply system in the iron and steel-making process. Applied Energy 2016; 164: 462-474.

[7] Zhao X, Bai H, Shi Q, Lu X, Zhang Z. Optimal scheduling of a byproduct gas system in a steel plant considering time-of-use electricity pricing. Applied Energy 2017; 195: 100-113.

[8] Zeng Y, Xiao X, Li J, Sun L, Floudas CA. A Novel Multi-Period Mixed-Integer Linear Optimization Model for Optimal Distribution of Byproduct Gases, Steam and Power in an Iron and Steel Plant. Energy 2018; 143: 881-899.

532968 Model Predictive Control with Active Learning Under Model Uncertainty

Abstract:

Optimal control relies on a model, which is generally uncertain as a result of incomplete knowledge of the system as well as possible changes in the dynamics over time. The most widely used approach to optimal control of multivariable systems with state and input constraints is model predictive control, or MPC [1,2]. Typical sources of uncertainty in a prediction model used in MPC include inaccurate estimates of model parameters and unknown aspect of the model structure itself, as in the case of an unknown kinetic mechanism in a physics-based model, or the appropriate order of a data-driven model. The uncertainty in the model structure may be further increased if the system also undergoes abrupt changes (such as faults in actuators and sensors [3]).

Feedback, through its corrective nature, ensures a certain degree of robustness to uncertainty in MPC, but control performance will degrade when feedback alone is inadequate in compensating for incomplete knowledge of the system. As a consequence, problems such as large offsets in setpoint tracking, excessive constraint violations, and even instability may arise. This has led to the development of robust and stochastic MPC, or RMPC and SMPC [4]. While these approaches both systematically account for model uncertainty, their control performance is highly dependent on the accuracy and precision of the uncertainty descriptions, which is generally not updated in real time. Adaptive control involves continuously improving the model through adjusting it under closed-loop control [5]. However, the data generated in closed loop must be sufficiently informative for the model adaptation to be beneficial and to avoid problems such as bursting and the loss of controllability. In optimal control of systems with reducible model uncertainty the inputs must have a *probing* effect that generates informative closed-loop data for model adaptation through *active learning*, in addition to their *directing* effect to control the system state [6].

We here discuss the problem of MPC with active learning for systems with probabilistic uncertainty descriptions [7]. Through illustrative case studies [8], we demonstrate the potential ability of active learning to maintain MPC performance in the presence of model uncertainty. The first case study involves a continuously-stirred tank reactor with uncertain reaction parameters in the model. A change in the process, represented by a drop in a reaction constant, leads to significant performance loss with three standard approaches to MPC. Adding active learning to the controller mitigates most of the performance loss and enables quick recovery of the productivity. The second case study considers a continuous bioreactor, in which a change in the growth conditions can be captured by a structural change in the model. Compared with nominal and passively adaptive MPC, the addition of active learning here enables quick reduction of model-structure uncertainty through high-confidence selection of the growth model corresponding to the new conditions. These simulation results show that active learning can be highly beneficial when a system undergoes abrupt changes, including the sudden occurrence of a fault or malfunction, that can compromise operational safety, reliability, and profitability.

References

[1] D.Q. Mayne, J.B. Rawlings, C.V. Rao, and P.O.M. Scokaert, "Constrained model predictive control: Stability and optimality," Automatica, vol. 36, no. 6, pp. 789–814, 2000.

[2] S.J. Qin and T.A. Badgwell, "A survey of industrial model predictive control technology," Control Engineering Practice, vol. 11, no. 7, pp. 733–764, 2003.

[3] M. Blanke, M. Kinnaert, J. Lunze, and M. Staroswiecki, Diagnosis and Fault-Tolerant Control. Berlin, Germany: Springer, 2nd ed., 2006.

[4] B. Kouvaritakis and M. Cannon, Model Predictive Control: Classical, Robust and Stochastic. London: Springer, 2016.

[5] K.J. Åström and B. Wittenmark, Adaptive Control. Reading: Addison-Wesley, 2nd ed., 1995.

[6] A.A. Feldbaum, "Dual-control theory. I," Automation and Remote Control, vol. 21, no. 9, pp. 874–880, 1961.

[7] A. Mesbah, "Stochastic model predictive control with active uncertainty learning: A survey on dual control," Annual Reviews in Control, 2017.

[8] T.A.N. Heirung, J.A. Paulson, S.J. Lee, and A. Mesbah, "Model predictive control with active learning under model uncertainty: why, when, and how," AIChE Journal, 2018.

534995 A Modified SQP Method for MPC of a Supercritical Pulverized Coal-Fired Power Plant during Cycling

Abstract:

In recent years, with rising interest in different sources of energy and focus on greenhouse gas emissions, renewable energy sources have been increasingly penetrated into the grid. As fossil fueled power plants will still play an important role in future power generation [1], optimal operations of such power plants during cycling will become more important. In particular, power plant cycling includes ramping the power generation rate up/down for load-following, as well as shut down or start up, to accommodate the renewable penetrations. To achieve the optimal profiles for cycling the power plant, the development and implementation of model predictive control (MPC) algorithms is critical for the improved plant operation under the challenging scenarios that arise during cycling.

In this work, the design and implementation of linear and nonlinear MPC strategies are considered for power plant cycling scenarios. The proposed MPC strategy is based on a modified sequential quadratic programming (SQP) method, which is designed for better acceptance of long steps and faster convergence. The modified SQP algorithm is built upon a backtracking line search framework and employs a group of relaxed step acceptance conditions. In the proposed algorithm, a step is first tested based on adapted Armijo conditions. If the first condition does not hold, then the second condition is tested, which may still have the step contribution for a sufficient decrease in the objective with a controlled infeasibility. If the second condition fails, then a defined third condition can enforce the step to have global convergence. The proposed MPC strategy is tested and compared to the following other linear and nonlinear MPC controllers developed: (i) a linear MPC control strategy based on dynamic matrix control (DMC); (ii) a nonlinear MPC based on a classical SQP method from the literature as benchmark [2]; and (iii) a nonlinear MPC based on the direct transcription method [3, 4, 5], which employs IPOPT [6] as nonlinear programming solver and ADOL-C [7] for improving derivative calculations.

The application system addressed corresponds to a supercritical pulverized coal-fired power plant with carbon capture. A number of plant-wide control scenarios will be discussed including: (i) power plant load-following, which simulates renewable penetrations into the grid; (ii) disturbance rejection in the coal-feed quality while maintaining the carbon capture rate; and (iii) trajectory tracking for power plant cycling over a relatively long period of operation, exploring the use of industrial power plant data. The closed-loop responses associated with different scenarios and different MPC algorithms will be analyzed. The MPC based on the modified SQP method will be compared to other MPC algorithms in terms of tracking errors and computational time performance.

References

1. Annual Energy Outlook 2015 with projection to 2040. Contract No.: DOE/EIA-0383, 2015.
2. F. Martinsen, L. T. Biegler and B. A. Foss. Application of optimization algorithms to nonlinear MPC. IFAC Proceedings 2002, 35:369-374.
3. F. V. Lima, R. Amrit, M. Tsapatsis, and P. Daoutidis. Nonlinear model predictive control of IGCC plants with membrane reactors for carbon capture. In Proceedings of the American Control Conference. Washington, DC, June 17-19, 2013.
4. F. V. Lima, X. He, R. Amrit and P. Daoutidis. Advanced control strategies for IGCC plants with membrane reactors for CO_2 capture. Process systems and materials for CO_2 capture: modeling, design, control and integration, A.I. Papadopoulos and P. Seferlis (eds.), Wiley, 2017.
5. X. He, Y. Wang, D. Bhattacharyya, F.V. Lima and R. Turton. Dynamic modeling and advanced control of post-combustion CO_2 capture plants. Chemical Engineering Research and Design, 2018, 131:430-439.
6. A. Wächter, L.T. Biegler. On the implementation of an interior-point filter line-search algorithm for large-scale nonlinear programming. Mathematical Programming, 2006,106(1):25-57.
7. A. Wälther, A. Griewank. Getting started with ADOL-C. Combinatorial Scientific Computing, 2012, 181-202.

521957 A Model-Based Algorithm for Subtyping Patients with Clotting Abnormalities Using Thromboelastogram Response

Abstract:

Objectives: Early diagnosis of clotting abnormalities, or coagulopathies, would allow more rapid, effective and personalized transfusion support for trauma patients. We present a method that leverages thromboelastogram (TEG) data, a point-of-care assessment of clotting functionality in the form of a tracing, to accomplish this goal. While the TEG returns test results within an hour, our method takes the initial minutes of data collection and assigns patients to coagulopathy subgroups based on a database of simulated TEGs and a deterministic model.

Methods: RapidTEGs were collected from the Study of Tranexamic Acid during Air Medical Prehospital transport (STAAMP) and Prehospital Air Medical Plasma (PAMPer) clinical trials. After eliminating incomplete or discontinuous TEGs, 76% of TEGs remained (104 patients, n=104 tracings). For each, an adjudication was made whether there was insufficient coagulation, defined by a maximum amplitude (MA) of the tracing less than 50 mm (n=15), or excessive lysis, defined by a decline in MA of at least 7.5 percent over 30 minutes (n=5). Parameters of a 3-state ordinary differential equation (ODE) model of TEG response based on mass conservation and reaction-engineering principles were fit to individual tracings. Estimated parameters include the effective initial platelet count, and rates of platelet activation, thrombus growth, and lysis (P0, k1, k2, and k3, respectively). Simulations over a mesh comprised of the four parameters, over their clinically relevant ranges, defined a set of 167,400 reference TEG trajectories. For each of the 104 patient TEGs, segments of the initial 5 or 10 minutes of data were used in the analysis. Sum of squared error (SSE) was computed between all simulated trajectories and individual patient trajectory over the specified time, and the 10 nearest simulated trajectories were retained. Therefore, the dataset for analysis includes 1040 parameter values for the top 10 trajectories for each timespan of the 104 patients. Four different logistic regression (LR) models were developed to predict either outcome (insufficient clotting or high lysis) using parameter sets from nearest trajectories of either 5 or 10 minutes of patient data. LR models were developed and tested on 700 and 340 parameter sets, respectively (each with balanced incidence of insufficient coagulation or high lysis). LR performance was described by and AUC of the ROC curve reported on the test sets.

Results: For parameter values identified from 5 minutes of patient data, LR shows that model parameters can be highly predictive of MA and moderately predictive of lysis. Multivariate LR identified low MA (AUC = 0.906). LR less accurately identified excessive lysis with all parameters (AUC = 0.857). For the parameter values identified using 10 minutes of data, LR shows that model parameters are more highly predictive of MA and less predictive of lysis. LR using all parameters identified low MA (AUC = 0.966) with high certainty, while there was no improvement in the ability to identify excessive lysis (AUC = 0.829). The variability in accuracy for predicting lysis may be associated with the lower incidence in the total population (n=5).

Conclusion: This simulated TEG database structure quickly analyzes subgroups of high risk. This method combines the use of deterministic modeling with data analytics to determine subgroup membership. Results show reliable prediction of problems with MA and a potential to predict lysis early-on. Applying this method could allow the point-of-care RapidTEG to provide preliminary readouts (for example, at 5 and 10 minutes) of coagulopathy; and thereby, decrease time to patient-tailored treatment. This is particularly applicable for use with shock and trauma patients, where timing is critical.

538079 Hemolysis Prediction from CFD Simulations of Turbulent Blood Flow in a Functioning and Malfunctioning Bi-Leaflet Artificial Heart Valve

Abstract:

Artificial heart valves are commonly used devices for treating valvular defects. However, these prosthetic devices may expose the blood to turbulent flow conditions leading to high stress that can damage blood cells. The purpose of this research is to simulate blood flow in both a functioning and malfunctioning bi-leaflet artificial heart valve and predict the damage caused to red blood cells (RBCs) using eddy analysis. The magnitude of the stress and exposure time blood experiences, as determined by the size and spatial distribution of turbulent flow eddies derived from computational fluid dynamics (CFD) simulations, is used to calculate the expected percent hemolysis. Using CFD software ANSYS DesignModeler, two prosthetic heart valve models were constructed. In one valve, both leaflets were kept open in a fully functioning position, while in the other, one leaflet was kept mostly closed to replicate a serious problem that can occur in an implanted valve. Blood flow simulations and calculations were done using ANSYS Fluent and validated with experimental findings available in the literature. Results from the CFD simulations provided the spatial distribution of Kolmogorov length scales that were used for eddy analysis in the heart valves. The Kolmogorov length scale (KLS), or the dissipative length scale, was used to characterize eddies in the flow field. The analysis is centered on the hypothesis that only some of the turbulent flow eddies â€“ those with size comparable to or smaller than the size of RBCs â€“ are the ones that contribute to cell damage. This is because smaller KLS reflect higher stress levels, which causes more damage to RBCs at the surface of an eddy. This CFD-based research utilized the total surface area of small eddies in the blood as a way to predict the amount of hemolysis experienced by the blood cells. The spatial distribution of these small eddies provides an indication of the extent of exposure of the RBCs to these stresses. The flow fields, stresses, and eddy distributions were then compared between the two valves. Early results indicate that hemolysis levels are low, suggesting the need for further study of subhemolytic damage. However, hemolysis predictions could be used as a comparative analysis to determine what situations may cause increased damages. For example, additional preliminary results have shown an increase in hemolysis levels in the malfunctioning heart valve when compared to a functioning heart valve under the same flow conditions of simulation.

531039 Dynamic Modeling of Cardiovascular System for Optimal Control of Ventricular Assist Devices

Abstract:

Approximately 5.7 million adults in U.S. have heart failure (HF), and the annual cost of health care services and medications to treat HF is about $30.7 billion [1]. An estimated 150,000 new patients are diagnosed with end-state HF every year. The traditional solution of HF is heart transplantation. However, only few patients are eligible for a transplant due to the limited supply of donor hearts. Ventricular assist device (VAD) has become an established life-saving therapeutic option for end-stage HF patients, which can support the blood circulation as a bridge-to-transplantation [2]. In this work, a lumped parameter model of human cardiovascular circulatory system is developed, which is further combined with a blood pump model to describe the acceleration of the fluid in a mixed flow blood pump. Based on the models of circulatory system and blood pump, an optimization problem is proposed to control the speed profile of VAD, which can find a trade-off between the maximization of the blood flow to satisfy the blood need of human body and the minimization of suction in VADs.

Previous studies have formulated lumped parameter models of human circulatory system, which comprise the left and right ventricle, active left and right atrium, systemic and pulmonary arterial load, and systemic and pulmonary venous return [3, 4]. While these models have shown their potential in clinical applications, such as the speed control of VADs, the application is still limited due to the incapability of accounting for model uncertainty and variability among individual patients. For example, the flow resistance in the systemic arterial system of the cardiac regulatory mechanisms was modeled based on the mean value of systemic arterial pressure. The step to quantify the effect of variations in arterial pressure on the flow resistance was omitted in previously reported works. In addition, to make models tractable and useful, assumption and simplification are generally used, which may consequently induce model mismatch. In order to improve the reliability and credibility of model, it is necessary to account for uncertainty and model mismatch in the models of circulatory system. Therefore, a stochastic lumped model of the human circulatory system is developed in this work using a polynomial chaos expansion theory [5].

Based on the models of circulatory system, feedback controller can be used to adapt the pump speed of VAD to meet the physiologic needs of HF patients. However, control of VAD presents major challenges, since it can create suctions when the impeller speed, and thus flow, exceeds the venous return to the left ventricle. In addition, most of the available control algorithms of VAD uses a pre-selected speed profile, which ignores different demands of oxygen and blood, due to various activities of patients such as exercise or resting [6, 7]. In this work, an adaptive control algorithm is proposed by formulating an optimization problem to tune the pump speed of VAD in a real-time fashion, which can maximally satisfy the physiologic need of patients, while minimizing suction in VADs. The efficiency of the proposed modelling and control algorithm is illustrated with simulations, which will be further studied with a mock human circulatory cardiovascular system.

Reference

[1] B. Ziaeian and G. C. Fonarow, "Epidemiology and aetiology of heart failure," *Nature Reviews Cardiology volume,* vol. 13, no. 6, p. 368â€"378, 2016.

[2] D. Timms, "A review of clinical ventricular assist devices," *Medical Engineering & Physics,* vol. 33, no. 9, pp. 1041-1047, 2011.

[3] R. Amacher, J. Asprion, G. Ochsner, H. Tevaearai, M. J. Wilhelm, A. Plass, A. Amstutz, S. Vandenberghe and M. S. Daners, "Numerical Optimal Control of Turbo Dynamic Ventricular Assist Devices," *Bioengineering,* vol. 1, no. 1, pp. 22-46, 2014.

[4] E. Lim, S. Dokos, S. L. Cloherty, R. F. Salamonsen, D. G. Mason, J. A. Reizes and N. H. Lovell, "Parameter-Optimized Model of Cardiovascularâ€"Rotary Blood Pump Interactions," *IEEE Transactions on Biomedical Engineering,* vol. 57, no. 2, pp. 254-266, 2010.

[5] D. Xiu, Numerical Methods for Stochastic Computation: A Spectral Method Approach, Princeton, New Jersey, USA: Princeton University Press, 2010.

[6] T. Pirbodaghi, S. Axiak, A. Weber, T. Gempp and S. Vandenberghe, "Pulsatile control of rotary blood pumps: Does the modulation waveform matter?," *The Journal of Thoracic and Cardiovascular Surgery,* vol. 144, no. 4, pp. 970-977, 2012.

[7] P. He, J. Bai and D. D. Xia, "Optimum control of the Hemopump as a left-ventricular assist device," *Medical & Biological Engineering & Computing,* vol. 43, no. 1, pp. 136-141, 2005.

538165 First Passage of Molecular Motors on Networks of Cytoskeletal Filaments

Abstract:

Molecular motors facilitate intracellular transport by actively transporting cargo along cytoskeletal filaments. The behavior of a motor traversing a single filament has been well-characterized for many motors. However, understanding how motors traverse networks of cytoskeletal filaments remains an open problem with applications in cell biology and the design of active biological materials.

In this work, we use kinetic Monte Carlo computer simulations to investigate the transport of molecular motors across a domain containing a network of filaments. The overall motion consists of passive cytosolic diffusion and directed transport along filaments; the first passage time describes the time for a motor to first traverse the simulation domain. To probe the link between the network structure and long-range motor transport, we systematically vary properties of the network and assess the effect on the distribution of first passage times. For networks of randomly oriented filaments, domains with moderate numbers of long filaments have the largest mean first passage time with the largest coefficient of variation. This is indicative of slow, unreliable transport. Additionally, randomly generated networks with the same number and lengths of filaments can lead to markedly different first passage times, indicating the importance of the underlying filament distribution. Some networks produce anomalously large first passage times. These have spatial regions where motors become trapped for long periods of time. Removal of a small fraction of filaments significantly reduces the mean first passage time, but surprisingly, the most impactful filaments are far removed from the trapping regions. These filaments serve as lynchpins connecting one region to another. We develop analytical theory to describe effective diffusion coefficients and to highlight the importance of long-lived traps.

Taken together, our results relate the first passage of a molecular motor across a domain to features of the underlying filament network. We conclude by discussing implications of our results for plant cell biology, where large cell sizes can lead to prohibitively large diffusive time scales for cell mixing.

Received: 5 March 2018 | Revised: 4 July 2018 | Accepted: 15 July 2018
DOI: 10.1002/cnm.3132

RESEARCH ARTICLE - APPLICATION

WILEY

Numerical accuracy comparison of two boundary conditions commonly used to approximate shear stress distributions in tissue engineering scaffolds cultured under flow perfusion

Olufemi E. Kadri[1] | Cortes Williams III[2] | Vassilios Sikavitsas[2] | Roman S. Voronov[1]

[1]Otto H. York Department of Chemical and Materials Engineering, New Jersey Institute of Technology, Newark, NJ 07102, USA

[2]Stephenson School of Biomedical Engineering, The University of Oklahoma Norman, OK 73019, USA

Correspondence
Roman S. Voronov, Otto H. York Department of Chemical and Materials Engineering, New Jersey Institute of Technology, Newark, NJ 07102, USA.
Email: rvoronov@njit.edu

Funding information
National Science Foundation, Grant/Award Number: ACI-1548562; Gustavus and Louise Pfeiffer Research Foundation, Grant/Award Number: Major Research Grant (No Award Number)

Abstract

Introduction: Flow-induced shear stresses have been found to be a stimulatory factor in pre-osteoblastic cells seeded in 3D porous scaffolds and cultured under continuous flow perfusion. However, due to the complex internal structure of the scaffolds, whole scaffold calculations of the local shear forces are computationally intensive. Instead, representative volume elements (RVEs), which are obtained by extracting smaller portions of the scaffold, are commonly used in literature without a numerical accuracy standard.

Objective: Hence, the goal of this study is to examine how closely the whole scaffold simulations are approximated by the two types of boundary conditions used to enable the RVEs: "wall boundary condition" (WBC) and "periodic boundary condition" (PBC).

Method: To that end, lattice Boltzmann method fluid dynamics simulations were used to model the surface shear stresses in 3D scaffold reconstructions, obtained from high-resolution microcomputed tomography images.

Results: It was found that despite the RVEs being sufficiently larger than 6 times the scaffold pore size (which is the only accuracy guideline found in literature), the stresses were still significantly under-predicted by both types of boundary conditions: between 20% and 80% average error, depending on the scaffold's porosity. Moreover, it was found that the error grew with higher porosity. This is likely due to the small pores dominating the flow field, and thereby negating the effects of the unrealistic boundary conditions, when the scaffold porosity is small. Finally, it was found that the PBC was always more accurate and computationally efficient than the WBC. Therefore, it is the recommended type of RVE.

KEYWORDS
computational fluid dynamics, lattice Boltzmann method, periodic boundary condition, representative volume element, surface shear stresses, tissue engineering, wall boundary condition

1 | INTRODUCTION

Incidences of bone disorders constitute a significant economic burden to societies globally. In the United States alone, over $213 billion is the total annual cost (direct and indirect) of treating the estimated 126.6 million people affected by musculoskeletal disorders.[1] Unfortunately, with an increasingly obese and aging population, this trend is expected to continue further. Current approaches for replacing the damaged bone tissues include the use of bone grafts (ie, autografts or allografts). However, these methods have several shortcomings, limited availability, and risk of disease transmission.[2-4] To address those disadvantages, bone tissue engineering has emerged as an alternative regenerative strategy.

In bone tissue engineering, a combination of osteo-inductive biological factors, mesenchymal stem cells obtained from patients' own bone marrow, and porous biodegradable scaffolds are used. Typically, the process involves seeding the cells within the 3-D scaffolds, followed by culturing under flow in perfusion bioreactors. The flow is a necessary part of the culture, because the stimulatory shear that it imposes on the stem cells mimics the natural microenvironment in bone canaliculi.[5,6] Moreover, it has been shown to promote tissue regeneration.[7-10] Thus, the applied shear stresses should be within the physiological range required for stimulation: 0.1 to 25 dynes/cm^2,[11-13] because excessive shear of 26 to 54 dynes/cm^2 can cause cell lysing and/or detachment from the scaffold.[14,15]

Therefore, the ability to predict the shear stress distribution in different scaffold micro-architectures can provide insight into whether or not a particular scaffold design will promote tissue growth. Moreover, when used in conjunction with the latest advances in 3D microfabrication technologies, such predictive capabilities can be used to create optimized scaffold geometries. Unfortunately, however, the complex internal structure of the porous scaffolds makes estimation of the required shear stresses via experimental or analytical techniques impractical. Hence, computational fluid dynamics models, based on either idealized pore geometries[6,8,16-21] or actual scaffold images,[12,22-34] are commonly utilized.

The latter is the more realistic approach since it is based on the actual microscopic pore structures, which are typically obtained via a 3D scanning technique such as micro-computed tomography (μCT). Yet, due to the computationally intensive nature of the scaffold reconstructions resulting from such high-resolution imaging, researchers are forced to resort to implementing approximations.[12,22-34] For example, rectangular "representative volume elements" (RVE) are cut from whole scaffolds and implemented in conjunction with various boundary conditions along the artificially created periphery. Two common types of boundary conditions that are typically implemented for this purpose are the "wall boundary condition" (WBC)[8,12,19,22,23,27,28,30-34] and the "periodic boundary condition" (PBC).[18,20,24-26,29] In the former case, the RVE is surrounded by solid walls in the non-flow directions, while the latter is an application of periodicity in all three dimensions.

Although these approaches save on computation time, it is not obvious how accurate the resulting shear stresses are, or which of the two boundary conditions yields the better results. Consequently, the RVE approach is commonly questioned by journal reviewers, as no standards or guidance regarding their use exist. We have found only one publication that investigated the accuracy of the RVE-WBC, as compared with the whole scaffold simulation.[22] Here, a guideline was provided stating that for scaffolds with a homogenous pore distribution *the domain size should be at least 6 times the average pore size*. However, this suggestion was made based on *average* wall stresses only, while in reality the *spatial distribution* of the stresses is also important to tissue growth. For example, the cells within the scaffold migrate around in a nonrandom manner[35] are therefore more likely to experience stresses at some preferred locations. Furthermore, only scaffolds prepared using the same fabrication technique were studied, though two different materials were used in their manufacturing. Nonetheless, the scaffold's structure depends more on the fabrication method than it does on the material.[24,25,36] Moreover, just a single scaffold sample was used for each type of the material. Hence, a more thorough investigation of the RVEs' accuracy is warranted, especially given that no PBC studies were found at all.

Therefore, in this work, we set out to quantify how the two relevant boundary conditions compare against each other, when applied to scaffolds manufactured using different fabrication methods, and for a large number of samples with varying porosities. To achieve this, an in-house lattice Boltzmann method (LBM) code is used to simulate fluid flow through reconstructions of salt-leached foam and nonwoven fiber mesh scaffolds, all of which are imaged using μCT. The overall computational approach used in this work is summarized in Figure 1. Ultimately, the accuracy of the spatial stress distributions is reported for both the RVE-WBC and the RVE-PBC. Finally, a descriptive statistical analysis is used to demonstrate the accuracy and computational efficiency differences between the two RVE approaches.

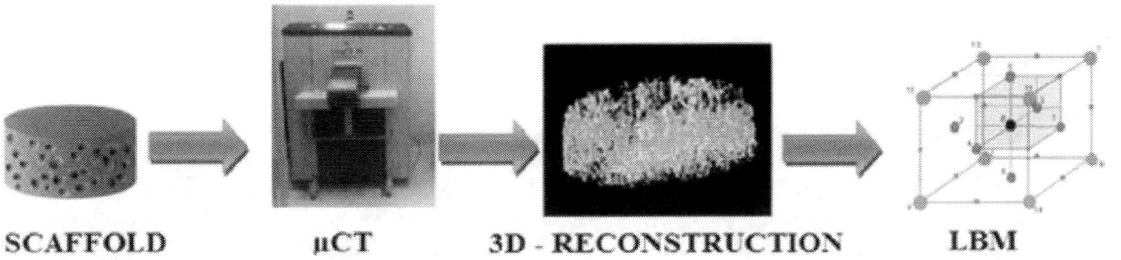

FIGURE 1 The image-based modeling methodology used in this work. The scaffolds were first manufactured, and then they were scanned using high-resolution µCT. Afterwards, their architecture was reconstructed virtually in 3D and imported into the LBM fluid flow solver. Finally, the LBM simulation results were used to compute the shear stresses on the scaffold surfaces

2 | MATERIALS AND METHODS

2.1 | Scaffold preparation and 3D imaging

Two types of porous scaffolds, salt-leached foam and nonwoven fiber scaffolds, were produced from poly-L-lactic acid using methods described in detail elsewhere.[24] The scaffolds were then scanned via µCT using a ScanCo VivaCT40 system (ScanCo Medical, Bassersdorf, Switzerland) to obtain 10-µm resolution, 2D intensity image slices (see Figure 2) at the optimum settings of 88 µA (intensity) and 45 kV (energy). The acquired X-ray images were filtered for noise reduction and assembled into 3D reconstructions of the scaffolds using a custom Matlab code (MathWorks Inc., Natick, MA). The scans were segmented using global thresholding. Threshold values were chosen such that the porosity of scaffolds from 3D reconstructions were within 1% of experimentally calculated porosities. Experimental porosities were obtained by measuring the solid volume (mass of the scaffolds divided by the density of scaffold materials) and comparing with total scaffold volume (assuming a cylindrical scaffold) as reported in VanGordon et al[24] and Voronov et al.[25,26]

2.2 | Simulation domain and boundary conditions

For the non-RVE calculations, each simulation domain was composed of a *whole* scaffold placed inside of a pipe (see Figure 3A). This is meant to mimic the cassette holder that typically fixes the scaffold in the perfusion bioreactors. The pipe's length was taken to be approximately 10 times greater than the scaffold thickness, in order to avoid periodicity artifacts and ensure that a uniform parabolic profile is developed before flow reaches the scaffolds. Simulations were

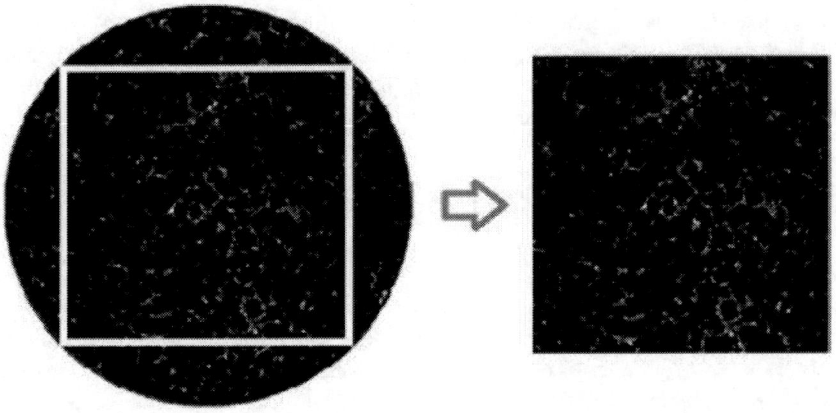

FIGURE 2 A 2D analogy of how an RVE simulation domain is cut out of a whole scaffold µCT image. LEFT—A slice representative of a cross section through a typical fiber mesh scaffold. The white box marks the largest rectangular area that could possibly be inscribed into the circular scaffold; RIGHT—The resulting RVE cutout from the rectangular region inscribed into the whole scaffold. Grayscale color is the X-ray radio-density of the scaffold material as imaged via µCT

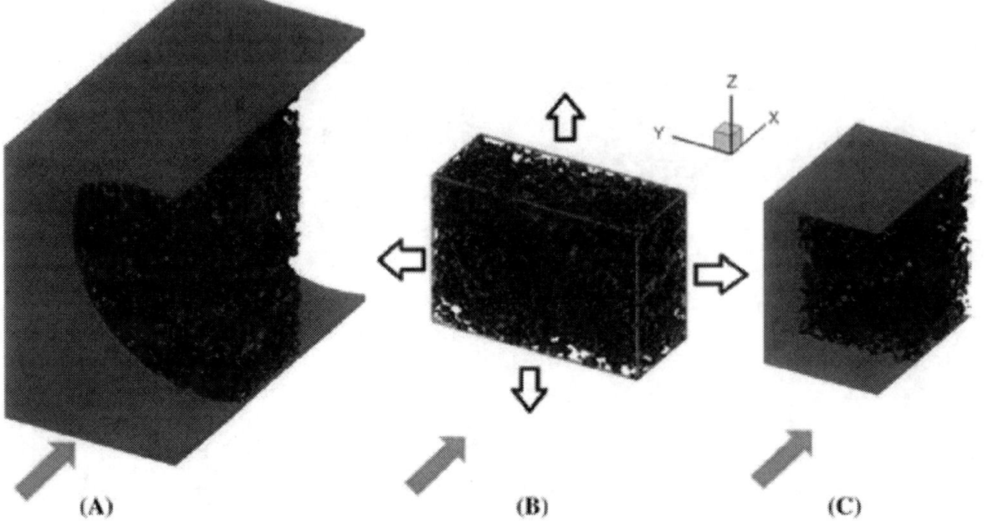

FIGURE 3 Simulation domain comparison between whole scaffold simulations and the RVEs. A, Cross-sectional view of the whole simulation domain (right half is omitted for clarity). The wall and the entrance length are shown in red. In all panes, the blue arrow is the flow direction, and the gray scale color is the X-ray radio-density of the scaffold material, as imaged via μCT. B, A representative RVE-PBC simulation domain. White arrows show the directions in which periodicity is applied. Note that periodicity is also applied in both x directions as well, but the arrows are omitted for clarity. C, A cross-sectional view of RVE-WBC simulation domain (right half is omitted for clarity). The wall and the entrance length are shown in red

performed for a flow rate of 0.15 mL/min. This is considered a suitable flow rate for mechanical stimulation in many perfusion bioreactors.[9,37]

For the RVE calculations, the simulation domain was obtained by extracting the *largest* rectangle that could possibly be inscribed into the circular whole scaffold (see Figure 2-Left**)**.

Generally, the size of the RVE (denoted below as "L") is taken to satisfy the so-called "separation of scales", $d < L$, where d is the average size of the porous media heterogeneity.[38] Specifically, the RVE domain size should be *at least 6 times the average pore size*.[22] We tried to stay well above this criterion, in order to maximize accuracy and examine the *best-case scenario* produced by the RVEs. Subsequently, the resulting domain sizes (see Figure 2-Right) were all between 8 and 14 times the scaffolds pore sizes, which is significantly greater than the suggested minimum for RVEs.[22]

For the RVE-PBC calculations, periodicity was applied in all three directions (see Figure 3B) to approximate an infinite domain representing the full scaffold. On the other hand, for the RVE-WBC, the scaffold was surrounded by solid walls in the non-flow directions (see Figure 3C). In addition, an entrance length equal to half of the scaffold's thickness in the flow direction was added, in order to stay consistent with the previous analysis of this boundary condition type.[22] In both implementations, the total flow rate through the RVE was decreased proportionally to the cutout size, in order to compensate for the reduction in the cross-sectional area available to flow relative to the whole scaffold.

2.3 | Fluid flow modeling: lattice Boltzmann method

LBM was chosen for the present application, because it is especially appropriate for modeling pore-scale flow through porous media (such as scaffolds) due to the simplicity with which it handles complicated boundaries.[24-26,29,39-42] This is because LBM uses structured meshes for complex geometries, unlike classical CFD approaches which will rather utilize unstructured meshes. Another advantage of LBM is that it uses a direct method based on first principles at the mesoscopic scale rather than modeling terms of the fluid flow governing equations at the macroscopic scale. In addition, the LBM method has gained popularity within the scientific computing community because of the ease with which it can be parallelized on supercomputers.[43]

A previously developed custom-written, in-house code was used in this work.[24-26,39,41,44-47] The D3Q15 lattice[48] in conjunction with the single-relaxation time Bhatnagar, Gross and Krook[49] collision term approximation was used to perform simulations. The no-slip boundary condition was applied at solid faces using the "bounce-back" technique.[50] To take advantage of the inherent LBM parallelizability, domains were decomposed using message passing interface.[39,41] Simulation convergence was defined as when average and highest velocities computed for the simulation domain vary by less than 0.01% for two consecutive time steps. The code has been validated for several flow cases for which analytical solutions are available: forced flow in a slit, flow in a pipe, and flow through an infinite array of spheres.[25,39]

2.4 | Surface stress calculations and error analysis

Shear stress on the surface of the scaffolds was calculated following a scheme suggested in Porter et al,[29] where the full shear stress tensor is calculated first, and then the maximum eigen value is evaluated using a Jacobi iteration technique. The cell culture media was assumed a Newtonian fluid, and the shear stresses at every location within the scaffolds were estimated using

$$\underline{\underline{\sigma}} \approx \mu \left(\frac{1}{2}\right) \left(\nabla U + \nabla U^T\right) \tag{1}$$

where $\underline{\underline{\sigma}}$ is the shear stress tensor, and U is local velocity vector. The fluid dynamic viscosity was 0.01 g/(cm s), which is close to that of α-MEM supplemented with 10% FBS typically used in cell culturing experiments.[51] Velocity vectors used in calculations were derived for the specified flow rate. Computed shear stress values are the largest eigenvalues of $\underline{\underline{\sigma}}$. Stress maps generated using Tecplot 360 EX 2017 (Tecplot Inc., Bellevue, WA USA) were used to visualize localized shear stress values on the scaffolds. For the accuracy comparisons, the error at every fluid surface node was calculated as follows:

$$\varepsilon_t = 100\% * \left| \frac{True\ (Whole\ Scaffold) Stress - Approximated\ (RVE)\ Stress}{True\ (Whole\ Scaffold)\ Stress} \right| \tag{2}$$

where the "True Stress" refers to the shear stress value obtained from the Whole scaffold simulations, while the "Approximated Stress" is obtained from the RVE scaffold simulations.

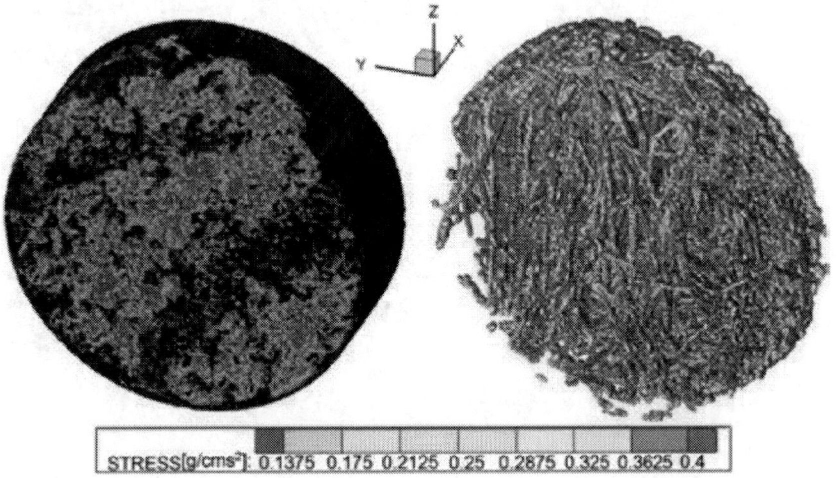

FIGURE 4 3D reconstructions of the two scaffold types used in this study, with the surface stress maps (color) calculated from LBM overlaid on the μCT images (gray scale). LEFT—porous foam scaffold. RIGHT—Fiber mesh scaffold. The pipe surrounding the scaffold, and stress values below 0.1dynes/cm^2 are omitted for clarity. Dimensions are 1.61 mm × 5.1 mm × 5.1 mm and 1.10 mm × 5.1 mm × 5.1 mm for salt-leached and nonwoven fiber-mesh scaffolds, respectively

3 | RESULTS

In order to compare the RVEs' performance relative to each other, as well as to that of the whole scaffold simulations, we performed image-based LBM modeling using a flow rate typically encountered during artificial tissue culturing conditions. Figure 4 shows representative results for *whole* scaffolds of two different architectures: salt-leached foam (Figure 4-Left) and nonwoven fiber mesh (Figure 4-Right). These very computationally intensive models exemplify the best stress estimates, because they are the most representative of the actual flow perfusion culturing conditions. However, they take approximately 30 000 to 40 000 LBM steps to converge. Unfortunately, for high-resolution scaffold images

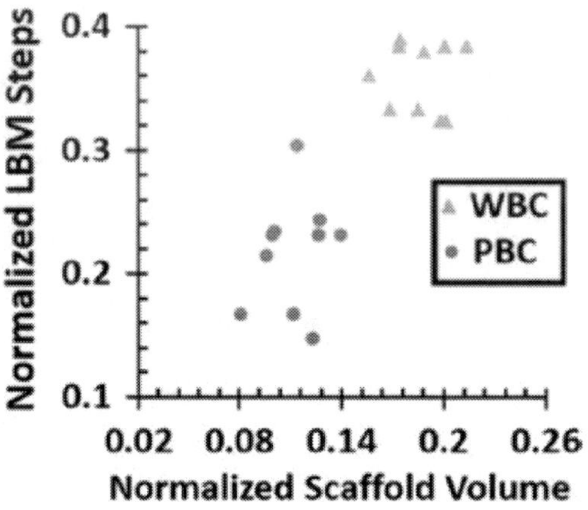

FIGURE 5 Total number of LBM steps required to reach convergence versus scaffold volume for RVE-PBC and RVE-WBC. The vertical and horizontal axes are normalized with respect to the whole scaffold values, respectively

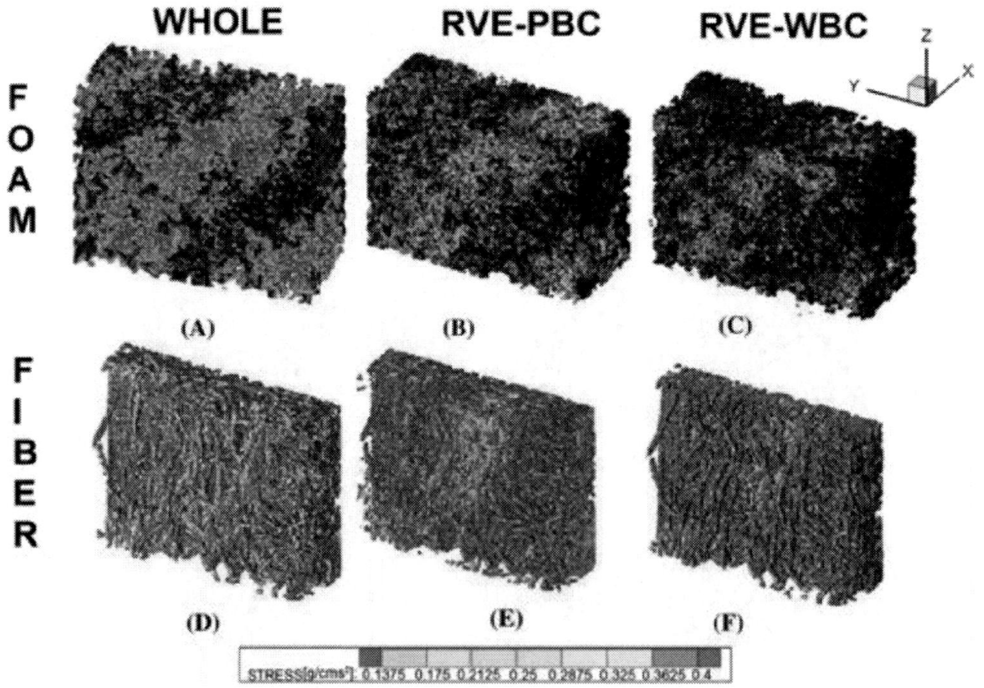

FIGURE 6 Surface *stress* maps overlays for the RVE-equivalents cutout from whole scaffold simulations and the two types of boundary conditions used for the RVEs. Top row: Salt-leached foam scaffold. Bottom row: Nonwoven fiber mesh. Left column: RVE-equivalent volumes cutout from a *whole* scaffold models. Center column: RVE-PBCs. Right column: RVE-WBCs. Dimensions are 1.61 mm × 3.91 mm × 2.81 mm and 1.10 mm × 3.79 mm × 2.99 mm for salt-leached and nonwoven fiber-mesh scaffolds, respectively

(eg, μCT), this can translate to weeks of waiting for results, even on a large supercomputer. The RVEs, on the other hand, take only a fraction of that to converge and, therefore, have a higher computational efficiency. This is shown in Figure 5, where both axes are normalized by the corresponding whole scaffold values. From this figure, it is apparent that the RVEs take roughly 3 to 6 times less LBM steps to converge, which translates into a significant waiting time reduction for the user. Hence, this makes the RVE approaches attractive, as long as some error can be tolerated.

Next, we visualized the stresses produced by the RVEs, in order to gauge the amount of error incurred from their use. The first question we wanted to answer was whether the stress approximations provided by the RVEs follow a *spatial* pattern similar to that observed in the whole scaffold simulations. To that end, Figure 6A,D shows stresses in RVE-equivalent domains cut out from the *whole* scaffold simulations in Figure 4. These are presented in order to provide a one-to-one comparison for the actual RVE-WBC and RVE-PBC cutouts shown in Figure 6B,E and Figure 6C,F, respectively. From Figure 6, it is apparent that the *spatial* stress patterns produced by the RVEs are indeed very similar to those obtained from the whole scaffold simulations. However, the former stresses are significantly lower when compared with the latter.

Consequently, we used Equation (2) to quantify the spatial *accuracy* of the results produced by the RVE approximations. This was done by calculating the amount by which they deviate from the *true* stress values at every surface voxel of the scaffolds. For the purposes of this comparison, the whole scaffold stresses are considered to be the *true* values in the Equation (2) calculations, while the stresses produced by the RVE are used as the *approximated* values in the same equation. Figure 7 is a 3D overlay of the resulting spatial error patterns in the two scaffold types from Figure 4. From left to right, the top row shows the salt-leached foam RVE-PBC and WBC, respectively, while the bottom row shows the fiber-mesh RVEs, in the same order. It can be immediately observed from this figure that the bulk error is similar for both RVEs,

FIGURE 7 Surface *error* maps overlays for the two RVE types, relative to the whole scaffold simulation RVE equivalents. Top row: Salt-leached foam scaffold. Bottom row: Nonwoven fiber mesh. Left column: RVE-PBC. Right column: RVE-WBC. The wall is omitted from the WBC reconstructions for clarity. Top right quarter of each scaffold is removed in order to provide a view inside. Dimensions are 1.61 mm × 3.91 mm × 2.81 mm and 1.10 mm × 3.79 mm × 2.99 mm for salt-leached and nonwoven fiber-mesh scaffolds, respectively

varying roughly between 40 and 60% (for these particular samples). However, there is also a spatial trend in the error: in the case of the RVE-PBC it appears to *decrease* towards the cutout's periphery, while for the RVE-WBC the opposite is the case. This makes sense, given that the presence of the wall in the latter alters the flow field near the boundary. Therefore, there are significant differences in accuracies between the two RVEs; with the RVE-PBC yielding the more superior results.

In addition to the spatial distributions of stresses, we analyzed the probability distributions of the wall shear stress magnitude. Figure 8A,C shows the histogram distributions of stresses in salt-leached foam and nonwoven fiber-mesh scaffolds, respectively. Stresses are normalized using (*stress* − *mean stress*)/(*standard deviation of stress distribution*), as is reported in our previous works.[24,25,36]

From the figures, it is observed that all the normalized distributions appear similar in shape but differ by frequency of occurrence of stresses. Hence, we quantified the differences, errors in probabilities for RVE-PBC and RVE-WBC are calculated as percentages of the whole scaffold simulation values using

$$Error = |True\ Probability - Approximated\ Probability| * 100\% / (True\ Probability) \tag{3}$$

where *True Probability value* is probability of occurrence of each stress histogram bin in *whole* scaffold simulation histogram and *Approximated Probability value* is probability of occurrence of each stress bin in *RVE* simulation histogram. Figure 8B,D shows error histograms for salt-leached foam and nonwoven fiber-mesh scaffold. The figures show that for both scaffold architectures, the RVE-PBC produced lower errors than RVE-WBC at low stresses, while higher errors resulted at high stress values. However, since each of the histogram bins corresponds to a different probability of stress occurrence, a *weighted* average is calculated for each of the error distributions to aid better comparison between them. The *weighted* average error is calculated using

$$W_{error} = \sum_{i=1}^{N} (p_i{}^{*}\mathrm{err}_i) \tag{4}$$

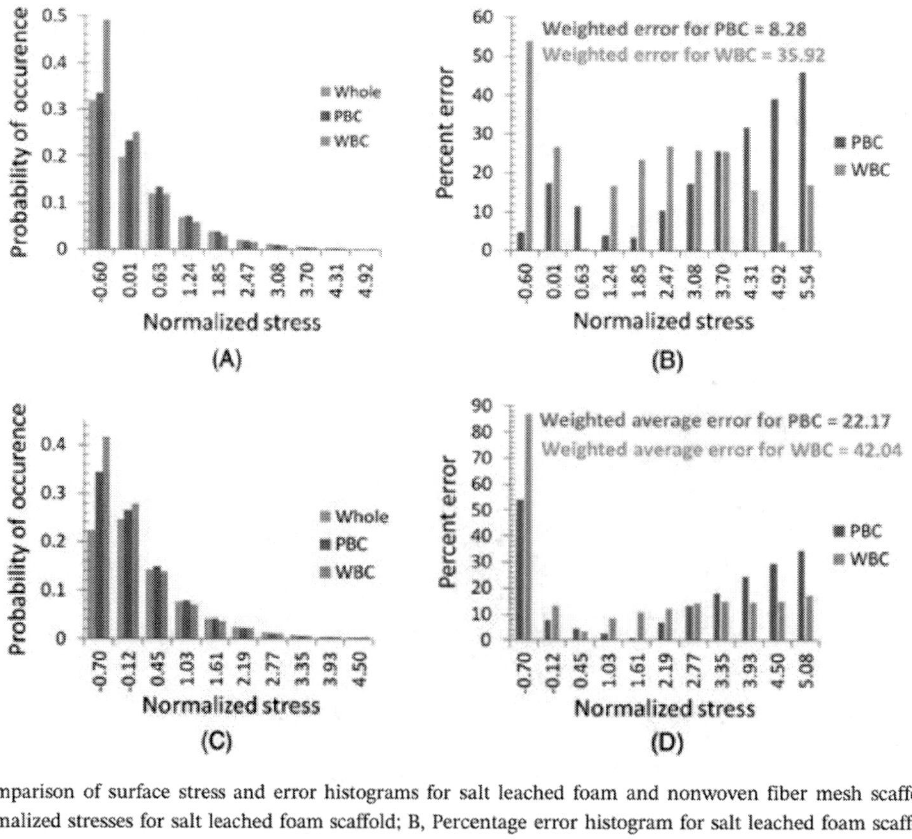

FIGURE 8 Comparison of surface stress and error histograms for salt leached foam and nonwoven fiber mesh scaffolds. Top row: A, Histogram of normalized stresses for salt leached foam scaffold; B, Percentage error histogram for salt leached foam scaffold. Bottom row: C, Histogram of normalized stresses for nonwoven fiber mesh scaffold; D, Percentage error histogram for nonwoven fiber mesh scaffold. Gray color—Whole. Orange color—WBC. Blue color—PBC

where W_{error} is *weighted* average error, i is an index for histogram bin i, N is total number of bins in histogram, p_i is the probability of occurrence of histogram bin i, and err_i is the error obtained for histogram bin i using Equation (3). As shown in Figure 8B,D, the weighted average errors are lower for the RVE-PBC for both scaffold architectures, thus, confirming its superiority over the RVE-WBC.

Further, we wanted to check whether the type of the scaffold architecture has a significant effect on the surface stress accuracy. To that end, Figure 9 plots the mean error in the RVE surface stress, relative to those from the whole scaffold simulations. We chose the abscissa to be porosity, since it also describes the scaffold morphology. From this figure, it is evident that the RVE-WBC always results in a greater error relative to that obtained from the corresponding RVE-PBC. Another observation is that the foam scaffolds always had a smaller error than the fiber scaffolds, although this is likely an effect of their lower porosity. Specifically, the error was found to increase proportionally to the latter, indicating that the porosity influences the simulation accuracy. The absolute stresses for the data in Figure 9 are given in Table 1.

Lastly, the validity of LBM simulation results presented is demonstrated. We compared permeability values calculated by LBM to those obtained from the well-established Kozeny-Carman (KC) equation for porous materials (given below):

$$k = \frac{\varepsilon^3}{(1-\varepsilon)^2}\left(\frac{1}{S^2 K}\right) \tag{5}$$

k is the permeability of material, ε is the porosity, S is the surface area-to-volume ratio, and K is the Kozeny constant. Based on experiment,[52] K is often assigned a value of 5, but recent studies have shown that its value varies with porosity.[53] The comparisons are given in Table 2, and they show that permeabilities obtained by LBM are of similar order of magnitude to those from the analytical KC equation. The percentage errors between estimates obtained from the KC equation and LBM are plotted in Figure 10. It shows that values obtained by RVE-PBCs are mostly more accurate than the RVE-WBC. The discrepancy observed for scaffold with porosity of 0.977 is likely because calculations were performed using K = 5, despite its dependence on porosity.

Additionally, we have validated the simulation results against a three-parameter gamma distribution, $\Gamma(\alpha, \gamma, \beta)$ which was shown to be representative of normalized wall stresses in flows through highly porous random porous media[36]:

$$\Gamma(\alpha, \gamma, \beta) = \frac{(\tau^* - \gamma)^{\alpha-1}}{\beta^\alpha \Gamma(\alpha)} \exp\left[-{(\tau^* - \gamma)}/{\beta}\right] \tag{6}$$

where τ^* is the normalized stress, $\Gamma(\alpha)$ is a complete gamma function, $\alpha = 2.91$, $\beta = 0.45$, and $\gamma = -1.43$. The comparison is given in Supplementary Figure 1. It is observed that RVE-PBC, RVE-WBC, and whole simulation results all show good agreement with $\Gamma(\alpha, \gamma, \beta)$; however, the PBC is closer to it than the WBC.

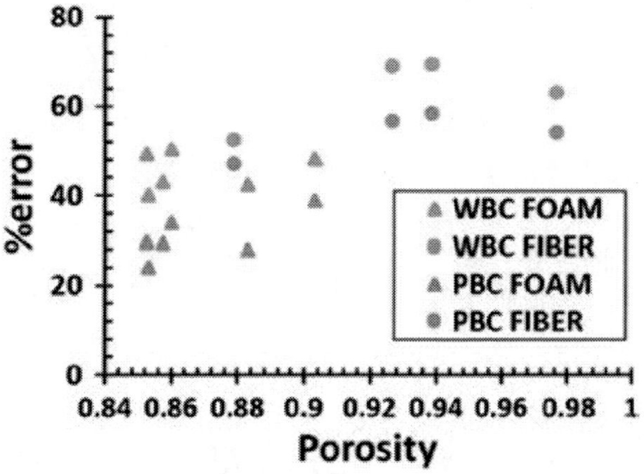

FIGURE 9 Average error in fluid-induced surface stress resulting from the RVE calculations plotted versus scaffold porosity and boundary condition type. Orange color—WBC. Blue color—PBC. Triangles—Salt leached foam. Circles—Nonwoven fiber mesh

TABLE 1 Summarized stress data and LBM steps to convergence for all the scaffolds studied

Scaffold type	Porosity	Shear Stress, dynes/cm² Average	Shear Stress, dynes/cm² Maximum	Mean % error	LBM steps
Foam 1					
Whole scaffold	0.8830	0.078	1.695	N/A	37 000
RVE-PBC		0.057	0.620	27.799	9 000
RVE-WBC		0.045	0.547	42.462	12 000
Foam 2					
Whole scaffold	0.9033	0.092	1.462	N/A	39 000
RVE-PBC		0.056	0.608	38.860	9 000
RVE-WBC		0.048	0.551	48.137	15 000
Foam 3					
Whole scaffold	0.8526	0.076	2.035	N/A	38 500
RVE-PBC		0.054	0.815	29.537	9 000
RVE-WBC		0.038	0.649	49.389	15 000
Foam 4					
Whole scaffold	0.8600	0.081	1.818	N/A	39 000
RVE		0.054	0.813	33.780	9 000
RVE-WBC		0.040	0.692	50.292	15 000
Foam 5					
Whole scaffold	0.8573	0.074	1.958	N/A	39 500
RVE		0.052	0.957	29.342	12 000
RVE-WBC		0.042	0.813	43.062	15 000
Foam 6					
Whole scaffold	0.853	0.068	1.263	N/A	39 000
RVE		0.052	0.790	24.059	9 000
RVE-WBC		0.041	0.565	40.097	15 000
Fiber 1					
Whole scaffold	0.8790	0.122	3.420	N/A	36 000
RVE		0.065	0.780	46.762	6 000
RVE-WBC		0.058	0.920	52.340	13 000
Fiber 2					
Whole scaffold	0.9268	0.134	2.318	N/A	34 000
RVE		0.059	0.835	56.404	5 000
RVE-WBC		0.042	0.576	68.975	10 000
Fiber 3					
Whole scaffold	0.9388	0.146	3.298	N/A	36 000
RVE		0.061	0.766	58.146	6 000
RVE-WBC		0.045	0.641	69.195	12 000
Fiber 4					
Whole scaffold	0.9770	0.127	4.345	N/A	42 000
RVE-PBC		0.058	0.721	54.117	9 000
RVE-WBC		0.047	0.382	62.909	14 000

4 | DISCUSSION

The ability to predict the shear stresses experienced by osteoblasts during culturing is crucial for optimizing in vitro bone tissue engineering experiments and scaffold designs. For example, such models could help reveal scaffold features that are key to inducing bone forming responses: increased nitric oxide, prostaglandin, and osteopontin production.[5,54,55]

TABLE 2 Permeability and interstitial velocity for scaffolds

		Permeability, cm^2	Error, %		Average Interstitial Velocity, cm/s	Error, %
Foam 1	KC	1.84E−5		Whole	1.44E−2	
	PBC	2.10E−5	13.85	PBC	1.22E−2	15.30
	WBC	2.12E−5	13.85	WBC	1.13E−2	21.14
Foam 2	KC	2.50E−5		Whole	1.40E−2	
	PBC	2.72E−5	8.76	PBC	1.20E−2	14.41
	WBC	2.25E−5	10.04	WBC	1.11E−2	20.77
Foam 3	KC	1.29E−5		Whole	1.57E−2	
	PBC	1.55E−5	20.22	PBC	1.24E−2	21.39
	WBC	1.57E−5	21.56	WBC	9.21E−3	41.49
Foam 4	KC	1.43E−5		Whole	1.52E−2	
	PBC	1.66E−5	15.64	PBC	1.23E−2	19.20
	WBC	1.78E−5	24.00	WBC	1.03E−2	32.38
Foam 5	KC	1.38E−5		Whole	1.52E−2	
	PBC	1.67E−5	21.24	PBC	1.23E−2	18.87
	WBC	1.80E−5	30.43	WBC	1.04E−2	31.79
Foam 6	KC	1.40E−5		Whole	1.55E−2	
	PBC	1.69E−5	20.95	PBC	1.24E−2	20.06
	WBC	1.78E−5	27.21	WBC	1.03E−2	33.18
Fiber 1	KC	1.16E−5		Whole	1.48E−2	
	PBC	1.63E−5	40.39	PBC	9.66E−3	34.57
	WBC	1.61E−5	38.94	WBC	9.56E−3	35.25
Fiber 2	KC	1.59E−5		Whole	1.35E−2	
	PBC	1.66E−5	4.16	PBC	9.42E−3	30.17
	WBC	1.72E−5	8.20	WBC	9.27E−3	31.29
Fiber 3	KC	2.41E−5		Whole	1.34E−2	
	PBC	1.77E−5	26.54	PBC	8.95E−3	33.05
	WBC	1.73E−5	28.32	WBC	8.69E−3	35.06
Fiber 4	KC	1.11E−4		Whole	1.31E−2	
	PBC	4.43E−5	60.08	PBC	8.85E−3	32.59
	WBC	3.71E−5	66.58	WBC	8.79E−3	33.07

However, there is a danger that the simpleton RVE models, commonly implemented by researchers to save on computation time, could potentially yield unrealistic results. Yet, the only guideline on accuracy that currently exists in literature[22] applies to the RVE-WBC only, while the RVE-PBC remains uncharacterized. Moreover, the guideline is based on *average* stress values, which is not truly representative of what the cells experience during an actual experiment, given that they are not uniformly distributed in the scaffolds. Hence, we explored the *spatial* variations in the fluid-induced shear stresses obtained via *both* the RVE-PBC and the RVE-WBC scaffold approximations. These were compared by benchmarking against the computationally intensive *whole* scaffold LBM simulations, which are considered to yield the *true* stress values. Moreover, we made sure that our cutout sizes were well beyond the "6 times the average pore size" guideline suggested by Maes et al,[22] in order to examine the *best-case scenario* produced by the RVEs.

Consequently, the RVE simulations were found to have a significantly lower computational overhead (see Figure 5), which ultimately translates to shorter waiting times for obtaining results. Furthermore, between the two boundary condition types commonly employed in the RVE simulations, the PBC was found to converge faster than the WBC. This is

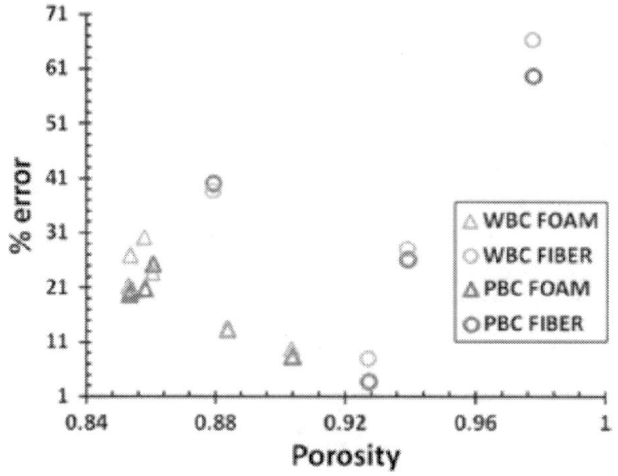

FIGURE 10 Percentage error in permeabilities calculated from LBM and analytical Kozeny-Carman equation plotted versus scaffold porosity and boundary condition type. Orange color—WBC. Blue color—PBC. Triangles—Salt leached foam. Circles—Nonwoven fiber mesh

largely attributed to (1) the smaller simulation domain size of the former, since the latter models typically include an entrance length that is added in order to avoid entrance effects; and (2) the flow field being considerably simpler, since there is no wall contributing to its complexity.

Moreover, the RVE-PBC was also found to be more accurate when compared with the RVE-WBC (see Figures 7 and 8). Specifically, the RVE-PBC produced less error towards the cut-outs' edges, while the RVE-WBC caused more. This makes sense, given that the RVE-PBC is not affected by its boundaries due the periodicity applied, while the RVE-WBC is essentially a flow bounded by a duct. The latter is also less representative of an equivalent cut-out from the whole scaffold. Furthermore, we found that for both RVE types, the error goes up with the scaffold porosity (see Figure 9). The likely reason for this observation is that when the scaffold pores are tight, they have a dominating effect on shaping the flow field and the boundary condition effects are negligible. Conversely, when the pores become less restrictive, the effect of the boundary condition becomes more evident and contributes to the error.

However, even for the best-case scenario (i.e., when the cut-out is >> 6 times larger than the average pore size), the deviation from the true stress values was still between 20% and 80%, for the 10 scaffolds that we tested. This is significant and can lead to misleading conclusions about the efficacy of the scaffold. Therefore, caution must be taken when using the RVE approximations. Overall, however, the fact that most of the spatial *stress* patterns were preserved, despite the use of the unrealistic boundary conditions, adds legitimacy to their use. Additionally, information about the variation of *error*, such as what is provided in this manuscript, could in theory help undo some of the inaccuracies associated with the RVEs. Until that becomes possible, however, the RVE-PBC was found to be the clear winner over the RVE-WBC, on both the accuracy and computational efficiency fronts. Therefore, it is the recommended boundary condition for large-scale simulations, especially when the porosity of the scaffold is high (which is typical for tissue engineering).

A limitation of this study is that it was performed on *empty* scaffolds, without any cells or tissues in them. As mentioned earlier, the cells are likely to build tissues in preferred locations within the scaffolds. Therefore, the effects of the RVEs should be quantified in greater detail in the locations favored by the cells. Finally, in addition to calculating the stimulatory stresses, the flow field produced by the RVEs is often also used to model the influences of metabolite transport on the tissue growth.[26,36] Therefore, although they are beyond the scope of this manuscript, the effects of the RVE cutouts and their boundary conditions on the mass transport within the scaffolds will be addressed in our future investigations.

Overall, the results of this study are expected to be useful to CFD applications beyond tissue engineering and LBM. For example, the RVEs are commonly used in porous media modeling[38,56,57] and with non-LBM methods.[20,21,58] The biggest difference for the latter is that the non-LBM CFD methods may implement a non-structured mesh instead of a structured one. In the case of PBC, "pointers" are used at each boundary mesh-node in order to re-address the flow to a location appropriate for achieving periodicity.[59] Hence, the conclusions of this work apply to RVE-PBCs implemented with CFD approaches other than LBM as well.

5 | CONCLUSIONS

In this work, we investigated the numerical differences between two types of boundary conditions, the PBC and the WBC, which are commonly employed for enabling RVE approximations in bone tissue engineering scaffold simulations. We found that, in general, both of the RVE types followed the same *spatial* surface stress patterns as the whole scaffold simulations. However, they under-predicted the absolute stress values by a considerable amount: 20% to 80%. Moreover, it was found that the error grew with higher porosity of the scaffold but did not depend significantly on its manufacturing method. Lastly, we found that the PBC always resulted in a better prediction (ie, lower error) than the WBC. It was also more computationally efficient, due to a smaller simulation domain size requirement. Therefore, the PBC is recommended as the boundary condition of choice for the RVE approximations. Overall, these findings fill an important knowledge gap in literature regarding the accuracy of the widely used RVE approximations. Therefore, it is our expectation that they will be used to help researchers decide whether the use of the RVE approximations is justified for their application.

ACKNOWLEDGEMENTS

Financial support from Gustavus and Louise Pfeiffer Research Foundation is gratefully acknowledged. We also acknowledge the support provided by the University of Oklahoma Supercomputing Center for Education and Research (OSCER) and Texas Advanced Computing Center (TACC) at The University of Texas at Austin under allocations for granting us access to their High-Performance Computing facilities. Both have contributed to the research results reported within this paper. URLs: http://www.ou.edu/oscer.html and http://www.tacc.utexas.edu, respectively. This work used the Extreme Science and Engineering Discovery Environment (XSEDE), which is supported by the National Science Foundation grant number ACI-1548562.[60] Allocations: TG-BCS170001 and TG-BIO160074.

CONFLICTS OF INTEREST

None declared.

ORCID

Roman S. Voronov http://orcid.org/0000-0001-6685-949X

REFERENCES

1. American Academy of Orthopaedic Surgeons, A. One in two Americans have a musculoskeletal condition. ScienceDaily [Science News] 2016 https://www.sciencedaily.com/releases/2016/03/160301114116.htm. Accessed 06/13/2017
2. Black CRM, Goriainov V, Gibbs D, Kanczler J, Tare RS, Oreffo ROC. Bone tissue engineering. *Curr Mol Biol Rep.* 2015;1(3):132-140.
3. Delloye C, Cornu O, Druez V, Barbier O. Bone allografts. *J Bone Joint Surg Br Vol.* 2007;89-B(5):574-579.
4. Graham SM, Leonidou A, Aslam-Pervez N, et al. Biological therapy of bone defects: the immunology of bone allo-transplantation. *Expert Opin Biol Ther.* 2010;10(6):885-901.
5. Verbruggen SW, Vaughan TJ, McNamara LM. Fluid flow in the osteocyte mechanical environment: a fluid-structure interaction approach. *Biomech Model Mechanobiol.* 2014;13(1):85-97.
6. Guyot Y, Luyten FP, Schrooten J, Papantoniou I, Geris L. A three-dimensional computational fluid dynamics model of shear stress distribution during neotissue growth in a perfusion bioreactor. *Biotechnol Bioeng.* 2015;112(12):2591-2600.
7. Miyashita S, Ahmed NEMB, Murakami M, et al. Mechanical forces induce odontoblastic differentiation of mesenchymal stem cells on three-dimensional biomimetic scaffolds. *J Tissue Eng Regen Med.* 2017;11(2):434-446.
8. Zhao F, Vaughan TJ, McNamara LM. Quantification of fluid shear stress in bone tissue engineering scaffolds with spherical and cubical pore architectures. *Biomech Model Mechanobiol.* 2016;15(3):561-577.
9. Sikavitsas VI, Bancroft GN, Holtorf HL, Jansen JA, Mikos AG. Mineralized matrix deposition by marrow stromal osteoblasts in 3D perfusion culture increases with increasing fluid shear forces. *Proc Natl Acad Sci.* 2003;100(25):14683-14688.
10. Yourek G, McCormick SM, Mao JJ, Reilly GC. Shear stress induces osteogenic differentiation of human mesenchymal stem cells. *Regen Med.* 2010;5(5):713-724.

11. Stolberg S, McCloskey KE. Can shear stress direct stem cell fate? *Biotechnol Prog*. 2009;25(1):10-19.

12. Raimondi MT, Moretti M, Cioffi M, et al. The effect of hydrodynamic shear on 3D engineered chondrocyte systems subject to direct perfusion. *Biorheology*. 2006;43(3-4):215-222.

13. Kim KM, Choi YJ, Hwang JH, et al. Shear stress induced by an interstitial level of slow flow increases the osteogenic differentiation of mesenchymal stem cells through TAZ activation. *PLOS One*. 2014;9(3):e92427.

14. Brindley D, Moorthy K, Lee JH, Mason C, Kim HW, Wall I. Bioprocess forces and their impact on cell behavior: implications for bone regeneration therapy. *J Tissue Eng*. 2011;2011:620247.

15. Alvarez-Barreto JF, Linehan SM, Shambaugh RL, Sikavitsas VI. Flow perfusion improves seeding of tissue engineering scaffolds with different architectures. *Ann Biomed Eng*. 2007;35(3):429-442.

16. Guyot Y, Papantoniou I, Luyten FP, Geris L. Coupling curvature-dependent and shear stress-stimulated neotissue growth in dynamic bioreactor cultures: a 3D computational model of a complete scaffold. *Biomech Model Mechanobiol*. 2016;15(1):169-180.

17. Hossain MS, Bergstrom DJ, Chen XB. Prediction of cell growth rate over scaffold strands inside a perfusion bioreactor. *Biomech Model Mechanobiol*. 2015;14(2):333-344.

18. Bergemann C, Elter P, Lange R, et al. Cellular nutrition in complex three-dimensional scaffolds: a comparison between experiments and computer simulations. *Int J Biomater*. 2015;2015:584362.

19. Olivares AL, Marsal È, Planell JA, Lacroix D. Finite element study of scaffold architecture design and culture conditions for tissue engineering. *Biomaterials*. 2009;30(30):6142-6149.

20. Chen Y, Schellekens M, Zhou S, et al. Design optimization of scaffold microstructures using wall shear stress criterion towards regulated flow-induced erosion. *J Biomech Eng*. 2011;133(8):081008. -081008-10

21. Ali D, Sen S. Finite element analysis of mechanical behavior, permeability and fluid induced wall shear stress of high porosity scaffolds with gyroid and lattice-based architectures. *J Mech Behav Biomed Mater*. 2017;75:262-270.

22. Maes F, Claessens T, Moesen M, van Oosterwyck H, van Ransbeeck P, Verdonck P. Computational models for wall shear stress estimation in scaffolds: a comparative study of two complete geometries. *J Biomech*. 2012;45(9):1586-1592.

23. Maes F, van Ransbeeck P, van Oosterwyck H, Verdonck P. Modeling fluid flow through irregular scaffolds for perfusion bioreactors. *Biotechnol Bioeng*. 2009;103(3):621-630.

24. VanGordon SB, Voronov RS, Blue TB, Shambaugh RL, Papavassiliou DV, Sikavitsas VI. Effects of scaffold architecture on preosteoblastic cultures under continuous fluid shear. *Ind Eng Chem Res*. 2011;50(2):620-629.

25. Voronov R, VanGordon S, Sikavitsas VI, Papavassiliou DV. Computational modeling of flow-induced shear stresses within 3D salt-leached porous scaffolds imaged via micro-CT. *J Biomech*. 2010;43(7):1279-1286.

26. Voronov RS, VanGordon SB, Sikavitsas VI, Papavassiliou DV. Efficient Lagrangian scalar tracking method for reactive local mass transport simulation through porous media. *Int J Numer Methods Fluids*. 2011;67(4):501-517.

27. Cioffi M, Boschetti F, Raimondi MT, Dubini G. Modeling evaluation of the fluid-dynamic microenvironment in tissue-engineered constructs: a micro-CT based model. *Biotechnol Bioeng*. 2006;93(3):500-510.

28. Cioffi M, Küffer J, Ströbel S, Dubini G, Martin I, Wendt D. Computational evaluation of oxygen and shear stress distributions in 3D perfusion culture systems: macro-scale and micro-structured models. *J Biomech*. 2008;41(14):2918-2925.

29. Porter B, Zauel R, Stockman H, Guldberg R, Fyhrie D. 3-D computational modeling of media flow through scaffolds in a perfusion bioreactor. *J Biomech*. 2005;38(3):543-549.

30. Hendrikson WJ, Deegan AJ, Yang Y, et al. Influence of additive manufactured scaffold architecture on the distribution of surface strains and fluid flow shear stresses and expected osteochondral cell differentiation. *Front Bioeng Biotechnol*. 2017;5(6).

31. Zermatten E, Vetsch JR, Ruffoni D, Hofmann S, Müller R, Steinfeld A. Micro-computed tomography based computational fluid dynamics for the determination of shear stresses in scaffolds within a perfusion bioreactor. *Ann Biomed Eng*. 2014;42(5):1085-1094.

32. Jungreuthmayer C, Donahue SW, Jaasma MJ, et al. A comparative study of shear stresses in collagen-glycosaminoglycan and calcium phosphate scaffolds in bone tissue-engineering bioreactors. *Tissue Eng Part a*. 2009;15(5):1141-1149.

33. Campos Marin A, Lacroix D. The inter-sample structural variability of regular tissue-engineered scaffolds significantly affects the micromechanical local cell environment. *Interface Focus*. 2015;5(2).

34. Liu B et al. Perfusion applied to a 3D model of bone metastasis results in uniformly dispersed mechanical stimuli. *Biotechnol Bioeng*. 2017;1-10.

35. Wu P-H, Giri A, Sun SX, Wirtz D. Three-dimensional cell migration does not follow a random walk. *Proc Natl Acad Sci U S a*. 2014;111(11):3949-3954.

36. Voronov RS, VanGordon SB, Sikavitsas VI, Papavassiliou DV. Distribution of flow-induced stresses in highly porous media. *Appl Phys Lett*. 2010;97(2):024101.

37. Simmons AD, Williams C III, Degoix A, Sikavitsas VI. Sensing metabolites for the monitoring of tissue engineered construct cellularity in perfusion bioreactors. *Biosens Bioelectron*. 2017;90:443-449.

38. Du, X. and M. Ostoja-Starzewski, On the size of representative volume element for Darcy law in random media. *Proceedings of the Royal Society A: Mathematical, Physical and Engineering Science*, 2006.462(2074): p. 2949.

39. Voronov, R., Fluid shear stress and nutrient transport effects via lattice-Boltzmann and Lagrangian scalar tracking simulations of cell culture media perfusion through artificial bone engineering constructs imaged with microCT [PhD Thesis], in *School of Chemical,Biological and materials Engineering*. 2010, University of Oklahoma. p. 131.

40. Shakhawath Hossain M, Bergstrom DJ, Chen XB. A mathematical model and computational framework for three-dimensional chondrocyte cell growth in a porous tissue scaffold placed inside a bi-directional flow perfusion bioreactor. *Biotechnol Bioeng*. 2015;112(12):2601-2610.

41. Papavassiliou DV, Pham NH, Kadri OE, Voronov RS. Chapter 23—lattice Boltzmann methods for bioengineering applications. In: *Numerical Methods and Advanced Simulation in Biomechanics and Biological Processes*. Academic Press; 2018:415-429.

42. Williams C, Kadri O, Voronov R, Sikavitsas V. Time-dependent shear stress distributions during extended flow perfusion culture of bone tissue engineered constructs. *Fluids*. 2018;3(2):25.

43. Wang J, Zhang X, Bengough AG, Crawford JW. Domain-decomposition method for parallel lattice Boltzmann simulation of incompressible flow in porous media. *Physical Review E*. 2005;72(1):016706.

44. Voronov RS, Stalker TJ, Brass LF, Diamond SL. Simulation of intrathrombus fluid and solute transport using in vivo clot structures with single platelet resolution. *Ann Biomed Eng*. 2013;41(6):1297-1307.

45. Alam TA, Pham QL, Sikavitsas VI, Papavassiliou DV, Shambaugh RL, Voronov RS. Image-based modeling: a novel tool for realistic simulations of artificial bone cultures. *Technology*. 2016;04(04):229-233.

46. Pham NH, Voronov RS, Tummala NR, Papavassiliou DV. Bulk stress distributions in the pore space of sphere-packed beds under Darcy flow conditions. *Phys Rev E Stat Nonlin Soft Matter Phys*. 2014;89(3):033016.

47. Pham NH, Voronov RS, Vangordon SB, Sikavitsas VI, Papavassiliou DV. Predicting the stress distribution within scaffolds with ordered architecture. *Biorheology*. 2012;49(4):235-247.

48. Qian YH, Dhumieres D, Lallemand P. Lattice Bgk models for Navier-Stokes equation. *Europhysics Letters*. 1992;17(6bis):479-484.

49. Bhatnagar PL, Gross EP, Krook M. A model for collision processes in gases.1. Small amplitude processes in charged and neutral one-component systems. *Phys Rev*. 1954;94(3):511-525.

50. Sukop MC, Thorne DT, NetLibrary Inc. *Lattice Boltzmann Modeling: An Introduction for Geoscientists and Engineers*. Berlin; New York. p. ix: Springer; 2006:172.

51. Lakhotia S, Papoutsakis ET. Agitation induced cell injury in microcarrier cultures—protective effect of viscosity is agitation intensity dependent—experiments and modeling. *Biotechnol Bioeng*. 1992;39(1):95-107.

52. Wyllie MRJ, Rose WD. Application of the Kozeny equation to consolidated porous media. *Nature*. 1950;165(4207):972.

53. Ozgumus T, Mobedi M, Ozkol U. Determination of Kozeny constant based on porosity and pore to throat size ratio in porous medium with rectangular rods. *Eng Appl Comp Fluid Mech*. 2014;8(2):308-318.

54. Bacabac RG, Smit TH, Mullender MG, Dijcks SJ, van Loon JJWA, Klein-Nulend J. Nitric oxide production by bone cells is fluid shear stress rate dependent. *Biochem Biophys Res Commun*. 2004;315(4):823-829.

55. Bakker AD, Soejima K, Klein-Nulend J, Burger EH. The production of nitric oxide and prostaglandin E_2 by primary bone cells is shear stress dependent. *J Biomech*. 2001;34(5):671-677.

56. Kim J-H, Ochoa JA, Whitaker S. Diffusion in anisotropic porous media. *Transp Porous Media*. 1987;2(4):327-356.

57. Efendiev Y, Hou T. Multiscale finite element methods. In: *Surveys and Tutorials in the Applied Mathematical Sciences*. 1st ed. Vol.4 New York: Springer-Verlag New York. XII; 2009:234.

58. Van Bael S, Chai YC, Truscello S, et al. The effect of pore geometry on the in vitro biological behavior of human periosteum-derived cells seeded on selective laser-melted Ti6Al4V bone scaffolds. *Acta Biomater*. 2012;8(7):2824-2834.

59. Habashi F, H.M. M. *Computational Fluid Dynamics Techniques*. CRC Press; 1995:872.

60. Towns J, Cockerill T, Dahan M, et al. XSEDE: accelerating scientific discovery. *Comput Sci Eng*. 2014;16(5):62-74.

SUPPORTING INFORMATION

Additional supporting information may be found online in the Supporting Information section at the end of the article.

How to cite this article: Kadri OE, Williams III C, Sikavitsas V, Voronov RS. Numerical accuracy comparison of two boundary conditions commonly used to approximate shear stress distributions in tissue engineering scaffolds cultured under flow perfusion. *Int J Numer Meth Biomed Engng*. 2018;e3132. https://doi.org/10.1002/cnm.3132

531879 An *in-Silico* study of Feedforward Predictive Control in Blood Glucose Concentration for People with Type 1 Diabetes

Abstract:

Artificial pancreas (AP) has the potential to tightly control glucose concentration for people with type 1 diabetes. AP has seen rapid development in semi-automatic control in recent years. To further advance AP, fully automatic/closed-loop control that can determine meal bolus infusion automatically is the next step. The core component in fully closed-loop AP is a control algorithm that can determine insulin infusion rate. Model-based control algorithms are wide used in various AP applications. However, a critical drawback in them is the need for accurate glucose prediction far into the future due to long time lags for system inputs (e.g. carbs, insulin). In this context, the proposed method, feedforward predictive control (FFPC), does not require future prediction of glucose and has the potential for perfect control, which are major advantages with great potential to significantly tighten glucose control in this application.

This work first demonstrates theoretical performance limits of FFPC in first order plus dead time (FOPDT) processes with two inputs under ideal conditions (i.e. manipulated variables (MVs) with the ability to both raise and reduce the controlled variable). This FOPDT study presents the proposed algorithm mathematically and demonstrates its strengths under ideal conditions. Next, the proposed FFPC is applied and evaluated using 30 virtual subjects from a FDA approved diabetes simulator in a comparison study against model predictive control (MPC), with insulin infusion rate as the MV that only has one-directional effects on glucose (i.e. insulin being only capable of reducing glucose level).

For the FOPDT study, FFPC shows it can reach perfect control under ideal conditions, while MPC failed to reach the same level of performance under the same conditions. For simulation study in diabetes simulator, in terms of the standard deviation (Stdev) of glucose about its mean, FFPC was about 4% smaller than MPC on the average, which indicates the potential of FFPC with restrictions on inputs (i.e. insulin as the only MV).

This study shows FFPC has the potential of perfect control given hormones that can raise glucose can be manipulated in the future (e.g. dual-hormone infusion systems). This prospect of perfect control and not using glucose predictions will be a major advantage in AP research.

533270 A Model-Based Investigation of Cytokine Storm for T-Cell Therapy

Abstract:

In the last decade, scientists developed a process to re-engineer T-cells from a patient which currently serve as a last resort against blood cancers, such as acute lymphoblastic leukemia. The genetically engineered T-cells are called Chimeric-Antigen Receptor (CAR-T) cells, where the T-cells are reproducing with the specific antigen against cancer cells (1) (2). Those treated have seen wildly successful rates of cancer remission (3). Introduction of CAR-T cells into a patient's blood system often leads to a moderate to severe cytokine release syndrome (CRS), and in its most lethal stage what is called a cytokine storm (3). Normally, CRS affects limbs, organs and other areas of the body, but can also affect the brain. This is referred to as cerebral CRS (4). Cytokines are regulated and produced in a number of different ways; from T-cells, within feedback loops of other cytokines, and with other signalizing and indicator molecules (5). Cytokine storm and cytokine release syndrome are characterized by an unregulated immune response to the CAR-T cell therapy or other foreign molecule, inducing a large quantity of cytokines to be released which can be potentially life-threatening to the patient receiving the therapy. The most infamous case of cytokine storm was the millions of healthy individuals who perished in 1918 thanks to the Spanish Pandemic flu. Thankfully today, there are several drugs on the market which serve to counteract this upregulation of cytokines, such as Tocilizumab, an antagonist of the IL6 receptor, as well as an IL2 receptor antagonist antibody (6) (3). In several cases, these two drugs have saved patients experiencing cytokine release and cytokine storm from near-death experiences (7).

In 2006, a Phase 1 clinical trial of a CD28 monoclonal antibody superagonist was introduced to 6 healthy patients, all male subjects (7). The drug had previously been tested in other animal studies and had been successful (8). In less than an hour, all the patients exposed to the superagonist had spiked cytokine levels, where some of the markers of cell death and inflammation, such as cytokines Interferon-γ and TNF-α were at rising at a dangerous rate (7). Doctors stepped in and delivered analgesic drugs and an IL2 receptor antagonist antibody to blunt the reaction. Although all 6 patients survived, all of them had permanent effects from only 8mg of the CD28 superagonist. Doctors captured frequent blood samples and measured cytokine levels using cytobeads, as well as levels of T-cells, macrophages and other indicators of inflammation (7). Half a decade later, Stengel's group at Princeton University developed the first human multi-cytokine computational model in an article entitled, *Dynamics of a Cytokine Storm (6)*. It is the only model to date which models human cytokine levels on this scale. The researchers developed a second order linear system of equations, both uncoupled and coupled for all nine cytokines monitored. This model illustrated the positive and negative relationship between cytokines; it was an 18 variable, 90 parameter model.

In this work, we further simulated Stengel's CRS model to: 1) study the variation ranges (i.e., uncertainties) of TNF-α, IFN-γ, IL1, IL2, IL4, IL6, IL8, IL10 and IL12 from 5,000 *in silico* patients treated by the T-Cell therapy; and 2) test the synergistic effect of simultaneous inhibitions of multiple cytokines on reducing CRS. Specifically, a local sensitivity analysis was conducted to determine the most sensitive parameters out of the 90 parameters in the model. Each parameter was changed -50% and 50% and the change of the peak values of all 9 cytokines was recorded. Parameters were then ranked based on their sensitivity values. It is found that 7 of 10 of the most sensitive parameters in the model were related to IL12 and IL12 coupling with other cytokines. This gave insight to look into IL12 interactions. From the learnings of the sensitivity analysis, a Monte Carlo simulation was conducted with variation in these 10 parameters to simulate 5,000 *in-silico* patients and obtain the corresponding peak cytokine levels. The results enabled the team to conduct a principal component analysis (PCA) as well as hierarchical clustering, which is different from the original model developed in the Yiu article (6). The original article looked at the wave function profile of just the patients from the clinical trial, while this PCA took a look at the variation in the *in-silico* patients generated in MATLAB. In addition, the PCA enabled the team to utilize current drugs out on the market to simulate *in-silico* individual cytokine knockouts using the model from the original Yiu paper. The results demonstrate the effect that IL2 and IL6 have on the relationship with other cytokines, which confirms the positive and negative relationships from other models and literature reviews (5).

References

1. Kinetics and biomarkers of severe cytokine release syndrome after CD19 chimeric antigen receptor-modified T-cell therapy. Hay, K., et al. 2017, Blood, pp. 2295-2306.

2. Engineered T Cells for cancer therapy. June, C, et al. s.l. : Cancer Immunology Immunotherapy, 2014.

3. Current Concepts in the diagnosis and management of cytokine release syndrome. Lee, D. W., et al. 2014, Blood , pp. 188-195.

4. Predominant Cerebral Cytokine Release syndrome in Cd19-directed Chimeric Antigen receptor-modified T-Cell Therapy. Hu, Y, et al. s.l. : Journal of Hematology and Oncology, 2016.

5. Different Subsets of T Cells, Memory, Effector Functions, and CAR-T Immunotherapy. Golubovskaya, V. and Wu, L. 2016, Cancers, pp. 1-12.

6. Dynamics of a Cytokine Storm. Yiu, H. H., Graham, A. L. and Stengel, R. F. 2012, PLOS One, pp. 1-15.

7. Cytokine Storm in a Phase 1 Trial of the Anti-CD28 Monoclonal Antibody TGN1412. Suntharalingam, G., et al. s.l. : The New England Journal of Medicine, 2006, Vol. 355.

8. Monoclonal antibody TGN1412 trial failure explained by species differences in CD28 expression on CD4+ Effector memory T-cells. Eastwood, D, et al. Hertfordshire, UK : British Journal of Pharmacology, 2010.

535781 Data-Driven Discovery of Novel Therapeutic Targets through Metabolic Modeling of *Staphylococcus Aureus*

Abstract:

Recent trends have recorded a significant rise in community-associated methicillin-resistant *Staphylococcus aureus*(CA-MRSA) infections. Additionally, most clinical isolates can develop into recalcitrant multiple-antibiotic-resistant and host immune-tolerant biofilm communities, necessitating novel therapeutic strategies for combating *S. aureus*infections. Therefore, there has been a steady interest and focus towards understanding how staphylococcal metabolism relates to antibiotic resistance and pathogenesis. In recent years, whole genome sequencing has accelerated the reconstruction of genome-scale metabolic models of several species. However, these models often do not accurately predict biological phenotypes due to inherent insufficiencies, including metabolic gaps, unbalanced reactions, and lack of key physiological information.

We reconstructed the genome-scale metabolic model of *S. aureus* by combining genome annotation data, reaction stoichiometry and thermodynamic information, and regulation information from biochemical databases and previous strain-specific models [1-7], which were validated though our experimental observations. We then employed our model to identify metabolic-model driven genetic intervention strategies to combat the pathogenic activity of *S. aureus*. Reactions in the model were elemental and charge balanced. Upon manual and automated gap-filling, the model has 850 metabolic genes, 1437 metabolites and 1718 reactions that include inter-compartment transport and exchange reactions. To resolve the growth and no-growth inconsistencies in the model, we used an automated procedure called GrowMatch to reconcile the inconsistency predictions by suppressing or adding functionalities in the model while maintaining the already identified correct growth and no-growth predictions [8].

The extensive manual curation performed on the draft genome-scale reconstruction resulted in improved prediction capabilities while successfully capturing experimental results on growth inhibition. Our synthetic lethality analyses identified both intuitive and novel multiple gene-knockout strategies for reducing growth of *Staphylococcus*to a specific predetermined threshold. We performed correlation analyses between the synthetic lethal gene pairs to pinpoint the most probable candidates for genetic manipulations that fight the pathogenicity of *S. aureus*. We identified some experimentally-observed metabolic bottlenecks in wild-type and mutant growth, and predicted additional growth-inhibiting single- and double-knockouts which can potentially provide a solution to the prevailing antibiotic resistance of this organism.The results serve as a foundation from which to build, modify, and constantly improve, through the incorporation of future â€œ-omicsâ€ data, the prediction and evaluation of novel therapeutic targets, thus, enhancing its functional utility and use as a resource to augment staphylococcal research worldwide.

References:

1. Becker SA, Palsson BO. Genome-scale reconstruction of the metabolic network in Staphylococcus aureus N315: an initial draft to the two-dimensional annotation. *BMC microbiology*. Mar 07 2005;5:8.
2. Bosi E, Monk JM, Aziz RK, Fondi M, Nizet V, Palsson BO. Comparative genome-scale modelling of Staphylococcus aureus strains identifies strain-specific metabolic capabilities linked to pathogenicity. *Proc Natl Acad Sci U S A*. Jun 28 2016;113(26):E3801-3809.
3. Henry CS, DeJongh M, Best AA, Frybarger PM, Linsay B, Stevens RL. High-throughput generation, optimization and analysis of genome-scale metabolic models. *Nature biotechnology*. Sep 2010;28(9):977-982.
4. Schellenberger J, Park JO, Conrad TM, Palsson BO. BiGG: a Biochemical Genetic and Genomic knowledgebase of large scale metabolic reconstructions. *BMC bioinformatics*. Apr 29 2010;11:213.
5. Kanehisa M, Goto S. KEGG: Kyoto Encyclopedia of Genes and Genomes. *Nucleic acids research*. Jan 1 2000;28(1):27-30.
6. Pence HE, Williams A. ChemSpider: An Online Chemical Information Resource. *J Chem Educ*. Nov 2010;87(11):1123-1124.
7. Ren Q, Chen K, Paulsen IT. TransportDB: a comprehensive database resource for cytoplasmic membrane transport systems and outer membrane channels. *Nucleic Acids Res*. // 2007;35:D274-D279.
8. Kumar VS, Maranas CD. GrowMatch: an automated method for reconciling in silico/in vivo growth predictions. *PLoS computational biology*. Mar 2009;5(3):e1000308.

533963 A Process Resilience Analysis Framework (PRAF) Application for Recoverability Assessment Model of Offshore Oil and Gas Platforms

Abstract:

Abstract

In the offshore oil and gas platforms, process safety incidents cause severe consequences such as fatalities, injuries, loss of property, financial losses, and poor reputation. The effectiveness of safety and emergency measures and their implementation is impacted due to the demanding environment and complex nature of the offshore operations [1,2]. Many studies in literature have introduced the recoverability or emergency assessment models [3, 4, 5]. These studies have addressed mostly the technical aspects or social aspects in a fragmented manner. The challenge is to integrate both the technical and social aspects in a comprehensive manner that utilizes the key metrics data.

In this work, a recoverability assessment model is described and an optimum combination and prioritization of mitigation barriers are determined. The overall purpose is to obtain a reduction in the response time culminating in reduced consequences and enhanced recovery. This model is based on the Process Resilience Analysis Framework (PRAF)[6,7] and considers the resilience metrics that integrate both technical (process parameters variations) and social (human and organizational) factors. Some examples of these process system resilience metrics are tests for emergency systems and procedures; mock drills for emergency situations; process safety required training sessions completed; process safety required emergency procedures reviewed or revised; and shift handover communication violations. It is a quantitative and data-driven model. An illustration of an offshore oil and gas platform with various emergency scenarios is applied to demonstrate the developed recoverability assessment model. The key elements of this model are â€" 1) a process simulation model that captures the details such as process variations, 2) a safety model that takes into account the impact scenarios and respective losses, 3) building model that covers the geometry of the platform, and 4) a cost objective function. This work is a significant contribution as it employs the data that is collected at the facility and models scenarios that are challenging to simulate. The results indicate that this recoverability assessment model based on PRAF can accurately simulate the emergency scenarios and aids in improving the response to such situations. This leads to decreased severity of consequences and hence betters recovery of the process system to transit to the normal operations. The established model is also applicable to other industries such as the pharmaceutical, construction, and nuclear industry.

References

1. Cheng, J. C., Tan, Y., Song, Y., Mei, Z., Gan, V. J., & Wang, X. (2018). Developing an evacuation evaluation model for offshore oil and gas platforms using BIM and agent-based model. Automation in Construction, 89, 214-224.
2. Norazahar, N., Smith, J., Khan, F., & Veitch, B. (2018). The use of a virtual environment in managing risks associated with human responses in emergency situations on offshore installations. Ocean Engineering, 147, 621-628.
3. Andersen, S., & Mostue, B. A. (2012). Risk analysis and risk management approaches applied to the petroleum industry and their applicability to IO concepts. Safety Science, 50(10), 2010-2019.
4. Deacon, T., Amyotte, P. R., & Khan, F. I. (2010). Human error risk analysis in offshore emergencies. Safety science, 48(6), 803-818.
5. Joo, J., Kim, N., Wysk, R. A., Rothrock, L., Son, Y. J., Oh, Y. G., & Lee, S. (2013). Agent-based simulation of affordance-based human behaviors in emergency evacuation. Simulation Modelling Practice and Theory, 32, 99-115.
6. Jain, P., Pasman, H. J., Waldram, S. P., Rogers, W. J., & Mannan, M. S. (2016). Did we learn about risk control since Seveso? Yes, we surely did, but is it enough? An historical brief and problem analysis. Journal of Loss Prevention in the Process Industries.
7. Jain, P., Pasman, H. J., Waldram, S., Pistikopoulos, E. N., & Mannan, M. S. (2017). Process Resilience Analysis Framework (PRAF): A systems approach for improved risk and safety management. Journal of Loss Prevention in the Process Industries.

532078 A Mixed-Integer Programming Framework for Combined Placement of Fire and Gas Detectors in Chemical Processing Facilities

Abstract:

In chemical processing plants, the placement of sensors to alert personnel of hazards is critical for public safety. However, given the large number and variety of chemical and physical sensors necessary to ensure process safety, this optimal placement problem becomes difficult to formulate and solve. A typical petrochemical plant includes many gas (e.g. ultrasonic, catalytic bead) and fire detectors (e.g. point infrared, open path infrared, optical). Gas detectors must be in contact with gas leaks to be effective, while fire detectors require a visual path to the fire, free of obstructions, for successful detection. Thus, it is important that both types of detectors be placed according to their respective characteristics for maximum effectiveness.

The optimal gas detector placement problem was formulated by [1] as a stochastic program based on the p-median facility location problem and relied on scenarios generated from CFD simulations within a petrochemical facility. Their results showed that coupling coverage with expected detection time using a scenario-based method outperformed the existing coverage-only approaches, and has inspired formulations based on Conditional-Value-at-Risk [2], minimization of false positives and false negative alarms with voting schemes [3], maximum risk reduction of scenarios [4], and MINLP approaches for non-uniform failure probabilities [5].

But while gas detectors must be placed in the path of gas leaks for effective detection, fire detectors must be placed within unobstructed visual distance of flames. The optimal placement of visual fire detectors in processing plants can be particularly difficult due to the large number of physical obstructions (e.g. piping, valves, tanks). Most relevantly, [6] has proposed a basic formulation for placing visual fire detectors on the walls of a room with cylindrical objects, weighted by pre-determined hazard levels.

This work focuses on a mixed-integer optimization formulation that integrates placement of both fire (visual) and gas (point) detectors. The scenario-based, p-median approach demonstrated by [1] for gas detectors is combined with a maximum coverage model for fire detectors in a 3D space within a multi-objective framework. We show that this approach allows for the simultaneous placement of both sensor types for a fixed sensor budget while under uncertainty. We demonstrate the effectiveness of this formulation with fire and gas leak simulations from a chemical facility. We further discuss how the framework is readily scalable for larger problem sizes and amenable to various objectives.

References:

[1] Legg, S. W., Benavides-Serrano, A. J., Siirola, J. D., Watson, J. P., Davis, S. G., Bratteteig, A., and Laird, C. D. (2012). A stochastic programming approach for gas detector placement using CFD-based dispersion simulations. Computers and Chemical Engineering, 47:194–201.

[2] Legg, S. W., Wang, C., Benavides-Serrano, A. J., and Laird, C. D. (2013). Optimal gas detector placement under uncertainty considering Conditional-Value-at-Risk. Journal of Loss Prevention in the Process Industries, 26(3):410–417.

[3] Benavides-Serrano, A. J., Legg, S. W., Vázquez-Román, R., Mannan, M. S., Laird, C. D., Vaquez-Roma, R., Mannan, M. S., Laird, C. D., and Kay, M. O. (2014). A stochastic programming approach for the optimal placement of gas detectors: Unavailability and voting strategies. Industrial and Engineering Chemistry Research, 53(13):5355–5365.

[4] Rad, A., Rashtchian, D., and Badri, N. (2017). A risk-based methodology for optimum placement of flammable gas detectors within open process plants. Process Safety and Environmental Protection, 105:175–183.

[5] Liu, J., and Laird, C.D., A Global Stochastic Programming Approach for the Optimal Placement of Gas Detectors with Nonuniform Unavailabilities, to appear in Journal of Loss Prevention in the Process Industries, 2017.

[6] Yang, T. P., Asirvadam, V. S., and Saad, N. B. (2012). Optimal placement of fire detector using dual-view. In 2012 4th International Conference on Intelligent and Advanced Systems (ICIAS2012), pages 599–603. IEEE.

537842 Optimal Design, Control and Scheduling of Multi-Product Systems Under Uncertainty: A Stochastic Back-Off Approach

Abstract:

Integration of design and control aims to specify the optimal design of a process while taking into account the dynamics of the system under different scenarios [1], [2]. In recent years, it has been shown that, for certain applications, process design and control decisions may also impact process scheduling [3], [4]. Although these tasks can be optimized individually, this approach is often found to be inadequate as it may result in sub-optimal or dynamically inoperable designs. Moreover, processes are subject to uncertainty and disturbances during operation thus adding an additional layer of complexity into the analysis. As a result, the explicit integration of these tasks results in the specification of overly challenging optimization problems, which are often found to be intractable due to the complex interactions between the tasks. Hence, there is a motivation to develop new frameworks that can perform optimal process integration.

The aim of this work is to present a decomposition algorithm for the optimal design, control and scheduling of multi-product systems subject to uncertainty and disturbances. The proposed decomposition framework involves the solution of a dynamic flexibility analysis followed by a dynamic feasibility analysis [5]. In the first (dynamic flexibility analysis), process design, control and scheduling decisions are sought using the expected values in the uncertain parameters and disturbances whereas in the latter (dynamic feasibility analysis), the solution obtained from the first is evaluated under stochastic realizations in the uncertain parameters and disturbances. The proposed framework assumes that the uncertain parameters and disturbances are stochastic random variables that follow specific probability distribution functions defined by the user. This a key feature considered in the present formulation since previous studies typically discretize the random uncertain variables [3], [5]. A back-off algorithm recently proposed in the literature is adopted to perform the optimal process integration under stochastic realizations process disturbances and uncertainty [6]. Back-off terms representing the variability of the process constraints due to stochastic realizations in uncertain parameters and disturbances are calculated using Monte Carlo (MC) sampling techniques. These back-off terms are updated at each iteration in the decomposition algorithm and added to each process constraint in the dynamic flexibility formulation to account for the stochastic realizations in the uncertain parameters and disturbances. The decomposition algorithm iterates in this fashion and eventually converges to a solution that can accommodate process variability due to uncertainty and disturbances. A case study involving the production of multiple products from a single equipment is examined to demonstrate the benefits and limitations of the proposed approach.

References

[1] Diangelakis N.A., Burnak. B., Katz J., Pistikopoulos E.N. Process design and control optimization: A simultaneous approach by multiparametric programming. AIChE Journal. 2017;63(8):4827–4846.

[2] Rafiei, M., Ricardez-Sandoval, L. Stochastic Back-Off Approach for Integration of Design and Control Under Uncertainty. Industrial & Engineering Chemistry Research. 2018;*57*(12):4351–4365.

[3] Patil B., Maia E., Ricardez-Sandoval L.A. Integration of Design, Scheduling and Control of Multi-Product Chemical Processes Under Uncertainty. AIChE Journal. 2015;61:2456-2470.

[4] Pistikopoulos, E.N., Diangelakis, N.A. Towards the integration of process design, control and scheduling: are we getting closer? Computers & Chemical Engineering. 2015;91:85–92.

[5] Koller, R., Ricardez-Sandoval, L. A Dynamic Optimization Framework for Integration of Design, Control and Scheduling of Multi-product Chemical Processes under Disturbance and Uncertainty. Computers & Chemical Engineering. 2017;106:147-159.

[6] Shi J., Biegler L.T., Hamdan I., Wassick J.M. Optimization of grade transitions in polyethylene solution polymerization process under uncertainty. Computers & Chemical Engineering. 2016;95:260–279.

533404 Risk-Averse Health-Aware Control of Subsea Plants

Abstract:

Optimal process operation is important in the subsea oil and gas industry, especially for fields with high break-even oil prices. In order to optimize production, it is desirable to maximize the throughput. However, it is at the same time necessary to consider plant reliability and availability, as maintenance costs can have a significant impact on the profitability of the plant. Maximizing throughput while simultaneously maximizing the availability of the plant is often not possible in reality, as the two objectives are conflicting. This is why it is necessary to find an acceptable trade-off between plant revenue and maintenance cost.

This trade-off is usually not defined in a systematic manner. Instead, a sequential approach is commonly employed. First, the safe operating regions are defined and imposed as back-offs from safety-critical constraints. Only after the allowable operating region has been defined it is attempted to optimize the operation. This method may be overly conservative, for at least two reasons:

- The back-off is often defined with a worst-case scenario in mind, which leads to sub-optimal performance under regular operating conditions.
- When the system is complex and many degradation modes are effecting the reliability simultaneously, it might be difficult to specify meaningful back-offs manually.

Our proposed approach is to integrate prognostic and diagnostic models into an advanced, optimization-based control strategy, either by imposing constraints on the system availability, or by penalizing system degradation in the objective function (Verheyleweghen & Jäschke, 2017). We thereby achieve proactive fault prevention rather than retroactive fault-tolerance, which is more in line with industrial needs (Escobet, Puig, & Nejjari, 2012). The problem formulation is made risk-averse by using a CVaR-formulation of the problem, similar to what is commonly done in financial decision making (Rockafellar & Uryasev, 2000).

By doing so, we get improved economic performance compared to traditional control strategies by exploiting the additional flexibility gained by reducing the large back-off from safety-critical constraints. Simultaneously, we ensure safe operation and minimize unexpected breakdowns and maintenance interventions. The degree of conservativeness of the control structure can be adjusted online to reflect the risk aversion of the decision maker. Our proposal is different from current industrial practice, where it is still common to treat reliability and optimal operation in a sequential manner.

References

Escobet, T., Puig, V., & Nejjari, F. (2012). Health aware control and model-based prognosis. *20th Mediterranean Conference on Control & Automation* (pp. 691 - 696). Barcelona: IEEE.

Rockafellar, R. T., & Uryasev, S. (2000). Optimization of conditional value-at-risk. *Journal of risk*, 21-42.

Verheyleweghen, A., & Jäschke, J. (2017). Framework for Combined Diagnostics, Prognostics and Optimal Operation of a Subsea Gas Compression System. *IFAC-PapersOnLine*, 15916-15921.

527867 Analytical and Triangular Representation of Flexibility Space

Abstract:

In process design, process parameters are usually uncertain with swift changes in the internal and external conditions. The uncertainties may lead to plant-model mismatch. A good process design should not only exhibit an optimal balance between capital and operating costs but also exhibit good operability in a practical operating environment [1]. One key process operability is the flexibility, which refers to the capability of the process to maintain feasible operation over the entire domain of uncertain parameters [2]. Evaluating the operational flexibility is an important issue that must be incorporated into design considerations. The flexibility analysis methods have been applied in many practical industries [3,4,5]. The main purpose of flexibility analysis is to determine and describe the flexibility space. Because of the inherent nonlinearity of the process design model, the flexibility space is commonly non-convex. In recent decades, the algorithms for describing non-convex flexibility spaces can be mainly divided into four types: (1) transforming the flexibility analysis model into a global or robust optimization model [6,7]; (2) iteratively constructing the boundary via the simplicial approximation method [8]; (3) generating a polygonal representation through properly sampled points [9]; (4) building the response surface of the flexibility function by using the surrogate models [10]. However, these existing methods are mainly developed by numerical calculation methods, resulting in the fact that they are used to estimate only the contour of the flexibility space, rather than providing an accurate description.

In this work, a novel method is proposed for analytically representing the flexibility space. This method can accurately describe the flexibility space and provide a triangular model to explicitly express the functional relationships between uncertain parameters. First, the original flexibility analysis model is viewed as an existential quantifier model. Then, a technique of quantifier elimination, cylindrical algebraic decomposition [11,12], is introduced to deduce the model to a set of explicitly triangular semi-algebraic systems, each of which is an analytical representation of a flexibility subspace, and any two subspaces are disjoint. Last, a logical combination of the semi-algebraic systems can be used to describe the whole flexibility space. In summary, the proposed method in flexibility analysis has four properties, that is, explicit property, nonconvex property, triangular property and operational property. In the case studies, the proposed method is successfully applied to represent non-convex flexibility spaces and disjoint flexibility spaces. The results show that the explicit expressions deduced by the proposed method can accurately and effectively describe the flexibility spaces and then guide the steady-state operations.

Keywords: process design, flexibility analysis, cylindrical algebraic decomposition.

Reference:

[1] Rooney WC, Biegler LT. Optimal process design with model parameter uncertainty and process variability. *AIChE J.* 2003;49(2):438-449.

[2] Swaney RE, Grossmann IE. An index for operational flexibility in chemical process design. Part II: Computational algorithms. *AIChE J.* 1985;31(4):631-641.

[3] Lima FV, Jia Z, Ierapetritou MG, Georgakis C. Similarities and differences between the concepts of operability and flexibility: the steady-state case. *AIChE J.* 2010;56(3):702-716.

[4] Sirdeshpande AR, Ierapetritou MG, Andrecovich MJ, Naumovitz JP. Process synthesis optimization and flexibility evaluation of air separation cycles. *AIChE J.* 2005;51(4):1190-1200.

[5] Wang H, Mastragostino R, Swartz CLE. Flexibility analysis of process supply chain networks. *Comput Chem Eng.* 2016;84:409-421.

[6] Bansal V, Perkins JD, Pistikopoulos EN. Flexibility analysis and design using a parametric programming framework. *AIChE J.* 2002;48(12):2851-2868.

[7] Zhang Q, Grossmann IE, Lima RM. On the Relation Between Flexibility Analysis and Robust Optimization for Linear Systems. *AIChE J.* 2016;62(9):3109-2132.

[8] Goyal V, Ierapetritou MG. Framework for evaluating the feasibility operability of nonconvex processes. *AIChE J.* 2003;49(5):1233-1240.

[9] Banerjee I, Ierapetritou MG. Feasibility evaluation of nonconvex systems using shape reconstruction techniques. *Ind Eng Chem Res.* 2005;44(10):3638-3647.

[10] Wang Z, Ierapetritou MG. A novel feasibility analysis method for black-box processes using a radial basis function adaptive sampling approach. *AIChE J.* 2017;63(2):532-550.

[11] Collins GE. Quantifier elimination for real closed fields by cylindrical algebraic decomposition. *Lec Notes Comp Sci.* 1975;33:134-183.

[12] Arnon DS, Collins GE, McCallum S. Cylindrical algebraic decomposition I: the basic algorithm. *SIAM J Comput.* 1984;13:865-877.

524832 Design of Flare Systems Under Uncertainty: A Chance-Constrained Nonlinear Programming Approach

Abstract:

Flare systems are used as safety (relief) devices to manage abnormal situations in infrastructure systems (natural gas and oil processing plants, and pipelines), manufacturing facilities (chemical plants, offshore rings), and power generation facilities. Flare design is influenced by several uncertain factors such as the amount and composition of the waste stream and the ambient conditions.
Here, we propose to use a chance-constrained nonlinear programming (CC-NLP) formulation to find designs that minimize cost and that mitigate the risk that the combustion radiation exceeds a certain safety threshold.
CCs are a powerful modeling paradigm that helps decision-makers control risk. Unfortunately, CCs cannot be handled directly by off-the-shelf optimization solvers. CCs can reformulated into a standard NLP when the quantile function of the constraint can be expressed in algebraic form. This property has been exploited extensively in the special case in which the constraint mapping is linear in both arguments and the random data vector is Gaussian [1]. Under these assumptions, the constraint mapping is Gaussian and its quantile can be expressed as a weighted sum of its two moments (the expectation and the variance). For more general settings it is possible to derive exact reformulations using integer variables, as originally proposed in [3]. Unfortunately, in the context of CC-NLP, integer reformulations would lead to large-scale and nonconvex MINLPs.
We review different strategies to tackle large-scale CC-NLPs arising in flare system design. In particular, we consider the special case in which the algebraic structure of the moments and of the quantile function of the constraint mapping is known or approximately known and propose to use moment matching (MM) to compute its parameters and to derive its quantile function. We argue that this procedure can be applied to a wide range of distributions that go beyond Gaussian such as uniform, log-normal, and generalized extreme value (Weibull, Frechet, Gumbel). The reason is that many random phenomena observed in science and engineering can be explained using asymptotic results such as the central limit and the extreme value (Fisher-Tippett) theorem or because basic transformations and fundamental relationships between distributions can be exploited. While this approach cannot be applied to general settings, we believe that there is value in studying the structure of the problem at hand to explore if suitable algebraic approximations emerge. To handle more general set tings, we consider the use of a recently proposed sigmoidal approximation approach (SigVaR), which can be used to solve large-scale NLPs with chance constraints by using powerful serial and parallel solvers such as Ipopt and PIPS-NLP [2, 4]. We demonstrate that this approach outperforms existing conservative approximations such as the conditional value-at-risk (CVaR) approximation that not offer a mechanism to enforce convergence to a solution of CC-P, and as the almost-surely (AS) approximation which enforces the constraint constraint with probability one (almost surely). This is equivalent to enforce the constraint for all possible realizations of the uncertainty. Our flare design study confirms that MM and SigVaR provide nearly exact solutions of the CC-NLP while AS and CVaR solutions exhibit extreme conservatism.

References

[1] D. Bienstock, M. Chertkov, and S. Harnett. Chance-constrained optimal power flow: Risk-aware network control under uncertainty. SIAM Review, 56(3):461–495, 2014.

[2] N.-Y. Chiang and V. M. Zavala. An inertia-free filter line-search algorithm for large-scale nonlinear programming. Computational Optimization and Applications, 64(2):327–354, 2016.

[3] J. Luedtke, S. Ahmed, and G. L. Nemhauser. An integer programming approach for linear programs with probabilistic constraints. Mathematical Programming, 122(2):247–272, 2010.

[4] A. W¨achter and L. T. Biegler. On the implementation of a primal-dual interior point filter line search algorithm for large-scale nonlinear programming. Mathematical Programming, 106:25–57, 2006.

536993 Multistage Stochastic Programming Using Hybrid Scenario and Decision Rule Formulation

Abstract:

As an important technique for addressing dynamic decision making problem under uncertainty, multistage stochastic programming (MSSP) has received lots of applications in process systems engineering, including planning, scheduling and supply chain management. In the literature, two of the most popular approaches for solving MSSP are the scenario based method and the linear decision rule based method. In scenario based method, the uncertainty is represented by scenarios (or scenario tree for time-dependent uncertainty). The decision is scenario dependent so the model complexity is affected by the dimension of the uncertainty [1]. On the other hand, in linear decision rule method, adjustable variables are modeled as affine functions of uncertain parameters [2]. Subsequently, constraints are converted to their deterministic counterparts using duality theorem. Compared to the scenario based formulation, the model complexity is significantly reduced. Furthermore, while scenario formulation guarantees the feasibility of the solution for scenarios considered in the model, linear decision rule based solution guarantee solution feasibility for any uncertainty realization in the predefined uncertainty set.

In practical applications, uncertainty often comes from multiple sources. They can appear as objective coefficient uncertainty, constraint left-hand-side coefficient uncertainty or right-hand-side uncertainty. In addition, those uncertainties can be static or time-dependent. Simultaneously addressing those uncertainties brings challenges to the traditional scenario based or linear decision rule based method [3]. First, using only scenario based formulation can result in a very large problem size due to the presence of multiple sources of uncertainty. Therefore, the solution can be very time consuming and computationally expensive. On the other hand, it is hard to only use a decision rule based method for problems that include multiple sources of uncertainty, especially when uncertain parameters are coefficients of adjustable variables. This results in non-convex constraints with respect to uncertain parameters and thus duality theorem cannot be used for reformulation.

In this work, we propose two strategies to combine the use of scenario and decision rule in MSSP to address the above challenges. In the first approach, time-dependent uncertainties are modeled using the scenario tree and the corresponding adjustable decisions are modeled as scenario dependent. Rest static uncertainties are modeled using uncertainty sets. The corresponding semi-infinite constraints with respect to static uncertainty can be reformulated using duality theorem. This results in a scenario dependent dynamic decision solution which is robust for the static uncertainty at any nodes of the tree. In the second approach, time-dependent uncertainty is modeled using uncertainty sets and adjustable decisions are modeled using linear decision rules with respect to those uncertainties, while static uncertain parameters are modeled as discrete scenarios in finite sets and the corresponding constraints are enforced for all scenarios. The resulting semi-infinite constraints with respect to time-dependent uncertainty can further be reformulated using duality theorem. This approach leads to an adaptive decision rule solution which is feasible for defined scenarios of the static uncertainty and any realization of the time-dependent uncertainty in the uncertainty set.

Computational studies were performed to study the performance of the two approaches using a capacity expansion planning problem. MSSP problem under multiple sources of uncertainty is studied. The problem addresses the capacity expansion of a chemical plant under demand and yield uncertainties. Both methods are implemented to solve the problem. The results demonstrate that the first approach results in a larger size problem while the second approach leads to a much smaller problem which can be solved more efficiently. In addition, while the first approach has more flexibility in adjusting the binary decisions, it only leads to slightly better solution quality than the second approach. Since the second approach has much better computational efficiency, it is recommended for multistage stochastic programming problem.

References

1. A. J. Conejo, M. Carrion, J. M. Morales, Decision Making under Uncertainty in Electricity Markets, 2010, International series in operations research and management science.
2. Ben-Tal, A., Goryashko, A., Guslitzer, E., & Nemirovski, A. (2004). Adjustable robust solutions of uncertain linear programs. Mathematical Programming, 99 (2), 351-376.
3. F. Babonneau, J.-P. Vial, R. Apparigliato, Uncertainty and Environmental Design Making, 2010, International series in operations research and management science.

535578 Optimization of Wastewater Treatment Plant Design Using an Early-Stage Techno-Economic Analysis Under Uncertainty

Abstract:

Selection of an optimal wastewater treatment plant (WWTP) layout is a formidable challenge that is growing in complexity due to increasingly stringent legal demands on effluent quality, the ever-growing number of competing treatment technologies, and the limited financial resources. Several environmental decision support tools (Bozkurt et al., 2015; Castillo et al., 2016) have been proposed to assist design engineers in their early stage decision making for building or retrofitting a WWTP network. These tools provide users with a number of candidate process flow diagrams for WWTP layout given a defined objective by making use of expert knowledge-based and mathematical programming-based process synthesis approaches. Further techno-economic evaluation of these candidate designs with a level of detail available at the early stage of the project development is needed to foster the decision making process. Although a variety of methodological approaches have been reported in the literature which account for different types of costs of wastewater treatment (i.e. energy costs, environmental externalities such as effluent tax, etc.) with a varying level of detail (Rodríguez-Garcia et al.,2011; Molinos-Senante et al., 2013), the comparability and applicability of these works at the early stage of decision making is limited. It is well-known that significant economic trade-offs exists between investment and operational costs which are subject to a large uncertainty (U-tapao et al., 2015). Therefore, there is a significant need for a comparative techno-economic analysis of the early-stage alternative candidate designs to further assist decision making under various uncertainties like variations in influent wastewater pollutant load and composition, cost of electricity used for aeration, and the price of biogas produced in sludge digesters, etc.

The main objective of this work is to develop an early-stage techno-economic assessment methodology for the selection of the optimal network among the alternative WWTP networks by taking into account available plant-wide simulation data to better estimate operational costs (sludge disposal, effluent tax, and energy consumption) along with major capital investment costs. To this end, as opposed to simplistic input-output type process models used at the design stage, a Monte Carlo sampling-based optimization under uncertainty framework, utilizing rigorous non-linear mechanistic models of individual technological treatment units able to handle industrially relevant problems and scales, has been developed to select candidate designs among alternative treatment technologies. In order also to incorporate uncertainties at the design selection stage, a techno-economic analysis methodology accounting for investment costs and operational costs with data generated from rigorous plant-wide simulations is used. Effluent quality index (EQI) and net present value (NPV) metrics are respectively used to evaluate technical and economic performances of the alternative designs. Uncertainty and sensitivity analyses are complementarily carried out in order to judge the robustness of the assessments against uncertainties in influent wastewater characteristics and technical performance metrics. The proposed methodology effectively quantifies the effect of uncertain information on the optimum design selection, further fostering decision-making in industrially relevant wastewater treatment design problems.

References

Bozkurt, H., Quaglia, A., Gernaey, K. V., & Sin, G. (2015). A mathematical programming framework for early stage design of wastewater treatment plants. *Environmental Modelling & Software*, *64*, 164–176. https://doi.org/10.1016/j.envsoft.2014.11.023

Castillo, A., Cheali, P., Gómez, V., Comas, J., Poch, M., & Sin, G. (2016). An integrated knowledge-based and optimization tool for the sustainable selection of wastewater treatment process concepts. *Environmental Modelling & Software*, *84*, 177–192. https://doi.org/10.1016/j.envsoft.2016.06.019

Molinos-Senante, M., Hernández-Sancho, F., Sala-Garrido, R., & Cirelli, G. (2013). Economic feasibility study for intensive and extensive wastewater treatment considering greenhouse gases emissions. *Journal of Environmental Management*, *123*, 98–104. https://doi.org/10.1016/j.jenvman.2013.02.044

Rodriguez-Garcia, G., Molinos-Senante, M., Hospido, A., Hernández-Sancho, F., Moreira, M. T., & Feijoo, G. (2011). Environmental and economic profile of six typologies of wastewater treatment plants. *Water Research*, *45*(18), 5997–6010. https://doi.org/10.1016/j.watres.2011.08.053

U-tapao, C., Gabriel, S. A., Peot, C., & Ramirez, M. (2015). Stochastic, Multiobjective, Mixed-Integer Optimization Model for Wastewater-Derived Energy. *Journal of Energy Engineering*, *141*(1), B5014001. https://doi.org/10.1061/(ASCE)EY.1943-7897.0000195

537220 Adaptive Scheduling of Steelmaking and Continuous Cast Process Under Uncertainty

Abstract:

Steelmaking and continuous casting (SCC) is widely used in industry to convert raw iron ore and scrap iron into various steel products. SCC comprises three major processing stages: steelmaking, refining and continuous casting. Each of them involves high-energy, high-temperature, complex processes which require heavy investment and operation costs. The optimal scheduling of these processes would result in reduced energy consumption and costs, increased profits and an improved refining-casting-delivery schedule. In practice, SCC process scheduling is complicated because of the interacting and sequential nature of the processes, and strict material, flow time, quality and delivery time constraints. Furthermore, additional complexity is introduced into the scheduling exercise when one accounts for the uncertainty associated with a large host of parameters involved, such as demand fluctuation and processing time uncertainty.

Tang, et al (1998) presented a linear programming model approach to develop a production schedule for SCC. Their technique consisted of a just-in-time approach based on meeting product demand as well as ensuring continuity in process operations. Atighehchian, et al (2009) proposed an algorithm that treats the SCC scheduling process as a hybrid flow shop scheduling exercise and solves it using a combination of ant colony optimization and nonlinear optimization. The stochastic SCC scheduling process can be dealt with in two ways: reactive scheduling and preventive scheduling. Reactive scheduling deals with uncertainties pertaining to unforeseen product cancellations or breakdowns during operation, whereas preventive scheduling deals with uncertainties pertaining to processing time, product demand and prices. Rodrigues, et al (1996) presented a mixed integer linear programming formulation for reactive scheduling using state task network representation and a rolling horizon approach. Ye, et al (2014) compared two approaches for dealing with product demand uncertainty during the SCC process: static robust optimization and two-stage stochastic optimization framework, and compared the quality of solutions obtained from both approaches.

The focus of this work is to generate adaptive scheduling policy for the SCC process when there is uncertainty in ladle processing time. The proposed method is based on multistage adaptive optimization using a linear decision rule based solution technique. Compared to static robust optimization techniques, the proposed method generates less conservative, yet robust schedule solution, which is adjustable based on the observed uncertainty realizations. The proposed solution framework handles two types of objectives: a risk-averse objective function which leads to adaptive robust optimization and a risk-neutral objective function which leads to adaptive stochastic optimization. The optimal solution is a set of linear decision rules for adaptive decision variables, rather than the exact values. Accordingly, realized uncertainties until a particular decision stage are plugged into the linear decision rules, and then, the values for decision variables are determined for the next stage. Moreover, since the decision rules are linear, the problem size is a polynomial function of the number of stages. Therefore, this approach is a good candidate for large scale SCC scheduling problems.

The proposed methods were tested through several case studies. Although the adaptive robust and adaptive stochastic solution give different objective values, the scheduling solutions given by both models are robust and feasible for all possible uncertainty realizations. However, the adaptive robust solution is more conservative than the adaptive stochastic solution. Hence, the adaptive stochastic optimization method is found to be preferable for SCC scheduling under uncertainty.

References

1. J. Li, X. Xiao, Q. Tang, C. A. Floudas (1998). Production Scheduling of a Large-Scale Steelmaking Continuous Casting Process via Unit-Specific Event-Based Continuous-Time Models: Short-Term and Medium-Term Scheduling. *Ind. Eng. Chem. Res.*, volume (51), 7300-7319.
2. L. Tang, J. Liu, A. Rong, and Z. Yang (1998). A mathematical programming model for scheduling steelmaking-continuous casting production. *European Journal of Operational Research*, volume (120), 423-435
3. A. Atighehchian, M. Bijari and H. Tarkesh (2009). A novel hybrid algorithm for scheduling steel-making continuous casting production. *Computers & Operations Research*, volume (36), no. 8, 2450-2461.
4. M. T. M. Rodrigues, L. Gimeno, C. A. S. Passos and M. D. Campos (1996). Reactive scheduling approach for multipurpose chemical batch plants. *Computers & Chemical Engineering*, volume (20), no. Supplement 2, S1215-S1220.
5. Y. Ye, J. Li, Z. Li, Q. Tang, X. Xiao and C. A. Floudas (2014). Robust optimization and stochastic programming approaches for medium-term production scheduling of a large-scale steelmaking continuous casting process under demand uncertainty. *Computers & Chemical Engineering*, volume (66), no. Supplement C, 165-185.

536744 A Rolling Horizon Scheduling Algorithm Considering Electricity Load Tracking and Future Load Preduction

Abstract:

The bill for an electricity consumer is dependent on many factors. One factor is the type of electricity purchase contract. These contracts include, long-term (up to a year) base load contracts, where a constant amount of electricity is supplied at a constant price. There are also medium-term (up to a few months) time-of-use load contracts where electricity is supplied at a multi-tiered pricing scheme (e.g. day and night). For shorter time periods, the day-ahead market can be used to purchase electricity loads at hourly prices, typically up to 12 hours in advance of the first delivery for the next day. Once commissioned, the total load from these contracts represents the total hourly committed load curve for a consumer for the next day (24 hours). This curve then must be followed to within a tolerance to avoid so-called deviation penalties. In the event that a consumer experiences a major shift in electricity demand (e.g. in the case of a unit breakdown) there also exists an intraday market in which consumers can modify their committed load in order to pay fewer deviation penalties. In addition to electricity consumed, grid operator fees need to be paid. One such fee is a peak-pricing scheme in which a company must pay a fee based on their highest simultaneous electricity consumption over a given time-interval. Due to the nature of these markets, it is evident that consumers can save considerable amounts on their electricity bills by shifting their production in response to these time-dependent electricity costs. This activity is known as Demand-Side Management (DSM).

Due to the strong time-dependence of these electricity-related concerns, effective scheduling is essential in DSM, especially when a complex manufacturing system is involved. Several publications have looked at DSM for scheduling including Hadera *et al.* (2015) who looked at the scheduling of steel manufacturing based on a multi-contract load commitment problem using a precedence-based model. Castro *et al.* (2013) also studied the load commitment problem using a discrete-time Resource-Task Network (RTN) formulation. The downside of studying only a 24-hour problem stems from two major factors. The first is that they unrealistically consider the intraday and day-ahead markets by either assuming both to be active at the same time or by ignoring one of the markets. The second is that the committed electricity load in such formulations drops off towards the end of the day as the considered jobs are completed. This represents an unrealistic demand curve as in actuality production is performed on a continuous basis.

We address this problem by considering a two-day problem, combining the current-day problem of load commitment (considering the intraday market) with the future problem of load-prediction (considering the day-ahead market). The downside to such an approach is that it results in very large models that are difficult to solve. In this work, a novel rolling-horizon algorithm is proposed based on the discrete-time RTN formulation. In this algorithm, a non-uniform grid is considered. In the detailed portion of the time-grid, a fine time discretization is used to capture the details of the scheduling model. In the remaining model a coarser grid is used in order to capture the necessary details of the scheduling and electricity models, while reducing the computational complexity of the overall problem. Results show that this algorithm is able to greatly reduce the computation time of such scheduling problems, while still capturing the key aspects of the formulation. In addition, the formulation naturally lends itself to an online scheduling environment.
Castro, P. M., Sun, L., & Harjunkoski, I. (2013). Resource-Task Network Formulations for Industrial Demand Side Management of a Steel Plant. *Industrial and Engineering Chemistry Research*(52), 13046-16058.

Hadera, H., Harjunkoski, I., Sand, G., Grossmann, I. E., & Engell, S. (2015). Optimization of steel production scheduling with complex time-sensitive electricity cost. *Computers and Chemical Engineering, 76*, 117-136.

533559 Comparison of Risk-Averse Stochastic Programming and Adaptive Robust Optimization: A Virtual Power Plant Scheduling Application

Abstract:

In this presentation, we compare two-stage risk-averse stochastic programming (SP) and two-stage adaptive robust optimization (RO) to address the self-scheduling and electricity market involvement of a virtual power plant (VPP). We investigate both approaches concerning formulations, uncertainty and risk, and decomposition algorithms and their computational performance. To quantify the risk in SP, we use the Conditional Value at Risk (CVaR) because it can resemble a worst-case measure, which naturally links to RO.

The VPP comprises a thermal plant, a pumped-storage hydro unit, and a wind farm. The objective of the VPP is to maximize the operating profit for a time horizon of one week. The self-scheduling of the VPP involves the commitment and dispatch of the thermal unit and the hydro plant. The determination of the commitment and dispatch of the thermal unit is subjected to minimum up and down times, limits to ramp up and down rates, startup costs depending on the previous state of the unit, and upper and lower limits on the output generation. Regarding the hydro plant, the commitment and dispatch include the determination of generation and consumption with its region of operation constrained by the power generation function, limits on the output generation, and limits on the volumes of water in the reservoir; see Lima et al. (2013) and Lima and Novais (2016) for further details on these constraints.

The electricity market involvement includes the selection of forward contracts to sell or buy electricity and the option to sell or buy electricity in an hourly pool.

The deterministic VPP problem is represented by a Mixed-Integer Linear Programming model. However, the decision-making problem of the VPP involves uncertainty in the wind speed and electricity price forecast, leading to an MILP stochastic problem with mixed-integer first-stage variables.

The electricity prices forecasts are sampled from an Auto-Regressive Integrated Moving Average model, and the wind speed forecasts are based on a 51-member wind ensemble obtained from weather forecast models from the European Centre for Medium-Range Weather Forecasts. Using the ensemble, we construct a Karhunen-Loève expansion for sampling the wind speed.

We use two efficient implementations of the decomposition algorithms for SP and RO (Lima et al. 2015, Lima et al. 2017) and we assess 1) their performance taking into consideration different sample sizes and risk management parameters; and 2) the operational results regarding first-stage decision variables. Furthermore, we use a Monte Carlo Sample Average Approximation methodology (Lima et al. 2018) to evaluate inference statistics on the expected profit and CVaR of the profit to compare both approaches.

The results show that the SP approach is computationally competitive with the RO approach in specific circumstances, and that there is a set of instances that poses convergence difficulties to both approaches. In these problematic instances, the RO decomposition algorithm cannot meet the stop criteria, whereas the SP decomposition method meets the stop criteria for a large percentage of the samples tested but at a high computational cost. A relevant insight is also discussed regarding the risk management: similar first-stage solutions are obtained with SP and RO depending on the risk parameterizations used in each formulation.

References

Lima, R. M., Marcovecchio, M. G., Novais, A. Q., Grossmann, I. E., 2013. On the Computational Studies of Deterministic Global Optimization of Head Dependent Short-Term Hydro Scheduling. IEEE Transactions on Power Systems 28 (4), 4336â€“4347

Lima, R. M., Novais, A. Q., Conejo, A. J., 2015. Weekly self-scheduling, forward contracting, and pool involvement for an electricity producer. An adaptive robust optimization approach. European Journal of Operational Research 240 (2), 457â€“475.

Lima, R. M., Novais, A. Q., 2016. Symmetry breaking in MILP formulations for Unit Commitment problems. Computers and Chemical Engineering 85, 162â€“176.

Lima, R. M., Conejo, A., Langodan, S., Hoteit, I., Knio, O., 2017. Risk-averse formulations and methods for a virtual power plant. Computers & Operations Research (Accepted).

Lima, R. M., Conejo, A. J., Giraldi, L., Le Maître, O., Hoteit, I., Knio, O., 2018. Sample average approximation for risk-averse problems: A virtual power plant scheduling application. (To be submitted).

525924 Novel Formulation for Optimal Schedule with Demand Side Management in Multi-Product Air Separation Processes

Abstract:

Since the deregulation of electricity markets and the integration of renewable energy systems into the electricity grid, the electricity market has more flexible clearing pricing strategies on customers, which significantly influence the profitability of energy-intensive industries, like air separation plants. In this context, it is critical for industries to optimally adjust their production in order to benefit from the variable energy cost.

In this work, we focus on the optimal scheduling of air separation in industrial gases processes with electricity purchased from spot market. We consider the electricity price forecast in the scheduling horizon as a model input, based on fixed time intervals. As for the continuous process, the variable production levels are achieved by properly adjusting the units to meet various production constraints, such as different product demands. This could be interpreted as the transition of operation states of the production system. In this work, the continuous space of plant operation is discretized into several operation states. Therefore, the main goal of the scheduling problem is to minimize the total production costs by selecting the optimal operation state for each unit in the production system for each time period. Besides the energy consumption, we also consider the setup costs and other related cost item.

To represent the production states and the transition between them, an advanced process state transition network (APSTN) is proposed. The formulation relies on a basic concept of previous work [1], where states with minimum processing time are disaggregated into sub states. However, the APSTN representation is more compact, especially when dealing with more complex production systems. Specifically, for multi-product air separation processes, a discrete-time mixed integer linear (MILP) model is developed based on the APSTN, which has proven to provide a tighter LP relaxation to handling industrial-scale instances.

Three main features characterize the air separation process: 1) minimum consecutive operating time for specific states; 2) setup costs of transition of states and finally, 3) pre-defined direction for some transitions. The proposed MILP model involves the constraints with the previously mentioned features, as well as other considerations, such as inventory level and product demand. The computational efficiency of the model is demonstrated with data provided by the company. Numerical experiments are performed in order to make comparisons between our model and previous ones [1-3]. The response of the scheduling and production level is also tested regarding to various interval's length for the electricity pricing and the time horizon for the scheduling.

References:

[1] Basán, N.P., Grossmann, I.E., Gopalakrishnan, A., Lotero, I., Mendez, C.A., 2018. Novel MILP Scheduling Model for Power-Intensive Processes under Time-Sensitive Electricity Prices. Ind. Eng. Chem. Res., 57 (5), 1581–1592.

[2] Mitra, S., Grossmann, I.E., Pinto, J.M., Arora, N., 2012. Optimal production planning under time-sensitive electricity prices for continuous power-intensive processes. Comput. Chem. Eng. 38, 171–184.

[3] Mitra, S.; Sun, L.; Grossmann, I. E. Optimal Scheduling of Industrial Combined Heat and Power Plants Under Time-sensitive Electricity Prices. Energy 2013, 54, 194.

524663 Multi-Operational Development Planning for Multi-System Shale Gas Production

Abstract:

Shale gas production is expected to increase by more than 50% over the next 34 years, from 14 trillion cubic feet (TCF) in 2016 to approximately 29 TCF in 2040 [1]. This increase in production will make shale gas a major player in the natural gas field, with shale gas production fulfilling 69% of the expected total dry gas natural gas demand [1]. With this expected growth in shale gas, informed decisions, as to where, when, and how to develop assets, must be made to keep the development and production economical. This includes development scheduling, such as when and where to drill, and strategic planning, such as when and which contracts should be made for gas transportation and resource utilization.

Due to the increased impact of shale gas on the energy sector, the optimization of the development and transportation of shale gas has grown in popularity during the past decade. This research has concentrated on adding breadth to the supply chain optimization problem [2,3], and depth to the development and production scheduling [2,4]. However, previous models lumped all pre-production operations (i.e., top-setting, horizontal drilling, fracturing, and turning-in-line) into the generic group of development operations. While this does provide a method of tracking well production as a function of development, it does not allow for the consideration of drilling rig and fracturing crew contracts or mobilization costs, which may significantly impact the development of a shale gas region.

Motivated by the above, in this paper, we present the planning of the multi-operational asset development and shale gas production for a set of prospective capacity systems in an expansive shale gas region. By explicitly accounting for the sequencing of development operations at the pad level, the planning considers the time to perform individual operations, the length of contracts and the number of resources (rigs, crews) needed for development, and costs for moving rigs from one pad to another. By initially formulating the multi-system development planning using General Disjunctive Programming (GDP), we can systematically derive mixed-integer inequalities for all disjunctions and logic propositions to obtain a discrete-time mixed integer programming (MILP) formulation [5,6]. The MILP can be used to optimize the planning of asset development and infrastructure commitments, considering the long-term impacts of gas production and transportation for multiple shale gas systems. To demonstrate the validity of this formulation and the impact of rig considerations, we apply the planning to an industrial case study based on real data, quantifying the value of optimizing shale gas development and production.

References:

[1] EIA (2016). *Most natural gas production growth is expected to come from shale gas and tight oil plays.* https://www.eia.gov/todayinenergy/detail.php?id=26552

[2] Drouven, M. G., & Grossmann, I. E. (2016). Multi period planning, design, and strategic models for long term, quality sensitive shale gas development. *AIChE Journal*, *62*(7), 2296-2323.

[3] Cafaro, D. C., & Grossmann, I. E. (2014). Strategic planning, design, and development of the shale gas supply chain network. *AIChE Journal*, *60*(6), 2122-2142.

[4] Gao, J., & You, F. (2015). Deciphering and handling uncertainty in shale gas supply chain design and optimization: Novel modeling framework and computationally efficient solution algorithm. *AIChE Journal*, *61*(11), 3739-3755.

[5] Grossmann, I. E., & Trespalacios, F. (2013). Systematic modeling of discrete continuous optimization models through generalized disjunctive programming. *AIChE Journal*, *59*(9), 3276-3295.

[6] Williams, H. P. "Mathematical building in mathematical programming." (1985).

533261 Continuous-Time Scheduling Formulation for Pipeline Systems with Branches

Abstract:

In the petroleum supply chain, refined products are frequently transported by pipeline from refineries to harbors and distribution centers serving local markets. Pipelines can carry large volumes, are more energy efficient than trucks and trains, leave a low carbon footprint, are less affected by weather conditions, and are safe. Multiproduct pipelines carry liquids (e.g. gasoline, diesel, jet fuel) one after the other, and so it may take days to cover distances that by truck would take just a few hours. The low speed of delivery coupled with a complex system configuration with multiple nodes, can make it challenging to meet demand on time.

Research on pipeline scheduling can be divided into product- [1-4] and batch-centric approaches [5-9]. The latter, convert product into batches when entering the pipeline, admitting two or more batches of a product inside a pipeline segment. Because multiple batches can enter/leave a segment during a slot [9], their other advantage is fewer time slots to represent a schedule. There are however, three disadvantages: (i) the way in which batch coordinates are computed makes the extension to generic systems with reversible flow quite difficult; (ii) there are more tuning parameters affecting solution quality, unless there is a single input node; (iii) constrains for enforcing forbidden sequences (very common in practice) are not rigorous, since they refer to the injected batch sequence, which can be altered by removing the batch at an intermediate output node or a branch (different path taken).

In this work, we extend our modular batch-centric continuous-time formulation, from straight pipelines [9] to treelike systems with a single input node and multilevel branching (our work in [10] features a more complex mixed-integer linear programming model and is only suitable for one-level branching). We also propose rigorous constraints for handling forbidden product sequences that are derived from Generalized Disjunctive Programming [11-13].

The proposed MILP model is used to tackle a real-life problem from the Iranian Oil Pipelines and Telecommunication Company (IOPTC), addressing, for the first time, the preference for schedules with a single product inside idle segments. The problem involves the first-week-of-November-2017 plan for transporting gasoline, kerosene and gasoil from the refinery in Tabriz to the cities of Maragheh, Miandoab and Urmia.

The rule-based schedule implemented by the company failed to deliver the full product demand, whereas our optimization model could do it more than half a day earlier, increasing pipeline capacity by 6.2%. When considering 3 benchmark problems from the literature, we were able to obtain new best schedules for all, due to a superior computational performance. In particular, when compared to the most general (product-centric) formulation from the literature [14] in terms of system configuration and directions of flow, we could prove global optimality for the problem with the highest number of segments, whereas the model in [14] could not even find a feasible solution.

Acknowledgments: Financial support from Fundação para a Ciência e Tecnologia (FCT) through projects IF/00781/2013 andUID/MAT/04561/2013.

References:

[1] Rejowski, R., Pinto, J.M., 2003. Scheduling of a multiproduct pipeline system. Comput. Chem. Eng. 27, 1229–1246.

[2] Magatão, L., Arruda, L.V.R., Neves, F.A., 2004. A mixed integer programming approach for scheduling commodities in a pipeline. Comput. Chem. Eng., 28, 171-185.

[3] Castro, P.M., 2010. Optimal scheduling of pipeline systems with a resource-task network continuous-time formulation. Ind. Eng. Chem. Res.49, 11491-11505.

[4] Castro, P.M., Mostafaei, H., 2017. Product-centric continuous-time formulation for pipeline scheduling. Comp. Chem. Eng. 104, 283-295.

[5] Cafaro, D.C., Cerdá, J., 2004. Optimal scheduling of multiproduct pipeline systems using a non-discrete MILP formulation. Comput. Chem. Eng. 28, 2053-2068.

[6] Relvas, S., Matos, H.M., Barbosa-Póvoa, A.P.F.D., Fialho, J., Pinheiro, A.S., 2006. Pipeline scheduling and inventory management of a multiproduct distribution oil system. Ind. Eng. Chem. Res. 45, 7841–7855.

[7] MirHassani, S.A., Jahromi, H.F., 2011. Scheduling multi-product tree-structure pipelines. Comput. Chem. Eng. 35, 165–176.

[8] Mostafaei, H., Ghaffari-Hadigheh, A., 2014. A General modeling framework for the long-term scheduling of multiproduct pipelines with delivery constraints. Ind. Eng. Chem. Res. 53, 7029-7042.

[9] Mostafaei, H., Castro, P.M., 2017. Continuous-time scheduling formulations for straight pipelines. AIChE J. 63, 1923-1936.

[10] Mostafaei, H., Castro, P.M., Ghaffari-Hadigheh, A., 2015. Novel monolithic MILP framework for lot-sizing and scheduling of multiproduct treelike pipeline networks, Ind. Eng. Chem. Res. 54, 9202-9221.

[11] Balas, E., 1979. Disjunctive programming. Annals of Discrete Mathematics 5, 3-51.

[12] Raman, R., Grossmann, I.E., 1994. Modeling and computational techniques for logic based integer programming. Comput. Chem. Eng. 18, 563-578.

[13] Castro, P.M., Grossmann, I.E., 2012. Generalized disjunctive programming as a systematic modeling framework to derive scheduling formulations. Ind. Eng. Chem. Res. 51, 5781-5792.

[14] Castro, P.M., 2017. Optimal scheduling of multiproduct pipelines in networks with reversible flow. Ind. Eng. Chem. Res. 56, 9638-9656.

531994 Preprocessing Algorithms and Tightening Constraints for Blend Scheduling Models

Abstract:

Preprocessing Algorithms and Tightening Constraints for Blend Scheduling Models

Yifu Chen, Christos T. Maravelias

Department of Chemical and Biological Engineering, University of Wisconsin-Madison

1415 Engineering Dr., Madison, WI 53706, USA

The multiperiod blend scheduling problem (MBSP) considers the scheduling in a production environment that includes blending processes, where streams with different properties are mixed (blended) to obtain products that satisfy given property specifications. MBSP can be considered as a scheduling extension of the well-studied pooling problem[1,2], or time-indexed pooling problem[3]. In general, MBSP is formulated as mixed integer nonlinear programming (MINLP) problem, where binary variables are used to model decisions on material transfers, and nonlinear constraints containing bilinear terms are employed to model the blending process.

Researchers have proposed a range of methods to address global optimization problem containing binary variables and bilinear terms in the context of pooling and blending problems over the past few decades, including decomposition, linear approximation, valid inequalities, as well as identifying parametric structures [4–9]. However, solving large MBSP remains challenging, partly due the weak models obtained after applying convex relaxation to the bilinear terms. Tightening methods based on strong valid inequalities and reformulations have been proven to be efficient to address instances of industrial relevance in chemical production scheduling problem[10,11] but have not been generalized for blending problems.

Accordingly, in this work we build upon the insights offered by tightening methods for mixed integer linear programming (MILP) scheduling models, to develop new tightening constraints for MBSP. The goal is to use instance-specific data, such as product demand and unit capacity, to tighten the formulation, and thereby obtain improvement in computational performance.

Specifically, we first introduce novel preprocessing algorithms, which can be executed in seconds, to calculate bounds on critical variables in MBSP. In particular, we propose lower bounds on demand for streams when given product demand and specifications, and upper bounds on production and streams utilization when stream availability is given. Second, we introduce new disaggregated variables for stream flows and define product dedicated flow variables to address product specific features involved in MBSP. Third, the bounds obtained from the preprocessing algorithms along with the new variables are used to generate tightening constraints. In some special cases, we show that some of the proposed tightening constraints coincide with a subset of the facet defining inequalities for fixed charge transportation problem with product blending[12].

The methods are applicable to both (1) final product blending and (2) crude oil blending, including downstream distillation units. Furthermore, they are applicable to all previously proposed MINLP models for MBSP, as well as MILP models obtained from the MINLP formulations using different linear relaxation/approximation methods. Finally, we discuss how they can be modified so they are applicable to both cost minimization for given demand, and profit maximization problems.

We close with a computational study including a wide range of (1) previously proposed MINLP and MILP models, (2) global optimization solvers, and (3) literature instances.

Reference

1. Kolodziej, S. P., Grossmann, I. E., Furman, K. C. & Sawaya, N. W. A discretization-based approach for the optimization of the multiperiod blend scheduling problem. *Comput. Chem. Eng.* **53,** 122–142 (2013).
2. Lotero, I., Trespalacios, F., Grossmann, I. E., Papageorgiou, D. J. & Cheon, M. S. An MILP-MINLP decomposition method for the global optimization of a source based model of the multiperiod blending problem. *Comput. Chem. Eng.* **87,** 13–35 (2016).
3. Gupte, A., Ahmed, S., Dey, S. S. & Cheon, M. S. Relaxations and discretizations for the pooling problem. *J. Glob. Optim.* **67,** 631–669 (2017).
4. Meyer, C. A. & Floudas, C. A. Global optimization of a combinatorially complex generalized pooling problem. *AIChE J.* **52,** 1027–1037 (2006).
5. D'Ambrosio, C., Linderoth, J. & Luedtke, J. Valid Inequalities for the Pooling Problem with Binary Variables. in *Integer Programming and Combinatoral Optimization* (eds. Günlük, O. & Woeginger, G. J.) 117–129 (Springer Berlin Heidelberg, 2011).
6. Li, J., Misener, R. & Floudas, C. A. Continuous-time modeling and global optimization approach for scheduling of crude oil operations. *AIChE J.* **58,** 205–226 (2012).
7. Kolodziej, S. P., Castro, P. M. & Grossmann, I. E. Global optimization of bilinear programs with a multiparametric disaggregation technique. *J. Glob. Optim.* **57,** 1039–1063 (2013).
8. Castillo, P. A. C., Mahalec, V. & Kelly, J. D. Inventory pinch algorithm for gasoline blend planning. *AIChE J.* **59,** 3748–3766 (2013).
9. Baltean-Lugojan, R. & Misener, R. Piecewise parametric structure in the pooling problem: from sparse strongly-polynomial solutions to NP-hardness. *J. Glob. Optim.* 1–36 (2017). doi:10.1007/s10898-017-0577-y
10. Velez, S., Sundaramoorthy, A. & Maravelias, C. T. Valid Inequalities Based on Demand Propagation for Chemical Production Scheduling MIP Models. *AIChE J.* **59,** 872–887 (2013).
11. Merchan, A. F., Lee, H. & Maravelias, C. T. Discrete-time mixed-integer programming models and solution methods for production scheduling in multistage facilities. *Comput. Chem. Eng.* **94,** 387–410 (2016).
12. Papageorgiou, D. J., Toriello, A., Nemhauser, G. L. & Savelsbergh, M. W. P. Fixed-Charge Transportation with Product Blending. *Transp. Sci.* **46,** 281–295 (2012).

533751 Long-Term Maintenance and Production Planning for the Integrated Chemical Enterprise

Abstract:

We address an enterprise-wide strategic maintenance turnaround and production planning problem. For networks of large, continuous plants, maintenance and production planning clearly affect each other and here we consider them together for the full network. It is clear that non-safety-related tasks performed for routine maintenance turnarounds and outages significantly disrupt production and profits, and that high production targets can prompt evaluation to defer routine maintenance during periods of high profitability. Consequently, the timing of maintenance turnarounds can affect both production and profitability significantly, especially when a facility is running at capacity. For example, production rates are commonly affected by fouling, catalyst activity, and overall unit reliability---and these depend on the cumulative production or time since the last scheduled maintenance turnaround. Separately from conversion rates and throughput, selectivities may also change with time and production rate and have an impact on production in integrated site networks. Further, deferring a scheduled outage/turnaround can increase the scope of work required for the next scheduled turnaround, and may also increase the cost and duration of the turnaround. There may also be restrictions and requirements on site imports/exports, material swaps, and mandatory inspections that prevent a deferral.

In order to create a rigorous enterprise production and profitability plan, these factors must be taken into account. They require the tracking of time and production since the previous turnaround, the timing of which is itself a decision variable. Additionally, the large time intervals between routine maintenance turnarounds necessitate very long planning horizons due to end effects, significantly adding to problem size. In this work, an MILP-based framework is presented that includes the above conditions, and, in addition, is able to consider long- and short-term financial performance by providing recommendations that not only maximize long-term NPV, but also balance quarter-to-quarter profits while respecting strict safety requirements and standards. Though profits from sales may be booked soon after delivery, costs for turnarounds are usually incurred several months prior to and several months after execution---this needs to be considered in the formulation when there are quarterly budget constraints or profitability targets.

To summarize, attention to safety, mandatory inspections, production technologies, reliability, maintenance, finance and accounting, and supply chain are all critical to production planning, and we afford these consideration in the current context. Effective reformulations, modeling strategies, and associated case studies are presented to address this real-world large-scale problem.

532401 Ensemble Models for Univariate Time Series Forecasting

Abstract:

Univariate time series (UTS) are stochastic processes and poorly modeled by conventional time-dependent regression due to their strong historical dependence and periodic behavior. Statistical models, like ARIMA (AutoRegressive Integrated Moving Average) methods, spectral modeling, and other non-linear techniques have long since dominated the UTS modeling literature; however, in 2010, Ahmed et al. demonstrated that machine learning models like multilayer perceptron neural networks and Gaussian Processes can compete with and outperform ARIMA models on business and economic forecasting problems [1]. This work first benchmarks the performance of many statistical and machine learning UTS models on datasets derived from scientific, natural, and business sources in order to establish a better understanding of each modelâ€™s relative accuracy and utility.

This work also leverages its benchmarking framework to identify prospective ensemble models and to guide modelers to an effective methodology for their problem class and topical domain. The use of ensembles has been shown to improve UTS forecasting accuracy already. Gheyas and Smith found success unifying multiple Generalized Regression Neural Networks (GRNN) with a Combiner GRNN [2]. Lai et al. implemented a deep learning framework coupled with a standard ARIMA model capture both long- and short-term trends in multivariate time series [3]. In 2017, Cerqueirla et al. implemented a dynamic ensemble capable of adapting to regime changes in the data [4]. This work extends upon the existing literature by applying best subset selection to the ensemble learning problem.

[1] Ahmed, N.K., Atiya, A.F., Gayar, N.E., El-Shishiny, H., An Empirical Comparison of Machine Learning Models for Time Series Forecasting, Econometric Reviews, 29:594â€“6215, 2010.

[2] Gheyas, I.A., Smith, L.S., A novel neural network ensemble architecture for time series forecasting, Neurocomputing, 74: 3855â€“3864, 2011.

[3] Lai, G., Chang, W.-C., Yang, Y., Liu, H., Modeling Long- and Short-Term Temporal Patterns with Deep Neural Networks, arXiv:1703.07015v2 [cs.LS], 2017.

[4] Cerqueira, V., Torgo, L., Oliveira, M., Pfahringer, B., Dynamic and Heterogeneous Ensembles for Time Series Forecasting, IEEE International Conference on Data Science and Advanced Analytics, DSAA, 2017.

Abstract:

Ontology engineering approach to support process of model and data integration

Linsey Koo, Edlira Kalemi, Franjo Cecelja,

Process and Information Systems Engineering, Faculty of Engineering and Physical Sciences, University of Surrey, Guildford, Surrey, U. K.

Process modelling and simulation is a vital tool to plan, evaluate, assess, and develop different alternatives for the design of products and processes. The complexity of problems as well as heterogeneity of modelling methods make process modelling and simulation challenging, time consuming and often tedious process requiring a wide range of expertise. Inconsistencies in model development are the main cause for redundant work. Models remain implicit to the engineers who have built them, which further limits the potential of reusability.

The only model integration framework in use, the CAPE-OPEN, addresses the issue of standardisation of interfaces to enable interoperability between simulator software components from different sources. It is the framework built around a middleware, the Common Object Request Broker Architecture (CORBA) that hosts communication between unit operations defined for a specific function and the process modelling environments. The standard specification is defined as a property package which is needed for a thermodynamic or physical property calculation. The interoperability of models, such as model selection, parameter identification, and experimental work is enabled through the connection related to the unit operations and physical properties. It is not necessary to match all parameters in order to facilitate Input-Output (I-O) matching. However, the shortcoming of the CAPE-OPEN is in the need for identifying key variables for each unit operation. Also, it does not facilitate data integration.

In this paper a new approach for model and data integration is proposed which builds upon the CAPE-OPEN framework and proposes the use of ontology and replaces the object bus with more flexible semantic repository (Koo, Trokanas and Cecelja, 2017). Models are described by Semantic Web Services (SWS) using Ontology Web Service Description (OWL-S) as an enabler of web services through service discovery, selection, composition, and execution stages (REF _Ref511651865 \h Figure 1
08D0C9EA79F9BACE118C8200AA004BA90B02000000080000000E0000005F0052006500660035003100310036003500310038003600350000000
). By The same token, datasets are also described by the OWL-S through their outputs and auxiliary precondition for their execution, i.e. extraction, modification and deletion (REF _Ref511651844 \h Figure 2
08D0C9EA79F9BACE118C8200AA004BA90B02000000080000000E0000005F0052006500660035003100310036003500310038003400340000000
). The discovery stage allows formation of an integrated model through matching requests from a public repository(ies). The best match that satisfy the requestor's functionality is selected in the model selection stage. The model composition stage then formulates the chain of integrated models and execution of integrated model takes place during the execution stage.

Figure 1 Semantically described model

Figure 2 Semantically described datasets

This paper focuses on the matching parameters related to the domain of process system engineering, with emphasis placed on the role of physical properties and unit operation. Each model and dataset representing a device (e.g. unit operations, flowsheets, subflowsheets etc.) is semantically described in domain ontology including domain assumptions and descriptions of the functionality of the model. The domain ontology guides the process of registering models and instantiation of ontology through ontology parsing, which makes the model and data explicit and sharing terminology in domain ontology improves consistency. The devices in a process are connected via streams that transmit information through multiple inlets and outlets from one unit to the other. The connection between devices are described in ontology by introducing the concept of 'ports' and 'connections'. The 'ports' generally describe inlets and outlets of devices and three different types of streams are distinguished: material, energy, and information, which are further described by objective properties. The 'connections' are the object that is responsible for establishing a link between two ports, which contains information regarding methods, types, quantities, and units of streams.

The whole process is verified by a case of model integration for a supply chain modelling in biorefining.

Koo, L., N. Trokanas, et al. (2017). "A semantic framework for enabling model integration for biorefining." Computers & Chemical Engineering 100: 219-231

534774 Comparison of Surrogate Modeling Techniques for Surrogate-Based Optimization

Abstract:

Surrogate models, also known as response surfaces, black-box models, metamodels, or emulators, are simplified approximations of more complex, higher order models (Wang et al., 2014). These models are used to map input data to output data when the actual relationship between the two is unknown or computationally expensive to evaluate (Han and Zhang, 2012). Surrogate models also allow the use of traditional derivative-based optimization techniques for optimization of systems defined by these complex, higher order models. Surrogate modeling techniques are of particular interest where high-fidelity, thus expensive, simulations are used, for example in computational fluid dynamics (CFD) or computational structural dynamics (CSD). These techniques are of great importance to the chemical engineering community as they have wide reaching related applications, including in process synthesis, process controls, and supply chain management. Some recent examples of applications of surrogate modeling approaches include several process synthesis applications, such as in optimization of multiphase flow networks (Grimstad et al., 2016), and process controls applications in the pharmaceutical production industry (Icten et al., 2015).

Construction of a surrogate model is comprised of three steps: (1) selection of the sample points, (2) optimization or "training" of the model parameters, and (3) evaluation of the accuracy of the surrogate model (Wang et al., 2014). Although several machine learning and regression techniques have been developed for surrogate model construction, there has been little work done on how to best select the appropriate model for a particular application for both surrogate modeling and surrogate based optimization. The majority of the studies comparing surrogate model performance only compare a few models on a limited number of functions or complex models. Davis et al. (2017) investigated a more extensive selection of surrogate modeling techniques for 35 challenge functions and concluded that Artificial Neural Networks, Automated Learning of Algebraic Models using Optimization, and Extreme Learning Machines yielded the most accurate predictions for the challenge functions tested. This work aims to build upon that study and to further address the knowledge gap by comparing the ability of eight different surrogate modeling techniques to both learn and accurately model the responses of a set of challenge functions and to locate the extrema of these functions using surrogate based optimization. The surrogate-modeling techniques considered include Artificial Neural Networks (ANN), Automated Learning of Algebraic Models using Optimization (ALAMO), Radial Basis Networks (RBN), Extreme Learning Machines (ELM), Gaussian Progress Regression (GPR), Random Forests (RF), Support Vector Regression (SVR), and Multivariate Adaptive Regression Splines (MARS). These techniques are used to construct surrogate models for the 47 optimization challenge functions from the Virtual Library of Simulation Experiments (Surjanovic, 2013). The effects of the challenge function characteristics, including function shape and number of inputs, and sampling methods on the surrogate model performance are evaluated. The sampling methods studied are Sobol sequence sampling and Latin Hypercube sampling (LHS). Four performance measures are used to evaluate the accuracy of the surrogate models: root mean squared error (RMSE), maximum percent error (MPE), the R-squared value, and the Akaike Information Criteria (AIC). The models' ability to locate the extrema of the functions are evaluated by calculating the distance between the extreme point(s) estimated by the model and the actual function extrema. The results provide guidance on selecting which surrogate modeling technique to use based on the specifics and characteristics of the function or data set being modeled.

References:

Davis, S., Cremaschi, S., Eden, M., 2017, "Efficient Surrogate Model Development: Optimum Model Form Based on Input Function Characteristics",Computer Aided Chemical Enginering 70.1, 457-462.

Grimstad, B., Foss, B., Heddle, R., Woodman, M., 2016, "Global optimization of multiphase flow networks using spline surrogate models", Computers and Chemical Engineering 84.1, 237-254.

Han, ZH. and Zhang, KH., 2012,"Surrogate-Based Optimization", Real-World Applications of Genetic Algorithms. InTech. 343-362.

Icten, E., Nagy, Z., Reklaits, G, 2015, "Process control of a dropwise additive manufacturing system for pharmaceuticals using polynomial chaos expansion based surrogate model", 2015, Computers and Chemical Engineering 83.1, 221-231.

Surjanovic, S., Bingham, D., 2013, "Virtual Library of Simulation Experiments: Test Functions and Datasets", http://www.sfu.ca/~ssurjano.

Wang, C., Duan, Q., Gong, W., Ye, A., Di, Z., Miao, C., 2014, "An evaluation of adaptive surrogate modeling based optimization with two benchmark problems", Environmental Modelling and Software 60.1, 167-179.

528035 Semantic Networking Facility for the Biorefining Community

Abstract:

Semantic Networking Facility for the Biorefining Community

Edlira KALEMI[1], Linsey KOO[1], Franjo CECELJA[1]

[1]Process & Information Systems Engineering Research Centre

University of Surrey

e.kalemi@surrey.ac.uk, f.cecelja@surrey.ac.uk, m.koo@surrey.ac.uk

Abstract

Biorefining is a dynamic field with ever growing number of computer models developed, heterogeneous data acquired and generally new knowledge generated in large volumes, all to serve functions at different scales and for different purposes. Sharing and reusing of these resources, especially models and data, inherently saves developing time, but also enables solving ever more complex problems and addressing ever more complex functions. This, in turn, necessitates efficient networking tool to allow for publishing, discovery, but also inferring new knowledge and integration of resources, dominantly computer models.

This paper introduces a semantic networking facility developed to enable an efficient reuse and sharing of knowledge and resources in the domain of biorefining. Built around model integration platform, this networking facility allows for knowledge and other resources in the domain of biorefining to be accessible by humans, but also by agents and services. It is organised as a collaborative, WiKi like platform, aiming to facilitate collaboration and to bring biorefining community together by sharing knowledge and resources. The operation of the proposed facility is supported by a biorefining domain ontology, but also by Semantic Media Wiki extension. Outstanding features of the Biorefining Semantic Wiki include semantic repository for biorefining models, model discovery and integration tool, supply chain networking and demonstration, documentation, including published article repository and discovery, forum for the community and with active social media of the biorefining community. The operation of Biorefining Semantic Wiki has been implemented as a semantic web service and its operation verified in practice.

532403 The Use of Asset-Oriented Data Models for Data Integration Enables Advanced Analytics in the Process Industry

Abstract:

Industrial organizations in resource extraction, continuous or discrete manufacturing, and power generation each generate large volumes of data during the operation of their plants and facilities. The effective use of this data to drive business outcomes is a central theme behind recent trends in IIoT or digitization of industry as well as cross-cutting efforts such as Industry 4.0. Machine- or sensor-generated process time series data, engineering design data, maintenance records, and business records are often separate data silos within large organizations. The existence of these disparate data silos has historically hindered the effective implementation of advanced analytics in areas such as asset benchmarking, process monitoring, fault detection, and machine learning for predictive maintenance or other applications.

This work shares practical lessons from the industrial application of advanced analytics in prototype and production phases. It also highlights real analytics use-cases and the transformation in data management and information architecture necessary to enable these activities. Several commonly observed challenges in enabling data scientists and quantitative analysts to be effective with process data are discussed. The utility of asset-oriented data models is shown in an example application with a realistic set of databases; including the data from instrumentation, control systems, and transactional business records. The data and data model requirements of business intelligence technologies and advanced analytical applications are discussed, with a focus on how a shared data layer must be flexible to support a variety of analytics activities.

A property graph structure stored in a graph database can effectively support the storage and construction of relationships relevant to an asset-oriented data model. The ingestion of structured information complying with established standards for data interchange can accelerate the construction of such a graph model. This is illustrated with the use of piping and instrumentation diagrams (P&IDs) to establish the process topology as well as link data streams from instrumentation with process units and higher-level views of a plant.

533083 Parameter Prediction for Stochastic Job Shop Scheduling Using Probabilistic Machine Learning

Abstract:

Modern industrial job shop processes generate significantly large amount of scheduling and sensor data; the number of sensors in these processes is in the order of hundreds or even thousands (Gungor and Hancke, 2009). By applying machine learning algorithms to the data, the operators have the potential to extract useful insight of their processes. This insight can be used to improve the scheduling algorithms by enhancing their solution quality and/or reducing their computing time.

The most common approach of applying machine learning algorithms to job shop scheduling is to learn dispatching rules from historical data. Li and Olafsson (2005) applied an inductive learning algorithm (C4.5) to train a decision tree model using historical data. Later, Olafsson and Li (2010) improved the performance of the model by training it only with a subset of the historical data, representing the best decisions, which were chosen using a genetic algorithm. Mouelhi-Chibani and Pierreval (2010) proposed a neural network-based approach to select, in real time, the best job for a resource, once it becomes available. Instead of using historical data, they trained their model using simulated scheduling data, generated by simulated annealing. As calling a trained model is nearly instantaneous, this type of approaches eliminate, or significantly reduce, the computational time of the scheduling procedure. This feature is advantageous to dynamic job shop scheduling, in which any computational overhead in the scheduling procedure may cause the decisions to be outdated. For a more extensive review of learning dispatching rules from historical data, the reader may wish to consult the review article by Priore et at. (2014).

An alternative approach of applying machine learning algorithms to job shop scheduling is to predict the key scheduling parameters (e.g., job processing times and power consumptions) for the actual scheduling algorithm. When using typical deterministic algorithms in dynamic scheduling, inaccuracies in scheduling parameters can significantly affect the optimality, or even the feasibility, of the solution. In the literature, several studies have been conducted in order to improve the quality of the scheduling via more accurate input parameter predictions. Berral et al. (2010) predicted power consumption levels, CPU loads and SLA timings of a data center using linear regression and the M5P algorithm. They used these predictions in the scheduling of computing tasks, with an objective of reducing the total power consumption of the data center. Jiang et al. (2016) applied Gaussian process regression in dynamic scheduling of continuous casting of steel to predict slack ratios of jobs. A slack ratio describes the trade-off between low production time and increased risk of a cast-break.

Regardless of the size and quality of the available historical data, the parameter predictions contain uncertainty due to noise and model limitations. By acknowledging this fact, the quality of scheduling solutions in a stochastic environment can be further improved by taking these uncertainties into account in the scheduling algorithm (e.g., by using stochastic programming (Birge and Louveaux, 2011)). Although the information of prediction uncertainty lies in any (extensive) historical data, this aspect of machine learning-aided job shop scheduling is rarely studied in the literature.

Thus, in this work, we, first, use probabilistic machine learning algorithms to predict the scheduling parameters and their uncertainties from an industrial job shop scheduling dataset. For simplicity, we narrow the predicted scheduling parameters to be the job processing times. Second, we use stochastic programming to define a scheduling algorithm, which objective is to minimize the make span of a set of products in a dynamic scheduling environment. Finally, we benchmark our approach to traditional deterministic scheduling methods, for which the job processing times are predicted from the same historical data using linear regression models.

References:

Berral, J. L., Goiri, Í., Nou, R., Julià, F., Guitart, J., Gavaldà, R., & Torres, J. 2010. Towards energy-aware scheduling in data centers using machine learning. *Proceedings of the 1st International Conference on energy-Efficient Computing and Networking*, ACM, 215-224.

Birge, J. R., & Louveaux, F. 2011. Introduction to stochastic programming. *Springer Science & Business Media.*

Gungor, V. C., & Hancke, G. P. 2009. Industrial wireless sensor networks: Challenges, design principles, and technical approaches. *IEEE Transactions on industrial electronics*, 56(10), 4258-4265.

Jiang, S. L., Liu, M., Lin, J. H., & Zhong, H. X. 2016. A prediction-based online soft scheduling algorithm for the real-world steelmaking-continuous casting production. *Knowledge-Based Systems,* 111, 159-172.

Li, X., & Olafsson, S. 2005. Discovering dispatching rules using data mining. *Journal of Scheduling*, *8*(6), 515-527.

Mouelhi-Chibani, W., & Pierreval, H. 2010. Training a neural network to select dispatching rules in real time. *Computers & Industrial Engineering*, *58*(2), 249-256.

Olafsson, S., & Li, X. 2010. Learning effective new single machine dispatching rules from optimal scheduling data. *International Journal of Production Economics*, *128*(1), 118-126.

Priore, P., Gómez, A., Pino, R., & Rosillo, R. 2014. Dynamic scheduling of manufacturing systems using machine learning: An updated review. *AI EDAM*, *28*(1), 83-97.

527041 Two Stage Stochastic Programming Modeling Approach for Energy Storage System Operation Under Uncertainty

Abstract:

Employing renewable energy sources has become an irreversible trend in the current energy systems of various sectors including process industries. Their decision-makers should cope with often occurring mismatches between varying energy supply and demand over multiple time periods. Considering the inherent variability of renewable output, it would be often witnessed that surplus energies from renewable sources is simply thrown away without being used. It is thus understandable to see the introduction of energy storage system (ESS) for the sake of the economic benefit of using the surplus again later. We can increase the penetration of renewable energies by installing the ESS. On the other hand, it is also right to point out that there is much to be explored in maximizing the potential of ESS with other entities in the energy system. In order to transform the expectation into reality, the corresponding decision-making framework should be rigorously designed. As an alternative, this paper investigates the impact of variation on the design and operation of ESS. Specifically, the design of ESS under uncertainty is transformed into a two-stage stochastic programming model. The stochastic model was formulated in terms of a mixed integer linear programming (MILP) problem. As illustrated in numerical examples, the proposed modeling approach allows us to gain the insight on renewable energy systems. More similar works are expected to establish the robust energy system in process industries.

527226 Theoretical and Experimental Investigation of the Microbial Degradation of Solitary Oil Microdroplets

Abstract:

Natural seeps and accidental releases of crude oil in the sea result in clouds of droplets that are carried away by underwater sea currents. The droplets may be created either at the sea surface during the breakup of an oil slick by sea waves, or at the seafloor during the atomization of live crude oil extruding at sufficiently high speed from a natural crack or a broken wellhead. Drifting droplet clouds disturb the established ecosystem dynamics and pose a high risk of toxic effects to many marine species. It is therefore of crucial importance to understand and quantify the factors that determine the retention of the dispersed droplets within the seawater column. Dissolution, biodegradation and aggregation with marine snow are the main physical processes that rule the fate of dispersed oil droplets in marine waters.

With regard to the biodegradation process, microbes have developed three fundamental strategies for accessing and assimilating oily substrates. Depending on their affinity for the oily phase and ability to proliferate in multicellular structures, microbes might either attach to the oil surface and directly uptake compounds from the oily phase, or grow suspended in the aqueous phase consuming solubilized oil, or form three-dimensional biofilms over the oil-water interface (Kapellos, 2017).

We have recently developed a compound particle model for the biodegradation of solitary oil microdroplets moving through a water column (Kapellos et al., 2018). The compound particle is of the core-shell type and consists of an oily core that is successively surrounded by a bioreactive skin of negligible thickness and another bioreactive shell of finite thickness. The bioreactive skin represents a thin layer of microbes that uptake oil directly from the oily phase, whereas the bioreactive shell represents a distinct biofilm phase. The new model accounts for all three modes of biodegradation: direct interfacial uptake, bioreaction in the bulk aqueous phase, and bioreaction in a biofilm formed around the droplet. A closed-form correlation has been established for the overall dissolution rate for a given particle configuration and two coupled ordinary differential equations have been derived for the evolution of the dimensions of the compound particle. Numerical analysis is currently used to extend the domain of validity of the model by relaxing key hypotheses. In this presentation, major findings of the theoretical analysis will be discussed in detail and in conjunction with preliminary results from the biodegradation of hexadecane droplets by Marinobacter sp. in a microfluidic setting.

Acknowledgements: This work has received funding from the European Unionâ€™s Horizon 2020 research and innovation programme under the Marie Sklodowska-Curie grant agreement No 741799 (project "OILY MICROCOSM").

References:

Kapellos, G.E. "Microbial strategies for oil biodegradation", In *Modeling of Microscale Transport in Biological Processes*; Becker, S.M., Ed.; Academic Press, pp. 19-39 (2017).

Kapellos, G.E., Paraskeva, C.A., Kalogerakis, N. and Doyle, P.S. "Theoretical insight into the biodegradation of solitary oil microdroplets moving through a water column", *Bioengineering*, 5:15, (2018).

530591 Modeling Faradaic Capacitive Deionization with Redox Active Porous Electrodes

Abstract:

Capacitive deionization (CDI) is an attractive technology for brackish water treatment, due to the low energy consumption at low source water salinity level compared to conventional water treatment technologies such as reverse osmosis or distillation.

It has been shown that the surface charge on the porous carbon electrodes has significant impact on CDI performance. By introducing immobilized chemical groups, the point of zero charge of carbon materials can be tuned, thus enhanced or inversed CDI performance could be obtained. Redox active electrodes have been investigated in CDI applications recently. The Faradaically induced chemical charges add to electronic charges within the electric double layers formed at the electrode surfaces to provide a much higher adsorption capacity than attained in purely capacitive processes, which is denoted as Faradaic capacitive deionization (FaCDI) here. The redox active ligand can be considered as variable chemical charges that depend on the cell voltage, as opposed to the fixed charges introduced by chemical modification.

Here, we developed a model which incorporates the effect of variable chemical charge due to general faradaic reactions of redox species immobilized on the electrode surfaces. The theory describes the variable chemical charge by the generalized Frumkin-Butler-Volmer equation across the stern layer as an extension of current electrosorption theory. As a first approximation, the model assumes the spacer channel is well-mixed with uniform electrosorption in each electrode. The model is able to predict the experimentally observed enhanced and inverted performance of CDI cells. Furthermore, the deionization performance of FaCDI cells is shown to be superior to that of CDI and enhanced-CDI systems with equilibrium adsorption capacities 50-100% higher than attained with CDI systems. By properly choosing the equilibrium potentials of the redox active material, FaCDI is able to achieve the same ion adsorption capacity at smaller voltages compared to conventional CDI, which indicates the promise of obtaining higher salt adsorption capacity with lower energy input through the use of FaCDI materials.

The model enables us to optimize the operation parameters of the process, such as flow rate, applied voltage as well as the design of CDI cells, such as the surface to volume ratio of the device, in order to obtain the optimal deionization performance.

532901 Finding the Ideal Feedstock for Biodiesel Production Via Generalized Kinetic Model for Transesterification and Saponification

Abstract:

The hunt for an assortment of energy sources is just about as old as humanity's need for energy. This search has been escalated by an expanding industrialization since the 19th century. The internal combustion engine is a prime example. Even though fuels obtained from fossil sources have been a prime choice since the onset of these engines, biofuels especially biodiesel have also found some interest. This interest has particularly gained momentum in the past few decades due to rapidly increasing energy demand and depletion in fossil fuel resources.

Biodiesel is a non-toxic, biodegradable, and environmentally benign fuel that can be used in diesel engines. The most commonly used method for the production of biodiesel is transesterification. It involves the reaction of oil with an alcohol in the presence of a catalyst to form fatty acid alkyl ester (biodiesel) and glycerol. Although, biodiesel is a promising alternative fuel, its production poses significant challenges due to its higher costs compared to that of petroleum-based diesel. Typically one of the major contributors to this issue is the cost of the feedstock, which accounts for approximately 60 to 80% of the total production cost.

Even though the literature is replete with the studies about modelling and optimization of transesterification [1, 2, 3, 4], very few researchers have focused their attention on the selection of feedstock for biodiesel production. Although, a few comparative studies have been reported in the literature [2, 5], all of them focus on static optimization, thereby neglecting the kinetics of biodiesel production. Therefore, in order to address these issues, we aim to achieve the following objectives:

1. Find optimal operating conditions of the reactor for biodiesel production using different vegetable oils as feedstock.
2. Develop meta-models depicting the effect of the different vegetable oils/their blends on the properties of the biodiesel produced using these feedstocks.
3. Use the developed meta-models to find the best possible feedstock for a particular country.

The purpose of this study is to find optimal feedstock (vegetable oil blend) for a particular country based on the standard property specifications and the availability of oils in that country. To this end, we use the generalized kinetic model, developed in our previous work [6] to find optimal conditions with 4,500 different blends of six major oils *viz.*, Jatropha, Linseed, Olive, Palm, Rapeseed, and Sunflower oil as feedstock. We employ a property based objective function (considering density, viscosity, cold point, pour point, cetane number, higher heating value, lower calorific value, and cold filter plugging point of biodiesel) to solve these optimization problems. We then employ correlations from the literature [3] to calculate properties of biodiesel formed using a particular feedstock and at corresponding optimal conditions. This is followed by developing meta-models with the computed properties of biodiesel as output variables and different characteristics of the feedstocks as the input variables. We use five different approaches to characterize the feedstock *viz.* a) average chain length and the unsaturation degree of the fatty acids that constitute the oil (adopted from [7], number of dimensions (N) = 2); b) average chain length, weighted unsaturation degree calculated based on the triglyceride composition of oil (proposed in this work, N = 2); c) triglyceride composition (proposed in this work, N = 34); d) weights on the considered oils for the different blends (proposed in this work, N = 6); and e) properties of oils, *i.e.* viscosity, density, heat capacity, enthalpy of vaporization, and vapour pressure calculated based on the triglyceride composition of oil (proposed in this work, N = 5). Finally, we use the best performing meta-models to obtain optimal feedstock for production of biodiesel in a particular country.

The predictions made using the developed meta-models are in accord with those made using the original models. We quantify this agreement, by computing the Percentage Average Absolute Error (PAAE) as follows,

PAAE = (Average Absolute Error/Average Response)x100

The error values are below 1% for most of the cases with the maximum value at 2.55%, which shows excellent prediction performance. Moreover, we use a normalized error metric to compare the performance of the meta-models developed using different data sets in this study. We observed that the meta-models developed using triglyceride composition and weight distribution of oils in the blend as input variables performed superior to the other meta-models. However, owing to the large number of dimensions in the former case, it is apt to use the latter. Finally, we used the meta-models to obtain optimal feedstock blend for a particular country. We analysed different scenarios based on the standard specification imposed on the biodiesel quality, availability and type of feedstock (*i.e.* edible or non-edible). We observed that the quality of biodiesel produced can change significantly based on the considered scenarios, thereby highlighting the importance of such a study. Besides, we also observed that it is possible to produce premium quality biodiesel even with just non-edible oils. Such a result can definitely bolster the efforts for commercialization of biodiesel production as the use of non-edible oils can significantly reduce the costs associated with the biodiesel production process.

REFERENCES

[1] V. Singh, K. Solanki, M. N. Gupta, Process optimization for biodiesel production, Recent patents on biotechnology 2 (2) (2008) 130{143.

[2] J.Yang, T. Astatkie, Q. S. He, A comparative study on the effect of unsaturation degree of camelina and canola oils on the optimization of bio-diesel production, Energy Reports 2 (2016) 211-217.

[3] S. Pinzi, D. Leiva, G. Arzamendi, L. Gandia, M. Dorado, Multiple response optimization of vegetable oils fatty acid composition to improve biodiesel physical properties, Bioresource technology 102 (15) (2011) 7280-7288.

[4] R. S. Kumar, K. Sureshkumar, R. Velraj, Optimization of biodiesel production

from manilkara zapota (l.) seed oil using taguchi method, Fuel 140 (2015) 90-96.

[5] T. Issariyakul, A. K. Dalai, Comparative kinetics of transesterication for biodiesel production from palm oil and mustard oil, The Canadian Journal of Chemical Engineering 90 (2) (2012) 342-350.

[6] P. Chhabra, S. Mosbach, M. Kraft, I. A. Karimi, A generalized kinetic model for transesterication and saponification (2017).

URL https://aiche.confex.com/aiche/2017/meetingapp.cgi/Paper/500597

[7] E. G. Giakoumis, A statistical investigation of biodiesel physical and chemical properties, and their correlation with the degree of unsaturation, Renewable Energy 50 (2013) 858-878.

530255 Accelerating the Generation of Coal Power Plant Property Models

Abstract:

Power plants are complex systems that are represented by first principles or empirical models. As power plants are developed with new technologies outside of well-studied thermodynamic regions, data-driven approaches have been used to model these complex systems [5]. For example, thermophysical properties or kinetic reaction models can be explained to an arbitrary level of complexity with an empirical model. The challenge is that these process models are not well understood, and it is not always intuitive what functional forms can be utilized to represent these systems. When optimizing proposed model systems, the efficiency of power plants relies on having accurate models. To maximize efficiency as well as to maintain safe operating conditions, it is essential that complex kinetic and thermophysical relationships are represented accurately, even though determining such a representation may be computationally expensive.

In this work we address the problem of speeding up the model generation process. In particular, we look at accelerating best subset selection algorithms with the use of GPU parallel computing. Best subset selection is a regression problem that seeks to determine a small set of features that best relates a set of inputs to an output [7]. There are many exact and heuristic approaches that can be used to solve the best subset selection problem, such as branch-and-bound [4], forward and backward selection [3], exhaustive search, the lasso [7], and mixed-integer optimization [6, 2, 1]. The proposed approach speeds up the model building process, allowing us to consider a larger feature space and generate more realistic models. The proposed methodology is compared against best subset selection techniques as well as the Automated Learning of Algebraic MOdels (ALAMO) methodology [2].

References

[1] D. Bertsimas, A. King, and R. Mazumder. Best subset selection via a modern optimization lens. The Annals of Statistics, 44:813–852, 2016.

[2] A. Cozad, N. V. Sahinidis, and D. C. Miller. Automatic learning of algebraic models for optimization. AIChE Journal, 60:2211–2227, 2014.

[3] M.A. Efroymson. Multiple regression analysis. Mathematical methods for digital computers, pages 191–203, 1960.

[4] C. Gatu and EJ Kontoghiorghes. Branch-and-bound algorithms for computing the best-subset regression models. Computational and Graphical Statistics, 15:139–156, 2006.

[5] S Lu and BW Hogg. Dynamic nonlinear modelling of power plant by physical principles and neural networks. International Journal of Electrical Power & Energy Systems, 22.

[6] A. Miller. Subset selection in regression. CRC Press, 2002.

[7] R. Tibshirani. Regression shrinkage and selection via the lasso. Royal Statistical Society, Methodological:267–288, 1996.

524382 Dynamic Optimisation of Water-Injection Wells Operation for Enhanced Oil Production from a Mature Oil and Gas Field

Abstract:

Field engineers in the oil and gas industry are constantly faced with the challenge of maintaining profitability amidst several operational constraints whose effects span a wide spectrum of timescales. Given the inevitably high environmental and financial stakes associated with the exploration and production of an oil and gas prospect, there is a strong incentive to enhance hydrocarbon recovery and production via systematic, model-based optimisation approaches [1-3]. In further response to this challenge, the application of sophisticated simulation methodologies to comprehensively ocapture the reservoir behaviour, multiphase flows (in wellbores and flowlines) [4-5] and gas-oil-water separation in the processing facilities is of exceptional industrial as well as academic interest. Liquid loading in gas wells and artificial lift design considerations, reservoir pressure maintenance via water injection, gas/water coning during production from deviated wells, pressure drop and liquid holdup of multiphase mixtures in highly deviated flowlines are some of the specific features of the problem which are frequently encountered and known to affect its complexity.

Simulation of these prevalent subsurface and surface phenomena does not always guarantee an accurate prediction of the onset of these problems, let alone the trouble-free operation of an existing petroleum field. In order to tackle the shortcomings and thus improve the fidelity of the currently employed models, it is necessary to also combine robust optimisation methods with these oil and gas CFD simulations [6-7]. This combination of simulation and optimisation algorithms increases the order of complexity due to model nonlinearity, non-convexity and the presence of discrete variables. There has been an increasing number of contributions in the application of optimisation techniques to field production, and this can be attributed to the advances in the development of specialised algorithms and accompanying computational power. To date, most of these attempts have focused on fields undergoing primary production through predominantly vertical wells [1-3].

This paper advances the state of the art in this field by formulating a production optimisation problem, via simultaneous consideration of production and injection wells with both vertical and deviated (inclined/horizontal) well geometries. A field undergoing secondary production (Enhanced Oil Recovery, EOR) is considered in this study. We use multiphase flow and reservoir simulation software in order to compute flowrates and pressure drops in wellbores and flowlines of the considered production network; extra measures are taken to ensure all flowlines are free from hydrate and wax at the prevalent temperature and pressure conditions. The problem is solved as a nonlinear program (NLP), comprising an economic objective function and several constraints to ensure operational feasibility. The adopted optimisation technique yields a reasonable increase in oil production compared to the usual and direct application of a black-box simulator. This also translates to an increased NPV with the current oil price, thus demonstrating the efficiency of the proposed method as a value addition tool.

REFERENCES

[1] Gunnerud, V., Conn, A. and Foss, B., 2013. Embedding structural information in simulation-based optimization. *Comput. Chem. Eng.* **53**: 35–43.
[2] Gerogiorgis, D.I., Georgiadis, M., Bowen, G., Pantelides, C.C. and Pistikopoulos, E.N., 2006. Dynamic oil and gas production optimization via explicit reservoir simulation. *Comput. Aided Chem. Eng.*, **21**: 179–184.
[3] Tavallali, M.S., Karimi, I.A., Teo, K.M., Baxendale, D. and Ayatollahi, S., 2013. Optimal producer well placement and production planning in an oil reservoir. *Comput. Chem. Eng.*, **55**: 109–125.
[4] Epelle, E.I. and Gerogiorgis, D.I., 2017. A multiparametric CFD analysis of multiphase annular flows for oil and gas drilling applications. *Comput. Chem. Eng.*, **106**: 645–661.
[5] Epelle, E.I. and Gerogiorgis, D.I., 2018. Transient and steady state analysis of drill cuttings transport phenomena under turbulent conditions. *Chem. Eng. Res. Des.* **131**: 520-544.
[6] Gunnerud, V. and Foss, B., 2010. Oil production optimization—A piecewise linear model, solved with two decomposition strategies. *Comput. Chem. Eng.*, **34**(11): 1803–1812.
[7] Kosmidis, V.D., Perkins, J.D. and Pistikopoulos, E.N., 2004. Optimization of well oil rate allocations in petroleum fields. *Ind. Eng. Chem. Res.* **43**(14): 3513–3527.

537394 Voluntary Oil Well Ignition As a Blowout Response: Analysis of Factors Influencing Viability

Abstract:

While most off-shore drilling use conventional steel rigs, for large reservoirs in shallow waters, artificial islands make an attractive alternative for offshore drilling because it offers more space, lower environmental impact, greater personnel safety, and lower cost. Drilling from an off-shore island may allow voluntary ignition as a blowout response if the oil contains sufficient gas. For voluntary ignition to be a viable blowout response, wellhead burning â€" that depends on a number of factors â€" needs to be nearly 100% efficient.

In this work, we investigate wellhead burning as a planned response to oilwell blowouts. The work presents a case study of proposed wellhead burning of 880 GOR, 27° API oil, should a blowout occur. The work takes into account some of the concerns regarding possible significant oil spill on Beaufort Sea as a result of incomplete combustion of the blowout effluent.

Burning of two-phase mixtures of oil and gas is a complex phenomenon that depends on the fraction of gas in the mixture, percentage of oil being entrained as liquid droplets in the gas stream, the size of the droplets, their residence time in the flame, among other parameters. Available models and correlations for the influence of these parameters on burning efficiency are examined in view of their relevance to wellhead burning.

This work will demonstrate that for some oil wells with high enough GOR, wellhead burning could be efficient enough for to be a reasonable response to a well blowout. In the studied case, a GOR of 880 combined with high total flowrate of the mixture, strongly implies nearly complete combustion of the well effluent during a blowout. The case study helped us formulate an algorithm to examine whether any particular well could be a possible candidate for wellhead burning as a planned response to a blowout. The approach suggests determining a wellâ€™s worst case discharge (WCD), estimating fluid velocities at well exit, estimating adiabatic flame temperature, and performing an energy balance to determine if all oil drops will evaporate.

532256 Multi-Objective and Dynamic Real-Time Optimization of Postcombustion Carbon Capture Processes for Cycling Applications

Abstract:

As the penetration of intermittent renewable energy sources, such as solar and wind, into the power grid increases[1], there is a growing need for baseload fossil-fueled power plants to cycle their load accordingly[2]. Cycling baseload plants may introduce unexpected inefficiencies into the system and increase costs, as the power plants are no longer operating at their optimal design point. Previously[3], a dynamic real-time optimization (DRTO) framework was implemented on a reduced-order model of a methylethylamine (MEA)-based carbon capture subsystem (CCS) from a supercritical coal-fired power plant[4]. This framework considered the cycling profile, electricity price market and *cap&trade* as forcing functions to determine the optimal output trajectories for different carbon capture landscapes. The obtained optimal dynamic trajectories can then ultimately be sent to lower-level advanced model-based controllers[5].

Although the implementation of the DRTO framework for the CCS was performed successfully, in general, real-life situations are guided by more than a stand-alone economic objective to determine the optimum point at a certain time. Particularly, for the cycling scenario, objectives just as important such as emission reduction or minimization of wear and tear may be conflicting with the economic cost. To fill this gap, it is proposed in this work the incorporation of a multi-objective optimization (MOO) component into the DRTO framework for the calculation of optimal output trajectories considering all the objectives in tandem. In this case, more reliable optimal trajectories can be obtained, as the framework considers a more comprehensive context. As a trade-off, there is an increase in algorithm complexity, as new definitions (e.g., dominance) need to be introduced, and different strategies (e.g. aggregation, Pareto-front) need to be assessed to achieve the solution.

The reduced-order model of the subsystem considered in this work is a nonlinear autoregressive with exogenous inputs (NARX) with wavelets networks. The DRTO calculates the output optimal trajectories considering a one day-ahead prediction with a 1 hour simulated resolution. For the MOO component, it is desirable to minimize the decision maker interaction with the optimizer, so that the optimal trajectories can be determined and implemented at every time interval in an online and automated fashion. For instance, the optimizer should not wait for the decision maker to choose a point among a set of solutions to send the trajectories to the controllers. Thus, a multi-objective framework that guarantees a Pareto-optimal operating point is implemented in this work to keep the optimization independent of the hourly input of the decision maker. Furthermore, an algorithm without the necessity for system/search space convexity is also explored. The augmented framework is applied to a case study of the cycling carbon capture subsystem and the results are generated considering different carbon capture landscapes.

References

[1] REN21. (2018). Renewable Energy Policy Network for 21st Century. Available at: http://www.ren21.net/status-of-renewables/global-status-report/. Accessed on April 16, 2018.

[2] BENTEK ENERGY LLC. (2010). How Less Became More: Wind, Power and Unintended Consequences in the Colorado Energy Market. *Prepared for Independent Petroleum Association of Mountain States.*

[3] KIM, R., LIMA F. V., (2017). Nonlinear System Identification and Dynamic Real Time Optimization of Postcombustion CO_2 Capture Processes for Cycling Applications. *AIChE Annual Meeting, Minneapolis, MN.*

[4] ZHANG Q., TURTON R., & BHATTACHARYYA D., (2016). Development of Model and Model-Predictive Control of an MEA-Based Postcombustion CO_2 Capture Process. Industrial and Engineering Chemistry Research. 55, 1292-1308.

[5] HE X, WANG Y, BHATTACHARYYA D, LIMA F. V, & TURTON R., (2018). Dynamic Modeling and Advanced Control of Post-Combustion CO_2 Capture Plants. Chemical Engineering Research and Design. 131, 430-439.

536204 Dynamic Modeling and Control of a Post-Combustion Solid-Sorbent Capture System with the Ccsi Models and Tools

Abstract:

Solid sorbent-based post-combustion CO_2 capture processes are a viable alternative to solvent-based capture that may provide better energy efficiency. Integrated dynamic models of these types of processes are essential for understanding transient system behavior and for the development of control strategies to efficiently reject process disturbances and optimize tracking of set points. In this work, a suite of tools and models from the Carbon Capture Simulation Initiative (CCSI) was used to develop a dynamic integrated solid sorbent, post-combustion capture model and implement Nonlinear Model Predictive Control (NMPC). The CCSI Toolset has been released under an open-source license with the goal of accelerating and reducing scale-up risk of carbon capture technologies.

The post-combustion carbon capture flowsheet was developed in Aspen Custom Molder (ACM) and was constructed using the CCSI bubbling fluidized bed (BFB) model[1]. The model is one-dimensional with two-phases and an integrated heat-exchanger that is flexible over a wide-range of operating conditions so that it can represent both adsorption and regeneration in either a steady-state or dynamic framework. A dynamic flowsheet was developed that is based on an optimal steady-state design that includes a three stage BFB for CO_2 adsorption, a two stage BFB regenerator, and includes a crossflow heat exchanger between the rich and lean solid streams. The flowsheet also accounts for the pressure flow dynamics and downcomers in the BFB.

The integrated model is computationally expensive and not suitable for direct use in an NMPC framework, hence a reduced order model is required for model-based control. The CCSI data-driven system identification tool, D-RM Builder,[2] is used to generate computationally efficient dynamic reduced-order models from a set of transient data. D-RM Builder allows for the selection of input/output variables, the generation of a sequence of step-changes, and interfaces with rigorous modeling platforms, such as ACM, to automatically run and collect step-test data. This tool was used to generate a computationally efficient nonlinear Decoupled A-B Net (DABNet) model[3] involving multiple time-scales corresponding to "fast" pressure-flow dynamics and "slower" temperature-composition transients. The DABNet model was used as the control model in CCSI's Advanced Process Control (APC) Framework[4,5]. The APC Framework solves the constrained NMPC problem in a computationally efficient manner utilizing "fast" analytical derivative evaluations, implementing the control moves on the rigorous ACM flowsheet model. The control strategy is tested and compared to traditional control for its servo control and disturbance rejection characteristics.

References

1. Lee, A., & Miller, D. C. (2012). A one-dimensional (1-d) three-region model for a bubbling fluidized-bed adsorber. *Industrial & Engineering Chemistry Research*, *52*(1), 469-484.
2. Ma, J., Mahapatra, P., Zitney, S. E., Biegler, L. T., & Miller, D. C. (2016). D-RM Builder: A software tool for generating fast and accurate nonlinear dynamic reduced models from high-fidelity models. *Computers & Chemical Engineering*, *94*, 60-74.
3. Sentoni, G. B., Biegler, L. T., Guiver, J. B., & Zhao, H. (1998). State space nonlinear process modeling: Identification and universality. *AIChE Journal*, *44*(10), 2229-2239.
4. Omell, B. P., Ma, J., Mahapatra, P., Yu, M., Lee, A., Bhattacharyya, D., Zitney, S., Biegler, L., & Miller, D. C. (2016). Advanced Modeling and Control of a Solid Sorbent-Based CO2 Capture Process. *IFAC-PapersOnLine*, *49*(7), 633-638.
5. Mahapatra, P., Ma, J. & Zitney, S.E. (2018). Nonlinear Model Predictive Control using Decoupled A-B Net Formulation for Carbon Capture Systems – Comparisons with Algorithmic Differentiation Approach. *American Control Conference*, Jun 27-29, Milwaukee, WI.

537271 Application of a Data-Driven Modeling Approach to a Large-Scale Power Plant

Abstract:

Most conventional fossil-fuel-based power plants are designed to operate in baseload operation with small variations in the power output [1]. With the increasing contribution of renewable energy sources to the power grid, the conventional power plants are experiencing more fluctuations in the grid. This is mainly because most of the renewable energy sources are intermittent in nature [2] and conventional power plants must ramp up/down to provide room for the renewable power to be utilized or to compensate for them when they become unavailable. The sharp ramps in the power output of conventional sources result in increased inefficiencies in the power plant operation, which results in increased generation of pollutants such as NOx. It is, therefore, critical to optimize the operation of the power plant when it undergoes transient circumstances.

In this research study, a large scale coal-fired power generation unit in the United States was considered to demonstrate the benefits of dynamic optimization, as opposed to the steady-state optimization approach that is currently implemented in this unit. Currently, power production from renewable sources as well as the electricity demand are forecasted a few hours ahead of time. This serves as an opportunity to plan the operation of the power plant ahead of time such that NOx generation in the plant is minimized while it undergoes fluctuations. The first step in implementing a dynamic optimization approach is to develop a dynamic model of the plant. This presentation provides the results of a dynamic model of the plant as opposed to a steady-state approach that is the currently utilized in the plant. The power plant considered in this analysis has hundreds of variables that makes it very complex to develop a first-principles model of the plant. Thus, a data-driven modeling approach was considered in this study. The method used to develop the dynamic model of the plant is based on a recurrent neural network (RNN) [3]. An RNN model uses neurons that have nonlinear activation functions, which is used to map inputs to outputs. An RNN model takes advantage of some of the past values of measurements (e.g. NOx in this study) that are fed back to the model; thus, making it dynamic in nature. Measurement feedback also serves as a recalibration of the model. An RNN model is also capable of predicting into the future of the plant for as many time steps ahead as there is forecasted data for power demand and renewable power sources.

To develop an RNN model of the plant, it should first be trained i.e. the different weighting and bias factors should be calibrated such that the model predictions for target outputs (i.e. NOx in this analysis) should have the minimum deviations with the corresponding historical data. To achieve this goal, historical data for all decision variables and target NOx values for over four months of the plant operation was used. Additionally, different numbers of neurons as well as model orders were investigated to balance the model prediction accuracy and its complexity. The order of an RNN model refers to the number of past measurements that are fed-back to the model. To compare the accuracy of the models with different neurons and orders, coefficient of correlation (R) was used as the performance metric.

The optimal number of neurons that can best represent the nonlinear system of this power plant was obtained after a sensitivity analysis. Subsequently, the training of the RNN was implemented for orders 1 to 3. Additionally, a static neural network was also trained with the same historical data to provide a comparison basis for the RNN models. In a static neural network, none of the past values of NOx is given to the model; thus, essentially making the neural network a steady-state model. Training of all models was implemented in the MATLAB programming language. The R-values associated with the static and RNN models of order 1 to 3 were observed to be 0.85962, 0.92919, 0.9295, and 0.92816, respectively. While all models show a strong relationship between the model prediction and historical data, the R-value associated with a recurrent neural network was always greater than 90 and greater than a static neural network. This demonstrates that a dynamic model can better represent the historical data than a static model. Among the different orders considered for the recurrent neural network, a marginal improvement was also seen by increasing the model order. Because increasing the model order increases the computational time to train and use the neural network model, a model order of 2 was selected for the further analysis of this study to balance the accuracy and computational complexity of running an RNN model.

The trained neural network was then used to predict NOx for future time steps. A sensitivity analysis was implemented on the length of the prediction horizon for both the static and RNN models. Consequently, it was observed that smaller prediction horizons resulted in higher and lower R- and root mean square error (RMSE) values, respectively, when compared to the historical NOx data; i.e. the model can more accurately predict the future operation of the plant with shorter horizons. This conclusion was expected because the further the model has to foresee into the future, it has to predict additional fluctuations that the actual plant has undergone through this period. A 3-hour prediction horizon was therefore selected for this analysis, which resulted in accurate prediction of NOx over the horizon while also enabling to demonstrate the benefits of dynamic simulation in comparison to a static model. In this study, historical data for the gross power output of the plant was used as a given input to the model for the desired prediction horizon and the NOx prediction of the model was compared with the historical NOx generation of the plant during the same period. While the static model does not require any measurement update for the 3-hour prediction horizon, this analysis assumed that the actual NOx data from the plant was available to the RNN model every 5 time steps (with 5-minute interval). For a 3-hour ahead prediction of NOx, the R-value for the static model was obtained to be 0.7796 while a value of 0.8298 was observed for the second order RNN model. This difference between the static and second order RNN models represents a 6.44% improvement in the accuracy of model prediction in favor of the RNN model. Improvement of the predictions with a RNN model became more obvious when a 24-hour prediction horizon was considered. With a 24-hour prediction horizon, it was observed that the R-value of the RNN model improved 26.92% over the static model. This is because the plant experiences more transient circumstances over a 24-hour horizon and the RNN model better encounters for these transient operations over this period. These results demonstrate the value of dynamic simulation in comparison to steady-state simulation, when the plant undergoes transient operations such as those encountered when renewable sources are being utilized in the power grid. The future research in this study includes developing a dynamic optimization algorithm for the transient operation of the plant as well as improving the accuracy of the second order RNN model.

References

[1] "Dynamic optimization of a hybrid system of energy-storing cryogenic carbon capture and a baseline power generation unit," Applied Energy, pp. 66-79, 2016.

[2] "Intermittency and Value of Renewable Energy," Journal of Political Economy, pp. 124, (4), 2016.

[3] Principles of Artificial Neural Networks, World Scientific Publishing Company, 2013.

531756 Model-Based Analysis of a Thermofluidic Engine for Low-Grade Heat Recovery: Accounting for Irreversible Thermal Losses

Abstract:

The Non-Inertive-Feedback Thermofluidic Engine (NIFTE) is a two-phase thermofluidic oscillator that can convert low-grade heat into useful mechanical work. It consists of one hot and one cold heat exchangers alongside several compartments as illustrated on Fig. 1. The working fluid undergoes cyclic condensation and vaporization phases by contacting with the heat exchangers, which induce sustained fluid oscillations. Most previous mathematical models proposed to analyse the dynamic behaviour of NIFTE have considered linear descriptions [1-4]. These linear models provide useful insights about the operation of NIFTE, e.g. in terms of the minimum temperature difference required, the oscillation frequency and exergetic efficiencies; but despite their ability to give accurate quantitative predictions, they fail to explain the sustained robust periodic oscillations that have been observed experimentally on a NIFTE prototype. This is the so-called limit-cycle behaviour, which is essentially a nonlinear phenomenon. Markides et al. [5] first proposed a nonlinear inertive NTP model for NIFTE, and this model has been validated experimentally. Later, Wang et al. [6] conducted an optimization-based analysis using this model and they predicted the co-existence of multiple cyclic steady states for certain NIFTE configurations. **Fig. 1.** A schematic diagram of the NIFTE device.

However, none of the previous (linear or nonlinear) models account for the thermal losses in the device, which can result in significant overestimation of the efficiencies reported in experiments. Solanki et al. [7,8] refined several linear models models by incorporating a dissipative thermal loss parameter to account for (i) the loss of exergy arising from irreversible heat transfer, and (ii) alternating parasitic condensation and re-evaporation within the engine. This refinement enables much closer predictions of the exergetic efficiencies of NIFTE, while having a limited effect on the predicted oscillation frequencies.

The main development in this work is a description of thermal losses in the nonlinear inertive NTP model of NIFTE to improve its predictive capability – namely, the inertive model with thermal loss (NTP-TL). We build on the electric-circuit analogy previously developed for the NTP model, by adding extra resistor and capacitor elements to describe the overall irreversible thermal losses. It is found that the NTP-TL model provides much smaller exergetic efficiencies in agreement with experimental observations, and also improves the oscillation frequency predictions compared with its linear counterparts. Finally, we apply an optimization-based approach based on NTP-TL to detect and characterize the cyclic steady states corresponding to various engine configurations, and we also conduct a (global) sensitivity analysis to determine those (combinations of) parameters having the greatest effect in terms of the exergetic efficiency. Overall, the NTP-TL model enables a more realistic understanding of the performance of the NIFTE prototype, and provides useful insights on how to design better prototypes.

References:

[1] Smith, T.C.B: Thermally driven oscillations in dynamic applications. PhD thesis. Cambridge: University of Cambridge, UK. 2006
[2] Markides, C.N. and Smith, T.C.B.: A dynamic model for the efficiency optimization of an oscillatory low grade heat engine. Energy, 36, 6967-6980, 2011
[3] Solanki, R., Galindo, A., and Markides, C.N.: Dynamic modelling of a two-phase thermofluidic oscillator for efficient low grade heat utilization: Effect of fluid inertia. Applied Energy, 89(1), 156-163, 2012
[4] Solanki, R., Galindo, A., and Markides, C.N.: The role of heat exchangeon the behaviour of an oscillatory two-phase low-grade heat engine. Applied Thermal Engineering, 53, 177-187, 2013.
[5] Markides, C.N., Osuosslale, A., Solanki, R., and Stan, G.B.V.: Nonlinear heat transfer processes in a two-phase thermofluidic oscillator. Applied Energy, 104, 958-977, 2013
[6] Wang, Y., Markides, C.N., and Chachuat, B.: Optimization-based Investigations of a Thermofluidic Engine for Low-grade Heat Recovery. Manuscript accepted by ADCHEM 2018
[7] Solanki, R., Mathie, R., Galindo, A., and Markides, C.N.: Modelling of a two-phase thermofluidic oscillator for low-grade heat utilisation: Accounting for irreversible thermal losses. Applied Energy, 106, 337-354, 2013
[8] Markides, C.N., Solanki, R., Galido, A.: Working fluid selection for a two-phase thermofluidic oscillator: Effect of thermodynamic properties. Applied Energy, 124, 167-185, 2014

537091 Model Predictive Control Designs to Achieve Uniform Growth of Simultaneously Propagating Multiple Fractures in Hydraulic Fracturing

Abstract:

Natural gas and oil resources are found in shale rock formations with ultra-low permeability. In ultra-low permeability formations, horizontal wells are typically drilled and each horizontal well is stimulated with massive fractures for enhanced recovery of oil and gas [1]. In multi-stage hydraulic fracturing treatments, simultaneously propagating multiple fractures with close spacing often induce non-uniform fracture development, resulting in one or two dominant fractures due to the uneven distribution of fracturing fluids [2]. One of the important contributing factors for the uneven distribution is the well-known phenomenon called ``stress shadow effects" [3]. In general, fracturing fluids are distributed to multiple fractures inversely proportional to their flow resistances, which is a function of perforation friction, wellbore friction and fracture propagation. The problem with the stress shadow effects is that it exerts extra compressional stresses on the interior fractures and increases the flow resistance within interior fractures, resulting in fluid diverting into the exterior fractures. Therefore, to achieve uniform growth of simultaneously propagating multiple fractures and facilitate more fracturing fluids entering into the interior fractures, the stress shadow effects should be balanced or mitigated.

In this regard, several hydraulic fracture simulators have been developed to describe the stress shadow effects in simultaneously propagating multiple fractures [4]. Typically, these high-fidelity hydraulic fracture simulators require several days, and sometimes over one week, to compute the growth of simultaneously propagating multiple fractures at real reservoir length and time scales. Hence, optimization of hydraulic fracturing to find the operating condition for uniform growth of hydraulic fractures, which may often require hundreds or thousands of simulation runs, were not practically viable; in these directions, very limited efforts have been made by selecting a few important parameters via sensitivity analysis [4, 5].

Motivated by our previous efforts on single fracture propagation [6, 7, 8], in this work, we focus on the development of a new model order-reduction technique for simultaneously propagating multiple fractures by integrating the analytical models to calculate the pressure drop due to perforation friction and wellbore friction and a data-based ROM to describe the pressure drop along the fractures due to stress shadow effects and fracture interaction. Then, we propose a model-based pumping schedule design technique by utilizing the ROM and the uniform limited entry design technique to compute the flow rate of fracturing fluids which will promote equal distribution of fracturing fluids to achieve uniform growth of multiple fractures while mitigating the undesired stress shadow effects. Simulation results are presented to compare the proposed technique with state-of-the-art techniques in the field.

References:

[1] Lecampion, B., & Desroches, J. (2015). Simultaneous initiation and growth of multiple radial hydraulic fractures from a horizontal wellbore. *Journal of the Mechanics and Physics of Solids*, 82, 235-258.

[2] Wu, K., & Olson, J. E. (2016). Mechanisms of simultaneous hydraulic-fracture propagation from multiple perforation clusters in horizontal wells. *SPE Journal*, 21(03), 1000-1008.

[3] Germanovich, L. N., Astakhov, D. K., Mayerhofer, M. J., Shlyapobersky, J., & Ring, L. M. (1997). Hydraulic fracture with multiple segments I. Observations and model formulation. *International Journal of Rock Mechanics and Mining Sciences*, 34(3-4), 97.e1-97.e19.

[4] Kan, W., Anusarn, S., & Jizhou, T. (2016). Numerical study of flow rate distribution for simultaneous multiple fracture propagation in horizontal wells. *In 50th US Rock Mechanics/Geomechanics Symposium. American Rock Mechanics Association (ARMA-2016-038),* Houston TX.

[5] Cheng, C., Bunger, A.P., Peirce, A.P., 2016. Optimal perforation location and limited entry design for promoting simultaneous growth of multiple hydraulic fractures. *In SPE Hydraulic Fracturing Technology Conference. Society of Petroleum Engineers (SPE 179158),* The Woodlands TX.

[6] Siddhamshetty, P., Yang, S., & Kwon, J. S. I. (2017). Modeling of hydraulic fracturing and designing of online pumping schedules to achieve uniform proppant concentration in conventional oil reservoirs. *Computers & Chemical Engineering. URL:* https://doi.org/10.1016/j.compchemeng.2017.10.032.

[7] Siddhamshetty, P., Kwon, J. S. I., Liu, S., & Valkó, P. P. (2017). Feedback control of proppant bank heights during hydraulic fracturing for enhanced productivity in shale formations. *AIChE Journal*, 64(05), 1638-1650.

[8] Narasingam, A., & Kwon, J. S. I. (2017). Development of local dynamic mode decomposition with control: Application to model predictive control of hydraulic fracturing. *Computers & Chemical Engineering*, 106, 501-511.

526564 Real-Time Control and Balancing of a Reformer Furnace

Abstract:

In a vision of carbon-emission-free and sustainable energy systems, the hydrogen economy, in which hydrogen is the major energy carrier of the renewable energy sources and is the main transportation fuel, is a viable option and has the potential to reduce greenhouse gas emissions and to resolve issues associated with climate change without halting economic growth. Therefore, solving challenges encountered in commercial-scale production of hydrogen and optimizing these production lines has become an issue of great interest to both academia and industry. Among a variety of commercial hydrogen production processes, steam methane reforming (SMR) is by far the most common. For instance, SMR was responsible for 80% to 85% of the world-wide hydrogen production in 2007 [1]. SMR is typically carried out in top-fired steam methane reforming furnaces which are the most expensive equipment in terms of the maintenance and operating costs compared to other major equipment such as the hydrotreating, prereforming, water-shift and purification units at centralized SMR-based facilities. For instance, the re-tubing cost of a furnace is 10% of the total capital investment, and the annual operating cost to procure fresh natural gas for a SMR-based hydrogen plant with a production rate of 2.7 million Nm^3 per day is $62 million [2]. In addition, the conversion of methane via SMR in the furnace can only be increased by increasing the total energy input, which is constrained by the extreme sensitivity of the reforming tube service life to its operating temperature [3]; consequently, the plant throughput is compromised to avoid premature failure of the reformer and to reduce the plant risk of suffering substantial capital and production losses. Hence, the need for a robust tool that identifies the optimized total fuel flow rate and the corresponding valve distribution such that the conversion of methane via SMR is maximized, and the maximum OTWT along the reforming tube length among all reforming tubes is strictly less than the design temperature of the reforming tube wall material becomes apparent.

Motivated by the above considerations, the present work utilizes the framework for the furnace-balancing scheme [3], the valve-to-flow-rate converter [3], the statistical-based model identification [4] and a heuristic search algorithm to create a real-time balancing procedure, which recursively tests different total fuel flow rates of which the respective spatial distribution to burners is optimized in real-time until key operational specifications are satisfied. The ability to adjust the total fuel flow rate of the balancing procedure is of special interest for the hydrogen manufacturing industry as it can potentially lead to substantial savings in the re-tubing cost of the reformer. Subsequently, a computational fluid dynamic (CFD) model of the furnace developed in [5] is used to characterize the previously unstudied dynamic behavior of the reformer, based on which we develop an optimal strategy to implement the optimized total fuel flow rate to maximize the reformer throughput. Finally, a case study in which the balancing procedure is implemented on the reformer initially operated under the nominal reformer input is proposed, and the results are used to demonstrate that the furnace-balancing scheme successfully determines the optimized reformer fuel input to increase the reformer throughput while meeting the OTWT limits.

[1] Simpson, A.P., Lutz, A.E., 2007. Exergy analysis of hydrogen production via steam methane reforming.

International Journal of Hydrogen Energy 32, 4811–4820.

[2] Latham, D.A., McAuley, K.B., Peppley, B.A., Raybold, T.M., 2011. Mathematical modeling of an industrial steam-methane reformer for on-line deployment. Fuel Processing Technology 92, 1574–1586.

[3] Tran, A., Aguirre, A., Crose, M., Durand, H., Christofides, P.D., 2017a. Temperature balancing in steam methane reforming furnace via an integrated CFD/databased optimization approach. Computers & Chemical Engineering 104, 185–200.

[4] Tran, A., Pont, M., Aguirre, A., Durand, H., Crose, M., Christofides, P.D., in press, 2018. Bayesian model averaging for estimating the spatial temperature distribution in a steam methane reforming furnace. Chemical Engineering Researchand Design.

[5] Tran, A., Aguirre, A., Durand, H., Crose, M., Christofides, P.D., 2017b. CFD modeling of a industrial-scale steam methane reforming furnace. Chemical Engineering Science 171, 576–598.

521550 Finite Element Modeling and Optimization of Heat Exchangers

Abstract:

The rise in energy demand throughout the globe calls for more energy efficient design using process intensification techniques. These designs require solving complicated steady state models describing the physical phenomenon happening inside the system. One of the most important and well-researched aspects of chemical plant design are heat exchangers, yet these are currently designed under the assumption that the physical properties of the streams remain constant inside the heat exchanger. This assumption breaks down significantly under phase change as liquid and vapor phase properties vary drastically and non-smoothly. Lack of *a priori* knowledge of where the phase transition occurs inside the heat exchanger posts a serious problem for the solution algorithms. This complexity is compensated by adding safety factors that lead to inefficient and over-design of heat exchangers thus decreasing the energy efficiency of the whole process.

In this work, a new optimization framework for heat exchanger design is presented that can solve the difficulty of non-smooth nature resulting from phase change inside the heat exchanger. With this in mind, we describe a discretized finite element structure [1], which can be used as a candidate heat exchanger model for large-scale process design problems including phase change. We model the non-smooth nature of phase-change using complementarity constraints [3] that are relaxed and solved iteratively until an acceptable tolerance level is achieved. The discontinuity in heat transfer coefficient arising from the huge difference in latent and sensible heat transfer is described by a smooth approximation. We apply the proposed model by designing heat exchangers in a single-stage and multi-stage Vapor Compression Refrigeration (VCR) cycle with R-152a (1,1-Difluoroethane) as the refrigerant. The VCR cycle subject to complementarity and process constraints with around 4000 variables and constraints is modeled as a non-linear program (NLP) in AMPL and solved using IPOPT [2]. The intricacies of the model along with initialization and the results from the optimization will be presented. Extensions to modeling the system dynamics and solving it using optimization solvers will be discussed.

[1] Ravikumaur, S.G., Seetharamu, K.N. and Aswatha Narayana, P.A., Finite element analysis of shell and tube heat exchanger, International Communications in Heat and Mass Transfer, Volume 15, Issue 2, Pages 151-163,(1986)

[2] Wächter Andreas, Biegler L.T., On the implementation of an interior-point filter line-search algorithm for large-scale nonlinear programming, Mathematical Programming, Volume 106, Issue1, Pages 25-57,(2006)

[3] B.T. Baumrucker, L.T. Biegler, MPEC strategies for optimization of a class of hybrid dynamic systems, Journal of Process Control, Volume 19, Issue 8, Pages 1248-1256(2009)

538305 Application of Paroc in the Optimization and Control of PEM Water Electrolysis Process

Abstract:

Proton Exchange Membrane Water Electrolyzer (PEMWE) has gained attention within the last decade due to its relative advantage over other water electrolysis technologies. However, its high operational energy intensity (losses) has been the major setback (1). Energy losses are categorized as faradaic losses, and overpotentials and they are caused by complex irreversible interactions within the PEMWE process. The bubble effect, which involves the accumulation of oxygen gas bubble on the surface of the Membrane Electrode Assembly (MEA), is one of the various phenomena that has a significant impact on the overpotential of the PEMWE process (2–4). Furthermore, the flow of the electrolyte has a substantial effect on the accumulation of bubbles on the surface of the MEA(3). Few researchers have considered the bubble effect in their models (5) and none has attempted to systematically optimize their model to obtain an optimal operating point.

In this work, we present an experimentally validated dynamic mathematical model and control optimization of a PEMWE system using the Parametric Optimization and Control (PAROC) framework (6). The main objectives of this work are to (i) formulate and validate a comprehensive model; (ii) to obtain optimal operating points that minimizes the energy losses in the PEMWE process; (iii) and to design advanced controller to maintain the optimal operating points obtained. The PAROC framework enables the representation and solution of demanding model-based operational optimization and control problems. The application of PAROC on PEMWE features: (i) a high-fidelity dynamic mathematical model of the PEMWE that captures the detailed electrochemical interaction, transport phenomenon, and other interaction associated with energy losses in the system; (ii) a model reduction step that develops a more tractable state-space model from the original complex model without losing its sense of the original model; (iii) a Model Predictive Control (MPC) design and its reformulation to a multi-parametric (mp) optimization problem (which becomes mp-MPC) (7). The solution of the mp-MPC provides a map of solution that spans the feasible operating region of the PEMWE. Unlike the MPC the mp-MPC avoids the online optimization procedure at every time step (7). The optimization is done once and offline to determine the control action at every realization of the parameters while simultaneously accounting for physical and operational constraints. For a system with fast dynamics (like the PEMWE), mp-MPC is ideal.

Reference

1. Grigoriev SA, Porembsky VI, Fateev VN. Pure hydrogen production by PEM electrolysis for hydrogen energy. Int J Hydrogen Energy [Internet]. 2006 Feb 1 [cited 2018 Jan 17];31(2):171–5. Available from: http://www.sciencedirect.com/science/article/pii/S036031990500145X
2. Qian K, Chen ZD, Chen JJJ. Bubble coverage and bubble resistance using cells with horizontal electrode. J Appl Electrochem [Internet]. 1998;28(10):1141–5. Available from: http://link.springer.com/10.1023/A:1003447913204
3. Balzer RJ, Vogt H. Effect of Electrolyte Flow on the Bubble Coverage of Vertical Gas-Evolving Electrodes. J Electrochem Soc [Internet]. 2003;150(1):E11. Available from: http://jes.ecsdl.org/cgi/doi/10.1149/1.1524185
4. Kiuchi D, Matsushima H, Fukunaka Y, Kuribayashi K. Ohmic Resistance Measurement of Bubble Froth Layer in Water Electrolysis under Microgravity. J Electrochem Soc [Internet]. 2006;153(8):E138. Available from: http://jes.ecsdl.org/cgi/doi/10.1149/1.2207008
5. Grigoriev SA, Kalinnikov AA, Millet P, Porembsky VI, Fateev VN. Mathematical modeling of high-pressure PEM water electrolysis. J Appl Electrochem. 2010;40(5):921–32.
6. Pistikopoulos EN, Diangelakis NA, Oberdieck R, Papathanasiou MM, Nascu I, Sun M. PAROC An integrated framework and software platform for the optimisation and advanced model-based control of process systems. Chem Eng Sci. 2015;136:115–38.
7. Bemporad A, Morari M, Dua V, Pistikopoulos EN. The explicit linear quadratic regulator for constrained systems. Automatica. 2002;38(1):3–20.

526054 Predictive Control of Solar Collector Energy System with Gaussian Process Priors of Uncertain Solar Irradiance

Abstract:

In this work, we study the application of the receding horizon optimal control for solar collector energy system governed by coupled nonlinear hyperbolic partial differential equations [1]. The optimal solution is calculated by using the calculus of variations approach. Solar irradiance is the only dominant source to provide heat to the system and suppose that current time solar irradiance can be measured by a solar irradiance measuring device such as pyranometer. In order to apply nonlinear model predictive control (MPC) techniques [2], the prediction of solar irradiance in future certain time horizon is necessary. In this work, we consider the multi-step ahead prediction of solar irradiance (time series) using non-parametric Gaussian process (GP) regression parametrized by mean and covariance function [3]. k-step ahead forecasting of a discrete-time non-linear dynamic system can be performed by doing repeated one-step ahead predictions. k-step ahead prediction method works as follows: it predicts only one time step ahead, using the estimate of the output of the current prediction, as well as previous outputs (up to the lag L), as the input to the prediction of the next time step, until the prediction k steps ahead is made. In the simulation part, the integration of GP model of solar irradiance and MPC is performed to achieve optimal outlet temperature trajectory tracking.

[1] Xu, Xiaodong, Yuan Yuan, and Stevan Dubljevic. "Receding horizon optimal operation and control of a solarâ €thermal district heating system." AIChE Journal (2017)

[2] Van Pham, Thang, Didier Georges, and Gildas Besancon. "Predictive control with guaranteed stability for water hammer equations." IEEE transactions on automatic control 59.2 (2014): 465-470.

[3] Rasmussen, Carl Edward. "Gaussian processes in machine learning." Advanced lectures on machine learning. Springer, Berlin, Heidelberg, 2004. 63-71.

523250 Fragrance Product Design/Screening Methods Using an Integrated Machine Learning and Camd Model

Abstract:

Abstract

Fragrances are used in a wide variety of daily products such as perfumes, cosmetics, toiletries and household cleaners [1]. The business of flavors and fragrances has become a multibillion dollar market with a great impact all over the world, but the design of fragrances remains mostly empirical, based on the experience and knowledge of experts. However, potentially better fragrance products could be missed when employing experience and knowledge-based design methods. Therefore, a systematic computer-aided design method is needed to assist in designing/screening fragrance molecules to meet various product needs as a first step before focused experiment-based verification can be performed. However, the lack of relevant property models prevents the computer-aided design methods being applied to the fragrance products. In chemical and pharmaceutical research, large amounts of experimental data are available. Knowledge may be extracted from such data, for example by deriving models that can predict properties for new compounds. Therefore, machine learning techniques are used to establish Quantitative Structure-Property Relationships (QSPR). A growing number of new potentially useful machine learning (ML) methods are being used in chemistry. Although many models for property prediction have been developed and many more are needed, very few have been developed by applying machine learning methods even though sufficient fragrance data can be found.

In this paper, an integrated model of ML and Computer-Aided Molecular Design (CAMD) is developed for the design of new fragrance molecules. The odors of the molecules are predicted using a data-driven ML approach. In the developed ML method, group-based representation is selected to represent the molecular structures. The database contains 480 different molecules, and 20 odor characters are used [2] to classify them. Convolutional Neural Network (CNN) is used in the modeling of odor of organic molecules. The established machine learning (CNN) model structure consists of a 50×64 embedding layer, two 47×128 convolutional layers (47×128 and 44×128), a 22×128 max-pooling layer, a 22×128 dropout layer, a 2816×1 flatten layer, a 128×1 dense layer, another 128×1 dropout layer and 20×1 dense layer. Finally, the properties of odor characters and odor pleasantness are predicted using this model structure, with regressed model parameters, using the available data for training and verification. The average correctness of odor characters is 92.9%, while the average prediction error of odor pleasantness is 18.4%. Next, a CAMD model is developed based on the established work-flow for molecular design and group-contribution based models for other needed properties. The optimization model includes an objective function, molecular structural constraints and property constraints including odor character, odor pleasantness, diffusion coefficient, vapor pressure, normal boiling and melting points, solubility parameter, viscosity, density and LC_{50}. A decomposed-based solution approach, in which the established MILP/MINLP model is decomposed into an ordered set of sub-problems, is used to obtain the optimal design results. Finally, two case studies are presented to demonstrate the method for optimal design of fragrance molecules.

References

[1] Teixeira, M.A., et al., Chapter 1 - A Product Engineering Approach in the Perfume Industry, in Perfume Engineering. 2013, Butterworth-Heinemann: Oxford. p. 1-13.

[2] Keller, A. and L.B. Vosshall, Olfactory perception of chemically diverse molecules. Bmc Neuroscience, 2016. 17(1): p. 55.

525804 Optimal Production of Power from Mid-Temperature Geothermal Sources: Scale and Safety Issues

Abstract:

Geothermal energy is a promising source from within the Earth's interior. In spite of its potential, it has not been extensively exploited yet. High-temperature reservoirs, with temperatures above 220 ºC, are the ones most suitable for commercial production of electricity. However, the medium- and low-temperature water-dominated systems, with temperatures between 110 and 160 ºC, are the most abundant [1]. So far most of the work focuses on evaluating the lay out of the Organic Rankine Cycles (ORC), to determine their capacity of extracting energy from low temperature sources [2] and the fluid that can be used [3]. However, either simplified cycles are used to compare differ fluids or details analysis of just a particular fluid are presented in the literature.

In this work we use a superstructure optimization approach for the mathematical optimal design and operation of Organic Rankine binary Cycles for the recovery of energy from medium and low geothermal sources. We focus on the three fluids, Benzene, toluene and cyclohexane, that our previous study identified as the most promising from environmental, economic and safety points of view [4]. In particular, we develop a superstructure for the heat exchanger network to use the hot brine from a geothermal well allowing double and single extractions to optimally design the best cycle, where the thermodynamics of the fluids is included through surrogate models for the enthalpies and entropies. Two solutions are presented, the first one optimizes the energy output and a second one where economic, environmental and safety considerations are included in a normalized objective function to evaluate their effect on the selection of the cycle and on the operating conditions. Finally, the effect of the scale on the investment and production costs is evaluated.

The economic optimization selects a two expansion cycle using toluene as organic fluid, producing 10.4 MW at 0.075€/kWh with an investment of 102M€. Safety considerations slightly change the operating conditions, reducing the pressures and temperatures, but not the selection of the working fluid. Sustainable and economic terms overcome safety issues. The results are competitive with other renewable-based technologies for thermal power production such as CSP or biomass.

[1] Barbier, E. Geothermal energy technology and current status: an overview. *Renew. Sust. Energy Reviews* 2002. 6 (1-2): 3-65

[2] DiPippo, R. Geothermal Power Plants. Principles, applications, case studies and environmental impact. Elsevier, 2008. ISBN: 978-00-8055-476-1

[3] Wang, J.; Yan, Z.; Wang, M.; Ma, S.; Dai, Y. Thermodynamic analysis and optimization of an organic Rankine Cycle (Orc) using low grade heat source. *Energy* 2013; 49: 356-365

[4] Martínez-Gómez, J.; Peña-Lamas, J., Martín M.; Ponce–Ortega, J.M. A Multi-objective optimization

approach for the selection of working fluids of geothermal facilities: Economic, environmental and social aspects. *J. Environ. Manag.* 2017; 203 (3): 962-972

527623 Computer-Aided Reaction Solvent Design Integrated with New Reaction Mechanism Model

Abstract:

With the development of process industry, solvents are widely used in chemical manufacturing processes, such as extraction, absorption, crystallization and reaction. When involved in liquid homogeneous-phase kinetic reactions, solvents can have significant impacts on reaction rate and selectivity to facilitate the existing synthetic routes or even develop new routes [1]. Thus, the choice of reaction solvents is particularly important to product yield and economic benefit. However, it is impossible to perform thousands of extensive and costly kinetic experiments for screening reaction solvents based on trial and error method or qualitative chemical knowledge. Therefore, efficient theoretical methods should be developed to help the selection of reaction solvents. A very promising avenue of research in this direction is the development of Computer-Aided Molecular Design (CAMD) technique, which has been successfully adopted for the optimal design of solvents in the last three decades [2]. It offers a systematic methodology that numerous of chemical species can be assembled by a predefined set of functional groups, and then Quantitative Structure-Property Relationship (QSPR) method, such as Group Contribution (GC) method, was constructed to formulate the molecular group structures with macroscopic solvent properties for establishing mathematical optimization models. Although CAMD had contributed a lot to the solvent design for separations, the work in reaction systems was still immature comparatively. On the one hand, pure Quantum Mechanics (QM) calculations integrated with CAMD have been performed for the reaction rate constant estimations without any experimental rate constant. However, large quantitative deviations often occurred due to the defect of Conventional Transition State Theory (CTST) and Continuum Solvation Model (CSMs). On the other hand, some empirical models, functioned as surrogate models, have been proposed to correlate some experimental reaction rate constants with solvent properties [3] or QM-based descriptors [4], but the model accuracy and universality can hardly be guaranteed simultaneously for the reaction solvent design. Therefore, reliable reaction kinetic models are urgently in need on the basis of reaction mechanisms.

In this paper, an optimization-based framework is developed for the reaction solvent design. A new reaction mechanism model has been established using the hybrid method of CTST derivation, knowledge-based selection and model evaluation, in which rigorously thermodynamic deduction was first performed to formulate the primary reaction mechanism model consisting of the infinite dilution activity coefficients and solvent molar volumes. The state-of-the-art COSMO-SAC (COnduct-like Screening MOdel-Segment Activity Coefficient) model is employed for the calculation of activity coefficients efficiently. Then, according to the solvatochromic equation [3], knowledge-based selection method was used to add the rate-related descriptors to our primary equation for model correction. Afterwards, optimal linear regression equation with the largest adjusted R square was identified through model evaluation, in which the significances of all model descriptors were tested again to overcome the overfitting problem. This hybrid method was demonstrated by two case studies, namely Diels-Alder and Menschutkin reaction, and impressive consistency of the results was found out that the infinite dilution activity coefficients, hydrogen-bond donor, hydrogen-bond acceptor and solvent surface tension are the most influential descriptors to the reaction rate constant. The multiple linear regression results of determination coefficient (R^2) and Average Absolute Percent Error (AAPE) between experimental logk and predictive logk are illustrated in Table 1, where the results of Zhou and FoliĬ's methods are also exhibited for comparisons. As we can see from Table 1, our method shows higher R^2 and lower AAPE, which indicates that the reaction mechanism model proposed in this paper exhibits appealing characteristics with wide universality and high accuracy simultaneously.

Table 1. Regression results of different methods for comparisons

Method	Diels-Alder reaction		Menschutkin reaction	
	R^2	AAPE	R^2	AAPE
Our method	**0.962**	**1.83%**	**0.911**	**8.07%**
Zhou's method [4]	0.923	2.57%	0.292	20.22%
FoliĬ's method [3]	-	-	0.654	18.75%

Later GC method was integrated for the prediction of above descriptors. CAMD technique was incorporated with the reaction mechanism model for the optimal reaction solvent design. With an Mixed-Integer Non-Linear Programming (MINLP) problem formulated, decomposition-based solution algorithm was employed to reduce the difficulty in dealing with the nonlinear COSMO-SAC equations. Ultimately, the promising solvents with double bond and carboxyl groups were identified for Diels-Alder reaction, while nitryl, azyl and nitrile groups for Menschutkin reaction. Experimental verification will be further carried out in our future research.

References:

[1] Struebing, H; Ganase, Z; Karamertzanis, P. G; Siougkrou, E; Haycock, P; Piccione, P. M; Armstrong, A; Galindo, A; Adjiman, C. S. Computer-aided molecular design of solvents for accelerated reaction kinetics. *Nature Chemistry*. **2013**, *5*(11), 952-957.

[2] Zhang, L.; Babi, D. K.; Gani, R. New Vistas in Chemical Product and Process Design. *Annual Review of Chemical and Biomolecular Engineering*. **2016**, *7*(1), 557-582.

[3] FoliÄ‡ M; Adjiman C. S; Pistikopoulos E. N. Computer-Aided Solvent Design for Reactions: Maximizing Product Formation. *Industrial & Engineering Chemistry Research*. **2008**, *47*(15), 5190-5202.

[4] Zhou T; Lyu, Z; Qi Z; Sundmacher, K. Robust design of optimal solvents for chemical reactionsâ€"A combined experimental and computational strategy. *Chemical Engineering Science*, **2015**, *137*(2), 613-625.

528775 Reactor Network Development for Rigid Polyols Production

Abstract:

Rigid polyols are defined as polyether polyols formed by alkoxylation of initiators with 3 to 8 active sites to form polyethers used with isocyanates to manufacture closed cell foams. Closed cell foams are used in the production of refrigerators, building and construction insulation. Demand for these products have increased significantly in the last 10 years and are expected to continue that growth in the next 10 years, and tight product specifications are imposed based on properties related to the molecular weight distribution (MWD).

In most cases, semi-batch reactors are used to produce rigid polyols, with only a few examples of continuous reactors in the patent literature. As the product demand of a polymer increases, larger or more semi-batch reactors are built to meet the demand, and the capital cost per production capacity is predictably lowered. However, this capital cost is still too high for re-investment, and alternative lower capital cost solutions, such as switching the production from semi-batch to continuous processes, could help sustain growth. Hence, continuous processes can reduce the capital cost and the cost required for heat transfer equipment, since the heat load is more evenly distributed over time. In addition, continuous reactors are easier to optimize at a steady state to maximize the overall production rate. However, continuous reactors are not as flexible regarding multiple products, product transition can be difficult and scale-up of the products need extra investment. The motivation of this study is to transform the production from the existing semi-batch reactors to continuous reactors, such as plug flow and continuous stir tank reactors, to lower the capital cost and to maintain product specifications based on properties of the MWD. Prior to developing an optimization superstructure, which predicts an optimal combination of CSTRs and/or PFRs in series/parallel, the reaction mechanism and the kinetic parameters also needs to be determined.

Here, the kinetic parameter estimation problem, which includes MWD information, can be written in a general form as a dynamic optimization problem. Moreover, the simultaneous collocation method can be adopted to deal with the dynamic optimization problem for the reactor network. This method discretizes the continuous time or length horizon into a finite element mesh, and then the differential-algebraic equation optimization problems are converted into nonlinear programming problems. The differential and algebraic state variables z and y, respectively can be denoted by the Runge-Kutta basis representation. Furthermore, a three-point Radau collocation is applied to parameter estimation model. The model is constructed in AMPL and solved by NLP solver IPOPT. This optimization is applied to synthesis an optimal reactor network for rigid polyol production, where we demonstrate significant performance improvements over previous studies.

532076 Mixed Integer Linear Programming Optimization Framework Applied to a Platinum Group Metals Flotation Circuit

Abstract:

This research focuses on determining an optimum flotation circuit in terms of operating variables such as mass flowrates of flotation streams and structural variable like flotation circuit configuration. Determination of optimal circuit involves the use of Mixed-Integer Linear Programming framework in which a constrained objective function is formulated. The objective function in this work is maximization of Platinum Group Metals (PGMs) recovery from flotation circuit constrained by the quality specification of the final concentrate, i.e. gram of PGMs per ton of concentrate.

A superstructure was employed to provide a set of alternative circuits out of which an optimal one is obtained. Optimization was carried out on the basis of a developed flotation model. The model formulation begins with mass balance for each flotation cell and then integrates the cells to build the overall optimization model of the flotation circuit. Optimum operating conditions and structural variables such as concentrate flowrate and flotation circuit configuration respectively, were determined. Flotation process parameters such solid density and rate constants were also fitted in this model.

The results generated by the optimization algorithm show that high recoveries occur at rougher stages while low recoveries at cleaner stages. This high recovery at rougher stage is mainly attributed to the fact that recovery at this stage takes place through two mechanisms, namely: True flotation and Entrainment. Whilst at cleaner stages entrainment is unlikely to occur due to high froth heights in the cells. The increase in concentrate mass flowrate increases recovery with the assumption that the process is at steady state in terms of feed grade whilst other operating variables such as reagent dosage, particle sizes, and aeration rate are at optimal values.

533954 Simultaneous Working Medium Selection, Process and Control System Design for Organic Rankine Cycles

Abstract:

Organic Rankine cycles (ORC) are efficient processes for the recovery of relatively low enthalpy waste heat from industrial streams and exhaust gas streams in combustion engines and its conversion to electricity. The decisions involved in the design of an ORC range from the selection of a suitable working medium for the available temperature levels of the hot and cold sources to the unit operation sizing and extending to the associated controller design. It is well known that mixtures of components exhibit a superior performance as they allow the reduction of the exergy losses in the evaporator and the condenser in the ORC. The closer tracking of the hot source temperature curves is the primary reason for such a behavior. Papadopoulos et al (2013) offered a comprehensive investigation of a large number of potential mixtures as working media in an ORC that also considered variability in the operating conditions. Later, Zarogiannis et al (2017) studied the influence the working medium has on the achieved control performance. However, no study has ever performed a simultaneous working medium selection with process and control system design.

The present work encompasses the design of the process equipment in an ORC with the control system design for a set of potentially promising working media under a unified framework. Working media selection involves the identification of the components forming the mixture and the associated concentration of each component. The set of mixtures that possess the desired properties and have the potential for a highly performing ORC system have been identified using a computer aided molecular design (CAMD) procedure (Papapdopoulos et al., 2013). The CAMD procedure targets specific mixture properties but it does not consider the dynamic behavior of the system under disturbance influence or variability in the operating conditions. Since CAMD is based on multi-objective optimization it identifies all mixtures that are included in the Pareto optimal front.

The integrated working medium selection with the process and control system design aims to investigate the optimal combination of the working mixture with the ORC configuration that satisfy an economic criterion under steady-state and closed loop conditions. Therefore, for a selected working mixture the steady-state design of the ORC is performed based on economic optimization, where the optimal configuration and sizing of the equipment is obtained. In addition, a set of disturbance scenarios is constructed that are representative of the most commonly occurring operating conditions for the process. A model predictive control (MPC) scheme is employed for the satisfaction of the process goals under process variability and disturbance influence. The use of a MPC enables the best handling of the interactions among the various processes in the cyclic process. The integrated design framework enables the determination of the most suitable input-output structure for the controller by allowing the selection of the controlled and manipulated variables from a pool of screened options. The performance of the MPC under the specified disturbance scenarios is calculated through dynamic simulations for the selected working medium and process flowsheet configuration, in each optimization iteration. The dynamic performance is converted to an economic equivalent term so that it can be combined with the steady-state design economic term to generate the overall objective function value.

Novel and conventional hydrocarbon- and halogenated hydrocarbon-based mixtures are considered for the integrated ORC process and control system design under the proposed unified framework. Variation in the hot and cold source streams is considered in the system where the employed MPC aims to maintain the evaporator temperature at the desired level and achieve the maximum work in the expander. The simultaneous problem of working mixture selection and process design under static and closed loop conditions is solved using a stochastic optimization technique. The results provide significant insights in the behavior of the ORC system under closed loop conditions. The ability of the ORC system with the MPC to maintain a high performance under variability is found to be influenced by the sensitivity of the working medium physical properties within the range of variation that is transferred through the system.

Cited References

Papadopoulos A.I., M.Z. Stijepovic, P. Linke, P. Seferlis, and S.S. Voutetakis, "Towards Optimum Working Fluid Mixtures for Organic Rankine Cycles using Molecular Design and Sensitivity Analysis", Industrial and Engineering Chemistry Research, 52(34),12116–12133, 2013.

Zarogiannis T., A. I. Papadopoulos, P. Seferlis, and P. Linke, "The Impact of Novel and Conventional Working Fluids on the Control Performance in Organic Rankine Cycles", Computer Aided Chemical Engineering, 40, 2443-2448, 2017.

537211 Process Optimization-Centric Design and Screening of Nanoporous Adsorbents for Gas Separations

Abstract:

There are several works on computational screening of nano-porous materials such as zeolites and metal-organic frameworks (MOFs) for gas separation applications including CO_2 capture and natural gas purification[1,2]. Other studies involve developing novel adsorbent screening metrics to compare adsorbents without performing process simulations[3-5]. While these are computationally inexpensive, they may not be enough to check the feasibility with respect to specified process constraints. Recent efforts have focused on obtaining optimal materials by performing material screening along with process optimization[6,7].

In this work, we determine the optimal material properties for a given process operation (e.g. pressure swing adsorption) through optimization of a completely discretized algebraic model of the process. This involves optimizing a non-linear programming problem with material properties as decision variables. The adsorbent properties such as isotherms, heat of adsorption and selectivity of materials (e.g. zeolites) in existing databases are calculated using molecular simulations and their range is provided as bounds for the optimization. The materials from the database with properties closest to the obtained optimal material properties can then be rigorously explored. This approach eliminates the need for including individual material alternatives in the optimization formulation. Revised optimal properties for any changes in process and feed conditions can also be determined easily. For example, in the case of an optimal zeolite (SBN) obtained through material screening for a simultaneous methane storage and separation application[8], a 1% change in isotherm parameters results in a 3.27% increase in optimal methane storage from 186.13 v(STP)/v to 192.23 v(STP)/v. The structural similarity[9] between materials possessing similar adsorption properties is used to then predict the geometric properties including pore size, channel diameter, void fraction and surface area corresponding to the optimal adsorbent properties by using a data-driven surrogate-model based approach[10]. Such a process optimization-centric methodology can help in the design of adsorbents tailored to the specific requirements of gas separation process applications.

References:

(1) Lin, L.-C.; Berger, A. H.; Martin, R. L.; Kim, J.; Swisher, J. A.; Jariwala, K.; Rycroft, C. H.; Bhown, A. S.; Deem, M. W.; Haranczyk, M.; et al. In silico Screening of Carbon-Capture Materials. *Nat. Mater.*, **2012**, 11 (7), 633â€"641.

(2) Krishna, R.; van Baten, J. M. In silico Screening of Metal-Organic Frameworks in Separation Applications. *Phys. Chem. Chem. Phys.* **2011**, 13 (22), 10593â€"10616.

(3) Wiersum, A. D., Chang, J. S., Serre, C. and Llewellyn, P. L. An Adsorbent Performance Indicator as a First Step Evaluation of Novel Sorbents for Gas Separations: Application to Metal-organic Frameworks. Langmuir, **2013**, 29(10), 3301-3309.

(4) Rajagopalan, A. K.; Avila, A. M. and Rajendran, A. Do Adsorbent Screening Metrics Predict Process Performance? A Process Optimisation based Study for Post-Combustion Capture of CO_2. *Int. J. Greenh. Gas Con.*, **2016**, 46, 76-85.

(5) Ga, S.; Jang, H. and Lee, J. H. New Performance Indicators for Adsorbent Evaluation Derived from a Reduced Order Model of an Idealized PSA Process for CO_2 capture. *Comput. Chem. Eng.* **2017**, 102, 188-212.

(6) Hasan, M. M. F.; First, E. L. and Floudas, C. A. Cost-effective CO_2 Capture based on *In silico* Screening of Zeolites and Process Optimization. *Phys. Chem. Chem. Phys.*, **2013**, 15 (40), 17601-17618.

(7) Hasan, M. M. F.; First, E. L. and Floudas, C. A. Discovery of Novel Zeolites and Multi-zeolite Processes for p-xylene Separation using Simulated Moving Bed (SMB) chromatography. *Chem. Eng. Sci.*, **2017**, 159, 3-17.

(8) Iyer, S. S.; Hasan, M. M. F. A Novel Plug-and-Store Technology for Natural Gas Purification and Strage. In CAMX 2015 - Composites and Advanced Materials Expo; **2015**.

(9) Martin, R. L.; Willems, T. F.; Lin, L.-C.; Kim, J.; Swisher, J. A.; Smit, B.; Haranczyk, M. Similarity-Driven Discovery of Zeolite Materials for Adsorption-Based Separations. *Chem. Phys. Chem A Eur. J. Chem. Phys. Phys. Chem.*, **2012**, 13 (16), 3595â€"3597.

(10) Bajaj, I.; Iyer, S. S. and Hasan, M. M. F. A Trust Region-based Two Phase Algorithm for Constrained Black-box and Grey-box Optimization with Infeasible Initial Point. *Comput. Chem. Eng.*, **2017**, doi: 10.1016/j.compchemeng.2017.12.011.

Universal Graph Structures for Property Estimation

Abstract ID: 538734

Author(s): Charles C. Solvason and Russell Burnett, Bryan Research & Engineering, Bryan, TX

Session Date: Friday, November 2, 2018

Session Time: 12:30 PM - 3:00 PM

Over the past century, physical-chemical properties have become an increasingly important element of industrial and engineering chemistry (O'Connell et al. 2009). In particular, the uses of physical-chemical property estimation have become more sophisticated, with a greater ability to describe the precise molecular structures needed to bring specific properties to the marketplace. The multitude of competing chemical mechanisms that control these properties has required chemical product designers to become more creative and, as a result, the number of property models available to today's chemical engineer can be overwhelming.

This paper highlights a technique to consolidate multiple molecular structure-based property methods into a single, common underlying structure domain from which property estimation can be efficiently facilitated. Each user-defined physical property model is first mapped to a set of universal graph structures and stored in a pre-configured molecular library database. Once mapped, new molecular species can be evaluated without changes to the underlying code. Furthermore, new structures can be added to each method, almost always with no additional development, giving the approach a desired expandability. The method makes extensive use of the IUPAC International Chemical Identifier (InChI) to map chemical species into molecular graphs and back again.

A case study on species relevant to the natural gas industry will be utilized to illustrate the development of the technique and its application via a molecular library database.

O'Connell, J. P., R. Gani, P. M. Mathias, G. Maurer, J. D. Olson and P. A. Crafts (2009). "Thermodynamic Property Modeling for Chemical Process and Product Engineering: Some Perspectives." Industrial & Engineering Chemistry Research 48(10): 4619.

525543 Estimating Parameters and Model Uncertainty in Fundamental Dynamic Models Using Historical Data

Abstract:

Nonlinear models based on fundamental mass and energy balances describe dynamic behavior of chemical processes in process monitoring, control and optimization applications. Predictions from these models are always (at least slightly) wrong due to inaccurate parameter values and model imperfections associated with unmodelled disturbances, inadequate process understanding, simplifying assumptions, and uncertain inputs. We have developed new methods to use batches of old dynamic process data to simultaneously estimate model parameters and the "badness" of the associated nonlinear dynamic model.

To account for model imperfections, stochastic terms are added to the right-hand sides of the dynamic model equations, resulting in stochastic differential equation (SDE) models of the form:

$$\dot{x}(t) = f(x(t), u(t), \theta) + \eta(t)$$

$$x(t_0) = x_0$$

$$y(t_j) = g(x(t_j), u(t_j), \theta) + \varepsilon(t_j)$$

where x is the vector of state variables, t is time, f is a vector of nonlinear functions (derived based on scientific knowledge and assumptions), u is the vector of input variables, θ is the parameter vector and $\eta(t)$ is a vector of stochastic independent continuous white-noise processes corresponding to a power spectral density matrix Q whose diagonal elements are disturbance intensities. These disturbance intensities indicate the level of imperfection of each corresponding ODEs (i.e., the differential equations above with $\eta(t)$ removed). Large disturbance intensities correspond to poor predictive power of the model and large mismatch due to stochastic disturbances or other model imperfections.

Our proposed SDE parameter estimation methods provide estimates of the model parameters θ (and their associated confidence intervals) along with estimates of the disturbance intensities in Q and the measurement noise variances, if they are unknown. These techniques readily accommodate situations where: *i*) some or all of the initial conditions in are unknown and must also be estimated from the data; *ii*) different types of measurements are made at different sampling frequencies or some measurements are made at irregular intervals; *iii*) the modeler is willing to specify prior knowledge about some or all of the model parameters and/or noise variances (e.g., initial guesses and corresponding initial uncertainty levels). Time-varying parameters and nonstationary disturbances can also be accommodated by augmenting the state vector with additional unmeasured parameter or disturbance states.

An Approximate Bayesian Expectation Maximization (ABEM) methodology and a Laplace Approximation Bayesian (LAB) methodology are proposed for estimating parameters and uncertainties associated with SDE models and data. These methodologies are more powerful than our previous maximum-likelihood methodologies for SDEs because they enable modelers to account for prior information about unknown parameters and initial conditions. The ABEM methodology is suitable for situations where the modeler can assume that measurement noise variances are well known, whereas LAB includes measurement noise variances among the parameters that require estimation. Both techniques estimate the magnitudes of stochastic terms included in the differential equations to account for model mismatch and unknown process disturbances. The proposed ABEM and LAB methodologies are illustrated using a two-state nonlinear continuous stirred tank reactor (CSTR) case study, with simulated data sets generated using a variety of scenarios. The ABEM and LAB objective functions derived for the case study have additional terms and result in improved estimates of model parameters and noise parameters compared to our previous maximum-likelihood objective functions, especially in situations where data available for parameter estimation are sparse. Benefits of including prior information about unknown parameters and noise variances are illustrated and advice is provided regarding which objective function should be used for parameter estimation in different situations. Because the proposed ABEM and LAB methodologies rely on B-spline basis functions rather than Markov Chain Monte Carlo techniques, they are straightforward to implement using available optimizers and modeling software and require only modest computational effort.

We hope that the proposed ABEM and LAB objective functions for SDE parameter estimation will be attractive to practitioners who develop dynamic models for online monitoring and control and for dynamic optimization. We anticipate that they will be able to readily adopt the proposed objective function structure in their own modeling and parameter estimation problems, thereby obtaining useful uncertainty information that can aid model improvement and/or implementation.

534732 Accommodating Missing, Non-Uniformly Sampled and Delayed Measurements for Modeling and Control of Variable Duration Batch Processes in a Subspace Identification Framework

Abstract:

Batch processes are an indispensable part of chemical manufacturing industry used for manufacturing of specialty chemical products such as polymers, pharmaceuticals etc. Further, large chemical plants in nominal continuous operation exhibit batch behavior during plant startups and shutdown phase. Acquisition of measurements from operation of batch processes is an important part of process control where important processes variables are sampled and stored in a database for process modeling and monitoring / control. Often some measurements may be corrupted, for instance, due to sensor faults leading to missing values in the database. Further, it is always not possible to measure all the variables at the same sampling time. For instance, measurement of certain variables may depend on offline analysis which may not arrive at a fixed time interval leading to delayed and non-uniform measurements. Therefore, a batch process modeling and control approach is required which can explicitly handle such cases.

A variety of approaches for development of data-driven models have been proposed in the literature. One excellent approach is partial least squares (PLS), which models the process in a projected latent space [3]. These models are essentially time-varying linear models, linearized around mean past trajectories, and therefore require the batches to be of same length, or to recognize an appropriate alignment variable. To account for these limitations, a multi-model approach was proposed in [2]. These models were based on the 'current measurements' of the process instead of the 'time'. These developments were followed by contributions in integration of these data-driven models with the advanced control formulations [1-2]. More recently a subspace identification based batch control approach was proposed in [4] where a LTI state-space model of the batch process is estimated. The subspace identification approach in [4] differed from conventional subspace identification approach for continuous process [5] in Hankel matrices construction to utilize data from multiple batches. A direct consequence of this approach is that it does not require alignment of different batch lengths. This approach will be referred to as batch subspace identification in this work.

The existing formulations of batch subspace identification method have dealt generally with all process variables being measured at a fixed rate [4,6,7]. Recently, [8] discussed the formulation for a multi rate process, to handle the case where some of the measurements are available at different rate than others. However, these results do not explicitly consider missing and delayed measurements in the batch subspace identification formulation. Further, the formulation of batch subspace identification requires generalization for the case of non-uniformly sampled measurements, that is, the rate at which variables are measured is not same during the entire batch duration.

Motivated by these considerations, this work presents a subspace identification based state-space modeling and control approach for batch process with missing, non-uniformly sampled and delayed measurements. The key idea is to adapt batch subspace identification using the error minimization principles of classical system identification through an iterative algorithm to explicitly handle missing and non-uniformly sampled measurements. The efficacy of the

proposed modeling approach is demonstrated using a simulated batch process. The historical batch database is first modeled using the proposed approach. The identified model is then validated on new batches to demonstrate its predictive quality. Finally, the identified model is deployed in an MPC to demonstrate its control ability.

[1] S. Aumi, P. Mhaskar, Integrating Data-Based Modeling and Nonlinear Control Tools For Batch Process Control, AIChE Journal 58 (2012) 2105-2119.

[2] S. Aumi, B. Corbett, P. Mhaskar, and T. Clarke-Pringle. Data-based modeling and control of nylon-6, 6 batch polymerization. Control Systems Technology, IEEE Transactions on, 21(1):94-106, Jan 2013.

[3] J. Flores-Cerrillo, J. F. MacGregor, Control of particle size distributions in emulsion semibatch polymerization using mid-course correction policies, Industrial and Engineering Chemistry Research 41 (7) (2002) 1805-1814.

[4] B. Corbett, P. Mhaskar, Subspace identification for data-driven modelling and quality control of batch processes, AIChE Journal 62 (2016) 1581-1601.

[5] L. Ljung, System Identification: Theory for the User, Pearson Education, 1998.

[6] Abhinav Garg, P. Mhaskar (2017), Subspace Identification Based Modeling and Control of Batch Particulate Processes, Industrial and Engineering Chemistry Research, 56 (26) 7491-7502.

[7] Abhinav Garg, Brandon Corbett, Prashant Mhaskar, Gangshi Hu, Jesus Flores-Cerrillo (2017), High Fidelity Model Development and Subspace Identification of a Hydrogen Plant Startup Dynamics, Computers and Chemical Engineering, 106 183-190.

[8] Rashid, M. M., Mhaskar, P., & Swartz, C. L. (2017). Handling multiâ€rate and missing data in variable duration economic model predictive control of batch processes, AIChE Journal *63*(7), 2705-2718.

536219 Investigation of CO_2 Sorption Mechanisms in Isothermal Columns Via Transient Material and Energy Flow PDE Models

Abstract:

In recent years, extensive research effort has been dedicated to improving CO_2 capture technology due to contribution of CO_2 to global climate change. Compared with ammonium solution, CO_2 capture sorbents are less caustic and consume less energy during CO_2 desorption. To understand and predict the dynamics of CO_2 sorption in a fixed bed column, many models have been proposed in the literature. In previous work [1], we proposed a PDE model to circumvent the unphysical assumptions of linear driving force and spatially uniform adsorption rates; unknown parameters were determined using breakthrough curves that were experimentally measured by mass spectrometry. However, the model could not accurately capture the temporal profiles of the heat released due to adsorption (experimentally measured by micro-calorimetry).

The present work attempts to improve the accuracy of our previous PDE model. It relaxes the assumptions of (a) constant concentration of accessible sites and (b) uniform superficial velocity down the column. Using an improved transport description, we consider the accuracy of five reaction mechanisms to describe the experimental observations. Physical quantities are estimated (including CO_2 sorption capacity, rate constants and heat of sorption) by minimizing the integral in time of the difference between model predictions and experimental measurements. In previous publication [1] by our group we only investigated the concentration of CO_2 while it was being sorbed, operating the column at one temperature. An elementary third order reaction was adequate to explain the column behavior. In this work, we also utilized the concentration profile of CO_2 during desorption and the associated calorimetry profiles to provide more data information for estimation. We then evaluated the different reaction mechanisms by comparing (a) the associated PDE modelsâ€™ predictions with the experimental profiles, (b) the computed accessible sites with the theoretical maximum, (c) heat of sorption as a function of temperature with the range of viable values reported in the literature. We also analyzed (d) the consistency of the rate constants to an Arrhenius model and (e) the model sensitivity to these rate constants.

We investigated five mechanisms. Based on the analysis of the optimization results, we conclude that the dual-site physisorption-chemisorption scheme is the most probable mechanism that describes the experimental data in a physically consistent manner. Since the model is insensitive to two reaction rate constants, design of new experiments and further study are needed to provide arguments for or against this mechanism.

REFERENCE

[1] Babaei Pourkargar, D.; Kamali Shahri, S. M.; Rioux, R. M.; Armaou, A. Spatiotemporal Modeling and Parametric Estimation of Isothermal CO_2 Adsorption Columns. Industrial and Engineering Chemistry Research 2016, 55, 6443-6453.

531813 Output Feedback Regulation Via Carleman Based Receding Horizon Estimation and Control

Abstract:

We present an algorithm that combines Carleman moving horizon estimation (CMHE) with Carleman model predictive control (CMPC) to design an output feedback receding horizon controller. CMHE identifies the system state and noise terms from noisy measurements, based on measurement history and process model. The identified process state is then provided as the initial condition to CMPC, which then makes optimal control decisions based on predications made via the process model. The first decision is then enacted on the process and the optimal control decisions made by the CMPC update the dynamic model used in the CMHE to increase the precision of the estimations. The novelty in our contribution lies in modeling the nonlinear dynamics with Carleman approximation, which allows us to estimate the system evolution for both CMHE and CMPC analytically. The Gradient vectors and Hessian matrices are then provided to facilitate the optimizations by these observer controller designs. To further reduce real-time computation, we adapt the advanced-step NMHE and advanced-step NMPC concepts [1,2] to our CMHE/CMPC pair to develop an asCMHE/asCMPC pair. The new design pre-estimates the states and pre-designs the manipulated input sequence one step in advance of real-time using the analytical solutions to the process model, and then it updates the estimation and control decisions almost in the real-time using pre-calculated analytical sensitivities of the estimation and control solutions to the system state.

A nonlinear open-loop unstable CSTR is studied as the illustration example. Any small perturbation in the operating condition may cause large oscillations on both the concentration and the reactor temperature. Under unmeasured process and sensor noises, both the concentration and the temperature go through a periodic orbit away from the desired steady state. With output feedback control, the closed-loop system identifies the noise and the state, regulates the system back to the desired temperature. Using standard nonlinear MHE/nonlinear MPC, the controller takes more than one sampling time to make a decision, which is **infeasible** in practice. With the CMHE/CMPC pair, computational time is decreased twentyfold to 7.48 % of one sampling time. With asCMHE/asCMPC pair, the computational time of estimation and control is further reduced to a negligible amount in terms of **on-time** calculation [3].

Reference:

1. Zavala V, Laird C, Biegler L. A fast moving horizon estimation algorithm based on nonlinear programming sensitivity. *J Process Control*. 2008;18(412):876-884. doi:10.1016/j.jprocont.2008.06.003.
2. Zavala V, Biegler L. The advanced-step NMPC controller: Optimality, stability and robustness. *Automatica*. 2009;45(1):86-93. doi:10.1016/j.automatica.2008.06.011.
3. Fang Y, Armaou A. Output Feedback Receding Horizon Regulation via Moving Horizon Estimation and Model Predictive Control. *Submitted to J Process Control*. 2018

535412 Adaptive Control of System with Unknown Inputs with Application to Chemical Reaction Control

Abstract:

Adaptive Control of System with Unknown Inputs with Application to Chemical Reaction Control

We present an approach to control a system composed of a deterministic nonlinear system plus an uncertain parts modeled as a time-varying parameter vector, also called unknown input or disturbance. These inputs may represent unknown reaction rates, heat or mass transfer rates. The adaptive controller is composed of two parts: a passivity-based input observer to estimate the uncertain, time-varying parameter and a passivity-based controller that updates its control decision based on the observer estimates and on-line measurements.

The main assumptions and the measurement vector has at least the same dimension as the unknown parameter vector, along with an observability condition [1], and that the input or disturbance is uniformly continuous [3].

The unknown input observer treats the estimation problem as a control problem. In this approach, the observer constructs the error between the observations and the output produced from the observer, and uses Lyapunov method to construct parameter adaptation law that will ensure both measurement and parameter tracking. The estimated parameters are used in a passivity-based back-stepping controller [2]. The back-stepping was necessary since we need to augment original dynamics with a first order filter to ensure the existence of a solution to both control and parameter estimation problems. Asymptotic control stability and tuning intuition of the observer and controller are presented based on control stability and trade-off between estimation convergence rate and amplification of measurements noise [3]. We show that the controller can be written in a PID with a filter.

The practical problem is motivated by the fact that reaction kinetics and other dynamic mechanisms usually have large uncertainty, and it can affect the effectiveness of model-based controller. The unknown input observer can be used to estimate the reaction rates, heat and mass transfer rates if their mechanisms are not clear. The passivity-based control follows naturally and uses estimates in the feedforward term to update control decisions.

[1] Moreno, Jaime A., and Denis Dochain. "Global observability and detectability analysis of uncertain reaction systems and observer design." International Journal of Control 81.7 (2008): 1062-1070.

[2] Zhao, Zixi, et al. "Passivity-based Backstepping Control of a Semi-batch Reactor." IFAC-PapersOnLine 50.1 (2017): 13741-13746.

[3] Levant, Arie. Robust exact differentiation via sliding mode technique. Automatica 34(3): 379-384 (1998).

527182 Optimization of Heterogeneous Batch Reactor Under Parameter Uncertainty

Abstract:

Heterogeneous batch processes are widely used in food, agrochemical and pharmaceutical industries to generate small amounts of high-priced fine chemicals. The optimal control of this type of processes is typically aimed to reduce the batch time, increase conversion and product purity, which ultimately enhance the profitability of the batch process. The optimal control highly depends on accurate mathematical models. A first principle model is usually developed based on fundamental physical and chemical mechanisms, carrying a set of unknown parameters. Parameter estimation and model discrimination are then conducted by fitting process data. However, due to model mismatch, measurement errors and even the absence of variable measurements, the parameters are always estimated with uncertainties. Exerting the optimal control designed at the nominal parameter values is likely to violate path and end-point constraints. In this work, we are going to quantify parameter uncertainty from estimation and formulate the robust optimal control to act against uncertainty and improve feasibility, while reducing the batch time.

We firstly introduce a dynamic model of a solid-liquid batch reactor, the parameters of which are estimated by limited experimental and plant data. The confidence region of estimated parameters is described by a hyper-ellipsoid or an arbitrary shape based on linear and nonlinear approximation [1].

The parameter uncertainty can be addressed in different ways. Two effective methods, multi-scenario approach and back-off approach, are implemented. In the multi-scenario approach, the optimal control problem is formulated initially at the estimated parameter set and extreme points in the confidence region [2]. Subsequently, the optimal control problem is iteratively augmented with worst case scenarios, where the largest constraint violations are observed by Monte Carlo (MC) simulations. In the back-off approach, a small term is added to the hard constraint to prevent violations in a high probability. The value of the back-off term can be approximated by MC simulations or sensitivities [3]. Both methods are implemented to the heterogeneous batch reactor case study to improve feasibility of the optimal control design.

References

[1] Albuquerque, João S., Lorenz T. Biegler, and Robert E. Kass. "Inference in dynamic error in variable measurement problems." *AIChE journal* 43.4 (1997): 986-996.

[2] Rooney, W. C. and Biegler, L. T. (2001), Design for model parameter uncertainty using nonlinear confidence regions. AIChE J., 47: 1794-1804.

[2] Diehl, Moritz, Hans Georg Bock, and Ekaterina Kostina. "An approximation technique for robust nonlinear optimization." Mathematical Programming 107.1 (2006): 213-230.

538636 Estimating the Drainage Area of "Frac-HIT" or Refractured Horizontal Well

Abstract:

Production from unconventional (low-permeability) plays has emerged as a major source of oil and gas production in North America in the past decade. Use of horizontal wells stimulated by multiple transverse hydraulic fractures has enabled this development. To extract the most amount of hydrocarbons from a given acreage, it has become common practice to space multiple horizontal wells very close to each other. This often results in so-called "frac-hits", i.e. a phenomenon where production from an existing well is considerably decreased due to interaction with a new adjacent well being fractured, as fractures from the new well may reach the original well's drainage area. Production from an existing well may also not be optimal due to the original fracturing resulting in limited drainage area or due to reduction of the original fractures' conductivity over time. This makes it appealing to re-stimulate the well via "re-fracturing" operations, to increase well productivity. For both cases – frac-hits or refracturing - it is important to quantify the evolution of the drainage area of such existing wells. Developing a model that accomplishes this calculation is the focus of this presentation. The objective is to develop a relatively simple methodology that can easily be used in the field. The approach taken is as follows:

- Develop computer simulations of scenarios that model drainage area varitions due to frac-hits or refractures,
- Conduct rate-normalized pressure data analysis to look for factors that control the alteration of production,
- Predict simple algebraic relationships/equations with the help of identified factors to quantitatively predict altered performance of a fractured horizontal well,
- Verify the predicted equations by application to data from real wells.

The approach presented above attempts to combine the advantages of sophisticated simulation employing fundamentals and of simplified algebraic equations collectively described as decline curves.

530871 Thin Falling Film Layer Monitoring and State Estimation Via Discrete Kuramoto-Sivashinsky Observer and Kalman Filter

Abstract:

Kuramoto-Sivashinsky equation (KSE) as a fourth order nonlinear partial differential equation (PDE) has been widely used as model for description of various complex phenomena, such as falling film fronts, unstable flame fronts, phase turbulence in Belousov-Zhabotinsky reaction diffusion and interfacial instabilities between multiple viscous phases [1-2]. Considering the four-order spatial derivative term and nonlinear multiplication term of the state and its first-order spatial derivative, solving this distributed fourth-order parabolic KSE is not a trivial task, which poses a challenge for its associated observer and state estimation design [3-5]. With the understanding that it is not feasible and practical to install distributed sensors to monitor the film layer propagation evolution due to given physical constraints and equipment costs, it is of significant interest to propose boundary applied discrete observer for state monitoring and estimation that can be easily realized in the state-of-the-art digital computing devices.

This work develops a discrete close-form solution for linearized infinite-dimensional Kuramoto-Sivashinsky partial differential equation (PDE) by solving for the four-by-four-matrix-form resolvent operator in this distributed parameter system setting. Furthermore, since the Euler discretization framework cannot avoid spatial and temporal approximation and/or induce numerical instability, energy preserving symmetric in time Crank-Nicolson integration method is utilized without any spatial approximation nor model reduction of the underlying KS linear PDE model [6-7]. Based on this, a four-by-four-matrix-form resolvent operator is derived [8], from which a discrete Luenberger [9] is designed for thin falling film layer monitoring. In order to consider the noise appearing in the state and output measurements, discrete Kalman filter is designed for linear Kuramoto-Sivashinsky PDE. Finally, different simulation studies which include different noise levels and/or input signals will be presented to verify performances of the proposed method.

[1] Kuramoto Y., Tsuzuki T. 1975. On the formation of dissipative structures in reactionâ€"diffusion systems, Progress of Theoretical Physics, 54: 687-699.

[2] Michelson D.M., Sivashinsky G.I. 1977. Nonlinear analysis of hydrodynamic instability in laminar flames. II. Numerical experiments, Acta Astronautica, 4: 1207-1221.

[3] Dubljevic S. 2010b. Model predictive control of Kuramotoâ€"Sivashinsky equation with state and input constraints, Chemical Engineering Science, 65(15): 4388-4396.

[4] Dubljevic, S. 2010a. Boundary model predictive control of kuramoto-sivashinsky equation with input and state constraints. Computers & Chemical Engineering, 34(10):1655-1661.

[5] Yang Y, Dubljevic S. 2013. Boundary model predictive control of thin film thickness modelled by the Kuramoto-Sivashinsky equation with input and state constraints, Journal of Process Control 23(9): 1362-1379.

[6] Xu Q., Dubljevic S., 2017. Linear model predictive control for transport-reaction processes. AIChE Journal 63 (7): 2644-2659.

[7] Havu V., Malinen J., 2007. The cayley transform as a time discretization scheme. Numerical Functional Analysis and Optimization 28 (7-8): 825-851.

[8] Curtain R. F. and Zwart H. An introduction to infinite-dimensional linear systems theory, Springer, 1995.

[9] Luenberger D. 1971. An Introduction to Observes, IEEE Transactions on Automatic Control, 16 (6): 596-602.

533995 A Generalized State-Space Model for Online Scheduling

Abstract:

Due to disruptions or arrival of new information, an incumbent schedule can become suboptimal or even infeasible, thus motivating the need for online (re)scheduling (Gupta et al., 2016). Subramanian et al. (2012) proposed the first scheduling state-space model, providing a systematic way to account for different uncertainties as disturbances, viz., demand fluctuation, task delays, unit breakdowns, and material handling losses. Many common applications, however, require a richer set of disturbances and counter-decisions. For example, in bio-manufacturing, the use of live systems such as bacteria, mammalian, or insect cells, introduces several operational challenges, including batch-to-batch variability, parallel growth of desired products and unwanted toxic byproducts, and possible random shocks that can lead to complete failure of a batch (Martagan et al. 2018). There is currently no scheduling formulation for modeling these operational features, and more importantly, no model that can be used for the online scheduling of such processes.

We have developed a general state-space model, particularly motivated by an online scheduling perspective, that allows modeling (1) task-delays and unit breakdowns with a new, more intuitive convention over that of Subramanian et al., 2012, (2) fractional delays and unit downtimes, when using discrete-time grid, (3) variable batch-sizes, (4) robust scheduling through the use of conservative yield estimates and processing times, (5) feedback on task-yield estimates before the task finishes, (6) task termination during its execution, (7) post-production storage of material in unit, and (8) unit capacity degradation and maintenance. Further, we propose new methods for updating the state of the process, as well as methods to enforce additional constraints, based on feedback information, on future decisions. We demonstrate the effectiveness of this model on a case study from the field of bio-manufacturing.

Through this new state-space model, we have enabled a natural way to handle routinely encountered processing features and disturbance information in online scheduling, in general. The proposed model, therefore, greatly extends and enables the possible application of mathematical programming based online scheduling solutions to diverse application settings. Finally, it is important to note, that although our model uses state task network based representation, these generalizations can also be adapted to resource task network based representation (Pantelides, 1994).

References

Gupta, D.; Maravelias, C.T.; Wassick, J.M. From rescheduling to online scheduling. *Chemical Engineering Research and Design*, 116 (2016), 83-97.

Subramanian, K.; Rawlings J.B.; Maravelias, C.T. A state-space model for chemical production scheduling. *Computers & Chemical Engineering*, 47 (2012), 97-110.

Martagan, T.; Krishnamurthy, A.; Leland, P.; Maravelias, C.T. Performance Guarantees and Optimal Purification

Decisions for Engineered Proteins. *Operations Research*, 6 (1), (2018), 18-41.

Pantelides, C. C. (1994). Unified frameworks for optimal process planning and scheduling. In *Proceedings of the second conference on foundations of computer aided operations* (pp. 253--274). New York: Publications, Cache.

536659 Integration of Automation Logic and Scheduling for Real-Time Batch Chemical Plant Optimization

Abstract:

In this work, we present a novel approach to incorporating a model of the low-level automation logic in a chemical plant into the online scheduling problem. We address general scheduling problems in network production environments with the following characteristics:

- Tasks can consume and produce multiple materials, batches can be mixed or split, and recycle streams may be present.
- The resolution of the scheduling problem is on the order of a single task (e.g., batch reaction).
- The plant's automation system operates at a much finer resolution, so that a single reaction may consist of 10 or more steps. ,
- The automation logic may restrict the dynamics of the plant in ways that are not accounted for in the scheduling problem; for example, a delay due to automation logic (not accounted for in scheduling) may lead to delays in the completion of a task.

Two of the challenges that arise from the mismatch between the scheduling model and the actual plant dynamics are that (1) during execution of a schedule, disturbances or delays that render the current schedule infeasible, but which can only be detected using the detailed model of the automation system, may occur; and (2) the scheduling solution, obtained using an optimization model, may lead to behavior that is prevented by the automation system. To address these challenges, we develop a methodology that integrates automation-derived information with the scheduling model.

First, building upon previous work (Gupta *et al.*, 2016; Subramanian *et al.*, 2012), we develop an iterative procedure that contains (1) a generalized state-space discrete-time resource task network (RTN) model that can be implemented online, and (2) the update equations that carry information from one online iteration to the next. The model includes features that have not been examined before, such as considering delays in processing time of tasks as optimization decisions, which may be triggered due to delays in other tasks and/or limited resource availability. In addition, we generalize the definition of tasks to account for the multiple steps involved. Furthermore, we develop a systematic way to introduce disturbances into the scheduling model based on measurements from the plant. The model accounts for new types of disturbances, including (1) processing delays detected before the completion of a task, (2) disruptions from nominal pump flow rates, (3) yield changes during the execution of a task, and (4) other disturbances during the execution of tasks (e.g., resource requirements) that can lead to task delays.

Second, to detect disturbances during the execution of a schedule, we construct a finite transition system that models the plant's automation logic (Rawlings *et al.*, 2017), and augment the model with historic data on the amount of time the system spends in each step. With this detailed model of the plant's dynamics, we treat the current schedule as a specification of the desired system behavior, and compute the set of states from which the plant could complete the schedule. If the plant ever reaches a state that is not included in this set (due to a disturbance), then we conclude that the schedule needs to be recomputed. Additional information can be extracted from the model to explain why the schedule was no longer feasible, and to add constraints to the scheduling problem when resolving to account for the detected disturbance. The same approach also applies to the case when infeasibility of the scheduling solution will be detected (and corrected), even in the absence of any disturbance.

To evaluate our methodology, we use SIMIT, a simulation platform developed by Siemens, to build a simulation model that mimics the process dynamics and automation logic of a real plant. We develop a Delay Prediction Module (DPM) that analyzes automation logic to determine step transitions. Delays are detected by DPM and the schedule is recomputed to account for these disturbances. Other disturbances such as yield changes or disruptions from nominal pump flow rates are communicated to the scheduling model from SIMIT. The three components (SIMIT, DPM, and RTN model) are interfaced through the â€œexecutive programâ€, which ensures proper data flow between these components.

We illustrate the applicability of our methods through a case study that involves both batch and continuous processes, multiple final products and tasks with various steps (including material filling and withdrawing, reaction, maintenance, and quality control steps). We study how the integration of scheduling and automation may lead to better quality closed-loop solutions, with higher throughput than in solutions obtained through iterative rescheduling.

References

1. Gupta, D., Maravelias, C. T., and Wassick, J.M. (2016) From rescheduling to online scheduling. *Chemical Engineering Research and Design*, **116**, 83-97.
2. Subramanian, K., Maravelias, C. T., and Rawlings, J.B. (2012) A state-space model for chemical production scheduling. *Computers & Chemical Engineering*, **47**, 97-110.
3. Rawlings, B.C., Wassick, J.M., and Ydstie,B.E. (2017) Application of formal verification and falsification to large-scale chemical plant automation systems. In *CPC 2017*.

This work is sponsored in part by the Department of the Army through the Digital Manufacturing and Design Innovation Institute. Any opinions, findings, and conclusions or recommendations expressed in this material are those of the author(s) and do not necessarily reflect the views of the Department of the Army.

532017 Dynamic Real-Time Optimization of a Coal-Fired Power Plant Using an Artificial Neural Network Model

Abstract:

Dynamic Real-Time Optimization of a Coal-Fired Power Plant Using an Artificial Neural Network Model

Load demand on fossil fuel power plants is more variable today than ever before. A major contributor to these variable load demands is the penetration of renewable energy sources into the grid. As intermittent renewable energy like wind and solar becomes available, the load demand to baseload units like coal power plants must decrease rapidly to allow room for renewable energy to enter the grid. Similarly, when these same renewables become unavailable, base load units must ramp up quickly to make up the difference in supply and demand[2]. The process of continual ramping is highly inefficient, costing between $150 million and $450 million in wear-and-tear costs annually, and in effect increases emissions from power generators as a whole[6]. Due to the process's overall complexity, techniques used in other industries to improve efficiency (such as first-principles modeling) are difficult to apply due to massive computational time and limited engineering resources[3,5]. A proven method of reducing emissions of coal-fired boilers is the use of neural networks to model the combustion process, and subsequent optimization of the input variables[4]. Due to the highly nonlinear nature of the process, swarm based optimization approaches have grown increasingly popular for combustion optimization[1] in place of more traditional gradient based optimizers or collocation methods. Traditional real-time optimization (RTO) methods worked well when base load power plants operated as they were originally designed, consistently at maximum or near-maximum loads. With today's variable load demands however, power plants operate less often in steady-state configurations, or more often in ramping conditions, attempting to meet these ever-changing demands, and as a result, traditional RTO methods are less-effective. In many systems, by the time the optimizer is able to operate and determine input conditions, the system has already changed configuration to where the current solution from the optimizer is no longer valid. The major cause of this is the increasingly dynamic nature of the process, while still attempting to apply methods that are meant for steady-state operation. This work focuses on improving methods of combustion optimization by applying dynamic real-time optimization techniques (DRTO) using a neural network model. To demonstrate the effectiveness of this configuration, a basic simulation model representing a coal-fired power generation unit is constructed. Treating the simulation as the power plant, a nonlinear autoregressive exogenous (NARX) neural network model is developed using data from the simulation as the training set. Upon development of a suitable model, constrained model predictive control (MPC) is implemented on the unit. Emissions values and unit efficiency from this application are compared to those produced from a traditional RTO configuration applied to the same process. Improvements are shown to both emissions levels and unit efficiency from the DRTO configuration over the RTO configuration.

[1] Song, J., et al., "Improved artificial bee colony-based optimization of boiler combustion considering NOx emissions, heat rate and fly ash recycling for on-line applications," *Fuel*, **172**, pp. 20–28, (2016)

[2] Gowrisankaran, G., et al., "Intermittency and Value of Renewable Energy," *Journal of Political Economy*, **124**, (4), (2016)

[3] Thornock, D. E., et al., "Evaluation of Carbon Conversion/NOx Emissions from Full Scale Boilers Using CFD-Based Combustion Simulations," presented at meeting of the American Flame Research Committee, Maui, HI (Oct 16-20, 1994)

[4] Chu, J. et al., "Constrained Optimization of Combustion in a Simulated Coal-Fired Boiler Using Artificial Neural Network Model and Information Analysis," *Fuel*, **82**, (6), pp. 693-703, (2003)

[5] Darby, M. L., et al., "RTO: An Overview and Assessment of Current Practice," *Journal of Process Control*, **21**, (6), pp 874–884, (2011)

[6] Moarefdoost, M. M., et al., "A Robust Model for the Ramp-Constrained Economic Dispatch Problem with Uncertain Renewable Energy," *Energy Economics*, **56**, pp. 310–325, (2016)

534118 Efficient Real Time Optimization Using Approximate Dynamic Programming

Abstract:

In this work, an efficient real time optimization framework is developed based on the approximate dynamic programming (ADP) with less computational demand and high robustness to the uncertainties.

Nowadays, due to serious competitions in the market and high price of energy, how to optimize the plant-wide operations to enhance the profitability is of course important to the entire chemical industry. The real time optimization (RTO), as a module to coordinate the operating conditions (setpoint) of different units, thus becomes more valuable in the process system engineering. The steady state RTO, even though requires less computational efforts, totally ignores the dynamics of the process and thus may lead to infeasible operations [1]. The dynamic RTO, even though takes dynamic model into account, may not be robust to the uncertainties and is computationally expensive because of the multistep prediction. Thus, it is necessary to consider some new methodologies to overcome all drawbacks mentioned above. The approximate dynamic programming (ADP) was introduced into process system engineering community for a decade [2]. It relies on the principle of optimality [3] to reduce online computational demand and makes use of simulation/historical data to find profitable operational patterns for plant-wide system.

We propose an ADP-RTO to reduce online computational work and achieve better economic objective values. First, through data analysis, the information of process states, setpoint and production is extracted from operational/historical data. Second, a surrogate model and approximate value function are constructed to characterize the transient and steady performances of a plant under certain control schemes. Third, in the online implementation, a small-scale mixed-integer linear program (MILP) is solved at each time instant to maximize the one-step ahead value function and then decide whether the setpoint should be changed such that the economic objective along the prediction horizon is optimized.

Several features of proposed ADP-RTO are shown in order. First, even though this ADP-RTO is based on the historical data, the real time measurement also can be used to continuously update the value function and surrogate model in order to reduce the impact of uncertainties. Second, this ADP-RTO can be integrated with any types of controller, not limited to the model predictive control (MPC). Third, this ADP-RTO avoid identifying nonlinear dynamic representation for the entire plant, because such modeling process can be time consuming, costly and very sensitive to the disturbances. Fourth, due to its less computational demand, the ADP-RTO can be triggered more frequently to optimize the dynamic behavior of the plant.

Reference

[1] Tosukhowong, T., Lee, J. M., Lee, J. H., Lu J. (2004). An introduction to a dynamic plant-wide optimization strategy for an integrated plant. Computers and Chemical Engineering, 29, 199-208.

[2] Lee, J. H., & Lee, J. M. (2006). Approximate dynamic programming based approach to process control and scheduling. Computers and Chemical Engineering, 30, 1603-1618.

[3] Bellman, R. E. (1957). Dynamic programming. Princeton University Press.

521893 An Overview and Evaluation of Approaches for Online Process Optimization

Abstract:

This paper discusses approaches to online process optimization, where the objective is usually to minimize an economic cost. In many process control application, traditional real-time optimization (RTO) uses nonlinear steady-state process models to compute the optimal setpoint at steady state operation. The most common approach to RTO is the so-called, two-step model-adaptation approach, where the static model used in the optimization is first updated using measurement data and the updated model is then used to re-compute the new optimal setpoints.

However, traditional RTO faces some challenges, which limits its industrial use, including (in expected order of importance):

1. Cost of developing and updating the model structure (offline update of the model),
2. Model uncertainty, including uncertain values of disturbances and parameters and structural uncertainty (online update of the model),
3. Frequent grade changes, which makes steady-state optimization less relevant,
4. Dynamic limitations, including infeasibility due to dynamic constraint violation.

It would seem that reason 2, 3, and 4 suggests towards dynamic RTO. However, except for processes with frequent grade changes (reason 3), static optimization may be close to economic optimum, at a much lower computational cost. In a recent review paper on current practice of RTO [1], it was noted that the fundamental limiting factor of RTO implementation is the steady-state wait time associated with the online update of the model (reason 2). Since only static models are used, the model adaptation step must be carried out using measurements that corresponds to steady-state operation. If the process is frequently subject to disturbances or if the settling times are too long, this can lead to the plant being operated in transients for significant periods. With the lack of sufficient steady-state measurements, the model is not updated frequently. Consequently, the plant is operated sub-optimally for long periods. Many commercially available steady-state detection methods may also falsely accept transient data as steady state data as demonstrated in [2] using a real industrial example.

On the other hand, dynamic RTO uses dynamic models for model adaptation and the updated dynamic model is used to solve a dynamic nonlinear optimization problem online. However, the main implementation issue with DRTO today is the computational cost, as noted in [3].

In this preseantaion, a number of alternative approaches are discussed, given here in the order from most complex (and model-based) to most simple (and data based):

1. Economic (nonlinear) model predictive control (EMPC)
2. Dynamic real-time optimization (DRTO)
3. Hybrid RTO - Steady-state RTO with dynamic model update *(new method)*
4. Feedback-based Hybrid RTO *(new method)*
5. Conventional steady-state RTO
6. Self-optimizing control
7. Extremum seeking control, and similar feedback approaches based on measuring the cost such as NCO tracking and hill-climbing.

Except possibly for EMPC, the optimizer sends setpoints to a control layer, which could be a PID layer or MPC. Dynamic stability issues (reason4) are then handled by the lower control layer.

In the conventional steady-state RTO, one must wait for the process to reach steady-state before updating the model using "data reconciliation". In the hybrid RTO approach, we solve the same steady-state optimization problem as in traditional Steady-state RTO, but instead of a steady-state model update, we use dynamic model adaptation with the use of transient measurements, for example, using an extended Kalman Filter. This avoids the steady-state wait time.

In the feedback-based hybrid RTO approach, we don't solve the steady-state optimization problem numerically as in conventional RTO, but instead the steady-state gradient is estimated by linearizing the nonlinear dynamic model around the current operating point. The gradient is controlled to zero using standard feedback controllers, for example, a PI-controller. In case of disturbances, the proposed method is able to adjust quickly to the new optimal operation. The advantage of the feedback RTO strategy compared to standard steady-state real-time optimization is that it reaches the optimum much faster and without the need to wait for steady-state to update the model. The advantage compared to dynamic RTO and the closely related economic NMPC, is that the computational cost is much reduced and the tuning is simpler. It is significantly faster than classical extremum-seeking control and does not require the measurement of the cost function and additional process excitation. We compare the performance of different RTO approaches using several simulation case studies.

References

1. Darby ML, Nikolaou M, Jones J, Nicholson D. RTO: An overview and assessment of current practice. Journal of Process Control. 2011 Jul 1;21(6):874-84.
2. Câmara MM, Quelhas AD, Pinto JC. Performance Evaluation of Real Industrial RTO Systems. Processes. 2016 Nov 22;4(4):44.
3. Campos MC, Teixeira H, Liporace F, Gomes M. Challenges and problems with advanced control and optimization technologies. IFAC Proceedings Volumes. 2009 Jan 1;42(11):1-8.

528146 Accelerated Parallel Alternating Method of Multipliers (ADMM) for Distributed Optimization

Abstract:

The real-time optimization in the operation of distributed large-scale plants demands effective distributed optimization algorithms, where the optimization problems are solved based on a system decomposition. That is, the variables in the entire optimization problem are partitioned into several blocks with interconnecting constraints, and the resulting sub-problems are optimized in a distributed manner, coordinated through the information change between them.

Alternating direction method of multipliers (ADMM) is a widely used algorithm of distributed optimization, especially for problems in machine learning and image processing [1]. For chemical engineering, its application to distributed real-time optimization has been recently introduced [2]. ADMM can also be used for distributed model predictive control (see e.g. [3]). A significant feature of ADMM is that it allows each optimizer (controller) to work with only local information, i.e. the model of the corresponding subsystem, its connections with upstream and downstream variables, and its own optimization objective and constraints. In this sense, ADMM is a localized distributed optimization algorithm suitable for local information availability.

However, the real-time implementation of the classical multi-block ADMM algorithm is largely prohibited by a high computational cost due to its linear convergence rate and the sequential optimization scheme of solving the sub-problems. Recently, there have been works that seek to parallelize the multi-block optimization in ADMM (see e.g. [4]), and accelerate the convergence of two-block ADMM to a quadratic rate (see e.g. [5]).

In this work, we incorporate the efforts of [4] and [5] to parallelize and accelerate ADMM in a novel *accelerated parallel ADMM algorithm* in the context of solving multi-block, nonconvex, constrained optimization problems, which arise in the distributed control or real-time optimization of nonlinear chemical process systems. We demonstrate with a case study that the ADMM formulation can be applied to the localized distributed control of a chemical plant, and that the accelerated parallel ADMM significantly improves the computational performance compared to the classical ADMM.

References

[1] Boyd, S., *et al.* (2011). Distributed optimization and statistical learning via the alternating direction method of multipliers. *Found. Trends Mach. Learn.*, *3*(1), 1-122.

[2] Harwood, S., *et al.* (2017). Analysis of the alternating direction method of multipliers for the optimization of distributed nonconvex systems. *2017 AIChE Annual Meeting*, 522f.

[3] Mota, J. F., *et al.* (2015). Distributed optimization with local domains: Applications in MPC and network flows. *IEEE Trans. Autom. Control*, *60*(7), 2004-2009.

[4] Deng, W., *et al.* (2017). Parallel multi-block ADMM with $o(1/k)$ convergence. *J. Sci. Comput.*, *71*(2), 712-736.

[5] Goldstein, T., *et al.* (2014). Fast alternating direction optimization methods. *SIAM J. Imaging Sci.*, *7*(3), 1588-1623.

529141 Dynamic Modeling for Improved Operation and Control of a Supercritical CO_2 Brayton Power Cycle

Abstract:

Supercritical carbon dioxide (sCO_2) Brayton power cycles are an attractive alternative to conventional sub/supercritical steam Rankine power plants. The high density of the sCO_2 working fluid allows for highly compact equipment and a power cycle with potentially higher thermal efficiencies and lower overall cost of electricity. Due to these advantages over traditional systems, the U.S. Department of Energy, under its Supercritical Transformational Electric Power (STEP) Program, is working with industry and research partners to design, build, and operate a 10 MWe (net) sCO_2 pilot plant test facility that uses a natural gas-fired furnace as the indirect heat source. The new facility will support the demonstration of systems integration, performance, operability, and controls for the future commercialization of utility-scale sCO_2 Brayton power cycles. In this presentation, a pressure-driven dynamic process model of an indirect-fired 10MWe sCO_2 recompression Brayton cycle is used to develop and evaluate operational and control strategies for improving cycle performance during transient, off-design conditions.

Considering Carnot's rule, the sCO_2 cycle efficiency is maximized by keeping the ratio between the hot turbine inlet temperature (TIT) and the cold main compressor inlet temperature (MCIT) as high as possible. To maintain the TIT at its maximum design point during cycle turndown and ramp-up operations, this study uses inventory control wherein sCO_2 is removed and added, respectively, from the closed cycle. The impacts of inventory tank capacity and initial pressure on maximum sCO_2 cycle turndown are investigated. An advanced controller logic is developed that regulates the inventory inlet and outlet valves in a fast and effective manner to achieve the TIT control objectives while satisfying process constraints. Similarly, cooling water flow is used to maintain the MCIT at its minimum design point slightly above the critical temperature of CO_2, thereby avoiding transition to the two-phase region. In addition, a controller is designed that aims to maintain an optimal flow split between the main compressor and bypass compressor streams during transient operations. The presentation will include results for various cycle turndown and ramp-up scenarios, as well as analysis of trade-offs between systems efficiency and control system performance.

AUTHOR INDEX

AUTHOR INDEX

AUTHOR INDEX

AUTHOR INDEX

AUTHOR INDEX

AUTHOR INDEX

AUTHOR INDEX

AUTHOR INDEX

AUTHOR INDEX

AUTHOR INDEX

AIChE
120 Wall Street, FL 23
New York, NY 10005-4020

ISBN 978-1-5108-7623-1
9 781510 876231

CURRAN ASSOCIATES INC.
proceedings.com